Quantum Communication, Quantum Networks, and Quantum Sensing

Quantum Communication, Quantum Networks, and Quantum Sensing

Ivan B. Djordjevic
Department of Electrical and Computer Engineering,
University of Arizona,
Tucson, AZ, United States

ACADEMIC PRESS
An imprint of Elsevier
ELSEVIER

Academic Press is an imprint of Elsevier
125 London Wall, London EC2Y 5AS, United Kingdom
525 B Street, Suite 1650, San Diego, CA 92101, United States
50 Hampshire Street, 5th Floor, Cambridge, MA 02139, United States
The Boulevard, Langford Lane, Kidlington, Oxford OX5 1GB, United Kingdom

ISBN: 978-0-12-822942-2

For information on all Academic Press publications visit our website at
https://www.elsevier.com/books-and-journals

Publisher: Mara Conner
Acquisitions Editor: Tim Pitts
Editorial Project Manager: Kyle Gravel
Production Project Manager: Prem Kumar Kaliamoorthi
Cover Designer: Miles Hitchen

Typeset by TNQ Technologies

Contents

Preface

Quantum information is related to using quantum mechanical concepts to perform information processing and information transmission. Quantum information processing (QIP) is an exciting research area with numerous applications, including quantum key distribution (QKD), quantum teleportation, quantum and distributed quantum computing, quantum sensing, quantum networking, quantum radar, and quantum machine learning. Quantum information science (QIS) is experiencing rapid development, as judged by the number of published books on QIP, quantum computing, and quantum communications and the numerous conferences devoted solely to QIS. Given the novelty of underlying QIS concepts, this topic is expected to be a subject of interest to many scientists, engineers, governmental agencies, industries, and investors, not just those involved in QIP research. The entanglement represents a unique QIP feature enabling quantum computers to solve problems that are numerically intractable for classical computers, enabling quantum-enhanced sensors with measurement sensitivities exceeding classical limits and providing certifiable data transmission security guaranteed by the laws of quantum mechanics rather than the unproven assumptions used in conventional cryptography for computational security. Preshared entanglement can be used to (1) improve classical channel capacity, (2) enable secure communications, (3) improve sensor sensitivity, (4) enable distributed quantum computing, (5) enable entanglement-assisted distributed sensing, and (6) enable provably secure quantum computer access, as well as perform many other tasks.

Because of the interdisciplinary nature of QIS, this book aims to provide the right balance among QIP, quantum error correction, quantum communication, quantum sensing, and quantum networking. The main *objectives* of the book can be summarized as follows:

(1) Book describes trends in QIP, quantum error correction, quantum communication, quantum sensing, quantum networking, and quantum machine learning.
(2) Book represents a self-contained introduction to quantum communication, quantum error correction, quantum sensing, and quantum networking.
(3) Book targets a wide range of readers: electrical engineers, communication engineers, optical engineers, applied mathematicians, computer scientists, and physicists.
(4) Book does not require prior knowledge of quantum mechanics or QIP. The basic concepts are provided in Chapter 3.
(5) The book does not require any prerequisite background, except for an understanding of the basic concepts of vector algebra at the undergraduate level.
(6) For readers not familiar with information theory, channel coding, and detection theory, Chapter 2 provides basic concepts and definitions in the respective fields. The chapter only covers information, coding, and detection theory concepts relevant in later chapters.
(7) Readers not interested in quantum error correction can still approach the book's other chapters.
(8) Readers solely interested in quantum error correction do not need to read other chapters, except Chapters 3 and 4, and can concentrate on just the quantum error correction chapters. Namely, an effort has been made to create independent chapters while ensuring a proper flow between chapters.
(9) The book starts with intuitive and basic concepts before moving to medium-difficulty topics and then to complex, mathematically involved topics.
(10) Several courses can be offered using this book. Examples include courses on (1) quantum error correction, (2) QIP and quantum machine learning, (3) quantum communication, (4) quantum

sensing, (5) quantum networking, and (6) integration of the concepts of quantum communication, quantum sensing, and quantum networking.
(11) The reader who completes this book will be able to perform independent research in various fields of QIS.

This book represents a self-contained introduction to quantum information, quantum communication, quantum error correction, quantum sensing, and quantum networking. The successful reader of the book will be ready for further study in these areas and will be prepared to perform independent research. The reader who has completed the book will be able to design QIP circuits, stabilizer codes, Calderbank–Shor–Steane (CSS) codes, subsystem codes, topological codes, surface codes, and entanglement-assisted quantum error correction codes. The reader will also be proficient in fault-tolerant quantum error correction. Further, the reader will be proficient in quantum communication, QKD, and quantum machine learning. The reader will have gained the required knowledge in quantum sensing and quantum radars to perform independent research in these fields. Readers of this book will also be able to propose quantum transceivers suitable for quantum communication and quantum sensing applications. Moreover, book readers will be able to propose quantum communication networking protocols and quantum networking architectures.

Some *unique features* of the book can be summarized as follows:

(1) This book integrates QIP, communication, error correction, sensing, networking, and machine learning.
(2) The book does not require knowledge of quantum mechanics.
(3) This book does not require any prerequisite material except for basic concepts of vector algebra at the undergraduate level.
(4) This book offers in-depth exposition on the design of QIP, error correction, communication and networking, and sensing circuits and modules.
(5) The successful student will be prepared for further study in these areas and qualified to perform independent research.
(6) The reader who has completed this book will be able to design information-processing circuits, stabilizer codes, CSS codes, topological codes, surface codes, subsystem codes, quantum low-density parity-check (LDPC) codes, and entanglement-assisted quantum error correction codes and will be proficient in fault-tolerant design.
(7) The successful student will be able to propose quantum transceivers for quantum communication and quantum sensing applications.
(8) The successful student will be able to propose quantum communication networking protocols and quantum networking architectures.
(9) Students who have completed the book will be proficient in quantum machine learning.
(10) Finally, students will be proficient in QKD and quantum teleportation.
(11) Extra material to support the book will be provided on an accompanying website to be developed after publication.

The author would like to acknowledge the support of the National Science Foundation.

Finally, special thanks are extended to Prem Kumar Kaliamoorthi, Kyle Gravel, and Tim Pitts of Elsevier for their tremendous efforts in organizing the logistics of the book, including editing and promotion, which are indispensable to making this book happen.

CHAPTER

1

Basics of quantum information, quantum communication, quantum sensing, and quantum networking

Chapter outline

1.0 Overview

This chapter is devoted to introducing quantum information processing (QIP), quantum communication, quantum networking, quantum sensing, and quantum error correction coding (QECC) [1−18]. *Quantum information* is related to quantum mechanical concepts used in information processing or the transmission of quantum information. QIP is an exciting research area with numerous applications, including quantum key distribution (QKD), quantum teleportation, quantum and distributed quantum computing, quantum sensing, quantum networking, and quantum radar.

Fundamental features of QIP are different from those of classical signal processing and can be summarized as three characteristics [1,2,6,7,10]: (1) linear superposition, (2) entanglement and (3) quantum parallelism. The following are some basic details of these features:

(1) Linear superposition. Contrary to the classical bit, a quantum bit or *qubit* can take not only two discrete values, 0 and 1, but also *all* possible *linear combinations* of them. This is a consequence of a fundamental property of quantum states—it is possible to construct a *linear superposition* of quantum state $|0\rangle$ and quantum state $|1\rangle$.

(2) Entanglement. At a quantum level, it appears that two quantum objects can form a single entity even when they are well separated from each other. Any attempt to consider this entity as a combination of two independent quantum objects given by the tensor product of quantum states

Quantum Communication, Quantum Networks, and Quantum Sensing. https://doi.org/10.1016/B978-0-12-822942-2.00002-9

fails unless the possibility of signal propagation at a superluminal speed is allowed. These quantum objects that cannot be decomposed into the tensor product of individual independent quantum objects are commonly referred to as *entangled* quantum objects. Given that arbitrary quantum states cannot be copied, which is a consequence of the no-cloning theorem, communication at superluminal speed is not possible, and as a consequence, entangled quantum states cannot be written as the tensor product of independent quantum states. Moreover, it can be shown that the amount of information contained in an entangled state of N qubits grows exponentially instead of linearly as is the case for classical bits.

(3) Quantum parallelism. *Quantum parallelism* is the possibility of performing a large number of operations in parallel, which represents its key difference from classical information processing and computing. Namely, in classical computing, it is possible to know the internal status of the computer. On the other hand, because of the no-cloning theorem, it is not possible to know the current state of a quantum computer. This property has led to the development of Shor factorization algorithm, which can be used to crack the Rivest–Shamir–Adleman (RSA) encryption protocol. Some other important quantum algorithms include the Grover search algorithm, which is used to search for an entry in an unstructured database; the quantum Fourier transform, which is the basis for a number of algorithms; and Simon's algorithm. The quantum computer can encode all input strings of length N simultaneously in a single computation step. In other words, the quantum computer can simultaneously pursue 2^N classical paths, indicating that the quantum computer is significantly more powerful than the classical one.

Although QIP has opened up some fascinating perspectives, certain limitations must be overcome before QIP becomes a commercial reality. The first is related to the number of existing quantum algorithms, which is significantly lower than that of classical algorithms. The second problem is related to physical implementation issues. Another problem, which can be considered problem number one, relates to *decoherence*. Decoherence is related to the interaction of qubits with the environment that blurs the fragile superposition states. Decoherence also introduces errors, indicating that the quantum register should be sufficiently isolated from the environment so that only a few random errors occasionally occur and can be corrected by QECC techniques. One of the most powerful applications of quantum error correction is protecting quantum information as it dynamically undergoes quantum computation. Imperfect quantum gates affect quantum computation by introducing errors in computed data. Moreover, the imperfect control gates introduce errors in processed sequences, since the wrong operations can be applied. The QECC scheme now must deal not only with errors introduced by the quantum channel but also with errors introduced by imperfect quantum gates during the encoding/decoding process.

Nevertheless, QIP opens new avenues for various applications, including high-performance computing, high-precision sensing, and secure communications [1,2]. The entanglement represents a unique QIP feature enabling quantum computers capable of solving the problems that are numerically intractable for classical computers [19], enabling quantum-enhanced sensors with measurement sensitivities exceeding the classical limit [20]; and providing certifiable security for data transmissions whose security is guaranteed by the laws of quantum mechanics, known as the quantum key distribution (QKD), rather than unproven assumptions used in cryptography based on computational security [1,2]. The preshared entanglement can be used to (1) improve the classical channel capacity

[21–27], (2) enable secure communications, (3) improve sensor sensitivity, (4) enable distributed quantum computing [28], (5) enable entanglement-assisted distributed sensing, and (6) enable provably secure quantum computer access [29], to mention a few.

The *quantum network* represents the network of nodes interconnected by quantum links. The quantum links could be heterogeneous, including fiber-optics links, free-space optical (FSO) links, and satellite links, to mention a few. The quantum communication network (QCN) can be defined as a system of quantum nodes in which an uninterrupted quantum channel can be established between any two nodes. When the quantum nodes are equipped with quantum computers capable of exchanging the qubits (or qudits) over the QCN by teleportation, the corresponding QCN is commonly referred to as the *quantum Internet*. Several QKD testbeds have been reported, including the DARPA QKD network [30], the Tokyo QKD network [31], and secure communication based on the quantum cryptography (SECOQC) network [32]. QKD can also be used to establish a QKD-based campus-to-campus virtual private network employing the IPsec protocol [33], as well as to establish a network setup to use transport-layer security based on QKD [34]. However, all these networks employ a dark fiber infrastructure. Classical communication networks have nodes connected by various channel types, including FSO, optical fiber, ground-satellite links, wireless RF, and coaxial cables. Such a heterogeneous architecture will be equally important for quantum nodes to access a QCN. Indeed, quantum communications (QuComs) are individually validated in free-space, optical fiber, and between a satellite and a ground station, but a heterogeneous QCN with multiple channel types remains currently elusive.

Quantum sensing is an application of QIP that can find practical realization in the near term. When we use nonclassical sources to improve the sensitivity of measuring a certain parameter in a sensing application, we refer to such a methodology as *quantum metrology* [20,35,36]. When we repeat the measurement of a certain parameter N times and average out the individual measurement results, we can achieve the precision that scales as $N^{-1/2}$, which is commonly referred to as the standard quantum limit (SQL). By employing nonclassical sources such as the entangled source or squeezed light, we can achieve better precision than SQL. As an illustration, in Laser Interferometer Gravitational-Wave Observatory (LIGO) experiments [35,36], squeezed light is injected into a Michelson interferometer to overcome the SQL (dictated by the shot noise). Alternatively, in distributed quantum sensing [20], multiple entangled sensors are employed to obtain a certain *global property* of an interrogated object and beat the SQL. In addition to LIGO and target detection/quantum radars [37,38], quantum metrology has also been applied in biological sensing [39], microscopy [40], and phase-tracking applications [41].

This introductory chapter is organized as follows. In Section 1.1, photon polarization is described as representing a simple and the most natural connection to the QuComs. In the same section, we introduce some basic quantum mechanics concepts, such as the state concept. We also introduce the Dirac notation, which will be used throughout the book. In Section 1.2, we formally introduce the concept of qubit and provide its geometric interpretation. Section 1.3 is devoted to basic quantum gates and QIP fundamentals. The basic concepts of quantum teleportation are introduced in Section 1.4. Section 1.5 is devoted to the basic QECC concepts. Section 1.6 introduces the quantum sensing concept. Section 1.7 is devoted to the quantum key distribution (QKD), also known as quantum cryptography. Section 1.8 is related to the quantum networks basics. The organization of the book is described in Section 1.9.

1.1 Photon polarization

The electric/magnetic field of plane linearly polarized waves is described as follows [42]:

$$\boldsymbol{A}(\boldsymbol{r},t) = \boldsymbol{p}A_0 \exp[j(\omega t - \boldsymbol{k}\cdot\boldsymbol{r})], A \in \{\boldsymbol{E},\boldsymbol{H}\} \tag{1.1}$$

where $\boldsymbol{E}$ ($\boldsymbol{H}$) denotes electric (magnetic) field, $\boldsymbol{p}$ denotes the polarization orientation, $\boldsymbol{r} = x\boldsymbol{e}_x + y\boldsymbol{e}_y + z\boldsymbol{e}_z$ is the position vector, and $\boldsymbol{k} = k_x\boldsymbol{e}_x + k_y\boldsymbol{e}_y + k_z\boldsymbol{e}_z$ denotes the wave propagation vector whose magnitude is $k = 2\pi/\lambda$, with λ being the operating wavelength. For the x-polarization waves ($\boldsymbol{p} = \boldsymbol{e}_x$, $\boldsymbol{k} = k\boldsymbol{e}_z$), Eq. (1.1) becomes

$$\boldsymbol{E}_x(z,t) = \boldsymbol{e}_x E_{0x} \cos(\omega t - kz), \tag{1.2}$$

while for y-polarization ($\boldsymbol{p} = \boldsymbol{e}_y$, $\boldsymbol{k} = k\boldsymbol{e}_z$), it becomes

$$\boldsymbol{E}_y(z,t) = \boldsymbol{e}_y E_{0y} \cos(\omega t - kz + \delta), \tag{1.3}$$

where δ is the relative phase difference between the two orthogonal waves. The resultant wave can be obtained by combining Eqs. (1.2) and (1.3) as follows:

$$\boldsymbol{E}(z,t) = \boldsymbol{E}_x(z,t) + \boldsymbol{E}_y(z,t) = \boldsymbol{e}_x E_{0x}\cos(\omega t - kz) + \boldsymbol{e}_y E_{0y}\cos(\omega t - kz + \delta). \tag{1.4}$$

The *linearly polarized* wave is obtained by setting the phase difference to an integer multiple of 2π, namely $\delta = m\cdot 2\pi$:

$$\boldsymbol{E}(z,t) = (\boldsymbol{e}_x E\cos\theta + \boldsymbol{e}_y E\sin\theta)\cos(\omega t - kz);\ E = \sqrt{E_{0x}^2 + E_{0y}^2},\ \theta = \tan^{-1}\frac{E_{0y}}{E_{0x}}. \tag{1.5}$$

We can represent the linear polarization by ignoring the time-dependent term as shown in Fig. 1.1A. On the other hand, if $\delta \neq m\cdot 2\pi$, the *elliptical polarization* is obtained. From Eqs. (1.2) and (1.3), by eliminating the time-dependent term, we obtain the following equation of ellipse:

$$\left(\frac{E_x}{E_{0x}}\right)^2 + \left(\frac{E_y}{E_{0y}}\right)^2 - 2\frac{E_x}{E_{0x}}\frac{E_y}{E_{0y}}\cos\delta = \sin^2\delta,\ \tan 2\phi = \frac{2E_{0x}E_{0y}\cos\delta}{E_{0x}^2 - E_{0y}^2}, \tag{1.6}$$

which is shown in Fig. 1.1B. By setting $\delta = \pm\frac{\pi}{2}, \pm 3\frac{\pi}{2}, \ldots$ the equation of the ellipse becomes

$$\left(\frac{E_x}{E_{0x}}\right)^2 + \left(\frac{E_y}{E_{0y}}\right)^2 = 1. \tag{1.7}$$

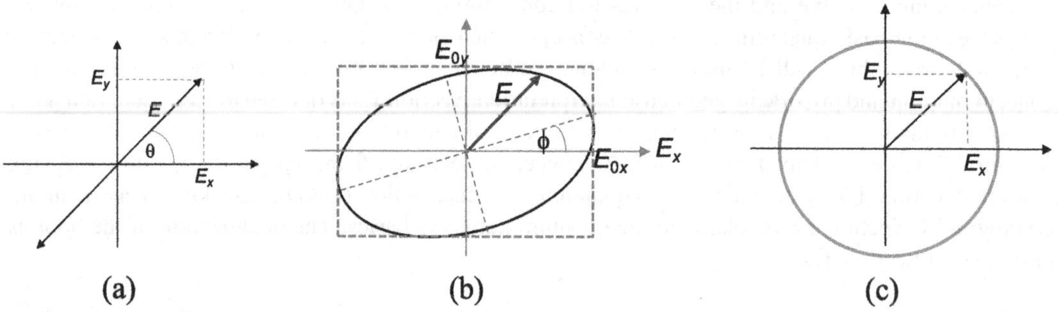

FIGURE 1.1

Various forms of polarization: (A) linear polarization, (B) elliptic polarization, and (C) circular polarization.

Further, by setting $E_{0x} = E_{0y} = E_0$; $\delta = \pm\frac{\pi}{2}, \pm 3\frac{\pi}{2}, \ldots$ the equation of the ellipse becomes the circle

$$E_x^2 + E_y^2 = 1, \tag{1.8}$$

and the corresponding polarization is known as *circular polarization* (see Fig. 1.1C). A *right circularly polarized* wave is obtained for $\delta = \pi/2 + 2m\pi$:

$$\boldsymbol{E} = E_0\left[\boldsymbol{e}_x \cos(\omega t - kz) - \boldsymbol{e}_y \sin(\omega t - kz)\right], \tag{1.9}$$

Otherwise, for $\delta = -\pi/2 + 2m\pi$, the polarization is known as left circularly polarized.

Quite often, the *Jones vector representation* of the polarization wave is used:

$$\boldsymbol{E}(t) = \begin{bmatrix} E_x(t) \\ E_y(t) \end{bmatrix} = E \begin{bmatrix} \sqrt{1-\kappa} \\ \sqrt{\kappa} e^{j\delta} \end{bmatrix} e^{j(\omega t - kz)}, \tag{1.10}$$

where κ is the power-splitting ratio between states of polarization (SOPs), with the complex phasor term typically omitted in practice.

Another interesting representation is the *Stokes vector representation*:

$$\boldsymbol{S}(t) = \begin{bmatrix} S_0(t) \\ S_1(t) \\ S_2(t) \\ S_3(t) \end{bmatrix}, \tag{1.11}$$

where the parameter S_0 is related to the optical intensity by

$$S_0(t) = |E_x(t)|^2 + |E_y(t)|^2. \tag{1.12a}$$

The parameter $S_1 > 0$ is related to the preference for horizontal polarization and is defined by

$$S_1(t) = |E_x(t)|^2 - |E_y(t)|^2. \tag{1.12b}$$

The parameter $S_2 > 0$ is related to the preference for the $\pi/4$ SOP:

$$S_2(t) = E_x(t)E_y^*(t) + E_x^*(t)E_y(t). \tag{1.12c}$$

Finally, the parameters $S_3 > 0$ is related to the preference for right-circular polarization, and it is defined by

$$S_3(t) = j\left[E_x(t)E_y^*(t) - E_x^*(t)E_y(t)\right]. \tag{1.12d}$$

The parameter S_0 is related to other Stokes parameters by

$$S_0^2(t) = S_1^2(t) + S_2^2(t) + S_3^2(t). \tag{1.13}$$

The *degree of polarization* is defined by

$$p = \frac{\left[S_1^2 + S_2^2 + S_3^2\right]^{1/2}}{S_0}, \quad 0 \le p \le 1 \tag{1.14}$$

For $p = 1$, the polarization does not change with time. The Stokes vector can be represented in terms of Jones vector parameters as

$$s_1 = 1 - 2\kappa, s_2 = 2\sqrt{\kappa(1-\kappa)}\cos\delta, s_3 = 2\sqrt{\kappa(1-\kappa)}\sin\delta. \tag{1.15}$$

After *normalization* with respect to S_0, normalized Stokes parameters are given by

$$s_i = \frac{S_i}{S_0}; i = 0, 1, 2, 3. \tag{1.16}$$

The polarization state can be represented as a point on a Poincaré sphere if normalized Stokes parameters are used, as shown in Fig. 1.2. The points at the opposite sides of the line crossing the center represent orthogonal polarizations.

The polarization ellipse is often represented by ellipticity and azimuth, which are illustrated in Fig. 1.3. Ellipticity is defined by the ratio of the half-axes lengths. The corresponding angle is called the ellipticity angle and is denoted ε. A small ellipticity means that the polarization ellipse is highly elongated, while for zero ellipticity, polarization is linear. For $\varepsilon = \pm\pi/4$, polarization is circular. For

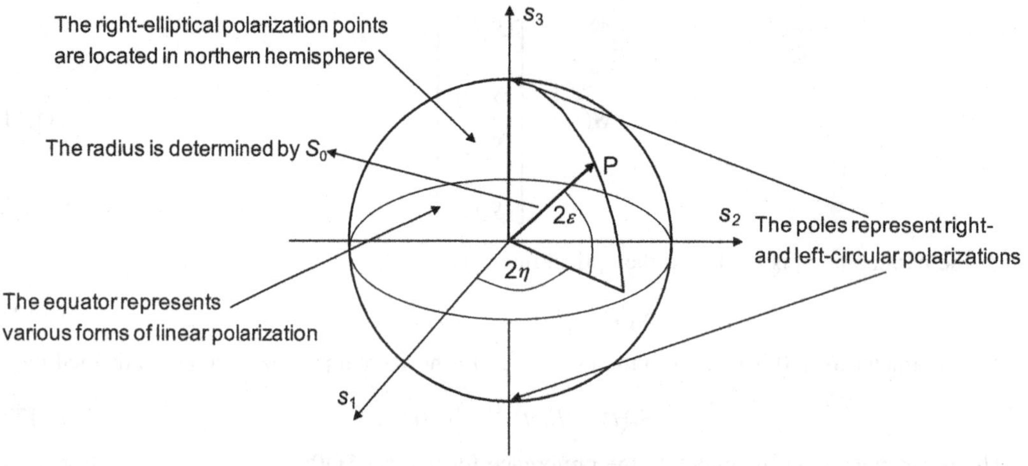

FIGURE 1.2

Representation of polarization state as a point on a Poincaré sphere.

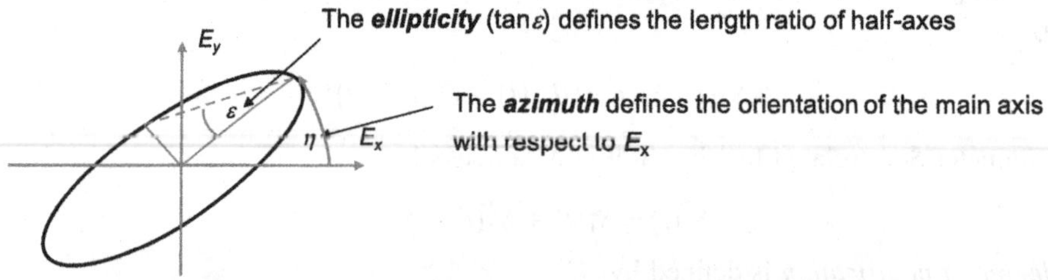

FIGURE 1.3

The ellipticity and azimuth of the polarization ellipse.

$\varepsilon > 0$, polarization is right-elliptical. On the other hand, the azimuth angle η defines the orientation of the main axis of the ellipse with respect to E_x.

The parameters of polarization ellipse can be related to the Jones vector parameters by

$$\sin 2\,\varepsilon = 2\sqrt{\kappa(1-\kappa)}\sin\delta,\ \tan 2\,\eta = \frac{2\sqrt{\kappa(1-\kappa)}\sin\delta}{1-2\kappa}. \tag{1.17}$$

Finally, the parameters of the polarization ellipse can be related to the Stokes vector parameters by

$$s_1 = \cos 2\eta \cos 2\varepsilon,\ s_2 = \sin 2\eta \cos 2\,\varepsilon,\ s_3 = \sin 2\varepsilon, \tag{1.18}$$

and the corresponding geometrical interpretation is provided in Fig. 1.2.

Let us now observe the *polarizer–analyzer ensemble*, shown in Fig. 1.4. When an electromagnetic wave passes through the polarizer, it can be represented as a vector in the xOy plane transversal to the propagation direction, as given by Eq. (1.5), where the angle θ depends on the filter orientation. By introducing the unit vector $\widehat{p} = (\cos\theta, \sin\theta)$, Eq. (1.5) can be rewritten as

$$\boldsymbol{E} = E_0\widehat{p}\cos(\omega t - kz). \tag{1.19}$$

If $\theta = 0$ rad, the light is polarized along the x-axis, while for $\theta = \pi/2$ rad, it is polarized along the y-axis. *Natural light* is *unpolarized*, as it represents an *incoherent superposition* of 50% of light polarized along the x-axis and 50% along the y-axis. After the analyzer, the polarization vector makes an angle ϕ with respect to the x-axis, which can be represented by a unit vector $\widehat{n} = (\cos\phi, \sin\phi)$, and the output electric field is given by

$$\begin{aligned}\boldsymbol{E}' &= (\boldsymbol{E}\cdot\widehat{n})\widehat{n} = E_0\cos(\omega t - kz)(\widehat{p}\cdot\widehat{n})\widehat{n} = E_0\cos(\omega t - kz)[(\cos\theta, \sin\theta)\cdot(\cos\phi, \sin\phi)]\widehat{n} \\ &= E_0\cos(\omega t - kz)[\cos\theta\cos\phi + \sin\theta\sin\phi]\widehat{n} \\ &= E_0\cos(\omega t - kz)\cos(\theta - \phi)\widehat{n}.\end{aligned} \tag{1.20}$$

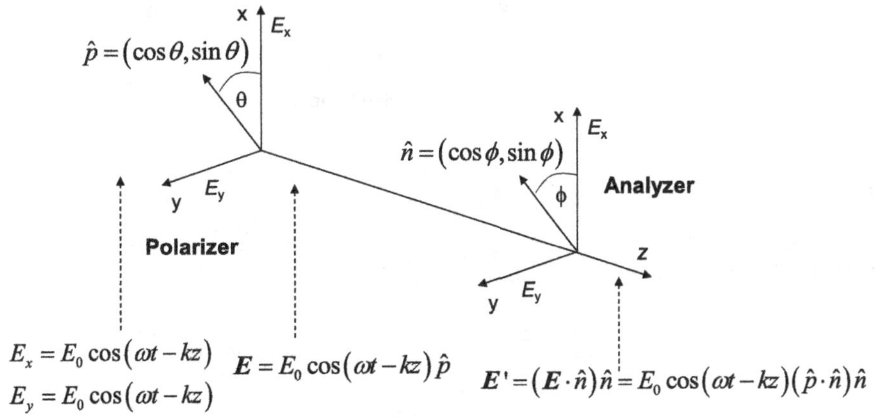

FIGURE 1.4

The polarizer–analyzer ensemble for the study of photon polarization.

The intensity of the analyzer output field is given by

$$I' = |\mathbf{E}'|^2 = I\cos^2(\theta - \phi), \tag{1.21}$$

which is commonly referred to as the *Malus law.*

Polarization decomposition by a birefringent plate is studied in Fig. 1.5. Experiments indicate that photodetectors PD_x and PD_y are never triggered simultaneously, meaning that an entire photon reaches either PD_x or PD_y (a photon never splits). Therefore, the corresponding probabilities that photodetector PDx and PDx detect a photon can be determined by

$$p_x = \Pr(PD_x) = \cos^2\theta, p_y = \Pr(PD_y) = \sin^2\theta. \tag{1.22}$$

If the total number of photons is N, the number of detected photons in x-polarization will be $N_x \cong N\cos^2\theta$, and the number of detected photons in y-polarization will be $N_y \cong N\sin^2\theta$. In the limit as $N \to \infty$, we would expect the Malus law to be obtained.

Let us now study the *polarization decomposition and recombination* by means of birefringent plates, as illustrated in Fig. 1.6. Classical physics prediction of total probability of a photon passing the polarizer-analyzer ensemble is given by

$$p_{\text{tot}} = \cos^2\theta\cos^2\phi + \sin^2\theta\sin^2\phi \neq \cos^2(\theta - \phi), \tag{1.23}$$

which is inconsistent with the Malus law, given by Eq. (1.21). To reconstruct the results from wave optics, it is necessary to introduce the concept into quantum mechanics of the *probability amplitude*

FIGURE 1.5

Polarization decomposition by a birefringent plate. *PD*: photodetector.

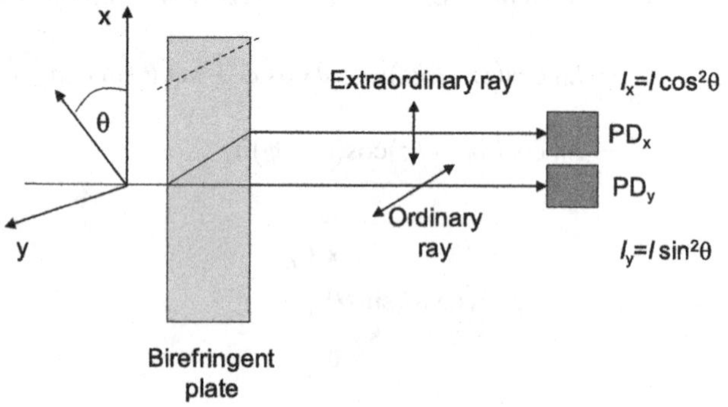

FIGURE 1.6

Polarization decomposition and recombination by a birefringent plate.

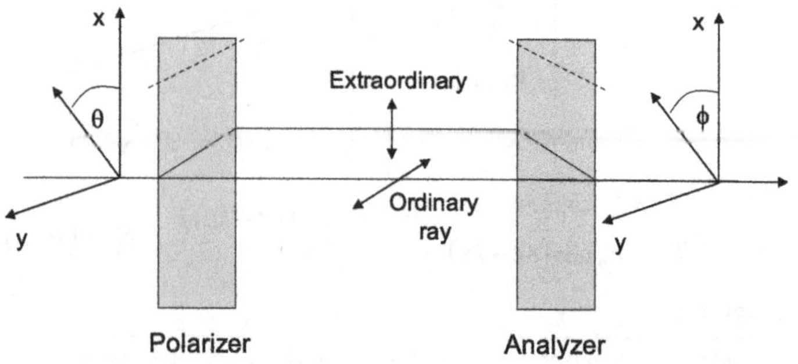

that α is detected as β, which is denoted $a(\alpha \rightarrow \beta)$, which is a complex number. The probability is obtained as the squared magnitude of probability amplitude:

$$p(\alpha \rightarrow \beta) = |a(\alpha \rightarrow \beta)|^2. \tag{1.24}$$

The relevant probability amplitudes related to Fig. 1.6 are summarized as

$$\begin{aligned} a(\theta \rightarrow x) &= \cos\theta \quad a(x \rightarrow \phi) = \cos\phi \\ a(\theta \rightarrow y) &= \sin\theta \quad a(x \rightarrow \phi) = \sin\phi \end{aligned} \tag{1.25}$$

The basic principle of quantum mechanics is to sum up the probability amplitudes for indistinguishable paths to obtain

$$a_{\text{tot}} = \cos\theta\cos\phi + \sin\theta\sin\phi = \cos(\theta - \phi). \tag{1.26}$$

The corresponding total probability is

$$p_{\text{tot}} = |a_{\text{tot}}|^2 = \cos^2(\theta - \phi), \tag{1.27}$$

and this result is consistent with the Malus law!

Based on the previous discussion, *the state vector* of the photon polarization is given by

$$|\psi\rangle = \begin{pmatrix} \psi_x \\ \psi_y \end{pmatrix}, \tag{1.28}$$

where ψ_x is related to x-polarization, and ψ_y is related to y-polarization, with normalization condition as follows:

$$|\psi_x|^2 + |\psi_y|^2 = 1. \tag{1.29}$$

In this representation, the x- and y-polarization photons can be represented by

$$|x\rangle = \begin{pmatrix} 1 \\ 0 \end{pmatrix}, |y\rangle = \begin{pmatrix} 0 \\ 1 \end{pmatrix}, \tag{1.30}$$

and the right and left circular polarization photons are represented by

$$|R\rangle = \frac{1}{\sqrt{2}}\begin{pmatrix} 1 \\ j \end{pmatrix}, |L\rangle = \frac{1}{\sqrt{2}}\begin{pmatrix} 1 \\ -j \end{pmatrix}. \tag{1.31}$$

In Eqs. (1.28)–(1.31), we used Dirac notation to denote the column-vectors (kets). In Dirac notation, with each column vector ("ket") $|\psi\rangle$, we associate a row vector ("bra") $\langle\psi|$ as follows:

$$\langle\psi| = \begin{pmatrix} \psi_x^* & \psi_y^* \end{pmatrix}. \tag{1.32}$$

The *scalar (dot) product* of ket $|\phi\rangle$ bra $\langle\psi|$ is defined by "bracket" as follows:

$$\langle\phi|\psi\rangle = \phi_x^*\psi_x + \phi_y^*\psi_y = \langle\psi|\phi\rangle^*. \tag{1.33}$$

The normalization condition can be expressed in terms of a scalar product by

$$\langle\psi|\psi\rangle = 1. \tag{1.34}$$

Based on Eqs. (1.30) and (1.31), it is evident that

$$\langle x|x\rangle = \langle y|y\rangle = 1, \quad \langle R|R\rangle = \langle L|L\rangle = 1.$$

Because the vectors $|x\rangle$ and $|y\rangle$ are *orthogonal*, their dot product is zero:

$$\langle x|y\rangle = 0, \tag{1.35}$$

and they form the *basis*. Any state vector $|\psi\rangle$ can be written as a *linear superposition* of basis kets as follows:

$$|\psi\rangle = \begin{pmatrix} \psi_x \\ \psi_y \end{pmatrix} = \psi_x|x\rangle + \psi_y|y\rangle. \tag{1.36}$$

We can now use Eqs. (1.34) and (1.36) to derive an important relation in quantum mechanics, known as *completeness relation*. The projections of state vector $|\psi\rangle$ along basis vectors $|x\rangle$ and $|y\rangle$ are given by

$$\langle x|\psi\rangle = \psi_x \underbrace{\langle x|x\rangle}_{1} + \psi_y \underbrace{\langle x|y\rangle}_{0} = \psi_x,\ \langle y|\psi\rangle = \psi_x\langle y|x\rangle + \psi_y\langle y|y\rangle = \psi_y. \tag{1.37}$$

By substituting Eq. (1.37) into Eq. (1.36), we obtain

$$|\psi\rangle = |x\rangle\langle x|\psi\rangle + |y\rangle\langle y|\psi\rangle = \underbrace{(|x\rangle\langle x| + |y\rangle\langle y|)}_{\hat{1}}|\psi\rangle, \tag{1.38}$$

and from the right side of Eq. (1.38), we derive the completeness relation:

$$|x\rangle\langle x| + |y\rangle\langle y| = \hat{1}, \tag{1.39}$$

where $\hat{1}$ denotes the identity operator.

The probability that the photon in state $|\psi\rangle$ will pass the x-polaroid is given by

$$p_x = \frac{|\psi_x|^2}{|\psi_x|^2 + |\psi_y|^2} = |\psi_x|^2 = |\langle x|\psi\rangle|^2, \tag{1.40}$$

and the probability amplitude of the photon in state $|\psi\rangle$ to pass the x-polaroid is

$$a_x = \langle x|\psi\rangle. \tag{1.41}$$

Let $|\phi\rangle$ and $|\psi\rangle$ be two physical states, the probability amplitude of finding ϕ in ψ, denoted as $a(\phi\rightarrow\psi)$, is given by

$$a(\phi\rightarrow\psi) = \langle\phi|\psi\rangle, \tag{1.42}$$

and the probability for ϕ to pass ψ test is given by

$$p(\phi\rightarrow\psi) = |\langle\phi|\psi\rangle|^2. \tag{1.43}$$

1.2 The concept of qubit

Based on the previous section, it can be concluded that the quantum bit, also known as the qubit, lies in a two-dimensional Hilbert space H isomorphic to the C^2-space, where C is the complex number space, and can be represented as

$$|\psi\rangle = \alpha|0\rangle + \beta|1\rangle = \begin{pmatrix}\alpha\\ \beta\end{pmatrix}; \quad \alpha, \beta \in C; \quad |\alpha|^2 + |\beta|^2 = 1, \tag{1.44}$$

where the $|0\rangle$ and $|1\rangle$ states are the computational basis (CB) states, and $|\psi\rangle$ is a superposition state. If we measure a qubit, we will obtain $|0\rangle$ with a probability $|\alpha|^2$ and $|1\rangle$ with a probability of $|\beta|^2$. Measurement changes the state of a qubit from a superposition of $|0\rangle$ and $|1\rangle$ to the specific state consistent with the measurement result. If we parametrize the probability amplitudes α and β as follows:

$$\alpha = \cos\left(\frac{\theta}{2}\right), \quad \beta = e^{j\phi}\sin\left(\frac{\theta}{2}\right), \tag{1.45}$$

where θ is a polar angle, and ϕ is an azimuthal angle. We can geometrically represent the qubit by Bloch sphere (or the Poincaré sphere for the photon) as illustrated in Fig. 1.7. (Note that the Bloch sphere from Fig. 1.7, commonly used in QIP, is different from that of the Poincaré sphere from Fig. 1.2, commonly used in optics.) Bloch vector coordinates are given by ($\cos\phi\ \sin\theta$, $\sin\phi\ \sin\theta$, $\cos\theta$). This Bloch vector representation is related to CB by

$$|\psi(\theta,\phi)\rangle = \cos(\theta/2)|0\rangle + e^{j\phi}\sin(\theta/2)|1\rangle \doteq \begin{pmatrix}\cos(\theta/2)\\ e^{j\phi}\sin(\theta/2)\end{pmatrix}, \tag{1.46}$$

where $0 \le \theta \le \pi$ and $0 \le \phi < 2\pi$. The north and south poles correspond to computational $|0\rangle$ ($|x\rangle$-polarization) and $|1\rangle$ ($|y\rangle$-polarization) basis kets, respectively. Other important bases are the *diagonal basis* $\{|+\rangle, |-\rangle\}$, very often denoted as $\{|\nearrow\rangle, |\searrow\rangle\}$, related to CB by

$$|+\rangle = |\nearrow\rangle = \frac{1}{\sqrt{2}}(|0\rangle + |1\rangle), \quad |-\rangle = |\searrow\rangle = \frac{1}{\sqrt{2}}(|0\rangle - |1\rangle). \tag{1.47}$$

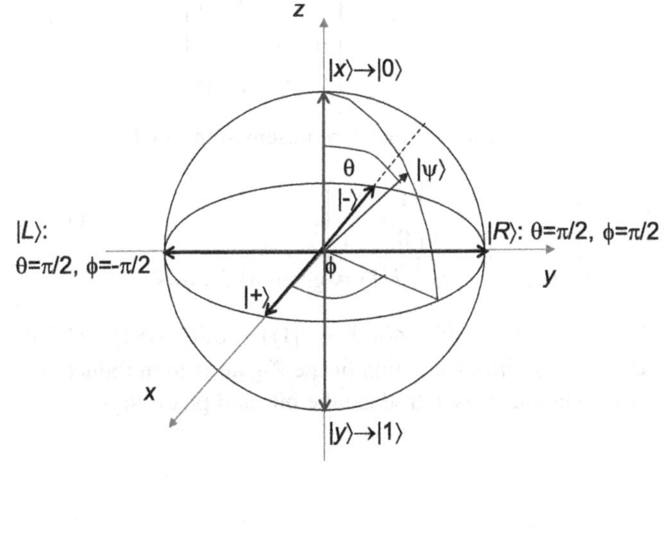

FIGURE 1.7

Bloch (Poincaré) sphere representation of a single qubit.

and the *circular basis* $\{|R\rangle, |L\rangle\}$, related to the CB as follows:

$$|R\rangle = \frac{1}{\sqrt{2}}(|0\rangle + j|1\rangle), \quad |L\rangle = \frac{1}{\sqrt{2}}(|0\rangle - j|1\rangle). \tag{1.48}$$

1.3 Quantum gates and quantum information processing

In quantum mechanics, the primitive, undefined concepts are *physical system, observable,* and *state.* The concept of state has been introduced in previous sections. An observable, such as momentum and spin, can be represented by an *operator,* such as A, in the vector space in question. An operator, or gate, acts on a ket from the left: $(A) \cdot |\alpha\rangle = A|\alpha\rangle$, and results in another ket. A linear operator (gate) B can be expressed in terms of eigenkets $\{|a^{(n)}\rangle\}$ of an Hermitian operator A. (An operator A is said to be *Hermitian* if $A^\dagger = A$, $A^\dagger = (A^T)^*$.) The *operator* X is associated with a *square matrix* (albeit infinite in extent), whose elements are

$$X_{mn} = \langle a^{(m)}|X|a^{(n)}\rangle, \tag{1.49}$$

and can explicitly be written as

$$X \doteq \begin{pmatrix} \langle a^{(1)}|X|a^{(1)}\rangle & \langle a^{(1)}|X|a^{(2)}\rangle & \cdots \\ \langle a^{(2)}|X|a^{(1)}\rangle & \langle a^{(2)}|X|a^{(2)}\rangle & \cdots \\ \vdots & \vdots & \ddots \end{pmatrix}, \tag{1.50}$$

where we use the notation $\doteq$ to denote that operator X is represented by the matrix above.

Very important single-qubit gates are the Hadamard gate H, the phase shift gate S, the $\pi/8$ (or T) gate, the controlled-NOT (or CNOT) gate, and Pauli operators X, Y, and Z. The Hadamard gate H, phase shift gate, T gate, and CNOT gate have the following matrix representation in computational basis (CB) $\{|0\rangle, |1\rangle\}$:

$$H \doteq \frac{1}{\sqrt{2}}\begin{bmatrix} 1 & 1 \\ 1 & -1 \end{bmatrix}, S \doteq \begin{bmatrix} 1 & 0 \\ 0 & j \end{bmatrix}, T \doteq \begin{bmatrix} 1 & 0 \\ 0 & e^{j\pi/4} \end{bmatrix}, CNOT \doteq \begin{bmatrix} 1 & 0 & 0 & 0 \\ 0 & 1 & 0 & 0 \\ 0 & 0 & 0 & 1 \\ 0 & 0 & 1 & 0 \end{bmatrix}. \tag{1.51}$$

The Pauli operators, on the other hand, have the following matrix representation in CB:

$$X \doteq \begin{bmatrix} 0 & 1 \\ 1 & 0 \end{bmatrix}, Y \doteq \begin{bmatrix} 0 & -j \\ j & 0 \end{bmatrix}, Z \doteq \begin{bmatrix} 1 & 0 \\ 0 & -1 \end{bmatrix}. \tag{1.52}$$

The action of Pauli gates on an arbitrary qubit $|\psi\rangle = a|0\rangle + b|1\rangle$ is given as follows:

$$X(a|0\rangle + b|1\rangle) = a|1\rangle + b|0\rangle, Y(a|0\rangle + b|1\rangle) = j(a|1\rangle - b|0\rangle), Z(a|0\rangle + b|1\rangle) = a|0\rangle - b|1\rangle. \tag{1.53}$$

So the action of the X-gate is to introduce the bit-flip, the action of the Z-gate is to introduce the phase-flip, and the action of the Y-gate is to simultaneously introduce the bit- and phase-flips.

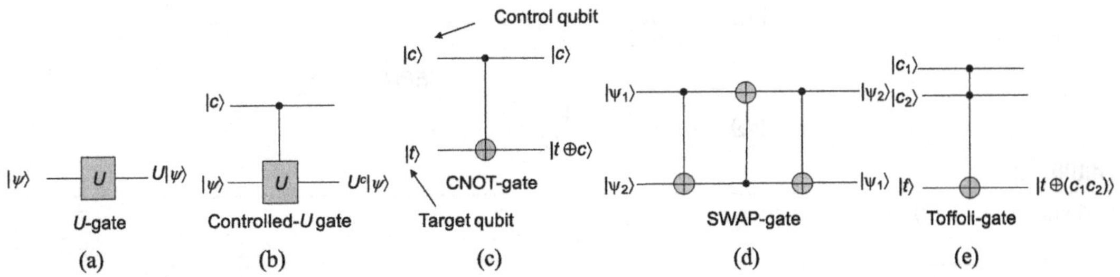

FIGURE 1.8

Important quantum gates and their action: (A) single-qubit gate, (B) controlled-U gate, (C) CNOT-gate, (D) SWAP-gate, and (E) Toffoli gate.

Several important single-, double-, and three-qubit gates are shown in Fig. 1.8. The action of the single-qubit gate is to apply the operator U on qubit $|\psi\rangle$, which results in another qubit. Controlled-U gate conditionally applies the operator U on target qubit $|\psi\rangle$, when the control qubit $|c\rangle$ is in $|1\rangle$-state. One particularly important controlled U-gate is the controlled-NOT (CNOT) gate. This gate flips the content of target qubit $|t\rangle$ when the control qubit $|c\rangle$ is in $|1\rangle$-state. The purpose of SWAP-gate is to interchange the positions of two qubits and can be implemented using three CNOT-gates, as shown in Fig. 1.8D. Finally, the Toffoli gate represents the generalization of CNOT-gate, where two control qubits are used.

The minimum set of gates that can perform an arbitrary quantum computation algorithm is known as the *universal set of gates* [43,44]. The most popular sets of universal quantum gates are {H, S, CNOT, Toffoli} gates, {H, S, $\pi/8$ (T), CNOT} gates, the Barenco gate [45], and the Deutsch gate [46]. With these universal quantum gates, more complicated operations can be performed. As an illustration, Fig. 1.10 shows the Bell state [Einstein–Podolsky–Rosen (EPR) pair] preparation circuit, which is highly important in quantum teleportation and QKD applications.

So far, single-, double-, and triple-qubit quantum gates have been considered. An arbitrary quantum state of K qubits has the form $\sum_s \alpha_s |s\rangle$, where s runs over all binary strings of length K. Therefore, there are 2^K complex coefficients, all independent except the normalization constraint:

$$\sum_{s=00\ldots00}^{11..11} |\alpha_s|^2 = 1. \tag{1.54}$$

For example, the state $\alpha_{00}|00\rangle + \alpha_{01}|01\rangle + \alpha_{10}|10\rangle + \alpha_{11}|11\rangle$ (with $|\alpha_{00}|^2 + |\alpha_{01}|^2 + |\alpha_{10}|^2 + |\alpha_{11}|^2 = 1$) is the general 2-qubit state (we use $|00\rangle$ to denote the tensor product $|0\rangle \otimes |0\rangle$). The multiple qubits can be **entangled**, so they cannot be decomposed into two separate states. For example, the Bell state or EPR pair $(|00\rangle + |11\rangle)/\sqrt{2}$ cannot be written in terms of tensor product $|\psi_1\rangle\ |\psi_2\rangle = (\alpha_1|0\rangle + \beta_1|1\rangle) \otimes (\alpha_2|0\rangle + \beta_2|1\rangle) = \alpha_1\alpha_2|00\rangle + \alpha_1\beta_2|01\rangle + \beta_1\alpha_2|10\rangle + \beta_1\beta_2|11\rangle$, because in that case, the following conditions need to be satisfied $\alpha_1\alpha_2 = \beta_1\beta_2 = 1/\sqrt{2}$ and $\alpha_1\beta_2 = \beta_1\alpha_2 = 0$, which a priori has no reason to be valid. This state can be obtained using the circuit shown in Fig. 1.9, for a two-qubit input state $|00\rangle$.

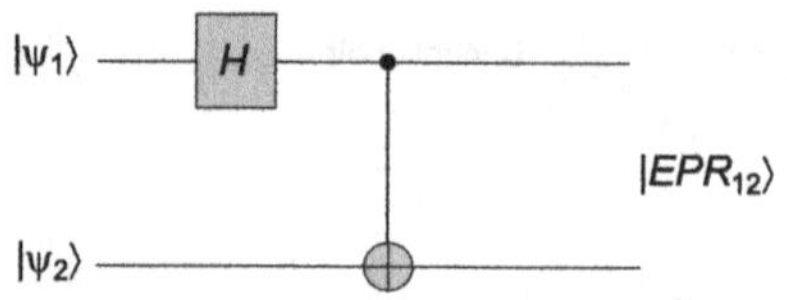

FIGURE 1.9

Bell state (EPR pair) preparation circuit.

1.4 Quantum teleportation

Quantum teleportation [47] is a technique to transfer quantum information from source to destination by employing the entangled states. Namely, in quantum teleportation, the entanglement in the Bell state (EPR pair) is used to transport arbitrary quantum state $|\psi\rangle$ between two distant observers A and B (often called Alice and Bob), as illustrated in Fig. 1.10. The quantum teleportation system employs three qubits, qubit 1 is an arbitrary state to be teleported, while qubits 2 and 3 are in Bell state $|B_{00}\rangle=(|00\rangle+|11\rangle)/\sqrt{2}$. Let the state to be teleported be denoted by $|\psi\rangle = a|0\rangle+b|1\rangle$. The input to the circuit shown in Fig. 1.10 is, therefore, $|\psi\rangle|B_{00}\rangle$, and can be rewritten as

$$|\psi\rangle|B_{00}\rangle = (a|0\rangle + b|1\rangle)(|00\rangle + |11\rangle)/\sqrt{2} = (a|000\rangle + a|011\rangle + b|100\rangle + b|111\rangle)/\sqrt{2}. \quad (1.55)$$

The CNOT-gate is then applied with the first qubit serving as the control and the second qubit as the target, which transforms Eq. (1.55) into

$$\begin{aligned} CNOT^{(12)}|\psi\rangle|B_{00}\rangle &= CNOT^{(12)}(a|000\rangle + a|011\rangle + b|100\rangle + b|111\rangle)/\sqrt{2} \\ &= (a|000\rangle + a|011\rangle + b|110\rangle + b|101\rangle)/\sqrt{2}. \end{aligned} \quad (1.56)$$

In the next stage, the Hadamard gate is applied to the first qubit, which maps $|0\rangle$ to $(|0\rangle+|1\rangle)/\sqrt{2}$ and $|1\rangle$ to $(|0\rangle-|1\rangle)/\sqrt{2}$, so the overall transformation of the quantum state of Eq. (1.55) is as follows:

$$H^{(1)}\, CNOT^{(12)}|\psi\rangle|B_{00}\rangle = \frac{1}{2}(a|000\rangle + a|100\rangle + a|011\rangle + a|111\rangle + b|010\rangle - b|110\rangle + b|001\rangle - b|101\rangle). \quad (1.57)$$

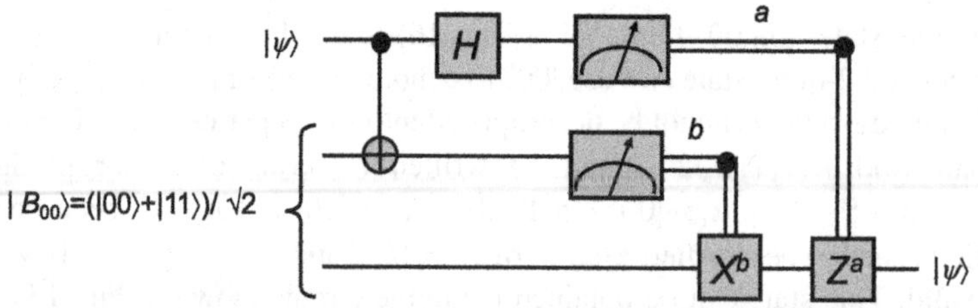

FIGURE 1.10

Illustration of quantum teleportation principle.

The measurements are performed on qubits 1 and 2, and based on the results of measurements, denoted respectively as *a* and *b*, the controlled-*X* (CNOT) and controlled-Z gates are applied conditionally to lead to the following content on qubit 3:

$$\frac{1}{2}(2a|0\rangle + 2b|1\rangle) = a|0\rangle + b|1\rangle = |\psi\rangle, \tag{1.58}$$

indicating that the arbitrary state $|\psi\rangle$ is teleported to the remote destination and can be found at the qubit 3 position.

1.5 Quantum error correction concepts

QIP relies on delicate superposition states that are sensitive to environmental interactions, resulting in decoherence. Moreover, quantum gates are imperfect, and quantum error correction coding (QECC) is necessary to enable fault-tolerant computing and deal with quantum errors [48–53]. QECC is also essential in quantum communication and quantum teleportation applications. The elements of quantum error correction codes are shown in Fig. 1.11A. The (N,K) QECC code performs encoding of the quantum state of K qubits specified by 2^K complex coefficients α_s into a quantum state of N qubits in such a way that errors can be detected and corrected, and all 2^K complex coefficients can be perfectly restored, up to the global phase shift. Namely, we know from quantum mechanics that two states $|\psi\rangle$ and $e^{j\theta}|\psi\rangle$ are equal up to a *global phase shift* as the results of measurement on both states are the same. A quantum error correction consists of four major steps: encoding, error detection, error recovery, and decoding. The sender (Alice) encodes quantum information in state $|\psi\rangle$ with the help of local ancilla qubits $|0\rangle$, and then sends the encoded qubits over a noisy quantum channel (say free-space optical channel or fiber-optics channel). The receiver (Bob) performs multi-qubit measurement on all qubits to diagnose the channel error and performs a recovery unitary operation R to reverse the action of the channel. The quantum error correction is essentially more complicated than classical error correction,

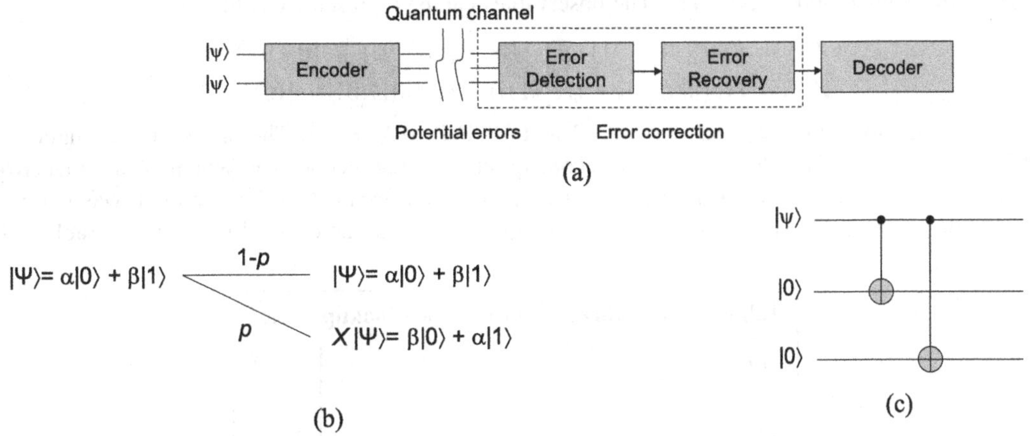

FIGURE 1.11

(A) A quantum error correction principle. (B) Bit-flipping channel model. (C) Three-qubit flip code encoder.

with key differences being summarized as follows: (1) the no-cloning theorem indicates that it is impossible to make a copy of an arbitrary quantum state, (2) quantum errors are continuous, and a qubit can be in any superposition of the two bases states, and (3) the measurements destroy the quantum information. The quantum error correction principles will be more evident after a simple example below.

Assume we want to send a single qubit $|\Psi\rangle = \alpha|0\rangle + \beta|1\rangle$ through the quantum channel in which during transmission the transmitted qubit can be flipped to $X|\Psi\rangle = \beta|0\rangle + \alpha|1\rangle$ with probability p. Such a quantum channel is called a *bit-flip channel,* and it can be described as shown in Fig. 1.11B. Three-qubit flip code sends the same qubit three times and therefore represents the *repetition code* equivalent. The corresponding codewords in this code are $|\overline{0}\rangle = |000\rangle$, $|\overline{1}\rangle = |111\rangle$. The three-qubit flip code encoder is shown in Fig. 1.11C. One input qubit and two ancillae are used at the input encoder, which can be represented by $|\psi_{123}\rangle = \alpha|000\rangle + \beta|100\rangle$. The first ancilla qubit (the second qubit at the encoder input) is controlled by the information qubit (the first qubit at encoder input), so its output can be represented by $CNOT_{12}(\alpha|000\rangle + \beta|100\rangle) = \alpha|000\rangle + \beta|110\rangle$ (if the control qubit is $|1\rangle$ the target qubit gets flipped, otherwise it stays unchanged). The output of the first CNOT gate is used as input to the second CNOT gate in which the second ancilla qubit (the third qubit) is controlled by the information qubit (the first qubit), so the corresponding encoder output is obtained as $CNOT_{13}(\alpha|000\rangle + \beta|110\rangle) = \alpha|000\rangle + \beta|111\rangle$, which indicates that basis codewords are indeed $|\overline{0}\rangle$ and $|\overline{1}\rangle$. With this code, we can correct a single qubit-flip error that occurs with probability $(1-p)^3 + 3p(1-p)^2 = 1 - 3p^2 + 2p^3$. Therefore, the probability of an error remaining uncorrected or wrongly corrected with this code is $3p^2 - 2p^3$. It is clear from Fig. 1.11C that a three-qubit bit-flip encoder is a *systematic encoder* in which the information qubit is unchanged, and the ancilla qubits are used to impose the encoding operation and create the parity qubits (the output qubits 2 and 3).

Let us assume that a qubit flip occurred on the first qubit, leading to received quantum word $|\Psi_r\rangle = \alpha|100\rangle + \beta|011\rangle$. To identify the error, measurements must be performed on the observables Z_1Z_2 and Z_2Z_3, where the subscript denotes the index of the qubit on which a given Pauli gate is applied. The measurement result is eigenvalue ±1, and the corresponding eigenvectors are two valid codewords, namely $|000\rangle$ and $|111\rangle$. The observables can be represented as follows:

$$\begin{aligned} Z_1Z_2 &= (|00\rangle\langle 11| + |11\rangle\langle 11|) \otimes I - (|01\rangle\langle 01| + |10\rangle\langle 10|) \otimes I \\ Z_2Z_3 &= I \otimes (|00\rangle\langle 11| + |11\rangle\langle 11|) - I \otimes (|01\rangle\langle 01| + |10\rangle\langle 10|) \end{aligned} \tag{1.59}$$

It can be shown that $\langle\psi_r|Z_1Z_2|\psi_r\rangle = -1$ and $\langle\psi_r|Z_2Z_3|\psi_r\rangle = +1$. The first syndrome indicates that an error occurred on either the first or second qubit while the second syndrome indicates that error is not located on the second or third qubits. The intersection reveals that the first qubit was in error. Using this approach, we can create the three-qubit look-up table (LUT), given as Table 1.1.

Table 1.1 The three-qubit flip-codc lookup table.

$Z_1 Z_2$	$Z_2 Z_3$	Error
+1	+1	I
+1	−1	X_3
−1	+1	X_1
−1	−1	X_2

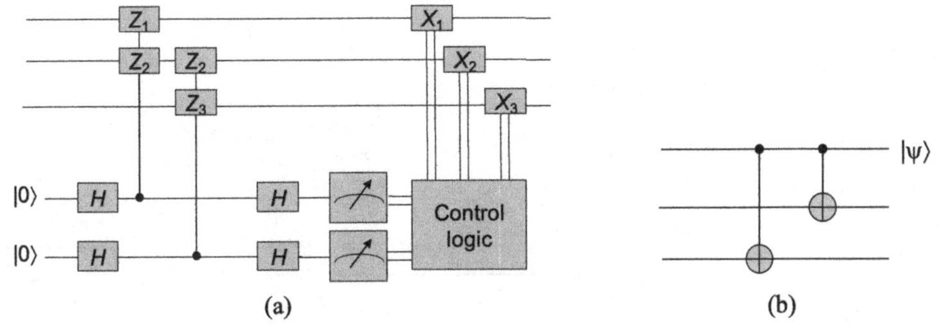

FIGURE 1.12

(A) Three-qubit flip code error detection and error correction circuit. (B) Decoder circuit configuration.

Three-qubit flip code error detection and error correction circuits are shown in Fig. 1.12. The results of measurements on ancillae, as illustrated in Fig. 1.12A, will determine the error syndrome $[\pm 1 \pm 1]$, and based on the LUT given by Table 1.1, we identify the error event and apply the corresponding X_i gate on the ith qubit being in error, and the error is corrected since $X^2 = I$. The control logic operation is described in Table 1.1. For example, if both outputs at the measurement circuits are -1, the operator X_2 is activated. The last step is decoding, as shown in Fig. 1.12B, by simply reversing the order of elements in the corresponding encoder.

1.6 Quantum sensing

As an illustration of quantum sensing applications, the conceptual schematic of the distributed quantum sensing (DQS) concept [20,54,55] is provided in Fig. 1.13. Based on initial quantum state $\rho_0 = |\psi_0\rangle\langle\psi_0|$, the source quantum module generates a multipartite entangled probe state that is shared by M sensors to scan an object of interest. The measurement results obtained by M sensors are (classically) postprocessed to determine a certain global parameter of the scanned object. On the other hand, in distributed classical sensing, the quantum state of M sensors can be separated; that is, it can be written as a tensor product $\rho_1 \otimes \cdots \otimes \rho_M$. The distributing sensing scenario can be described by a unitary operator being the product of M unitary operators, that is, $U\left(\widetilde{\boldsymbol{\theta}}\right) = \otimes_{m=1}^{M} U\left(\widetilde{\theta}_m\right)$, wherein each unitary measures one parameter. The output quantum state can be described by the following transformation:

$$\rho_M(\boldsymbol{\theta}) = U(\boldsymbol{\theta})\rho_M U^{\dagger}(\boldsymbol{\theta}),\ \boldsymbol{\theta} = [\theta_1, \cdots, \theta_M], \tag{1.60}$$

where ρ_M is the input entangled probe state. In DQS, we are interested in estimating the global parameter rather than estimating all unknown parameters. This can be achieved by calculating the weighted average as follows:

$$\overline{\theta} = \sum_{m=1}^{M} w_m \theta_m = \boldsymbol{w} \cdot \boldsymbol{\theta}^{\mathrm{T}},\ \boldsymbol{w} = [w_1, \cdots, w_M],\ w_m \geq 0,\ \sum_{m=1}^{M} w_m = 1. \tag{1.61}$$

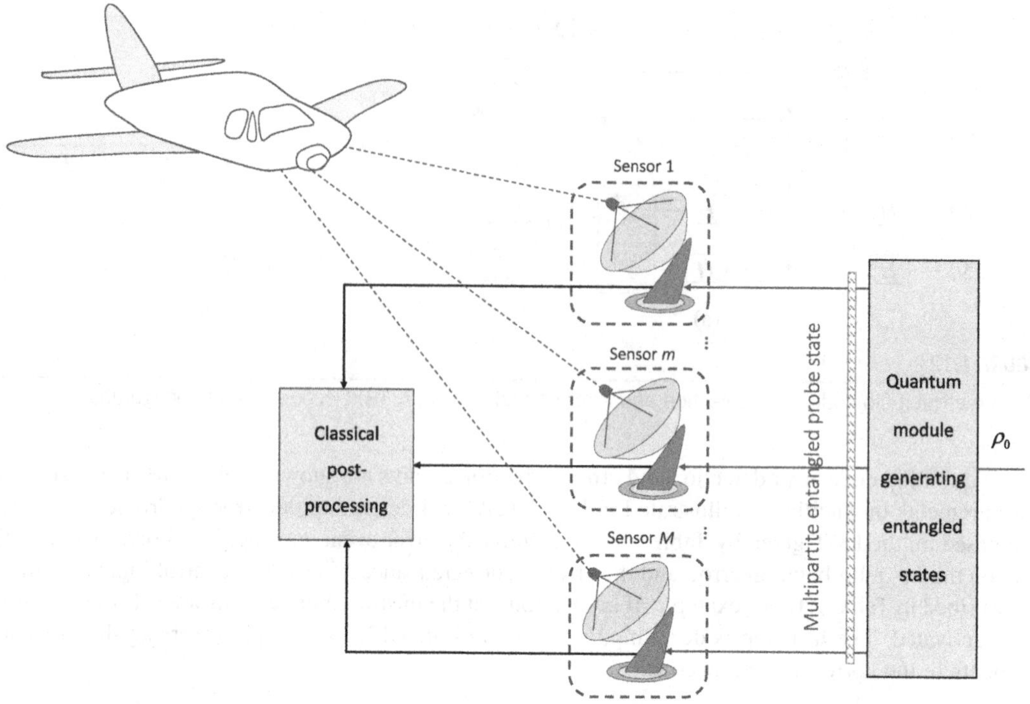

FIGURE 1.13

Illustrating the distributed quantum sensing concept for entangled inputs.

In the special case where $w_m = 1/M$ and $\theta_m = \theta$, using the entangled probe state, we can beat the SQL and achieve the Heisenberg limit. For additional details, an interested reader is referred to Chapter 11.

1.7 Quantum key distribution

QKD exploits the principle of quantum mechanics to enable the provably secure distribution of a private key between remote destinations. Private key cryptography is much older than the public key cryptosystems commonly used today. In a private key cryptosystem, Alice (sender) must have an *encoding key*, while Bob (receiver) must have a matching *decoding key* to decrypt the encoded message. The simplest private key cryptosystem is the *Vernam cipher* (*one-time pad*), which operates as follows [8]: (1) Alice and Bob share n-bit key strings, (2) Alice encodes her n-bit message by adding the message and the key together, and (3) Bob decodes the information by subtracting the key from the received message. The are several *drawbacks* of this scheme: (1) secure distribution of the key as well as the key length must be at least as long as the message length, (2) the key bits cannot be reused, and (3) the keys must be delivered in advance, securely stored until use, and destroyed after use.

On the other hand, if the classical information, such as key, is transmitted over the quantum channel, thanks to the no-cloning theorem, which states that a device cannot be constructed to produce an exact copy of an arbitrary quantum state, the eavesdropper cannot get an exact copy of the key in a QKD system. Namely, in an attempt to distinguish between two nonorthogonal states, information gain is only possible at the expense of introducing the disturbance to the signal. This observation can be proved as follows. Let $|\psi_1\rangle$ and $|\psi_2\rangle$ be two nonorthogonal states Eve is interested to learn about and $|\alpha\rangle$ be the standard state prepared by the Eve. The Eve tries to interact with these states without disturbing them, and as the result of this interaction, the following transformation is performed [8]:

$$|\psi_1\rangle \otimes |\alpha\rangle \rightarrow |\psi_1\rangle \otimes |\alpha\rangle, \ |\psi_2\rangle \otimes |\alpha\rangle \rightarrow |\psi_2\rangle \otimes |\beta\rangle. \tag{1.62}$$

The Eve hopes that the states $|\alpha\rangle$ and $|\beta\rangle$ would be different, in an attempt to learn something about the states. However, any unitary transformation must preserve the dot product, so from Eq. (1.60), we obtain

$$\langle\alpha|\beta\rangle\langle\psi_1|\psi_2\rangle = \langle\alpha|\alpha\rangle\langle\psi_1|\psi_2\rangle, \tag{1.63}$$

which indicates that $\langle\alpha|\alpha\rangle = \langle\alpha|\beta\rangle = 1$, and consequently $|\alpha\rangle = |\beta\rangle$. Therefore, an attempt to distinguish between $|\psi_1\rangle$ and $|\psi_2\rangle$, Eve will disturb them. The key idea of QKD, therefore, is to transmit nonorthogonal qubit states between Alice and Bob, and by checking for disturbances in the transmitted state, they can establish an upper bound on the noise/eavesdropping level in their communication channel.

One of the simplest QKD protocols is the *BB84 protocol*, named after Bennett and Brassard, who proposed it in 1984 [56,57]. The BB84 protocol can be implemented using different degrees of freedom (DOF), including the polarization DOF [58] or the phase of the photons [59]. Experimentally, the BB84 protocol has been demonstrated over both fiber-optics and free-space optical (FSO) channels [58,60]. The polarization-based BB84 protocol over the fiber-optics channel is affected by polarization mode dispersion (PMD), polarization-dependent loss (PDL), and fiber loss, which affect the transmission distance. To extend the transmission distance, phase encoding is employed as in Ref. [60]. Unfortunately, the secure key rate over 405 km of ultra-low-loss fiber is extremely low, only 6.5 b/s. In an FSO channel, the polarization effects are minimized; however, the atmospheric turbulence can introduce the wavefront distortion and random phase fluctuations. Previous QKD demonstrations include a satellite-to-ground FSO link demonstration and a demonstration over an FSO link between two locations in the Canary Islands [58].

Two bases that are used in the BB84 protocol are the computational basis CB = $\{|0\rangle, |1\rangle\}$ and the diagonal basis:

$$DB = \{|+\rangle = (|0\rangle + |1\rangle)/\sqrt{2}, \ |-\rangle = (|0\rangle - |1\rangle)/\sqrt{2}\}. \tag{1.64}$$

These bases belong to the class of *mutually unbiased bases* (MUBs) [61–64]. The keyword *unbiased* means that if a system is prepared in a state belonging to one of the bases, all measurement outcomes with respect to the other basis are equally likely. Alice randomly selects the MUB, followed by random selection of the basis state. The logical 0 is represented by $|0\rangle$, $|+\rangle$; while the logical 1 is represented by $|1\rangle$, $|-\rangle$. Bob measures each qubit by randomly selecting the basis, computational or diagonal. In a sifting procedure, Alice and Bob announce the bases used for each qubit and keep only instances when they have used the same basis. The *polarization-based BB84 QKD protocol* employs four corresponding states as shown in Fig. 1.14. The bulky optics implementation of the BB84 protocol is illustrated in Fig. 1.15.

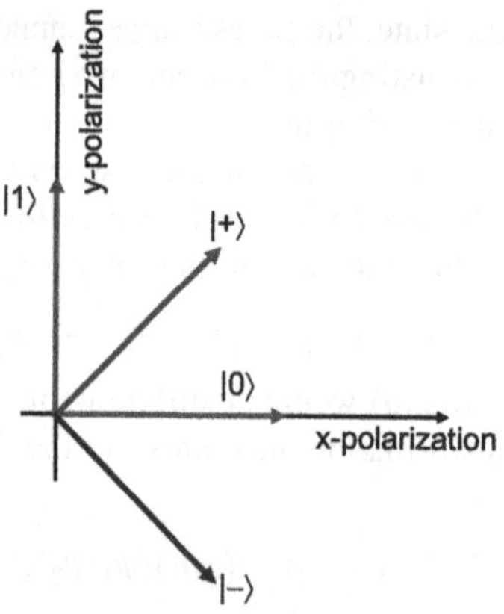

FIGURE 1.14

The four states being employed in the BB84 protocol.

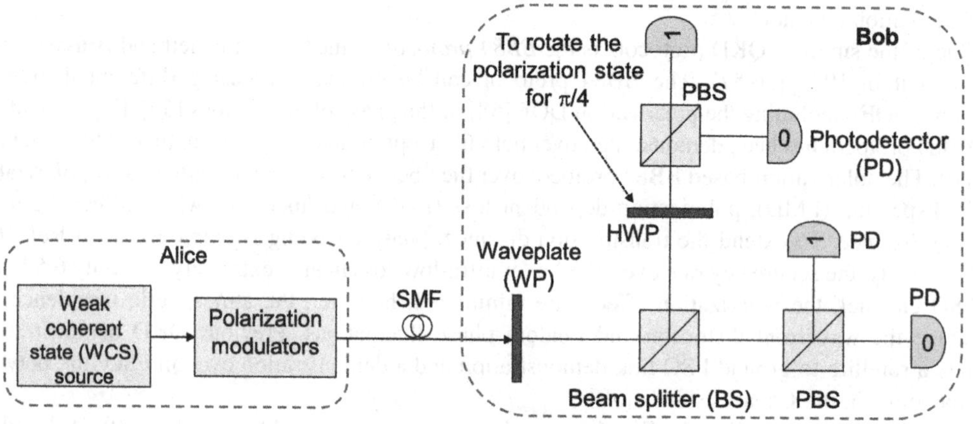

FIGURE 1.15

Illustrating the polarization-based BB84 protocol. *PBS*: polarization beam splitter. *PD*: single-photon detector.

Once the QKD protocol is completed, transmitter A and receiver B perform a series of classical steps, as illustrated in Fig. 1.16. As the transmission distance increases and for higher key distribution speeds, error correction becomes increasingly important. By performing the information reconciliation by low-density parity-check (LDPC) codes, the higher input bit-error rates (BERs) can be tolerated compared to other coding schemes. The *privacy amplification* is further performed to eliminate any information obtained by an eavesdropper. Privacy amplification must be performed between Alice and Bob to distill a smaller set of bits from the generated key whose correlation with Eve's string is below the desired threshold. One way to accomplish privacy amplification is through universal hash functions. The simplest family of hash functions is based on multiplication with a random element of the Galois field $GF(2^m)$ $(m > n)$, where n denotes the number of bits remaining after information

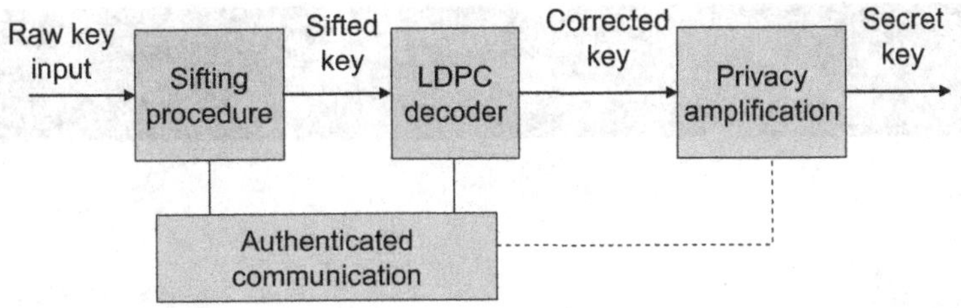

FIGURE 1.16

Classical postprocessing steps.

reconciliation is completed. The threshold for the maximum tolerable error rate is dictated by the efficiency of the implemented information reconciliation and privacy amplification protocols. The reader interested in learning about QKD with more details is referred to Chapter 6.

1.8 Quantum networking

The QuCom is the cornerstone to fully exploit the power of QIP to distribute entanglement at a distance to interconnect quantum devices. However, the distribution of entanglement over long distances has been an outstanding challenge due to photon losses—i.e., quantum signals cannot be amplified without introducing additional noise that degrades or destroys the transmitted entanglement. Hence, QuCom calls for fundamentally distinct loss-mitigation mechanisms to establish long-range entanglement. In this regard, quantum repeaters have been pursued over the last decade to overcome the exponentially low entanglement distribution rate versus transmission distance in optical fiber. Despite encouraging advances of using silicon vacancy (SiV) defects in diamond quantum memory to surpass the repeaterless key-generation bound [65], several technological hurdles, including the scalability of quantum devices, the indistinguishability of emitted photons, and quantum error correction, have to be overcome before the development of fully functional quantum repeaters for long-distance QuCom. As an alternate solution, satellites have been leveraged as relays for QuCom over thousands of kilometers by virtue of the more favorable quadratic scaling of photon loss versus the distance in FSO links. Remarkably, QKD [59], quantum teleportation [66], and entanglement distribution [67] have been demonstrated in a ground-to-satellite QuCom testbed across several thousand kilometers, showing tremendous potential for building up satellite-mediated wide-area QuCom networks via FSO links. As another ubiquitous use case, FSO links would grant ad hoc quantum devices access to QuCom networks, akin to the internet connecting billions of mobile devices. As such, the envisioned future quantum internet will interconnect various genres of quantum devices through heterogeneous fiber, ground-to-satellite links, and FSO links as illustrated in Fig. 1.17.

At present, while QuCom is individually validated over a specific type of quantum link, heterogeneous quantum interconnect that enables full integration of diverse quantum modules into a unified QuCom network remains elusive. To implement such a quantum Internet infrastructure, various quantum interconnects and quantum interfaces are needed. The *quantum interface* can be defined as a quantum module or subsystem that connects the stationary qubits, typically located in quantum memories, and flying (mobile) qubits to establish a quantum channel between any two remote nodes.

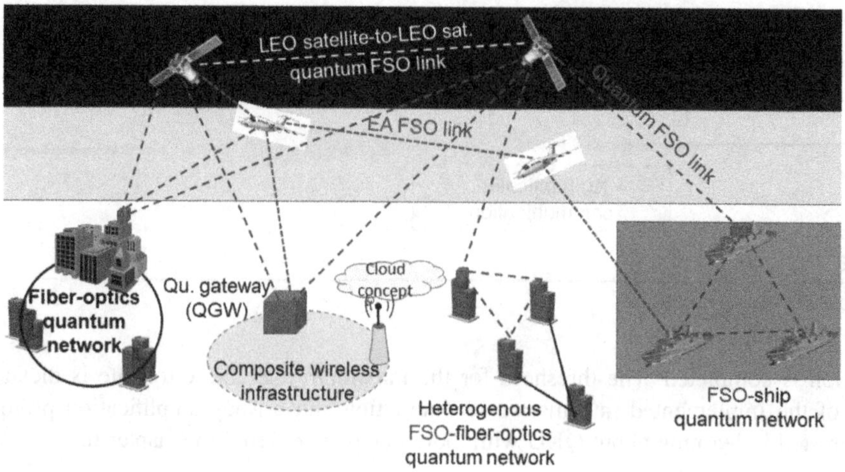

FIGURE 1.17

Envisioned wide-area quantum communication network empowered by heterogeneous quantum interconnects. EA: entanglement assisted communication.

There are two outstanding challenges to be solved before such a global QCN (or quantum Internet) becomes realizable. First, photon loss fundamentally limits the entanglement distribution rate because quantum signals cannot be amplified without introducing additional noise that degrades or destroys the transmitted entanglement. Second, QIP devices rest upon stationary quantum information carried by, e.g., solid-state spin states, superconducting qubits, and ultracold atoms, whereas entanglement distribution hinges on flying quantum information encoded on photons. To interconnect QIP devices, the quantum interconnect modules must convert back and forth between stationary and flying quantum information. Some possible solutions to the various challenges that current heterogeneous QCNs are facing are described in Chapter 10.

1.9 Organization of the book

This section is related to the organization of the book. This book covers various topics of quantum information processing, quantum communication, quantum error correction, quantum sensing, and quantum networking. The chapters of the book are organized as described below.

Chapter 2 is devoted to the basic concepts of information theory, coding theory, and detection and estimation theory. Section 2.1 provides the definitions of entropy, joint entropy, conditional entropy, relative entropy, mutual information, and channel capacity, followed by the information capacity theorem. We also briefly describe the source coding and data compaction concepts. Further, we discuss the channel capacity of discrete memoryless, continuous, and optical channels. The standard FEC schemes that belong to the class of hard-decision codes are described as well. More powerful FEC schemes belong to the class of soft iteratively decodable codes, but their description is beyond the scope of this chapter. Iteratively decodable codes, such as LDPC codes, are described in Chapter 9. In Section 2.2 we introduce classical channel coding preliminaries. Section 2.3 is devoted to the basics of linear block codes (LBCs), such as the definition of the generator and parity-check matrices, syndrome decoding, distance properties of LBCs, and some important coding bounds. In Section 2.4, cyclic codes are introduced. BCH codes are described in Section 2.5. RS, concatenated, and product codes are described in Section 2.6. The fundamentals of detection and estimation theory are provided in Section 2.7.

Chapter 3 is devoted to QIP fundamentals. The basic QIP features are discussed in Section 3.1. The basic concepts of quantum mechanics such as state vectors, operators, projection operators, and density operators are introduced in Section 3.2. In Section 3.3, the measurements, uncertainty relations, and dynamics of a quantum system are discussed. The superposition principle and quantum parallelism concept are introduced in Section 3.4, together with QIP basics. The no-cloning theorem is formulated and proved in Section 3.5. Further, the theorem of the inability to distinguish nonorthogonal quantum states is formulated and proved in Section 3.6. Section 3.7 is devoted to various aspects of entanglement, including the Schmidt decomposition theorem, purification, superdense coding, and entanglement swapping. Various entanglement measures are introduced, including the entanglement of formation and concurrence. How entanglement can be used in detecting and correcting quantum errors is described as well. Further, in Section 3.8, the operator-sum representation of a quantum operation is introduced and applied to the bit-flip and phase-flip channels. In Section 3.9, quantum errors and decoherence effects are discussed, and phase damping, depolarizing, and amplitude damping channels are described.

Chapter 4 is devoted to the quantum information theory fundamentals. After introductory remarks in Section 4.1, the chapter introduces classical and von Neumann entropies definitions and properties, followed by the quantum representation of classical information in Section 4.2. Next, in Section 4.3, we introduce accessible information as the maximum mutual information over all possible measurement schemes. In the same section, we also introduce an important bound, the Holevo bound, and prove it. In Section 4.4, we introduce the typical sequence, typical state, typical subspace, and projector on the typical subspace. Further, we formulate and prove Schumacher's source coding theorem, the quantum equivalent of Shannon's source coding theorem. In the same section, we introduce the concept of quantum compression and decompression as well as the concept of reliable compression. In Section 4.5, we describe various quantum channel models including bit-flip, phase-flip, depolarizing, and amplitude damping channel models. In Section 7.6, we formulate and prove the Holevo–Schumacher–Westmoreland theorem, the quantum equivalent of Shannon's channel capacity theorem.

Chapter 5 is devoted to quantum detection theory, quantum communications, and Gaussian quantum information theory. The density operators are briefly reviewed in Section 5.1 before describing the quantum detection theory concepts in Section 5.2. This theory is then applied to the binary detection problem in Section 5.3. The coherent states are then described in Section 5.4, followed by the introduction of quadrature operators and derivation of the uncertainty principle for continuous-variable systems. In Section 5.5, the binary quantum optical communications in the absence of noise are discussed, including photon-counting receiver (Subsections 5.5.1 and 5.5.2), OOK receiver (Subsection 5.5.3), BPSK receiver (Subsection 5.5.4), Kennedy receiver (Subsection 5.5.5), and Dolinar receiver (Subsection 5.5.6). The P-representation is further introduced in Section 5.6 and applied to represent the thermal noise and the coherent state in the presence of thermal noise. Binary optical communication in the presence of noise is discussed in Section 5.7, with a focus on OOK and BPSK. The Gaussian and squeezed states are then introduced in Section 5.8, followed by

Wigner function definition as well as the definition of covariance matrices in Subsection 5.8.1. The Gaussian transformation and Gaussian channels are discussed in Subsection 5.8.2, with beam splitter operation, phase rotation operation, and squeezing operator as the representative examples. The thermal decomposition of Gaussian states is discussed in Subsection 5.8.3, and the von Neumann entropy for thermal states is derived. The covariance matrices for two-mode Gaussian states are discussed in Subsection 5.8.4, and how to calculate the symplectic eigenvalues, relevant in von Neumann entropy calculation. The Gaussian states measurements and detection are discussed in Subsection 5.8.5, emphasizing homodyne detection, heterodyne detection, and partial measurements. In Section 5.9, the focus moves to nonlinear quantum optics fundamentals, particularly three-wave and four-wave mixing. In the same section, the generation of the Gaussian states is discussed, particularly the two-mode squeezed state generation is discussed in detail. Section 5.10 is devoted to multilevel quantum optical communication, wherein the scenarios in the absence and presence of noise are described. The square root measurements approach to analyze the performance of these schemes is described in detail in Subsection 5.10.1. This method is then applied to the signal constellations with the geometrically uniform symmetry such as M-ary PSK in Subsection 5.10.2. Finally, Subsection 5.10.3 is devoted to the M-ary quantum communications in the presence of noise.

Chapter 6 is related to the fundamental concepts of QKD. The main differences between conventional cryptography and QKD are described in Section 6.1. In the section on QKD basics (Section 6.2), after a historical overview, different QKD types are briefly introduced. Two fundamental theorems on which QKD relies on, namely the no-cloning theorem and the theorem of inability to unambiguously distinguish nonorthogonal quantum states, are described in Section 6.3. In Section 6.4, the DV-QKD systems, the BB84, B92, Ekert (E91), EPR, and time-phased encoding protocols are described in corresponding Subsections 6.4.1–6.4.4. Section 6.5 is devoted to QKD security, wherein the secret-key rate (SKR) is represented as the product of raw key rate and fractional rate. Moreover, the generic expression for the fractional rate is provided, followed by a description of different eavesdropping strategies, including individual attacks (Section 6.5.1), collective attacks (Section 6.5.2), and quantum hacking/side-channel attacks (Section 6.5.3). For individual and coherent attacks, the corresponding secrete fraction expressions are provided, including the secret fraction for the BB84 protocol in Section 6.5.4. After that, in Section 6.6, the decoy-state protocol is described together with the corresponding SKR calculation. Next, the key concepts for measurement-device-independent (MDI)-QKD protocols are introduced in Section 6.7, including photonic Bell state measurements (Section 6.7.1), polarization-based MDI-QKD protocol (Section 6.7.2), and time-phased encoding-based MDI-QKD protocol (Section 6.7.3), while the secrecy fraction calculation is provided in Section 6.7.4. The twin-field QKD protocols are further described in Section 6.8, whose performance is evaluated against decoy-state and MDI-QKD protocols. The focus then moves to classical processing steps in Section 6.9, including information reconciliation (Section 6.9.1) and privacy amplification (Section 6.9.2) steps. To facilitate the description of continuous-variable (CV)-QKD protocols, the fundamentals of quantum optics and Gaussian information theory are introduced in Section 6.10. Among different topics, quadrature operators, Gaussian states, and squeezed states are described in Section 6.10.1, Gaussian transformation and generation of quantum states are introduced in Section 6.10.2, while the thermal decomposition of Gaussian states in Section 6.10.3. Finally, the two-mode Gaussian states and the measurements on Gaussian states are discussed in Section 6.10.4. Section 6.11 is devoted to the CV-QKD protocols, wherein the homodyne and heterodyne detection schemes are described in Section 6.11.1, following the brief description of the squeezed states-based protocols in Section 6.11.2. Given that coherent states are much easier to generate and manipulate, coherent state-based protocols are described in detail in Section 6.11.3. For lossy transmission channels, the corresponding covariance matrices are derived for both homodyne and coherent detections schemes, followed by the secret-key rate derivation for prepare-and-measure Gaussian modulation (GM)-based CV-QKD in Section 6.11.4. Some illustrative reconciliation results are provided in Section 6.11.5, related to the GM-based CV-QKD schemes.

Chapter 7 is devoted to the quantum error correction concepts, ranging from an intuitive description to a rigorous mathematical framework. After the introduction of Pauli operators, we describe basic quantum codes, such as three-qubit flip code, three-qubit phase flip code, Shor's nine-qubit code, stabilizer codes, and CSS codes. We then formally introduce the quantum error correction, including the quantum error correction mapping, quantum error representation, stabilizer group definition, quantum-check matrix representation, and quantum syndrome equation. We then provide the necessary and sufficient conditions for quantum error correction, discuss the distance properties and error correction capability, and revisit the CSS codes. Section 7.4 is devoted to important quantum coding bounds including the quantum Hamming bound, Gilbert–Varshamov bound, and Singleton bound. In the same section, we discuss the quantum weight enumerators and quantum MacWilliams identities. We further discuss the quantum superoperators and various quantum channels, including quantum depolarizing, amplitude damping, and generalized amplitude damping channels.

Chapter 8 is related to the stabilizer codes and their related relatives. The stabilizer codes are introduced in Section 8.1. Their basic properties are discussed in Section 8.2. Encoded operations are introduced in the same section (Section 8.2). In Section 8.3, finite geometry interpretation of stabilizer codes is introduced. This representation is used in Section 8.4 to introduce the so-called standard form of stabilizer code. The standard form is the basic representation for the efficient encoder and decoder implementations, which is discussed in Section 8.5. Section 8.6 is devoted to nonbinary stabilizer code, generalizing the previous sections. The subsystem codes are introduced in Section 8.7. In the same section, efficient encoding and decoding of subsystem codes are also discussed. An important class of quantum codes, namely the topological codes is discussed in Section 8.8. Section 8.9 is devoted to the surface codes. The entanglement-assisted codes are described in Section 8.10.

Chapter 9 is devoted to the quantum LDPC codes. In Section 9.1, we describe classical LDPC codes, their design, and decoding algorithms. In Section 9.2, we describe dual-containing quantum LDPC codes. Various design algorithms are described. Section 9.3 is devoted to entanglement-assisted quantum LDPC codes. Various classes of finite geometry codes are described. In Section 9.4, we describe the probabilistic sum-product algorithm based on the quantum-check matrix. The quantum spatially coupled LDPC codes are introduced in Section 9.5.

Chapter 10 is devoted to discussing how to build a future quantum network. The basic concepts of quantum communication networks and quantum Internet are introduced in Section 10.1, followed by quantum teleportation and quantum relay concepts in Section 10.2. Entanglement distribution, a key technique enabling quantum networking by employing the quantum teleportation concept, is described in detail in Section 10.3. The section starts with a brief description of how to generate the entangled states. The entanglement swapping, Bell state measurements, and Hong–Ou–Mendel effect are further described in Subsections 10.3.1 and 10.3.2. Continuous variable quantum teleportation is described in Subsection 10.3.3, followed by the basics of quantum network coding in Subsection

10.3.4. Section 10.4 is devoted to describing how to engineer the entangled states by photon addition and photon subtraction. The hybrid CV-DV networking concept is introduced in Subsection 10.4.1. After describing photon subtraction and addition concepts in Subsection 10.4.2, we describe how to apply them to build a hybrid CV−DV quantum network. The following topics on hybrid quantum networks are described: generation of hybrid DV-CV entangled states (in Subsection 10.4.3), hybrid CV-DV states teleportation (Subsection 10.4.4), and entanglement swapping through entangling measurements (Subsection 10.4.4). Next, the generation of entangled macroscopic light states is discussed in Subsection 10.4.5, given that macroscopic light states are much more tolerant than DV states to channel loss. The section ends with Subsection 10.4.6 on noiseless amplification, which represents a possible approach to extend the transmission distance between quantum nodes. The focus of the chapter then moves to cluster state-based quantum networking in Section 10.5. The cluster states concept is introduced first, followed by cluster state processing, both in Subsection 10.5.1. The cluster state-based quantum networking concept is described in Subsection 10.5.3 as both quantum Internet and distributed quantum computing enabling technology. In Section 10.6, surface code-based and quantum LDPC code-based quantum networking concepts are described. The final section of the chapter (Section 10.7) is devoted to entanglement-assisted communication networks.

Chapter 11 is devoted to quantum sensing and quantum radar. In Section 11.1 the quantum phase estimation problem is addressed, and the physical limits on phase measurements are described. The quantum interferometry is described in Section 11.1.1 and applied to the sparable quantum probe to derive the standard quantum limit. In Section 11.1.2, the highly entangled NOON states are used as the quantum probe to achieve the Heisenberg limit. The quantum Fisher information is introduced in Section 11.2 and used to derive the quantum Cramér-Rao bound, representing the lower bound for the precision of quantum sensing. In the same section, the classical Cramér-Rao bound is derived as well. Section 11.3 is devoted to distributed quantum sensing in which multiple quantum sensors have been used to measure a global parameter. It is demonstrated that the optimum entangled probes can beat the standard quantum limit and achieve the Heisenberg limit. Distributed quantum sensing (DQS) is employed to determine quadrature displacements. The performance limits for DQS are derived as well. Section 11.4 is focused on quantum radars. After a brief description of common quantum radar architectures, the following promising quantum radar technologies are described: interferometric quantum radars (Section 11.4.1), quantum illumination-based quantum radars (Section 11.4.2), and entanglement assisted target detection (Section 11.4.3). The quantum radar equation is discussed in Section 11.4.4.

Chapter 12 is devoted to the relevant both classical and quantum machine learning algorithms. In Section 12.1.1, the relevant classical machine learning algorithms, categorized as supervised, unsupervised, and reinforcement learning algorithms, are introduced briefly. In the same section, the concept of the feature space is introduced as well as the corresponding generalization performance of machine learning algorithms. The following topics from classical machine learning are described: principal component analysis (PCA), support vector machines, clustering, boosting, regression analysis, and neural networks. In the PCA section (Section 12.1.2), we describe how to determine the principal components from the correlation matrix, followed by a description of the singular value decomposition-based PCA as well as the PCA scoring phase. Section 12.1.3 is related to the support vector machines (SVM)-based algorithms, which are described from both geometric and Lagrangian method-based points of view. For classification problems that are not linearly separable, the concept of soft margins is introduced. For classification problems with nonlinear decision boundaries, the kernel method is described. Section 12.1.4 is devoted to the clustering problems. The following clustering algorithms are described: K-means (Section 12.1.4.1), expectation-maximization (Section 12.1.4.3), and K-nearest neighbors (Section 12.1.4.4). In the same section, the clustering quality evaluation (Section 12.1.4.2) is discussed. In boosting section (Section 12.1.5), the procedure how to build the strong learner from weak learners is described, with special attention devoted to the AdaBoost algorithm. In the regression analysis Section (12.1.6), the least square estimation, pseudoinverse approach, and ridge regression methods are described. Section 12.1.7 is related to artificial neural networks. The following topics are described in detail: the perceptron (Section 12.1.7.1), activation functions (Section 12.1.7.1), and feedforward networks (Section 12.1.7.2). The focus is then moved to the quantum machine learning (QML) algorithms. In Section 12.2, the Ising model is introduced, followed by the adiabatic computing and quantum annealing. In Subsection 12.2.1, the Ising model is related to the quadratic unconstrained binary optimization (QUBO) problem. The focus is moved then to solving the QUBO problem by adiabatic quantum computing and quantum annealing in corresponding Subsections 12.22 and 12.2.3, respectively. The variational quantum eigensolver and quantum approximate optimization algorithm (QAOA) are discussed in Section 12.3 as approaches to performing QML using imperfect and noisy quantum circuits. To illustrate its impact, QAOA is applied in the same section to combinatorial optimization and MAX-CUT problems. The quantum boosting is discussed in Section 12.4, and it is related to the QUBO problem. Section 12.5 is devoted to the quantum random access memory, allowing us to address the superposition of memory cells with the help of a quantum register. The quantum matrix inversion, also known as the Harrow–Hassidim–Lloyd (HHL) algorithm, is described in Section 12.6, which can be used as a basic ingredient for other QML algorithms such as the quantum PCA, which is described in Section 12.7. In the quantum optimization-based clustering section (Section 12.8), we describe how the MAX-CUT problem can be related to clustering and thus be solved by adiabatic computing, quantum annealing, and QAOA. In Section 12.9, the Grover algorithm-based quantum optimization is discussed. In the quantum K-means section (Section 12.10), we describe how to calculate the dot product and quantum distance and how to use them in the Grover search-based K-means algorithm. In the quantum SVM section (Section 12.11), we formulate the SVM problem using least-squares and describe how to solve it with the help of the HHL algorithm. In the quantum neural networks (QNNs) section (Section 12.12), we describe feedforward QNNs, quantum perceptron, and quantum convolutional networks.

Chapter 13 is devoted to the fault-tolerant quantum error correction. After the introduction of fault-tolerance basics and traversal operations in Section 13.1, we provide the basics of fault-tolerant quantum information processing concepts and procedures in Section 13.2 by employing Steane's code as an illustrative example. Section 13.3 is devoted to a rigorous description of fault-tolerant quantum error correction. After a short review of some basic concepts from stabilizer codes, in Subsection 13.3.1, the fault-tolerant syndrome extraction is described, followed by the description of fault-tolerant encoded operations in Subsection 13.3.2. Subsection 13.3.3 is concerned with the application of the quantum gates on a quantum register by means of a measurement protocol. This method, when applied transversally, is used in Subsection 13.3.4 to enable fault-tolerant error correction based on an arbitrary quantum stabilizer code. The fault-tolerant stabilizer codes are described in Subsection 13.3.4. As an illustrative example, the [1,3,5] fault-tolerant stabilizer code is described in Subsection 13.3.5.

References

[1] I.B. Djordjevic, Quantum Information Processing, Quantum Computing, and Quantum Error Correction: An Engineering Approach, second ed., Elsevier/Academic Press, 2021.

[2] I.B. Djordjevic, Physical-Layer Security and Quantum Key Distribution, Springer International Publishing, Switzerland, 2019.

[3] C.W. Helstrom, Quantum Detection and Estimation Theory, Academic, New York, 1976.

[4] C.W. Helstrom, J.W.S. Liu, J.P. Gordon, Quantum mechanical communication theory, Proc. IEEE 58 (1970) 1578–1598.

[5] V. Vilnrotter, C.-W. Lau, Quantum detection theory for the free-space channel, IPN Progress Report 42–146 (Aug. 15, 2001) 1–34.

[6] G. Cariolaro, Quantum Communications, Springer International Publishing Switzerland, Cham: Switzerland, 2015.

[7] R. van Meter, Quantum Networking, ISTE/Wiley, London-Hoboken, 2014.

[8] M.A. Neilsen, I.L. Chuang, Quantum Computation and Quantum Information, Cambridge University Press, Cambridge, 2000.

[9] M. Le Bellac, An Introduction to Quantum Information and Quantum Computation, Cambridge University Press, 2006.

[10] F. Gaitan, Quantum Error Correction and Fault Tolerant Quantum Computing, CRC Press, 2008.

[11] G. Jaeger, Quantum Information: An Overview, Springer, 2007.

[12] D. Petz, Quantum Information Theory and Quantum Statistics, Theoretical and Mathematical Physics, Springer, Berlin, 2008.

[13] P. Lambropoulos, D. Petrosyan, Fundamentals of Quantum Optics and Quantum Information, Springer-Verlag, Berlin, 2007.

[14] G. Johnson, A Shortcut Thought Time: The Path to the Quantum Computer, Knopf, New York, 2003.

[15] J. Preskill, Quantum Computing, 1999 available at: http://www.theory.caltech.edu/~preskill/.

[16] J. Stolze, D. Suter, Quantum Computing, Wiley, New York, 2004.

[17] R. Landauer, Information is physical, Phys. Today 44 (5) (May 1991) 23–29.

[18] R. Landauer, The physical nature of information, Phys. Lett. 217 (1991) 188–193.

[19] P.W. Shor, Polynomial-time algorithms for prime factorization and discrete logarithms on a quantum computer, SIAM J. Comput. 26 (1997) 1484.

[20] Z. Zhang, Q. Zhuang, Distributed quantum sensing, Quantum Sci. Technol. 6 (4) (2021) 043001.

[21] A.S. Holevo, R.F. Werner, Evaluating capacities of Bosonic Gaussian channels, Phys. Rev. 63 (3) (2001) 032312.

[22] A.S. Holevo, On entanglement assisted classical capacity, J. Math. Phys. 43 (9) (2002) 4326–4333.

[23] C.H. Bennett, P.W. Shor, J.A. Smolin, A.V. Thapliyal, Entanglement-assisted capacity of a quantum channel and the reverse Shannon theorem, IEEE Trans. Inf. Theor. 48 (2002) 2637–2655.

[24] M. Sohma, O. Hirota, Capacity of a channel assisted by two-mode squeezed states, Phys. Rev. 68 (2) (2003) 022303.

[25] H. Shi, Z. Zhang, Q. Zhuang, Practical route to entanglement-assisted communication over noisy Bosonic channels, Phys. Rev. Appl. 13 (3) (2020) 034029.

[26] I.B. Djordjevic, On entanglement assisted classical optical communications, IEEE Access 9 (March 17 , 2021) 42604–42609.

[27] I.B. Djordjevic, Quantum receivers for entanglement assisted classical optical communications, IEEE Photonics J. (April 27 , 2021), https://doi.org/10.1109/JPHOT.2021.3075881.

[28] R.V. Meter, S.J. Devitt, The path to scalable distributed quantum computing, IEEE Computer 49 (2016) 9.

[29] A.M. Childs, Secure assisted quantum computation, Quant. Inf. Comput. 5 (6) (2005) 456–466.
[30] C. Elliott, A. Colvin, D. Pearson, O. Pikalo, J. Schlafer, H. Yeh, Current status of the DARPA quantum network (Invited Paper), in: Proc. SPIE 5815, Quantum Information and Computation III, 25 May 2005, https://doi.org/10.1117/12.606489.
[31] M. Sasaki, et al., Field test of quantum key distribution in the Tokyo QKD Network, Opt Express 19 (11) (2011) 10387–10409.
[32] R. Alléaume, et al., Using quantum key distribution for cryptographic purposes, J. Theoretical Computer Science 560 (P1) (Dec. 2014) 62–81.
[33] S. Nagayama, R. Van Meter, Internet-Draft: IKE for IPsec with QKD, 2009. https://tools.ietf.org/html/draft-nagayama-ipsecme-ipsec-with-qkd-01.
[34] A. Mink, S. Frankel, R. Perlner, Quantum key distribution (QKD) and commodity security protocols: introduction and integration, Int. J. Netw. Secur. Appl. 1 (2) (2009) 101–112.
[35] J. Abadie, et al., A gravitational wave observatory operating beyond the quantum shot-noise limit, Nat. Phys. 7 (2011) 962–965.
[36] M. Tse, et al., Quantum-enhanced advanced LIGO detectors in the era of gravitational-wave astronomy, Phys. Rev. Lett. 123 (December 5 , 2019) 231107.
[37] Z. Zhang, S. Mouradian, F.N.C. Wong, J.H. Shapiro, Entanglement-enhanced sensing in a lossy and noisy environment, Phys. Rev. Lett. 114 (11) (2015) 110506.
[38] M. Lanzagorta, Quantum Radar, Morgan and Claypool Publishers, 2012.
[39] M.A. Taylor, J. Janousek, V. Daria, J. Knittel, B. Hage, H.A. Bachor, W.P. Bowen, Biological measurement beyond the quantum limit, Nat. Photonics 7 (2013) 229–233.
[40] T. Ono, R. Okamoto, S. Takeuchi, An entanglement-enhanced microscope, Nat. Commun. 4 (2013) 2426.
[41] H. Yonezawa, et al., Quantum-enhanced optical-phase tracking, Science 337 (6101) (2012) 1514–1517.
[42] G. Keiser, Optical Fiber Communications, McGraw Hill, 2000.
[43] D.P. DiVincenzo, Two-bit gates are universal for quantum computation, Phys. Rev. 51 (2) (1995), 1015–1022.
[44] A. Barenco, C.H. Bennett, R. Cleve, D.P. DiVincenzo, N. Margolus, P. Shor, T. Sleator, J.A. Smolin, H. Weinfurter, Elementary gates for quantum computation, Phys. Rev. 52 (5) (1995), 3457–3467.
[45] A. Barenco, A universal two-bit quantum computation, Proc. Roy. Soc. Lond. A 449 (1937) (1995) 679–683.
[46] D. Deutsch, Quantum computational networks, Proc. Roy. Soc. Lond. A 425 (1868) (1989) 73–90.
[47] C.H. Bennett, G. Brassard, C. Crépeau, R. Jozsa, A. Peres, W.K. Wootters, Teleporting an unknown quantum state via dual classical and Einstein-Podolsky-Rosen channels, Phys. Rev. Lett. 70 (13) (1993) 1895–1899.
[48] D.J.C. MacKay, G. Mitchison, P.L. McFadden, Sparse-graph codes for quantum error correction, IEEE Trans. Inf. Theor. 50 (2004) 2315–2330.
[49] I.B. Djordjevic, Photonic implementation of quantum relay and encoders/decoders for sparse-graph quantum codes based on optical hybrid, IEEE Photon. Technol. Lett. 22 (19) (2010) 1449–1451.
[50] I.B. Djordjevic, Photonic entanglement-assisted quantum low-density parity-check encoders and decoders, Opt. Lett. 35 (9) (2010) 1464–1466.
[51] I.B. Djordjevic, Quantum LDPC codes from balanced incomplete block designs, IEEE Commun. Lett. 12 (2008) 389–391.
[52] I.B. Djordjevic, Cavity quantum electrodynamics (CQED) based quantum LDPC encoders and decoders. IEEE Photonics J. (accepted for publication).
[53] I.B. Djordjevic, Photonic quantum dual-containing LDPC encoders and decoders, IEEE Photon. Technol. Lett. 21 (13) (July 1, 2009) 842–844.
[54] Q. Zhuang, Z. Zhang, J.H. Shapiro, Distributed quantum sensing using continuous-variable multipartite entanglement, Phys. Rev. 97 (3) (March 21 , 2018) 032329.

[55] Q. Zhuang, J. Preskill, L. Jiang, Distributed quantum sensing enhanced by continuous-variable error correction, New J. Phys. 22 (2020) 022001.
[56] C.H. Bennet, G. Brassard, Quantum Cryptography: public key distribution and coin tossing, in: Proc. IEEE International Conference on Computers, Systems, and Signal Processing, 1984, pp. 175–179. Bangalore, India.
[57] C.H. Bennett, Quantum cryptography: uncertainty in the service of privacy, Science 257 (August 7, 1992) 752–753.
[58] R. Ursin, et al., Entanglement-based quantum communication over 144 km, Nat. Phys. 3 (7) (2007) 481–486.
[59] S.-K. Liao, et al., Satellite-to-ground quantum key distribution, Nature 549 (2017) 43–47.
[60] A. Boaron, et al., Secure quantum key distribution over 421 km of optical fiber, Phys. Rev. Lett. 121 (2018) 190502.
[61] W.K. Wootters, B.D. Fields, Optimal state-determination by mutually unbiased measurements, Ann. Phys. 191 (1989) 363–381.
[62] I.D. Ivanovic, Geometrical description of quantal state determination, J. Physiol. Anthropol. 14 (1981) 3241–3245.
[63] I. Bengtsson, Three ways to look at mutually unbiased bases, AIP Conf. Proc. 889 (1) (2007) 40.
[64] I.B. Djordjevic, FBG-based weak coherent state and entanglement assisted multidimensional QKD, IEEE Photonics J. 10 (4) (2018) 7600512.
[65] M.K. Bhaskar, et al., Experimental demonstration of memory-enhanced quantum communication, Nature 580 (2020) 60.
[66] J.-G. Ren, et al., Ground-to-satellite quantum teleportation, Nature 549 (2017) 70.
[67] J. Yin, et al., Satellite-based entanglement distribution over 1200 kilometers, Science 356 (2017) 1140–1144.

CHAPTER

Information theory, error correction, and detection theory

2

Chapter outline

Quantum Communication, Quantum Networks, and Quantum Sensing. https://doi.org/10.1016/B978-0-12-822942-2.00013-3

2.1 Classical information theory fundamentals

In this section, we attempt to answer two fundamental questions [1–13]: (1) What is the lowest irreducible complexity below which the signal cannot be compressed? (2) What is the highest reliable transmission rate over a given noisy communication channel? To answer the first question, we introduce the concept of *entropy*, which represents a basic measure of information defined in terms of the probabilistic behavior of the source. We later show that the average number of bits per symbol representing a discrete source can be made as small as possible but not smaller than the source entropy, representing the claim of the so-called *source-coding theorem*. To answer the second question, we introduce the concept of *channel capacity* as the largest possible amount of information that can be reliably transmitted over the channel. As long as the transmission rate is lower than the channel capacity, we can design a channel coding scheme that allows for reliable transmission and reconstruction of the transmitted information with an arbitrarily small probability of error.

2.1.1 Entropy, conditional entropy, relative entropy, and mutual information

Let us observe a *discrete memoryless source* (DMS) characterized by a finite alphabet $S=\{s_0,s_1,\ldots,s_{K-1}\}$, wherein each symbol is generated with probability $P(S=s_k)=p_k$, $k=0,1,\ldots,K-1$. At a given instance, the DMS must generate one symbol from the alphabet, so we can write $\sum_{k=0}^{K-1} p_k=1$. The generation of a symbol at a given instance is independent of previously generated symbols. The amount of information a given symbol carries is related to the surprise when it occurs and is, therefore, reversely proportional to the probability of its occurrence. Since there is uncertainty about which symbol will be generated by the source, the terms uncertainty, surprise, and amount of information appear interrelated. Given that certain symbols can occur with very low probability, the information value will be huge if the reverse probability is used to determine the amount of information. In practice, we use the logarithm of the reverse probability of occurrence as the amount of information to solve this problem:

$$I(s_k)=\log\left(\frac{1}{p_k}\right)=-\log p_k;\ \ k=0,1,\cdots,K-1 \tag{2.1}$$

The most common logarithm base is 2, and the unit for the amount of information is a *bi*nary un*it* (bit). When $p_k=1/2$, the amount of information is $I(s_k)=1$ bit, indicating that 1 bit is the amount of information gained when one of two equally likely events occurs. It is straightforward to show that the amount of information is nonnegative—that is, $I(s_k)\geq 0$. Further, when $p_k>p_i$ then $I(s_i)>I(s_k)$. Finally, when two symbols s_i and s_k are independent, the joint amount of information is additive—that is, $I(s_k s_i)=I(s_k)+I(s_i)$.

The average information content per symbol is commonly referred to as the *entropy*:

$$H(S)=E_P\ I(s_k)=E\left[\log\left(\frac{1}{p_k}\right)\right]=\sum_{k=0}^{K-1} p_k\log\left(\frac{1}{p_k}\right) \tag{2.2}$$

It is easily shown that the entropy is upper- and lower-bounded as follows:

$$0\leq H(S)\leq |S|. \tag{2.3}$$

The entropy of binary source $S=\{0,1\}$ is given by

$$H(S)=p_0 \log\left(\frac{1}{p_0}\right)+p_1 \log\left(\frac{1}{p_1}\right)=-p_0 \log p_0-(1-p_0)\log(1-p_0)=H(p_0), \tag{2.4}$$

where $H(p_0)$ is known as the *binary entropy function*.

The entropy definition is also applicable to any random variable X; namely, we can write $H(X)=-\sum_k p_k \log p_k$. The *joint entropy* of a pair of random variables (X,Y), denoted as $H(X,Y)$, is defined as

$$H(X,Y)=-E_{p(x,y)} \log p(X,Y)=-\sum_{x\in X}\sum_{x\in Y} p(x,y)\log p(x,y). \tag{2.5}$$

When a pair of random variables (X,Y) has the joint distribution $p(x,y)$, the *conditional entropy* $H(Y|X)$ is defined as

$$H(Y|X)=-E_{p(x,y)} \log p(Y|X)=-\sum_{x\in X}\sum_{y\in Y} \underbrace{p(x,y)}_{p(y|x)p(x)} \log p(y|x)$$

$$=-\sum_{x\in X} p(x)\sum_{y\in Y} p(y|x)\log p(y|x). \tag{2.6}$$

Since $p(x,y)=p(x)p(y|x)$, by taking logarithm, we obtain

$$\log p(X,Y)=\log p(X)+\log p(Y|X), \tag{2.7}$$

and now, by applying the expectation operator, we obtain

$$\underbrace{E \log p(X,Y)}_{H(X,Y)}=\underbrace{E \log p(X)}_{H(X)}+\underbrace{E \log p(Y|X)}_{H(Y|X)} \tag{2.8}$$
$$\Leftrightarrow H(X,Y)=H(X)+H(Y|X)$$

The equation above is commonly referred to as the *chain rule*.

The *relative entropy* is a measure of the distance between two distributions $p(x)$ and $q(x)$ and is defined as follows:

$$D(p\|q)=E_p \log\left[\frac{p(X)}{q(X)}\right]=\sum_{x\in X} p(x)\log\left[\frac{p(X)}{q(X)}\right]. \tag{2.9}$$

The relative entropy is also known as the Kullback–Leibler distance and can be interpreted as measuring the inefficiency of assuming that the distribution is $q(x)$ when the true distribution is $p(x)$. Now, by replacing $p(X)$ with $p(X,Y)$ and $q(X)$ with $p(X)q(Y)$, the corresponding relative entropy between the joint distribution and product of distributions, well known as *mutual information*, is obtained:

$$D(p(X,Y)\|p(X)q(Y))=E_{p(X,Y)} \log\left[\frac{p(X,Y)}{p(X)q(Y)}\right]$$

$$=\sum_{x\in X}\sum_{y\in Y} p(x,y)\log\left[\frac{p(X,Y)}{p(X)q(Y)}\right]\doteq I(X,Y). \tag{2.10}$$

2.1.2 Source coding and data compaction

An important problem in communication systems and data storage is efficiently representing the data generated by a source. The source coding represents the process by which this goal is achieved. In particular, when the statistics of the source are known, we can assign short codewords to frequently occurring symbols and long codewords to rarely occurring symbols, which results in a *variable-length source code*. Two basic requirements must be satisfied: (1) the codewords generated by a source encoder are binary, and (2) the source code must be *uniquely decodable*—that is, the original sequence can be perfectly recovered from the binary-encoded sequence.

The average number of bits per symbol can be determined by

$$L = \sum_i p_k L_k, \quad L_k = \text{length}(s_k), \tag{2.11}$$

where L_k is the length of the k-th source symbol codeword s_k, which occurs with probability p_k. The source coding efficiency can be defined by $\eta = L_{\min}/L$, $L_{\min} = \min L$.

The *source coding theorem*, also known as Shannon's first theorem, can be formulated as follows [1–6]—the average source codeword length for distortionless transmission for a given memoryless source S of entropy $H(S)$ is lower-bounded by

$$L \geq H(S). \tag{2.12}$$

In other words, we can make the average number of bits per symbol for a DMS as small as possible but not smaller than the entropy of the source. Now we can redefine the *source coding efficiency* as follows:

$$\eta = \frac{H(S)}{L}. \tag{2.13}$$

For efficient transmission, redundant information should be removed before transmission takes place. This can be done by quantization (lossy compression) or *data compaction* (lossless data compression), which is considered here. An efficient way to do so is by prefix coding. The *prefix code* is source code for which not any codeword can be used as a prefix of a longer codeword. It can be shown that the prefix code is uniquely decodable, but the converse is not necessarily true. Prefix codes can be decoded as soon as a source symbol codeword is fully received, and as such, they can be called instantaneous codes. To verify whether a given source code is uniquely decodable, we can use the *Kraft–McMillan inequality* [1–6]:

$$\sum_{k=0}^{K-1} 2^{-L_k} \leq 1, \tag{2.14}$$

where K is the cardinality of the source.

Because the entropy of the DMS is $H(S)$, we can construct the prefix code such that

$$p_k = 2^{-l_k} \Rightarrow \sum_{k=0}^{K-1} 2^{-l_k} = \sum_{k=0}^{K-1} p_k = 1. \tag{2.15}$$

The average codeword length will then be

$$L = \sum_{k=0}^{K-1} p_k L_k = \sum_{k=0}^{K-1} 2^{-L_k} L_k. \tag{2.16}$$

The corresponding entropy of this prefix code will be

$$H(S) = \sum_{k=0}^{K-1} p_k \log\left(\frac{1}{p_k}\right) = \sum_{k=0}^{K-1} 2^{-L_k} \log_2\left(2^{L_k}\right) = \sum_{k=0}^{K-1} \frac{L_k}{2^{L_k}} = L. \tag{2.17}$$

Clearly, this prefix code is matched to the source since $L = \text{H}(S)$, and thus, the code efficiency is $\eta = 1$. The key question now arises, how to match the prefix code to an arbitrary DMS? The answer to this question is with the help of so-called *extended source.* In extended code, we consider the blocks of symbols rather than the individual symbols obtained by the Cartesian product of symbols from the original source. If the symbols from the source are independent, the probability of the n-th-order extended source will be n times larger than the entropy of the original source; in other words, we can write

$$H(S^n) = nH(S). \tag{2.18}$$

The average codeword length of the extended source, denoted as $\overline{L}_n$, will then be limited as

$$H(S^n) \le \overline{L}_n \le H(S^n) + 1 \Rightarrow nH(S) \le \overline{L}_n \le nH(S) + 1. \tag{2.19}$$

By dividing both sides of inequalities by n, we obtain

$$H(S) \le \frac{\overline{L}_n}{n} \le H(S) + \frac{1}{n}. \tag{2.20}$$

In the limit as n tends to infinity, we obtain

$$\lim_{n->\infty} \frac{\overline{L}_n}{n} = H(S), \tag{2.21}$$

indicating that we can faithfully represent the source by sufficiently extending the order of the source.

2.1.2.1 Huffman coding

An important class of prefix codes is a class of *Huffman codes* [14]. The key idea behind the Huffman code is to represent a symbol from a source alphabet by a sequence of bits of a length proportional to the amount of information conveyed by the symbol under consideration—that is, $L_k \cong -\log(p_k)$. Clearly, Huffman codes require knowledge of the source statistics and attempt to represent DMS statistics more simply. The *Huffman encoding algorithm* has the following steps:

i. List the symbols of the source in decreasing order of probability of occurrence. The two symbols with the lowest probability are assigned 0 or 1.
ii. The two symbols with the lowest probability are combined into a new symbol (a supersymbol), with its probability being the sum of individual symbols within the supersymbol. The location of the supersymbol in the list in the next stage is according to combined probability.
iii. The procedure is repeated until we are left with only two symbols, to which we assign bits 0 and 1.

By reading out the bits assigned until we reach the root, we obtain the codeword assigned to each symbol from the source. The Huffman code is applicable to both binary and nonbinary source codes.

Example. A DMS has an alphabet of eight symbols whose probabilities of occurrence are as follows:

Symbols:	s_1	s_2	s_3	s_4	s_5	s_6	s_7	s_8
Probabilities:	0.28	0.18	0.15	0.13	0.10	0.07	0.05	0.04

We are concerned with designing the Huffman code for this source by moving a supersymbol as high as possible and assuming *ternary* transmission with symbols {0, 1, 2}. The Huffman procedure for this ternary code is summarized in Fig. 2.1.

2.1.2.2 Ziv–Lempel algorithm

The *Ziv–Lempel algorithm*, on the other hand, does not require knowledge about the source statistics and is therefore easier to implement. The key idea behind this algorithm is to parse the source data stream into the shortest segments not previously encountered. As an illustration, let us consider the following binary sequence: 110100111011110... We would like to use the Ziv–Lempel algorithm to encode this sequence. We assume that the binary symbols 0 and 1 are already in the book at addresses 1 and 2, respectively. By parsing this sequence into segments not encountered before, we obtain 11 | 01| 00| 111| 011| 110 ... The corresponding Ziv–Lempel code is provided in Fig. 2.2. The first row represents the numerical positions of segments, assuming that the symbols 0 and 11 are already in the code book. The second row represents the segments of the parsed sequence. The third row provides the numerical representation of segments. Each segment is represented by the address of the segment already encountered previously and the bit in the last position representing the new bit from the sequence, known as the *innovation bit*. The final row provides binary-encoded blocks, wherein each address is represented in binary form followed by the innovation bit.

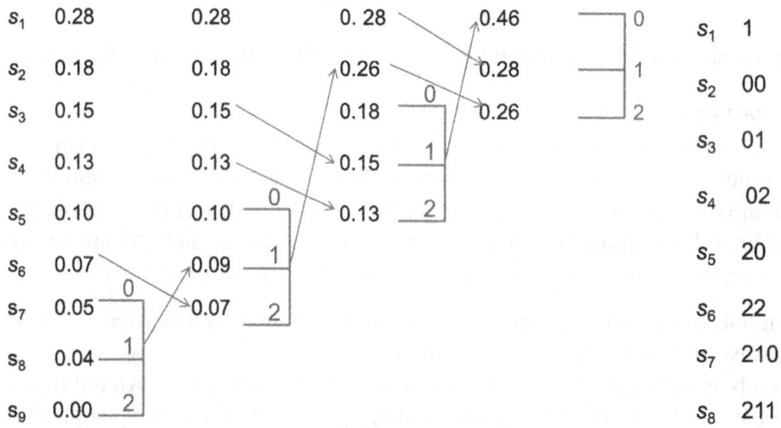

FIGURE 2.1

Illustration of the Huffman procedure (left) and corresponding ternary code (right).

FIGURE 2.2

Ziv–Lempel algorithm on binary sequence 110100111011110....

Addresses:	1	2	3	4	5	6	7	8
Segments:	0	1	11	01	00	111	011	110
Numerical representations:			21	11	10	31	41	30
Binary encoded blocks:			**0101**	**0011**	**0010**	**0111**	**1001**	**0110**

2.1.3 Mutual information, channel capacity, channel coding theorem, and information capacity theorem

2.1.3.1 Mutual information and information capacity

Fig. 2.3 shows an example of a discrete memoryless channel (DMC), which is characterized by channel (transition) probabilities. If $X=\{x_0, x_1, \ldots, x_{I-1}\}$ and $Y=\{y_0, y_1, \ldots, y_{J-1}\}$ denote the channel input alphabet and the channel output alphabet, respectively, the channel is characterized completely by the following set of transition probabilities:

$$p\left(y_j|x_i\right)=P\left(Y=y_j|X=x_i\right),\ 0\leq p\left(y_j|x_i\right)\leq 1, \tag{2.22}$$

where $i\in\{0,1,\ldots,I-1\}$, $j\in\{0,1,\ldots,J-1\}$, while I and J denote the sizes of the input and output alphabets, respectively. The transition probability $p(y_j|x_i)$ represents the conditional probability that $Y=y_j$ for a given input $X=x_i$.

One of the most important characteristics of the transmission channel is *information capacity*, which is obtained by maximization of mutual information $I(\boldsymbol{X},\boldsymbol{Y})$ over all possible input distributions:

$$C=\max_{\{p(x_i)\}} I(\boldsymbol{X},\boldsymbol{Y}),\ \ I(\boldsymbol{X},\boldsymbol{Y})=H(\boldsymbol{X})-H(\boldsymbol{X}|\boldsymbol{Y}), \tag{2.23}$$

where $H(\boldsymbol{U})=E\{-\log_2 P(\boldsymbol{U})\}$ denotes the entropy of a random variable $\boldsymbol{U}$.

For a DMC, the mutual information can be determined as

$$I(\boldsymbol{X};\boldsymbol{Y})=H(\boldsymbol{X})-H(\boldsymbol{X}|\boldsymbol{Y})$$

$$=\sum_{i=1}^{M}p(x_i)\log_2\left[\frac{1}{p(x_i)}\right]-\sum_{j=1}^{N}p\left(y_j\right)\sum_{i=1}^{M}p\left(x_i|y_j\right)\log_2\left[\frac{1}{p\left(x_i|y_j\right)}\right]. \tag{2.24}$$

In the equation above, $H(\boldsymbol{X})$ represents the uncertainty about the channel input before observing the channel output—also known as *entropy*—while $H(\boldsymbol{X}|\boldsymbol{Y})$ denotes the conditional entropy or amount of uncertainty remaining about the channel input after the channel output has been received. (The log function from the above equation relates to base 2, and it will be like that throughout this chapter.)

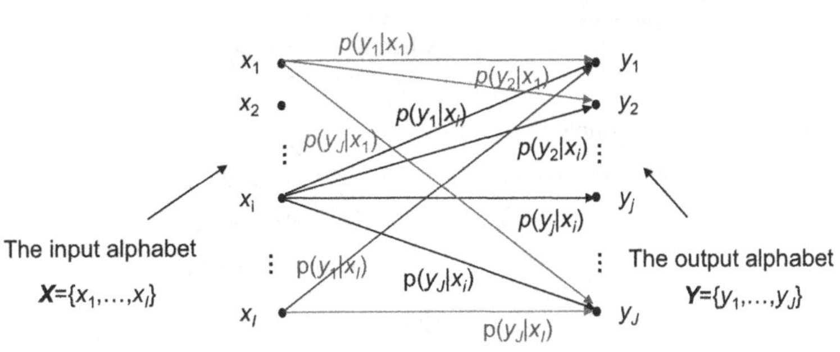

FIGURE 2.3

Discrete memoryless channel.

Therefore, mutual information represents the amount of information (per symbol) conveyed by the channel, representing the uncertainty about channel input that is resolved by observing channel output. Mutual information can be interpreted with a Venn diagram [1–13], as shown in Fig. 2.4A. The left and right circles represent the entropy of channel input and output, respectively, while the mutual information is obtained as the intersection area of the two circles. Another interpretation is illustrated

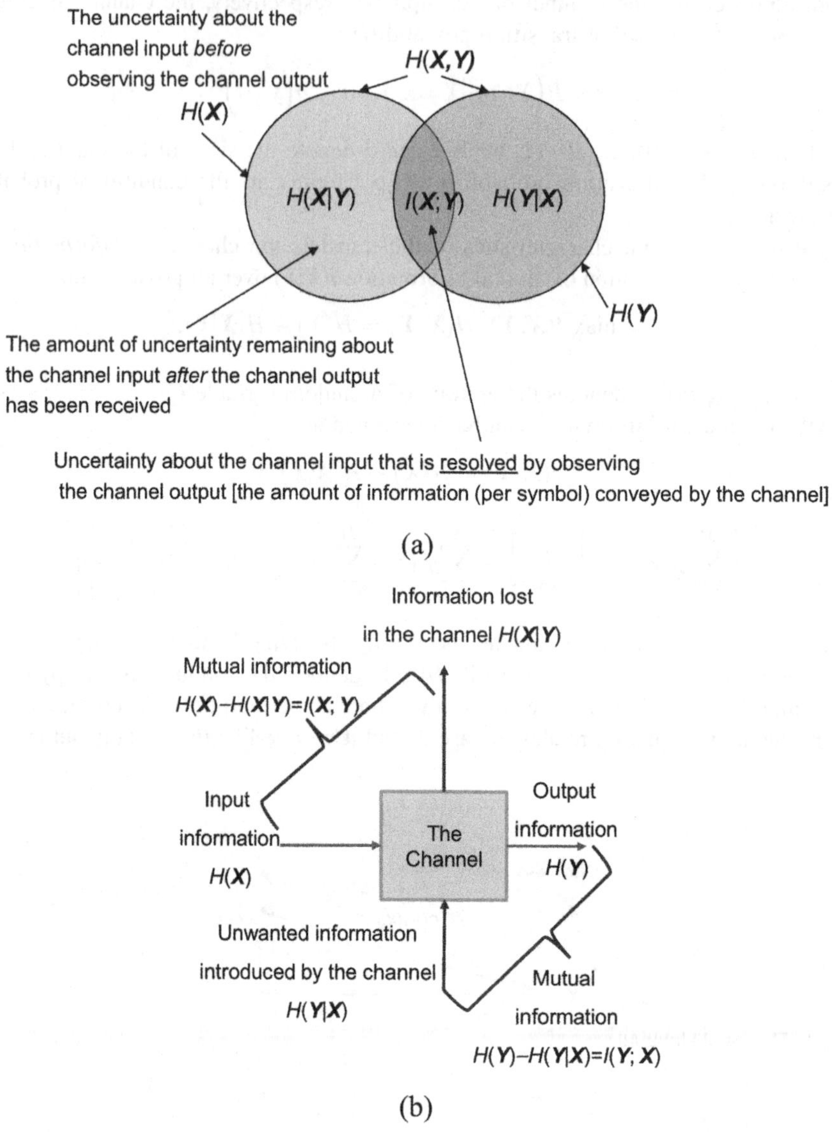

FIGURE 2.4

Interpretation of mutual information using (A) Venn diagram and (B) the approach of Ingels.

in Fig. 2.4B [3]. The mutual information, i.e., the information conveyed by the channel, is obtained as the output information minus unwanted information introduced by the channel.

Since for an M-ary input and M-ary output symmetric channel (MSC), we have $p(y_j|x_i) = P_s/(M-1)$ and $p(y_j|x_j) = 1-P_s$, where P_s is symbol error probability, the channel capacity in bits/symbol can be determined by

$$C = \log_2 M + (1 - P_s)\log_2(1 - P_s) + P_s \log_2\left(\frac{P_s}{M-1}\right). \tag{2.25}$$

Channel capacity represents an important bound on data rates achievable by any modulation and coding schemes. It can also be used to compare different coded modulation schemes for their distance to the maximum channel capacity curve. In Fig. 2.5, we show the channel capacity for a Poisson M-ary pulse-position modulation (PPM) channel against the average number of signal photons per slot, expressed in dB scale, for an average number of background photons of $K_b = 1$.

2.1.3.2 Channel coding theorem

Now we have gained sufficient knowledge to introduce a very important concept, the *channel coding theorem* [1–13], which is formulated as follows. Let a DMS with an alphabet S have entropy $H(S)$ and emit symbols every T_s seconds. Let a DMC have capacity C and be used once in T_c seconds. Then, if

$$H(S)/T_s \leq C/T_c \tag{2.26}$$

a coding scheme exists for which the source output can be transmitted over the channel and reconstructed with an arbitrarily small probability of error. The parameter $H(S)/T_s$ is related to the average information rate, while the parameter C/T_c is related to channel capacity per unit time.

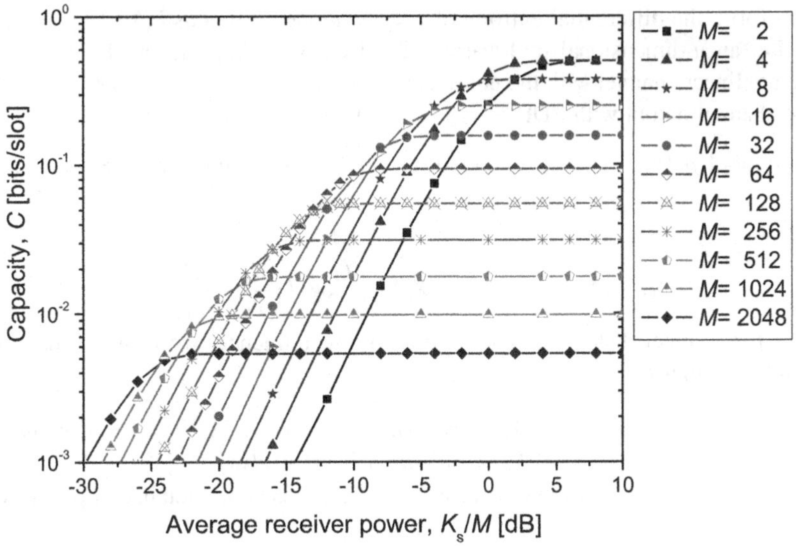

FIGURE 2.5

Channel capacity for M-ary pulse-position modulation on a Poisson channel for $K_b = 1$.

For a binary symmetric channel ($N = M = 2$), the inequality is reduced to $R \le C$, where R is the code rate. Since the proof of this theorem can be found in any textbook on information theory, such as [2–5], the proof of this theorem is omitted here.

2.1.3.3 Capacity of continuous channels

In this section, we discuss the channel capacity of continuous channels. Let $\boldsymbol{X} = [X_1, X_2, \ldots, X_n]$ denote an n-dimensional multivariate with a probability density function (PDF) $p_1(x_1, x_2, \ldots, x_n)$ representing the channel input. The corresponding *differential entropy* is defined by [2–6]

$$\begin{aligned} h(X_1, X_2, \ldots, X_n) &= -\underbrace{\int_{-\infty}^{\infty} \cdots \int_{-\infty}^{\infty}}_{n} p_1(x_1, x_2, \ldots, x_n) \log p_1(x_1, x_2, \ldots, x_n)\, dx_1 dx_2 \ldots dx_n \\ &= \langle -\log p_1(x_1, x_2, \ldots, x_n) \rangle, \end{aligned} \tag{2.27}$$

where for brevity we use $\langle \cdot \rangle$ to denote the expectation operator. To simplify explanations, we use the compact form of Eq. (2.27), namely $h(\boldsymbol{X}) = \langle -\log p_1(\boldsymbol{X}) \rangle$, which was introduced in Ref. [4]. In similar fashion, the channel output can be represented as an m-dimensional random variable $Y = [Y_1, Y_2, \ldots, Y_m]$ with a PDF $p_2(y_1, y_2, \ldots, y_m)$, while the corresponding differential entropy is defined by

$$\begin{aligned} h(Y_1, Y_2, \ldots, Y_m) &= -\underbrace{\int_{-\infty}^{\infty} \cdots \int_{-\infty}^{\infty}}_{m} p_2(y_1, y_2, \ldots, y_m) \log p_1(y_1, y_2, \ldots, y_m)\, dy_1 dy_2 \ldots dy_m \\ &= \langle -\log p_2(y_1, y_2, \ldots, y_m) \rangle. \end{aligned} \tag{2.28}$$

In compact form, the differential entropy of output can be written as $h(\boldsymbol{Y}) = \langle -\log p_1(\boldsymbol{Y}) \rangle$.

Example. Let an n-dimensional multivariate $\boldsymbol{X} = [X_1, X_2, \ldots, X_n]$ with a PDF $p_1(x_1, x_2, \ldots, x_n)$ be applied to the nonlinear channel with the nonlinear characteristic $\boldsymbol{Y} = g(\boldsymbol{X})$, where $\boldsymbol{Y} = [Y_1, Y_2, \ldots, Y_n]$ represents the channel output with PDF $p_2(y_1, y_2, \ldots, y_n)$. Since the corresponding PDFs are related by the Jacobian symbol $p_2(y_1, \cdots, y_n) = p_1(x_1, \cdots, x_n) \left| J\left(\frac{X_1, \cdots, X_n}{Y_1, \cdots, Y_m}\right) \right|$, the output entropy can be determined as

$$h(Y_1, \cdots, Y_m) \cong h(X_1, \cdots, X_n) - \left\langle \log \left| J\left(\frac{X_1, \cdots, X_n}{Y_1, \cdots, Y_m}\right) \right| \right\rangle.$$

To account for channel distortions and additive noise influence, we observe the corresponding conditional and joint PDFs:

$$\begin{aligned} &P(y_1 < Y_1 < y_1 + dy_1, \ldots, y_m < Y_m < y_m + dy_m | X_1 = x_1, \ldots, X_n = x_n) = p(\tilde{y}|\tilde{x}) d\tilde{y} \\ &P(y_1 < Y_1 < y_1 + dy_1, \ldots, x_n < X_n < x_n + dx_n) = p(\tilde{x}, \tilde{y}) d\tilde{x} d\tilde{y} \end{aligned} \tag{2.29}$$

The mutual information (also known as information rate) can be written in compact form as follows [4]:

$$I(\boldsymbol{X}; \boldsymbol{Y}) = \left\langle \log \frac{p(\boldsymbol{X}, \boldsymbol{Y})}{p(\boldsymbol{X}) P(\boldsymbol{Y})} \right\rangle. \tag{2.30}$$

Notice that various differential entropies $h(X)$, $h(Y)$, $h(Y|X)$ do not have direct interpretation as far as the information processing in the channel is concerned compared with their discrete counterparts from Subsection 2.1.3.1. Some authors, such as Gallager in Ref. [13], prefer to define the mutual information directly by Eq. (2.30) without considering the differential entropies at all. The mutual information, however, has theoretical meaning and represents the average information processed in the channel (or amount of information conveyed by the channel). The mutual information has the following important properties [2–6]: (1) it is symmetric: $I(X;Y)=I(Y;X)$; (2) it is nonnegative; (3) it is finite; (4) it is invariant under linear transformation; (5) it can be expressed in terms of the differential entropy of channel output by $I(X;Y) = h(Y)-h(Y|X)$; and (6) it is related to channel input differential entropy by $I(X;Y) = h(X)-h(X|Y)$.

The *information capacity* can be obtained by maximization of Eq. (2.30) under all possible input distributions, which is

$$C = \max I(\boldsymbol{X};\boldsymbol{Y}). \tag{2.31}$$

Let us now determine the mutual information of two random vectors $\boldsymbol{X} = [X_1,X_2, \ldots,X_n]$ and $\boldsymbol{Y} = [Y_1,Y_2, \ldots,Y_m]$ that are normally distributed. Let $\boldsymbol{Z} = [\boldsymbol{X};\boldsymbol{Y}]$ be the random vector describing the joint behavior. Without loss of generality, we further assume that $\overline{X}_k = 0 \;\forall\; k$ and $\overline{Y}_k = 0 \;\forall\; k$ (the mean values are zero). The corresponding PDFs for $\boldsymbol{X}$, $\boldsymbol{Y}$, and $\boldsymbol{Z}$ are, respectively, given as [4]

$$\begin{gathered} p_1(\boldsymbol{x}) = \frac{1}{(2\pi)^{n/2}(\det \boldsymbol{A})^{1/2}}\exp\left(-0.5\left(\boldsymbol{A}^{-1}\boldsymbol{x},\boldsymbol{x}\right)\right), \\ \boldsymbol{A} = \left[a_{ij}\right], a_{ij} = \int x_i x_j p_1(\boldsymbol{x})d\boldsymbol{x} \end{gathered} \tag{2.32}$$

$$\begin{gathered} p_2(\boldsymbol{y}) = \frac{1}{(2\pi)^{n/2}(\det \boldsymbol{B})^{1/2}}\exp\left(-0.5\left(\boldsymbol{B}^{-1}\boldsymbol{y},\boldsymbol{y}\right)\right), \\ \boldsymbol{B} = \left[b_{ij}\right],\; b_{ij} = \int y_i y_j p_2(\boldsymbol{y})d\boldsymbol{y} \end{gathered} \tag{2.33}$$

$$\begin{gathered} p_3(\boldsymbol{z}) = \frac{1}{(2\pi)^{(n+m)/2}(\det \boldsymbol{C})^{1/2}}\exp\left(-0.5\left(\boldsymbol{C}^{-1}\boldsymbol{z},\boldsymbol{z}\right)\right), \\ \boldsymbol{C} = \left[c_{ij}\right],\; c_{ij} = \int z_i z_j p_3(\boldsymbol{z})d\boldsymbol{z} \end{gathered} \tag{2.34}$$

where $(\bullet,\bullet)$ denotes the dot-product of two vectors. By substitution of Eqs. (2.32)–(2.34) into Eq. (2.30), we obtain [4]

$$I(\boldsymbol{X};\boldsymbol{Y}) = \frac{1}{2}\log\frac{\det \boldsymbol{A} \det \boldsymbol{B}}{\det \boldsymbol{C}}. \tag{2.35}$$

The mutual information between two Gaussian random vectors can also be expressed in terms of their correlation coefficients [4]:

$$I(\boldsymbol{X};\boldsymbol{Y}) = -\frac{1}{2}\log\left[\left(1-\rho_1^2\right)..\left(1-\rho_l^2\right)\right],\; l = \min(m,n) \tag{2.36}$$

where ρ_j is the correlation coefficient between X_j and Y_j.

To obtain the information capacity for additive Gaussian noise, we make the following assumptions: (1) input X, output Y, and noise Z are n-dimensional random variables; (2) $\overline{X}_k = 0$, $\overline{X_k^2} = \sigma_{x_k}^2$ $\forall$ k and $\overline{Z}_k = 0$, $\overline{Z_k^2} = \sigma_{z_k}^2$ $\forall$ k; and (3) the noise is additive: $Y = X + Z$. Since we have

$$p_X(\boldsymbol{y}|\boldsymbol{x}) = p_X(\boldsymbol{x}+\boldsymbol{z}|\boldsymbol{x}) = \prod_{k=1}^{n}\left[\frac{1}{(2\pi)^{1/2}\sigma_{z_k}} e^{-z_k^2/2\sigma_{z_k}^2}\right] = p(\boldsymbol{z}), \tag{2.37}$$

the conditional differential entropy can be obtained as

$$H(Y|X) = H(Z) = -\int_{-\infty}^{\infty} p(\boldsymbol{z})\log p(\boldsymbol{z})d\boldsymbol{z}. \tag{2.38}$$

The mutual information is then

$$I(X;Y) = h(Y) - h(Y|X) = h(Y) - h(Z) = h(Y) - \frac{1}{2}\sum_{k=1}^{n}\log 2\,\pi e \sigma_{z_k}^2. \tag{2.39}$$

The information capacity expressed in bits-per-channel use is therefore obtained by maximizing $h(Y)$. Because the distribution maximizing the differential entropy is Gaussian, the information capacity is obtained as

$$C(X;Y) = \frac{1}{2}\sum_{k=1}^{n}\log 2\,\pi e \sigma_{y_k}^2 - \frac{1}{2}\sum_{k=1}^{n}\log 2\,\pi e \sigma_{z_k}^2 = \frac{1}{2}\sum_{k=1}^{n}\log\left(\frac{\sigma_{y_k}^2}{\sigma_{z_k}^2}\right)$$

$$= \frac{1}{2}\sum_{k=1}^{n}\log\left(\frac{\sigma_{x_k}^2 + \sigma_{z_k}^2}{\sigma_{z_k}^2}\right) = \frac{1}{2}\sum_{k=1}^{n}\log\left(1 + \frac{\sigma_{x_k}^2}{\sigma_{z_k}^2}\right), \tag{2.40}$$

For $\sigma_{x_k}^2 = \sigma_x^2$, $\sigma_{z_k}^2 = \sigma_z^2$ we obtain the following expression for information capacity:

$$C(X;Y) = \frac{n}{2}\log\left(1 + \frac{\sigma_x^2}{\sigma_z^2}\right), \tag{2.41}$$

where σ_x^2/σ_z^2 presents the signal-to-noise ratio (SNR). The expression above represents the maximum amount of information that can be transmitted per symbol.

From a practical point of view it is important to determine the amount of information conveyed by the channel per second, which is the information capacity per unit time, also known as the *channel capacity*. For bandwidth-limited channels and the Nyquist signaling employed, there will be $2BW$ samples per second (BW is the channel bandwidth), and the corresponding channel capacity becomes

$$C = W\log\left(1 + \frac{P}{N_o BW}\right) \quad \text{[bits/s]}, \tag{2.42}$$

where P is the average transmitted power and $N_0/2$ is the noise power spectral density (PSD). Eq. (2.42) represents the well-known *information capacity theorem*, commonly referred to as Shannon's third theorem [15]. Since the Gaussian source clearly has maximum entropy, it maximizes the mutual information. Therefore, the equation above can be derived as follows. Let the n-dimensional multivariate $\boldsymbol{X} = [X_1, \ldots, X_n]$ represent the Gaussian channel input with samples generated from zero-mean Gaussian distribution with variance σ_x^2. Let the n-dimensional multivariate $\boldsymbol{Y} = [Y_1, \ldots, Y_n]$ represent the Gaussian channel output with samples spaced $1/(2BW)$ apart. The channel is additive with noise samples generated from zero-mean Gaussian distribution with variance σ_z^2. Let the PDFs of input and output be denoted by $p_1(\boldsymbol{x})$ and $p_2(\boldsymbol{y})$, respectively. Finally, let the joint PDF of channel input and output be denoted by $p(\boldsymbol{x},\boldsymbol{y})$. The maximum mutual information can be calculated from

$$I(\boldsymbol{X};\boldsymbol{Y}) = \iint p(\boldsymbol{x},\boldsymbol{y}) \log \frac{p(\boldsymbol{x},\boldsymbol{y})}{p(\boldsymbol{x})p(\boldsymbol{y})} d\boldsymbol{x}d\boldsymbol{y}. \tag{2.43}$$

By following a procedure similar to that above, we obtain the corresponding expression for Gaussian channel capacity given by Eq. (2.42).

Using Eq. (2.42) and Fano's inequality [2], we obtain

$$H(\boldsymbol{X}|\boldsymbol{Y}) \leq H(P_e) + P_e \log_2(M-1), \quad H(P_e) = -P_e \log_2 P_e - (1-P_e)\log_2(1-P_e) \tag{2.44}$$

for an optical channel amplified spontaneous emission (ASE) noise-dominated scenario and binary phase-shift keying (BPSK) at 40 Gb/s in Fig. 2.6A. We report the minimum BERs against the optical SNR for different code rates. In Fig. 2.6B, we show the minimum channel capacity BERs for different code rates for 64-PPM over a Poisson channel in the presence of background radiation with an average number of background photons of $K_b = 0.2$. For example, it has been shown in Ref. [16] that for 64-PPM over a Poisson channel with serial concatenated turbo code (8184,4092), we are about 1 dB away from channel capacity (for $K_b = 0.2$ at BER of 10^{-5}), while with low-density parity-check (LDPC) (8192,4096) codes of fixed column-weight 6 and irregular row-weight, we are 1.4 dB away from channel capacity.

2.2 Channel coding preliminaries

Two key system parameters—transmitted power and channel bandwidth—together with additive noise sources determine the SNR and correspondingly the BER. In practice, we often come into situations where the target BER cannot be achieved with a given modulation format. For a fixed SNR, the only practical option to change data transmission quality from unacceptable to acceptable is through *channel coding*. Another practical motivation of introducing channel coding is to reduce the required SNR for a given target BER. The energy that can be saved by coding is commonly described as coding gain. *Coding gain* refers to the savings attainable in energy per information bit to the noise spectral density ratio (E_b/N_0) required to achieve a given BER when coding is used, compared with no coding. A typical digital communication system employing channel coding is shown in Fig. 2.7. The discrete source generates information as a sequence of symbols. The channel encoder accepts message symbols and adds redundant symbols according to a corresponding prescribed rule. Channel coding is the act of transforming a length-k sequence into a length-n codeword. The set of rules specifying this transformation is called the channel code and is represented by the following mapping:

$$C: M \rightarrow X, \tag{2.45}$$

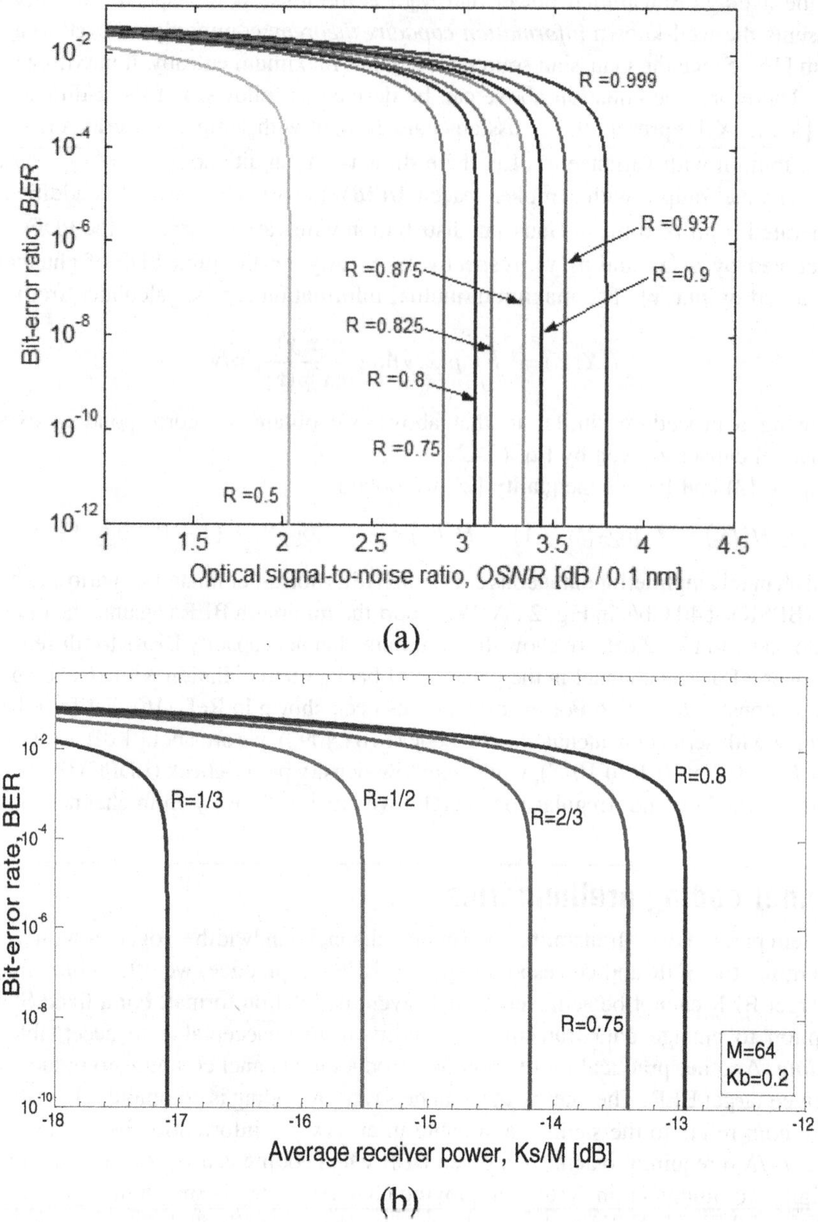

FIGURE 2.6

(A) Minimum bit error rate (BER) against optical signal-to-noise ratio for different code rate values (for binary phase-shift keying at 40 Gb/s). (B) Channel capacity BER curves against average receiver power for 64-pulse-position modulation over a Poisson channel with background radiation of $K_b = 0.2$ and different code rates.

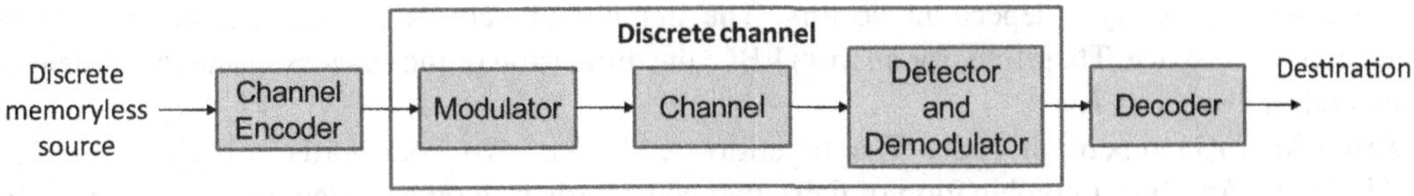

FIGURE 2.7

Block diagram of a point-to-point digital communication system.

where C is the channel code, M is the set of information sequences of length k, and X is the set of codewords of length n. The decoder exploits these redundant symbols to determine which message symbol was actually transmitted. The encoder and decoder consider the whole digital transmission system as a discrete channel. Various classes of channel codes can be categorized into three broad categories: (1) *error detection*, in which we are concerned only with detecting errors during transmission (examples include automatic requests for transmission-ARQs), (2) *forward error correction* (FEC), where we are interested in correcting errors during transmission, and (3) hybrid channel codes that combine the previous two approaches. In this chapter, we are concerned only with FEC.

The key idea behind FEC codes is adding extra redundant symbols to the message to be transmitted and using those redundant symbols in the decoding procedure to correct errors introduced by the channel. Redundancy can be introduced in time, frequency, or space domains. For example, redundancy in the time domain is introduced if the same message is transmitted at least twice—this technique is known as *repetition code*. Space redundancy is used to achieve high spectrally efficient transmission in which modulation is combined with error control.

The codes commonly considered in digital communications and storage applications can be either *block codes* or *convolutional codes*. In (n,k) *block code*, the channel encoder accepts information in successive k-symbol blocks and adds $n-k$ redundant symbols algebraically related to the k message symbols, thereby producing an overall encoded block of n symbols ($n > k$), known as a *codeword*. If the block code is *systematic*, the information symbols stay unchanged during the encoding operation, and the encoding operation is considered as adding $n-k$ generalized parity checks to k information symbols. Since the information symbols are statistically independent (a consequence of source coding or scrambling), the next codeword is independent of the content of the current codeword. The *code rate* of an (n,k) block code is defined as $R = k/n$ and *overhead* by $OH=(1/R-1)\cdot 100\%$. In *convolutional code*, however, the encoding operation is considered the discrete-time convolution of the input sequence with the impulse response of the encoder. Therefore, the $n-k$ generalized parity checks are functions of not only k information symbols but also m previous k-tuples, with $m+1$ being the encoder impulse response length. Statistical dependence is introduced to the window of length $n(m+1)$, the parameter known as the *constraint length* of convolutional codes.

We now provide an elementary introduction to linear block, cyclic, Reed–Solomon (RS), concatenated, and product codes. For a detailed treatment of different error-control coding schemes, the interested reader is referred to Refs. [10,16–22].

2.3 Linear block codes

A *linear block code* (n,k), or LBC, using the language of vector spaces, can be defined as a subspace of a vector space over a finite field GF(q), with q as the prime power. Every space is described by its *basis*—a set of linearly independent vectors. The number of vectors in the basis determines the dimension of the space. Therefore, for an (n,k) LBC, the dimension of the space is n, and the dimension of the code subspace is k.

Example: $(n,1)$ repetition code. The repetition code has two codewords, $\boldsymbol{x}_0=(00 \ldots 0)$ and $\boldsymbol{x}_1=(11 \ldots 1)$. Any linear combination of these two codewords is another codeword, as shown below:

$$\boldsymbol{x}_0+\boldsymbol{x}_0=\boldsymbol{x}_0$$

$$\boldsymbol{x}_0+\boldsymbol{x}_1=\boldsymbol{x}_1+\boldsymbol{x}_0=\boldsymbol{x}_1$$

$$\boldsymbol{x}_1+\boldsymbol{x}_1=\boldsymbol{x}_0$$

The set of codewords from an LBC forms a group under the addition operation, because the all-zero codeword serves as the identity element, and the codeword itself serves as the inverse element. This is the reason why LBCs are also called group codes. The LBC (n,k) can be observed as a k-dimensional subspace of the vector space of all n-tuples over the binary file $GF(2)=\{0,1\}$, with addition and multiplication rules given in Table 2.1. All n-tuples over GF(2) form the vector space. The sum of two n-tuples $\boldsymbol{a}=(a_1\ a_2 \ldots a_n)$ and $\boldsymbol{b}=(b_1\ b_2 \ldots b_n)$ is clearly an n-tuple and commutative rule is valid because $\boldsymbol{c}=\boldsymbol{a}+\boldsymbol{b}=(a_1+b_1\ a_2+b_2 \ldots a_n+b_n)=(b_1+a_1\ b_2+a_2 \ldots b_n+a_n)=\boldsymbol{b}+\boldsymbol{a}$. The all-zero vector $\boldsymbol{0}=(0\ 0 \ldots 0)$ is the identity element, while n-tuple $\boldsymbol{a}$ itself is the inverse element $\boldsymbol{a}+\boldsymbol{a}=\boldsymbol{0}$. Therefore, the n-tuples form the Abelian group with respect to the addition operation. The scalar multiplication is defined by: $\alpha\boldsymbol{a}=(\alpha\ a_1\ \alpha a_2 \ldots \alpha a_n)$, $\alpha\in GF(2)$. The distributive laws

$$\alpha(\boldsymbol{a}+\boldsymbol{b})=\alpha\boldsymbol{a}+\alpha\boldsymbol{b}$$

$$(\alpha+\beta)\boldsymbol{a}=\alpha\boldsymbol{a}+\beta\boldsymbol{a},\ \forall\alpha,\beta\in GF(2)$$

are also valid. The associate law $(\alpha\cdot\beta)\boldsymbol{a}=\alpha\cdot(\beta\boldsymbol{a})$ is clearly satisfied. Therefore, the set of all n-tuples is a vector space over GF(2). The set of all codewords from an (n,k) LBC forms an Abelian group under the addition operation. It can be shown, in a fashion similar to that above, that all codewords of (n,k) LBCs form the vector space of dimensionality k. There exist k basis vectors (codewords) such that every codeword is a linear combination of these codewords.

Example: $(n,1)$ repetition code of $C=\{(0\ 0 \ldots 0),\ (1\ 1 \ldots 1)\}$. Two codewords in C can be represented as a linear combination of an all-ones basis vector: $(11 \ldots 1)=1\cdot(11 \ldots 1)$, $(00 \ldots 0)=1\cdot(11 \ldots 1)+1\cdot(11 \ldots 1)$.

Table 2.1 Addition (+) and multiplication (·) rules.

+	0	1	·	0	1
0	0	1	0	0	0
1	1	0	1	0	1

2.3.1 Generator matrix for linear block code

Any codeword x from the (n,k) LBC can be represented as a linear combination of k basis vectors g_i $(i = 0,1,\ldots,k-1)$, as given below:

$$x = m_0 g_0 + m_1 g_1 + \ldots + m_{k-1} g_{k-1} = m \begin{bmatrix} g_0 \\ g_1 \\ \cdots \\ g_{k-1} \end{bmatrix} = mG; \quad G = \begin{bmatrix} g_0 \\ g_1 \\ \cdots \\ g_{k-1} \end{bmatrix}, \quad m = (m_0 \quad m_1 \quad \ldots \quad m_{k\text{-}1}) \tag{2.46}$$

where m is the message vector, G is the generator matrix (of dimensions $k \times n$), and every row represents a vector from the coding subspace. Therefore, to encode, the message vector $m(m_0,m_1, \ldots,m_{k-1})$ must be multiplied with a generator matrix G to obtain $x = mG$, where $x(x_0,x_1, \ldots,x_{n-1})$ is a codeword.

Example: Generator matrices for repetition $(n,1)$ code G_{rep} and $(n,n-1)$ single-parity-check code G_{par} are given, respectively, as

$$G_{\text{rep}} = [11\ldots1] \quad G_{\text{par}} = \begin{bmatrix} 100\ldots01 \\ 010\ldots01 \\ \cdots \\ 000\ldots11 \end{bmatrix}.$$

By elementary operations on rows in the generator matrix, the code may be transformed into systematic form:

$$G_s = [I_k|P], \tag{2.47}$$

where I_k is unity matrix of dimensions $k \times k$, and P is the coefficient matrix of dimensions $k \times (n-k)$ with columns denoting the positions of parity checks:

$$P = \begin{bmatrix} p_{00} & p_{01} & \cdots & p_{0,n-k-1} \\ p_{10} & p_{11} & \cdots & p_{1,n-k-1} \\ \cdots & & \cdots & \cdots \\ p_{k-1,0} & p_{k-1,1} & \cdots & p_{k-1,n-k-1} \end{bmatrix}.$$

The codeword of a systematic code is obtained by

$$x = [m|b] = m[I_k|P] = mG, \quad G = [I_k|P], \tag{2.48}$$

and the structure of the systematic codeword is shown in Fig. 2.8.

Therefore, during encoding, the message vector is unchanged, and the elements of vector of parity checks b are obtained by

$$b_i = p_{0i} m_0 + p_{1i} m_1 + \ldots + p_{k-1,i} m_{k-1}, \tag{2.49}$$

FIGURE 2.8

Structure of systematic codeword.

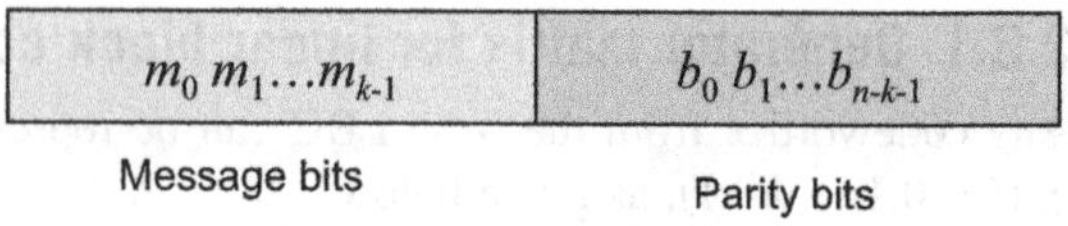

where

$$p_{ij} = \begin{cases} 1, & \text{if } b_i \text{ depends on } m_j \\ 0, & \text{otherwise} \end{cases}.$$

During transmission, the channel introduces errors, so the received vector $\boldsymbol{r}$ can be written as $\boldsymbol{r} = \boldsymbol{x} + \boldsymbol{e}$, where $\boldsymbol{e}$ is the error vector (pattern) with components determined by

$$e_i = \begin{cases} 1 & \text{if an error occured in the} i\text{th location} \\ 0 & \text{otherwise} \end{cases}$$

To determine whether the received vector $\boldsymbol{r}$ is a codeword vector, we introduce the concept of a *parity-check matrix.*

2.3.2 Parity-check matrix for linear block code

A useful matrix associated with LBCs is the parity-check matrix. Let us expand the matrix equation $\boldsymbol{x} = \boldsymbol{mG}$ in scalar form as follows:

$$\begin{aligned}
x_0 &= m_0 \\
x_1 &= m_1 \\
&\ldots \\
x_{k-1} &= m_{k-1} \\
x_k &= m_0 p_{00} + m_1 p_{10} + \ldots + m_{k-1} p_{k-1,0} \\
x_{k+1} &= m_0 p_{01} + m_1 p_{11} + \ldots + m_{k-1} p_{k-1,1} \\
&\ldots \\
x_{n-1} &= m_0 p_{0,n-k-1} + m_1 p_{1,n-k-1} + \ldots + m_{k-1} p_{k-1,n-k-1}
\end{aligned} \tag{2.50}$$

Using the first k equalities, the last $n-k$ equations can be rewritten as follows:

$$\begin{aligned}
&x_0 p_{00} + x_1 p_{10} + \ldots + x_{k-1} p_{k-1,0} + x_k = 0 \\
&x_0 p_{01} + x_1 p_{11} + \ldots + x_{k-1} p_{k-1,0} + x_{k+1} = 0 \\
&\ldots \\
&x_0 p_{0,n-k+1} + x_1 p_{1,n-k-1} + \ldots + x_{k-1} p_{k-1,n-k+1} + x_{n-1} = 0
\end{aligned} \tag{2.51}$$

The matrix representation of (2.51) is given below:

$$\begin{bmatrix} x_0 & x_1 & \cdots & x_{n-1} \end{bmatrix} \begin{bmatrix} p_{00} & p_{10} & \cdots & p_{k-1,0} & 1 & 0 & \ldots & 0 \\ p_{01} & p_{11} & \cdots & p_{k-1,1} & 0 & 1 & \ldots & 0 \\ \cdots & \cdots & & \cdots & & \cdots & & \cdots \\ p_{0,n-k-1} & p_{1,n-k-1} & \cdots & p_{k-1,n-k-1} & 0 & 0 & \ldots & 1 \end{bmatrix}^T = \boldsymbol{x}\begin{bmatrix} \boldsymbol{P}^T & \boldsymbol{I}_{n-k} \end{bmatrix}$$

$$= \boldsymbol{x}\boldsymbol{H}^T = \boldsymbol{0}, \quad \boldsymbol{H} = \begin{bmatrix} \boldsymbol{P}^T & \boldsymbol{I}_{n-k} \end{bmatrix}_{(n-k)xn} \tag{2.52}$$

The $\boldsymbol{H}$-matrix in (2.52) is known as the *parity-check matrix*. We can easily verify that

$$\boldsymbol{G}\boldsymbol{H}^T = [\boldsymbol{I}_k \quad \boldsymbol{P}]\begin{bmatrix} \boldsymbol{P} \\ \boldsymbol{I}_{n-k} \end{bmatrix} = \boldsymbol{P} + \boldsymbol{P} = \boldsymbol{0}, \tag{2.53}$$

meaning that the parity check matrix of an (n,k) LBC $\boldsymbol{H}$ is a matrix of rank $n-k$ and dimensions $(n-k) \times n$ whose null-space is a k-dimensional vector with its basis forming the generator matrix $\boldsymbol{G}$.

Example: Parity-check matrices for $(n,1)$ repetition code $\boldsymbol{H}_{\text{rep}}$ and $(n,n-1)$ single-parity check code $\boldsymbol{H}_{\text{par}}$ are given respectively as

$$\boldsymbol{H}_{\text{rep}} = \begin{bmatrix} 100\ldots01 \\ 010\ldots01 \\ \ldots \\ 000\ldots11 \end{bmatrix} \qquad \boldsymbol{H}_{\text{par}} = [11\ldots1].$$

Example: For Hamming (7,4) code, the generator $\boldsymbol{G}$ and parity check $\boldsymbol{H}$ matrices are given, respectively, as

$$\boldsymbol{G} = \begin{bmatrix} 110|1000 \\ 011|0100 \\ 111|0010 \\ 101|0001 \end{bmatrix}, \quad \boldsymbol{H} = \begin{bmatrix} 100|1011 \\ 010|1110 \\ 001|0111 \end{bmatrix}$$

Every (n,k) LBC with generator matrix $\boldsymbol{G}$ and parity-check matrix $\boldsymbol{H}$ has a dual code with generator matrix $\boldsymbol{H}$ and parity check matrix $\boldsymbol{G}$. For example, $(n,1)$ repetition and $(n,n-1)$ single-parity check codes are dual.

2.3.3 Distance properties of linear block codes

To determine the *error correction capability* of the code, we have to introduce the concept of Hamming distance and Hamming weight. The *Hamming distance* between two codewords $\boldsymbol{x}_1$ and $\boldsymbol{x}_2$, $d(\boldsymbol{x}_1,\boldsymbol{x}_2)$ is defined as the number of locations in which their respective elements differ. The *Hamming weight*, $w(\boldsymbol{x})$, of a codeword vector $\boldsymbol{x}$ is defined as the number of nonzero elements in the vectors. The *minimum distance*, $d_{\min}$, of an LBC is defined as the smallest Hamming distance between any pair of code vectors in the code. Since the zero-vector is a codeword, the minimum distance of an LBC can be determined simply as the smallest Hamming weight of the nonzero code vectors in the code. Let the parity check matrix be written as $\boldsymbol{H} = [\boldsymbol{h}_1\ \boldsymbol{h}_2\ \ldots\ \boldsymbol{h}_n]$, where $\boldsymbol{h}_i$ is the ith column in $\boldsymbol{H}$. Since every codeword $\boldsymbol{x}$ must satisfy the syndrome equation $\boldsymbol{x}\boldsymbol{H}^{\mathrm{T}} = \boldsymbol{0}$ (see Eq. 2.52), the minimum distance of an LBC is determined by the minimum number of columns of the $\boldsymbol{H}$-matrix whose sum equals the zero vector. For example, the (7,4) Hamming code in the example above has a minimum distance $d_{\min} = 3$ since adding the first, fifth, and sixth columns leads to a zero vector. The codewords can be represented as points in n-dimensional space, as shown in Fig. 2.9. The decoding process can be visualized by creating spheres of radius t around codeword points. The received word vector $\boldsymbol{r}$ in Fig. 2.9A will be decoded as a codeword $\boldsymbol{x}_i$ because its Hamming distance $d(\boldsymbol{x}_i,\boldsymbol{r}) \leq t$ is closest to the codeword $\boldsymbol{x}_i$. On the other hand, in the example shown in Fig. 2.9B, the Hamming distance $d(\boldsymbol{x}_i,\boldsymbol{x}_j) \leq 2t$ and received vector $\boldsymbol{r}$ that fall in the intersection area of two spheres cannot be uniquely decoded.

Therefore, an (n,k) LBC of minimum distance $d_{\min}$ can correct up to t errors if, and only if, $t \leq \lfloor 1/2(d_{\min}-1) \rfloor$ (where $\lfloor\rfloor$ denotes the largest integer less than or equal to the enclosed quantity) or equivalently, $d_{\min} \geq 2t + 1$. If we are only interested in detecting e_d errors, then $d_{\min} \geq e_d + 1$. Finally, if we are interested in detecting e_d errors and correcting e_c errors, $d_{\min} \geq e_d + e_c + 1$. The Hamming (7,4) code is therefore a single-error-correcting and double-error-detecting code. More generally, (n,k) LBCs with the following parameters:

- Block length: $n = 2^m - 1$
- Number of message bits: $k = 2^m - m - 1$
- Number of parity bits: $n - k = m$
- $d_{\min} = 3$

where $m \geq 3$, are in the family of *Hamming* codes. Hamming codes belong to the class of perfect codes, which satisfy the Hamming inequality below with equality sign [2,11]:

$$2^{n-k} \geq \sum_{i=0}^{t} \binom{n}{i}. \tag{2.54}$$

This bound gives how many errors t can be corrected with an (n,k) LBC using syndrome decoding (described below).

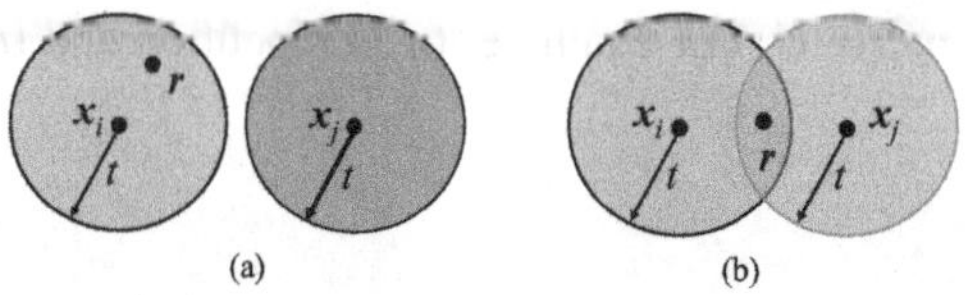

FIGURE 2.9

Illustration of Hamming distance: (A) $d(\boldsymbol{x}_i,\boldsymbol{x}_j) \geq 2t+1$, and (B) $d(\boldsymbol{x}_i,\boldsymbol{x}_j) < 2t+1$.

2.3.4 Coding gain

A very important characteristics of an (*n*,*k*) LBC is the so-called *coding gain*, which was introduced above as the savings attainable in the energy per information bit to noise spectral density ratio (E_b/N_0) required to achieve a given bit error probability when coding is used compared with no coding. Let E_c denote the transmitted bit energy and E_b denote the information bit energy. Since the total information word energy kE_b must be the same as the total codeword energy nE_c, we obtain the following relationship between E_c and E_b:

$$E_c = (k/n)E_b = RE_b. \tag{2.55}$$

The probability of error for BPSK on an additive white Gaussian noise (AWGN) channel, when a coherent hard decision (bit-by-bit) demodulator is used, can be obtained as follows:

$$p = \frac{1}{2}\operatorname{erfc}\left(\sqrt{\frac{E_c}{N_0}}\right) = \frac{1}{2}\operatorname{erfc}\left(\sqrt{\frac{RE_b}{N_0}}\right), \tag{2.56}$$

where the erfc(x) function is defined by

$$\operatorname{erfc}(x) = \frac{2}{\sqrt{\pi}}\int_x^{+\infty} e^{-z^2}dz.$$

For high SNRs, the word error probability (remaining upon decoding) of a t-error-correcting code is dominated by a $t+1$ error event:

$$P_w(e) \approx \binom{n}{t+1}p^{t+1}(1-p)^{n-t+1} \approx \binom{n}{t+1}p^{t+1}. \tag{2.57}$$

The bit error probability P_b is related to the word error probability by

$$P_b \approx \frac{2t+1}{n}P_w(e) \approx c(n,t)p^{t+1}, \tag{2.58}$$

because $2t+1$ or more errors per codeword cannot be corrected, and they can be located anywhere on n codeword locations, and $c(n,t)$ is a parameter dependent on error-correcting capability t and codeword length n. Using the upper bound on erfc(x), we obtain

$$P_b \approx \frac{c(n,t)}{2}\left[\exp\left(\frac{-RE_b}{N_0}\right)\right]^{t+1}. \tag{2.59}$$

The corresponding approximation for the uncoded case is

$$P_{b,\text{uncoded}} \approx \frac{1}{2}\exp\left(-\frac{E_b}{N_0}\right). \tag{2.60}$$

By equating Eqs. (2.59) and (2.60) and ignoring the parameter $c(n,t)$, we obtain the following expression for hard decision decoding coding gain:

$$\frac{(E_b/N_0)_{\text{uncoded}}}{(E_b/N_0)_{\text{coded}}} \approx R(t+1). \tag{2.61}$$

The corresponding soft decision asymptotic coding gain of convolutional codes is [23–27]:

$$\frac{(E_b/N_0)_{\text{uncoded}}}{(E_b/N_0)_{\text{coded}}} \approx Rd_{\min}, \tag{2.62}$$

and is about 3 dB better than hard decision decoding (because $d_{\min} \geq 2t + 1$).

In optical communications, it is very common to use the Q-factor as the figure of merit instead of SNR, which is related to the BER on an AWGN, as follows:

$$\text{BER} = \frac{1}{2}\text{erfc}\left(\frac{Q}{\sqrt{2}}\right). \tag{2.63}$$

Let BER_{in} denote the BER at the input of the FEC decoder, let BER_{out} denote the BER at the output of the FEC decoder, and let BER_{ref} denote the target BER (such as 10^{-12} or 10^{-15}). The corresponding coding gain (CG) and net coding gain (NCG) are respectively defined as [28]

$$\text{CG} = 20\log_{10}\left[\text{erfc}^{-1}(2\text{BER}_{\text{ref}})\right] - 20\log_{10}\left[\text{erfc}^{-1}(2\text{BER}_{\text{in}})\right] \ [\text{dB}], \tag{2.64}$$

$$\text{NCG} = 20\log_{10}\left[\text{erfc}^{-1}(2\text{BER}_{\text{ref}})\right] - 20\log_{10}\left[\text{erfc}^{-1}(2\text{BER}_{\text{in}})\right] + 10\log_{10}R \ [\text{dB}]. \tag{2.65}$$

In fact, all coding gains reported in this chapter are NCGs, although they are sometimes simply called coding gains because that is the common practice in the coding theory literature [23–27].

2.3.5 Syndrome decoding and standard array

The received vector $\boldsymbol{r} = \boldsymbol{x} + \boldsymbol{e}$ ($\boldsymbol{x}$ is the codeword, and $\boldsymbol{e}$ is the error pattern introduced above) is a codeword if the following *syndrome equation* is satisfied: $\boldsymbol{s} = \boldsymbol{rH}^{\text{T}} = \boldsymbol{0}$. The syndrome has the following important properties:

1. The syndrome is only a function of the error pattern. This property can easily be proved from the definition of syndrome as follows: $\boldsymbol{s} = \boldsymbol{rH}^{\text{T}} = (\boldsymbol{x} + \boldsymbol{e})\boldsymbol{H}^{\text{T}} = \boldsymbol{xH}^{\text{T}} + \boldsymbol{eH}^{\text{T}} = \boldsymbol{eH}^{\text{T}}$.
2. All error patterns that differ by a codeword have the same syndrome. This property can also be proved from the syndrome definition. Let $\boldsymbol{x}_i$ be the ith ($i = 0,1, \ldots,2^{k-1}$) codeword. The set of error patterns that differ by a codeword is known as coset: $\{\boldsymbol{e}_i = \boldsymbol{e} + \boldsymbol{x}_i;\ i = 0,1, \ldots,2^{k-1}\}$. The syndrome corresponding to ith error pattern from this set $\boldsymbol{s}_i = \boldsymbol{r}_i\boldsymbol{H}^{\text{T}} = (\boldsymbol{x}_i + \boldsymbol{e})\boldsymbol{H}^{\text{T}} = \boldsymbol{x}_i\boldsymbol{H}^{\text{T}} + \boldsymbol{eH}^{\text{T}} = \boldsymbol{eH}^{\text{T}}$ is only a function of the error pattern, and therefore all error patterns from the coset have the same syndrome.
3. The syndrome is a function of only those columns of the parity-check matrix corresponding to error locations. The parity-check matrix can be written in the following form: $\boldsymbol{H} = [\boldsymbol{h}_1 \ldots \boldsymbol{h}_n]$, where the ith element $\boldsymbol{h}_i$ denotes the ith column of $\boldsymbol{H}$. Based on the syndrome definition for an error pattern $\boldsymbol{e} = [e_1 \ldots e_n]$, the following is valid:

$$\boldsymbol{s} = \boldsymbol{eH}^T = [e_1 \quad e_2 \quad \ldots \quad e_n]\begin{bmatrix} \boldsymbol{h}_1^T \\ \boldsymbol{h}_2^T \\ \ldots \\ \boldsymbol{h}_n^T \end{bmatrix} = \sum_{i=1}^{n} e_i \boldsymbol{h}_i^T, \tag{2.66}$$

which proves the claim of property 3.

4. With syndrome decoding, an (n,k) LBC can correct up to t errors, provided that the Hamming bound (2.54) is satisfied. (This property will be proved in next subsection.)

Using property 2, 2^k codewords partition the space of all received words into 2^k disjoint subsets. Any received word within the subset will be decoded as a unique codeword. A *standard array* is a technique by which this partition can be achieved and can be constructed using the following two steps [11,12,16,17,19,20,22]:

1. Write 2^k codewords as elements of the first row, with the all-zero codeword as the leading element.
2. Repeat the following steps 2(a) and 2(b) until all 2^n words are exhausted.
   - **(a)** Of the remaining unused n-tuples, select the one with the least weight for the leading element of the next row.
   - **(b)** Complete the current row by adding the leading element to each nonzero codeword appearing in the first row and writing the resulting sum in the corresponding column.

The standard array for an (n, k) block code obtained by this algorithm is illustrated in Fig. 2.10. The columns represent 2^k disjoint sets, and every row represents the coset of the code, with leading elements being called the coset leaders:

$$\begin{array}{ccccccc}
\mathbf{x}_1 = 0 & \mathbf{x}_2 & \mathbf{x}_3 & \cdots & \mathbf{x}_i & \cdots & \mathbf{x}_{2^k} \\
\mathbf{e}_2 & \mathbf{x}_2+\mathbf{e}_2 & \mathbf{x}_3+\mathbf{e}_2 & \cdots & \mathbf{x}_i+\mathbf{e}_2 & \cdots & \mathbf{x}_{2^k}+\mathbf{e}_2 \\
\mathbf{e}_3 & \mathbf{x}_2+\mathbf{e}_3 & \mathbf{x}_3+\mathbf{e}_3 & \cdots & \mathbf{x}_i+\mathbf{e}_2 & \cdots & \mathbf{x}_{2^k}+\mathbf{e}_3 \\
\cdots & \cdots & & \cdots & & \cdots & \cdots \\
\mathbf{e}_j & \mathbf{x}_2+\mathbf{e}_j & \mathbf{x}_3+\mathbf{e}_j & \cdots & \mathbf{x}_i+\mathbf{e}_j & \cdots & \mathbf{x}_{2^k}+\mathbf{e}_j \\
\cdots & \cdots & & \cdots & & & \cdots \\
\mathbf{e}_{2^{n-k}} & \mathbf{x}_2+\mathbf{e}_{2^{n-k}} & \mathbf{x}_3+\mathbf{e}_{2^{n-k}} & \cdots & \mathbf{x}_i+\mathbf{e}_{2^{n-k}} & \cdots & \mathbf{x}_{2^k}+\mathbf{e}_{2^{n-k}}
\end{array}$$

FIGURE 2.10

The standard array architecture.

Example: The standard array of (5,2) code $C = \{(00000), (11010), (10101), (01111)\}$ is given in Table 2.2. The parity-check matrix of this code is given by

$$\mathbf{H} = \begin{bmatrix} 1 & 0 & 0 & 1 & 1 \\ 0 & 1 & 0 & 1 & 0 \\ 0 & 0 & 1 & 0 & 1 \end{bmatrix}.$$

Because the minimum distance of this code is 3 (first, second, and fourth columns add to zero), this code can correct all single errors. For example, if the word 01010 is received, it will be decoded to the topmost codeword 11010 of the column in which it lies. In the same table, corresponding syndromes are also provided.

Table 2.2 Standard array of (5,2) code and corresponding decoding table.

	Codewords				Syndrome s	Error pattern
Coset leader	00000	11010	10101	01111	000	00000
	00001	11011	10100	01110	101	00001
	00010	11000	10111	01101	110	00010
	00100	11110	10001	01011	001	00100
	01000	10010	11101	00111	010	01000
	10000	01010	00101	11111	100	10000
	00011	11001	10110	01100	011	00011
	00110	11100	10011	01001	111	00110

The syndrome decoding procedure is a three-step procedure [11,12,23–27]:

1. For the received vector $\boldsymbol{r}$, compute the syndrome $\boldsymbol{s}=\boldsymbol{r}\boldsymbol{H}^{\mathrm{T}}$. From property 3, we can establish one-to-one correspondence of the syndromes and error patterns (see Table 2.2), leading to the lookup table (LUT) containing the syndrome and corresponding error pattern (the coset leader).
2. Within the coset characterized by the syndrome $\boldsymbol{s}$, identify the coset leader, say, $\boldsymbol{e}_0$. The coset leader corresponds to the error pattern with the largest probability of occurrence.
3. Decode the received vector as $\boldsymbol{x}=\boldsymbol{r}+\boldsymbol{e}_0$.

Example: Let the received vector for the (5,2) code example above be $\boldsymbol{r}$=(01010). The syndrome can be computed as $\boldsymbol{s}=\boldsymbol{r}\boldsymbol{H}^{\mathrm{T}}$=(100), and the corresponding error pattern from LUT is found to be $\boldsymbol{e}_0$=(10000). The decoded word is obtained by adding the error pattern to the received word $\boldsymbol{x}=\boldsymbol{r}+\boldsymbol{e}_0$=(11010), and the error on the first bit position is corrected.

The standard array can be used to determine the probability of word error as follows:

$$P_w(e)=1-\sum_{i=0}^{n}\alpha_i p^i(1-p)^{n-i}, \tag{2.67}$$

where α_i is the number of coset leaders of weight i (the distribution of weights is also known as the *weight distribution* of the coset leaders), and p is the crossover probability of BSC. Any error pattern that is not a coset leader will result in a decoding error. For example, the weight distribution of coset leaders in (5,2) code is $\alpha_0=1$, $\alpha_1=5$, $\alpha_2=2$, $\alpha_i=0$, $i=3,4,5$, which leads to the following word error probability:

$$P_w(e)=1-(1-p)^5-5p(1-p)^4-2p^2(1-p)^3\big|_{p=10^{-3}}=7.986\cdot 10^{-6}.$$

We can use Eq. (2.67) to estimate the coding gain of a given LBC. For example, the word error probability for Hamming (7,4) code is

$$P_w(e)=1-(1-p)^7-7p(1-p)^6=\sum_{i=2}^{7}\binom{7}{i}p^i(1-p)^{7-i}\approx 21p^2.$$

In the previous section, we established the following relationship between bit and word error probabilities: $P_b \approx P_w(e)(2t+1)/n = (3/7)P_w(e) \approx (3/7)21p^2 = 9p^2$. Therefore, the crossover probability can be evaluated as

$$p = \sqrt{P_b}/3 = (1/2)\text{erfc}\left(\sqrt{\frac{RE_b}{N_0}}\right).$$

From this expression, we can easily calculate the required SNR to achieve the target P_b. By comparing such an obtained SNR with a corresponding SNR for the uncoded BPSK, we can evaluate the corresponding coding gain.

To evaluate the probability of undetected error, we must determine the other codeword weights as well. Because undetected errors are caused by error patterns identical to the nonzero codewords, the undetected error probability can be evaluated by

$$P_u(e) = \sum_{i=1}^{n} A_i p^i (1-p)^{n-i} = (1-p)^n \sum_{i=1}^{n} A_i \left(\frac{p}{1-p}\right), \tag{2.68}$$

where p is the crossover probability and A_i denotes the number of codewords of weight i. The codeword weight can be determined by the *McWilliams identity*, which establishes the connection between codeword weights A_i and the codeword weights of corresponding dual code B_i, by [23]

$$A(z) = 2^{-(n-k)}(1+z)^n B\left(\frac{1-z}{1+z}\right), \; A(z) = \sum_{i=0}^{n} A_i z^i, \; B(z) = \sum_{i=0}^{n} B_i z^i \tag{2.69}$$

where $A(z)$ ($B(z)$) is the polynomial representation of codeword weights (dual codeword weights). By substituting $z = p/(1\text{-}p)$ in (2.69) and knowing that $A_0 = 1$, we obtain

$$A\left(\frac{p}{1-p}\right) - 1 = \sum_{i=1}^{n} A\left(\frac{p}{1-p}\right)^i. \tag{2.70}$$

Substituting Eq. (2.70) into Eq. (2.68), we obtain

$$P_u(e) = (1-p)^n \left[A\left(\frac{p}{1-p}\right) - 1\right]. \tag{2.71}$$

An alternative expression for $P_u(e)$ in terms of $B(z)$ can be obtained from Eq. (2.69), which is more suitable for use when $n-k < k$, as follows:

$$P_u(e) = 2^{-(n-k)} B(1-2p) - (1-p)^n. \tag{2.72}$$

For large n, k, $n-k$, the McWilliams identity is impractical; in this case, an upper bound on the average probability of the undetected error of an (n,k) systematic code should be used instead:

$$\overline{P_u}(e) \le 2^{-(n-k)}[1-(1-p)^n]. \tag{2.73}$$

For a q-ary maximum-distance separable (MDS) code, which satisfies the Singleton bound with equality introduced in the next section, we can determine a closed formula for weight distribution [23]:

$$A_i = \binom{n}{i}(q-1)\sum_{j=0}^{i-d_{\min}} (-1)^j \binom{i-1}{j} q^{i-d_{\min}-j}, \tag{2.74}$$

where $d_{\min}$ is the minimum distance of the code, $A_0 = 1$, and $A_i = 0$ for $i \in [1, d_{\min}-1]$.

2.3.6 Important coding bounds

In this section, we describe several important coding bounds including Hamming, Plotkin, Gilbert–Varshamov, and Singleton [11,12,23–27]. The *Hamming* bound has already been introduced for binary LBCs in Eq. (2.54). The Hamming bound for a q-ary (n,k) LBC is given by

$$\left[1+(q-1)\binom{n}{1}+(q-1)^2\binom{n}{2}+\ldots+(q-1)^i\binom{n}{i}+\ldots+(q-1)^t\binom{n}{t}\right]q^k \le q^n, \quad (2.75)$$

where t is the error correction capability and $(q-1)^i\binom{n}{i}$ is the number of received words that differ from a given codeword in i symbols. Namely, there are n chooses i ways in which symbols can be chosen from n and $(q\text{-}1)^i$ possible choices for symbols. The codes satisfying the Hamming bound with equality sign are known as *perfect codes*. Hamming codes are perfect codes because $n=2^{n-k}-1$ (which is equivalent to $(1+n)2^k=2^n$), so the above inequality is satisfied with equality. The $(n,1)$ repetition code is also a perfect code. The 3-error-correcting (23,12) Golay code is another example of perfect code because

$$\left[1+\binom{23}{1}+\binom{23}{2}+\binom{23}{3}\right]2^{12}=2^{23}.$$

The *Plotkin* bound is the bound on the minimum distance of a code:

$$d_{\min} \le \frac{n2^{k-1}}{2^k-1}. \quad (2.76)$$

Namely, if all codewords are written as the rows of a $2^k \times n$ matrix, each column will contain 2^{k-1} 0s and 2^{k-1} 1s, with the total weight of all codewords being $n2^{k-1}$.

The *Gilbert–Varshamov* bound is based on the property that the minimum distance $d_{\min}$ of an (n,k) LBC can be determined as the minimum number of columns in an $\boldsymbol{H}$-matrix that sum to zero:

$$\binom{n-1}{1}+\binom{n-1}{2}+\ldots+\binom{n-1}{d_{\min}-2} < 2^{n-k}-1. \quad (2.77)$$

Another important bound is the *Singleton bound*:

$$d_{\min} \le n-k+1. \quad (2.78)$$

This bound is straightforward to prove. Let only 1 bit, of value 1, be present in an information vector. If it is involved in $n-k$ parity checks, then the total number of 1's in the codeword cannot be more than $n-k+1$. The codes satisfying the Singleton bound with equality sign are MDS codes (e.g., RS codes are MDS codes).

2.4 Cyclic codes

The most commonly used class of LBCs is cyclic codes. Examples of cyclic codes include BCH, Hamming, and Golay codes. RS codes are also cyclic but nonbinary. Even LDPC codes can be designed in cyclic or quasi-cyclic fashion.

Let us observe a vector space of dimension n. The subspace of this space is *cyclic code* if for any codeword $\boldsymbol{c}(c_0,c_1,\ldots,c_{n-1})$, an arbitrary cyclic shift $\boldsymbol{c}_j(c_{n-j},c_{n-j+1},\ldots,c_{n-1},c_0,c_1,\ldots,c_{n-j-1})$ is another codeword. With every codeword $\boldsymbol{c}(c_0,c_1,\ldots,c_{n-1})$ from a cyclic code, we associate the *codeword polynomial*:

$$c(x) = c_0 + c_1x + c_2x^2 + \cdots + c_{n-1}x^{n-1}. \quad (2.79)$$

The jth cyclic shift, observed mod(x^n-1), is also a codeword polynomial:

$$c^{(j)} = x^j c(x)\mathrm{mod}(x^n - 1). \quad (2.80)$$

It is straightforward to show that the observed subspace is cyclic if composed from polynomials divisible by a polynomial $g(x) = g_0 + g_1x + \ldots + g_{n-k}x^{n-k}$ that divides x^n-1 at the same time. The polynomial $g(x)$ of degree $n-k$ is called the *generator polynomial* of the code. If $x^n-1 = g(x)h(x)$, the polynomial of degree k is called the *parity-check polynomial*. The generator polynomial has the following three important properties [11,12,23–27]:

1. The generator polynomial of an (n,k) cyclic code is unique (usually proved by contradiction),
2. Any multiple of a generator polynomial is a codeword polynomial, and
3. The generator polynomial and parity-check polynomial are factors of x^n-1.

The generator polynomial $g(x)$ and parity-check polynomial $h(x)$ serve the same role as the generator matrix $\boldsymbol{G}$ and parity check matrix $\boldsymbol{H}$ of an LBC. The n-tuples related to the k polynomials $g(x), xg(x), \ldots, x^{k-1}g(x)$ may be used in rows of the $k\times n$ generator matrix $\boldsymbol{G}$, while n-tuples related to the $(n-k)$ polynomials $x^k h(x^{-1}), x^{k+1}h(x^{-1}), \ldots, x^{n-1}h(x^{-1})$ may be used in rows of the $(n-k)\times n$ parity-check matrix $\boldsymbol{H}$.

To encode, we simply multiply the message polynomial $m(x) = m_0 + m_1x + \ldots + m_{k-1}x^{k-1}$ with the generator polynomial $g(x)$, i.e., $c(x) = m(x)g(x)\mathrm{mod}(x^n-1)$, where $c(x)$ is the codeword polynomial. To encode in *systematic* form, we must find the reminder of $x^{n-k}m(x)/g(x)$ and add it to the shifted version of the message polynomial $x^{n-k}m(x)$, i.e., $c(x) = x^{n-k}m(x) + \mathrm{rem}[x^{n-k}m(x)/g(x)]$, where rem[ ] is denoted the remainder of a given entity. The general circuit for generating the codeword polynomial in systematic form is given in Fig. 2.11A. The encoder operates as follows. When switch S is in position 11 and the gate is closed (on), the information bits are shifted into the shift register and simultaneously transmitted onto the channel. Once all information bits are shifted into the register in k shifts, with the gate open (off), switch S is moved into position 2, and the content of the $(n-k)$-shift register is transmitted onto the channel.

To determine whether the received word polynomial is the codeword polynomial $r(x) = r_0 + r_1x + \ldots + r_{n-1}x^{n-1}$, we can simply determine the *syndrome polynomial* $s(x) = \mathrm{rem}[r(x)/g(x)]$. If $s(x)$ is zero, there is no error introduced during transmission. The corresponding circuit is shown in Fig. 2.11B.

For example, the encoder and syndrome calculator for the (7,4) Hamming code are given in Fig. 2.12A and B, respectively. The generating polynomial is given by $g(x) = 1 + x + x^3$. The polynomial $x^7 + 1$ can be factorized as follows: $x^7 + 1 = (1 + x)(1 + x^2 + x^3)(1 + x + x^3)$. If we select $g(x) = 1 + x + x^3$ as the generator polynomial, based on property 3 of the generator polynomial, the corresponding parity-check polynomial will be $h(x) = (1 + x)(1 + x^2 + x^3) = 1 + x + x^2 + x^4$. The message sequence 1001 can be represented in polynomial form by $m(x) = 1 + x^3$. For representation in systematic form, we must multiply $m(x)$ by x^{n-k} to obtain $x^{n-k}m(x) = x^3m(x) = x^3 + x^6$. The codeword polynomial is obtained by $c(x) = x^{n-k}m(x) + \mathrm{rem}[x^{n-k}m(x)/g(x)] = x + x^3 + \mathrm{rem}[(x + x^3)/(1 + x + x^3)] = x + x^2 + x^3 + x^6$. The

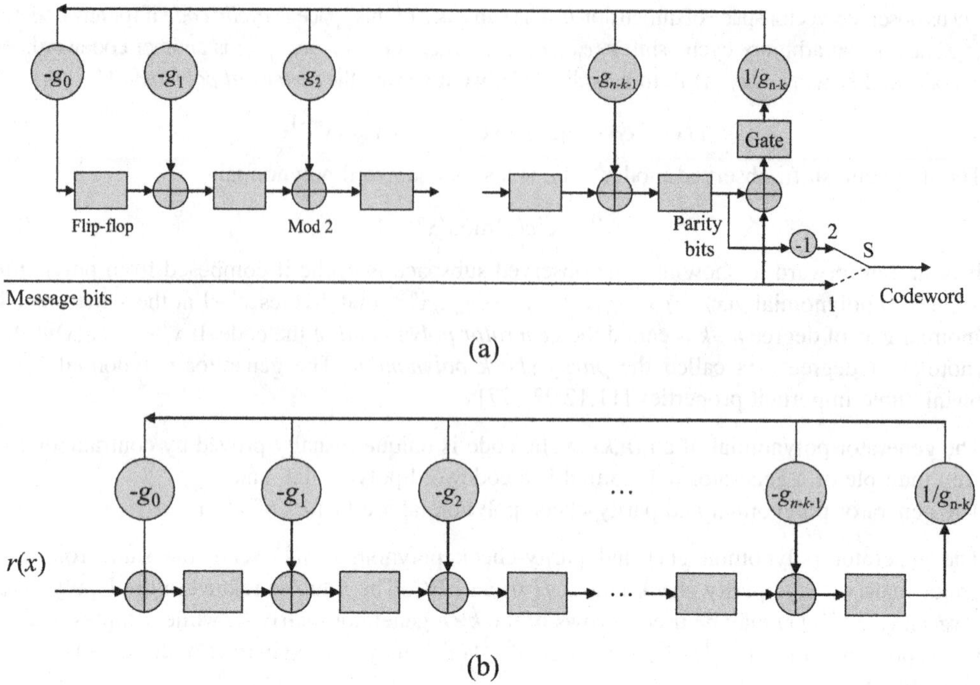

FIGURE 2.11

(A) Systematic cyclic encoder, (B) syndrome calculator.

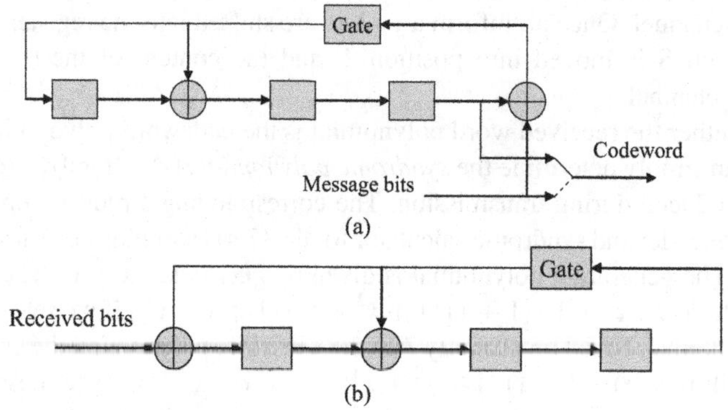

FIGURE 2.12

Hamming (7,4) encoder (A) and syndrome calculator (B).

corresponding codeword is 0111001. To obtain the generator matrix of this code, we can use the following polynomials: $g(x) = 1 + x + x^3$, $xg(x) = x + x^2 + x^4$, $x^2g(x) = x^2 + x^3 + x^5$, and $x^3g(x) = x^3 + x^4 + x^6$; we write the corresponding n-tuples in matrix form as follows:

$$G' = \begin{bmatrix} 1101000 \\ 0110100 \\ 0011010 \\ 0001101 \end{bmatrix}.$$

By Gaussian elimination, we can put the generator matrix into systematic form:

$$G = \begin{bmatrix} 1101000 \\ 0110100 \\ 1110010 \\ 1010001 \end{bmatrix}.$$

The parity-check matrix can be obtained from the following polynomials:

$$x^4h\left(x^{-1}\right) = 1 + x^2 + x^3 + x^4, x^5h\left(x^{-1}\right) = x + x^3 + x^4 + x^5, x^6h\left(x^{-1}\right) = x^2 + x^3 + x^5 + x^6,$$

by writing the corresponding n-tuples in the form of a matrix,

$$H' = \begin{bmatrix} 1011100 \\ 0101110 \\ 0010111 \end{bmatrix}.$$

The H-matrix can be put in systematic form by Gaussian elimination:

$$H = \begin{bmatrix} 1001011 \\ 0101110 \\ 0010111 \end{bmatrix}.$$

The syndrome polynomial $s(x)$ has the following three important properties, that can be used to simplify decoder implementation [11,12,23–27]:

1. The syndrome of received word polynomial $r(x)$ is also the syndrome of corresponding error polynomial $e(x)$,
2. The syndrome of a cyclic shift of $r(x)$, $xr(x)$ is determined by $xs(x)$, and
3. The syndrome polynomial $s(x)$ is identical to the error polynomial $e(x)$ if the errors are confined to the $(n-k)$ parity-check bits of the received word polynomial $r(x)$.

Maximal-length codes $(n = 2^m-1,m)$ $(m \geq 3)$ are dual Hamming codes and have a minimum distance of $d_{min} = 2^m-1$. The parity-check polynomial for (7,3) maximal-length codes is therefore $h(x) = 1 + x + x^3$. The encoder for (7,3) maximum-length code is given in Fig. 2.13. The generator

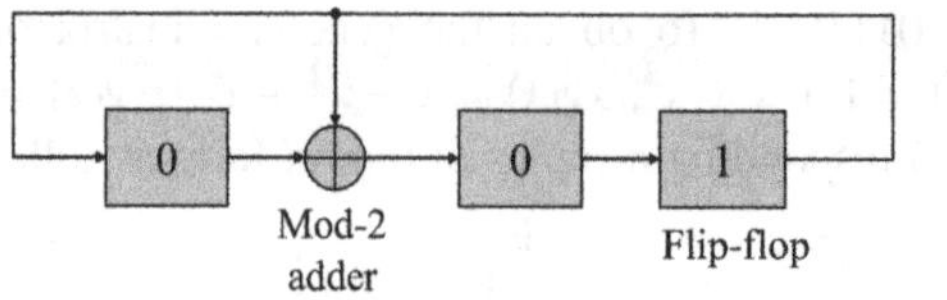

FIGURE 2.13

Encoder for the (7, 3) maximal-length code.

polynomial gives one period of maximum-length code provided that the encoder is initialized to 0 … 01. For example, the generator polynomial for the (7,3) maximum-length code above is $g(x) = 1 + x + x^2 + x^4$, and the output sequence is given by

$$\underbrace{100}_{\text{initial state}} \quad \underbrace{1110100}_{g(x)=1+x+x^2+x^4}$$

Cyclic redundancy check (CRC) codes are very popular codes for error detection. The (n,k) CRC codes are capable of detecting [11,12,23–27]

- All error bursts of length $n-k$, with an error burst of length $n-k$ being defined as a contiguous sequence of $n-k$ bits in which the first and last bits or any other intermediate bits are received in error;
- A fraction of error bursts of length equal to $n-k+1$; the fraction equals $1-2^{-(n-k-1)}$;
- A fraction of error of length greater than $n-k+1$; the fraction equals $1-2^{-(n-k-1)}$;
- All combinations of $d_{\min}-1$ (or fewer) errors;
- All error patterns with an odd number of errors if the generator polynomial $g(x)$ for the code has an even number of nonzero coefficients.

Table 2.3 lists the generator polynomials of several CRC codes currently used in various communication systems.

Decoding cyclic codes uses the same three steps as in LBC decoding, namely syndrome computation, error pattern identification, and error correction [11,12,23–27]. The **Meggitt decoder** configuration, which is implemented based on syndrome property 2 (also known as the Meggitt theorem), is shown in Fig. 2.14. The syndrome is calculated by dividing the received word by the generating polynomial $g(x)$, and at the same time, the received word is shifted into the buffer register. Once the last bit of a received word enters the decoder, the gate is turned off. The syndrome is further

Table 2.3 Generator polynomials of several cyclic redundancy check (CRC) codes.

CRC codes	Generator polynomial	$n-k$
CRC-8 code (IEEE 802.16, WiMax)	$1+x^2+x^8$	8
CRC-16 code (IBM CRC-16, ANSI, USB, SDLC)	$1+x^2+x^{15}+x^{16}$	16
CRC-ITU (X25, V41, CDMA, Bluetooth, HDLC, PPP)	$1+x^5+x^{12}+x^{16}$	16
CRC-24 (WLAN, UMTS)	$1+x+x^5+x^6+x^{23}+x^{24}$	24
CRC-32 (Ethernet)	$1+x+x^2+x^4+x^5+x^7+x^8+x^{10}+x^{11}+x^{12}+x^{16}+x^{22}+x^{23}+x^{26}+x^{32}$	32

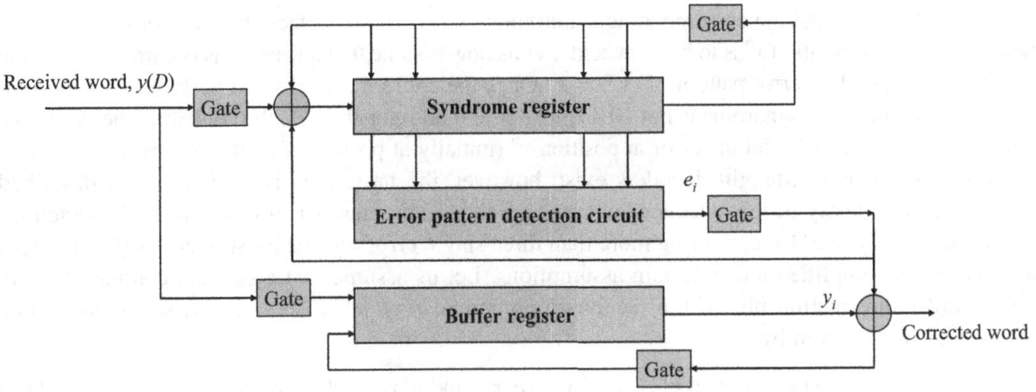

FIGURE 2.14

A Meggitt decoder configuration.

read into the error pattern detection circuit and implemented as the combinational logic circuit, which generates 11 if and only if the content of the syndrome register corresponds to a correctible error pattern at the highest-order position x^{n-1}. By adding the output of the error pattern detector to the bit in error, the error can be corrected, and at the same time, the syndrome must be modified. If an error occurs on position x^l, by cyclically shifting the received word $n-l-1$ times, the erroneous bit will appear in position x^{n-1} and can be corrected. The decoder, therefore, corrects the errors in a bit-by-bit fashion until the entire received word is read out from the buffer register.

The Hamming (7,4) cyclic decoder configuration for generating polynomial $g(x) = 1 + x + x^3$ is shown in Fig. 2.15. Because we expect only single errors, the error polynomial corresponding to the highest-order

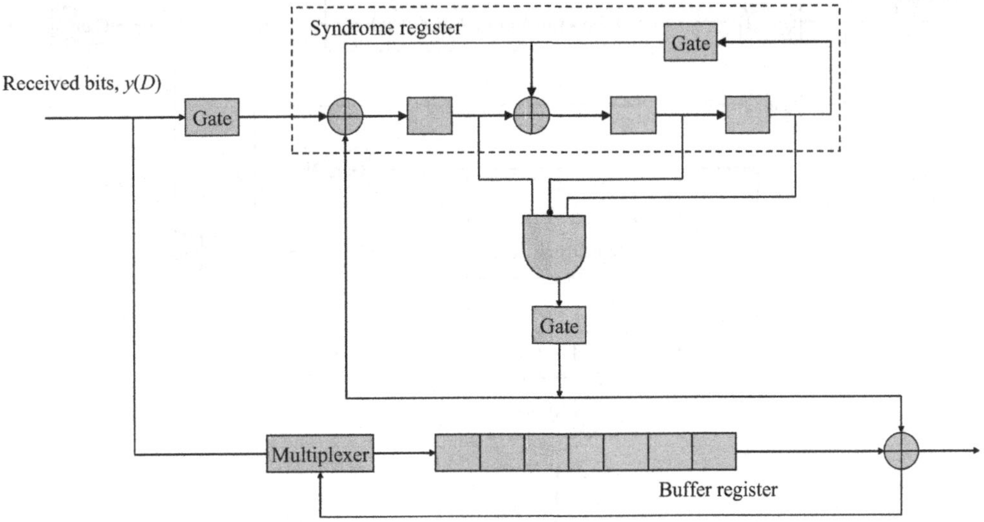

FIGURE 2.15

Hamming (7,4) decoder configuration.

position is $e(x) = x^6$, and the corresponding syndrome is $s(x) = 1 + x^2$. Once this syndrome is detected, the erroneous bit at position x^6 is to be corrected. Let us now assume that an error has occurred on position x^i with a corresponding error pattern of $e(x) = x^i$. Once the entire received word is shifted into the syndrome register, the error syndrome is not 101. But after $6-i$ additional shifts, the content of the syndrome register will become 101, and the error at position x^6 (initially at position x^i) will be corrected.

Several versions of Meggitt decoders exist; however, the basic idea is similar to that described above. The complexity of this decoder increases quickly as the number of errors to be corrected increases; it is rarely used for correcting more than three single errors or one burst of errors. The Meggitt decoder can be simplified under certain assumptions. Let us assume that errors are confined to only highest-order information bits of the received polynomial $r(x)$: x^k, x^{k+1}, …,x^{n-1}, so the syndrome polynomial $s(x)$ is given by

$$s(x) = r(x)\text{mod } g(x) = [c(x) + e(x)]\text{mod } g(x) = e(x)\text{mod } g(x), \tag{2.81}$$

where $r(x)$ is the received polynomial, $c(x)$ is the codeword polynomial, and $g(x)$ is the generator polynomial. The corresponding error polynomial can be estimated by

$$e'(x) = e'(x)\text{mod } g(x) = s'(x) = x^{n-k}s(x)\text{mod } g(x) = e_k + e_{k+1}x + \ldots + e_{n-2}x^{n-k-2} + e_{n-1}x^{n-k-1}. \tag{2.82}$$

The error polynomial will be most at the degree $n-k-1$, because $\deg[g(x)] = n-k$. Therefore, the syndrome register content is identical to the error pattern, and we say that the error pattern is *trapped* in the syndrome register. The corresponding decoder, known as the *error-trapping decoder*, is shown in Fig. 2.16 [23,24]. If t or fewer errors occur in $n-k$ consecutive locations, it can be shown that the error pattern is trapped in the syndrome register only when the weight of the syndrome $w(s)$ is less than or

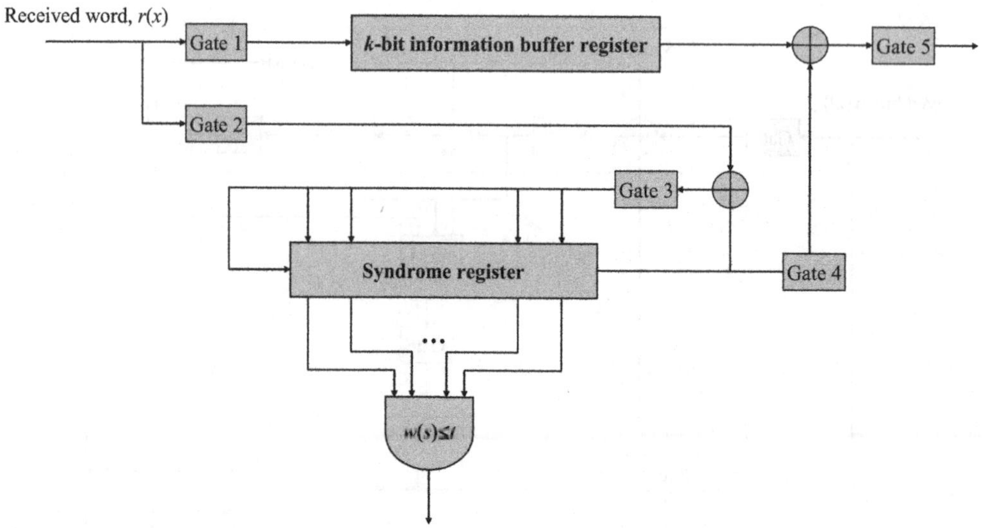

FIGURE 2.16

Error-trapping decoding architecture.

equal to t [23,24]. Therefore, the test for the error-trapping condition is to determine whether the weight of the syndrome is t or less. With gates 1, 2, and 3 closed (4 and 5 open), the received word is shifted into the syndrome register from the right end, which is equivalent to premultiplying the received polynomial $r(x)$ with x^{n-k}. Once the highest-order k bits (corresponding to information bits) are shifted into the information buffer, gate 11 is opened. Gate 2 is opened once all n bits from the received word are shifted into the syndrome register. The syndrome register at this point contains the syndrome corresponding to $x^{n-k}r(x)$. If its weight is t or less, gates 4 and 5 are closed (all others are opened), and the corrected information is shifted out. If $w(s)>t$, the errors are not confined to the $n-k$ higher-order positions of $s(x)$, and we keep shifting with gate 3 being on (the other gates are switched off) until $w(s)\leq t$. If $w(s)$ never goes $\leq t$ and the syndrome register is shifted k times, either an error pattern with errors confined to $n-k$ consecutive end-around locations has occurred or an uncorrectable error pattern has occurred. For additional explanation on this and other types of cyclic decoders, the interested reader is referred to Ref. [23].

2.5 Bose–Chaudhuri–Hocquenghem codes

The Bose–Chaudhuri–Hocquenghem (BCH) codes, the most famous of the cyclic codes, were discovered by Hocquenghem in 1959 and by Bose and Chaudhuri in 1960 [23–26]. Among the many decoding algorithms, the most important are the Massey–Berlekamp algorithm and Chien's search algorithm. An important subclass of the BCH is the class of RS codes proposed in 1960. Before we continue further with the study of BCH codes, we must introduce some properties of finite fields [19,20,23,26].

2.5.1 Galois fields

Before we can proceed further with finite fields, we introduce the concepts of ring, field, and congruencies from abstract algebra [19,20]. A *ring* is defined as a set of elements R with two operations—addition "+" and multiplication "•"—satisfying the following three properties: (1) R is an Abelian (commutative) group under addition, (2) the multiplication operation is associative, and (3) multiplication is associative over addition. A *field* is defined as a set of elements F with two operations, addition "+" and multiplication "•", satisfying the following properties: (1) F is an Abelian group under the addition operation, with 0 being the identity element, (2) the nonzero elements of F form an Abelian group under multiplication, with 11 being the identity element, and (3) the multiplication operation is distributive over the addition operation.

The quantity a is *congruent* to quantity b observed per modulus n, denoted as $a \equiv b \pmod{n}$, if $a-b$ is divisible by n. If $x \equiv a \pmod{n}$, then a is called a *residue* to x to modulus n. A *class of residues* to modulus n is the class of all integers congruent to a given residue (mod n), and every member of the class is called a representative of the class. There are n classes, represented by (0), (1), (2), …, ($n-1$), and the representative of these classes are called a *complete system* of incongruent residues to modulus n. If i and j are two members of a complete system of incongruent residues to modulus n, then addition and multiplication between i and j can be introduced by

1. $i+j=(i+j) \pmod{n}$
2. $i\cdot j=(i\cdot j) \pmod{n}$

A complete system of residues (mod n) forms a *commutative ring* with a unity element. Let s be a nonzero element of these residues. Then s possesses an inverse element if and only if n is prime, p. When p is a prime, a complete system of residues (mod p) forms a *Galois* (finite) *field* and is commonly denoted by GF(p).

Let P(x) be any given polynomial in x of degree m with coefficients belonging to GF(p), and let $F(x)$ be any polynomial in x with integral coefficients. Then $F(x)$ may be expressed as [19]

$$F(x) = f(x) + p \cdot q(X) + P(x) \cdot Q(x),$$

where $f(x) = a_0 + a_1x + a_2x^2 + \; ... \; + a_{m\text{-}1}x^{m-1}$, $a_i \in$ GF(p). This relationship may be written as $F(x) \equiv f(x) \bmod \{p, P(x)\}$, and we say that $f(x)$ is the *residue* of $F(x)$ modulus p and $P(x)$. If p and $P(x)$ are kept fixed but $f(x)$ varied, p^m classes can be formed (because each coefficient of $f(x)$ may take p values of GF(p)). The classes defined by $f(x)$ form a commutative (Abelian) ring, which will be a field if and only if $P(x)$ is *irreducible* over GF(p) (not divisible with any other polynomial of degree $m-1$ or less) [23,24,26]. The finite field formed by p^m classes of residues is called a *Galois field of order* p^m and is denoted by GF(p^m). Two important *properties of GF(q)*, $q = p^m$ are given below [23,24,26]:

1. The roots of polynomial $x^{q-1}-1$ are all nonzero elements of GF(q).
2. Let $P(x)$ be an irreducible polynomial of degree m with coefficients from GF(p) and β be a root from the extended field GF($q = p^m$). All possible roots of $P(x)$ are then given by $\beta, \beta^p, \beta^{p^2}, ..., \beta^{p^{m-1}}$.

The nonzero elements of GF(p^m) can be represented as polynomials of degree at most $m-1$ or as powers of a *primitive root* α, such that [23,24,26]

$$\alpha^{p^m-1} = 1, \quad \alpha^d \neq 1 (\text{for } d \text{ dividing } p^m - 1).$$

Therefore, the *primitive element* (root) is a field element that generates all nonzero file elements as its successive powers. An irreducible polynomial with a primitive element as its root is called a *primitive polynomial.*

The function $P(x)$ is said to be the *minimum polynomial* for generating the elements of GF(p^m) and represents the smallest-degree polynomial over GF(p) with field element $\beta \in$ GF(p^m) as a root. To obtain a minimum polynomial, we must divide x^q-1 ($q = p^m$) by the least common multiple (LCM) of all factors of the form x^d-1, where d is a divisor of p^m-1; and obtain so-called *cyclotomic equation* (that is, the equation having all primitive roots of the equation $x^{q-1}-1 = 0$ for its roots). The order of this equation is $O(p^m-1)$, where $O(k)$ is the number of all positive integers less than k and relatively prime to it. By substituting each coefficient in this equation by least nonzero residue to modulus p, we obtain the cyclotomic polynomial of order $O(p^m-1)$. Let $P(x)$ be an irreducible factor of this polynomial, then $P(x)$ is a minimum polynomial, which generally is not the unique one.

Example: Let us determine the minimum polynomial for generating the elements of GF(2^3). The cyclotomic polynomial is $(x^7-1)/(x-1) = x^6+x^5+x^4+x^3+x^2+x+1 = (x^3+x^2+1)(x^3+x+1)$. Hence, $P(x)$ can be either x^3+x^2+1 or x^3+x+1. Let us choose $P(x) = x^3+x^2+1$. The degree of this polynomial is deg[$P(x)$] = 3. Now we explain how we can construct GF(2^3) using the $P(x)$. The construction always starts with elements from the basic field (in this case, GF(2) = {0,1}). All nonzero elements can be obtained as successive powers of primitive root α until no new

Table 2.4 Three representations of GF(2^3) generated by $x^3 + x^2 + 1$.

Power of α	Polynomial	3-tuple
0	0	000
α^0	1	001
α^1	α	010
α^2	α^2	100
α^3	α^2+1	101
α^4	$\alpha^2+\alpha+1$	111
α^5	$\alpha+1$	011
α^6	$\alpha^2+\alpha$	110
α^7	1	001

element is generated, which is given in the first column of Table 2.4. The second column is obtained exploiting the primitive polynomial $P(x)=x^3+x^2+1$. α is the root of $P(x)$ and therefore $P(\alpha)=\alpha^3+\alpha^2+1=0$, and α^3 can be expressed as α^2+1. α^4 can be expressed as $\alpha\alpha^3=\alpha(\alpha^2+1)=\alpha^3+\alpha=\alpha^2+1+\alpha$. The third column in Table 2.4 is obtained by reading off coefficients in the second column, with the leading coefficient multiplying the α^2.

2.5.2 The structure and encoding of Bose–Chaudhuri–Hocquenghem codes

Equipped with this knowledge of Galois fields, we can continue our description of BCH code structures. Let the finite field GF(q) (*symbol field*) and extension field GF(q^m) (*locator field*), $m \geq 1$, be given. For every $m_0 \geq 1$ and Hamming distance d, there exists a BCH code with the generating polynomial $g(x)$ if and only if it is of smallest degree with coefficients from GF(q) and with roots from the extension field GF(q^m), as follows [23]:

$$\alpha^{m_0}, \alpha^{m_0+1}, \cdots, \alpha^{m_0+d-2}. \tag{2.83}$$

where α is from GF(q^m). The codeword length is determined as the least common multiple of orders of roots. (The order of an element β from a finite field is the smallest positive integer j such that $\beta^j=1$.)

It can be shown that for any positive integer m ($m \geq 3$) and t ($t < 2^{m-1}$), there exists a *binary* BCH code having the following properties [23,24]:

- Codeword length: $n=2^m-1$
- Number of parity bits: $n-k \leq mt$
- Minimum Hamming distance: $d \geq 2t+1$.

This code can correct up to t errors. The generator polynomial can be found as the LCM of the minimal polynomials of α^i [23,24]:

$$g(x)=\mathrm{LCM}[P_{\alpha^1}(x), P_{\alpha^3}(x), \ldots, P_{\alpha^{2t-1}}(x)] \tag{2.84}$$

where α is a primitive element in GF(2^m), and $P_{\alpha^i}(x)$ is the minimal polynomial of α^i.

Let $c(x) = c_0 + c_1x + c_2x^2 + \ldots + c_{n-1}x^{n-1}$ be the codeword polynomial, and let the roots of the generator polynomial be $\alpha, \alpha^2, \ldots, \alpha^{2t}$, where t is the error correction capability of the BCH code. Because the generator polynomial $g(x)$ is the factor of the codeword polynomial $c(x)$, the roots of $g(x)$ must also be the roots of $c(x)$:

$$c(\alpha^i) = c_0 + c_1\alpha^i + \ldots + c_{n-1}\alpha^{(n-1)i} = 0; 1 \le i \le 2t \tag{2.85}$$

This equation can also be written as the inner (scalar) product of the codeword vector $\boldsymbol{c} = [c_0\ c_1 \ldots c_{n-1}]$ and the following vector $[1\ \alpha^i\ \alpha^{2i} \ldots \alpha^{2(n-1)i}]$:

$$[c_0\ c_1 \ldots c_{n-1}] \begin{bmatrix} 1 \\ \alpha^i \\ \ldots \\ \alpha^{(n-1)i} \end{bmatrix} = 0; 1 \le i \le 2t \tag{2.86}$$

Eq. (2.86) can also be written as the following matrix:

$$[c_0\ c_1 \ldots c_{n-1}] \begin{bmatrix} \alpha^{n-1} & \alpha^{n-2} & \ldots & \alpha & 1 \\ (\alpha^2)^{n-1} & (\alpha^2)^{n-2} & \ldots & \alpha^2 & 1 \\ (\alpha^3)^{n-1} & (\alpha^3)^{n-2} & \ldots & \alpha^3 & 1 \\ \ldots & \ldots & \ldots & \ldots & \ldots \\ (\alpha^{2t})^{n-1} & (\alpha^{2t})^{n-2} & \ldots & \alpha^{2t} & 1 \end{bmatrix}^T = \boldsymbol{c}\boldsymbol{H}^T = \boldsymbol{0},$$

$$\boldsymbol{H} = \begin{bmatrix} \alpha^{n-1} & \alpha^{n-2} & \ldots & \alpha & 1 \\ (\alpha^2)^{n-1} & (\alpha^2)^{n-2} & \ldots & \alpha^2 & 1 \\ (\alpha^3)^{n-1} & (\alpha^3)^{n-2} & \ldots & \alpha^3 & 1 \\ \ldots & \ldots & \ldots & \ldots & \ldots \\ (\alpha^{2t})^{n-1} & (\alpha^{2t})^{n-2} & \ldots & \alpha^{2t} & 1 \end{bmatrix}, \tag{2.87}$$

where $\boldsymbol{H}$ is the parity-check matrix of BCH code. Using property 2 of GF(q) from the previous section, we conclude that α^i and α^{2i} are the roots of the same minimum polynomial, so the even rows in $\boldsymbol{H}$ can be omitted to obtain the final version of the parity-check matrix of BCH codes:

$$\boldsymbol{H} = \begin{bmatrix} \alpha^{n-1} & \alpha^{n-2} & \ldots & \alpha & 1 \\ (\alpha^3)^{n-1} & (\alpha^3)^{n-2} & \ldots & \alpha^3 & 1 \\ (\alpha^5)^{n-1} & (\alpha^5)^{n-2} & \ldots & \alpha^5 & 1 \\ \ldots & \ldots & \ldots & \ldots & \ldots \\ (\alpha^{2t-1})^{n-1} & (\alpha^{2t-1})^{n-2} & \ldots & \alpha^{2t-1} & 1 \end{bmatrix} \tag{2.88}$$

For example, (15,7) 2-error-correcting BCH code has the generator polynomial [29]

$$g(x) = LCM[P_{\alpha}(x), P_{\alpha^3}(x)]$$
$$= LCM\left[x^4 + x + 1, (x+\alpha^3)(x+\alpha^6)(x+\alpha^9)(x+\alpha^{12})\right]$$
$$= x^8 + x^7 + x^6 + x^4 + 1,$$

and the parity check matrix [29]

$$\boldsymbol{H} = \begin{bmatrix} \alpha^{14} & \alpha^{13} & \alpha^{12} & \cdots & \alpha & 1 \\ \alpha^{42} & \alpha^{39} & \alpha^{36} & \cdots & \alpha^3 & 1 \end{bmatrix}$$
$$= \begin{bmatrix} \alpha^{14} & \alpha^{13} & \alpha^{12} & \alpha^{11} & \alpha^{10} & \alpha^9 & \alpha^8 & \alpha^7 & \alpha^6 & \alpha^5 & \alpha^4 & \alpha^3 & \alpha^2 & \alpha & 1 \\ \alpha^{12} & \alpha^9 & \alpha^6 & \alpha^3 & 1 & \alpha^{12} & \alpha^9 & \alpha^6 & \alpha^3 & 1 & \alpha^{12} & \alpha^9 & \alpha^6 & \alpha^3 & 1 \end{bmatrix}.$$

In the previous expression, we used the fact that in GF(2^4), $\alpha^{15} = 1$. The primitive polynomial used to design this code was $p(x) = x^4 + x + 1$. Every element in GF(2^4) can be represented as a 4-tuple, as shown in Table 2.5.

To create the second column, we have used the relation $\alpha^4 = \alpha + 1$, and the 4-tuples are obtained reading off coefficients in the second column. By replacing the powers of α in the parity-check matrix above by corresponding 4-tuples, the parity-check matrix can be written in the following binary form:

Table 2.5 GF(2^4) generated by $x^4 + x + 1$.

Power of α	Polynomial of α	4-tuple
0	0	0000
α^0	1	0001
α^1	α	0010
α^2	α^2	0100
α^3	α^3	1000
α^4	$\alpha + 1$	0011
α^5	$\alpha^2 + \alpha$	0110
α^6	$\alpha^3 + \alpha^2$	1100
α^7	$\alpha^3 + \alpha + 1$	1011
α^8	$\alpha^2 + 1$	0101
α^9	$\alpha^3 + \alpha$	1010
α^{10}	$\alpha^2 + \alpha + 1$	0111
α^{11}	$\alpha^3 + \alpha^2 + \alpha$	1110
α^{12}	$\alpha^3 + \alpha^2 + \alpha + 1$	1111
α^{13}	$\alpha^3 + \alpha^2 + 1$	1101
α^{14}	$\alpha^3 + 1$	1001

$$H = \begin{bmatrix} 1 & 1 & 1 & 1 & 0 & 1 & 0 & 1 & 1 & 0 & 0 & 1 & 0 & 0 & 0 \\ 0 & 1 & 1 & 1 & 1 & 0 & 1 & 0 & 1 & 1 & 0 & 0 & 1 & 0 & 0 \\ 0 & 0 & 1 & 1 & 1 & 1 & 0 & 1 & 0 & 1 & 1 & 0 & 0 & 1 & 0 \\ 1 & 1 & 1 & 0 & 1 & 0 & 1 & 1 & 0 & 0 & 1 & 0 & 0 & 0 & 1 \\ 1 & 1 & 1 & 1 & 0 & 1 & 1 & 1 & 1 & 0 & 1 & 1 & 1 & 1 & 0 \\ 1 & 0 & 1 & 0 & 0 & 1 & 0 & 1 & 0 & 0 & 1 & 0 & 1 & 0 & 0 \\ 1 & 1 & 0 & 0 & 0 & 1 & 1 & 0 & 0 & 0 & 1 & 1 & 0 & 0 & 0 \\ 1 & 0 & 0 & 0 & 1 & 1 & 0 & 0 & 0 & 1 & 1 & 0 & 0 & 0 & 1 \end{bmatrix}$$

In Table 2.6, we listed the parameters of several BCH codes generated by primitive elements of an order less than 2^5-1 that are of interest for optical communications. The complete list can be found in Appendix C of [23].

Generally speaking, there is no need for q to be a prime; it could be a prime power. However, the symbols must be taken from GF(q) and the roots from GF(q^m). From the nonbinary BCH codes, RS codes are the most famous, and these codes are briefly explained in Section 2.6.

Table 2.6 A set of primitive binary Bose–Chaudhuri–Hocquenghem codes.

n	k	t	Generator polynomial (in octal form)
15	11	1	23
63	57	1	103
63	51	2	12471
63	45	3	1701317
127	120	1	211
127	113	2	41567
127	106	3	11554743
127	99	4	3447023271
255	247	1	435
255	239	2	267543
255	231	3	156720665
255	223	4	75626641375
255	215	5	23157564726421
255	207	6	16176560567636227
255	199	7	7633031270420722341
255	191	8	2663470176115333714567
255	187	9	52755313540001322236351
255	179	10	22624710717340432416300455

2.5.3 Decoding of Bose–Chaudhuri–Hocquenghem codes

The BCH codes can be decoded as with any other cyclic code class. For example, in Fig. 2.17, we provide the error-trapping decoder for BCH (15,7) double-error-correcting code. The operation principle of this circuit is already explained in previous section. Here we explain the decoding process by employing algorithms developed especially for decoding BCH codes. Let $g(x)$ be the generator polynomial with corresponding roots $\alpha, \alpha^2, \ldots, \alpha^{2t}$. Let $c(x) = c_0 + c_1x + c_2x^2 + \ldots + c_{n-1}x^{n-1}$ be the codeword polynomial, $r(x) = r_0 + r_1x + r_2x^2 + \ldots + r_{n-1}x^{n-1}$ be the received word polynomial, and $e(x) = e_0 + e_1x + e_2x^2 + \ldots + e_{n-1}x^{n-1}$ be the error polynomial. The roots of the generator polynomial are also the roots of the codeword polynomial—that is,

$$c(\alpha^i) = 0; \quad i = 0, 1 \ldots 2t. \tag{2.89}$$

For binary BCH codes, the only nonzero element is 1; therefore, the indices i of coefficients $e_i \neq 0$ (or $e_i = 1$) determine the error locations. For nonbinary BCH codes, the error magnitudes are important in addition to error locations. By evaluating the received word polynomial $r(x)$ for α^i, we obtain

$$r(\alpha^i) = c(\alpha^i) + e(\alpha^i) = e(\alpha^i) = S_i, \tag{2.90}$$

where S_i is the ith component of the syndrome vector, defined by

$$\boldsymbol{S} = [S_1 S_2 \ldots S_{2t}] = \boldsymbol{r}\boldsymbol{H}^T. \tag{2.91}$$

The BCH code can correct up to t errors. Let us assume that the error polynomial $e(x)$ does not have more than t errors, which can then be written as

$$e(x) = e_{j_1}x^{j_1} + e_{j_2}x^{j_2} + \ldots + e_{j_l}x^{j_l} + \ldots + e_{j_v}x^{j_v}; \quad 0 \le v \le t \tag{2.92}$$

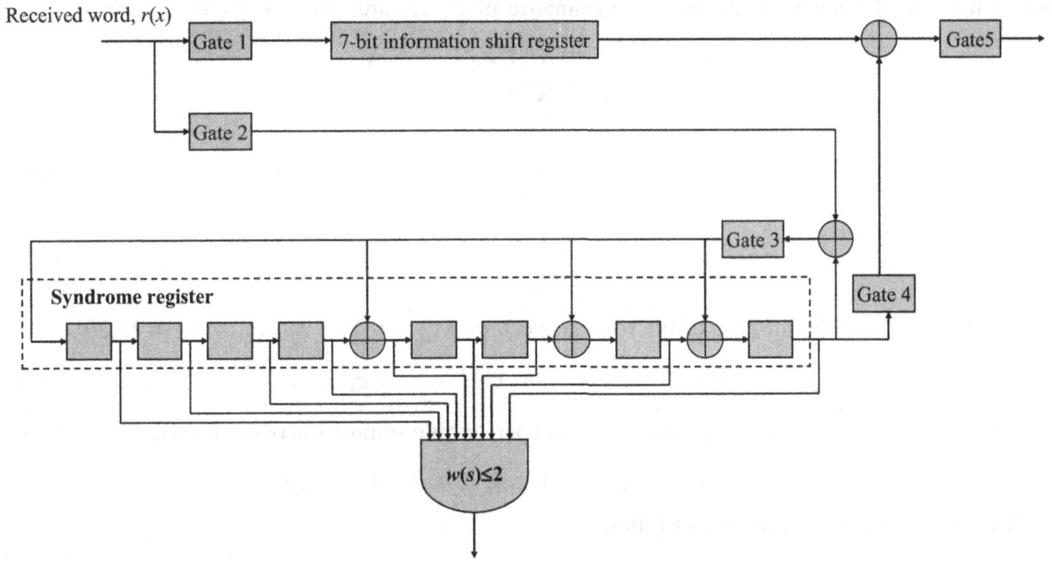

FIGURE 2.17

The error-trapping decoder for (15,7) Bose–Chaudhuri–Hocquenghem code generated by $g(x) = 1 + x^4 + x^6 + x^7 + x^8$.

where e_{j_l} is the error magnitude and j_l is the error location. The corresponding syndrome components can be obtained from Eqs. (2.90) and (2.92) as follows:

$$S_i = e_{j_1}(\alpha^i)^{j_1} + e_{j_2}(\alpha^i)^{j_2} + \ldots + e_{j_l}(\alpha^i)^{j_l} + \ldots + e_{j_v}(\alpha^i)^{j_v};\quad 0 \le v \le t \tag{2.93}$$

where α^{j_l} is the error-location number. Note that the error magnitudes are from the symbol field, while the error location numbers are from the extension field. To avoid double-indexing, let us introduce the following notation [6]: $X_l = \alpha^{j_l}$, $Y_l = e_{j_l}$. The pairs (X_l, Y_l) completely identify the errors ($l \in [1,v]$). We must solve the following set of equations [6]:

$$\begin{aligned} S_1 &= Y_1X_1 + Y_2X_2 + \ldots + Y_vX_v \\ S_2 &= Y_1X_1^2 + Y_2X_2^2 + \ldots + Y_vX_v^2 \\ &\ldots \\ S_{2t} &= Y_1X_1^{2t} + Y_2X_2^{2t} + \ldots + Y_vX_v^{2t}. \end{aligned} \tag{2.94}$$

The procedure to solve this system of equations represents the corresponding decoding algorithm. Directly solving this system of equations is impractical. Many algorithms exist to solve the system of Eq. (2.94), ranging from iterative to Euclidean algorithms [23–26]. A very popular decoding algorithm of BCH codes is the *Massey–Berlekamp algorithm* [6,16,17,19,20]. In this algorithm, the BCH decoding is observed as a shift register synthesis problem: given the syndromes S_i, we must find the minimal-length shift register that generates the syndromes. Once we determine the coefficients of this shift register, we construct the *error locator* polynomial [6,16,20]:

$$\sigma(x) = \prod_{i=1}^{v}(1 + X_ix) = \sigma_vx^v + \sigma_{v-1}x^{v-1} + \ldots + \sigma_1x + 1, \tag{2.95}$$

where the σ_i's, also known as elementary symmetric functions, are given by Viète's formulas:

$$\begin{aligned} \sigma_1 &= X_1 + X_2 + \ldots + X_v \\ \sigma_2 &= \sum_{i<j} X_iX_j \\ \sigma_3 &= \sum_{i<j<k} X_iX_jX_k \\ &\ldots \\ \sigma_v &= X_1X_2\ldots X_v \end{aligned} \tag{2.96}$$

Because $\{X_l\}$ are the inverses of the roots of $\sigma(x)$, $\sigma(1/X_l) = 0\ \forall\ l$, we can write [6,16,20]

$$X_l^v\sigma(X_l^{-1}) = X_l^v + \sigma_1X_l^{v-1} + \ldots + \sigma_v. \tag{2.97}$$

By multiplying the previous equation by X_l^j and performing summation over l for fixed j, we obtain

$$S_{v+j} + \sigma_1S_{v+j-1} + \ldots + \sigma_vS_j = 0;\ j = 1, 2, \ldots, v \tag{2.98}$$

This equation can be rewritten as follows:

$$S_{v+j} = -\sum_{i=1}^{v}\sigma_iS_{v+j-i};\ j = 1, 2, \ldots, v \tag{2.99}$$

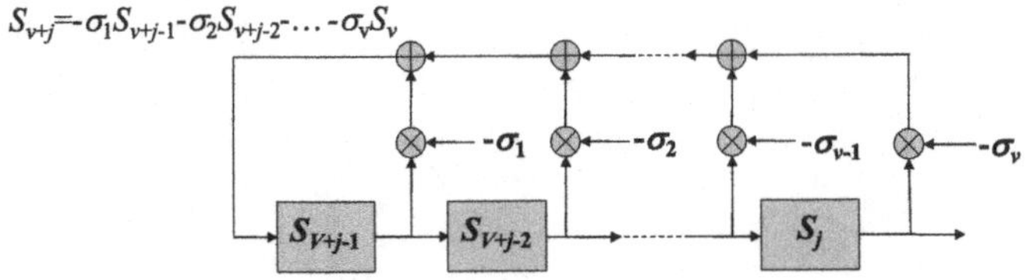

FIGURE 2.18

A shift register that generates the syndromes S_j.

S_{v+j} represents the output of the shift register shown in Fig. 2.18. The Massey–Berlekamp algorithm is summarized by the flowchart shown in Fig. 2.19, which is self-explanatory. Once the locator polynomial is determined, we must find the roots and invert them to obtain the error locators.

To determine the error magnitudes, we must define another polynomial, known as the *error-evaluator* polynomial, as follows [6,16]:

$$\varsigma(x) = 1 + (S_1 + \sigma_1)x + (S_2 + \sigma_1 S_1 + \sigma_2)x^2 + \ldots + (S_v + \sigma_1 S_{v-1} + \ldots + \sigma_v)x^v. \quad (2.100)$$

The error magnitudes are then obtained from Refs. [6,16]:

$$Y_l = \frac{\varsigma(X_l^{-1})}{\prod_{i=1,\, i\neq l}^{v} (1 + X_i X_l^{-1})}. \quad (2.101)$$

For binary BCH codes, we set $Y_l = 1$, so we do not need to evaluate the error magnitudes.

2.6 Reed–Solomon, concatenated, and product codes

RS codes were discovered in 1960 and represent a special class of nonbinary BCH codes [30,31]. RS codes represent the most commonly used nonbinary codes. Both the code symbols and the roots of generating polynomial are from the locator field. In other words, the symbol field and locator field are the same ($m = 1$) for RS codes. The codeword length of RS codes is determined by $n = q^m - 1 = q - 1$, so RS codes are relatively short. The minimum polynomial for some element β is $P_\beta(x) = x - \beta$. If α is the primitive element of GF(q) (q is a prime or prime power), the generator polynomial for t-error-correcting RS code is given by [6,16,17,19,20]

$$g(x) = (x - \alpha)(x - \alpha^2)\cdots(x - \alpha^{2t}). \quad (2.102)$$

The generator polynomial degree is $2t$ and is the same as the number of parity symbols $n - k = 2t$, while the block length of the code is $n = q - 1$. Since the minimum distance of BCH codes is $2t + 1$, the minimum distance of RS codes is $d_{\min} = n - k + 1$, therefore satisfying the Singleton bound ($d_{\min} \leq n - k + 1$) with equality and belonging to the class of MDS *codes*. When $q = 2^m$, the RS code parameters are $n = m(2^m - 1)$, $n - k = 2\,mt$, and $d_{\min} = 2\,mt + 1$. Therefore, the minimum distance of RS codes, when observed as binary codes, is large. The RS codes may be considered as burst error-correcting codes, and as such are suitable for bursty-errors prone channels. This binary code can correct up to t bursts of length m. Equivalently, this binary code can correct a single burst of length $(t - 1)m + 1$.

The weight distribution of RS codes can be determined by [23]

$$A_i = \binom{n}{i}(q-1)\sum_{j=0}^{i-d_{\min}} (-1)^j \binom{i-1}{j} q^{i-d_{\min}-j}, \quad (2.103)$$

and using this expression, we can evaluate the undetected error probability by Eq. (2.68).

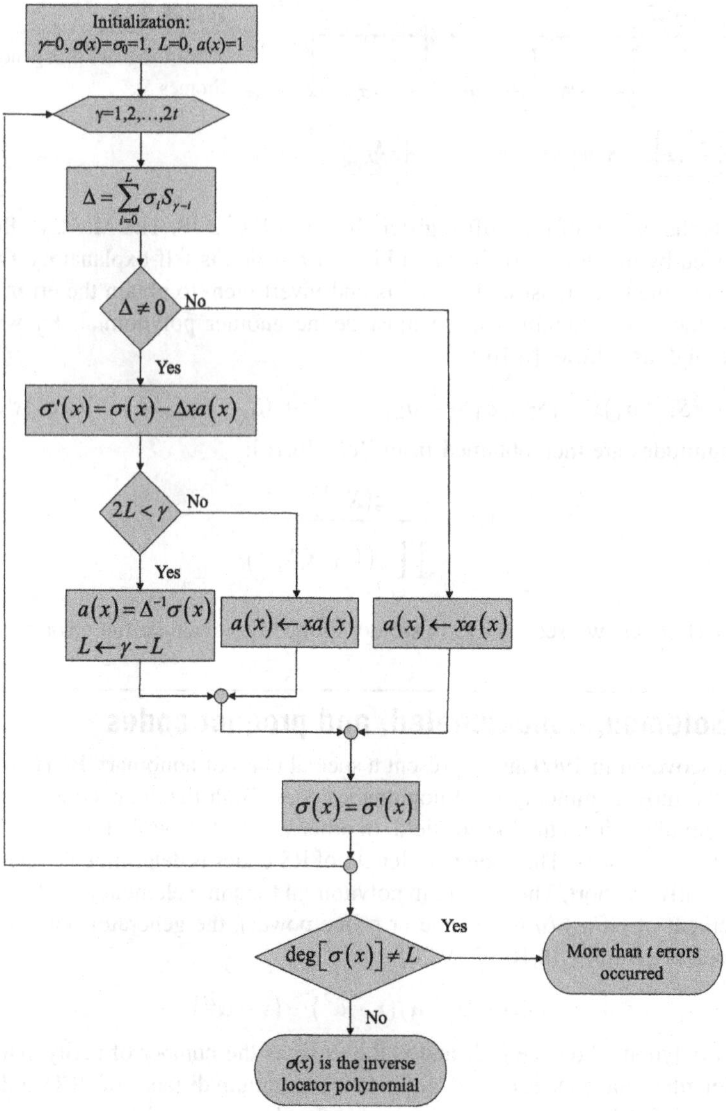

FIGURE 2.19

Flowchart of Massey–Berlekamp algorithm. Δ denotes the error (discrepancy) between the syndrome and the shift register output, and $a(x)$ stores the content of shift register (normalized by Δ^{-1}) prior to lengthening.

Example. Let GF(4) be generated by $1+x+x^2$, as explained in Section 2.4.1. The symbols of GF(4) are 0, 1, α, and α^2. The generator polynomial for the RS(3,2) code is given by $g(x)=x-\alpha$. The corresponding codewords are 000, 101, $\alpha 0\alpha$, $\alpha^2 0\alpha^2$, 011, 110, $\alpha 1\alpha^2$, $\alpha^2 1\alpha$, $0\alpha\alpha$, $1\alpha\alpha^2$, $\alpha\alpha 0$, $\alpha^2\alpha 1$, $0\alpha^2\alpha^2$, $1\alpha^2\alpha$, $\alpha\alpha^2 1$, and $\alpha^2\alpha^2 0$. This code is essentially the even parity-check code ($\alpha^2+\alpha+1=0$).

The generator polynomial for RS(3,1) is $g(x)=(x-\alpha)(x-\alpha^2)=x^2+x+1$, while the corresponding codewords are 000, 111, $\alpha\alpha\alpha$, and $\alpha^2\alpha^2\alpha^2$. Therefore, this code is in fact the repetition code.

Since RS codes are a special class of nonbinary BCH codes, they can be decoded using the same decoding algorithm explained in the previous section.

To improve the burst error correction capability of RS codes, RS code can be combined with an inner binary block code in a *concatenation* scheme as shown in Fig. 2.20. The key idea behind the concatenation scheme can be explained as follows [24]. Consider the codeword generated by inner (n,k,d) code (with d being the minimum distance of the code) and transmitted over the bursty channel. The decoder processes the erroneously received codeword and decodes it correctly. However, occasionally, the received codeword is decoded incorrectly. Therefore, the inner encoder, the channel, and the inner decoder may be considered a super channel whose input and output alphabets belong to GF(2^k). The outer encoder (N,K,D) (D-the minimum distance of outer code) encodes input K symbols and generates N output symbols transmitted over the super channel. The length of each symbol is k information digits. The resulting scheme, known as concatenated code and proposed initially by Forney [32], is an $(Nn,Kk,\geq Dd)$ code with a minimum distance of at least Dd. For example, RS (255,239,8) code can be combined with the (12,8,3) single parity check code in the concatenation scheme $(12\cdot 255,239\cdot 8,\geq 24)$. The concatenated scheme from Fig. 2.20 can be generalized to q-ary channels, the inner code operating over GF(q) and the outer over GF(q^k).

Two RS codes can be combined in a concatenated scheme by interleaving. An *interleaved code* is obtained by taking L codewords (of length N) of a given code $\mathbf{x}_j=(x_{j1},x_{j2},\ldots,x_{jN})$ $(j=1,2,\ldots,L)$, and forming the new codeword by interleaving the L codewords as follows $\mathbf{y}_i=(x_{11},x_{21},\ldots,x_{L1},x_{12},x_{22},\ldots,x_{L2},\ldots,x_{1N},x_{2N},\ldots,x_{LN})$. The process of interleaving can be visualized as the process of forming an $L\times N$ matrix of L codewords written row by row and transmitting the matrix column by column, as given below:

$$\begin{matrix} x_{11} & x_{12} & \cdots & x_{1N} \\ x_{21} & x_{22} & \cdots & x_{2N} \\ \cdots & \cdots & \cdots & \cdots \\ x_{L1} & x_{L2} & \cdots & x_{LN} \end{matrix}$$

The parameter L is known as the interleaving *degree*. The transmission must be postponed until L codewords are collected. To transmit a column whenever a new codeword becomes available, the codewords should be arranged down diagonals as given below, and the interleaving scheme is known as the *delayed interleaving* (1-frame delayed interleaving):

$$\begin{matrix} x_{i-(N-1),1} & \cdots & x_{i-2,1} & x_{i-1,1} & x_{i,1} \\ x_{i-(N-1),2} & \cdots & x_{i-2,2} & x_{i-1,2} & x_{i,2} \\ \cdots & \cdots & \cdots & \cdots & \cdots \\ & & x_{i-(N-1),N-1} & & x_{i-(N-2),N-1} \\ & & x_{i-(N-1),N} & & x_{i-(N-2),N} \end{matrix}$$

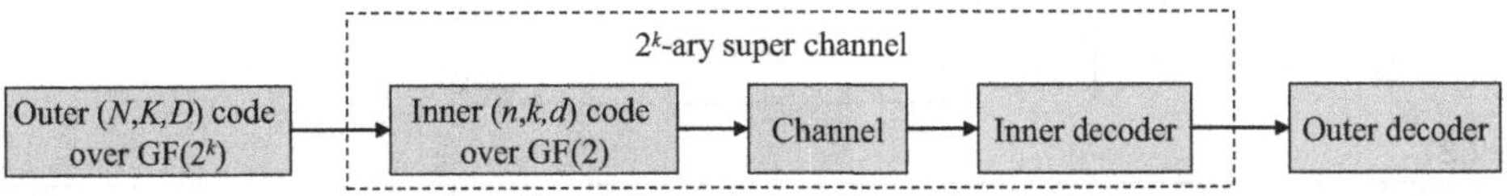

FIGURE 2.20

The concatenated $(Nn,Kk,\geq Dd)$ code.

Each new codeword completes one column of this array. In the example above the codeword $\boldsymbol{x}_i$ completes the column (frame) $x_{i,1}, x_{i-1,2}, \ldots, x_{i\text{-}(N-1),N}$. A generalization of this scheme, in which the components of the i-th codeword $\boldsymbol{x}_i$, say $x_{i,j}$ and $x_{i,j+1}$, are spaced λ frames apart, is known as λ-frame delayed interleaver.

Another way to deal with burst errors is to arrange two RS codes in a *product* manner as shown in Fig. 2.21. A product code [6,9,16] is an (n_1n_2,k_1k_2,d_1d_2) code in which codewords form an $n_1 x n_2$ array such that each row is a codeword from an (n_1,k_1,d_1) code C_1, and each column is a codeword from an (n_2,k_2,d_2) code C_2; with n_i, k_i and d_i $(i=1,2)$ being the codeword length, dimension and minimum distance, respectively, of ith component code. The product codes were proposed by Elias [33]. Both binary (such as binary BCH codes) and nonbinary codes (such as RS codes) may be arranged in the product code manner. It is possible to show [24] that the minimum distance of a product codes is the product of minimum distances of component codes. It is straightforwardly to show that the product code is able to correct the burst error of length $b=\max(n_1b_2,n_2b_1)$, where b_i is the burst error capability of component code $i=1,2$.

The results of Monte Carlo simulations for different RS concatenation schemes on an optical on-off keying AWGN channel are shown in Fig. 2.22. Interestingly, the concatenation scheme RS(255,239) + RS(255,223) of code rate $R=0.82$ outperforms the concatenation scheme RS(255,223) + RS(255,223) of lower code rate $R=0.76$, as well as the concatenation scheme RS(255,223) + RS(255,239) of the same code rate.

More powerful FEC schemes belong to the class of iteratively decodable codes [17,18,27,29,34–37], including turbo, turbo product, and LDPC codes.

2.7 Detection and estimation theory fundamentals

We conclude this chapter by providing fundamentals of detection and estimation theory, based on Refs. [9,12,21,22].

2.7.1 Geometric representation of received signals

In digital wireless/optical communication system there is a source of the message that at any time instance generates a symbol m_i $(i=1,2,\ldots,M)$ from the set of symbols $\{m_1,m_2,\ldots,m_M\}$. These

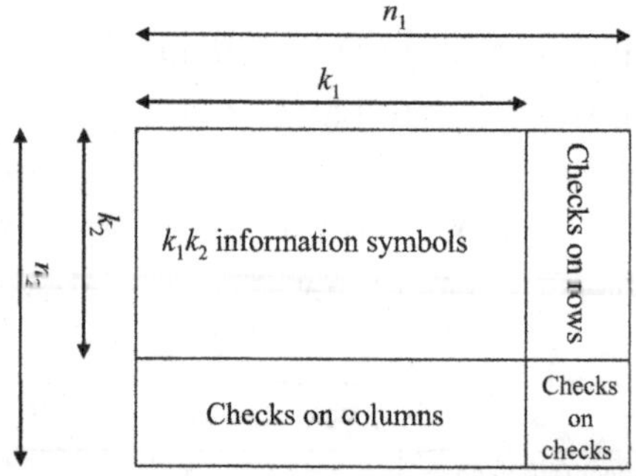

FIGURE 2.21

The structure of a codeword of a product code.

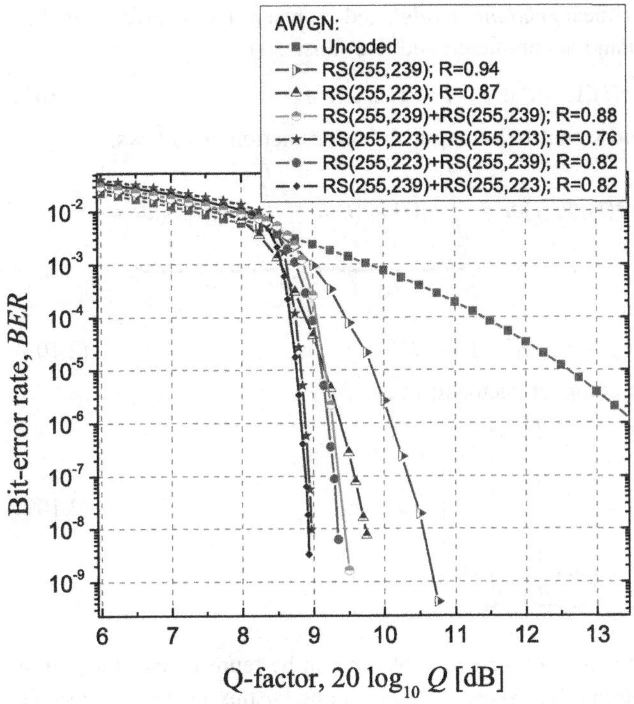

FIGURE 2.22

BER performance of concatenated RS codes on an optical on-off keying channel. The Q-factor is defined by $Q=(\mu_1-\mu_0)/(\sigma_1+\sigma_0)$, where μ_i and σ_i are the mean value and standard deviation corresponding to symbol i ($i=0,1.$).

symbols are generated with a priori probabilities: $p_1, p_2, \ldots, p_M$, where $p_i = P(m_i)$. Transmitter then converts the output m_i from the message source into a *distinct* real signal $s_i(t)$ of duration T, which is suitable for transmission over the wireless/optical channel. This set of signals $\{s_i(t)\}$ can be represented as properly weighted orthonormal basis functions $\{\Phi_n\}$ ($n = 1, \ldots, N$) (wherein $N \leq M$), so

$$s_i(t) = \sum_{n=1}^{N} s_{in}\Phi_n(t), 0 \leq t < T; \quad i = 1, 2, \cdots, M \tag{2.104}$$

where s_{in} is the projection of i-th signal along n-th basis function. The received signal $r(t)$ can also be represented in terms of projections along the same basis functions as follows:

$$r(t) = \sum_{j=1}^{N} r_n\Phi_n(t), r_n = \int_0^T r(t)\Phi_n(t)dt \ \ (n = 1, 2, \ldots, N). \tag{2.105}$$

Therefore, both received and transmitted signals can be represented as vectors $\boldsymbol{r}$ and $\boldsymbol{s}_i$ in the signal space, respectively, as

$$\boldsymbol{r} = [r_1 \ \ r_2 \ \ \cdots \ \ r_N]^{\mathrm{T}}, \quad \boldsymbol{s}_i = |s_i\rangle = [s_{i1} \ \ s_{i2} \ \ \cdots \ \ s_{iN}]^{\mathrm{T}}. \tag{2.106}$$

For Gaussian channels, we can use the Euclidean distance receiver to decide in favor of a signal constellation point closest to the received signal vector. In the linear regime for both wireless and optical communications, use the *additive* (linear) *channel model*, and represent the received signal as the addition of the transmitted signal $s_i(t)$ and accumulated additive noise $z(t)$:

$$r(t) = s_i(t) + z(t), 0 \le t < T; \ \ i = 1, 2, ..., M \tag{2.107}$$

We can represent projection of received signal $r(t)$ along n-th basis function as follows:

$$r_n = \int_0^T r(t)\Phi_n(t)dt = \int_0^T [s_i(t) + z(t)]\Phi_n(t)dt = \underbrace{\int_0^T s_i(t)\Phi_n(t)dt}_{s_{in}} + \underbrace{\int_0^{T_s} n(t)\Phi_n(t)dt}_{z_n} =$$

$$= s_{in} + z_n; \ \ n = 1, \cdots, N. \tag{2.108}$$

We can also rewrite Eq. (2.108) into a compact vector form as

$$\boldsymbol{r} = \begin{bmatrix} r_1 \\ r_2 \\ \vdots \\ r_N \end{bmatrix} = \underbrace{\begin{bmatrix} s_{i1} \\ s_{i2} \\ \vdots \\ s_{iN} \end{bmatrix}}_{\boldsymbol{s}_i} + \underbrace{\begin{bmatrix} z_1 \\ z_2 \\ \vdots \\ z_N \end{bmatrix}}_{\boldsymbol{z}} = \boldsymbol{s}_i + \boldsymbol{z}. \tag{2.109}$$

Therefore, the received vector, also known as *observation vector*, can be represented as a sum of transmitted vector and noise vector. The observation vector can be obtained using one of two possible alternative detection methods: matched filter based detector, and correlation detector, as summarized in Fig. 2.23.

The output of n-th matched filter having impulse response $h_n(t) = \Phi_n(T - t)$ is given as the convolution of the matched filter input and impulse response:

$$r_n(t) = \int_{-\infty}^{\infty} r(\tau)h_n(t - \tau)d\tau = \int_{-\infty}^{\infty} r(\tau)\Phi_n(T - t + \tau)d\tau. \tag{2.110}$$

By performing a sampling at the time instance $t = T$, we obtain

$$r_n(T) = \int_{-\infty}^{\infty} r(\tau)h_n(T - \tau)d\tau = \int_0^T r(\tau)\Phi_n(\tau)d\tau, \tag{2.111}$$

which has the same form as the n-th output of the correlation receiver. Therefore, the correlation and matched filter-based receivers are equivalent to each other.

Let us observe the random process obtained as a difference between received signal and signal obtained by projecting the received signal along the basis functions. This difference, in fact, is the "reminder" of the received signal:

$$r'(t) = r(t) - \sum_{n=1}^{N} r_n\Phi_n(t) = s_i(t) + z(t) - \sum_{n=1}^{N}(s_{in} + z_n)\Phi_n(t)$$

$$= z(t) - \sum_{n=1}^{N} z_n\Phi_n(t) = z'(t). \tag{2.112}$$

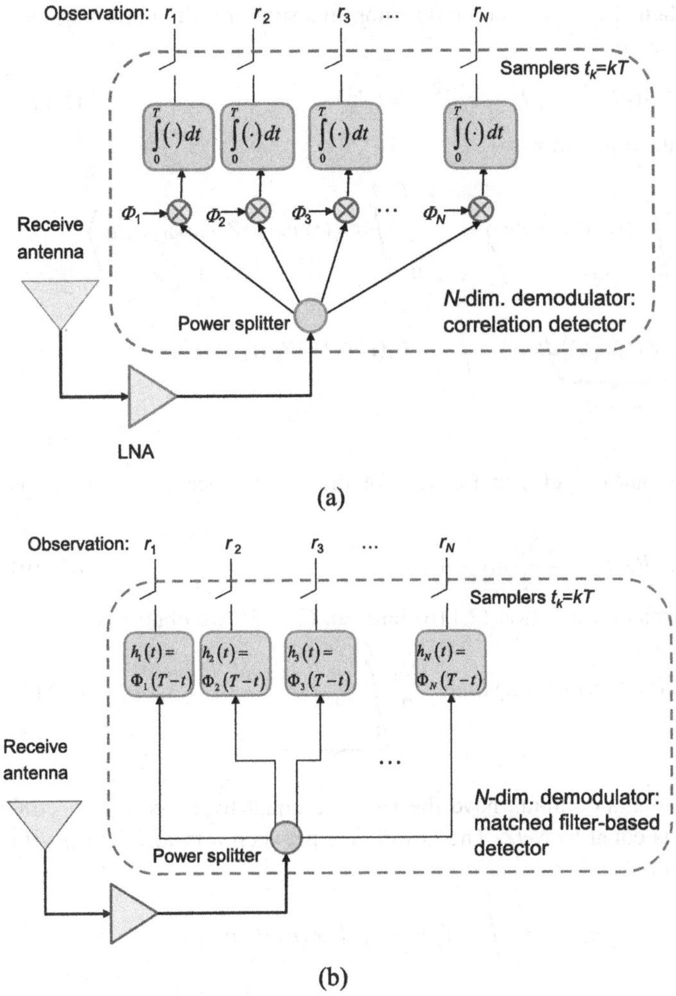

FIGURE 2.23

Two equivalent implementations of detector (or demodulator) for wireless communications: (A) correlation detector, and (B) matched filter-based detector.

We can notice that the reminder of the received signal is only noise dependent. In all equations above, we have represented the noise in terms of the same basis functions used to represent the transmitted signal. However, it is more appropriate to represent the noise using Karhunen-Loève expansion [21], by taking the correlation properties of noise process into account. However, for additive white Gaussian noise (AWGN), the noise projections along the basis functions of transmitted signals are uncorrelated and independent, as we will show shortly. Consequently, the basis used to represent noise is irrelevant in selection of the decision strategy in detection process.

Let $R(t)$ denote the random process established by sampling of the received signal $r(t)$. The projection R_n of $R(t)$ along the n-th basis function will be a random variable represented by the correlator output r_n, ($n = 1,2, \ldots,N$.) The corresponding mean value of R_n is given by

$$m_{R_n} = \langle R_n \rangle = \langle s_{in} + Z_n \rangle = s_{in} + \langle Z_n \rangle = s_{in}, \tag{2.113}$$

and it is dependent only on the n-th coordinate of transmitted signal s_{in}. With Z_n we denoted the random variable at the output of the n-th correlator (matched filter) originating from the noise component $z(t)$, and with $\langle\cdot\rangle$ the mathematical expectation operator. In a similar fashion, the variance of R_n is also only noise dependent since it is

$$\sigma_{R_n}^2 = \mathrm{Var}[R_n] = \langle (R_n - s_{in})^2 \rangle = \langle Z_n^2 \rangle. \tag{2.114}$$

The variance of R_n can be determined as follows:

$$\begin{aligned}\sigma_{R_n}^2 &= \langle Z_n^2 \rangle = \left\langle \int_0^T Z(t)\Phi_n(t)dt \int_0^T Z(u)\Phi_n(u)du \right\rangle = \left\langle \int_0^T \int_0^T \Phi_n(t)\Phi_n(u)Z(t)Z(u)dtdu \right\rangle \\ &= \int_0^T \int_0^T \Phi_n(t)\Phi_n(u)\underbrace{\langle Z(t)Z(u)\rangle}_{R_Z(t,u)}dtdu = \int_0^T \int_0^T \Phi_n(t)\Phi_n(u)R_Z(t,u)dtdu.\end{aligned} \tag{2.115}$$

In Eq. (2.115), $R_Z(\cdot)$ presents the autocorrelation function of the noise process $Z(t)$, which is otherwise given as [12,21]

$$R_Z(t,u) = \frac{N_0}{2}\delta(t-u), \tag{2.116}$$

Now, after substitution of autocorrelation function (2.116) into Eq. (2.115), we obtain

$$\sigma_{R_n}^2 = \frac{N_0}{2}\int_0^T \int_0^T \Phi_n(t)\Phi_n(u)\delta(t-u)dtdu = \frac{N_0}{2}\underbrace{\int_0^T \Phi_n^2(t)dt}_{1} = \frac{N_0}{2}. \tag{2.117}$$

Therefore, all matched filter (correlator) outputs have the variance equal to the power spectral density of noise process $Z(t)$, which is equal to $N_0/2$. The covariance between n-th and k-th ($n \neq k$) outputs of the matched filters is given by

$$\begin{aligned}\mathrm{Cov}[R_n R_k] &= \langle (R_n - s_{in})(R_k - s_{ik}) \rangle = \langle R_n R_k \rangle = \left\langle \int_0^T Z(t)\Phi_j(t)dt \int_0^T Z(u)\Phi_k(u)dt \right\rangle \\ &= \left\langle \int_0^T \int_0^T \Phi_n(t)\Phi_k(u)Z(t)Z(u)dtdu \right\rangle = \int_0^T \int_0^T \Phi_n(t)\Phi_k(u)\underbrace{\langle Z(t)Z(u)\rangle}_{R_Z(t-u)}dtdu \\ &= \int_0^T \int_0^T \Phi_n(t)\Phi_k(u)\underbrace{R_Z(t-u)}_{\frac{N_0}{2}\delta(t-u)}dtdu = \frac{N_0}{2}\int_0^T \int_0^T \Phi_n(t)\Phi_k(u)\delta(t-u)dtdu = \frac{N_0}{2}\underbrace{\int_0^T \Phi_n(t)\Phi_k(t)dt}_{0,n\neq k} = 0.\end{aligned} \tag{2.118}$$

Consequently, the outputs R_n of correlators (matched filters) are mutually uncorrelated. In conclusion, the outputs of correlators, given by Eq. (2.110), are Gaussian random variables. These outputs are statistically independent, so the joint conditional PDF of the observation vector $\boldsymbol{R}$ can be written as a product of conditional PDFs of its individual outputs from correlators, as follows:

$$f_{\boldsymbol{R}}(\boldsymbol{r}|m_i) = \prod_{n=1}^{N} f_{R_n}(r_n|m_i); i = 1, \cdots, M. \tag{2.119}$$

The component R_n's are independent Gaussian random variables with mean values s_{in} and variance equal to $N_0/2$, so the corresponding conditional PDF of R_n is given as

$$f_{R_n}(r_n|m_i) = \frac{1}{\sqrt{\pi N_0}} \exp\left[-\frac{1}{N_0}(r_n - s_{in})^2\right]; \begin{matrix} n = 1, 2, \ldots, N \\ i = 1, 2, \ldots, M \end{matrix} \tag{2.120}$$

By substituting the conditional PDFs into the joint PDF, after simple rearrangement, we obtain

$$f_{\boldsymbol{R}}(\boldsymbol{r}|m_i) = (\pi N_0)^{-N/2} \exp\left[-\frac{1}{N_0}\sum_{n=1}^{N}(r_n - s_{in})^2\right]; i = 1, \cdots, M. \tag{2.121}$$

We can conclude that the elements of a random vector given by Eq. (2.109) completely characterize the term $\sum_{n=1}^{N} r_n \Phi_n(t)$. We now characterize the reminder $z'(t)$ of the received signal given by Eq. (2.112), which is only noise dependent. It is easily shown that the sample of noise process $Z'(t_m)$ is statistically independent on outputs R_j of the correlators—in other words,

$$\langle R_n Z'(t_m)\rangle = 0, \begin{cases} n = 1, 2, \ldots, N \\ 0 \le t_m \le T_s \end{cases}. \tag{2.122}$$

Therefore, the outputs of correlators (matched filters) are only relevant in the statistics of the decision-making process. In other words, as long as the signal detection in AWGN is concerned, only the projections of the noise onto basis functions used to represent the signal set $\{s_i(t)\}$ $(i = 1, \ldots, M)$ affect the statistical process in the detection circuit; the remainder of the noise is irrelevant. This claim is sometimes called the *theorem of irrelevance* [9],[12,21,22]. According to this theorem, the channel model given by Eq. (2.107) can be represented by an equivalent N-dimensional vector given by Eq. (2.109).

2.7.2 Optimum and log-likelihood ratio receivers

This section concerns the receivers minimizing symbol error probability, commonly referred to as the *optimum receivers*. The joint conditional PDF given by $f_{\boldsymbol{R}}(\boldsymbol{r}|m_i)$ can be used in a decision-making process. It is often called the *likelihood function* and denoted $L(m_i)$. For Gaussian channels, the likelihood function contains exponential terms that can lead to numerical instabilities. It is more convenient to work with its logarithmic version, which is known as a *log-likelihood function*:

$$l(m_i) = \log L(m_i); \quad i = 1, 2, \ldots, M. \tag{2.123}$$

For the AWGN channel, the log-likelihood function is given by

$$l(m_i) = -\frac{1}{N_0}\sum_{j=1}^{N}(r_j - s_{ij})^2; i = 1, 2, \ldots, M. \tag{2.124}$$

One can notice that the log-likelihood function is similar to the Euclidean distance squared.

The log-likelihood function is quite useful in the decision-making procedure. Since we are particularly concerned with *optimum receiver* design, we can use the log-likelihood function as a design tool. Namely, given the observation vector $\boldsymbol{r}$, the optimum receiver performs the mapping from observation vector $\boldsymbol{r}$ to the estimated value $\hat{m}$ of the transmitted symbol (say m_i), so the symbol error probability is minimized. The probability P_e of an erroneous decision ($\hat{m} = m_i$) can be calculated by the following relation:

$$P_e(m_i|\boldsymbol{r}) = P(m_i \text{ is not sent } |\boldsymbol{r}) = 1 - P(m_i \text{ is sent } |\boldsymbol{r}). \tag{2.125}$$

where $P(\cdot|\cdot)$ denotes conditional probability.

The optimum decision strategy, also known as the *maximum* a posteriori *probability (MAP) rule*, can be formulated as the following decision rule:

$$\begin{aligned}&\text{Set } \hat{m} = m_i \text{if}\\ &P(m_i \text{ is sent } |\boldsymbol{r}) \geq P(m_k \text{ is sent } |\boldsymbol{r}), \ \forall k \neq i\end{aligned} \tag{2.126}$$

At this point, it is useful to provide the geometric interpretation of the optimum detection problem. Let D denote the N-dimensional signal space of all possible observation vectors (D, also known as the *observation space*). The total observation space is partitioned into M nonoverlapping N-dimensional decision regions $D_1, D_2, \ldots, D_M$, which are defined as

$$D_i = \{\boldsymbol{r}: P(m_i \text{ sent } |\boldsymbol{r}) > P(m_k \text{ sent } |\boldsymbol{r}) \forall \ k \neq i\}. \tag{2.127}$$

In other words, the i-th decision region D_i is defined as the set of all observation vectors for which the probability that m_i is sent is larger than any other probability. As an illustration, in Fig. 2.24 we show decision regions for QPSK (also known as 4-QAM), assuming equal transmission probability. Whenever the observation vector falls within decision region D_i, we can choose symbol m_i.

The optimum decision strategy given by Eq. (2.126) is not quite suitable from a practical point of view. Let us consider an alternative practical representation for $M = 2$. The average error probability for $M = 2$ is given by

$$P_e = p_1 \int_{\overline{D}_1} f_{\boldsymbol{R}}(r|m_1)dr + p_2 \int_{\overline{D}_2} f_{\boldsymbol{R}}(r|m_2)dr, \tag{2.128}$$

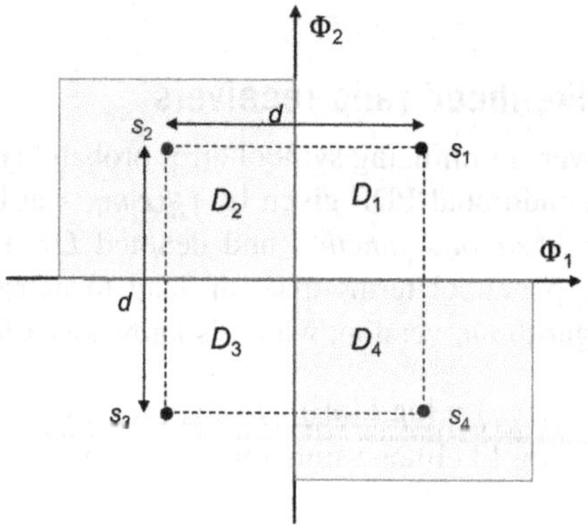

FIGURE 2.24

Decision regions for QPSK.

where $p_i = P(m_i)$ represents the a priori probability that symbol m_i is transmitted, while $\overline{D}_i = D - D_i$ denotes the complement of D_i ("not D_i decision region").

Since $D = D_1 + D_2$, we can rewrite Eq. (2.128) as

$$P_e = p_1 \int_{D-D_1} f_{\boldsymbol{R}}(\boldsymbol{r}|m_1)d\boldsymbol{r} + p_2 \int_{D_1} f_{\boldsymbol{R}}(\boldsymbol{r}|m_2)d\boldsymbol{r} =$$

$$= p_1 + \int_{D_1} [p_2 f_{\boldsymbol{R}}(\boldsymbol{r}|m_2) - p_1 f_{\boldsymbol{R}}(\boldsymbol{r}|m_1)]d\boldsymbol{r}. \tag{2.129}$$

To minimize the average error probability given by Eq. (2.129), we can apply the following decision rule: if $p_1 f_{\boldsymbol{R}}(\boldsymbol{r}|m_1)$ is greater than $p_2 f_{\boldsymbol{R}}(\boldsymbol{r}|m_2)$, then assign $\boldsymbol{r}$ to D_1 and accordingly decide in favor of m_1. Otherwise, we assign $\boldsymbol{r}$ to D_2 and decide in favor of m_2. This decision strategy can be generalized for $M > 2$ as follows:

$$\begin{aligned} &\text{Set } \widehat{m} = m_i \text{ if} \\ &p_k\, f_{\boldsymbol{R}}(\boldsymbol{r}|m_k) \text{ is maximum for } k = i. \end{aligned} \tag{2.130}$$

When all symbols occur with the same probability $p_i = 1/M$, then the corresponding decision rule is known as the *maximum likelihood (ML) rule* and can be formulated as

$$\begin{aligned} &\text{Set } \widehat{m} = m_i \text{ if} \\ &l(m_k) \text{is maximum for } k = i. \end{aligned} \tag{2.131}$$

In terms of the observation space, the ML rule can also be formulated as

$$\begin{aligned} &\text{Observation vector } \boldsymbol{r} \text{ lies in decision region } D_i \text{ when} \\ &l(m_k) \text{is maximum for } k = i. \end{aligned} \tag{2.132}$$

For the AWGN channel, using the likelihood function given by Eq. (2.124) (which is related to Euclidean distance squared), we can formulate the ML rule as

$$\begin{aligned} &\text{Observation vector } \boldsymbol{r} \text{ lies in region } Z_i \text{ if} \\ &\text{the Euclidean distance} \|\boldsymbol{r}\text{-}\boldsymbol{s}_k\| \text{is minimum for } k = i. \end{aligned} \tag{2.133}$$

From Euclidean distance in signal space, we know that

$$\|\boldsymbol{r} - \boldsymbol{s}_k\|^2 = \sum_{n=1}^{N} |r_n - s_{kn}|^2 = \sum_{n=1}^{N} r_n^2 - 2\sum_{n=1}^{N} r_n s_{kn} + \underbrace{\sum_{n=1}^{N} s_{kn}^2}_{E_k}, \tag{2.134}$$

and by substituting this equation into Eq. (2.133), we obtain the final version of the ML rule as follows:

$$\begin{aligned} &\text{Observation vector } \boldsymbol{r} \text{ lies in decision region } \mathrm{D}_i \text{ if} \\ &\sum_{n=1}^{N} r_n s_{kn} - \frac{1}{2} E_k \text{ is maximum for } k = i. \end{aligned} \tag{2.135}$$

The receiver configuration shown in Fig. 2.25, derived from Eq. (2.135), is optimum only for an AWGN channel (such as a wireless channel and optical channel dominated by ASE noise), whereas it

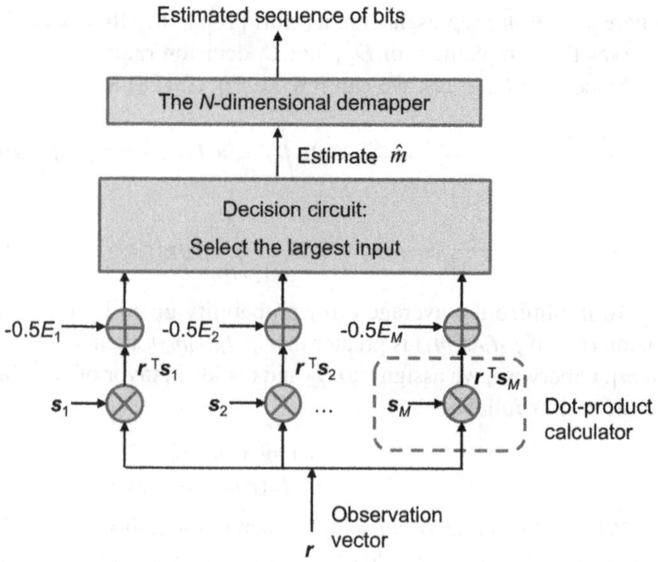

FIGURE 2.25

The signal transmission detector/decoder configuration for an optimum receiver.

is suboptimum for Gaussian-like channels. For non-Gaussian channels, we need to use the ML rule given by Eq. (2.131).

As an example, for modulation schemes with $M = 2$, we can use the decision strategy based on the following rule:

$$LLR(\mathbf{r}) = \log\left[\frac{f_{\mathbf{R}}(\mathbf{r}|m_1)}{f_{\mathbf{R}}(\mathbf{r}|m_2)}\right] \underset{m_2}{\overset{m_1}{\gtrless}} \log\left(\frac{p_2}{p_1}\right), \tag{2.136}$$

where $LLR(\bullet)$ denotes the log-likelihood ratio. Eq. (2.136) is also known as *Bayes' test* [21], while factor $\log(p_2/p_1)$ is known as a threshold of the test. The corresponding log-likelihood ratio receiver scheme is shown in Fig. 2.26. When $p_1 = p_2 = 1/2$, the test threshold is 0, while the corresponding receiver is the ML one.

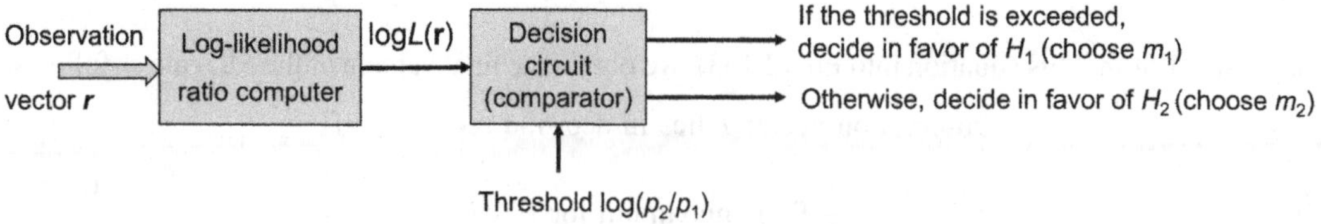

FIGURE 2.26

The log-likelihood ratio receiver.

2.7.3 Estimation theory fundamentals

So far, our focus has been on detecting signals in the presence of additive noise, such ASE or Gaussian noise. In this section, we will pay attention to the estimation of a certain signal parameter, such as phase, frequency offset, and so on, important from the detection point of view. Let the received signal be denoted by $r(t)$ while the transmitted signal is by $s(t,p)$, where p is the parameter to be estimated. The received signal in the presence of additive noise $z(t)$ can be written as

$$r(t) = s(t,p) + z(t), \quad 0 \leq t \leq T, \tag{2.137}$$

where T is the observation interval (symbol duration for modulation schemes). The operation of assigning a value $\widehat{p}$ to an unknown parameter p is known as *parameter estimation*; the value assigned is called *estimate*, and the algorithm used to perform this operation is called an *estimator.*

There are different criteria used to evaluate estimator, including minimum mean-square (MMS) estimate, MAP estimate, and ML estimate. We can define the cost function in MMS estimate as an integral of the quadratic form [21]:

$$C_{\mathrm{MMSE}}(\varepsilon) = \int \underbrace{(p - \widehat{p})^2}_{\varepsilon^2} f_P(p|\boldsymbol{r})dp, \tag{2.138}$$

where $\varepsilon = \widehat{p} - p$, while $f_P(p|\boldsymbol{r})$ is the a posteriori PDF of random variable P. The observation vector $\boldsymbol{r}$ in Eq. (2.138) is defined by its r_i components, where the i-th component is determined by

$$r_i = \int_0^T r(t)\Phi_i(t)dt; \quad i = 1, 2, \cdots, N \tag{2.139}$$

$\{\Phi_i(t)\}$ denotes the set of orthonormal basis functions. The estimate that minimizes the average cost function given by Eq. (2.138) is known as Bayes' MMS estimate [21,22]. The estimate that maximizes the a posteriori probability density function $f_P(p|\boldsymbol{r})$ is known as the MAP estimate. Finally, the estimate that maximizes the conditional PDF $f_r(\boldsymbol{r}|p)$, also known as a likelihood function, is called the ML estimate. The MAP estimate can be obtained as a special case of Bayes estimation using the uniform cost function $C(\varepsilon)=(1/\delta)[1\text{-rect}(\varepsilon/\delta)]$ as follows:

$$\begin{aligned}\langle \mathcal{C}(\varepsilon)\rangle &= \int \mathcal{C}(\varepsilon) f_P(p|\boldsymbol{r})dp = \frac{1}{\delta}\left[\int_{-\infty}^{\hat{p}-\delta/2} + \int_{\hat{p}+\delta/2}^{\infty}\right] f_P(p|\boldsymbol{r})dp \\ &= \frac{1}{\delta}\left[1 - \int_{\hat{p}-\delta/2}^{\hat{p}+\delta/2} f_P(p|\boldsymbol{r})dp\right] \cong (1/\delta)\left[1 - \delta f_P(p|\boldsymbol{r})\right] = \frac{1}{\delta} - f_P(p|\boldsymbol{r})\end{aligned} \tag{2.140}$$

where the approximation above is valid for sufficiently small δ. The quality of an estimate is typically evaluated in terms of the expected mean value $\langle\widehat{p}\rangle$ and variance of the estimation error. We can recognize several cases with respect to the expected value. If the expected value of the estimate equals the true value of the parameter $\langle\widehat{p}\rangle = p$, we say that estimate is *unbiased*. If the expected value of estimate differs from true parameter by fixed value b, i.e., if $\langle\widehat{p}\rangle = p + b$, we say that estimate has a *known bias*. Finally, if the expected value of the estimate is different from the true parameter by variable amount $b(p)$, i.e., if $\langle\hat{p}\rangle = p + b(p)$, we say that estimate has a *variable bias*. The estimation error ε is defined as the difference between true value p and estimate $\hat{p}$, namely, $\varepsilon = p - \hat{p}$. A good estimate is one that simultaneously provides a small bias and a small variance of estimation error. The lower bound on the variance of estimation error can be obtained from Cramér-Rao inequality [12,21], which is

$$\langle (p-\hat{p})^2 \rangle \geq \frac{-1}{\left\langle \frac{\partial^2 \log f_r(\mathbf{r}|p)}{\partial p^2} \right\rangle}, \tag{2.141}$$

where we assumed that the second partial derivative of conditional PDF exists and that it is absolutely integrable. Any estimate that satisfies the Cramér-Rao bound defined by the right side of Eq. (2.141) is known as an *efficient estimate*.

To explain ML estimation, we can assume that transmitted signal $s(t,p)$ can also be presented as vector $\mathbf{s}$ in signal space, with n-th component given as

$$s_n(p) = \int_0^T s(t,p)\Phi_n(t)dt; \quad n = 1,2,\cdots,N \tag{2.142}$$

The noise vector z can be presented similarly:

$$z = \begin{bmatrix} z_1 \\ z_2 \\ \vdots \\ z_N \end{bmatrix}, \quad z_n = \int_0^T z(t)\Phi_n(t)dt; n = 1,2,\cdots,N \tag{2.143}$$

Since the components z_n of the noise are zero-mean Gaussian with PSD equal $N_0/2$, the components of the observation vector will also be Gaussian with mean values $s_n(p)$. The joint conditional PDF for a given parameter p will be

$$f_{\mathbf{R}}(\mathbf{r}|p) = \frac{1}{(\pi N_0)^{N/2}} \prod_{n=1}^{N} e^{-[r_n - s_n(p)]^2/N_0}. \tag{2.144}$$

The likelihood function can be defined as

$$L[r(t),p] = \lim_{N\to\infty} \frac{f_r(\mathbf{r}|p)}{f_r(\mathbf{r})} = \frac{2}{N_0}\int_0^T r(t)s(t,p)dt - \frac{1}{N_0}\int_0^T s^2(t,p)dt. \tag{2.145}$$

The ML estimate is derived by maximizing the likelihood function. By differentiating the likelihood function with respect to estimate $\hat{p}$ and setting the result to zero, we obtain

$$\int_0^T [r(t) - s(t,\hat{p})]dt \frac{\partial s(t,\hat{p})}{\partial \hat{p}} = 0, \tag{2.146}$$

And this condition is known as the *likelihood equation*.

Example. Let us consider the following sinusoidal signal whose amplitude a and frequency ω are known, while the phase is unknown: $a\cos(\omega t + \theta)$. Based on likelihood Eq. (2.146), we obtain the following ML estimate of phase in the presence of AWGN:

$$\widehat{\theta} = \tan^{-1}\left[\frac{\int_0^T r(t)\cos(\omega t)dt}{\int_0^T r(t)\cos(\omega t)dt}\right].$$

2.8 Concluding remarks

This Chapter has been devoted to information theory, coding theory, and detection and estimation theory. In Section 2.1, we have provided the definitions of entropy, joint entropy, conditional entropy, relative entropy, mutual information, and channel capacity, followed by the information capacity theorem. We have also briefly described the source coding and data compaction concepts. Further, we have discussed the channel capacity of DMCs, continuous channels, and some optical channels.

The standard FEC schemes belonging to the class of hard-decision codes have been described in several sections of this chapter. Descriptions of more powerful FEC schemes belonging to the class of soft iteratively decodable codes lie outside the scope of this chapter (see, for instance, Refs. [9,10,18]). Iteratively decodable codes, such as LDPC codes, are described in Chapter 9. In Section 2.2, we have introduced the classical channel coding preliminaries. Section 2.3 has been devoted to the basics of LBCs, such as the definition of generator and parity-check matrices, syndrome decoding, distance properties of LBCs, and some important coding bounds. In Section 2.4, cyclic codes have been introduced. The BCH codes have been described in Section 2.5. The RS codes, concatenated codes, and product codes have been described in Section 2.6. The fundamentals of detection and estimation theory have been introduced in Sec. 2.7.

References

[1] S. Haykin, Digital Communication Systems, John Wile & Sons, 2014.

[2] T.M. Cover, J.A. Thomas, Elements of Information Theory, John Wiley & Sons, Inc., New York, 1991.

[3] F.M. Ingels, Information and Coding Theory, Intext Educational Publishers, Scranton, 1971.

[4] F.M. Reza, An Introduction to Information Theory, Dover Publications, New York, 1994.

[5] R.E. Blahut, Principles and Practice of Information Theory, Addison-Wesley Publishing Company, Reading, 1990.

[6] D. Drajic, P. Ivanis, Introduction to Information Theory and Coding, third ed., Academic Misao, Belgrade, 2009 (In Serbian).

[7] I.B. Djordjevic, Physical-Layer Security and Quantum Key Distribution, Springer International Publishing, Switzerland, 2019.

[8] I.B. Djordjevic, Quantum Biological Information Theory, Springer, Nov, 2015.

[9] I.B. Djordjevic, Advanced Optical and Wireless Communications Systems, Springer International Publishing, Switzerland, Dec. 2017.

[10] I.B. Djordjevic, W. Ryan, B. Vasic, Coding for Optical Channels, Springer, March 2010.
[11] S. Haykin, Communication Systems, John Wiley & Sons, Inc., 2004.
[12] J.G. Proakis, Digital Communications, McGaw-Hill, Boston, MA, 2001.
[13] R.G. Gallager, Information Theory and Reliable Communication, John Wiley and Sons, New York, 1968.
[14] D. Huffman, A method for the construction of minimum-redundancy codes, Proc. IRE 40 (9) (1952) 1098–1101.
[15] C.E. Shannon, A mathematical theory of communication, Bell Syst. Tech. J. 27 (1948), 379–423 and 623–656.
[16] M.F. Barsoum, B. Moision, M. Fitz, D. Divsalar, J. Hamkins, Itrative coded pulse-position-modulation for deep-space optical communications, in: Proc. ITW 2007, September 2–6, 2007, pp. 66–71. Lake Tahoe, California.
[17] D. Divsalar, F. Pollara, Turbo Codes for Deep-Space Communications, vols. 42–120, TDA Progress Report, February 15, 1995, pp. 29–39.
[18] W. Shieh, I. Djordjevic, OFDM for Optical Communications, Elsevier, Oct. 2009.
[19] D. Raghavarao, Constructions and Combinatorial Problems in Design of Experiments, Dover Publications, Inc., New York, 1988 (reprint).
[20] C.C. Pinter, A Book of Abstract Algebra, Dover Publications, Inc., New York, 2010 (reprint).
[21] R.N. McDonough, A.D. Whalen, Detection of Signals in Noise, second ed., Academic Press, San Diego, 1995.
[22] S. Haykin, Digital Communications, John Wiley & Sons, 1988.
[23] S. Lin, D.J. Costello, Error Control Coding: Fundamentals and Applications, Prentice-Hall, Inc., USA, 1983.
[24] J.B. Anderson, S. Mohan, Source and Channel Coding: An Algorithmic Approach, Kluwer Academic Publishers, Boston, MA, 1991.
[25] F.J. MacWilliams, N.J.A. Sloane, The Theory of Error-Correcting Codes, North Holland, Amsterdam, the Netherlands, 1977.
[26] S.B. Wicker, Error Control Systems for Digital Communication and Storage, Prentice-Hall, Inc., Englewood Cliffs, NJ, 1995.
[27] R.H. Morelos-Zaragoza, The Art of Error Correcting Coding, John Wiley & Sons, Boston, MA, 2002.
[28] T. Mizuochi, Recent progress in forward error correction and its interplay with transmission impairments, IEEE J. Sel. Top. Quant. Electron. 12 (4) (2006) 544–554.
[29] J.K. Wolf, Efficient maximum likelihood decoding of linear block codes using a trellis, IEEE Trans. Inf. Theor. IT-24 (1) (1978) 76–80.
[30] I.S. Reed, G. Solomon, Polynomial codes over certain finite fields, SIAM J. Appl. Math. 8 (1960) 300–304.
[31] S.B. Wicker, V.K. Bhargva (Eds.), Reed-Solomon Codes and Their Applications, IEEE Press, New York, 1994.
[32] G.D. Forney Jr., Concatenated Codes, MIT Press, Cambridge, MA, 1966.
[33] P. Elias, Error-free coding, IRE Trans. Inform. Theory IT-4 (1954) 29–37.
[34] L.R. Bahl, J. Cocke, F. Jelinek, J. Raviv, Optimal decoding of linear codes for minimizing symbol error rate, IEEE Trans. Inf. Theor. IT-20 (2) (1974) 284–287.
[35] W.E. Ryan, Concatenated convolutional codes and iterative decoding, in: J.G. Proakis (Ed.), Wiley Encyclopedia in Telecommunications, John Wiley and Sons, 2003.
[36] B. Vucetic, J. Yuan, Turbo Codes-Principles and Applications, Kluwer Academic Publishers, Boston, 2000.
[37] M. Ivkovic, I.B. Djordjevic, B. Vasic, Calculation of achievable information rates of long-haul optical transmission systems using instanton approach, IEEE/OSA J. Lightwave Technol. 25 (2007) 1163–1168.

Further reading

[1] C. Berrou, A. Glavieux, P. Thitimajshima, Near Shannon limit error-correcting coding and decoding: turbo codes, in: Proc. 1993 Int. Conf. Comm., ICC, 1993, pp. 1064–1070.
[2] C. Berrou, A. Glavieux, Near optimum error correcting coding and decoding: turbo codes, IEEE Trans. Commun. (1996) 1261–1271.
[3] R.M. Pyndiah, Near optimum decoding of product codes, IEEE Trans. Commun. 46 (1998) 1003–1010.
[4] R.G. Gallager, Low Density Parity Check Codes, MIT Press, Cambridge, MA, 1963.
[5] I.B. Djordjevic, S. Sankaranarayanan, S.K. Chilappagari, B. Vasic, Low-density parity-check codes for 40 Gb/s optical transmission systems, IEEE/LEOS J. Sel. Top. Quant. Electron. 12 (4) (2006) 555–562.
[6] I.B. Djordjevic, O. Milenkovic, B. Vasic, Generalized low-density parity-check codes for optical communication systems, IEEE/OSA J. Lightwave Technol. 23 (2005) 1939–1946.
[7] B. Vasic, I.B. Djordjevic, R. Kostuk, Low-density parity check codes and iterative decoding for long haul optical communication systems, IEEE/OSA J. Lightwave Technol. 21 (2003) 438–446.
[8] I.B. Djordjevic, et al., Projective plane iteratively decodable block codes for WDM high-speed long-haul transmission systems, IEEE/OSA J. Lightwave Technol. 22 (2004) 695–702.
[9] O. Milenkovic, I.B. Djordjevic, B. Vasic, Block-circulant low-density parity-check codes for optical communication systems, IEEE J. Sel. Top. Quant. Electron. 10 (2004) 294–299.
[10] B. Vasic, I.B. Djordjevic, Low-density parity check codes for long haul optical communications systems, IEEE Photon. Technol. Lett. 14 (2002) 1208–1210.
[11] I. B. Djordjevic, M. Arabaci, L. Minkov, Next generation FEC for high-capacity communication in optical transport networks, IEEE/OSA J. Lightwave Technol., Accepted for publication. (Invited Paper).
[12] I.B. Djordjevic, On advanced FEC and coded modulation for ultra-high-speed optical transmission, IEEE Commun. Surv. Tutorials 18 (3) (2016) 1920–1951.
[13] S. Chung, et al., On the design of low-density parity-check codes within 0.0045 dB of the Shannon Limit, IEEE Commun. Lett. 5 (2001) 58–60.
[14] I.B. Djordjevic, L.L. Minkov, H.G. Batshon, Mitigation of linear and nonlinear impairments in high-speed optical networks by using LDPC-coded turbo equalization, IEEE J. Sel. Areas Commun., Opt. Commun. Netw. 26 (6) (2008) 73–83.
[15] M.E. van Valkenburg, Network Analysis, third ed., Prentice-Hall, Englewood Cliffs, 1974.

CHAPTER

Quantum information processing fundamentals

3

Chapter outline

3.1 Quantum information processing features

The *fundamental features* of QIP are different from those of classical computing and can be summarized as the following three [1–4]: (i) linear superposition, (ii) quantum parallelism, and (iii) entanglement. We provide some basic details of these features below.

(i) *Linear superposition.* Contrary to the classical bit, a quantum bit or *qubit* can take not only the two discrete values 0 and 1 but also *all* possible *linear combinations* of them. This characteristic results from a fundamental property of quantum states—it is possible to construct a *linear superposition* of quantum state $|0\rangle$ and quantum state $|1\rangle$.

(ii) *Quantum parallelism.* Quantum parallelism allows many operations to be performed in parallel, which is a key difference from classical computing. Namely, in classical computing, it is

Quantum Communication, Quantum Networks, and Quantum Sensing. https://doi.org/10.1016/B978-0-12-822942-2.00008-X

possible to know the internal status of the computer. On the other hand, because of the no-cloning theorem, the current state of a quantum computer cannot be known. This property has led to the development of Shor's factorization algorithm, which can be used to crack the Rivest–Shamir–Adleman (aka RSA) encryption protocol. Other important quantum algorithms include the Grover search algorithm, which is used to search for an entry in an unstructured database; the quantum Fourier transform, which is the basis for several algorithms; and Simon's algorithm. The quantum computer can encode all input strings of length N simultaneously into a single computational step. In other words, the quantum computer can simultaneously pursue 2^N classical paths, indicating that the quantum computer is significantly more powerful than the classical one.

(iii) *Entanglement.* At a quantum level, it appears that two quantum objects can form a single entity even when they are well separated from each other. Any attempt to consider this entity a combination of two independent quantum objects given by the tensor product of quantum states will fail unless signal propagation at superluminal speed is allowed. These quantum objects that cannot be decomposed into the tensor product of independent quantum objects are called *entangled* quantum objects. Given that arbitrary quantum states cannot be copied, which is a consequence of the no-cloning theorem, communication at superluminal speed is not possible. Therefore, entangled quantum states cannot be written as the tensor product of independent quantum states. Moreover, it can be shown that the amount of information contained in an entangled state of N qubits grows exponentially instead of linearly as it does for classical bits.

In the upcoming sections, we describe these fundamental features in greater detail. First, let us briefly review the roots of QIP, as outlined in Refs. [1,2]. The roots of QIP lie with information theory [9–13], computer science [14–16], and quantum mechanics [1–8,17–25]. Shannon's development of information theory [9,10] allowed us to formally define "information," while Turing's seminal work [14–16] created the foundation of computer science and provided us with a computational mechanism to process information. Together, their work has led to breakthroughs in efficient and secure data communication over communication channels, such as a memory storage device and a wire (or wireless) communication link. These breakthroughs, however, are based on the classical unit of information, a binary digit or bit, that can be either 0 or 1. In this classical setup, to increase the computational speed of a classical computing machine, we must proportionally increase the processing power. For example, for a classical computer that can perform a certain number of computations per second, the number of computations possible can be approximately doubled by adding a similar computer. This picture dramatically changes when the computational consequences of quantum mechanics are introduced [26]. Now, rather than working with bits, we work with quantum bits or qubits representing quantum states (such as photon polarization or electron spin) in a two-dimensional Hilbert space.

3.2 State vectors, operators, projection operators, and density operators

In quantum mechanics, the primitive undefined concepts are the *physical system, observables,* and *state* [1–8,17–25]. A physical system is any sufficiently isolated quantum object, say an electron, a photon, or a molecule. An observable is associated with a measurable property of a physical system, say energy or a z-component of the spin. The state of a physical system is a trickier concept in quantum mechanics than in classical mechanics. The problem arises when considering composite physical systems. In particular, states exist, known as *entangled states*, for bipartite physical systems in which neither subsystem is in a definite state. Even in cases where physical systems can be described as being in a state, two classes of states are possible: pure and mixed.

3.2.1 State vectors and operators

The condition of a quantum-mechanical system is completely specified by its *state vector* $|\psi\rangle$ in a Hilbert space H (a vector space on which a positive-definite scalar product is defined) over the field of complex numbers. Any state vector $|\alpha\rangle$, also known as a **ket**, can be expressed in terms of basis vectors $|\phi_n\rangle$ by

$$|\alpha\rangle = \sum_{n=1}^{\infty} a_n|\phi_n\rangle \tag{3.1}$$

An **observable**, such as momentum or spin, can be represented by an **operator**, such as A, in the vector space in question. Quite generally, an operator acts on a ket from the left: $(A) \cdot |\alpha\rangle = A\,|\alpha\rangle$, which results in another ket. An operator A is said to be *Hermitian* if

$$A^\dagger = A, \quad A^\dagger = \left(A^{T}\right)^{*}. \tag{3.2}$$

Suppose that the Hermitian operator A has a discrete set of eigenvalues $a^{(1)}, \ldots, a^{(n)}, \ldots$ The associated eigenvectors (eigenkets) $|a^{(1)}\rangle, \ldots, |a^{(n)}\rangle, \ldots$ can be obtained from

$$A|a^{(n)}\rangle = a^{(n)}|a^{(n)}\rangle \tag{3.3}$$

The Hermitian conjugate of a ket $|\alpha\rangle$ is denoted by $\langle\alpha|$ and called the "bra." The space dual to ket space is known as **bra** space. There exists a one-to-one correspondence, dual-correspondence (D.C.), between a ket space and a bra space:

$$\begin{aligned}
&|\alpha\rangle \overset{\text{D.C.}}{\leftrightarrow} \langle\alpha| \\
&|a^{(1)}\rangle, |a^{(2)}\rangle, \ldots \overset{\text{D.C.}}{\leftrightarrow} \langle a^{(1)}|, \langle a^{(2)}|, \ldots \\
&|\alpha\rangle + |\beta\rangle \overset{\text{D.C.}}{\leftrightarrow} \langle\alpha| + \langle\beta| \\
&c_\alpha|\alpha\rangle + c_b|\beta\rangle \overset{\text{D.C.}}{\leftrightarrow} c_\alpha^*\langle\alpha| + c_\beta^*\langle\beta|
\end{aligned} \tag{3.4}$$

The *scalar (inner) product* of two state vectors $|\phi\rangle = \sum_n a_n\,|\phi_n\rangle$ and $|\psi\rangle = \sum_n b_n\,|\phi_n\rangle$ is defined by

$$\langle\beta|\alpha\rangle = \sum_{n=1}^{\infty} a_n b_n^* \tag{3.5}$$

3.2.2 Projection operators

The eigenkets $\{|\xi^{(n)}\rangle\}$ of operator Ξ form the basis so that arbitrary ket $|\psi\rangle$ can be expressed in terms of eigenkets by

$$|\psi\rangle = \sum_{n=1}^{\infty} c_n |\xi^{(n)}\rangle \tag{3.6}$$

By multiplying Eq. (3.6) by $\langle\xi^{(n)}|$ from the left, we obtain

$$\langle\xi^{(n)}|\psi\rangle = \sum_{j=1}^{\infty} c_j \langle\xi^{(n)}|\xi^{(j)}\rangle = c_n \langle\xi^{(n)}|\xi^{(n)}\rangle + \sum_{j=1,j\neq n}^{\infty} c_j \langle\xi^{(n)}|\xi^{(j)}\rangle. \tag{3.7}$$

Since the eigenkets $\{|\xi^{(n)}\rangle\}$ form the basis, the principle of orthonormality is satisfied, $\langle\xi^{(n)}|\xi^{(j)}\rangle = \delta_{nj}, \delta_{nj} = \begin{cases} 1, & n=j \\ 0, & n\neq j \end{cases}$, so that Eq. (3.7) becomes

$$c_n = \langle\xi^{(n)}|\psi\rangle. \tag{3.8}$$

By substituting Eqs. (3.8) into (3.6), we obtain

$$|\psi\rangle = \sum_{n=1}^{\infty} \langle\xi^{(n)}|\psi\rangle |\xi^{(n)}\rangle = \sum_{n=1}^{\infty} |\xi^{(n)}\rangle\langle\xi^{(n)}|\psi\rangle. \tag{3.9}$$

Because $|\psi\rangle = I|\psi\rangle$ (with I being the identity operator) from Eq. (3.9), it is clear that

$$\sum_{n=1}^{\infty} |\xi^{(n)}\rangle\langle\xi^{(n)}| = I, \tag{3.10}$$

and the relation above is known as the *completeness relation*. The operators under summation in Eq. (3.10) are known as *projection* operators P_n:

$$P_n = |\xi^{(n)}\rangle\langle\xi^{(n)}|, \tag{3.11}$$

which satisfy the relationship $\sum_{n=1}^{\infty} P_n = I$. It is easy to show that the ket in Eq. (3.6) with c_n determined with Eq. (3.8) is of unit length

$$\langle\psi|\psi\rangle = \sum_{n=1}^{\infty} \langle\psi|\xi^{(n)}\rangle\langle\xi^{(n)}|\psi\rangle = \sum_{n=1}^{\infty} |\langle\psi|\xi^{(n)}\rangle|^2 = 1. \tag{3.12}$$

The following is an important theorem used a great deal throughout the chapter. It can be shown that the eigenvalues of a Hermitian operator A are real, and the eigenkets are orthogonal:

$$\langle a^{(m)}|a^{(n)}\rangle = \delta_{nm}. \tag{3.13}$$

For the proof of this claim, the interested reader is referred to Refs. [3,7].

3.2.3 Photon, spin-1/2 systems, and Hadamard gate

Photon. The x- and y-polarizations of the photon can be represented by

$$|E_x\rangle = \begin{pmatrix} 1 \\ 0 \end{pmatrix} \quad |E_y\rangle = \begin{pmatrix} 0 \\ 1 \end{pmatrix}$$

On the other hand, the right- and left-circular polarizations can be represented by

$$|E_R\rangle = \frac{1}{\sqrt{2}} \begin{pmatrix} 1 \\ j \end{pmatrix} \quad |E_L\rangle = \frac{1}{\sqrt{2}} \begin{pmatrix} 1 \\ -j \end{pmatrix}.$$

The 45 degree polarization ket can be represented as follows:

$$|E_{45^o}\rangle = \cos\left(\frac{\pi}{4}\right)|E_x\rangle + \sin\left(\frac{\pi}{4}\right)|E_y\rangle = \frac{1}{\sqrt{2}}(|E_x\rangle + |E_y\rangle) = \frac{1}{\sqrt{2}} \begin{pmatrix} 1 \\ 1 \end{pmatrix}$$

The bras corresponding to the left- and right-polarizations can be written as

$$\langle E_R| = \frac{1}{\sqrt{2}}(1 \quad -j) \quad \langle E_L| = \frac{1}{\sqrt{2}}(1 \quad j)$$

It is easy to verify that the left and right states are orthogonal and that the right-polarization state is of unit length

$$\langle E_R|E_L\rangle = \frac{1}{2}(1 - j) \begin{pmatrix} 1 \\ -j \end{pmatrix} = 0 \quad \langle E_R|E_R\rangle = \frac{1}{2}(1 - j) \begin{pmatrix} 1 \\ j \end{pmatrix} = 1$$

The completeness relation is clearly satisfied because

$$|E_x\rangle\langle E_x| + |E_y\rangle\langle E_y| = \begin{pmatrix} 1 \\ 0 \end{pmatrix}(1 \quad 0) + \begin{pmatrix} 0 \\ 1 \end{pmatrix}(0 \quad 1) = \begin{pmatrix} 1 & 0 \\ 0 & 1 \end{pmatrix} = I_2$$

An arbitrary polarization state can be represented by

$$|E\rangle = |E_R\rangle\langle E_R|E\rangle + |E_L\rangle\langle E_L|E\rangle$$

For example, for $E = E_x$, we obtain

$$|E_x\rangle = |E_R\rangle\langle E_R|E_x\rangle + |E_L\rangle\langle E_L|E_x\rangle$$

For the photon spin operator S matrix representation, we must solve the following eigenvalue equation:

$$S|\psi\rangle = \lambda|\psi\rangle$$

The photon spin operator satisfies $S^2 = I$, so we can write

$$|\psi\rangle = S^2|\psi\rangle = S(S|\psi\rangle) = S(\lambda|\psi\rangle) = \lambda S|\psi\rangle = \lambda^2|\psi\rangle$$

It is clear from the previous equation that $\lambda^2 = 1$, so the corresponding eigenvalues are $\lambda = \pm 1$. By substituting the eigenvalues into the eigenvalue equation, we find that the corresponding eigenkets are the left and right polarization states:

$$S|E_R\rangle = |E_R\rangle \quad S|E_L\rangle = -|E_L\rangle$$

The photon spin represented on an $\{|E_x\rangle, |E_y\rangle\}$ basis can be obtained by

$$S \doteq \begin{pmatrix} S_{xx} & S_{xy} \\ S_{yx} & S_{yy} \end{pmatrix} = \begin{pmatrix} \langle E_x|S|E_x\rangle & \langle E_x|S|E_y\rangle \\ \langle E_y|S|E_x\rangle & \langle E_y|S|E_y\rangle \end{pmatrix} = \begin{pmatrix} 0 & -j \\ j & 0 \end{pmatrix}$$

Spin-1/2 systems. The S_z-basis in spin-1/2 systems can be written as $\{|E_z; +\rangle, |E_z; -\rangle\}$, where the corresponding basis kets represent the spin-up and spin-down states. The eigenvalues are $\{\hbar/2, -\hbar/2\}$, and the corresponding eigenket–eigenvalue relation is

$$S_z|S_z; \pm\rangle = \pm\frac{\hbar}{2}|S_z; \pm\rangle,$$

where S_z is the spin operator that can be represented in the basis above as follows:

$$S_z = \sum_{i=+,-}\sum_{j=+,-} |i\rangle\langle j|\underbrace{S_z|i\rangle}_{i\frac{\hbar}{2}|i\rangle}\langle j| = \sum_{i=+,-} i\frac{\hbar}{2}|i\rangle\langle i| = \frac{\hbar}{2}(|+\rangle\langle +| - |-\rangle\langle -|)$$

The matrix representation of spin-½ systems is obtained by

$$|S_z; +\rangle = \begin{pmatrix} \langle S_z; +|S_z; +\rangle \\ \langle S_z; -|S_z; +\rangle \end{pmatrix} \doteq \begin{pmatrix} 1 \\ 0 \end{pmatrix} \quad |S_z; -\rangle \doteq \begin{pmatrix} 0 \\ 1 \end{pmatrix}$$

$$S_z \doteq \begin{pmatrix} \langle S_z; +|S_z|S_z; +\rangle & \langle S_z; +|S_z|S_z; -\rangle \\ \langle S_z; -|S_z|S_z; +\rangle & \langle S_z; -|S_z|S_z; -\rangle \end{pmatrix} = \frac{\hbar}{2}\begin{pmatrix} 1 & 0 \\ 0 & -1 \end{pmatrix}$$

Hadamard gate. The matrix representation of the Hadamard operator (gate) is given by

$$H = \frac{1}{\sqrt{2}}\begin{bmatrix} 1 & 1 \\ 1 & -1 \end{bmatrix}$$

It is easily shown that the Hadamard gate is Hermitian and unitary as follows:

$$H^\dagger = \frac{1}{\sqrt{2}}\begin{bmatrix} 1 & 1 \\ 1 & -1 \end{bmatrix} = H$$

$$H^\dagger H = \frac{1}{\sqrt{2}}\begin{bmatrix} 1 & 1 \\ 1 & -1 \end{bmatrix}\frac{1}{\sqrt{2}}\begin{bmatrix} 1 & 1 \\ 1 & -1 \end{bmatrix} = \begin{bmatrix} 1 & 0 \\ 0 & 1 \end{bmatrix} = I$$

The eigenvalues for the Hadamard gate can be obtained from $\det(H - \lambda I) = 0$ as $\lambda_{1,2} = \pm 1$.

3.2.4 Density operators

Let a large number of quantum systems of the same kind be prepared, each in one of a set of orthonormal states $|\phi_n\rangle$, and let the fraction of the system in state $|\phi_n\rangle$ be denoted by probability P_n ($n = 1, 2, \ldots$):

$$\langle\phi_m|\phi_n\rangle = \delta_{mn}, \quad \sum_n P_n = 1 \tag{3.14}$$

Therefore, this ensemble of quantum states represents a classical *statistical mixture* of kets. The probability of obtaining ξ_n from the measurement of Ξ will be

$$\Pr(\xi_k) = \sum_{n=1}^{\infty} P_n |\langle\xi_k|\phi_n\rangle|^2 = \sum_{n=1}^{\infty} P_n \langle\xi_k|\phi_n\rangle\langle\phi_n|\xi_k\rangle = \langle\xi_k|\rho|\xi_k\rangle, \tag{3.15}$$

where the operator ρ is known as a *density operator* and defined by

$$\rho = \sum_{n=1}^{\infty} P_n |\phi_n\rangle\langle\phi_n|. \tag{3.16}$$

The expected value of operator Ξ is given by

$$\langle\Xi\rangle = \sum_{k=1}^{\infty} \xi_k \Pr(\xi_k) = \sum_{k=1}^{\infty} \xi_k \langle\xi_k|\rho|\xi_k\rangle = \sum_{k=1}^{\infty} \langle\xi_k|\rho\Xi|\xi_k\rangle = \mathrm{Tr}(\rho\Xi) \tag{3.17}$$

The density operator *properties* can be summarized as follows:

1. The density operator is Hermitian ($\rho^+ = \rho$), with the set of orthonormal eigenkets $|\phi_n\rangle$ corresponding to the nonnegative eigenvalues P_n and $\mathrm{Tr}(\rho) = 1$.
2. Any Hermitian operator with nonnegative eigenvalues and trace 1 may be considered a density operator.
3. The density operator is positive definite: $\langle\psi|\rho|\psi\rangle \geq 0$ for all $|\psi\rangle$.
4. The density operator has the property $\mathrm{Tr}(\rho^2) \leq 1$, with equality if one of the prior probabilities is 1 and all the rest 0: $\rho = |\phi_n\rangle\langle\phi_n|$; the density operator is then a projection operator.
5. The eigenvalues of a density operator satisfy $0 \leq \lambda_i \leq 1$

The proof of these properties is quite straightforward and is left as a homework problem. When ρ is the projection operator, we say it represents the system in a *pure state*; otherwise, with $\mathrm{Tr}(\rho^2) < 1$, it represents a *mixed state*. A mixed state in which all eigenkets occur with the same probability is known as a *completely mixed state* and can be represented by

$$\rho = \sum_{k=1}^{\infty} \frac{1}{n} |\phi_n\rangle\langle\phi_n| = \frac{1}{n} I \Rightarrow \mathrm{Tr}\left(\rho^2\right) = \frac{1}{n} \Rightarrow \frac{1}{n} \leq \mathrm{Tr}\left(\rho^2\right) \leq 1 \tag{3.18}$$

If the density matrix has off-diagonal elements different from zero, we say it exhibits *quantum interference*, which means that state terms can interfere with each other. Let us observe the following pure state:

$$|\psi\rangle = \sum_{i=1}^{n} \alpha_i |\xi_i\rangle \Rightarrow \rho = |\psi\rangle\langle\psi| = \sum_{i=1}^{n} |\alpha_i|^2 |a_i\rangle\langle a_i| + \sum_{i=1}^{n} \sum_{j=1, j\neq i}^{n} \alpha_i \alpha_j^* |\xi_i\rangle\langle\xi_j|$$

$$= \sum_{i=1}^{n} \langle\xi_i|\rho|\xi_i\rangle |\xi_i\rangle\langle\xi_i| + \sum_{i=1}^{n} \sum_{j=1, j\neq i}^{n} \langle\xi_i|\rho|\xi_j\rangle |\xi_i\rangle\langle\xi_j|. \tag{3.19}$$

The first term in Eq. (3.19) relates to the probability of the system being in state $|a_i\rangle$, and the second term relates to quantum interference. It appears that the off-diagonal elements of a mixed state will be zero, while those of a pure state will be nonzero. Notice that the existence of off-diagonal elements is base-dependent—therefore, to check for purity, it is a good idea to compute $\mathrm{Tr}(\rho^2)$ instead.

In quantum information theory, the density matrix can be used to determine the amount of information conveyed by the quantum state, i.e., to compute the von Neumann entropy:

$$S = -\mathrm{Tr}(\rho \log \rho) = -\sum_i \lambda_i \log_2 \lambda_i, \tag{3.20}$$

where λ_i are the eigenvalues of the density matrix. The corresponding Shannon entropy can be calculated by

$$H = -\sum_i p_i \log_2 p_i. \tag{3.21}$$

Now suppose that S is a *bipartite composite system* with component subsystems A and B. For example, subsystem A can represent the quantum register Q and subsystem B the environment E. The composite system can be represented by $AB = A \otimes B$, where $\otimes$ stands for the tensor product. If the dimensionality of Hilbert space H_A is m, and the dimensionality Hilbert space H_B is n, the dimensionality of Hilbert space H_{AB} will be mn. Let $|\alpha\rangle \in A$ and $|\beta\rangle \in B$; then, $|\alpha\rangle|\beta\rangle = |\alpha\rangle \otimes |\beta\rangle \in AB$. If operator A acts on kets from H_A and operator B on kets from H_B, the action of AB on $|\alpha\rangle|\beta\rangle$ can be described as follows:

$$(AB)|\alpha\rangle|\beta\rangle = (A|\alpha\rangle)(B|\beta\rangle). \tag{3.22}$$

The norm of state $|\psi\rangle = |\alpha\rangle|\beta\rangle \in AB$ is determined by

$$\langle\psi|\psi\rangle = \langle\alpha|\alpha\rangle\langle\beta|\beta\rangle. \tag{3.23}$$

Let $\{|\alpha_i\rangle\}$ ($\{|\beta_i\rangle\}$) be a basis for the Hilbert space H_A (H_B), and let E be an ensemble of physical systems S described by the density operator ρ. The *reduced density operator* ρ_A for subsystem A is defined as the partial trace of ρ over B:

$$\rho_A = \mathrm{Tr}_B(\rho) = \sum_j \langle\beta_j|\rho|\beta_j\rangle \tag{3.24}$$

Similarly, the *reduced density operator* ρ_B for subsystem B is defined as the partial trace of ρ over A:

$$\rho_B = \mathrm{Tr}_A(\rho) = \sum_i \langle\alpha_j|\rho|\alpha_j\rangle \tag{3.25}$$

3.3 Measurements, uncertainty relations, and dynamics of quantum systems

3.3.1 Measurements and generalized measurements

Each measurable physical quantity—observable (such as position, momentum, or angular momentum) is associated with a Hermitian operator with a complete set of eigenkets. According to P. A. Dirac, "A measurement always causes the system to jump into an eigenstate of the dynamical variable that is being measured" [27]. Dirac's statement can be formulated as the following *postulate*: exact measurement of an observable with operator A always yields one of the eigenvalues $a^{(n)}$ of A. Thus, the measurement changes the state, with the measurement system "thrown into" one of its eigenstates, which can be represented by $|\alpha\rangle \xrightarrow{A \text{ measurement}} |a^{(j)}\rangle$. If the system was in state $|\alpha\rangle$before measurement, the probability that a measurement result will be the eigenvalue $a^{(i)}$ is given by

$$\Pr(a^{(i)}) = |\langle a^{(i)}|\alpha\rangle|^2, \tag{3.26}$$

and this rule is commonly referred to as the *Born rule*. Since at least one eigenvalue must occur as the result of the measurements, these probabilities satisfy

$$\sum_i \Pr(a^{(i)}) = \sum_i |\langle a^{(i)}|\alpha\rangle|^2 = 1. \tag{3.27}$$

The expected value of the outcome of the measurement of A is given by

$$\langle A\rangle = \sum_i a^{(i)}\Pr(a^{(i)}) = \sum_i a^{(i)}|\langle a^{(i)}|\alpha\rangle|^2 = \sum_i a^{(i)}\langle\alpha|a^{(i)}\rangle\langle a^{(i)}|\alpha\rangle. \tag{3.28}$$

By applying the eigenvalue equation $a^{(i)}|a^{(i)}\rangle = A|a^{(i)}\rangle$, Eq. (3.28) becomes

$$\langle A\rangle = \sum_i \langle\alpha|A|a^{(i)}\rangle\langle a^{(i)}|\alpha\rangle. \tag{3.29}$$

Further, using the completeness relation $\sum_i |a^{(i)}\rangle\langle a^{(i)}| = I$, we obtain the expected value of the measurement of A as simply

$$\langle A\rangle = \langle\alpha|A|\alpha\rangle. \tag{3.30}$$

In various situations, such as the initial state preparations for quantum information processing applications, we must select one particular measurement outcome. This procedure is known as *selective measurement* (or filtration) and can be conducted as shown in Fig. 3.1.

The result of the selective measurement can be interpreted as applying the *projection operator* $P_{a'}$ to $|\alpha\rangle$ to obtain

$$P_{a'}|\alpha\rangle = |a'\rangle\langle a'|\alpha\rangle. \tag{3.31}$$

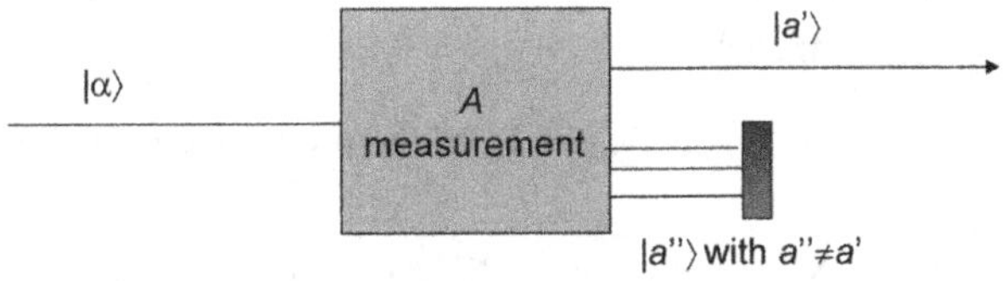

FIGURE 3.1

Illustration of the concept of selective measurement (filtration).

The probability that the outcome of the measurement of observable Ξ with eigenvalues $\xi^{(n)}$ lies between (a,b) is given by

$$\Pr(\xi \in R(a,b)) = \sum_{\xi^{(n)}\in R(a,b)} |\langle \xi^{(n)}|\alpha\rangle|^2 = \sum_{\xi^{(n)}\in R(a,b)} \langle \alpha|\xi^{(n)}\rangle\langle \xi^{(n)}|\alpha\rangle = \langle \alpha|P_{ab}|\alpha\rangle = \langle P_{ab}\rangle, \tag{3.32}$$

where with P_{ab}, we denote the following projection operator:

$$P_{ab} = \sum_{\xi^{(n)}\in R(a,b)} |\xi^{(n)}\rangle\langle \xi^{(n)}|. \tag{3.33}$$

It is straightforward to see that the projection operator P_{ab} satisfies

$$P_{ab}^2 = P_{ab} \Leftrightarrow P_{ab}(P_{ab} - I) = \mathbf{0}. \tag{3.34}$$

Therefore, the eigenvalues of projection operator P_{ab} are either 0 (corresponding to the "false proposition") or 1 (corresponding to the "true proposition"), which has great importance in *quantum detection theory* [5].

In terms of projection operators, the system state following the measurement is given by

$$|\alpha\rangle \xrightarrow{A \text{ measurement}} \frac{1}{\sqrt{\langle \alpha|P_j|\alpha\rangle}} P_j|\alpha\rangle, \quad P_j = |a^{(j)}\rangle\langle a^{(j)}| \tag{3.35}$$

In case operator A has the same eigenvalue a_i for the following eigenkets $\left\{\left|a_i^{(j)}\right\rangle\right\}_{j=1}^{d_i}$, with the corresponding characteristic equation,

$$A\left|a_i^{(j)}\right\rangle = a_i\left|a_i^{(j)}\right\rangle; \quad j = 1, \cdots, d_i \tag{3.36}$$

we say that eigenvalue a_i is *degenerate* of order d_i. The corresponding probability of obtaining measurement result a_i can be found by

$$\Pr(a_i) = \sum_{j=1}^{d_i} \left|\left\langle a_i^{(j)}\middle|\alpha\right\rangle\right|^2. \tag{3.37}$$

Generalized measurements. The projective measurements can be generalized as follows. Let the set of measurement operators be given by $\{M_m\}$, where index m stands for the possible measurement result satisfying the property $\sum_m M_m^\dagger M_m = I$. The probability of finding the measurement result m given state $|\psi\rangle$ is given by

$$\Pr(m) = \langle \psi|M_m^\dagger M_m|\psi\rangle \tag{3.38}$$

After the measurement, the system will be in the following state:

$$|\psi_f\rangle = \frac{M_m|\psi\rangle}{\sqrt{\langle \psi|M_m^\dagger M_m|\psi\rangle}}. \tag{3.39}$$

For projective measurements, clearly $M_m = P_m = |a^{(m)}\rangle\langle a^{(m)}|$, and from the property above, we obtain

$$\sum\nolimits_m M_m^\dagger M_m = \sum\nolimits_m |a^{(m)}\rangle \underbrace{\langle a^{(m)}|a^{(m)}\rangle}_{1} \langle a^{(m)}| = \sum\nolimits_m |a^{(m)}\rangle\langle a^{(m)}| = \sum\nolimits_m P_m = I, \tag{3.40}$$

which is the completeness relationship. The probability of obtaining m—the result of the measurement—is then

$$\Pr(m) = \mathrm{Tr}\left(P_m^\dagger P_m \rho\right) = \mathrm{Tr}(P_m \rho) = \mathrm{Tr}(|a^{(m)}\rangle\langle a^{(m)}|\rho) = \langle a^{(m)}|\rho|a^{(m)}\rangle. \tag{3.41}$$

Another important measurement is known as the *positive operator-valued measure* (POVM). A POVM consists of the set of operators $\{E_m\}$, where each operator E_m is positive semidefinite, i.e., $\langle\psi|E_m|\psi\rangle \geq 0$, satisfying the relationship

$$\sum_m E_m = I. \tag{3.42}$$

The POVM can be constructed from generalized measurement operators $\{M_m\}$ by setting $E_m = M_m^\dagger M_m$. The probability of obtaining the m-th result of the measurements is given by $\mathrm{Tr}(E_m\rho)$. The POVM concept is particularly suitable for situations when measurements are not repeatable. For instance, a photon can be destroyed during measurement, in which case repeated measurements are not possible.

3.3.2 Uncertainty principle

Let A and B be two operators that generally do not commute, i.e., $AB \neq BA$. The quantity $[A, B] = AB - BA$ is called the *commutator* of A and B, while the quantity $\{A, B\} = AB + BA$ is called the *anticommutator.* Two observables A and B are said to be **compatible** when their corresponding operators commute: $[A, B] = 0$. Two observables A and B are said to be *incompatible* when $[A, B] \neq 0$. If in the set of operators $\{A,B,C,\ldots\}$, all operators commute in pairs, namely $[A, B] = [A, C] = [B,C] = \ldots = 0$, we say the set is a *complete set of commuting observables.*

If two observables, say A and B, are to be measured simultaneously and exactly on the same system, the system after measurement must be left in the state $|a^{(n)};b^{(n)}\rangle$, which is an eigenstate of both observables:

$$A|a^{(n)};b^{(n)}\rangle = a^{(n)}|a^{(n)};b^{(n)}\rangle, \quad B|a^{(n)};b^{(n)}\rangle = b^{(n)}|a^{(n)};b^{(n)}\rangle \tag{3.43}$$

This is true only if $AB = BA$, or equivalently, the commutator $[A, B] = AB - BA = 0$—that is, when two operators *commute* as shown below:

$$\begin{aligned} AB|a^{(n)};b^{(n)}\rangle &= A(B|a^{(n)};b^{(n)}\rangle) = Ab^{(n)}|a^{(n)};b^{(n)}\rangle = b^{(n)}\cdot A|a^{(n)};b^{(n)}\rangle = a^{(n)}b^{(n)}|a^{(n)};b^{(n)}\rangle \\ BA|a^{(n)};b^{(n)}\rangle &= a^{(n)}b^{(n)}|a^{(n)};b^{(n)}\rangle \Rightarrow AB = BA \end{aligned} \tag{3.44}$$

When two operators do not commute, they cannot be simultaneously measured with complete precision. Given an observable A, we define the operator $\Delta A = A - \langle A\rangle$ and the corresponding expectation value of $(\Delta A)^2$, which is known as the *dispersion* of A, is defined by:

$$\langle(\Delta A)^2\rangle = \left\langle A^2 - 2A\langle A\rangle + \langle A\rangle^2\right\rangle = \langle A^2\rangle - \langle A^2\rangle \tag{3.45}$$

Then for any state, the following inequality is valid:

$$\langle(\Delta A)^2\rangle\langle(\Delta B)^2\rangle \geq \frac{1}{4}|\langle[\Delta A, \Delta B]\rangle|^2, \tag{3.46}$$

which is known as the **Heisenberg uncertainty principle.**

Example. The commutation relation for coordinate X and momentum P observables is $[X,P] = j\hbar$, as shown above. By substituting this commutation relation into Eq. (3.46), we obtain

$$\langle X^2\rangle\langle P^2\rangle \geq \frac{\hbar^2}{4}.$$

Assume we observe a large ensemble of N independent systems, all in the state $|\psi\rangle$. On some systems, X is measured, while on others, P is measured. The uncertainty principle asserts that the product of dispersions (variances) cannot be less than $\hbar^2/4$ for any state.

3.3.3 Time evolution—Schrödinger equation

The time-evolution operator $U(t,t_0)$ transforms the initial ket at time instance t_0, $|\alpha,t_0\rangle$, into the final ket at time instance t by

$$|\alpha, t_0; t\rangle = U(t, t_0)|\alpha, t_0\rangle. \tag{3.47}$$

This time-evolution operator must satisfy the following two properties:

1. *Unitary property*: $U^{+}(t,t_0)\, U(t,t_0) = I$ and
2. *Composition property*: $U(t_2,t_0) = U(t_2,t_1)\, U(t_1,t_0)$, $t_2 > t_1 > t_0$.

Following Eq. (3.47), the action of infinitesimal time-evolution operator $U(t_0 + \mathrm{d}t,t_0)$ can be described by

$$|\alpha, t_0; t_0+\mathrm{d}t\rangle = U(t_0+\mathrm{d}t, t_0)|\alpha, t_0\rangle. \tag{3.48}$$

The following operator satisfies all the propositions above when $dt\rightarrow 0$:

$$U(t_0+\mathrm{d}t, t_0) = 1 - j\Omega \mathrm{d}t, \quad \Omega^{\dagger} = \Omega \tag{3.49}$$

where the operator Ω is related to the Hamiltonian H by $H = \hbar\Omega$, and the Hamiltonian eigenvalues correspond to the energy $E = \hbar\omega$. For the infinitesimal time-evolution operator $U(t_0 + dt,t_0)$, we can derive the time-evolution equation as follows. The starting point in the derivation is the composition property:

$$U(t+dt, t_0) = U(t+dt, t)U(t, t_0) = \left(1 - \frac{j}{\hbar}Hdt\right)U(t, t_0). \tag{3.50}$$

Eq. (3.50) can be rewritten in the following form:

$$\lim_{dt\rightarrow 0}\frac{U(t+\mathrm{d}t, t_0) - U(t, t_0)}{\mathrm{d}t} = -\frac{j}{\hbar}HU(t, t_0), \tag{3.51}$$

Accounting for the partial derivative definition, Eq. (3.51) becomes

$$j\hbar\frac{\partial}{\partial t}U(t, t_0) = HU(t, t_0), \tag{3.52}$$

and this equation is known as the **Schrödinger equation for the time-evolution operator.**

The Schrödinger equation for a state ket can be obtained by applying the time-evolution operator on the initial ket:

$$j\hbar\frac{\partial}{\partial t}U(t,t_0)|\alpha,t_0\rangle = HU(t,t_0)|\alpha,t_0\rangle, \tag{3.53}$$

Based on Eq. (3.52), this can be rewritten as

$$j\hbar\frac{\partial}{\partial t}|\alpha,t_0;t\rangle = H|\alpha,t_0;t\rangle. \tag{3.54}$$

For *conservative systems*, for which the Hamiltonian is time-invariant, we can easily solve Eq. (3.54) to obtain

$$U(t,t_0) = e^{-\frac{j}{\hbar}H(t-t_0)}. \tag{3.55}$$

The time-evolution of kets in conservative systems can therefore be described by applying Eqs. (3.55) in (3.47), which yields

$$|\alpha(t)\rangle = e^{-\frac{j}{\hbar}H(t-t_0)}|\alpha(t_0)\rangle. \tag{3.56}$$

Therefore, the operators do not explicitly depend on time—this concept is known as the *Schrödinger picture*.

In the *Heisenberg picture*, on the other hand, the state vector is independent of time, but the operators are time-dependent:

$$A(t) = e^{\frac{j}{\hbar}H(t-t_0)}Ae^{-\frac{j}{\hbar}H(t-t_0)}. \tag{3.57}$$

The time-evolution equation in the Heisenberg picture is given by

$$j\hbar\frac{dA(t)}{dt} = [A(t),H] + j\hbar\frac{\partial A(t)}{\partial t}. \tag{3.58}$$

The density operator ρ, representing the statistical mixture of states, is independent of time in the Heisenberg picture. The expectation value of a measurement of an observable $\Xi(t)$ at time instance t is given by

$$E_t[\Xi] = \mathrm{Tr}[\rho\Xi(t)], \tag{3.59}$$

$$E_t[\Xi] = \mathrm{Tr}[\rho(t)\Xi],\ \ \rho(t) = e^{\frac{j}{\hbar}H(t-t_0)}\rho e^{-\frac{j}{\hbar}H(t-t_0)}. \tag{3.60}$$

Example. The Hamiltonian for a two-state system is given by

$$H = \begin{bmatrix} \omega_1 & \omega_2 \\ \omega_2 & \omega_1 \end{bmatrix}.$$

The basis for this system is given by $\{|0\rangle = [1\ 0]^{\mathrm{T}},\ |1\rangle = [0\ 1]^{\mathrm{T}}\}$.

(a) Determine the eigenvalues and eigenkets of H and express the eigenkets in terms of basis.
(b) Determine the time evolution of the system described by the Schrödinger equation:

$$j\hbar\frac{\partial}{\partial t}|\psi\rangle = H|\psi\rangle,\ \ |\psi(0)\rangle = |0\rangle.$$

To determine the eigenkets of H, we begin with the characteristic equation $\det(H-\lambda I)=0$ and find that the eigenvalues are $\lambda_{1,2}=\omega_1\pm\omega_2$. The corresponding eigenvectors are

$$|\lambda_1\rangle=\frac{1}{\sqrt{2}}\begin{bmatrix}1\\1\end{bmatrix}=\frac{1}{\sqrt{2}}(|0\rangle+|1\rangle)\quad |\lambda_1\rangle=\frac{1}{\sqrt{2}}\begin{bmatrix}1\\-1\end{bmatrix}=\frac{1}{\sqrt{2}}(|0\rangle-|1\rangle).$$

We now must determine the time-evolution of arbitrary ket $|\psi(t)\rangle=[\alpha(t)\beta(t)]^{\mathrm{T}}$. The starting point is the Schrödinger equation:

$$j\hbar\frac{\partial}{\partial t}|\psi\rangle=j\hbar\begin{bmatrix}\dot{\alpha}(t)\\ \dot{\beta}(t)\end{bmatrix},\quad H|\psi\rangle=\begin{bmatrix}\omega_1 & \omega_2\\ \omega_2 & \omega_1\end{bmatrix}\begin{bmatrix}\alpha(t)\\ \beta(t)\end{bmatrix}=\begin{bmatrix}\omega_1\alpha(t)+\omega_2\beta(t)\\ \omega_2\alpha(t)+\omega_1\beta(t)\end{bmatrix}\Rightarrow$$

$$j\hbar\begin{bmatrix}\dot{\alpha}(t)\\ \dot{\beta}(t)\end{bmatrix}=\begin{bmatrix}\omega_1\alpha(t)+\omega_2\beta(t)\\ \omega_2\alpha(t)+\omega_1\beta(t)\end{bmatrix}.$$

By substitution with $\alpha(t)+\beta(t)=\gamma(t)$ and $\alpha(t)-\beta(t)=\delta(t)$, we obtain the ordinary set of differential equations,

$$j\hbar\frac{d\gamma(t)}{dt}=(\omega_1+\omega_2)\gamma(t)\quad j\hbar\frac{d\delta(t)}{dt}=(\omega_1-\omega_2)\delta(t),$$

whose solution are $\gamma(t)=C\exp\left(\frac{\omega_1+\omega_2}{j\hbar}\right)$ and $\delta(t)=D\exp\left(\frac{\omega_1-\omega_2}{j\hbar}t\right)$. From the initial state $|\psi(0)\rangle=|0\rangle=[1\quad 0]^{\mathrm{T}}$, we obtain the unknown constants $C=D=1$ so that state time-evolution is given by

$$|\psi(t)\rangle=\exp\left(-\frac{j}{\hbar}\omega_1 t\right)\begin{bmatrix}\cos\left(\frac{\omega_2 t}{\hbar}\right)\\ -j\sin\left(\frac{\omega_2 t}{\hbar}\right)\end{bmatrix}.$$

3.4 Superposition principle, quantum parallelism, and quantum information processing basics

We say that the allowable states $|\mu\rangle$ and $|\nu\rangle$ of the quantum system satisfy the *superposition principle* if their linear combination $\alpha\,|\mu\rangle+\beta\,|\nu\rangle$—where α and β are the complex numbers ($\alpha,\beta\in C$) is also the allowable quantum state. Without loss of generality, we typically observe the computational basis composed of the orthogonal canonical states $|0\rangle=\begin{bmatrix}1\\0\end{bmatrix}$, $|1\rangle=\begin{bmatrix}0\\1\end{bmatrix}$, so the quantum bit, also known as a *qubit*, lies in a two-dimensional Hilbert space H isomorphic to C^2-space and can be represented as

$$|\psi\rangle=\alpha|0\rangle+\beta|1\rangle=\begin{pmatrix}\alpha\\ \beta\end{pmatrix};\quad \alpha,\beta\in C;\quad |\alpha|^2+|\beta|^2=1. \tag{3.61}$$

If we measure a qubit, we will get $|0\rangle$ with probability $|\alpha|^2$ and $|1\rangle$ with probability $|\beta|^2$. Measurement changes the state of a qubit from a superposition of $|0\rangle$ and $|1\rangle$ to the specific state consistent with the measurement result. If we parametrize the probability amplitudes α and β as follows,

$$\alpha = \cos\left(\frac{\theta}{2}\right), \quad \beta = e^{j\phi}\sin\left(\frac{\theta}{2}\right), \tag{3.62}$$

where θ is a polar angle, and ϕ is an azimuthal angle, we can geometrically represent the qubit by Bloch sphere (or the Poincaré sphere for the photon), as illustrated in Fig. 3.2. Bloch vector coordinates are given by ($\cos\phi \sin\theta$, $\sin\phi \sin\theta$, $\cos\theta$). This Bloch vector representation is related to the computational basis (CB) by

$$|\psi(\theta,\phi)\rangle = \cos(\theta/2)|0\rangle + e^{j\phi}\sin(\theta/2)|1\rangle \doteq \begin{pmatrix} \cos(\theta/2) \\ e^{j\phi}\sin(\theta/2) \end{pmatrix}, \tag{3.63}$$

where $0 \le \theta \le \pi$ and $0 \le \phi < 2\pi$. The north and south poles correspond to computational $|0\rangle$ ($|x\rangle$-polarization) and $|1\rangle$ ($|y\rangle$-polarization) basis kets, respectively. Other important bases are the *diagonal basis* $\{|+\rangle, |-\rangle\}$, very often denoted as $\{|\nearrow\rangle, |\searrow\rangle\}$, related to the CB by

$$|+\rangle = |\nearrow\rangle = \frac{1}{\sqrt{2}}(|0\rangle + |1\rangle), \quad |-\rangle = |\searrow\rangle = \frac{1}{\sqrt{2}}(|0\rangle - |1\rangle). \tag{3.64}$$

and the *circular basis* $\{|R\rangle, |L\rangle\}$, related to the CB as follows:

$$|R\rangle = \frac{1}{\sqrt{2}}(|0\rangle + j|1\rangle), \quad |L\rangle = \frac{1}{\sqrt{2}}(|0\rangle - j|1\rangle). \tag{3.65}$$

The pure qubit states lie on the Bloch sphere, while the mixed qubit states lie within the interior of the Bloch sphere. The maximally mixed state $I/2$ (I denotes the identity operator) lies in the center of the Bloch sphere. The orthogonal states are antipodal. From Fig. 3.2, we observe that the CB, diagonal basis, and circular basis are 90 degrees apart, and we often say that these three bases are mutually

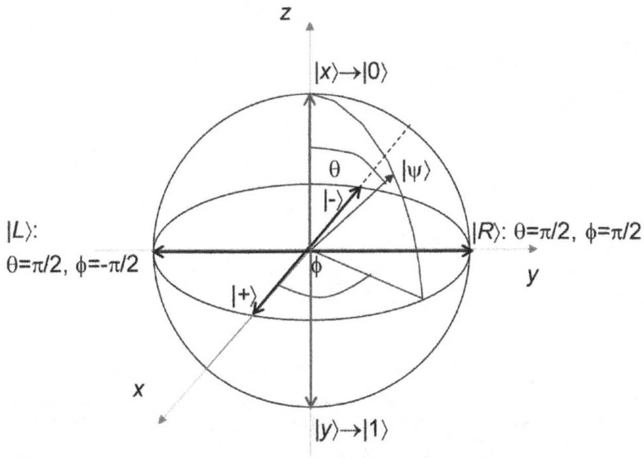

FIGURE 3.2

Bloch (Poincaré) sphere representation of the single qubit.

conjugate. These bases are used as three pairs of signal states for the six-state quantum key distribution (QKD) protocol. Another important basis used in QKD and eavesdropping is the *Breitbart basis* given by $\{\cos(\pi/8)|0\rangle + \sin(\pi/8)|1\rangle, -\sin(\pi/8)|0\rangle + \cos(\pi/8)|1\rangle\}$.

The superposition principle is the key property that makes quantum parallelism possible. To see this, let us juxtapose n qubits lying in n distinct two-dimensional Hilbert spaces $H_0, H_1, \ldots, H_{n-1}$ that are isomorphic to each other. In practice, this means the qubits have been prepared separately without any interaction, which can be mathematically described by the tensor product:

$$|\psi\rangle = |\psi_0\rangle \otimes |\psi_1\rangle \otimes \cdots \otimes |\psi_{n-1}\rangle \in H_0 \otimes H_1 \otimes \cdots \otimes H_{n-1} \tag{3.66}$$

Any arbitrary basis can be selected as the computational basis for H_i, $i = 0, 1, \ldots, n-1$. However, to facilitate the explanation, we assume the computational basis to be $|0_i\rangle$ and $|1_i\rangle$. Consequently, we can represent the i-th qubit as $|\psi_i\rangle = \alpha_i|0_i\rangle + \beta_i|1_i\rangle$. Introducing a further assumption, $\alpha_i = \beta_i = 2^{-1/2}$, without loss of generality, we now have

$$|\psi\rangle = \prod_{i=0}^{n-1} \frac{1}{\sqrt{2}}(|0_i\rangle + |1_i\rangle) = 2^{-n/2}\sum\nolimits_x |x\rangle, x = x_0x_1\cdots x_{n-1}, x_j \in \{0,1\} \tag{3.67}$$

This composite quantum system is called the ***n*-qubit register,** and as can be seen from the equation above, it represents a superposition of 2^n quantum states that exist simultaneously! This is an example of quantum parallelism. In the classical realm, a linear increase in size corresponds roughly to a linear increase in processing power. In the quantum world, due to the power of quantum parallelism, a linear increase in size corresponds to an exponential increase in processing power. The downside, however, is accessibility to this parallelism. Remember that superposition collapses the moment we attempt to measure it. The quantum circuit to create the superposition state above—in other words, the Walsh–Hadamard transform—is shown in Fig. 3.3. Therefore, the Walsh–Hadamard transform on n ancilla qubits in state $|00\ldots0\rangle$ can be implemented by applying the Hadamard operators (gates) H, whose action is described in Fig. 3.2, on ancillary qubits.

More generally, a linear operator (gate) B can be expressed in terms of the eigenkets $\{|a^{(n)}\rangle\}$ of a Hermitian operator A. The *operator B* is associated with a *square matrix* (albeit infinite in extent) whose elements are

$$B_{mn} = \langle a^{(m)}|B|a^{(n)}\rangle, \tag{3.68}$$

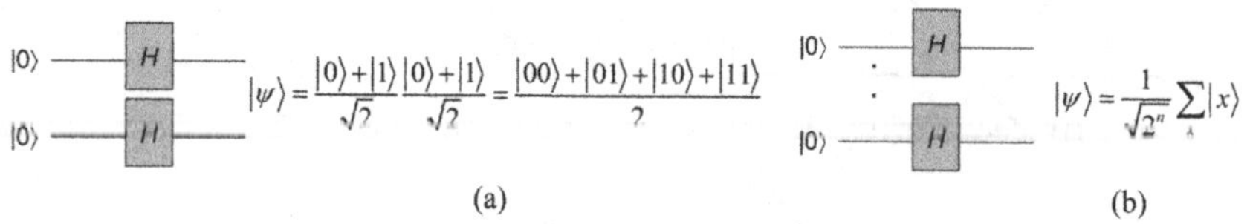

FIGURE 3.3

The Walsh–Hadamard transform (A) on two qubits and (B) on n qubits. The action of Hadamard gate H on the computational basis kets is given by $H|0\rangle = 2^{-1/2}(|0\rangle + |1\rangle)$ and $H|1\rangle = 2^{-1/2}(|0\rangle - |1\rangle)$.

and can be written explicitly as

$$B \doteq \begin{pmatrix} \langle a^{(1)}|B|a^{(1)}\rangle & \langle a^{(1)}|B|a^{(2)}\rangle & \cdots \\ \langle a^{(2)}|B|a^{(1)}\rangle & \langle a^{(2)}|B|a^{(2)}\rangle & \cdots \\ \vdots & \vdots & \ddots \end{pmatrix}, \tag{3.69}$$

using the notation $\doteq$ to denote that operator B is represented by the matrix above. The especially important single-qubit gates are the Hadamard gate H, phase shift gate S, $\pi/8$ (or T) gate, controlled-NOT (CNOT) gate, and Pauli operators X, Y, Z. The Hadamard gate H, phase shift gate, T gate, and $CNOT$ gate have the following matrix representation in CB $\{|0\rangle, |1\rangle\}$:

$$H \doteq \frac{1}{\sqrt{2}}\begin{bmatrix} 1 & 1 \\ 1 & -1 \end{bmatrix}, \quad S \doteq \begin{bmatrix} 1 & 0 \\ 0 & j \end{bmatrix}, \quad T \doteq \begin{bmatrix} 1 & 0 \\ 0 & e^{j\pi/4} \end{bmatrix}, \quad CNOT \doteq \begin{bmatrix} 1 & 0 & 0 & 0 \\ 0 & 1 & 0 & 0 \\ 0 & 0 & 0 & 1 \\ 0 & 0 & 1 & 0 \end{bmatrix}; \tag{3.70}$$

whereas the Pauli operators have the following matrix representation in the CB:

$$X \doteq \begin{bmatrix} 0 & 1 \\ 1 & 0 \end{bmatrix}, \quad Y \doteq \begin{bmatrix} 0 & -j \\ j & 0 \end{bmatrix}, \quad Z \doteq \begin{bmatrix} 1 & 0 \\ 0 & -1 \end{bmatrix}. \tag{3.71}$$

The action of Pauli gates on an arbitrary qubit $|\psi\rangle = \alpha|0\rangle + \beta|1\rangle$ is as follows:

$$\begin{aligned} &X(\alpha|0\rangle + \beta|1\rangle) = \alpha|1\rangle + \beta|0\rangle, \quad Y(\alpha|0\rangle + \beta|1\rangle) = j(\alpha|1\rangle - \beta|0\rangle), \\ &Z(\alpha|0\rangle + \beta|1\rangle) = \alpha|0\rangle - \beta|1\rangle. \end{aligned} \tag{3.72}$$

So the action of X-gate is to introduce the bit-flip, the action of Z-gate is to introduce the phase-flip, and the action of Y-gate is to simultaneously introduce the bit- and phase-flips.

Several important single-, two-, and three-qubit gates are shown in Fig. 3.4. The action of the single-qubit gate is to apply the operator U on qubit $|\psi\rangle$, which results in another qubit. The controlled-U gate conditionally applies the operator U on target qubit $|\psi\rangle$ when the control qubit $|c\rangle$ is in the $|1\rangle$

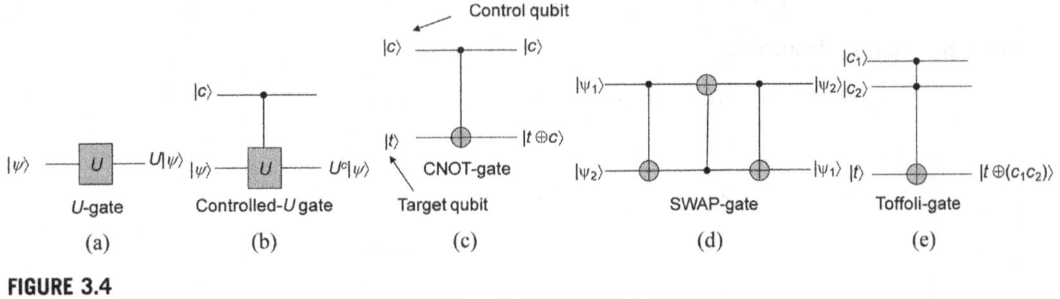

FIGURE 3.4

Important quantum gates and their action: (A) single-qubit gate, (B) controlled-U gate, (C) CNOT-gate, (D) SWAP-gate, and (E) Toffoli gate.

state. One particularly important controlled U-gate is the CNOT gate. This gate flips the content of target qubit $|t\rangle$ when the control qubit $|c\rangle$ is in the $|1\rangle$state. The purpose of the SWAP-gate is to interchange the positions of the two qubits and can be implemented using three CNOT gates, as shown in Fig. 3.3D. Finally, the Toffoli gate represents a generalization of the CNOT gate using two control qubits.

The minimum set of gates that can perform an arbitrary quantum computation algorithm is known as the *universal set of gates*. The most popular sets of universal quantum gates are {H, S, CNOT, Toffoli} gates, {H, S, π/8 (T), CNOT} gates, Barenco gate, and Deutsch gate. Using these universal quantum gates, more complicated operations can be performed. As an illustration, Fig. 3.5 shows the Bell state [Einstein–Podolsky–Rosen (EPR) pairs] preparation circuit, which is especially important in quantum teleportation and QKD applications.

Thus far, single-, two-, and three-qubit quantum gates have been considered. An arbitrary quantum state of K qubits has the form $\sum_s \alpha_s |s\rangle$, where s runs over all binary strings of length K. Therefore, there are 2^K complex coefficients, all of which are independent except from the normalization constraint:

$$\sum_{s=00...00}^{11..11} |\alpha_s|^2 = 1. \tag{3.73}$$

For example, the state $\alpha_{00}|00\rangle + \alpha_{01}|01\rangle + \alpha_{10}|10\rangle + \alpha_{11}|11\rangle$ (with $|\alpha_{00}|2 + |\alpha_{01}|^2 + |\alpha_{10}|^2 + |\alpha_{11}|^2 = 1$) is the general two-qubit state (we use $|00\rangle$ to denote the tensor product $|0\rangle \otimes |0\rangle$). The multiple qubits can be **entangled** so they cannot be decomposed into two separate states. For example, the Bell state or EPR pair $(|00\rangle + |11\rangle)/\sqrt{2}$ cannot be written in terms of tensor product $|\psi_1\rangle|\psi_2\rangle = (\alpha_1|0\rangle + \beta_1|1\rangle) \otimes (\alpha_2|0\rangle + \beta_2|1\rangle) = \alpha_1\alpha_2|00\rangle + \alpha_1\beta_2|01\rangle + \beta_1\alpha_2|10\rangle + \beta_1\beta_2|11\rangle$, because to do so would require $\alpha_1\alpha_2 = \beta_1\beta_2 = 1/\sqrt{2}$ and $\alpha_1\beta_2 = \beta_1\alpha_2 = 0$, which a priori has no reason to be valid. This state can be obtained using the circuit shown in Fig. 3.5 for the two-qubit input state $|00\rangle$. We will return to the concept of entanglement later in the chapter.

Quantum parallelism can now be introduced more formally. *QIP* implemented on a quantum register maps the input string $i_1 \ldots i_N$ to the output string $O_1(i), \ldots, O_N(i)$:

$$\begin{pmatrix} O_1(i) \\ \vdots \\ O_N(i) \end{pmatrix} = U(QIP) \begin{pmatrix} i_1 \\ \vdots \\ i_N \end{pmatrix}; \quad (i)_{10} = (i_1 \cdots i_N)_2 \tag{3.74}$$

The CB states are denoted by

$$|i_1 \cdots i_N\rangle = |i_1\rangle \otimes \cdots \otimes |i_N\rangle; \quad i_1, \cdots, i_N \in \{0, 1\} \tag{3.75}$$

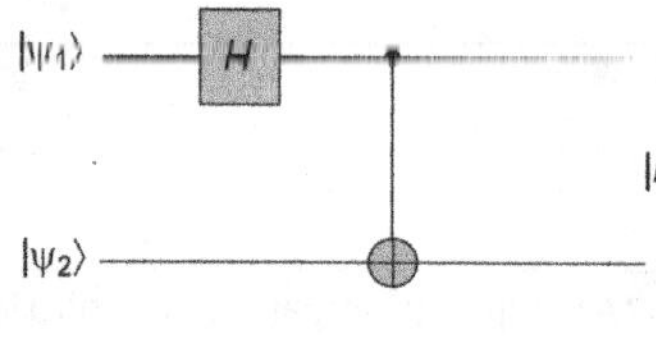

FIGURE 3.5

Bell state (EPR pairs) preparation circuit.

Linear superposition allows us to form the following $2N$-qubit state:

$$|\psi_{\text{in}}\rangle = \left[\frac{1}{\sqrt{2^N}}\sum_i |i_1 \cdots i_N\rangle\right] \otimes |0 \cdots 0\rangle, \tag{3.76}$$

and upon the application of quantum operation $U(QIP)$, the output can be represented by

$$|\psi_{\text{out}}\rangle = U(QIP)|\psi_{in}\rangle = \frac{1}{\sqrt{2^N}}\sum_i |i_1 \cdots i_N\rangle \otimes |O_1(i) \cdots O_N(i)\rangle. \tag{3.77}$$

The QIP circuit (or a quantum computer) encodes all input strings generated by QIP into $|\psi_{\text{out}}\rangle$; in other words, it simultaneously pursues 2^N classical paths. This ability of a QIP circuit to encode multiple computational results into a quantum state in a single quantum computational step is known as *quantum parallelism*, a concept that was introduced earlier in the chapter.

Within the classical realm, a linear increase in size corresponds to roughly a linear increase in processing power. In the quantum world, due to the power of quantum parallelism, a linear increase in size corresponds to an exponential increase in processing power. The downside, however, is accessibility to this parallelism. Remember that the superposition state collapses the same moment we attempt to measure it. This raises the question of whether quantum parallelism has practical uses. The first insight into this question was provided by the seminal works of Deutsch and Jozsa [28]. In 1985, Deutsch proposed the first quantum algorithm to determine whether a Boolean function is constant by exploiting quantum parallelism [29]. In 1992, Deutsch and Jozsa generalized the Deutsch algorithm to n qubits [30]. These algorithms showed the feasibility of quantum computing but did not provide any practical applications. Shor's groundbreaking work [31] changed that when he showed how to exploit quantum parallelism to achieve a polynomial-time quantum algorithm for prime factorization that achieved the promise of exponential speed-up over classical factorization algorithms. This was followed by Grover, who proposed a quantum algorithm [32] that achieved a quadratic speed-up over classical algorithms to perform unstructured database searches.

3.5 No-cloning theorem

Just as in quantum parallelism, quantum superposition is the key concept behind our inability to clone arbitrary quantum states. To see this, let us think of a quantum copier that takes as input an arbitrary quantum state and outputs two copies of that state, resulting in a clone of the original state. For example, if the input state is $|\psi\rangle$, the copier output is $|\psi\rangle|\psi\rangle$. For an arbitrary quantum state, such a copier raises a fundamental contradiction. Consider two arbitrary states $|\psi\rangle$ and $|\chi\rangle$ that are input to the copier. When they are input individually, we expect to receive $|\psi\rangle|\psi\rangle$ and $|\chi\rangle|\chi\rangle$. Now consider a superposition of these two states given by

$$|\varphi\rangle = \alpha|\psi\rangle + \beta|\chi\rangle. \tag{3.78}$$

Based on the above description that the quantum copier clones the original state, we expect the output

$$\begin{aligned}|\varphi\rangle|\varphi\rangle &= (\alpha|\psi\rangle + \beta|\chi\rangle)(\alpha|\psi\rangle + \beta|\chi\rangle) \\ &= \alpha^2|\psi\rangle|\psi\rangle + \alpha\beta|\psi\rangle|\chi\rangle + \alpha\beta|\chi\rangle|\psi\rangle + \beta^2|\chi\rangle|\chi\rangle\end{aligned} \tag{3.79}$$

On the other hand, the linearity of quantum mechanics, as evidenced by the Schrödinger wave equation, tells us that the quantum copier can be represented by a unitary operator that performs the cloning. If such a unitary operator were to act on the superposition state $|\varphi\rangle$, the output would be a superposition of $|\psi\rangle|\psi\rangle$ and $|\chi\rangle|\chi\rangle$ that is,

$$|\varphi'\rangle = \alpha|\psi\rangle|\psi\rangle + \beta|\chi\rangle|\chi\rangle \tag{3.80}$$

As is clearly evident, the difference between the previous two equations leads to the contradiction mentioned above. Consequently, there is no unitary operator that can clone $|\varphi\rangle$. We therefore formulate the no-cloning theorem as follows:

No-cloning Theorem. No quantum copier exists that can clone an arbitrary quantum state.

This result raises a related question: Do some specific states exist for which cloning is possible? The answer to this question is (surprisingly) yes. Remember, a key result of quantum mechanics is that unitary operators preserve probabilities. This implies that inner (dot) products $\langle\varphi \mid \varphi\rangle$ and $\langle\varphi' \mid \varphi'\rangle$ should be identical. The inner products $\langle\varphi |\varphi\rangle$ and $\langle\varphi' \mid \varphi'\rangle$ are respectively given by

$$\begin{aligned}
\langle\varphi|\varphi\rangle &= (\langle\psi|\alpha^* + \langle\chi|\beta^*)(\alpha|\psi\rangle + \beta|\chi\rangle) \\
&= |\alpha|^2\langle\psi|\psi\rangle + |\beta|^2\langle\chi|\chi\rangle + \alpha^*\beta\langle\psi|\chi\rangle + \alpha\beta^*\langle\chi|\psi\rangle \\
\langle\varphi'|\varphi'\rangle &= (\langle\psi|\langle\psi|\alpha^* + \langle\chi|\langle\chi|\beta^*)(\alpha|\psi\rangle|\psi\rangle + \beta|\chi\rangle|\chi\rangle) \\
&= |\alpha|^2|\langle\psi|\psi\rangle|^2 + |\beta|^2|\langle\chi|\chi\rangle|^2 + \alpha^*\beta|\langle\psi|\chi\rangle|^2 + \alpha\beta^*|\langle\chi|\psi\rangle|^2
\end{aligned} \tag{3.81}$$

We know that $\langle\psi|\psi\rangle = \langle\chi|\chi\rangle = 1$. Therefore, the discrepancy lies in the cross terms. Specifically, to avoid the contradiction that resulted in the no-cloning theorem, we require that $|\langle\psi|\chi\rangle|^2 = \langle\psi|\chi\rangle$. This condition can only be satisfied when the states are orthogonal. Thus, cloning is possible only for mutually orthogonal states. However, it is important here to remember a subtle point. Even if we have a mutually orthogonal set of states, we need a quantum copier (or unitary operator) specifically for those states. If the unitary operator is specific to a different set of mutually orthogonal states, cloning will fail. It seems that the no-cloning theorem would prevent us from exploiting the richness of quantum mechanics. It turns out that this is not the case. A key example is QKD, which has a very high probability of guaranteeing secure communication.

3.6 Distinguishing quantum states

Not only can nonorthogonal quantum states not be cloned, but also they cannot be reliably distinguished. There is no measurement device we can create to reliably distinguish nonorthogonal states. This fundamental result plays an important role in quantum cryptography. Its proof is based on contradiction. Let us assume that the measurement operator M is the Hermitian operator (with corresponding eigenvalues m_i and corresponding projection operators P_i) of an observable M, which allows unambiguous distinction between two nonorthogonal states $|\psi_1\rangle$ and $|\psi_2\rangle$. The eigenvalue m_1 (m_2) unambiguously identifies the state $|\psi_1\rangle$ ($|\psi_2\rangle$) as the premeasurement state. We know that for projection operators, the following properties are valid:

$$\begin{aligned}
&\langle\psi_1|P_1|\psi_1\rangle = 1 \quad \langle\psi_2|P_2|\psi_2\rangle = 1 \\
&\langle\psi_1|P_2|\psi_1\rangle = 0 \quad \langle\psi_2|P_1|\psi_2\rangle = 0
\end{aligned} \tag{3.82}$$

Since $|\psi_1\rangle$ and $|\psi_2\rangle$ are nonorthogonal states, $\langle\psi_1|\psi_2\rangle \neq 0$, and $|\psi_2\rangle$ can be represented in terms of $|\psi_1\rangle$ and another state $|\chi\rangle$orthogonal to $|\psi_1\rangle$ ($\langle\psi_1|\chi\rangle = 0$) as follows:

$$|\psi_2\rangle = \alpha|\psi_1\rangle + \beta|\chi\rangle. \tag{3.83}$$

From the projection operator properties listed above, we can conclude the following:

$$\begin{aligned} 0 &= \langle\psi_1|P_2|\psi_1\rangle \overbrace{=}^{P_2^2 = P} \langle\psi_1|P_2P_2|\psi_1\rangle = \|P_2|\psi_1\rangle\|^2 \Rightarrow P_2|\psi_1\rangle = 0 \\ 1 &= \langle\psi_2|P_2|\psi_2\rangle = \langle\psi_2|P_2P_2|\psi_2\rangle = (\alpha^*\langle\psi_1| + \beta^*\langle\chi|)(\alpha P_2|\psi_1\rangle + \beta P_2|\chi\rangle) = |\beta|^2\langle\chi|P_2|\chi\rangle \end{aligned} \tag{3.84}$$

Now we use the completeness relationship:

$$1 = \langle\chi|\chi\rangle = \langle\chi|\overbrace{\sum_i P_i}^{I}|\chi\rangle = \sum_i \langle\chi|P_i|\chi\rangle \overbrace{\geq}^{\langle\chi|P_i|\chi\rangle \geq 0} \langle\chi|P_2|\chi\rangle \tag{3.85}$$

By combining the previous two equations, we obtain

$$1 = \langle\psi_2|P_2|\psi_2\rangle \leq |\beta|^2 \Rightarrow |\beta|^2 = 1, \tag{3.86}$$

indicating that the probability of finding $|\psi_2\rangle$ in $|\chi\rangle$ is 1. Therefore, we conclude that $|\psi_2\rangle = |\chi\rangle$, which is a contradiction. Therefore, indeed, it is impossible to unambiguously distinguish nonorthogonal quantum states.

3.7 Quantum entanglement

Let $|\psi_0\rangle, \ldots, |\psi_{n-1}\rangle$ be n qubits lying in the Hilbert spaces $H_0, \ldots, H_{n-1}$, respectively, and let the state of the joint quantum system lying in $H_0 \otimes \ldots \otimes H_{n-1}$ be denoted by $|\psi\rangle$. The qubit $|\psi\rangle$ is then said to be entangled if it cannot be written in the product state form

$$|\psi\rangle = |\psi_0\rangle \otimes |\psi_1\rangle \otimes \cdots \otimes |\psi_{n-1}\rangle. \tag{3.87}$$

Important examples of two-qubit states are *Bell states*, also known as *EPR* states (pairs):

$$\begin{aligned} |B_{00}\rangle &= \frac{1}{\sqrt{2}}(|00\rangle + |11\rangle), \quad |B_{01}\rangle = \frac{1}{\sqrt{2}}(|01\rangle + |10\rangle), \\ |B_{10}\rangle &= \frac{1}{\sqrt{2}}(|00\rangle - |11\rangle), \quad |B_{11}\rangle = \frac{1}{\sqrt{2}}(|01\rangle - |10\rangle). \end{aligned} \tag{3.88}$$

The entangled states introduced above are famously called EPR pairs after Einstein, Podolsky, and Rosen, who raised doubts about quantum mechanics itself [33]. Their fundamental objection was regarding the entanglement of qubits and their apparent nonlocal interaction, which intrigued the quantum physics community for quite some time. The objection they raised is referred to as the hidden variable theory. To put the debate to rest, Bell developed a set of inequalities that, if violated, would validate quantum mechanics [34,35]. Bell's theoretical predictions were experimentally validated

[36–39], leading to the establishment of quantum mechanics as we know it today, representing the best physical description of nature thus far.

The n-qubit ($n > 2$) analogs of Bell states will now be briefly reviewed. One popular family of entangled multiqubit states is the Greenberger–Horne–Zeilinger (GHZ) states:

$$|\mathrm{GHZ}\rangle = \frac{1}{\sqrt{2}}(|00\cdots0\rangle \pm |11\cdots1\rangle). \tag{3.89}$$

Another popular family of multiqubit entangled states is known as the W-states:

$$|\mathrm{W}\rangle = \frac{1}{\sqrt{N}}(|00\cdots01\rangle + |00\cdots10\rangle + \cdots + |01\cdots00\rangle + |10\cdots00\rangle). \tag{3.90}$$

The W-state of n-qubits represents a superposition of single-weighted CB states, each occurring with a probability amplitude of $N^{-1/2}$.

The two-qubit quantum system density operator can be expanded in terms of Bell states as follows:

$$\rho = \sum\nolimits_{i,j} c_{ij}|B_{ij}\rangle\langle B_{ij}|, \tag{3.91}$$

where the corresponding outer products are given by $|B_{00}\rangle\langle B_0| = .25(II + XX - YY + ZZ)$, $|B_{01}\rangle\langle B_{01}| = 25(II + XX + YY - ZZ)$, $|B_{10}\rangle\langle B_1| = .25(II - XX + YY + ZZ)$, and $|B_{11}\rangle\langle B_{11}| = .25(II - XX - YY - ZZ)$. The state represented by density operator ρ is only separable when $c_{00} < 0.5$.

For a bipartite system, we can elegantly verify whether the qubit $|\psi\rangle$ is a product state or an entangled state by the Schmidt decomposition. The *Schmidt decomposition theorem* states that a pure state $|\psi\rangle$ of the composite system $H_A \otimes H_B$ can be represented by

$$|\psi\rangle = \sum\nolimits_i c_i|i_A\rangle|i_B\rangle, \tag{3.92}$$

where $|i_A\rangle$ and $|i_B\rangle$ are orthonormal bases of the subsystems H_A and H_B, respectively, and $c_i \in \mathscr{R}^+$ ($\mathscr{R}^+$ is the set of nonnegative real numbers) are *Schmidt coefficients* that satisfy the condition $\sum_i c_i^2 = 1$. The Schmidt coefficients can be calculated from the partial density matrix $\mathrm{Tr}_B(|\psi\rangle\langle\psi|)$, whose eigenvalues are c_i^2. The number of nonzero eigenvalues is the *Schmidt rank* (number). A corollary of the Schmidt decomposition theorem is that a pure state in a composite system can be a product state only if the Schmidt rank is 1. On the other hand, it can be an entangled state only if the Schmidt rank is greater than 1.

To illustrate, let us verify whether the Bell state $|B_{11}\rangle$ is entangled. We first determine the density matrix:

$$\begin{aligned}\rho = |B_{11}\rangle\langle B_{11}| &= \frac{1}{\sqrt{2}}(|01\rangle - |10\rangle)\frac{1}{\sqrt{2}}(\langle 01| - \langle 10|)\\ &= \frac{1}{2}(|01\rangle\langle 01| - |01\rangle\langle 10| - |10\rangle\langle 01| + |10\rangle\langle 10|)\end{aligned}$$

By tracing out the subsystem B, we obtain:

$$\begin{aligned}\rho_A = \mathrm{Tr}_B(|B_{11}\rangle\langle B_{11}|) &= \langle 0_B|B_{11}\rangle\langle B_{11}|0_B\rangle + \langle 1_B|B_{11}\rangle\langle B_{11}|1_B\rangle\\ &= \frac{1}{2}(|1\rangle\langle 1| + |0\rangle\langle 0|) = \frac{1}{2}I\end{aligned}$$

The eigenvalues of ρ are $c_1^2 = c_2^2 = 1/2$, and the Schmidt rank is 2, indicating that the Bell state $|B_{11}\rangle$ is entangled.

The *proof of the Schmidt decomposition theorem* is straightforward. Let $|j\rangle$ and $|k\rangle$ be fixed orthonormal bases for subsystems H_A and H_B; $|\psi\rangle$ can then be represented as

$$|\psi\rangle = \sum_{j,k} a_{jk}|j\rangle|k\rangle, \tag{3.93}$$

where the a_{jk} complex coefficients can be interpreted as the matrix $A=(a_{jk})$, which can be decomposed using the singular value decomposition as

$$A = U\Sigma V^{\dagger}, \tag{3.94}$$

where the $m \times n$ diagonal $\sum$-matrix corresponds to scaling (stretching) operation and is given by $\sum = \text{diag}(\sigma_1,\sigma_2, \ldots)$, with $\sigma_1 \geq \sigma_2 \geq \ldots \geq 0$ being the *singular values* of A. In fact, the singular values are the square roots of eigenvalues of $AA^{\dagger}$. The $m\times m$ matrix $U=(u_{ji})$ and $n \times n$ matrix $V=(v_{jk})$ are unitary matrices that describe the rotation operations. By substituting Eqs. (3.94) into (3.93), we obtain

$$|\psi\rangle = \sum_{i,j,k} u_{ji}\sigma_i v_{jk}|j\rangle|k\rangle. \tag{3.95}$$

Now by substituting $\sum_j u_{ji}\,|j\rangle = |i_A\rangle$, $\sum_j v_{jk}\,|k\rangle = |i_B\rangle$, and $\sigma_i = c_i$, we obtain Eq. (3.92). The set $\{|i_A\rangle\}$ ($\{|i_B\rangle\}$) forms an orthonormal set due to the orthonormality of $|j\rangle$ ($|k\rangle$) and unitarity of U (V).

An important concept in QIP is the *purification concept*, in which we create the reference system H_B such that given the original system H_A, the state $|\psi\rangle \in H_A \otimes H_B$ is a pure state. As with the Schmidt decomposition theorem, let $|i_A\rangle$ and $|i_B\rangle$ be orthonormal bases of the systems H_A and H_B, respectively. The state $|\psi\rangle$ is a purification of ρ_A if

$$\rho_A = \text{Tr}_B(|\psi\rangle\langle\psi|), \tag{3.96}$$

where we use $\text{Tr}_B(\bullet)$ to denote the partial trace over B introduced in Section 3.2.4. If ρ_A is the mixed state, it can be represented by

$$\rho_A = \sum_i p_i|i_A\rangle\langle i_A|. \tag{3.97}$$

The purification is then given by

$$|\psi\rangle = \sum_i \sqrt{p_i}|i_A\rangle|i_B\rangle. \tag{3.98}$$

The direct application of the definition of trace gives

$$\begin{aligned}\text{Tr}_B(|\psi\rangle\langle\psi|) &= \text{Tr}_B\left(\sum_i \sqrt{p_i}|i_A\rangle|i_B\rangle\sum_j\sqrt{p_j}\langle j_A|\langle j_B|\right) = \sum_{i,j}\sqrt{p_ip_j}\underbrace{\langle j_B|i_A\rangle}_{\delta_{i_A j_B}}(|i_B\rangle\langle j_A|)\\ &= \sum_i p_i|i_A\rangle\langle i_A| = \rho_A,\end{aligned} \tag{3.99}$$

thus proving the claim of Eq. (3.96).

The Schmidt rank is a relevant approach to determine whether a certain two-qubit state is entangled. For multiqubit states, we must use other entanglement measures such as concurrence, entanglement of formation, and negativity [40–43]. The amount of information conveyed by the quantum state can be computed using the *von Neumann entropy*:

$$S(\rho) = -\operatorname{Tr}(\rho \log \rho) = -\sum_i \lambda_i \log_2 \lambda_i, \tag{3.100}$$

where λ_i are the eigenvalues of the density matrix. To determine the level of resources needed to form a given entanglement state, the *entanglement of formation* is used, which, for a given mixed density operator ρ_{AB} of a bipartite system composed of two subsystems H_A and H_B, is defined as

$$E(\rho_{AB}) = \min \sum_i p_i E(|\psi_i\rangle), \tag{3.101}$$

where $E(|\psi_k\rangle)$ denotes the von Neumann entropy of either subsystem A or B:

$$E(|\psi_i\rangle) = -\operatorname{Tr}(\rho_{i,A} \log \rho_{i,A}) = -\operatorname{Tr}(\rho_{i,B} \log \rho_{i,B}), \tag{3.102}$$

for reduced density operators $\rho_{i,A} = \operatorname{Tr}_B(|\psi_i\rangle\langle\psi_i|)$ and $\rho_{i,B} = \operatorname{Tr}_B(|\psi_i\rangle\langle\psi_i|)$; and p_i denotes the probability of occurrence of state $|\psi_i\rangle$. The minimization in Eq. (3.101) is performed over all possible pure-state decompositions:

$$\rho_{AB} = \sum_i p_i |\psi_i\rangle\langle\psi_i|. \tag{3.103}$$

For the two-qubit case, we can come up with a closed-form expression for the entanglement of formation [43]:

$$E(\rho_{AB}) = h_2\left(\frac{1+\sqrt{1-C^2(\rho_{AB})}}{2}\right), h_2(x) = -p \log p - (1-p)\log(1-p), \tag{3.104}$$

where $h_2(x)$ is the binary entropy function introduced in Chapter 2. We use $C(\rho_{AB})$ to denote the *concurrence*, the measure of entanglement, determined by

$$C(\rho_{AB}) = \max(0, \lambda_1 - \lambda_2 - \lambda_3 - \lambda_4), \tag{3.105}$$

where λ_i are eigenvalues of the Hermitian matrix

$$R = \left(\sqrt{\rho_{AB}}\,\tilde{\rho}_{AB}\sqrt{\rho_{AB}}\right)^{1/2}, \tilde{\rho}_{AB} = XY\rho_{AB}^{*}XY, \tag{3.106}$$

arranged in decreasing order ($\lambda_1 \geq \lambda_2 \geq \lambda_3 \geq \lambda_4$). We used $\tilde{\rho}_{AB}$ to denote the density operator after the nonunitary *spin-flip transformation*, which maps each component qubit in its orthogonal state. In other words, the spin-flipped version of $|\psi\rangle$ is defined by

$$\left|\tilde{\psi}\right\rangle = YY|\psi^*\rangle. \tag{3.107}$$

The concurrence can also be defined as

$$C(|\psi\rangle) = \left|\left\langle\psi\middle|\tilde{\psi}\right\rangle\right|, \tag{3.108}$$

and thus represents the amount of overlap between the observed state and its spin-flipped version, and clearly, $0 \leq C \leq 1$.

To illustrate, let us again observe Bell state $|B_{11}\rangle$ and determine both the concurrence and entanglement of formation. Using the definition expression of Eq. (3.108), we find that $C(|B_{11}\rangle) = 1$. Using Eq. (3.104), we now determine that the entanglement of formation is $E(|B_{11}\rangle) = h_2(1/2) = 1$.

Entanglement can be used *to detect and correct errors* arising during a quantum computation [1–3]. At the beginning, the quantum computer Q, being in initial state $|q\rangle$, and its environment E, being in initial state $|e\rangle$, are not entangled, but they become entangled by the coupling interaction:

$$|e\rangle|q\rangle \rightarrow \sum_s |e_s\rangle\{E_s|q\rangle\}, \tag{3.109}$$

where $\{E_s\}$ are the errors introduced by the environment. The states of the environment $\{|e_s\rangle\}$ are not necessarily orthonormal. To deal with errors, we introduce the ancilla qubits $|a\rangle$coupled to qubits in quantum computer Q. The unitary interaction U to couple ancillae to the quantum computer is used to introduce the following effect:

$$U[|a\rangle\{E_s|q\rangle\}] = |s\rangle\{E_s|q\rangle\}, \tag{3.110}$$

indicating that the final ancilla state $|s\rangle$ depends on the error operator induced by the environment E_s, but not on the quantum state of the quantum computer $|q\rangle$. Further, the final ancilla states form the orthonormal set. The operator U produces the following transformation of the tripartite system environment-ancilla-quantum computer:

$$U\left[\sum_s |e_s\rangle|a\rangle\{E_s|q\rangle\}\right] = \sum_s |e_s\rangle|s\rangle\{E_s|q\rangle\}. \tag{3.111}$$

Given the orthonormality of states $\{|s\rangle\}$, they can be distinguished, and by measuring the ancilla in this basis, we obtain the nonentangled pure state as a postmeasurement state:

$$|\psi_{\text{post-measurement}}\rangle = |e_s\rangle|S\rangle\{E_s|q\rangle\}, \tag{3.112}$$

where S is the *syndrome error* identifying the error operator E_s. This process of determining the error syndrome is commonly referred to as *syndrome extraction*. Because the error operator E_s is identified from syndrome S, we can apply E_s^{-1} to quantum computer Q to undo the action of the environment:

$$|\psi_{\text{final}}\rangle = |e_s\rangle|S\rangle|q\rangle. \tag{3.113}$$

Therefore, the unwanted entanglement between the quantum computer Q and its environment E has been undone by further entangling Q with the corresponding set of ancillae, together with appropriate measurement on ancillae to disentangle the Q and E and identify the most probable error operator.

Entanglement can also find application in so-called *superdense coding* [44], in which Alice sends two classical bits even though she has only one qubit to work with. Let the Bell states $|B_{ab}\rangle$ correspond to the two classical bits ab to be transmitted, wherein $ab = 00$, 01, 10, or 11. The starting state is the Bell state $|B_{00}\rangle$:

$$|\Phi\rangle = |B_{00}\rangle = 2^{-1/2}(|00\rangle + |11\rangle), \tag{3.114}$$

and we assume that Alice is in possession of the first qubit and Bob possesses the second one. Depending on which classical bits she wants to send, she will act on her qubit by performing the corresponding single-qubit gate. If she wants to send 00 bits, she leaves her qubit untouched. If she applies the X-gate on initial qubit $|\Phi\rangle$, she will get

$$(X \otimes I)|\Phi\rangle = 2^{-1/2}(|10\rangle + |01\rangle) = |B_{01}\rangle, \tag{3.115}$$

and thus will be able to transmit classical bits $ab = 01$. On the other hand, if she applies single-qubit gate Z to her qubit before sending it to Bob, she will transform the $|\Phi\rangle$-state to

$$(Z \otimes I)|\Phi\rangle = 2^{-1/2}(|00\rangle - |11\rangle) = |B_{10}\rangle, \tag{3.116}$$

and thus will be able to transmit classical bits $ab = 10$. Finally, if she applies single-qubit gate jY to her qubit before sending to Bob, she will transform $|\Phi\rangle$-state to

$$(jY \otimes I)|\Phi\rangle = 2^{-1/2}(|01\rangle - |10\rangle) = |B_{11}\rangle, \tag{3.117}$$

and thus will be able to transmit classical bits $ab = 11$. On receiver side, once Bob gets the qubit from Alice he will perform the measurements in Bell basis and will be able to find one of the Bell states $\{|B_{00}\rangle, |B_{01}\rangle, |B_{10}\rangle, |B_{11}\rangle\}$, representing two classical bits 00, 01, 10, or 11.

The entanglement has applications in quantum teleportation, which has been discussed already in Introduction. Here we discuss the *entanglement swapping* instead. Entanglement swapping starts with two EPR pairs, labeled as $|B_{00}\rangle_{12}$ and $|B_{00}\rangle_{34}$, that is,

$$|B_{00}\rangle_{12} = 2^{-1/2}(|00\rangle_{12} + |11\rangle_{12}),\ |B_{00}\rangle_{34} = 2^{-1/2}(|00\rangle_{34} + |11\rangle_{34}). \tag{3.118}$$

Alice is in possession of qubits 1 and 4, while Bob is in position of qubits two and 3. The product state will be

$$|B_{00}\rangle_{12}|B_{00}\rangle_{34} = 2^{-1}(|00\rangle_{12}|00\rangle_{34} + |00\rangle_{12}|11\rangle_{34} + |11\rangle_{12}|00\rangle_{34} + |11\rangle_{12}|11\rangle_{34}). \tag{3.119}$$

This product state can be rearranged as follows:

$$|B_{00}\rangle_{12}|B_{00}\rangle_{34} = 2^{-1}(|B_{00}\rangle_{14}|B_{00}\rangle_{23} + |B_{01}\rangle_{14}|B_{01}\rangle_{23} + |B_{10}\rangle_{14}|B_{10}\rangle_{23} + |B_{11}\rangle_{14}|B_{11}\rangle_{23}). \tag{3.120}$$

Alice performs the Bell state measurement on qubits one and 4 with possible outcomes being $\{|B_{00}\rangle_{14}, |B_{01}\rangle_{14}, |B_{10}\rangle_{14}, |B_{11}\rangle_{14}\}$, each occurring with 1/4 probability. Depending on Alice measurement result, Bob's subsystem collapses into one of the Bell states $|B_{00}\rangle_{23}$, $|B_{01}\rangle_{23}$, $|B_{10}\rangle_{23}$, or $|B_{11}\rangle_{23}$ indicating that the qubits two and three get entangled even though they have been independent initially.

3.8 Operator-sum representation

Let the composite system C be composed of quantum register Q and environment E. This kind of system can be modeled as a closed quantum system. Because the composite system is closed, its dynamic is unitary, and final state is specified by a unitary operator U as follows $U(\rho \otimes \varepsilon_0)U^\dagger$, where ρ is a density operator of initial state of quantum register Q, and ε_0 is the initial density operator of the environment E. The reduced density operator of Q upon interaction ρ_f can be obtained by tracing out the environment:

$$\rho_f = \text{Tr}_E\left[U(\rho \otimes \varepsilon_0)U^\dagger\right] \equiv \xi(\rho) \tag{3.121}$$

The transformation (mapping) of initial density operator ρ to the final density operator ρ_f, denoted as $\xi : \rho \rightarrow \rho_f$, given by Eq. (3.121) is often called the *superoperator* or *quantum operation.* The final density operator can be expressed in so called *operator-sum representation* as follows:

$$\rho_f = \sum_k E_k \rho E_k^\dagger, \tag{3.122}$$

where E_k are the operation elements for the superoperator. Clearly, in the absence of environment, the superoperator becomes: $U\rho U^\dagger$, which is nothing else but a conventional time-evolution quantum operation.

The operator-sum representation can be used in classification of quantum operations into two categories: (i) *trace-preserving* when $\mathrm{Tr}\xi(\rho) = \mathrm{Tr}\, \rho = 1$ and (ii) *non-trace-preserving* when $\mathrm{Tr}\xi(\rho) < 1$. Starting from the trace-preserving condition,

$$\mathrm{Tr}\rho = \mathrm{Tr}\xi(\rho) = \mathrm{Tr}\left[\sum_k E_k \rho E_k^\dagger\right] = \mathrm{Tr}\left[\rho \sum_k E_k E_k^\dagger\right] = 1, \tag{3.123}$$

we obtain

$$\sum_k E_k E_k^\dagger = I. \tag{3.124}$$

For non-trace-preserving quantum operation, Eq. (3.124) is not satisfied, and informally we can write $\sum_k E_k E_k^\dagger < I$.

If the environment dimensionality is large enough it can be found in pure state, $\varepsilon_0 = |\phi_0\rangle\langle\phi_0|$, and the corresponding superoperator becomes

$$\xi(\rho) = \mathrm{Tr}_E\left[U(\rho \otimes \varepsilon)U^\dagger\right] = \sum_k \langle\phi_k|\left(U\rho \otimes \varepsilon U^\dagger\right)|\phi_k\rangle$$

$$= \sum_k \langle\phi_k|\left(U\rho \otimes \left(\underbrace{|\phi_0\rangle\langle\phi_0|}_{\varepsilon}\right)U^\dagger\right)|\phi_k\rangle$$

$$= \sum_k \underbrace{\langle\phi_k|U|\phi_0\rangle}_{E_k}\rho\underbrace{\langle\phi_0|U^\dagger|\phi_k\rangle}_{E_k^\dagger} =$$

$$= \sum_k E_k \rho E_k^\dagger, E_k = \langle\phi_k|U|\phi_0\rangle \tag{3.125}$$

The E_k operators in operator-sum representation are known as *Kraus operators*.

As an illustration, let us consider *bit-flip* and *phase-flip* channels, which are illustrated in Fig. 3.6. (The simultaneous bit-and-phase flip operation is described by the Pauli Y-operator.) Let the composite system be given by $|\phi_E\rangle|\psi_Q\rangle$, wherein the initial state of environment is $|\phi_E\rangle = |0_E\rangle$. Further, let the quantum subsystem Q interact with environment E by Pauli-X operator:

$$U = \sqrt{1-p} I \otimes I + \sqrt{p} X \otimes X, \;\; 0 \leq p \leq 1 \tag{3.126}$$

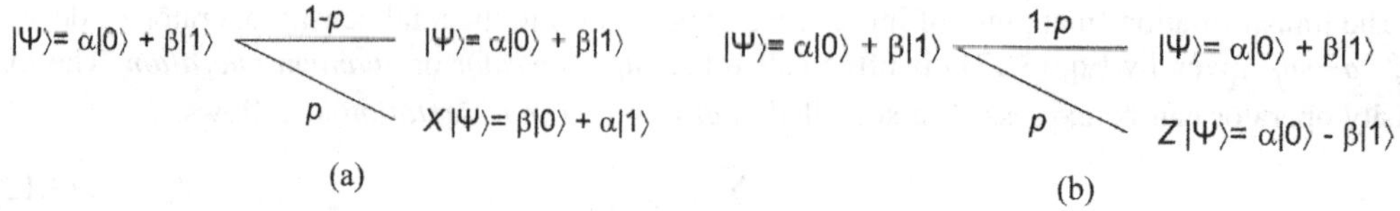

FIGURE 3.6

(A) Bit-flip channel model. (B) Phase-flip channel model.

Therefore, with probability $1-p$ we leave the quantum system untouched, while with probability p we apply Pauli-X operator to both quantum subsystem and the environment. By applying the operator U on environment state, we obtain

$$U|\phi_E\rangle=\sqrt{1-p}I\otimes I|0_E\rangle+\sqrt{p}X\otimes X|0_E\rangle=\sqrt{1-p}|0_E\rangle I+\sqrt{p}|1_E\rangle X, \tag{3.127}$$

The corresponding Kraus operators are given by

$$E_0=\langle 0_E|U|\phi_E\rangle=\sqrt{1-p}I,\ \ E_1=\langle 1_E|U|\phi_E\rangle=\sqrt{p}X. \tag{3.128}$$

Finally, the operator sum representation is given by

$$\xi(\rho)=E_0\rho E_0^\dagger+E_1\rho E_1^\dagger=(1-p)\rho+pX\rho X \tag{3.129}$$

In similar fashion, the Kraus operators for the phase-flip channel are given by

$$E_0=\langle 0_E|U|\phi_E\rangle=\sqrt{1-p}I,\ \ E_1=\langle 1_E|U|\phi_E\rangle=\sqrt{p}Z \tag{3.130}$$

and the corresponding operator sum representation is

$$\xi(\rho)=E_0\rho E_0^\dagger+E_1\rho E_1^\dagger=(1-p)\rho+pZ\rho Z \tag{3.131}$$

The Pauli operators play a key role in quantum computing and quantum communication applications. Let us prove that Pauli matrices along with the identity operator form an orthogonal basis for a complex Hilbert space of 2×2 matrices, denoted as $H_{2\times 2}$. Toward that end we employ the *Hilbert-Schmidt inner product*, defined as

$$\langle \boldsymbol{A},\boldsymbol{B}\rangle_{HS}=\mathrm{Tr}\left(\boldsymbol{A}^\dagger\boldsymbol{B}\right), \tag{3.132}$$

where $\boldsymbol{A},\boldsymbol{B}\in H_{2\times 2}$, to first show that all the four matrices are orthogonal:

$$\frac{1}{2}\mathrm{Tr}\left(\boldsymbol{I}^\dagger\boldsymbol{X}\right)=\frac{1}{2}\mathrm{Tr}\left[\begin{pmatrix}1&0\\0&1\end{pmatrix}\begin{pmatrix}0&1\\1&0\end{pmatrix}\right]=\frac{1}{2}\mathrm{Tr}\begin{pmatrix}0&1\\1&0\end{pmatrix}=0=\frac{1}{2}\mathrm{Tr}\left(\boldsymbol{I}^\dagger\boldsymbol{Z}\right)=\frac{1}{2}\mathrm{Tr}\left(\boldsymbol{I}^\dagger\boldsymbol{Y}\right), \tag{3.133.1}$$

$$\frac{1}{2}\mathrm{Tr}\left(\boldsymbol{X}^\dagger\boldsymbol{Y}\right)=\frac{1}{2}\mathrm{Tr}\left[\begin{pmatrix}0&1\\1&0\end{pmatrix}\begin{pmatrix}0&-j\\j&0\end{pmatrix}\right]=\frac{1}{2}\mathrm{Tr}\begin{pmatrix}j&0\\0&-j\end{pmatrix}=0=\frac{1}{2}\mathrm{Tr}\left(\boldsymbol{Y}^\dagger\boldsymbol{X}\right), \tag{3.133.2}$$

$$\frac{1}{2}\mathrm{Tr}\left(\boldsymbol{X}^\dagger\boldsymbol{Z}\right)=\frac{1}{2}\mathrm{Tr}\left[\begin{pmatrix}0&1\\1&0\end{pmatrix}\begin{pmatrix}1&0\\0&-1\end{pmatrix}\right]=\frac{1}{2}\mathrm{Tr}\begin{pmatrix}0&-1\\1&0\end{pmatrix}=0=\frac{1}{2}\mathrm{Tr}\left(\boldsymbol{Z}^\dagger\boldsymbol{X}\right), \tag{3.133.3}$$

$$\frac{1}{2}\mathrm{Tr}(\boldsymbol{Y}^{\dagger}\boldsymbol{Z}) = \frac{1}{2}\mathrm{Tr}\left[\begin{pmatrix} 0 & -j \\ j & 0 \end{pmatrix}\begin{pmatrix} 1 & 0 \\ 0 & -1 \end{pmatrix}\right] = \frac{1}{2}\mathrm{Tr}\begin{pmatrix} 0 & j \\ j & 0 \end{pmatrix} = 0 = \frac{1}{2}\mathrm{Tr}(\boldsymbol{Z}^{\dagger}\boldsymbol{Y}). \tag{3.133.4}$$

Based on the Hilbert-Schmidt inner product we can go a step further and show that ***I***, ***X***, ***Y***, ***Z*** are not only orthogonal but also orthonormal, which is left to prove as an exercise for the reader. Next, we need to prove that ***I***, ***X***, ***Y***, ***Z*** span $\mathrm{H}_{2\times 2}$, that is we can express any matrix $\boldsymbol{P} \in H_{2\times 2}$ as follows:

$$\boldsymbol{P} = c_1\boldsymbol{I} + c_2\boldsymbol{X} + c_3\boldsymbol{Y} + c_4\boldsymbol{Z}, \tag{3.134}$$

such that $c_i \in \mathcal{C}$. To show this let us first define ***P*** as

$$\boldsymbol{P} = \begin{pmatrix} r & s \\ t & u \end{pmatrix}, \tag{3.135}$$

where $r, s, t, u \in \mathcal{C}$. Due to the orthonormality of ***I***, ***X***, ***Y***, ***Z***, the coefficients can be easily calculated using the Hilbert-Schmidt inner product by

$$c_1 = \frac{1}{2}\mathrm{Tr}(\boldsymbol{I}^{\dagger}\boldsymbol{P}) = \frac{1}{2}\mathrm{Tr}\left[\begin{pmatrix} 1 & 0 \\ 0 & 1 \end{pmatrix}\begin{pmatrix} r & s \\ t & u \end{pmatrix}\right] = \frac{1}{2}(r+u) \in \mathcal{C}, \tag{3.136.1}$$

$$c_2 = \frac{1}{2}\mathrm{Tr}(\boldsymbol{X}^{\dagger}\boldsymbol{P}) = \frac{1}{2}\mathrm{Tr}\left[\begin{pmatrix} 0 & 1 \\ 1 & 0 \end{pmatrix}\begin{pmatrix} r & s \\ t & u \end{pmatrix}\right] = \frac{1}{2}(t+s) \in \mathcal{C}, \tag{3.136.2}$$

$$c_3 = \frac{1}{2}\mathrm{Tr}(\boldsymbol{Y}^{\dagger}\boldsymbol{P}) = \frac{1}{2}\mathrm{Tr}\left[\begin{pmatrix} 0 & -j \\ j & 0 \end{pmatrix}\begin{pmatrix} r & s \\ t & u \end{pmatrix}\right] = \frac{j}{2}(s-t) \in \mathcal{C}, \tag{3.136.3}$$

$$c_4 = \frac{1}{2}\mathrm{Tr}(\boldsymbol{Z}^{\dagger}\boldsymbol{P}) = \frac{1}{2}\mathrm{Tr}\left[\begin{pmatrix} 1 & 0 \\ 0 & -1 \end{pmatrix}\begin{pmatrix} r & s \\ t & u \end{pmatrix}\right] = \frac{1}{2}(r-u) \in \mathcal{C}. \tag{3.136.4}$$

Substituting Eqs. (3.136.1)–(3.136.4) in (3.134) verifies our claim, thereby completing the proof.

3.9 Decoherence effects, depolarization, and amplitude damping channel models

Quantum computation works by manipulating quantum interference effect. The quantum interference, a manifestation of coherent superposition of quantum states, is the cornerstone behind all quantum information tasks such as quantum computation and quantum communication. A major source of problem is our inability to prevent our quantum system of interest from interacting with the surrounding environment. This interaction results in an entanglement between the quantum system and the environment leading to decoherence. To understand this system-environment entanglement and decoherence better let us consider a qubit described by density state (matrix) $\rho = \begin{bmatrix} a & b \\ c & d \end{bmatrix}$ interacting with the environment, described by the following three states: $|0_E\rangle$, $|1_E\rangle$, and $|2_E\rangle$. Without loss of generality, we assume that environment was initially in state $|0_E\rangle$. The unitary operator introducing the entanglement between the quantum system and the environment is defined as

$$
\begin{aligned}
U|0\rangle|0_E\rangle &= \sqrt{1-p}|0\rangle|0_E\rangle + \sqrt{p}|0\rangle|1_E\rangle \\
U|0\rangle|0_E\rangle &= \sqrt{1-p}|1\rangle|0_E\rangle + \sqrt{p}|1\rangle|2_E\rangle.
\end{aligned}
\tag{3.137}
$$

The corresponding Kraus operators are given by

$$E_0 = \sqrt{1-p}I,\ \ E_1 = \sqrt{p}|0\rangle\langle 0|,\ \ E_2 = \sqrt{p}|1\rangle\langle 1| \tag{3.138}$$

The operator sum representation is given by

$$\xi(\rho) = E_0\rho E_0^\dagger + E_1\rho E_1^\dagger + E_2\rho E_2^\dagger = \begin{bmatrix} a & (1-p)b \\ (1-p)c & d \end{bmatrix} \tag{3.139}$$

By applying these quantum operation n-times, the corresponding final state would be

$$\rho_f = \begin{bmatrix} a & (1-p)^n b \\ (1-p)^n c & d \end{bmatrix}. \tag{3.140}$$

If the probability p is expressed as $p = \gamma\Delta t$ we can write $n = t/\Delta t$ and in limit we obtain

$$\lim_{\Delta t \to 0}(1-p)^n = (1-\gamma\Delta t)^{t/\Delta t} = e^{-\gamma t}. \tag{3.141}$$

Therefore, the corresponding operator sum representation as n tends to plus infinity is given by

$$\xi(\rho) = \begin{bmatrix} a & e^{-\gamma t}b \\ e^{-\gamma t}c & d \end{bmatrix}. \tag{3.142}$$

Clearly, the terms b and c go to zero as t increases, indicating that the relative phase in the original state of the quantum system is lost, and the corresponding channel model is known as the *phase damping* channel model.

In the above example, we have considered the coupling between a single qubit quantum system and the environment and discussed the resulting loss of interference or coherent superposition. In general for multiple qubit systems decoherence also results in loss of coupling between the qubits. In fact, with increasing complexity and size of the quantum computer the decoherence effect becomes worse. Additionally, the quantum system lies in some complex Hilbert space where there are infinite variations of errors that can cause decoherence.

A more general example of dephasing is the depolarization. The *depolarizing channel*, as shown in Fig. 3.7, with probability $1-p$ leaves the qubit as it is, while with probability p moves the initial state

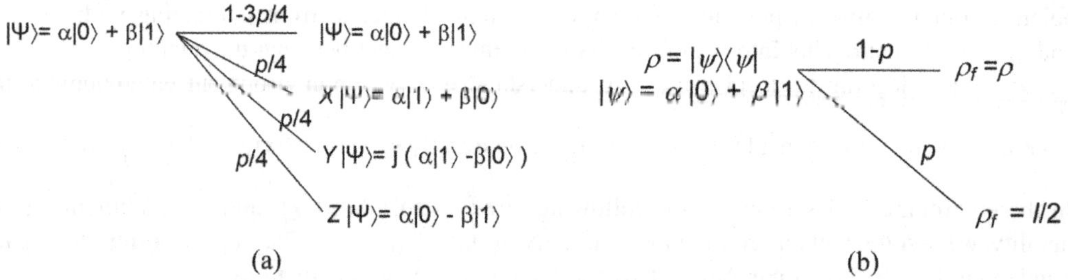

FIGURE 3.7

Depolarizing channel model: (A) Pauli operator description and (B) density operator description.

into $\rho_f = I/2$ that maximizes the von Neumann entropy $S(\rho) = -\,\mathrm{Tr}\;\rho\;\log\;\rho = 1$. The properties describing the model can be summarized as follows:

1. Qubit errors are independent,
2. Single-qubit errors (X, Y, Z) are equally likely,
3. All qubits have the same single-error probability $p/4$.

The Kraus operators E_i of the channel should be selected as follows:

$$E_0 = \sqrt{1-3p/4}I;\;\; E_1 = \sqrt{p/4}X;\;\; E_2 = \sqrt{p/4}Y;\;\; E_3 = \sqrt{p/4}Z \tag{3.143}$$

The action of depolarizing channel is to perform the following mapping: $\rho \rightarrow \xi(\rho) = \sum_i E_i \rho E_i^\dagger$, where ρ is the initial density operator. Without loss of generality we will assume that initial state was pure $|\psi\rangle = a|0\rangle + b|1\rangle$ so that

$$\begin{aligned}\rho &= |\psi\rangle\langle\psi| = (a|0\rangle + b|1\rangle)(\langle 0|a^* + \langle 1|b^*)\\
&= |a|^2\begin{bmatrix}1 & 0\\ 0 & 0\end{bmatrix} + ab^*\begin{bmatrix}0 & 1\\ 0 & 0\end{bmatrix} + a^*b\begin{bmatrix}0 & 0\\ 1 & 0\end{bmatrix} + |b|^2\begin{bmatrix}0 & 0\\ 0 & 1\end{bmatrix}.\\
&= \begin{bmatrix}|a|^2 & ab^*\\ a^*b & |b|^2\end{bmatrix}\end{aligned} \tag{3.144}$$

The resulting quantum operation can be represented using operator sum-representation as follows:

$$\begin{aligned}\xi(\rho) &= \sum_i E_i \rho E_i^\dagger = \left(1-\frac{3p}{4}\right)\rho + \frac{p}{4}(X\rho X + Y\rho Y + Z\rho Z)\\
&= \left(1-\frac{3p}{4}\right)\begin{bmatrix}|a|^2 & ab^*\\ a^*b & |b|^2\end{bmatrix} + \frac{p}{4}\begin{bmatrix}0 & 1\\ 1 & 0\end{bmatrix}\begin{bmatrix}|a|^2 & ab^*\\ a^*b & |b|^2\end{bmatrix}\begin{bmatrix}0 & 1\\ 1 & 0\end{bmatrix}\\
&\quad + \frac{p}{4}\begin{bmatrix}1 & 0\\ 0 & -1\end{bmatrix}\begin{bmatrix}|a|^2 & ab^*\\ a^*b & |b|^2\end{bmatrix}\begin{bmatrix}1 & 0\\ 0 & -1\end{bmatrix} + \frac{p}{4}\begin{bmatrix}0 & -j\\ j & 0\end{bmatrix}\begin{bmatrix}|a|^2 & ab^*\\ a^*b & |b|^2\end{bmatrix}\begin{bmatrix}0 & -j\\ j & 0\end{bmatrix}\\
&= \left(1-\frac{3p}{4}\right)\begin{bmatrix}|a|^2 & ab^*\\ a^*b & |b|^2\end{bmatrix} + \frac{p}{4}\begin{bmatrix}|b|^2 & a^*b\\ ab^* & |a|^2\end{bmatrix} + \frac{p}{4}\begin{bmatrix}|a|^2 & -ab^*\\ -a^*b & |b|^2\end{bmatrix} + \frac{p}{4}\begin{bmatrix}|b|^2 & -a^*b\\ -ab^* & |a|^2\end{bmatrix}\\
&= \begin{bmatrix}\left(1-\frac{p}{2}\right)|a|^2 + \frac{p}{2}|b|^2 & (1-p)ab^*\\ (1-p)a^*b & \frac{p}{2}|a|^2 + \left(1-\frac{p}{2}\right)|b|^2\end{bmatrix}\\
&= \begin{bmatrix}(1-p)|a|^2 + \frac{p}{2}\left(|a|^2 + |b|^2\right) & (1-p)ab^*\\ (1-p)a^*b & \frac{p}{2}\left(|a|^2 + |b|^2\right) + (1-p)|b|^2\end{bmatrix}\end{aligned} \tag{3.145.1}$$

Since $|a|^2 + |b|^2 = 1$, the operator-sum representation cab be written as

$$\xi(\rho) = \sum_i E_i \rho E_i^\dagger = (1-p)\begin{bmatrix} |a|^2 & ab^* \\ a^*b & |b|^2 \end{bmatrix} + \frac{p}{2}I = (1-p)\rho + \frac{p}{2}I. \tag{3.145.2}$$

The first line in Eq. (3.145.1) corresponds to the model shown in Fig. 3.7A and Eq. (3.145.2) corresponds to model shown in Fig. 3.7B. It is clear from Eqs. (3.145.1) and (3.145.2) that $\text{Tr}\xi(\rho) = \text{Tr}(\rho) = 1$, meaning that the superoperator is trace-preserving. Notice that the depolarizing channel model in some other books/papers can be slightly different from the one described here.

We now describe the *amplitude damping channel* model. In certain quantum channels the errors X, Y, and Z do not occur with the same probability. In amplitude damping channel, the operation elements are given by

$$E_0 = \begin{pmatrix} 1 & 0 \\ 0 & \sqrt{1-\varepsilon^2} \end{pmatrix}, \quad E_1 = \begin{pmatrix} 0 & \varepsilon \\ 0 & 0 \end{pmatrix} \tag{3.146}$$

The *spontaneous emission* is an example of a physical process that can be modeled using the amplitude damping channel model. If $|\psi\rangle = a\ |0\rangle\ +\ b\ |1\rangle$ is the initial qubit state $\left(\rho = \begin{bmatrix} |a|^2 & ab^* \\ a^*b & |b|^2 \end{bmatrix}\right)$, the effect of amplitude damping channel is to perform the following mapping:

$$\begin{aligned}
\rho \to \xi(\rho) = E_0 \rho E_0^\dagger + E_1 \rho E_1^\dagger &= \begin{pmatrix} 1 & 0 \\ 0 & \sqrt{1-\varepsilon^2} \end{pmatrix} \begin{bmatrix} |a|^2 & ab^* \\ a^*b & |b|^2 \end{bmatrix} \begin{pmatrix} 1 & 0 \\ 0 & \sqrt{1-\varepsilon^2} \end{pmatrix} \\
&\quad + \begin{pmatrix} 0 & \varepsilon \\ 0 & 0 \end{pmatrix} \begin{bmatrix} |a|^2 & ab^* \\ a^*b & |b|^2 \end{bmatrix} \begin{pmatrix} 0 & 0 \\ \varepsilon & 0 \end{pmatrix} \\
&= \begin{pmatrix} |a|^2 & ab^*\sqrt{1-\varepsilon^2} \\ a^*b\sqrt{1-\varepsilon^2} & |b|^2(1-\varepsilon^2) \end{pmatrix} + \begin{pmatrix} |b|^2\varepsilon^2 & 0 \\ 0 & 0 \end{pmatrix} \\
&= \begin{pmatrix} |a|^2 + \varepsilon^2|b|^2 & ab^*\sqrt{1-\varepsilon^2} \\ a^*b\sqrt{1-\varepsilon^2} & |b|^2(1-\varepsilon^2) \end{pmatrix}
\end{aligned} \tag{3.147}$$

Probabilities $P(0)$ and $P(1)$ that E_0 and E_1 occur are given by

$$\begin{aligned}
P(0) &= \text{Tr}\left(E_0 \rho E_0^\dagger\right) = \text{Tr}\begin{pmatrix} |a|^2 & ab^*\sqrt{1-\varepsilon^2} \\ a^*b\sqrt{1-\varepsilon^2} & |b|^2(1-\varepsilon^2) \end{pmatrix} = 1 - \varepsilon^2|b|^2 \\
P(1) &= \text{Tr}\left(E_1 \rho E_1^\dagger\right) = \text{Tr}\begin{pmatrix} |b|^2\varepsilon^2 & 0 \\ 0 & 0 \end{pmatrix} = \varepsilon^2|b|^2
\end{aligned} \tag{3.148}$$

The corresponding amplitude damping channel model is shown in Fig. 3.8.

To verify how much the entanglement is preserved during interaction with the quantum channel, the *entanglement fidelity* is typically used. Let us assume that the bipartite system $H_A \otimes H_B$ is prepared in general pure state $\rho_{AB} = |\psi_{AB}\rangle\langle\psi_{AB}|$. Under the assumption that the second subsystem undergoes the interaction with the channel, the *entanglement fidelity* can be defined as

$$F_e = \langle \rho_{AB}, \rho_{AB,final} \rangle_{HS} = \mathrm{Tr}(\rho_{AB}\rho_{AB,final}) = \langle \psi_{AB} | \rho_{AB,final} | \psi_{AB} \rangle, \tag{3.149}$$

which represents the Hilbert-Schmidt inner product [see Eq. 3.132] between initial and final density operators thus describing the variation in entanglement due to interaction with the quantum channel. The values of F_e close to one indicate that entanglement is well preserved, while the values close to 0 imply that the entanglement is destroyed during interaction with the quantum channel. If we trace out the first subsystem, the reduced density operator for subsystem H_B will be

$$\rho_{B,final} = \mathrm{Tr}_A \rho_{AB,final}. \tag{3.150}$$

Now we can define the fidelity between initial and final states of subsystem H_B as follows:

$$F(\rho_B, \rho_{B,final}) = \langle \rho_B, \rho_{B,final} \rangle_{HS} = \mathrm{Tr}(\rho_B \rho_{B,final}), \rho_B = Tr_A \rho_{AB}. \tag{3.151}$$

This definition of fidelity is an upper bound for the entanglement fidelity:

$$F_e \leq F(\rho_B, \rho_{B,final}), \tag{3.152}$$

because it is a measure of similarity (distance) between two states, while entanglement fidelity is the measure of preservation of entanglement after the interaction with the quantum channel is over. Mathematically speaking, the trace operation and the super operator do not commute in general. A given Bell state can be converted to a different Bell state by the quantum channel, which will result in 0 entanglement fidelity, while the final state is maximally entangled. This is the reason why other entanglements figures of merit should be considered as well, such as the concurrence and the entanglement of formation introduced in Section 3.7 already.

Before concluding this section, we provide another definition of fidelity. Let ρ and σ be two density operators. Then the fidelity can be defined as

$$F(\rho, \sigma) = Tr\left(\sqrt{\rho^{1/2}\sigma\rho^{1/2}}\right). \tag{3.153}$$

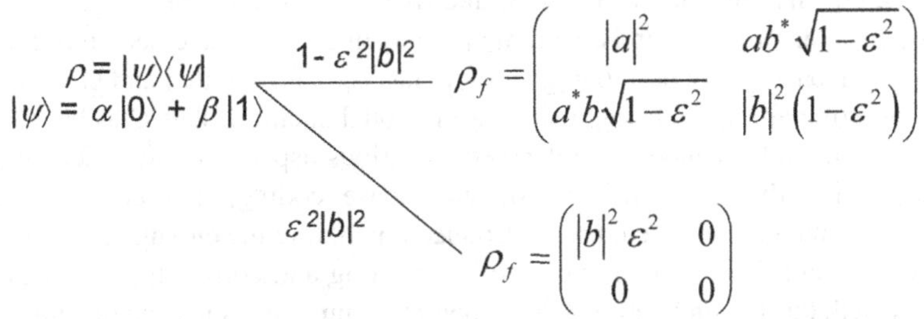

FIGURE 3.8

Amplitude damping channel model.

For two pure states with density operators $\rho = |\psi\rangle\langle\psi|, \sigma = |\phi\rangle\langle\phi|$, the corresponding fidelity will be

$$F(\rho,\sigma) = Tr\left(\sqrt{\rho^{1/2}\sigma\rho^{1/2}}\right) \overset{\rho^2=\rho}{=} Tr(\sqrt{|\psi\rangle\langle\psi|(|\phi\rangle\langle\phi|)|\psi\rangle\langle\psi|})$$

$$= Tr(\underbrace{|\psi\rangle\langle\psi|\phi\rangle\langle\phi|\psi\rangle}_{|\langle\phi|\psi\rangle|^2}\langle\psi|)^{1/2} = |\langle\phi|\psi\rangle|\underbrace{Tr(|\psi\rangle\langle\psi|)}_{1} = |\langle\phi|\psi\rangle| \tag{3.154}$$

where we used the property of the density operators for pure states $\rho^2 = \rho$ (or equivalently $\rho = \rho^{1/2}$). Clearly, for pure states the fidelity corresponds to the square root of probability of finding the system in state $|\phi\rangle$ if it is known to be prepared in state $|\psi\rangle$ (and vice versa). Since fidelity is related to the probability it ranges between 0 and 1, with 0 indicating there is no overlap and one meaning that the states are identical. The following *properties* of fidelity hold:

1. The symmetry property:

$$F(\rho,\sigma) = F(\sigma,\rho), \tag{3.155}$$

2. The fidelity is invariant under unitary operations:

$$F(U\rho U^{\dagger}, U\sigma U^{\dagger}) = F(\rho,\sigma), \tag{3.156}$$

3. If ρ and σ commute, the fidelity can be expressed in terms of eigenvalues of ρ, denoted as r_i, and σ, denoted as s_i, as follows:

$$F(\rho,\sigma) = \sum\nolimits_i (r_i s_i)^{1/2}. \tag{3.157}$$

3.10 Summary

This chapter has been devoted to quantum information processing (QIP) fundamentals. The basic QIP features have been discussed in Section 3.1. The basic concepts of quantum mechanics such as state vectors, operators, projection operators, and density operators have been introduced in Section 3.2. In Section 3.3 measurements, uncertainty relation, and dynamics of a quantum system have been discussed. The superposition principle and quantum parallelism concept have been introduced in Section 3.4, together with QIP basics. The no-cloning theorem has been formulated and proved in Section 3.5. Further, the theorem of inability to distinguish nonorthogonal quantum states has been formulated and proved in Sections 3.6. and 3.7 have been devoted to various aspects of entanglement including the Schmidt decomposition theorem, purification, superdense coding, and entanglement swapping. Various measures of entanglement have been introduced including the entanglement of formation and concurrence. How the entanglement can be used in detecting and correcting the quantum errors has been described as well. Further, in Section 3.8, the operator-sum representation of a quantum operation has been introduced and applied to the bit-flip and phase-flip channels. In Section 3.9, the quantum errors and decoherence effects have been discussed and the phase damping, depolarizing, and amplitude damping channels have been described.

References

[1] I.B. Djordjevic, Physical-Layer Security and Quantum Key Distribution, Springer International Publishing, Switzerland, 2019.

[2] I.B. Djordjevic, Quantum Information Processing, Quantum Computing, and Quantum Error Correction: An Engineering Approach, second ed., Elsevier/Academic Press, 2021.

[3] F. Gaitan, Quantum Error Correction and Fault Tolerant Quantum Computing, CRC Press, 2008.

[4] D. McMahon, Quantum Computing Explained, John Wiley & Sons, 2008.

[5] C.W. Helstrom, Quantum Detection and Estimation Theory, Academic, New York, 1976.

[6] C.W. Helstrom, J.W.S. Liu, J.P. Gordon, Quantum mechanical communication theory, Proc. IEEE 58 (1970) 1578–1598.

[7] J.J. Sakurai, Modern Quantum Mechanics, Addison-Wisley, 1994.

[8] G. Baym, Lectures on Quantum Mechanics, Westview Press, 1990.

[9] C.E. Shannon, A mathematical theory of communication, Bell Syst. Techn. J. 27 (1948) 379–423, 623–656.

[10] C.E. Shannon, W. Weaver, The Mathematical Theory of Communication, The University of Illinois Press, Urbana, Illinois, 1949.

[11] T. Cover, J. Thomas, Elements of Information Theory, Wiley-Interscience, 1991.

[12] R. Landauer, Irreversibility and heat generation in the computing process, IBM J. Res. Dev. 5 (3) (1961).

[13] A.N. Kolmogorov, Three approaches to the quantitative definition of information, Probl. Inf. Transm. 1 (1) (1965) 1–7.

[14] A. Turing, On computable numbers, with an application to the Entscheidungsproblem, Proc. Lond. Math. Soc. 2 (42) (1936) 230–265.

[15] A. Turing, Computability and lambda-Definability, J. Symbolic Logic 2 (1937) 153–163.

[16] A. Turing, Systems of logic based on ordinals, Proc. Lond. Math. Soc. 3 (45) (1939) 161–228.

[17] M. Planck, On the law of distribution of energy in the normal spectrum, Ann. Phys. 4 (1901) 553.

[18] A. Einstein, On a heuristic viewpoint concerning the production and transformation of light, Ann. Phys. 17 (1905) 132–148.

[19] L. de Broglie, Waves and quanta, Compt. Ren. 177 (1923) 507.

[20] W. Pauli, On the connection between the completion of electron groups in an atom with the complex structure of spectra, Z. Phys. 31 (1925) 765.

[21] W. Heisenberg, Quantum-theoretical re-interpretation of kinematic and mechanical relations, Z. Phys. 33 (1925) 879.

[22] M. Born, P. Jordan, On quantum mechanics, Z. Phys. 34 (1925) 858.

[23] E. Schrödinger, Quantization as a problem of proper values. Part I, Ann. Phys. 79 (1926) 361.

[24] E. Schrödinger, On the relation between the quantum mechanics of Heisenberg, Born, and Jordan, and that of schrödinger, Ann. Phys. 79 (1926) 734.

[25] W. Heisenberg, The actual content of quantum theoretical kinematics and mechanics, Z. Phys. 43 (1927) 172.

[26] R.P. Feynman, Simulating physics with computers, Int. J. Theor. Phys. 21 (1982) 467–488.

[27] P.A.M. Dirac, Quantum Mechanics, fourth ed., Oxford Univ. Press, London, 1958.

[28] D. Deutsch, The Fabric of Reality, Penguin Press, New York, 1997.

[29] D. Deutsch, Quantum theory, the church-turing principle and the universal quantum computer, Proc. Roy. Soc. Lond. A 400 (1985) 97–117.

[30] D. Deutsch, R. Jozsa, Rapid solution of problems by quantum computation, Proc. Math. Phys. Sci. 439 (1907) (1992) 553–558.
[31] P. Shor, Algorithms for Quantum Computation: Discrete Logarithms and Factoring, IEEE Comput. Soc. Press, November 1994, pp. 124–134.
[32] L.K. Grover, A fast quantum mechanical algorithm for database search, in: Proceedings 28th Annual ACM Symposium on the Theory of Computing, May 1996, p. 212.
[33] A. Einstein, B. Podolsky, N. Rosen, Can quantum-mechanical description of physical reality be considered complete? Phys. Rev. 47 (May 15, 1935) 777–780.
[34] J.S. Bell, On the Einstein-Podolsky-Rosen paradox, Physics 1 (1964) 195–200.
[35] J.S. Bell, Speakable and Unspeakable in Quantum Mechanics, *Cambridge University Press*, 1987.
[36] J.F. Clauser, A. Shimony, Bell's theorem experimental tests and implications, Rep. Prog. Phys. 41 (1978) 1981.
[37] A. Aspect, P. Grangier, G. Roger, Experimental tasks of realistic local theories by Bell's theorem, Phys. Rev. Lett. 47 (1981) 460.
[38] P.G. Kwiat, et al., New high-intensity source of polarization-entangled photon pairs, Phys. Rev. Lett. 75 (1995) 4337.
[39] R. Horodecki, P. Horodecki, M. Horodecki, K. Horodecki, Quantum entanglement, Rev. Mod. Phys. 81 (2009) 865.
[40] V. A. Mousolou, "Entanglement fidelity and measure of entanglement," Available at: https://arxiv.org/abs/1911.10854.
[41] C.H. Bennett, D.P. DiVincenzo, J. Smolin, W.K. Wootters, Mixed-state entanglement and quantum error correction, Phys. Rev. 54 (1996) 3824.
[42] S. Hill, W.K. Wootters, Entanglement of a pair of quantum bits, Phys. Rev. Lett. 78 (1997) 5022.
[43] W.K. Wootters, Entanglement of formation of an arbitrary state of two qubits, Phys. Rev. Lett. 80 (1998) 2245.
[44] A. Harrow, Superdense coding of quantum states, Phys. Rev. Lett. 92 (18) (May 7 , 2004) 187901-1–187901-187904.

Further reading

[1] R. LaRose, Overview and comparison of gate level quantum software platforms, Quantum 3 (2019) 130. Available at: https://arxiv.org/abs/1807.02500v2.
[2] J.D. Hidary, Quantum Computing: An Applied Approach, Springer Nature Switzerland AG, Cham, Switzerland, 2019.
[3] B. Abdo, et al., IBM Q 16 Rueschlikon V1.x.x, Sept. 2018. Available at: https://ibm.biz/qiskit-ibmqx5.
[4] Rigetti Forest: An API for quantum computing in the cloud, Available at: https://www.welcome.ai/tech/technology/rigetti-forest .
[5] R.S. Smith, M.J. Curtis, W.J. Zeng, A Practical Quantum Instruction Set Architecture, 2016. Available at: https://arxiv.org/abs/1608.03355.
[6] H. Zhang, et al., Quantized majorana conductance, Nature 556 (April 05 , 2018) 74–79.

CHAPTER

Quantum information theory fundamentals

4

Chapter outline

4.1 Introductory remarks

Quantum information theory, like its classical counterpart, studies the meanings and limits of communicating classical and quantum information over quantum channels. In this chapter, we introduce the basic concepts underlying this vast and fascinating area, which is currently the subject of intense research [1,2]. In the classical context, information theory provides answers to two fundamental questions [3–5]: (1) To what extent can information be compressed? (2) What is the maximum rate for reliable communication over a noisy channel? The former is the basis of source coding and data compression in communication theory, whereas the latter forms the basis for channel coding. Information theory also overlaps many areas in other disciplines. Since its inception—with the seminal 1948 paper by Claude Shannon titled, "A Mathematical Theory of Communication" [3]—information theory has established links with many areas, including Kolmogorov complexity in computer science [6], large deviation theory within probability theory [7], hypothesis testing (e.g., Fisher information [8]) in statistics, and Kelly's rule in investment and portfolio theory [9].

Over the last several decades, classical information theory has provided the foundation for developing its quantum analog. This is not to say that they are the same. In fact, as we show in this chapter, quantum information theory reveals consequences for the quantum world that classical

Quantum Communication, Quantum Networks, and Quantum Sensing. https://doi.org/10.1016/B978-0-12-822942-2.00005-4

information theory cannot predict. Nevertheless, classical information theory provides a logical structure for introducing the ideas of quantum information theory, and we exploit this link throughout the chapter. It is challenging to do justice to quantum information theory within a single chapter. It is a comprehensive area with many consequences, and whole books could be and have been devoted to the subject. Since our space is limited, we focus on telling a coherent story that we hope provides readers with a helpful introduction to the subject. The references at the end of this chapter are resources for readers who decide to pursue this field further. References [3–9] provide a background to classical information theory and its relations to computer science, statistics, probability, estimation and detection, and finance. References [10–25] list important papers that have led to breakthroughs in quantum information theory. Finally, references [14–25] detail interesting theoretical work done in quantifying the capacity of quantum optical channels.

4.2 Von Neumann entropy

Let us observe a classical discrete memoryless source with the alphabet $X=\{x_1,x_2,\ldots,x_N\}$. The symbols from the alphabet are emitted by the source with probabilities $P(X=x_n)=p_n, n=1,2,\ldots,N$. The amount of information carried by the k-th symbol is related to the uncertainty that is resolved when this symbol occurs, and it is defined as $I(x_n)=\log(1/p_n)=-\log(p_n)$, where the logarithm is to the base 2.

The *classical (Shannon) entropy* is defined as the measure of the average amount of information per source symbol:

$$H(X)=E[I(x_n)]=\sum_{n=1}^{N}p_nI(x_n)=\sum_{n=1}^{N}p_n\log_n\left(\frac{1}{p_n}\right). \tag{4.1}$$

Shannon's entropy satisfies the following inequalities:

$$0\leq H(X)\leq\log N=\log|X|. \tag{4.2}$$

In the research literature, $H(X)$ is interchangeably denoted $H(P)$ to highlight that Shannon's entropy is a functional of the probability mass function.

Let Q be a quantum system with the state described by the density operator $\boldsymbol{\rho}_x$. The probability that output $\boldsymbol{\rho}_x$ is obtained is given by $p_x=P(x)$. The quantum source, therefore, is described by the ensemble $\{\boldsymbol{\rho}_x, p_x\}$ characterized by the mixed density operator $\boldsymbol{\rho}=\sum_{x\in X}p_x\boldsymbol{\rho}_x$. Then, the *quantum (von Neumann) entropy* is defined as

$$S(\boldsymbol{\rho})=-Tr(\boldsymbol{\rho}\log\boldsymbol{\rho})=-\sum_{\lambda_i}\lambda_i\log\lambda_i,\quad \boldsymbol{\rho}=\sum_i\lambda_i|\lambda_i\rangle\langle\lambda_i|, \tag{4.3}$$

where λ_i are eigenvalues of $\boldsymbol{\rho}$, and $|\lambda_i\rangle$ are the corresponding eigenkets. When all quantum states are pure and mutually orthogonal, the von Neumann entropy equals the Shannon entropy, as in that case, $p_i=\lambda_i$. As an illustration, Fig. 4.1 provides an interpretation of the quantum representation of classical information. Two preparations $P^{(i)}$ and $P^{(j)}$, with $H(X^{(i)})\neq H(X^{(j)})$ $(i\neq j)$, can generate the same $\boldsymbol{\rho}$ and hence have the same von Neumann entropy $S(\rho)$ because the states of the two preparations may not be physically distinguishable from each other.

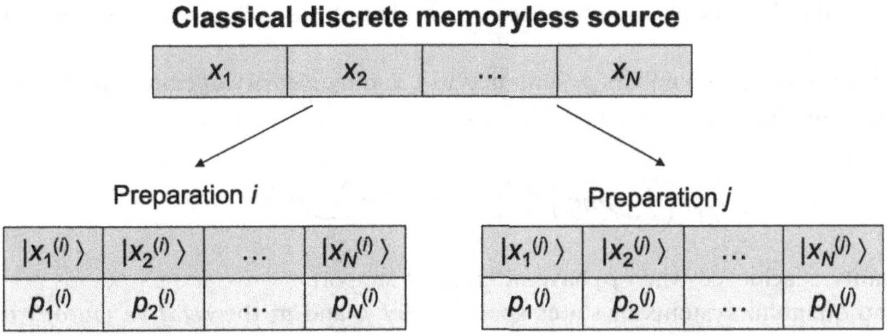

FIGURE 4.1

Quantum representation of classical information.

We should mention that although $S(\boldsymbol{\rho}) = H(P)$ for pure and mutually orthogonal states, they do not have the same meaning. $S(\boldsymbol{\rho})$ represents the measure of quantum uncertainty before any measurement takes place, whereas $H(P)$ measures the classical uncertainty of the outcomes after the quantum measurements. To clarify this point, let us assume that we measure the observable A denoted by

$$A = \sum_i a_i |a_i\rangle\langle a_i|. \tag{4.4}$$

This measurement for the quantum system Q in a state described by $\boldsymbol{\rho}$ yields a_i with the probability $p_i = \text{Tr}(|a_i\rangle\langle a_i|\boldsymbol{\rho})) = \langle a_i|\boldsymbol{\rho}|a_i\rangle$. If the observable A commutes with $\boldsymbol{\rho}$, that is, $[A,\boldsymbol{\rho}] = 0$, the same basis diagonalizes both A and $\boldsymbol{\rho}$, and the classical and quantum entropies are equal. This is the case for pure orthogonal states. In general, however, the two do not commute, and $S(\boldsymbol{\rho}) < H(P_{A,\boldsymbol{\rho}})$, where $P_{A,\boldsymbol{\rho}}$ denotes the probability of the measurement outcomes when the observable is A and the density operator is $\boldsymbol{\rho}$. Given that Shannon's entropy is upper limited by $\log N$ while $S(\rho)$ is upper limited by Shannon's entropy, we would expect $S(\rho) \leq \log d$, where d is the dimensionality of the Hilbert space. This situation corresponds to the completely mixed state for which $p_i = 1/d$. Because any density operator is nonnegative Hermitian with trace 1, we have

$$\sum_i \lambda_i = 1, \quad 0 \leq \lambda_i \leq 1. \tag{4.5}$$

When $\boldsymbol{\rho}$ is the pure state one of its eigenvalues λ_i equals 1, so $S(\boldsymbol{\rho}) = 0$. Otherwise, based on Eqs. (4.3) and (4.5), we conclude that $S(\boldsymbol{\rho}) > 0$. To summarize, the von Neumann entropy is bounded as follows:

$$0 \leq S(\boldsymbol{\rho}) \leq \log d. \tag{4.6}$$

Another relevant property of the von Neumann entropy is *invariance under unitary transformation*; in other words, for any unitary transformation U, the following is valid:

$$S\left(U\boldsymbol{\rho}U^{\dagger}\right) = S(\boldsymbol{\rho}). \tag{4.7}$$

If we choose U to diagonalize $\boldsymbol{\rho}$, we can simplify the calculation of the von Neumann entropy.

The von Neumann entropy is *concave*; that is, given the set of positive numbers p_i summing to 1 and the density operators $\boldsymbol{\rho}_i$, we have

$$S\left(\sum_i p_i \boldsymbol{\rho}_i\right) \geq \sum_i p_i S(\boldsymbol{\rho}_i). \tag{4.8}$$

In other words, the von Neumann entropy is higher if we are ignorant of how the state was prepared.

Given the set of positive numbers p_i summing to 1 and the density operators $\boldsymbol{\rho}_i$, the von Neumann entropy satisfies the following bound:

$$S\left(\sum_i p_i \boldsymbol{\rho}_i\right) \leq \sum_i p_i S(\boldsymbol{\rho}_i) - \sum_i p_i \log p_i, \tag{4.9}$$

wherein equality is achieved when $\boldsymbol{\rho}_i$ have orthogonal support.

Given two quantum systems in states specified by $\boldsymbol{\rho}$ and $\boldsymbol{\sigma}$, the *relative entropy* of $\boldsymbol{\rho}$ to $\boldsymbol{\sigma}$ is given by

$$S(\boldsymbol{\rho}\|\boldsymbol{\sigma}) = \mathrm{Tr}(\boldsymbol{\rho}\log\boldsymbol{\rho}) - \mathrm{Tr}(\boldsymbol{\rho}\log\boldsymbol{\sigma}). \tag{4.10}$$

This definition is analogous to the definition of classical relative entropy between two probability distributions $p(x)$ and $q(x)$ that was introduced in Chapter 2. Classical *relative entropy* is a measure of the distance between two distributions $p(x)$ and $q(x)$, and it is defined as follows:

$$D(p\|q) = E_p \log\left[\frac{p(X)}{q(X)}\right] = \sum_{x\in X} p(x)\log\left[\frac{p(X)}{q(X)}\right]. \tag{4.11}$$

Relative entropy is also known as the Kullback–Leibler (KL) distance. It can be interpreted as a measure of the inefficiency of assuming that distribution is $q(x)$ when the true distribution is $p(x)$. From the definition of classical relative entropy, it is clear that it can be infinite for certain cases—for example, when the support of $p(x)$ exceeds that of $q(x)$. A similar scenario arises for quantum relative entropy. Let us define the kernel of any density matrix as the eigenspace spanned by the eigenkets of the density operator corresponding to the eigenvalue 0. Additionally, let the support of any density operator refer to the eigenspace spanned by the eigenvectors of the density operator corresponding to the nonzero eigenvalues. Then, for cases for which the nontrivial intersection between the support of $\boldsymbol{\rho}$ and the kernel of $\boldsymbol{\sigma}$ is empty and the support of $\boldsymbol{\rho}$ is a subspace of the support of $\boldsymbol{\sigma}$, the relative entropy $S(\boldsymbol{\rho}\|\boldsymbol{\sigma})$ is finite. Otherwise, it is infinite. An important property of relative entropy, known as the *Klein inequality*, states that

$$S(\boldsymbol{\rho}\|\boldsymbol{\sigma}) \geq 0, \tag{4.12}$$

with equality being satisfied if and only if $\boldsymbol{\rho} = \boldsymbol{\sigma}$. By setting $\boldsymbol{\sigma} = \boldsymbol{I}/d$, from Eq. (4.12), we obtain

$$S(\boldsymbol{\rho}\,\|\boldsymbol{I}/d) = \mathrm{Tr}(\boldsymbol{\rho}\log\boldsymbol{\rho}) - \mathrm{Tr}(\boldsymbol{\rho}\log(\boldsymbol{I}/d)) \geq 0 \Rightarrow S(\boldsymbol{\rho}) = -\mathrm{Tr}(\boldsymbol{\rho}\log\boldsymbol{\rho}) \leq \log d, \tag{4.13}$$

thus proving the upper bound in Eq. (4.6).

To prove the Klein inequality, we perform the eigendecomposition of $\boldsymbol{\rho}$ and $\boldsymbol{\sigma}$:

$$\rho = \sum_i \rho_i |\rho_i\rangle\langle\rho_i|, \sigma = \sum_i \sigma_i |\sigma_i\rangle\langle\sigma_i|; 0 \leq \rho_i, \sigma_i \leq 1 \tag{4.14}$$

and apply the definition formula for relative entropy:

$$S(\boldsymbol{\rho}\|\boldsymbol{\sigma}) = \mathrm{Tr}(\boldsymbol{\rho}\log\boldsymbol{\rho}) - \mathrm{Tr}(\boldsymbol{\rho}\log\boldsymbol{\sigma}) = \sum_i \rho_i \log \rho_i - \mathrm{Tr}\left[\sum_i \rho_i |\rho_i\rangle\langle\rho_i| \log\boldsymbol{\sigma}\right]$$

$$= \sum_i \rho_i \log \rho_i - \sum_i \rho_i \langle\rho_i| \underbrace{\log\boldsymbol{\sigma}}_{\sum_j |\sigma_j\rangle \log \sigma_j \langle\sigma_j|} |\rho_i\rangle$$

$$= \sum_i \rho_i \left[\log \rho_i - \sum_j \log \sigma_j \underbrace{\langle\rho_i|\sigma_j\rangle\langle\sigma_j|\rho_i\rangle}_{|\langle\rho_i|\sigma_j\rangle|^2}\right] \tag{4.15}$$

Since

$$\sum_j \log \sigma_j |\langle\rho_i|\sigma_j\rangle|^2 \le \log\left[\underbrace{\sum_j \sigma_j |\langle\rho_i|\sigma_j\rangle|^2}_{\gamma_i}\right], \tag{4.16}$$

from Eq. (4.15), we obtain

$$S(\boldsymbol{\rho}\|\boldsymbol{\sigma}) \ge \sum_i \rho_i \log \frac{\rho_i}{\gamma_i} = D(\{\rho_i\} \parallel \{\gamma_i\}), \tag{4.17}$$

and since classical relative entropy is nonnegative, we find that the quantum relative entropy is also nonnegative, thus proving the Klein inequality of Eq. (4.12). In the upcoming subsection, we turn our attention to bipartite and multipartite quantum systems.

4.2.1 Composite systems

Let us consider the bipartite quantum system Q composed of uncorrelated subsystems A and B described by the product state density operator $\boldsymbol{\rho}_{AB} = \boldsymbol{\rho}_A \otimes \boldsymbol{\rho}_B$. This system has the interesting property that uncorrelated subsystems have the same quantum entropy—that is, we can write

$$S(\boldsymbol{\rho}_A) = S(\boldsymbol{\rho}_B). \tag{4.18}$$

To prove this claim, we employ the Schmidt decomposition theorem (proved in Chapter 3), in which a pure state $|\psi\rangle$ of a bipartite system AB can be decomposed as follows:

$$|\psi\rangle = \sum_i c_i |i_A\rangle |i_B\rangle, \tag{4.19}$$

where $|i_A\rangle$ and $|i_B\rangle$ are orthonormal bases of the subsystems H_A and H_B, respectively, and $c_i \in R^+$ (R^+ is the set of nonnegative real numbers) are *Schmidt coefficients* that satisfy the condition

$\sum_i c_i^2 = 1$. By employing the Schmidt decomposition theorem, we can express the density operator ρ_{AB} as follows:

$$\rho_{AB} = |\psi\rangle\langle\psi| = \sum_i \sum_j c_i c_j |i_A\rangle\langle j_A| \otimes |i_B\rangle\langle j_B|. \tag{4.20}$$

By taking the partial trace of ρ_{AB} over subsystem A, we obtain

$$\mathrm{Tr}_A(\rho_{AB}) = \sum_i \sum_j c_i c_j \underbrace{\langle j_A | i_A\rangle}_{\delta_{ji}} |i_B\rangle\langle j_B| = \sum_j c_j^2 |j_B\rangle\langle j_B| = \rho_B. \tag{4.21}$$

In a similar fashion, by taking the partial trace of ρ_{AB} over subsystem B, we obtain

$$\mathrm{Tr}_B(\rho_{AB}) = \sum_i \sum_j c_i c_j \underbrace{\langle i_B | j_B\rangle}_{\delta_{ij}} |i_A\rangle\langle j_A| = \sum_i c_i^2 |i_A\rangle\langle i_A| = \rho_A. \tag{4.22}$$

From Eqs. (4.21) and (4.22), we conclude that for a pure joint state given by $\rho_{AB} = \rho_A \otimes \rho_B$, the partial traces ρ_A and ρ_B have the same expansion coefficients c_i^2 and consequently, the same von Neumann entropy, i.e., $S(\rho_A) = S(\rho_B)$.

For a general bipartite system AB, however, the relation between the quantum entropies of the subsystem and composite system is described by the *subadditivity inequality,*

$$S(\rho_{AB}) \leq S(\rho_A) + S(\rho_B), \tag{4.23}$$

with the equality sign satisfied when $\rho_{AB} = \rho_A \otimes \rho_B$. Thus, the quantum entropy of the composite system is bounded from above by the sum of the entropies of the constituent subsystems except when the subsystems are uncorrelated, in which case it is purely additive. The subadditivity inequality can be thought of as the quantum analog of the Shannon inequality for classical sources X and Y,

$$H(X,Y) \leq H(X) + H(Y), \tag{4.24}$$

indicating that the joint entropy of XY with a joint distribution $p(x,y)$ is lower than the sum of the entropy of sources X and Y with marginal distributions $p(x)$ and $p(y)$, respectively. This applies except for the case when the outputs of the two sources are independent of each other, when the equality sign is satisfied. This result, however, should not lead the reader to draw a false equivalence between the classical and quantum worlds. In fact, the difference between them is starkly demonstrated by considering the lower bound on $S(\rho_{AB})$ given by the *Araki–Lieb triangle inequality*:

$$|S(\rho_A) - S(\rho_B)| \leq S(\rho_{AB}), \tag{4.25}$$

with the equality satisfied when ρ_{AB} is a pure state. Let us look at a consequence of this inequality through an example. The bounds of Eqs. (4.23) and (4.25) can be combined as follows:

$$|S(\rho_A) - S(\rho_B)| \leq S(\rho_{AB}) \leq S(\rho_A) + S(\rho_B). \tag{4.26}$$

Example. Here we consider the entangled subsystems of a bipartite system. Entanglement implies that the subsystems have nonlocal quantum correlations. Let us consider one such system whose state is specified by the Bell state:

$$|B_{10}\rangle = 2^{-1/2}(|00\rangle - |11\rangle).$$

The corresponding density operator is given by

$$\rho_{AB} = |B_{10}\rangle\langle B_{10}| = 2^{-1}(|00\rangle - |11\rangle)(\langle 00| - \langle 11|) = \frac{1}{2}\begin{bmatrix} 1 & 0 & 0 & -1 \\ 0 & 0 & 0 & 0 \\ 0 & 0 & 0 & 0 \\ -1 & 0 & 0 & 1 \end{bmatrix}.$$

Clearly, $S(\boldsymbol{\rho}_{AB}) = -\log 1 = 0$. The reduced density matrix for Bob will be

$$\rho_B = \mathrm{Tr}_A(\rho_{AB}) = \frac{1}{2}\begin{bmatrix} 1 & 0 \\ 0 & 1 \end{bmatrix},$$

while the corresponding von Neumann entropy for Bob will be $S(\boldsymbol{\rho}_B) = -\log_2(1/2) = 1$, and similarly for Alice. Thus, both inequalities in Eq. (4.26) are satisfied. This result indicates that the quantum entropy of a composite system with entangled subsystems is *less* than that of either subsystem. Thus, even though the composite system is in a pure state, measurements of the observables of the subsystems result in outcomes with associated uncertainties. This result has considerable application in quantum cryptography and quantum error correction codes. It is important to note that this observation has no classical counterpart. In fact, classical considerations provide exactly the opposite result, $H(X,Y) \geq H(X)$ and $H(X,Y) \geq H(Y)$, implying that the joint Shannon's entropy of the classical composite system is lower bounded by the Shannon's entropy of either subsystem.

To prove the *Araki–Lieb triangle inequality*, let us consider a purifying quantum system C such that the resulting tripartite system ABC is in a pure state. Employing the subadditivity inequality for a composite system, we have

$$S(\rho_{AC}) \leq S(\rho_A) + S(\rho_C), S(\rho_{BC}) \leq S(\rho_B) + S(\rho_C). \tag{4.27}$$

Furthermore, because ABC is in a pure state, using Eq. (4.18), we have

$$S(\rho_{AB}) = S(\rho_C), S(\rho_{AC}) = S(\rho_B), S(\rho_{BC}) = S(\rho_A). \tag{4.28}$$

Using Eqs. (4.27) and (4.28), we obtain

$$S(\rho_A) - S(\rho_B) = \underbrace{S(\rho_{BC})}_{\leq S(\rho_B)+S(\rho_C)} - S(\rho_B) \leq S(\rho_C) = S(\rho_{AB}), \tag{4.29}$$

which proves the Araki–Lieb triangle inequality. The equality is satisfied only when the bipartite system AB is in a pure state.

Before concluding this section, let us state, without proof, a quite important inequality in quantum information theory known as the *strong subadditivity*: For a tripartite system ABC in a state specified by $\boldsymbol{\rho}_{\mathrm{ABC}}$, the following inequalities are satisfied:

$$S(\rho_A) + S(\rho_B) \leq S(\rho_{AC}) + S(\rho_{BC}), \tag{4.30.1}$$

$$S(\rho_{ABC}) + S(\rho_B) \leq S(\rho_{AB}) + S(\rho_{BC}). \tag{4.30.2}$$

Strong subadditivity has important consequences in quantum information theory related to the conditioning of quantum entropy, bounds on quantum mutual information, and the effect of quantum operations on quantum mutual information.

4.3 Holevo information, accessible information, and Holevo bound

Let us go back to the KL distance. By replacing $p(X)$ with $p(X,Y)$ and $q(X)$ with $p(X)q(Y)$, the corresponding classical relative entropy between the joint distribution $p(X,Y)$ and the product of distributions $p(X)$ and $q(X)$ is well known as the *mutual information*:

$$D(p(X,Y)\|p(X)q(Y)) = E_{p(X,Y)}\log\left[\frac{p(X,Y)}{p(X)q(Y)}\right] = \sum_{x\in X}\sum_{y\in Y}p(x,y)\log\left[\frac{p(X,Y)}{p(X)q(Y)}\right] \doteq I(X,Y). \quad (4.31)$$

In this case, the mutual information between random variables X and Y is related to the amount of information that Y has about X on average, namely $I(X;Y)=H(X)-H(X|Y)$. In other words, it represents the amount of uncertainty about X that is resolved given that we know Y.

Let $\{\boldsymbol{\rho}_x\}$ be a set of mutually orthogonal mixed states occurring with probabilities $\{p_x\}$, and let Alice prepare a quantum state given by the density operator

$$\boldsymbol{\rho} = \sum_x p_x \boldsymbol{\rho}_x. \quad (4.32)$$

If all that is available to Bob is $\boldsymbol{\rho}$, the associated uncertainty is given by $S(\boldsymbol{\rho})$. However, if Bob also knows how Alice prepared the state, that is, he knows the probability p_x associated with each mixed state $\boldsymbol{\rho}_x$, the uncertainty is given by $\sum_x p_x S(\boldsymbol{\rho}_x)$. We would intuitively expect that

$$S\left(\sum_x p_x \boldsymbol{\rho}_x\right) \geq \sum_x p_x S(\boldsymbol{\rho}_x), \quad (4.33)$$

because ignorance about how $\boldsymbol{\rho}$ was prepared is reduced in the latter case. This inequality represents the special case of the *concavity property* of the von Neumann entropy given by Eq. (4.8). The difference between the two represents the average reduction in quantum entropy given that we know how $\boldsymbol{\rho}$ was prepared, namely $\boldsymbol{\rho} = \Sigma_x p_x \boldsymbol{\rho}_x$, and is commonly referred to as the *Holevo information*:

$$\chi = S(\boldsymbol{\rho}) - \sum_x p_x S(\boldsymbol{\rho}_x). \quad (4.34)$$

Let us now consider the ensemble $\{\boldsymbol{\rho}_x, p_x\}$ of mutually orthogonal mixed states, for which $\mathrm{Tr}(\boldsymbol{\rho}_{x_1}\boldsymbol{\rho}_{x_2}) = 0, x_1 \neq x_2$. By choosing the basis for which $\boldsymbol{\rho} = \sum_x p_x \boldsymbol{\rho}_x$ is block-diagonal, we have

$$S(\boldsymbol{\rho}) = -\mathrm{Tr}(\boldsymbol{\rho}\log\boldsymbol{\rho}) = -\sum_x \mathrm{Tr}(p_x\boldsymbol{\rho}_x)\log(p_x\boldsymbol{\rho}_x) = -\sum_x p_x\left[\underbrace{\mathrm{Tr}(\boldsymbol{\rho}_x)}_{=1}\log(p_x) + \mathrm{Tr}\boldsymbol{\rho}_x\log(\boldsymbol{\rho}_x)\right]$$

$$= \underbrace{-\sum_x p_x\log(p_x)}_{H(X)} + \sum_x p_x \underbrace{[-\mathrm{Tr}\boldsymbol{\rho}_x\log(\boldsymbol{\rho}_x)]}_{S(\boldsymbol{\rho}_x)} = H(X) + \sum_x p_x S(\boldsymbol{\rho}_x). \quad (4.35)$$

By rearrangement, we obtain

$$\underbrace{S(\rho) - \sum_x p_x S(\rho_x)}_{\chi} = H(X) \tag{4.36}$$

indicating that as long as ρ_x have mutually orthogonal support, the Holevo information equals the Shannon entropy of a classical source X with a probability distribution given by $P(X = x) = p_x$. In general, however, when the ensemble of mixed states does not satisfy the mutually orthogonal condition, Shannon's entropy upper bounds the Holevo information. To summarize, the Holevo information is upper bounded by Shannon's entropy; that is,

$$\chi \leq H(X). \tag{4.37}$$

Now suppose that Alice wants to send classical information to Bob over a quantum channel. Alice has a classical source X that outputs the character x drawn from an alphabet X with probability p_x. For each x, Alice prepares the state ρ_x and sends it over the channel. For the sake of simplicity, we assume that the quantum operation of the channel is characterized by an identity operator. When Bob receives ρ_x, he performs the generalized POVM M_y, resulting in outcome y with probability

$$p(y|x) = \text{Tr}\left(M_y \rho_x\right). \tag{4.38}$$

Bob's goal in performing this measurement is to maximize the mutual information $I(X,Y)$. This maximized mutual information is the best that Bob can hope for and represents the information accessible to Bob. Since Bob can perform the optimization over all possible measurement strategies to maximize the mutual information, the *accessible information* is defined as

$$H(X:Y) = \max_{M_y} I(X;Y) \tag{4.39}$$

If the quantum states are pure and mutually orthogonal, instead of POVM, we consider projective measurements such that $p(y|x) = \text{Tr}(M_y \rho_x) = \text{Tr}(P_y \rho_x) = 1$ if $x = y$ and zero otherwise. In that case, $H(X:Y) = H(X)$, and Bob (receiver) can accurately estimate the information sent by Alice (transmitter). If the states are nonorthogonal, the accessible information is bounded by

$$H(X:Y) \leq S(\rho) \leq H(X), \rho = \sum_x p_x \rho_x, \tag{4.40}$$

and this result is a consequence of the nonorthogonal states not being completely distinguishable. An important interpretation of Eq. (4.40) arises from the no-cloning theorem. To see this, suppose two nonorthogonal states $|\psi_1\rangle$ and $|\psi_2\rangle$ can be cloned. We can then perform repeated cloning of the states to obtain the states $|\psi_1\rangle \otimes \ldots \otimes |\psi_1\rangle$ and $|\psi_2\rangle \otimes \ldots \otimes |\psi_2\rangle$. As the number of clones tends to infinity, these new states start to become orthogonal, so Bob can unambiguously determine them with the result that $H(X:Y) = \text{H}(X)$. Since arbitrary quantum states cannot be cloned, we reach the desired result of Eq. (4.40).

When ρ_x states are mixed states, the accessible information is bounded by the Holevo information:

$$H(X:Y) \leq \chi, \tag{4.41}$$

and this generalized result is known as the *Holevo bound*. The equality holds when the mixed states have mutually orthogonal support. Since the Holevo information is upper bounded by Shannon's entropy, we can write

$$H(X:Y) \le \chi \le H(X), \tag{4.42}$$

implying that apart from the case when $\{\rho_x\}$ have mutually orthogonal support, Bob has no possibility of completely recovering the classical information, characterized by $H(X)$, that Alice sent him over the quantum channel!

Example. Let us consider two orthogonal pure states $|0\rangle$ and $|1\rangle$ with the corresponding density operators

$$\rho_0 = \begin{bmatrix} 1 & 0 \\ 0 & 0 \end{bmatrix}, \rho_1 = \begin{bmatrix} 0 & 0 \\ 0 & 1 \end{bmatrix}.$$

Clearly, $S(\rho_0) = S(\rho_1) = -\log 1 = 0$. Furthermore, let us assume that Alice selects each state with probability $\frac{1}{2}$, resulting in

$$\rho = \frac{1}{2}\begin{bmatrix} 1 & 0 \\ 0 & 1 \end{bmatrix}.$$

As a result, $S(\rho) = -2\times(1/2)\log_2(1/2) = 1$, while $\sum_x p_x S(\rho_x) = 0.5S(\rho_0) + 0.5S(\rho_1) = 0$. Consequently, $\chi = 1-0 = 1$, and as expected, equals the Shannon entropy $H(X) = -2\times(1/2)\log_2(1/2) = 1$. Now let us suppose we still have the same pure orthogonal states, but Alice is uncertain whether a given preparation is $|0\rangle$ and $|1\rangle$, resulting in the mixed density operators

$$\rho_0 = \rho_1 = \frac{1}{2}\begin{bmatrix} 1 & 0 \\ 0 & 1 \end{bmatrix}.$$

As in the previous case, Alice selects each with probability 1/2, resulting in

$$\rho = \frac{1}{2}\begin{bmatrix} 1 & 0 \\ 0 & 1 \end{bmatrix}.$$

and

$$\chi = S(\rho) - \sum_x p_x S(\rho_x) = 1 - 1 = 0.$$

Thus, the Holevo information is zero, and Bob has no hope of recovering any of the information sent to him by Alice. Note that although ρ_0 and ρ_1 are mixed states, they are not mutually orthogonal. In fact, they are identical, and therefore, Bob's measurement outcomes are entirely uncertain.

4.4 Data compression and Schumacher's noiseless quantum coding theorem

Having studied the basic building blocks of quantum information theory, we now turn to the quantum version of the first central issue in information theory: To what extent can a quantum message be compressed without significantly affecting the quantum message fidelity? The answer to this question, given by Schumacher's quantum noiseless channel coding theorem, is the topic of this section. Schumacher's theorem is the quantum analog of Shannon's noiseless channel coding theorem that establishes the conditions under which data compression of classical information can be reliably performed. Therefore, we begin with a discussion and outline of its main ideas. This discussion forms the basis for our development of Schumacher's theorem and quantum data compression.

4.4.1 Shannon's noiseless source coding theorem

Consider a classical source X that generates symbols 0 and 1 with probabilities p and $1-p$, respectively. The probability of output sequence $x_1, \ldots, x_N$ is given by

$$p(x_1,\ldots,x_N) = p^{\sum_i x_i}(1-p)^{\sum_i (1-x_i)} \xrightarrow[N \to \infty]{} p^{Np}(1-p)^{N(1-p)} \tag{4.43}$$

By taking the logarithm of this probability, we obtain

$$\log p(x_1,\ldots,x_N) \approx Np \log p + N(1-p)\log(1-p) = -NH(X) \tag{4.44}$$

Therefore, the probability of occurrence of the so-called *typical sequence* is $p(x_1, \ldots, x_N) \approx 2^{-NH(X)}$, and on average, $NH(X)$ bits are needed to represent any typical sequence, as illustrated in Fig. 4.2A. We denote the typical set by $T(N,\varepsilon)$. Clearly, not much information will be lost if we consider $2^{NH(X)}$ typical sequences instead of 2^N possible sequences, particularly for a sufficiently large N. This observation can be used in *data compression*. Namely, with $NH(X)$ bits, we can enumerate all typical sequences. If the source output is a typical sequence, it will be represented by $NH(X)$ bits. On the other hand, when an atypical

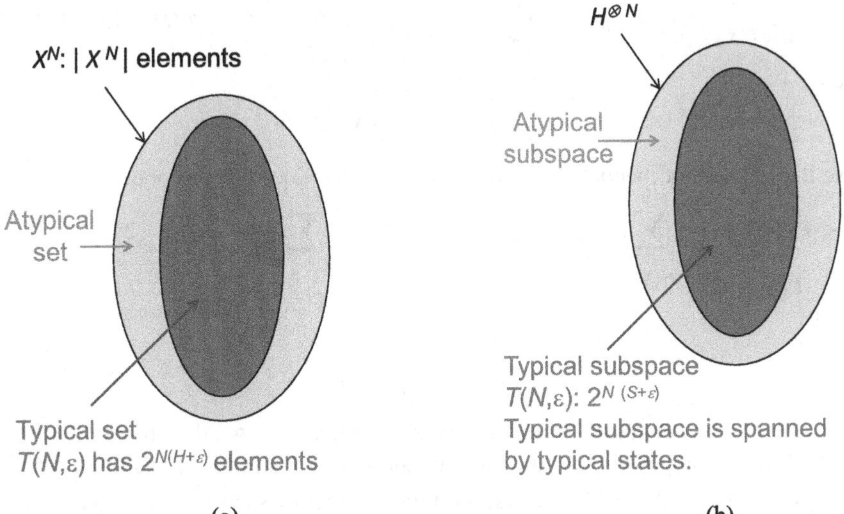

FIGURE 4.2

Typical set (A) and typical subspace (B).

sequence is generated by the source, we assign it a fixed index, resulting in compression loss. However, as $N \to \infty$, the probability of occurrence of an atypical sequence will tend to zero, resulting in an arbitrarily small information loss.

Let us now extend this concept to nonbinary sources. The law of large numbers applied to a nonbinary source with i.i.d. outputs $X_1, \ldots, X_N$ claims that as $N \to \infty$, the expected value of the source approaches the true value $E(X)$ in a probability sense. In other words,

$$P\left(\left|\frac{1}{N}\sum_i X_i - E(X)\right| \leq \varepsilon\right) > 1 - \delta \tag{4.45}$$

for sufficiently large N and ε, $\delta > 0$. By replacing the random variable X_i with the function of the random variable, namely $-\log P(X_i)$, we obtain

$$-N[H(X) + \varepsilon] \leq \log P(x_1, \ldots, x_N) \leq -N[H(X) - \varepsilon]. \tag{4.46}$$

The sequences $\{x_1, x_2, \ldots, x_N\}$ that satisfy Eq. (4.46) are typical sequences, and the set of these typical sequences is the typical set $T(N,\varepsilon)$. Thus, as a consequence of Eq. (4.45), for sufficiently large N, we have

$$P(T(N, \varepsilon)) > 1 - \delta. \tag{4.47}$$

Since $P[T(N,\varepsilon)] > 1 - \delta$ and given by Eq. (4.46), the size of the typical set, denoted $|T(N,\varepsilon)|$, is bounded by

$$(1 - \delta)2^{N[H(X) - \varepsilon]} \leq |T(N, \varepsilon)| \leq 2^{N[H(X) + \varepsilon]} \tag{4.48}$$

To prove the right side of the inequality, we note that the sum of the probabilities of all typical sequences cannot be more than 1, so the following is valid:

$$1 \geq \sum_{(x_1, \cdots, x_n) \in T(N,\varepsilon)} P(x_1, \cdots, x_n) \geq \sum_{(x_1, \cdots, x_n) \in T(N,\varepsilon)} 2^{-N[H(X)+\varepsilon]} = |T(N, \varepsilon)| 2^{-N[H(X)+\varepsilon]}. \tag{4.49}$$

resulting in

$$|T(N, \varepsilon)| \leq 2^{N[H(X)+\varepsilon]} \tag{4.50}$$

To prove the left side of inequality Eq. (4.48), we invoke Eq. (4.47) to obtain

$$\begin{aligned} 1 - \delta < P(T(N, \varepsilon)) &\leq \sum_{(x_1, \cdots, x_n) \in T(N,\varepsilon)} P(x_1, \cdots, x_n) \leq \sum_{(x_1, \cdots, x_n) \in T(N,\varepsilon)} 2^{-N[H(X)-\varepsilon]} \\ &= |T(N, \varepsilon)| 2^{-N[H(X)-\varepsilon]}, \end{aligned} \tag{4.51}$$

resulting in

$$|T(N, \varepsilon)| > (1 - \delta)2^{N[H(X)-\varepsilon]}. \tag{4.52}$$

Therefore, we obtain a couple of powerful results. First, as $N \to \infty$, the probability that an output sequence will be typical approaches 1, and the probability of occurrence of any one particular typical sequence is approximately $2^{-NH(X)}$. Second, the total number of such typical sequences, that is, the size of the typical set, is approximately $2^{NH(X)}$.

Now suppose that the compression rate R is larger than $H(X)$, say $R > H(X)+\varepsilon$. Then based on the discussion above, the total number of typical sequences for a sufficiently large N is bounded by

$$|T(N,\varepsilon)| \leq 2^{N[H(X)+\varepsilon]} \leq 2^{NR}, \tag{4.53}$$

and the probability of occurrence of the typical sequence is lower bounded by $1-\delta$, where δ is arbitrarily small.

The source encoding strategy can be described as follows. Let us divide the set of all sequences generated by the source into typical and atypical sets, as shown in Fig. 4.2A. The equation above clarifies that sequences in the typical set can be represented by NR bits at most. If an atypical sequence occurs, we assign it a fixed index. As N tends to infinity, the probability for this event to occur tends to zero, indicating that we can encode and decode typical sequences as reliable.

On the other hand, when $R < H(X)$, we can encode a maximum of 2^{NR} typical sequences, labeled as $T_R(N,\varepsilon)$. Since the probability of occurrence of a typical sequence is upper-bounded by $2^{-N[H(X)-\varepsilon]}$, the typical sequences in $T_R(N,\varepsilon)$ will occur with probability $2^{N[R-H(X)+\varepsilon]}$, which tends to zero as N tends to infinity, indicating that a reliable compression scheme does not exist in this case. With this, we have just proved Shannon's source coding theorem, which can be formulated as follows:

Shannon's Source Coding Theorem. For an i.i.d. source X, a reliable compression method exists for rate $R > H(X)$. If, on the other hand, rate $R < H(X)$, no reliable compression scheme exists.

4.4.2 Schumacher's noiseless quantum source coding theorem

Now consider a quantum source emitting a pure state $|\psi_x\rangle$ with probability p_x described in terms of the mixed density operator:

$$\rho = \sum_x p_x |\psi_x\rangle\langle\psi_x|. \tag{4.54}$$

The quantum message comprises N quantum source outputs, independent of each other, so

$$\rho_{\otimes N} = \rho \otimes \cdots \otimes \rho. \tag{4.55}$$

Schumacher's theorem gives us the condition under which there exists a reliable compression scheme that can compress and decompress the quantum message with high *fidelity*. The development of Schumacher's noiseless quantum channel coding theorem closely parallels Shannon's classical version.

For this purpose, let us express the mixed density operator in terms of its orthonormal eigenkets as

$$\rho = \sum_\alpha \lambda_\alpha |\lambda_\alpha\rangle\langle\lambda_\alpha|. \tag{4.56}$$

Then the quantum source can effectively be thought of as a classical source, and we say that the sequence $\{\lambda_1 \ldots \lambda_N\}$ is the typical sequence if it satisfies the following inequality for a sufficiently large N:

$$P\left(\left|\frac{1}{N}\log\left(\frac{1}{P(\lambda_1)P(\lambda_2)\cdots P(\lambda_N)}\right) - S(\rho)\right| \leq \varepsilon\right) > 1 - \delta, \tag{4.57}$$

and this inequality is a direct consequence of the law of large numbers in a manner identical to Eqs. (4.45) and (4.46). We can now define the *typical state* as a state $|\lambda_1\rangle \ldots |\lambda_N\rangle$ for which the eigenvalue sequence $\boldsymbol{\lambda} = \{\lambda_1, \ldots, \lambda_N\}$ is a typical sequence. Similarly to the concept of the set of typical sequences—the typical set—for a classical information source, we introduce the notion of a typical subspace spanned by the typical basis states $|\lambda_1\rangle \ldots |\lambda_N\rangle$. We can then define the projection operator onto the typical subspace as

$$\boldsymbol{P}_T = \sum\nolimits_{\lambda} |\lambda_1\rangle\langle\lambda_1| \otimes \cdots \otimes |\lambda_N\rangle\langle\lambda_N|. \tag{4.58}$$

The *projection* of the quantum message on the typical subspace is determined by $\boldsymbol{P}_T\boldsymbol{\rho}_{\otimes N}$. The probability of the projection is given by its trace. Using the projection operator, we can separate the total Hilbert space into typical and atypical subspaces, as illustrated in Fig. 4.2B. The probability that the state of the quantum message lies in the typical subspace is given by

$$P(\boldsymbol{P}_T\boldsymbol{\rho}_{\otimes N}) > 1 - \delta. \tag{4.59}$$

Earlier, we showed that for a sufficiently large N, the probability of the typical set approaches unity. Analogously, we find that the probability that the quantum message lies in the typical subspace approaches unity for a sufficiently large N. In the last section, we were able to bound the size of the typical set. Following a similar procedure, we can bound the dimensionality of the typical subspace, characterized by $\mathrm{Tr}(\boldsymbol{P}_T)$. Denoting the typical subspace by $T_\lambda(N,\varepsilon)$, the bounds are given by

$$(1-\delta)2^{N[S(\rho)-\varepsilon]} \le |T_\lambda(N,\varepsilon)| \le 2^{N[S(\rho)+\varepsilon]}. \tag{4.60}$$

Analogous to the classical development of Shannon's noiseless coding theorem, we have a couple of powerful results. First, as $N \to \infty$, the probability that the state of the quantum message will lie in the typical subspace approaches unity, and the dimensionality of the typical subspace approximates $2^{NS(\rho)}$, with the probability of the quantum message being in any one typical state approaching $2^{-NS(\rho)}$.

We are now in a position to formulate the corresponding *compression procedure*. Define the projector on the typical subspace $\boldsymbol{P}_T$ and its complement projecting on orthogonal subspace $\boldsymbol{P}_T^{\perp} = \boldsymbol{I} - \boldsymbol{P}_T$ ($\boldsymbol{I}$ is the identity operator) with corresponding outcomes 0 and 1. For compression purposes, we perform the measurements using two orthogonal operators with the outputs denoted 1 and 0, respectively. If the outcome is 1, we know that the message is in a typical state, and we do nothing further. If, on the other hand, the outcome is 0, we know that it belongs to an atypical subspace, and we numerate it as a *fixed* state from the typical subspace. Given that the probability of this happening can be made as small as possible for a large N, we can compress the quantum message without loss of information.

Schumacher showed that when rate $R > S(|\psi_x\rangle)$, the above-described compression scheme is reliable. Reliability implies that the compressed quantum message can be decompressed with high fidelity. In other words, we say that the *compression is reliable* if the corresponding *entanglement fidelity* tends to 1 for a large N:

$$F_e(\boldsymbol{\rho}_{\otimes N}; D \circ C) \xrightarrow[N \to \infty]{} 1; \boldsymbol{\rho}_{\otimes N} = |\psi_{\otimes N}\rangle\langle\psi_{\otimes N}|, |\psi_{\otimes N}\rangle = |\psi_{x_1}\rangle \otimes \ldots \otimes |\psi_{x_N}\rangle, \tag{4.61}$$

where we use D and C to denote the decompression and compression operations, respectively, defined as the following mappings:

$$D: H_c^N \to H^N, \quad C: H^N \to H_c^N, \tag{4.62}$$

with H_c^N being the 2^{NR}-dimensional subspace of H^N.

Let us reconsider our i.i.d. quantum source defined by the ensemble $\{|\psi_x\rangle, p_x\}$, where $|\psi_x\rangle$ are pure states. Given this ensemble, the kth instance of a length N quantum message is given by

$$|\psi_{\otimes N}^k\rangle = |\psi_{x_1}^k\rangle \otimes \cdots \otimes |\psi_{x_N}^k\rangle, \tag{4.63}$$

and the corresponding density operator is $\rho_{\otimes N}^k = |\psi_{\otimes N}^k\rangle\langle\psi_{\otimes N}^k|$. Alice encodes the message by performing a fuzzy measurement whose outcome will be 0_{typical} with probability $\text{Tr}(P_T\rho_{\otimes N}^k) > 1-\delta$ and 1_{atypical} with probability $\text{Tr}(P_T^\perp\rho_{\otimes N}^k) \le \delta$. As mentioned above, the atypical state is coded as some fixed state $|\varphi\rangle$ in the typical subspace, with

$$E = \sum_j \underbrace{|\varphi\rangle\langle j|}_{E_j} = \sum_j E_j, \; E_j = |\varphi\rangle\langle j|, \tag{4.64}$$

where $|j\rangle$ is the orthonormal basis of the atypical subspace. The compressed state is therefore given by

$$\rho_C = P_T\rho_{\otimes N}^k P_T + \sum_j E_j \rho_{\otimes N}^k E_j^\dagger. \tag{4.65}$$

By substituting Eq. (4.64) into Eq. (4.65), we obtain

$$\begin{aligned} \rho_C &= P_T\rho_{\otimes N}^k P_T + \sum_j |\varphi\rangle\langle j|\psi_{\otimes N}^k\rangle\langle\psi_{\otimes N}^k|j\rangle\langle\varphi| = P_T\rho_{\otimes N}^k P_T + |\varphi\rangle\langle\varphi|\sum_j \langle\psi_{\otimes N}^k|j\rangle\langle j|\psi_{\otimes N}^k\rangle \\ &= P_T\rho_{\otimes N}^k P_T + |\varphi\rangle\langle\varphi|\langle\psi_{\otimes N}^k|P_T^\perp|\psi_{\otimes N}^k\rangle. \end{aligned} \tag{4.66}$$

The decoding operator for the compressed Hilbert space of the typical subspace is the identity operator. Therefore, the decoding operation results in the state $\rho_{D\circ C}$ given by Eq. (4.66). The question we need to answer is what the fidelity of the decoding is when N is sufficiently large. For the kth quantum message we have considered, the fidelity $F_k = \langle\psi_{\otimes N}^k|\rho_{D\circ C}|\psi_{\otimes N}^k\rangle$ equals unity if $\rho_{D\circ C} = \rho_{\otimes N}^k$. Thus, fidelity is a measure of how close the decompressed quantum message is to the original quantum message. Because the quantum state of the source output is derived from an ensemble of quantum messages, we compute the average fidelity as

$$F = \sum_k p_k F_k = \sum_k p_k \langle\psi_{\otimes N}^k|\rho_{D\circ C}|\psi_{\otimes N}^k\rangle, \; p_k = P(\rho_{\otimes N}^k). \tag{4.67}$$

By substituting Eq. (4.66) into Eq. (4.67), we obtain

$$\begin{aligned} F &= \sum_k p_k F_k = \sum_k p_k \langle\psi_{\otimes N}^k|P_T\rho_{\otimes N}^k P_T|\psi_{\otimes N}^k\rangle + \underbrace{\sum_k p_k \langle\psi_{\otimes N}^k|\varphi\rangle\langle\varphi|\langle\psi_{\otimes N}^k|P_T^\perp|\psi_{\otimes N}^k\rangle|\psi_{\otimes N}^k\rangle}_{F_{\text{atypical}}} \\ &= \sum_k p_k \langle\psi_{\otimes N}^k|P_T|\psi_{\otimes N}^k\rangle\langle\psi_{\otimes N}^k|P_T|\psi_{\otimes N}^k\rangle + F_{\text{atypical}} = \sum_k p_k \underbrace{\text{Tr}(P_T\rho_{\otimes N}^k)\text{Tr}(P_T\rho_{\otimes N}^k)}_{\text{Tr}(P_T\rho_{\otimes N}^k)^2} + F_{\text{atypical}} \\ &\ge \sum_k p_k \text{Tr}(P_T\rho_{\otimes N}^k)^2 = \text{Tr}(P_T\rho_{\otimes N})^2. \end{aligned} \tag{4.68}$$

Given Eq. (4.59), from Eq. (4.68), we obtain the following inequality:

$$F > (1-\delta)^2 \geq 1-2\delta. \tag{4.69}$$

Thus, we have shown that we can compress the quantum message to $N(S(\boldsymbol{\rho})+\varepsilon)$ qubits with an arbitrarily small loss of fidelity for a sufficiently large N.

What happens if the quantum message is encoded to $R < S(\boldsymbol{\rho})$ qubits? In this case, the dimension of the Hilbert subspace Ω in which the compressed quantum message lies is 2^{NR}. Let the projection operator onto this subspace be V, and let the decompressed state be $\boldsymbol{\rho}_{D\circ C}$. This decompressed state lies in Ω and can consequently be expanded in the basis of this subspace as follows:

$$\boldsymbol{\rho}_{D\circ C} = \sum_i |v_i\rangle\langle v_i|\boldsymbol{\rho}_{D\circ C}|v_i\rangle\langle v_i|. \tag{4.70}$$

The average fidelity of reconstruction can now be calculated as

$$F = \sum_k p_k F_k = \sum_k p_k \langle \psi^k_{\otimes N}|\boldsymbol{\rho}_{D\circ C}|\psi^k_{\otimes N}\rangle = \sum_{k,i} p_k \langle \psi^k_{\otimes N}||v_i\rangle \underbrace{\langle v_i|\boldsymbol{\rho}_{D\circ C}|v_i\rangle}_{\geq 0} \langle v_i||\psi^k_{\otimes N}\rangle$$
$$\geq \sum_{k,i} p_k \langle \psi^k_{\otimes N}||v_i\rangle\langle v_i||\psi^k_{\otimes N}\rangle = \sum_k p_k \mathrm{Tr}\left(V\boldsymbol{\rho}^k_{\otimes N}\right). \tag{4.71}$$

Following a discussion similar to that for Shannon's noiseless coding theorem, the probability that a typical state lies in Ω is $2^{N[R-S(\rho)+\varepsilon]}$. For $R < S(\boldsymbol{\rho})$, as $N \to \infty$, this probability approaches 0. Thus, the fidelity of a quantum message reconstructed after being compressed to $R < S(\boldsymbol{\rho})$ qubits is not good at all. Therefore, we have the following statement for Schumacher's quantum noiseless coding theorem:

Schumacher's Source Coding Theorem. Let $\{|\psi_x\rangle, p_x\}$ be an i.i.d. quantum source. If $R > S(\rho)$, a reliable compression scheme of rate R exists for this source. Otherwise, if $R < S(\rho)$, no reliable compression scheme of rate R exists.

Example [1]. Let us consider a quantum source that produces state $|\psi_1\rangle$ and $|\psi_2\rangle$ with probabilities p_1 and p_2, respectively. Let us define the states as

$$|\psi_1\rangle = \frac{1}{\sqrt{2}}(|0\rangle + |1\rangle), \quad |\psi_2\rangle = \frac{\sqrt{3}}{2}|0\rangle + \frac{1}{2}|1\rangle,$$

with $p_1 = p_2 = 0.5$. Thus, the resulting density matrix is given by

$$\boldsymbol{\rho} = p_1|\psi_1\rangle\langle\psi_1| + p_2|\psi_2\rangle\langle\psi_2| = \begin{pmatrix} 0.6250 & 0.4665 \\ 0.4665 & 0.3750 \end{pmatrix}.$$

The eigendecomposition of $\boldsymbol{\rho}$ is

$$\boldsymbol{\rho} = \begin{pmatrix} -0.7934 & -0.6088 \\ -0.6088 & 0.7934 \end{pmatrix}\begin{pmatrix} 0.9830 & 0 \\ 0 & 0.0170 \end{pmatrix}\begin{pmatrix} -0.7934 & -0.6088 \\ -0.6088 & 0.7934 \end{pmatrix}^\dagger,$$

and therefore, the eigenvectors are

$$|\lambda_1\rangle = \begin{pmatrix} -0.7934 \\ -0.6088 \end{pmatrix}, \ |\lambda_2\rangle = \begin{pmatrix} -0.6088 \\ 0.7934 \end{pmatrix},$$

with the corresponding eigenvalues given by $\lambda_1 = 0.9830$ and $\lambda_2 = 0.0170$. With this setup in mind, let us consider the compression of a three-qubit quantum message. With our source as defined above, the possible quantum messages are

$$|\psi_{111}\rangle = |\psi_1\rangle \otimes |\psi_1\rangle \otimes |\psi_1\rangle, |\psi_{112}\rangle = |\psi_1\rangle \otimes |\psi_1\rangle \otimes |\psi_2\rangle,$$
$$|\psi_{121}\rangle = |\psi_1\rangle \otimes |\psi_2\rangle \otimes |\psi_1\rangle, \ |\psi_{122}\rangle = |\psi_1\rangle \otimes |\psi_2\rangle \otimes |\psi_2\rangle,$$
$$|\psi_{211}\rangle = |\psi_2\rangle \otimes |\psi_1\rangle \otimes |\psi_1\rangle, \ |\psi_{212}\rangle = |\psi_2\rangle \otimes |\psi_1\rangle \otimes |\psi_2\rangle,$$
$$|\psi_{221}\rangle = |\psi_2\rangle \otimes |\psi_2\rangle \otimes |\psi_1\rangle, \ |\psi_{222}\rangle = |\psi_2\rangle \otimes |\psi_2\rangle \otimes |\psi_2\rangle,$$

Let us also define the eigenvector-based basis states:

$$|\lambda_{111}\rangle = |\lambda_1\rangle \otimes |\lambda_1\rangle \otimes |\lambda_1\rangle, |\lambda_{112}\rangle = |\lambda_1\rangle \otimes |\lambda_1\rangle \otimes |\lambda_2\rangle,$$
$$|\lambda_{121}\rangle = |\lambda_1\rangle \otimes |\lambda_2\rangle \otimes |\lambda_1\rangle, |\lambda_{122}\rangle = |\lambda_1\rangle \otimes |\lambda_2\rangle \otimes |\lambda_2\rangle,$$
$$|\lambda_{211}\rangle = |\lambda_2\rangle \otimes |\lambda_1\rangle \otimes |\lambda_1\rangle, |\lambda_{212}\rangle = |\lambda_2\rangle \otimes |\lambda_1\rangle \otimes |\lambda_2\rangle,$$
$$|\lambda_{221}\rangle = |\lambda_2\rangle \otimes |\lambda_2\rangle \otimes |\lambda_1\rangle, |\lambda_{222}\rangle = |\lambda_2\rangle \otimes |\lambda_2\rangle \otimes |\lambda_2\rangle,$$

These states can be used to define the projective operators onto the Hilbert space $\mathbb{H}^{\otimes 3}$ spanned by $|\lambda_{ijk}\rangle$, $i, j = 1, 2$. Using these measurement states, let us see if a subspace of the Hilbert space exists where the quantum messages lie. In other words, can we find the typical subspace? To do this, we first note that

$$\mathrm{Tr}\big((|\psi_i\rangle\langle\psi_i|)(|\lambda_j\rangle\langle\lambda_j|)\big) = |\langle\psi_i||\lambda_j\rangle|^2 = \lambda_j; i, j = 1, 2$$

Furthermore,

$$\mathrm{Tr}\Big((|\psi_{ijk}\rangle\langle\psi_{ijk}|)\left(\left|\lambda_{i'j'k'}\right\rangle\left\langle\lambda_{i'j'k'}\right|\right)\Big) = |\langle\psi_i||\lambda_{i'}\rangle|^2 |\langle\psi_j||\lambda_{j'}\rangle|^2 |\langle\psi_k||\lambda_{k'}\rangle|^2 = \lambda_{i'}\lambda_{j'}\lambda_{k'},$$

where $i, j, k = 1, 2; i', j', k' = 1, 2$. We now compute $\lambda_{i'}\lambda_{j'}\lambda_{k'}$:

$$\lambda_1\lambda_1\lambda_1 = 0.9499, \lambda_1\lambda_1\lambda_2 = 0.0164, \lambda_1\lambda_2\lambda_1 = 0.0164, \lambda_1\lambda_2\lambda_2 = 2.8409E(-4),$$
$$\lambda_2\lambda_1\lambda_1 = 0.0164, \lambda_2\lambda_1\lambda_2 = 2.8409E(-4), \lambda_2\lambda_2\lambda_1 = 2.8409E(-4), \lambda_2\lambda_2\lambda_2 = 4.9130E(-6),$$

From these values, we can see that the probability that the quantum message lies in the subspace spanned by $|\lambda_{111}\rangle, |\lambda_{112}\rangle, |\lambda_{121}\rangle, |\lambda_{211}\rangle$ is $0.9499 + 0.0164 + 0.0164 + 0.0164 = 0.9991$, while the probability that the quantum message lies in the subspace spanned by $|\lambda_{122}\rangle, |\lambda_{212}\rangle, |\lambda_{221}\rangle, |\lambda_{212}\rangle$ is $2.8409\mathrm{E}(-4) + 2.8409\mathrm{E}(-4) + 2.8409\mathrm{E}(-4) + 4.9130\mathrm{E}(-6) = 8.5717\mathrm{E}(-4)$. Thus, with a probability close to 1, the quantum message will lie in the subspace spanned by the first set of the quantum state, whereas the probability δ of the quantum message lying in the subspace spanned by the second set of quantum states lies close to 0. This becomes even more apparent if n, which currently equals 3, increases. Thus, $|\lambda_{111}\rangle, |\lambda_{112}\rangle, |\lambda_{121}\rangle, |\lambda_{211}\rangle$ span the typical subspace, while $|\lambda_{122}\rangle, |\lambda_{212}\rangle, |\lambda_{221}\rangle, |\lambda_{212}\rangle$ span the complement subspace.

We can now perform data compression using unitary encoding. Unitary encoding implies that a unitary operator $\boldsymbol{U}$ acts on the Hilbert space such that any quantum message state $|\psi_{ijk}\rangle$ that lies in the typical subspace goes to $|\xi_1\rangle \otimes |\xi_2\rangle \otimes |fixed_{\text{typical}}\rangle$, whereas the message that lies in the atypical subspace goes to $|\xi_1\rangle \otimes |\xi_2\rangle \otimes |fixed_{\text{atypical}}\rangle$. The unitary operator can be chosen such that, for example, $|fixed_{\text{typical}}\rangle = |0\rangle$ and $|fixed_{\text{atypical}}\rangle = |1\rangle$. Thus, to encode the message, Alice measures the third qubit. If the output is 0, the quantum message is in the typical subspace, and Alice sends $|\xi_1\rangle \otimes |\xi_2\rangle$ to Bob. If the output is 1, the quantum message is in the atypical subspace. In this case, Alice assigns a fixed quantum state $|\varphi\rangle$ in the typical subspace whose unitary transformation is the state $|a\rangle \otimes |b\rangle \otimes |0\rangle$ Alice then sends $|a\rangle \otimes |b\rangle$ to Bob.

To decompress the message, Bob first appends $|0\rangle$ to the quantum state he receives and applies the transform $\boldsymbol{U}^{-1}$. The possible resulting states are $|\psi_{ijk}\rangle$ or $|\varphi\rangle$. Thus, Bob can decompress the message as desired. For the cases where we obtain $|\varphi\rangle$, we lose information. However, as shown above, the probability of this happening is very small and can be made smaller with a larger n.

Let us also look at the fidelity of the retrieved message. For a given quantum message m, the fidelity is given by

$$F_m = \langle \psi^m_{\otimes n} | \boldsymbol{P}_T \rho^m_{\otimes n} \boldsymbol{P}_T | \psi^m_{\otimes n} \rangle + \text{atypical contribution.}$$

Here, $\boldsymbol{P}_T$ is the projection onto the typical subspace. For our example, for any quantum message, fidelity reduces to

$$F = (1-\delta)^2 + \delta |\langle \varphi | \psi_{ijk}|^2$$

We have already computed $\delta = 8.5717\text{E}(-4)$, and this gives us fidelity greater than $(1-\delta)^2 = 0.9983$. We have thus achieved a rate of $R = \frac{2}{3} = 0.667$ with high reconstruction fidelity. Let us compare this rate with the quantum entropy of the source. The quantum entropy is determined to be

$$S(\boldsymbol{\rho}) = -\lambda_1 \log \lambda_1 - \lambda_2 \log \lambda_2 = 0.1242.$$

Therefore, we have $R > S$, thus satisfying Schumacher's condition. The equation above hints at another interesting fact: we can do better than two qubits and go down to a single qubit. For this case, $R = \frac{1}{3} = 0.333$ is still greater than $S(\boldsymbol{\rho})$. We note that quantum entropy here represents the incompressible information in the message generated by a quantum source similarly to Shannon's entropy, representing the incompressible information of a classical source.

4.5 Quantum channels

The quantum channel performs the transformation (mapping) of the initial density operator ρ to the final density operator ρ_f, denoted $\xi : \rho \to \rho_f$, which is known as the *superoperator* or *quantum operation*. The final density operator can be expressed in *operator-sum representation* as follows:

$$\rho_f = \sum_k E_k \rho E_k^\dagger, \tag{4.72}$$

where E_k are the operation elements for the superoperator. To describe the actions of the quantum channel, we must determine the Kraus operators E_k.

The operator-sum representation can be used to classify the quantum channels into two categories: (1) *trace-preserving* when $\mathrm{Tr}\xi(\rho) = \mathrm{Tr}\ \rho = 1$ and (2) *non-trace-preserving* when $\mathrm{Tr}\xi(\rho)<1$. Starting from the trace-preserving condition,

$$\mathrm{Tr}\rho = \mathrm{Tr}\xi(\rho) = \mathrm{Tr}\left[\sum_k E_k \rho E_k^\dagger\right] = \mathrm{Tr}\left[\rho \sum_k E_k E_k^\dagger\right] = 1, \tag{4.73}$$

and we obtain

$$\sum_k E_k E_k^\dagger = I. \tag{4.74}$$

The corresponding proof was provided in Chapter 3. For a non-trace-preserving quantum operation, Eq. (4.74) is not satisfied, and we can informally write $\sum_k E_k E_k^\dagger < I$.

As an illustration, let us consider *bit-flip* and *phase-flip* channels, which are illustrated in Fig. 4.3. The corresponding Kraus operators for the bit-flip (b.f.) channel are given by

$$E_0^{(\mathrm{b.f.})} = \sqrt{1-p}I, E_1^{(\mathrm{b.f.})} = \sqrt{p}X. \tag{4.75}$$

and the operator sum representation is given by

$$\xi^{(\mathrm{b.f.})}(\rho) = E_0^{(\mathrm{b.f.})}\rho\left(E_0^{(\mathrm{b.f.})}\right)^\dagger + E_1^{(\mathrm{b.f.})}\rho\left(E_1^{(\mathrm{b.f.})}\right)^\dagger = (1-p)\rho + pX\rho X \tag{4.76}$$

In a similar fashion, the Kraus operators for the phase-flip channel are given by

$$E_0^{(\mathrm{p.f.})} = \sqrt{1-p}I, E_1^{(\mathrm{p.f.})} = \sqrt{p}Z \tag{4.77}$$

and the corresponding operator sum representation is

$$\xi^{(\mathrm{p.f.})}(\rho) = E_0^{(\mathrm{p.f.})}\rho\left(E_0^{(\mathrm{p.f.})}\right)^\dagger + E_1^{(\mathrm{p.f.})}\rho\left(E_1^{(\mathrm{p.f.})}\right)^\dagger = (1-p)\rho + pZ\rho Z \tag{4.78}$$

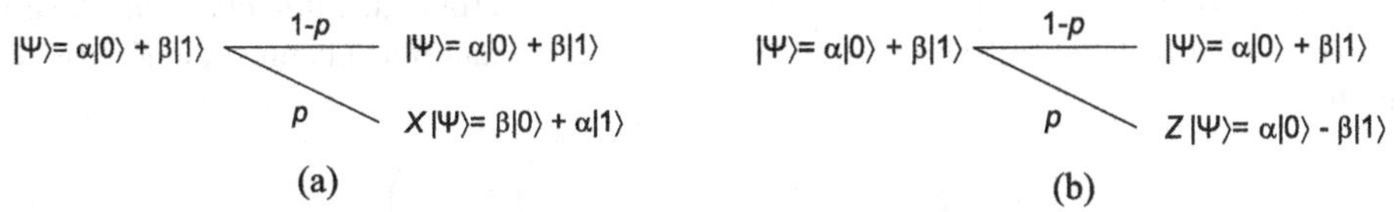

FIGURE 4.3

(A) Bit-flip channel model. (B) Phase-flip channel model.

Another relevant quantum channel model is the depolarizing channel. The *depolarizing channel*, as shown in Fig. 4.4, with probability $1-p$ leaves the qubit as it is, while probability p moves the initial state into $\rho_f = I/2$ that maximizes the von Neumann entropy $S(\rho) = -\mathrm{Tr}\ \rho \log \rho = 1$. The properties describing the model can be summarized as follows:

1. Qubit errors are independent,
2. Single-qubit errors (X, Y, Z) are equally likely,
3. All qubits have the same single-error probability, $p/4$.

The Kraus operators E_i of the channel should be selected as follows:

$$E_0 = \sqrt{1 - 3p/4}I; E_1 = \sqrt{p/4}X; E_2 = \sqrt{p/4}Y; E_3 = \sqrt{p/4}Z \qquad (4.79)$$

$|\Psi\rangle = \alpha|0\rangle + \beta|1\rangle$

$1-3p/4$ → $|\Psi\rangle = \alpha|0\rangle + \beta|1\rangle$

$p/4$

$X|\Psi\rangle = \alpha|1\rangle + \beta|0\rangle$

$p/4$

$Y|\Psi\rangle = j(\alpha|1\rangle - \beta|0\rangle)$

$Z|\Psi\rangle = \alpha|0\rangle - \beta|1\rangle$

(a)

$\rho = |\psi\rangle\langle\psi|$ $1-p$ $\rho_f = \rho$

$|\psi\rangle = \alpha|0\rangle + \beta|1\rangle$

p

$\rho_f = I/2$

(b)

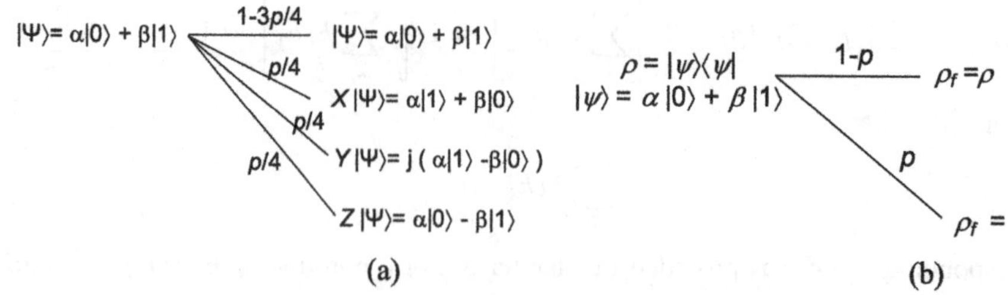

FIGURE 4.4

Depolarizing channel model: (A) Pauli operator description and (B) density operator description.

The action of the depolarizing channel is to perform the mapping $\rho \rightarrow \xi(\rho) = \sum_i E_i\rho E_i^\dagger$, where ρ is the initial density operator. The operator-sum representation for the depolarization channel can be written as

$$\xi(\rho) = \sum_i E_i\rho E_i^\dagger = \left(1 - \frac{3p}{4}\right)\rho + \frac{p}{4}(X\rho X + Y\rho Y + Z\rho Z)$$

$$= (1-p)\rho + \frac{p}{2}I. \qquad (4.80)$$

The first line in Eq. (4.80) corresponds to the model shown in Fig. 4.4A, and the second line in Eq. (4.80) corresponds to the model shown in Fig. 4.4B. It is clear from Eq. (4.80) that $\text{Tr}\xi(\rho) = \text{Tr}(\rho) = 1$, meaning that the superoperator is trace-preserving. Note that the depolarizing channel model in some other books/papers can differ slightly from the one described here.

We now describe the *amplitude damping channel* model. In certain quantum channels, the errors X, Y, and Z do not occur with the same probability. In the amplitude damping channel, the Kraus operators are given by

$$E_0 = \begin{pmatrix} 1 & 0 \\ 0 & \sqrt{1 - \varepsilon^2} \end{pmatrix}, \quad E_1 = \begin{pmatrix} 0 & \varepsilon \\ 0 & 0 \end{pmatrix} \qquad (4.81)$$

Spontaneous emission is an example of a physical process that can be modeled with the amplitude damping channel model. If $|\psi\rangle = a|0\rangle + b|1\rangle$ is the initial qubit state, the effect of amplitude damping channel is to perform the following mapping:

$$\rho \rightarrow \xi(\rho) = E_0\rho E_0^\dagger + E_1\rho E_1^\dagger = \begin{pmatrix} |a|^2 + \varepsilon^2|b|^2 & ab^*\sqrt{1 - \varepsilon^2} \\ a^*b\sqrt{1 - \varepsilon^2} & |b|^2(1 - \varepsilon^2) \end{pmatrix} \qquad (4.82)$$

The probabilities $P(0)$ and $P(1)$ that E_0 and E_1 occur are given by

$$P(0) = \mathrm{Tr}\left(E_0 \rho E_0^\dagger\right) = \mathrm{Tr}\begin{pmatrix} |a|^2 & ab^*\sqrt{1-\varepsilon^2} \\ a^*b\sqrt{1-\varepsilon^2} & |b|^2\left(1-\varepsilon^2\right) \end{pmatrix} = 1 - \varepsilon^2|b|^2$$
$$P(1) = \mathrm{Tr}\left(E_1 \rho E_1^\dagger\right) = \mathrm{Tr}\begin{pmatrix} |b|^2\varepsilon^2 & 0 \\ 0 & 0 \end{pmatrix} = \varepsilon^2|b|^2 \tag{4.83}$$

The corresponding amplitude damping channel model is shown in Fig. 4.5.

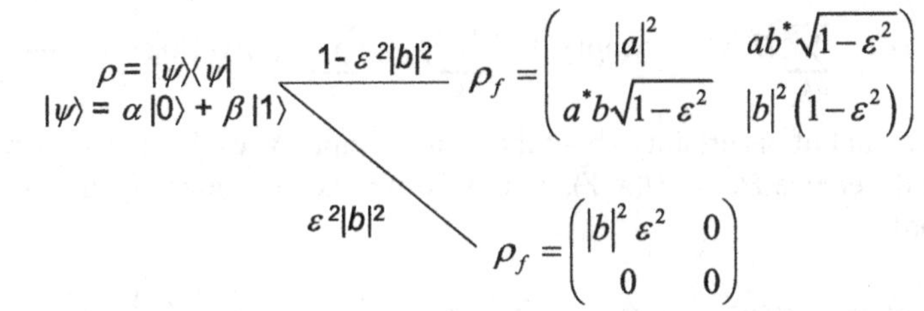

FIGURE 4.5

Amplitude damping channel model.

Every quantum channel error E (whether discrete or continuous) can be described as a linear combination of elements from the discrete set $\{I,X,Y,Z\}$, given by $E = e_1 I + e_2 X + e_3 Y + e_4 Z$, proved in Chapter 3. For example, the error $E = \begin{bmatrix} 1 & 0 \\ 0 & 0 \end{bmatrix}$ can be represented in terms of Pauli operators as $E = (I + Z)/2$. An error operator that affects several qubits can be written as a weighted sum of Pauli operators $\sum c_i P_i$ acting on ith qubits.

4.6 Quantum channel coding and Holevo–Schumacher–Westmoreland theorem

We now turn to the second question we raised at the beginning of this chapter: What is the maximum rate at which we can transfer information over a noisy communication channel? Specifically, we will attempt to answer this question for classical information over quantum channels. As in the previous section, we begin by looking at the classical counterpart, given the similar derivations.

4.6.1 Classical error correction and Shannon's channel coding theorem

The classical discrete memoryless channel (DMC) for input X and output Y is described by the set of transition probabilities:

$$p(y_k|x_j) = P(Y = y_k|X = x_j); 0 \le p(y_k|x_j) \le 1 \tag{4.84}$$

satisfying the condition $\sum_k p(y_k|x_j) = 1$. The conditional entropy of X, given $Y = y_k$, is related to the uncertainty of the channel input X by observing the channel output y_k and is defined as

$$H(X|Y=y_k)=\sum_j p(x_j|y_k)\log\left[\frac{1}{p(x_j|y_k)}\right], \tag{4.85}$$

where

$$p(x_j|y_k)=p(y_k|x_j)p(x_j)/p(y_k), p(y_k)=\sum_j p(y_k|x_j)p(x_j). \tag{4.86}$$

The amount of uncertainty remaining about channel input X after observing channel output Y can then be determined by

$$H(X|Y)=\sum_k H(X|Y=y_k)p(y_k)=\sum_k p(y_k)\sum_j p(x_j|y_k)\log\left[\frac{1}{p(x_j|y_k)}\right] \tag{4.87}$$

Therefore, the amount of uncertainty about the channel input X resolved by observing the channel input would be the difference $H(X)-H(X|Y)$, and this difference is known as the *mutual information* $I(X,Y)$. In other words,

$$I(X,Y)=H(X)-H(X|Y)=\sum_j p(x_j)\sum_k p(y_k|x_j)\log\left[\frac{p(y_k|x_j)}{p(y_k)}\right]. \tag{4.88}$$

The $H(X)$ represents the uncertainty about the channel input X before observing the channel output Y. In contrast, $H(X|Y)$ denotes the conditional entropy or the amount of uncertainty remaining about the channel input after the channel output has been received. Therefore, mutual information represents the amount of information (per symbol) conveyed by the channel. In other words, it represents the uncertainty about the channel input resolved by observing the channel output. Mutual information can be interpreted using the Venn diagram shown in Fig. 4.6A. The left

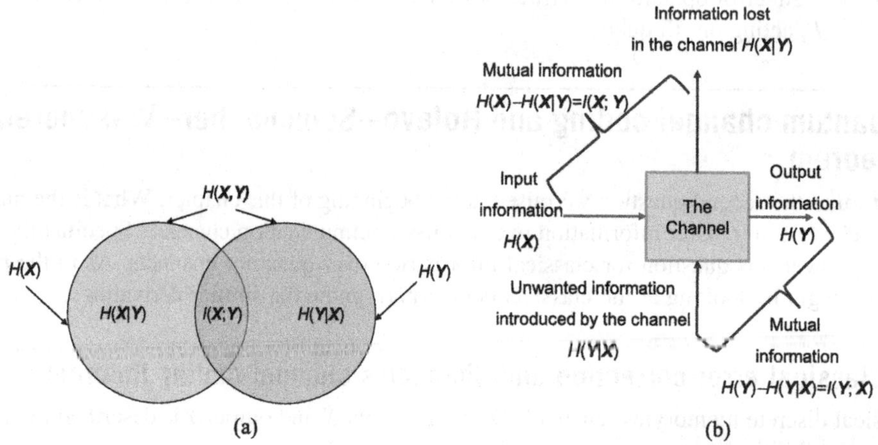

FIGURE 4.6

Interpretation of mutual information (A) using Venn diagrams and (B) using the approach of Ingels.

circle represents the entropy of channel input, the right circle represents the entropy of channel output, and the mutual information is obtained within the intersection of these two circles. Another interpretation, due to Ingels [26], is shown in Fig. 4.6B. The mutual information, i.e., the information conveyed by the channel, is obtained as the output information minus information introduced by the channel. In other words, the mutual information can be determined as the input information minus information lost in the channel. One important figure of merit of the classical channel is the *channel capacity*, which is obtained by maximization of mutual information *I*(*X*,*Y*) over all possible input distributions:

$$C = \max_{\{p(x_i)\}} I(X;Y). \tag{4.89}$$

The classical channel encoder accepts the message symbols and adds redundant symbols according to a corresponding prescribed rule. Channel coding is the act of transforming a length-k sequence into a length-n codeword. The set of rules specifying this transformation is commonly referred to as the *channel code*, which can be represented as the mapping C: $M \rightarrow X$, where C is the channel code, M is the set of information sequences of length k, and X is the set of codewords of length n. The decoder exploits these redundant symbols to determine which message symbol was actually transmitted. The concept of classical channel coding is introduced in Fig. 4.7. An important class of channel codes is the class of *block codes*. In an (*n*,*k*) *block code*, the channel encoder accepts information in successive k-symbol blocks. It adds $n-k$ redundant symbols that are algebraically related to the k message symbols, thus producing an overall encoded block of n symbols ($n > k$) known as a *codeword*. If the block code is *systematic*, the information symbols stay unchanged during the encoding operation, and the encoding operation may be considered as adding $n-k$ generalized parity checks to k information symbols. The code rate of the code is defined as $R = k/n$.

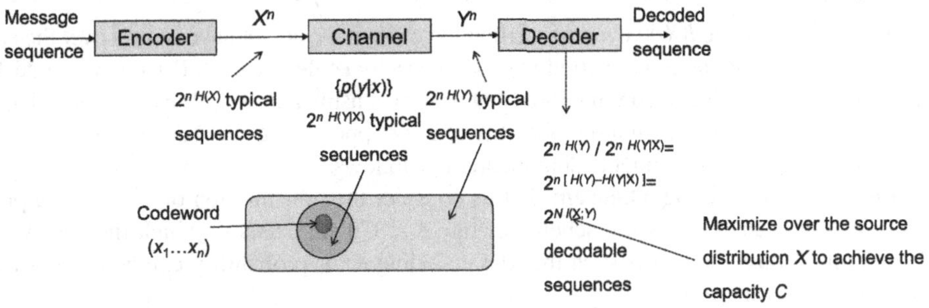

FIGURE 4.7

Classical error correction and Shannon's capacity theorem.

Shannon has shown that if $R < C$, we can construct 2^{NR} length-N codewords that can be sent over the (classical) channel with the maximum probability of error approaching zero for a large N. To prove this claim, we introduce the concept of jointly typical sequences. Two length-n sequences $\boldsymbol{x}$ and $\boldsymbol{y}$ are jointly typical sequences if they satisfy the following set of inequalities:

$$P\left(\left|\frac{1}{n}\log\left(\frac{1}{P(\boldsymbol{x})}\right) - H(X)\right| \leq \varepsilon\right) > 1 - \delta \tag{4.90.1}$$

$$P\left(\left|\frac{1}{n}\log\left(\frac{1}{P(\boldsymbol{y})}\right) - H(Y)\right| \leq \varepsilon\right) > 1 - \delta \tag{4.90.2}$$

$$P\left(\left|\frac{1}{n}\log\left(\frac{1}{P(\boldsymbol{x},\boldsymbol{y})}\right) - H(X,Y)\right| \leq \varepsilon\right) > 1 - \delta \tag{4.90.3}$$

where $P(\boldsymbol{x},\boldsymbol{y})$ denotes the joint probability of the two sequences, and $H(X,Y)$ is their joint entropy.

For an n-length input codeword randomly generated according to the probability distribution of a source X, the number of random input sequences is approximately $2^{nH(X)}$, and the number of output typical sequences is approximately $2^{nH(Y)}$. Furthermore, the total number of input and output sequences that are jointly typical is $2^{nH(X,Y)}$. Therefore, the total pairs of sequences that are simultaneously $\boldsymbol{x}$-typical, $\boldsymbol{y}$-typical, and jointly typical are $2^{n\underbrace{[H(X)+H(Y)-H(X,Y)]}_{I(X,Y)}} = 2^{nI(X,Y)}$, where $I(X,Y)$ is the mutual information between X and Y, as illustrated in Fig. 4.7. These are the maximum numbers of codeword sequences that can be distinguished. One way of seeing this is to consider a single codeword. For this codeword, the action of the channel characterized by the conditional probability $P(\boldsymbol{y}|\boldsymbol{x})$ defines the Hamming sphere in which this codeword can lie after the action of the channel. The size of this Hamming sphere is approximately $2^{nH(Y|X)}$. Given that the total number of output typical sequences is approximately $2^{nH(Y)}$, if we desire to have no overlap between two Hamming spheres, the maximum number of codewords we can consider is given by $2^{nH(X)}/2^{nH(Y|X)} = 2^{nI(X,Y)}$. To increase this number, we need to maximize $I(X,Y)$ over the distribution of X, as we do not have control over the channel. This maximal mutual information is referred to as the capacity of the channel. If we have a rate $R < C$, Shannon's noisy channel coding theorem tells us that we can construct 2^{nR} length-n codewords that can be sent over the channel with the maximum probability of error approaching zero for a large n. Now we can formally formulate Shannon's channel coding theorem as follows:

Shannon's Channel Coding Theorem. Let us consider the transmission of $2^{N(C-\varepsilon)}$ equiprobable messages. A classical channel coding scheme of rate $R < C$ then exists in which the codewords are selected from all 2^N possible words such that the decoding error probability can be made arbitrarily small for a sufficiently large N.

4.6.2 Quantum error correction and Holevo–Schumacher–Westmoreland theorem

Now we consider the communication over the quantum channel, as illustrated in Fig. 4.8. The quantum encoder for each message m out of $M = 2^{NR}$ messages generates a product state codeword $\boldsymbol{\rho}^m$ drawn from the ensemble $\{p_x, \boldsymbol{\rho}_x\}$ as follows:

$$\boldsymbol{\rho}^m = \boldsymbol{\rho}_{m_1} \otimes \cdots \otimes \boldsymbol{\rho}_{m_N} = \boldsymbol{\rho}_{\otimes N}. \tag{4.91}$$

This quantum codeword is sent over the quantum channel described by the trace-preserving quantum operation ξ, resulting in the received quantum word:

$$\sigma_m = \xi(\rho_{m_1}) \otimes \xi(\rho_{m_2}) \otimes \cdots \otimes \xi(\rho_{m_N}) = \sigma_{\otimes N}. \tag{4.92}$$

Bob performs measurements on σ_m to decode Alice's message. Our goal is to find the maximum rate at which Alice can communicate over the quantum channel such that the probability of decoding error is negligibly small. This maximum rate is the capacity $C(\xi)$ of the quantum channel for transmitting classical information. The key idea behind the Holevo–Schumacher–Westmoreland (HSW) theorem is that for $R < C(\xi)$ quantum codes with codewords of length N exist such that the probability of incorrectly decoding the quantum codeword can be made arbitrarily small.

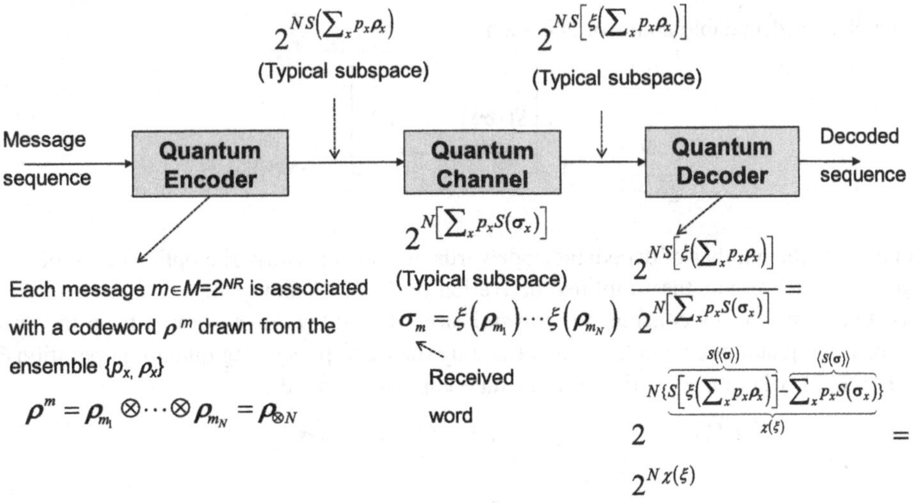

FIGURE 4.8

Quantum error correction and Holevo–Schumacher–Westmoreland theorem.

The von Neumann entropy associated with the quantum encoder is

$$S(\rho) = S\left(\sum\nolimits_x p_x\rho_x\right), \tag{4.93}$$

while the dimensionality of the typical subspace of quantum encoder is given by

$$2^{NS(\rho)} = 2^{NS(\sum_x p_x\rho_x)}. \tag{4.94}$$

On the other hand, the entropy associated with σ will be

$$\langle S(\sigma)\rangle = \sum\nolimits_x p_x S(\sigma_x), \tag{4.95}$$

while the dimensionality of the quantum subspace characterizing the quantum channel is given by

$$2^{N\langle S(\sigma)\rangle} = 2^{N\sum_x p_x S(\sigma_x)}. \tag{4.96}$$

The quantum channel perturbs the quantum codeword transmitted over the quantum channel by performing the trace-preserving quantum operation, so entropy at the channel output can be written as

$$S(\langle\sigma\rangle) = S\left(\xi\left(\sum_x p_x \rho_x\right)\right), \tag{4.97}$$

while the dimensionality of the corresponding subspace is given by

$$2^{NS(\langle\sigma\rangle)} = 2^{NS(\xi(\sum_x p_x \rho_x))}. \tag{4.98}$$

The number of decodable codewords is then

$$\frac{2^{NS(\langle\sigma\rangle)}}{2^{N\langle S(\sigma)\rangle}} = 2^{N\left[\underbrace{S(\langle\sigma\rangle) - \langle S(\sigma)\rangle}_{\chi(\xi)}\right]} = 2^{N\chi(\xi)}. \tag{4.99}$$

To maximize the number of decodable codewords, we must perform the optimization of $\chi(\xi)$ over p_x and ρ_x, which represents the simplified derivation of the HSW theorem.

HSW Theorem. Let us consider the transmission of a codeword ρ_m drawn from the ensemble $\{p_x, \rho_x\}$ over the quantum channel, characterized by the trace-preserving quantum operation ξ, with the rate $R < C(\xi)$, where $C(\xi)$ is the product state capacity defined as

$$\begin{aligned} C(\xi) &= \max_{\{p_x,\rho_x\}} \chi(\xi) = \max_{\{p_x,\rho_x\}} [S(\langle\sigma\rangle) - \langle S(\sigma)\rangle] \\ &= \max_{\{p_x,\rho_x\}} \left[S\left[\xi\left(\sum_x p_x\rho_x\right)\right] - \sum_x p_x S(\xi(\rho_x))\right], \end{aligned} \tag{4.100}$$

where the maximization of $\chi(\xi)$ is performed over p_x and ρ_x. A coding scheme then exists that allows reliable error-free transmission over the quantum channel.

To make the idea of capacity concrete, we compute the capacity of two example quantum channels [1], the *depolarizing channel* and the *phase-flip* channel. We assume quantum source outputs ρ_0 with probability p_0, and ρ_1 with probability p_1. We further assume that both the states are orthogonal pure states given by

$$\rho_0 = \begin{pmatrix} 1 & 0 \\ 0 & 0 \end{pmatrix}, \ \rho_1 = \begin{pmatrix} 0 & 0 \\ 0 & 1 \end{pmatrix}.$$

Depolarizing channel. The depolarizing channel is defined by the relation

$$\mathscr{E}(\rho) = q\frac{I}{2} + (1-q)\rho,$$

where $\mathscr{E}$ denotes the superoperator, which either transforms the input state to a uniform mixed state—denoted by the identity operator $\boldsymbol{I}$—with probability q or leaves the state unchanged with probability $1-q$. The transformation to the mixed state is referred to as the depolarization action of the channel. Applying the quantum operation from above to both $\boldsymbol{\rho}_0$ and $\boldsymbol{\rho}_1$, we obtain

$$\boldsymbol{\sigma}_0=\mathscr{E}(\boldsymbol{\rho}_0)=\frac{q}{2}\begin{pmatrix}1&0\\0&1\end{pmatrix}+(1-q)\begin{pmatrix}1&0\\0&0\end{pmatrix}=\begin{pmatrix}1-\frac{q}{2}&0\\0&\frac{q}{2}\end{pmatrix},$$

$$\boldsymbol{\sigma}_1=\mathscr{E}(\boldsymbol{\rho}_1)=\frac{q}{2}\begin{pmatrix}1&0\\0&1\end{pmatrix}+(1-q)\begin{pmatrix}0&0\\0&1\end{pmatrix}=\begin{pmatrix}\frac{q}{2}&0\\0&1-\frac{q}{2}\end{pmatrix}.$$

The corresponding quantum entropies can easily be seen as

$$S(\boldsymbol{\sigma}_0)=\left(1-\frac{q}{2}\right)\log\left(1-\frac{q}{2}\right)+\frac{q}{2}\log\frac{q}{2},\ S(\boldsymbol{\sigma}_1)=\frac{q}{2}\log\frac{q}{2}+\left(1-\frac{q}{2}\right)\log\left(1-\frac{q}{2}\right)=S(\boldsymbol{\sigma}_0).$$

Consequently, we have

$$S(\boldsymbol{\sigma})=p_0\,S(\boldsymbol{\sigma}_0)+p_1S(\boldsymbol{\sigma}_1)=(p_0+p_1)S(\boldsymbol{\sigma}_1)=\frac{q}{2}\log\frac{q}{2}+\left(1-\frac{q}{2}\right)\log\left(1-\frac{q}{2}\right).$$

This gives us the second term of Holevo information, χ. To find the first term, we compute

$$\langle\boldsymbol{\sigma}\rangle=\left\langle\mathscr{E}\left(\sum_i p_i\boldsymbol{\rho}_i\right)\right\rangle=\frac{q}{2}\begin{pmatrix}1&0\\0&1\end{pmatrix}+(1-q)\begin{pmatrix}p_0&0\\0&p_1\end{pmatrix}=\begin{pmatrix}\frac{q}{2}+(1-q)p_0&0\\0&\frac{q}{2}+(1-q)p_1\end{pmatrix},$$

and therefore, we have

$$S(\langle\boldsymbol{\sigma}\rangle)=\left(\frac{q}{2}+(1-q)p_0\right)\log\left(\frac{q}{2}+(1-q)p_0\right)+\left(\frac{q}{2}+(1-q)p_1\right)\log\left(\frac{q}{2}+(1-q)p_1\right).$$

Under the constraint that $p_0+p_1=1$, $S(\langle\boldsymbol{\sigma}\rangle)$ is maximized for $p_0=p_1=\frac{1}{2}$, and this maximum value is 1 bit. Therefore, the Holevo information reduces to

$$\chi=1-\left(1-\frac{q}{2}\right)\log\left(1-\frac{q}{2}\right)+\frac{q}{2}\log\frac{q}{2}.$$

We can see that χ equals 1 for $q=0$ and is 0 when $q=1$; thus, the capacity $C(\mathscr{E})=1$ bit. It is interesting to note that for $q=1$, χ vanishes, and the channel becomes useless. This result can be deduced from the definition of the depolarization channel. When we set $q=1$, the channel output is a mixed state $\frac{I}{2}$, no matter the input. Consequently, we know deterministically that the channel output is a mixed state (maximum uncertainty), and the channel has no use!

Phase-flip channel. As the next example, let us consider the phase-flip channel. If the input to the channel is a state given by $\alpha|0\rangle+\beta|1\rangle$, the action of the phase-flip channel results in $\alpha|0\rangle-\beta|1\rangle$ with probability q, or leaves the state unchanged with probability $1 - q$. The phase-flip action is performed using the Pauli matrix:

$$\mathbf{Z} = \begin{pmatrix} 1 & 0 \\ 0 & -1 \end{pmatrix}.$$

Therefore, we define the phase-flip channel mathematically by

$$\mathscr{E}(\boldsymbol{\rho}) = q\mathbf{Z}\boldsymbol{\rho}\mathbf{Z} + (1-q)\boldsymbol{\rho},$$

Following the steps in the previous example, we have

$$\boldsymbol{\sigma}_0 = \mathscr{E}(\boldsymbol{\rho}_0) = q\begin{pmatrix} 1 & 0 \\ 0 & -1 \end{pmatrix}\begin{pmatrix} 1 & 0 \\ 0 & 0 \end{pmatrix}\begin{pmatrix} 1 & 0 \\ 0 & -1 \end{pmatrix} + (1-q)\begin{pmatrix} 1 & 0 \\ 0 & 0 \end{pmatrix} = \begin{pmatrix} 1 & 0 \\ 0 & 0 \end{pmatrix},$$

$$\boldsymbol{\sigma}_1 = \mathscr{E}(\boldsymbol{\rho}_1) = q\begin{pmatrix} 1 & 0 \\ 0 & -1 \end{pmatrix}\begin{pmatrix} 0 & 0 \\ 0 & 1 \end{pmatrix}\begin{pmatrix} 1 & 0 \\ 0 & -1 \end{pmatrix} + (1-q)\begin{pmatrix} 0 & 0 \\ 0 & 1 \end{pmatrix} = \begin{pmatrix} 0 & 0 \\ 0 & 1 \end{pmatrix}.$$

The corresponding quantum entropies are

$$S(\boldsymbol{\sigma}_0) = (1)\log(1) + 0\log 0, S(\boldsymbol{\sigma}_1) = 0\log 0 + (1)\log(1) = 0 = S(\boldsymbol{\sigma}_0),$$

and therefore, the second term of the Holevo information is given by

$$\langle S(\boldsymbol{\sigma})\rangle = p_0 S(\boldsymbol{\sigma}_0) = p_1 S(\boldsymbol{\sigma}_1) = 0,$$

To calculate the first term, as in the previous example, we first find the expression for

$$\langle\boldsymbol{\sigma}\rangle = \left\langle \mathscr{E}\left(\sum_i p_i \boldsymbol{\rho}_i\right)\right\rangle = q\begin{pmatrix} 1 & 0 \\ 0 & -1 \end{pmatrix}\begin{pmatrix} p_0 & 0 \\ 0 & p_1 \end{pmatrix}\begin{pmatrix} 1 & 0 \\ 0 & -1 \end{pmatrix} + (1-q)\begin{pmatrix} p_0 & 0 \\ 0 & p_1 \end{pmatrix} = \begin{pmatrix} p_0 & 0 \\ 0 & p_1 \end{pmatrix}$$

This gives us

$$S(\langle\boldsymbol{\sigma}\rangle) = p_0 \log p_0 + p_1 \log p_1.$$

Therefore,

$$x = S(\langle\boldsymbol{\sigma}\rangle) - \langle S(\boldsymbol{\sigma})\rangle = p_0 \log p_0 + p_1 \log p_1.$$

Holevo information is maximized over $\{p_0, p_1\}$, and this maximization can be performed by noting that the capacity of any channel is upper bounded by the source entropy:

$$S(\langle\boldsymbol{\rho}\rangle) = S\begin{pmatrix} p_0 & 0 \\ 0 & p_1 \end{pmatrix} = p_0 \log p_0 + p_1 \log p_1,$$

which is the same as χ! Under the constraint $p_0 + p_1 = 1$, the maximum χ is 1 bit for $p_0 = p_1 = \frac{1}{2}$. Therefore, the capacity of the phase-flip channel is 1 bit!

Lossy bosonic channel. As the final example, we consider a continuous-variable quantum channel, i.e., a lossy bosonic channel. In quantum optical communication, bosonic channels typically refer to photon-based communication channels like free-space optical and fiber-optics transmission links. These are important research areas because they allow us to systematically explain and study quantum optical communication by exploiting system linearity coupled with certain nonlinear devices such as parametric downconverters and downsqueezers [20]. For a lossy channel, the effect of the channel on the input quantum states can be modeled by a quantum noise source. Taking this noise source into account, the evolution of a multimode lossy bosonic channel can be described as

$$A_{o_k} = \sqrt{T_k} A_{i_k} + \sqrt{1 - T_k} B_k, \tag{4.101}$$

where A_{i_k} are A_{o_k}, the annihilation ladder operators of the kth mode of the input and output electromagnetic field, B_k is the annihilation operator associated with the kth environmental noise mode, and T_k is the kth mode channel transmissivity or quantum efficiency. For such a channel, the capacity is given by

$$C(\xi) = \sum\nolimits_k g(T_k N_k(\beta)), \tag{4.102}$$

with $N_k(\beta)$ being the optimal photon-count distribution given by

$$N_k(\beta) = \frac{T_k^{-1}}{e^{\beta T_k^{-1} \hbar \omega_k} - 1}, \tag{4.103}$$

where $\hbar$ is the reduced Planck constant, ω_k is the frequency of the kth mode, and β is the Lagrange multiplier determined by the average transmitted energy. A detailed derivation of this result is given in Ref. [21] and the references therein. The function $g(\bullet)$,

$$g(x) = (x+1)\log(x+1) - x \log x, \tag{4.104}$$

is known as the Shannon's entropy of the Bose–Einstein probability distribution [24].

An interesting consequence of Eq. (4.102) is that by constructing the encoded product state

$$\rho_{\otimes k} = \otimes_k \int p_k(u) |u\rangle_k \langle u| du, \tag{4.105}$$

where the kth modal state $\int p_k(u)|u\rangle_k\langle u|du$ is a mix of coherent states $|u\rangle_k$, and $p_k(u)$ is the Gaussian probability distribution,

$$p_k(u) = \frac{1}{\pi N_k} e^{-|u|^2/N_k}, \tag{4.106}$$

with N_k being the modal photon count with u as the continuous variable, the channel capacity can be realized with a single use of the channel without employing entanglement. More details can be found in references [20–25].

4.7 Summary

This chapter was devoted to quantum information theory fundamentals. After introductory remarks in Section 4.1, the chapter introduced classical and von Neumann entropy definitions and properties, followed by quantum representation of classical information in Section 4.2. Section 4.3 introduced accessible information as the maximum mutual information over all possible measurement schemes. The same section introduced an important bound, the Holevo bound, and its proof. In Section 4.4, we introduced the typical sequence, state, and subspace, as well as the projector on the typical subspace. We also formulated and proved Schumacher's source coding theorem, which is the quantum equivalent of Shannon's source coding theorem. In the same section, we introduced the concepts of quantum compression, quantum decompression, and reliable compression. Section 4.5 described various quantum channel models, including bit-flip, phase-flip, depolarizing, and amplitude damping. In Section 4.6, we formulated and proved the HSW theorem, which is the quantum equivalent of Shannon's channel capacity theorem.

References

[1] I.B. Djordjevic, Quantum Information Processing and Quantum Error Correction: An Engineering Approach, Elsevier/Academic Press, 2012.

[2] I.B. Djordjevic, Physical-Layer Security and Quantum Key Distribution, Springer International Publishing, Switzerland, 2019.

[3] C.E. Shannon, A mathematical theory of communication, Bell Syst. Tech. J. 27 (1948), 379–423, 623–656.

[4] C.E. Shannon, W. Weaver, The Mathematical Theory of Communication, The University of Illinois Press, Urbana, Illinois, 1949.

[5] T. Cover, J. Thomas, Elements of Information Theory, Wiley-Interscience, 1991.

[6] A.N. Kolmogorov, Three approaches to the quantitative definition of information, Probl. Inf. Transm. 1 (1) (1965) 1–7.

[7] H. Touchette, The large deviation approach to statistical mechanics, Phys. Rep. 478 (1–3) (2009) 1–69.

[8] B.R. Frieden, Science from Fisher Information: A Unification, Cambridge Univ. Press, 2004.

[9] J.L. Kelly Jr., A new interpretation of information rate, Bell Syst. Tech. J. 35 (1956) 917–926.

[10] M.A. Nielsen, I.L. Chuang, Quantum Computation and Quantum Information, Cambridge Univ. Press, 2000.

[11] J. Preskill, Reliable quantum computers, Proc. R. Soc. Lond. A 454 (1998) 385–410.

[12] A.S. Holevo, Problems in the mathematical theory of quantum communication channels, Rep. Math. Phys. 12 (1977) 273–278.

[13] A.S. Holevo, Capacity of a quantum communications channel, Probl. Inf. Transm. 15 (1979) 247–253.

[14] R. Jozsa, B. Schumacher, A new proof of the quantum noiseless coding theorem, J. Mod. Opt. 41 (1994) 2343–2349.

[15] B. Schumacher, M. Westmoreland, W.K. Wootters, Limitation on the amount of accessible information in a quantum channel, Phys. Rev. Lett. 76 (1997) 3452–3455.

[16] A.S. Holevo, The capacity of the quantum channel with general signal states, IEEE Trans. Inf. Theor. 44 (1998) 269–273.

[17] M. Sassaki, K. Kato, M. Izutsu, O. Hirota, Quantum channels showing superadditivity in channel capacity, Phys. Rev. A 58 (1998) 146–158.

[18] C.A. Fuchs, Nonorthogonal quantum states maximize classical information capacity, Phys. Rev. Lett. 79 (1997) 1162–1165.

[19] C.H. Bennett, P.W. Shor, J.A. Smolin, A.V. Thapliyal, Entanglement enhanced classical capacity of noisy quantum channels, Phys. Rev. Lett. 83 (1999) 3081–3084.

[20] A.S. Holevo, R.F. Werner, Evaluating capacities of boson Gaussian channels, Phys. Rev. A 63 (03) (2001) 2312.

[21] V. Giovannetti, S. Lloyd, L. Maccone, P.W. Shor, Broadband channel capacities, Phys. Rev. A 68 (2003) 062323.

[22] V. Giovannetti, S. Lloyd, L. Maccone, P.W. Shor, Entanglement assisted capacity of the broadband lossy channel, Phys. Rev. Lett. 91 (2003) 047901.
[23] V. Giovannetti, S. Guha, S. Lloyd, L. Maccone, J.H. Shapiro, H.P. Yuen, Classical capacity of the lossy bosonic channel: the exact solution, Phys. Rev. Lett. 92 (2004) 027902.
[24] J.H. Shapiro, S. Guha, B.I. Erkmen, Ultimate channel capacity of free-space optical communications, J. Opt. Netw. 4 (2005) 501–516.
[25] S. Guha, J.H. Shapiro, B.I. Erkmen, Classical capacity of bosonic broadcast communication and a minimum output entropy conjecture, Phys. Rev. A 76 (2007) 032303.
[26] F.M. Ingels, Information and Coding Theory, Intext Educational Publishers, Scranton, 1971.

CHAPTER

Quantum detection and quantum communication

5

Chapter outline

Quantum Communication, Quantum Networks, and Quantum Sensing. https://doi.org/10.1016/B978-0-12-822942-2.00005-4

This chapter is devoted to quantum detection theory [1,2], quantum communication [3–6], and Gaussian quantum information theory [7]. Before introducing quantum detection theory, we briefly review basic concepts related to density operators. The chapter then describes quantum detection theory concepts. This theory is then applied to the binary detection problem. We then introduce coherent states and describe their basic properties, followed by quadrature operators and derivation of the uncertainty principle for continuous-variable (CV) systems. We then discuss binary quantum optical communication in the absence of noise, including the photon-counting, on–off keying (OOK), binary phase-shift keying (BPSK), Kennedy, and Dolinar receivers. The *P*-representation is introduced and applied to represent thermal noise and the coherent state in the presence of thermal noise. We then discuss binary optical communication in the presence of noise, focusing on OOK and BPSK. Gaussian and squeezed states are introduced, followed by definitions of the Wigner function and covariance matrices. We then discuss Gaussian transformation and Gaussian channels, using beam splitter (BS) operation, phase rotation operation, and squeezing operator as examples. The thermal decomposition of Gaussian states is discussed next, and the von Neumann entropy for thermal states is derived. The covariance matrices for two-mode Gaussian states are then discussed, as well as how to calculate symplectic eigenvalues, which are relevant for the von Neumann entropy calculation. Gaussian state measurements and detection are then discussed, emphasizing homodyne detection, heterodyne detection, and partial measurements. The focus then moves to nonlinear (NL) quantum optics fundamentals—in particular, detailed descriptions of three-wave mixing (TWM) and four-wave mixing (FWM). The generation of Gaussian states is described, and two-mode squeezed (TMS) state generation is discussed in detail. Finally, we discuss the multilevel quantum optical communication problem, wherein two scenarios are described—in the absence of noise and the presence of noise. Using square root measurements (SRMs) to analyze the performance of these schemes is described in detail.

5.1 Density operators (revisited)

Let a large number of quantum systems of the same kind be prepared, each in one of a set of states $|\phi_n\rangle$, and let the fraction of the system being in state $|\phi_n\rangle$ be denoted by probability p_n $(n = 1,2,\ldots)$:

$$\langle\phi_m|\phi_n\rangle = \delta_{mn}, \quad \sum_n p_n = 1 \tag{5.1}$$

Therefore, this ensemble of quantum states represents a classical *statistical mixture* of kets. The probability of obtaining measurement result ξ_n from the measurement of Ξ is

$$\Pr(\xi_k) = \sum_{n=1}^{\infty} p_n|\langle\xi_k|\phi_n\rangle|^2 = \sum_{n=1}^{\infty} p_n\langle\xi_k|\phi_n\rangle\langle\phi_n|\xi_k\rangle = \langle\xi_k|\rho|\xi_k\rangle, \tag{5.2}$$

where the operator ρ is known as a *density operator* and is defined by

$$\rho = \sum_{n=1}^{\infty} p_n|\phi_n\rangle\langle\phi_n|. \tag{5.3}$$

The expected value of the operator Ξ is given by

$$\langle\Xi\rangle = \sum_{k=1}^{\infty} \xi_k \Pr(\xi_k) = \sum_{k=1}^{\infty} \xi_k\langle\xi_k|\rho|\xi_k\rangle = \sum_{k=1}^{\infty}\langle\xi_k|\rho\Xi|\xi_k\rangle = \mathrm{Tr}(\rho\Xi) \tag{5.4}$$

The density operator *properties* can be summarized as follows:

1. The density operator is Hermitian ($\rho^+ = \rho$) with the set of orthonormal eigenkets $|\phi_n\rangle$ corresponding to the nonnegative eigenvalues p_n and $\mathrm{Tr}(\rho) = 1$.
2. Any Hermitian operator with nonnegative eigenvalues and trace 1 may be considered a density operator.
3. The density operator is positive definite: $\langle\psi|\rho|\psi\rangle \geq 0$ for all $|\psi\rangle$.
4. The density operator has the property $\mathrm{Tr}(\rho^2) \leq 1$, with equality if one of the prior probabilities is 1 and all the rest 0, that is, $\rho = |\phi_n\rangle\langle \phi_n|$; the density operator is then a projection operator.
5. The eigenvalues of a density operator satisfy $0 \leq p_n \leq 1$.

When ρ is the projection operator, we say it represents the system in a *pure state*; otherwise, with $\mathrm{Tr}(\rho^2)<1$, it represents a *mixed state.* A mixed state in which all eigenkets occur with the same probability is known as a *completely mixed state* and can be represented by

$$\rho = \sum_{k=1}^{\infty} \frac{1}{n} |\phi_n\rangle\langle\phi_n| = \frac{1}{n}\hat{1} \Rightarrow \mathrm{Tr}\left(\rho^2\right) = \frac{1}{n} \Rightarrow \frac{1}{n} \leq \mathrm{Tr}\left(\rho^2\right) \leq 1 \tag{5.5}$$

The density matrix can be used to determine the amount of information conveyed by the quantum state, i.e., to compute the *von Neumann entropy*:

$$S = -\mathrm{Tr}(\rho \log \rho) = -\sum_i \lambda_i \log_2 \lambda_i, \tag{5.6}$$

where λ_i are the eigenvalues of the density matrix. When the states $\{|\phi_n\rangle\}$ are orthogonal, we can use the *Shannon entropy* to determine the uncertainty of the quantum source:

$$H = -\sum_n p_n \log_2 p_n. \tag{5.7}$$

When the states $\{|\phi_n\rangle\}$ are orthonormal and pure, the von Neumann entropy is equal to the Shannon entropy.

The probability that the outcome of the measurement of observable Ξ with eigenvalues $\xi^{(n)}$ lies between (a,b) is given by

$$\Pr(\xi \in R(a,b)) = \sum_{\xi^{(n)} \in R(a,b)} \langle \xi^{(n)}|\psi\rangle^2 = \sum_{\xi^{(n)} \in R(a,b)} \langle\psi|\xi^{(n)}\rangle\langle\xi^{(n)}|\psi\rangle = \langle\psi|P_{ab}|\psi\rangle = \langle P_{ab}\rangle. \tag{5.8}$$

The operator P_{ab} is defined by

$$P_{ab} = \sum_{\xi^{(n)} \in R(a,b)} |\xi^{(n)}\rangle\langle\xi^{(n)}|, \tag{5.9}$$

which is projection operator since $P_{ab}^2 = P_{ab}$, and this property can be rearranged as

$$P_{ab}\left(P_{ab} - \hat{1}\right) = 0, \tag{5.10}$$

with $\hat{1}$ representing the identity operator, indicating that eigenvalues are either 0, corresponding to the *false proposition*, or 1, corresponding to the *true proposition.*

5.2 Quantum detection theory fundamentals

Let the transmitter input be one message of duration T of M possible messages $m_1,\ldots,m_M$. On the receiver side, the corresponding received electromagnetic state is in a statistical mixture of states denoted by a density operator ρ_i ($i = 1,\ldots,M$). The receiver chooses one of M possible hypotheses H_i ($i = 1,\ldots,M$) by which the symbol was transmitted to minimize the probability of error P_e, defined by

$$P_e = \Pr(\hat{m} \neq m). \tag{5.11}$$

Let $\boldsymbol{X}=(X_1,\ldots,X_L)$ denote the L-tuple of Hermitian operators corresponding to observable measured by the receiver. When these operators commute, a simultaneous measurement yields the L-tuple of eigenvalues $\boldsymbol{x}_n=(x_{n1},\ldots,x_{nL})$, with x_{ni} being an eigenvalue of operator X_i ($i = 1,\ldots,L$). Further, let $|\boldsymbol{x}_n\rangle$ be a simultaneous eigenvector of commuting operators $\{X_i\}$ corresponding to eigenvalue $\boldsymbol{x}_n$. The conditional probability that the outcome of the measurement of $\boldsymbol{X}$ is $\boldsymbol{x}_n$ providing that the message m_j is transmitted, denoted as $P(\boldsymbol{x}_n|m_j)$, can be determined by

$$P(\boldsymbol{x}_n|m_j) = \langle \boldsymbol{x}_n|\rho_j|\boldsymbol{x}_n\rangle. \tag{5.12}$$

By denoting with p_{jn} the probability that the receiver chooses the hypothesis H_j when the outcome of the measurement is x_n and denoting the *a priori* probability with ζ_j, the *probability of a correct decision* can be determined by

$$P_c = \sum_n \sum_{j=1}^{M} \zeta_j p_{jn} \langle \boldsymbol{x}_n|\rho_j|\boldsymbol{x}_n\rangle, \quad \sum_{j=1}^{M} p_{jn} = 1; \tag{5.13}$$

while the *probability of error* P_e is simply

$$P_e = 1 - P_c = 1 - \sum_n \sum_{j=1}^{M} \zeta_j p_{jn} \langle \boldsymbol{x}_n|\rho_j|\boldsymbol{x}_n\rangle. \tag{5.14}$$

The maximum *a posteriori* probability (MAP) receiver chooses the hypothesis H_j maximizing the *a posteriori* probability:

$$P(m_j|\boldsymbol{x}_n) = \zeta_j P(\boldsymbol{x}_n|m_j) / \sum_{j=1}^{M} \zeta_j P(\boldsymbol{x}_n|m_j) \tag{5.15}$$

In other words, we set $p_{jn} = 1$ for all n such that

$$\zeta_j \langle \boldsymbol{x}_n|\rho_j|\boldsymbol{x}_n\rangle \geq \zeta_i \langle \boldsymbol{x}_n|\rho_i|\boldsymbol{x}_n\rangle, \quad i \neq j \tag{5.16}$$

When the prior probabilities are equally likely, that is, $\zeta_j = 1/M$, the MAP receiver becomes the *maximum likelihood* receiver.

Let us introduce the projection operators Π_j:

$$\Pi_j = \sum_n p_{jn} |\boldsymbol{x}_n\rangle\langle \boldsymbol{x}_n|, \quad p_{jn} \in \{0, 1\}. \tag{5.17}$$

Because the probabilities p_{jn} are either zero or one, that is, $p_{in}p_{jn} = \delta_{ij}$, we can write

$$\Pi_i \Pi_j = \delta_{ij} \Pi_i, \tag{5.18}$$

and the projection operators satisfy the completeness relationship:

$$\sum_{j=1}^{M} \Pi_j = \hat{1}. \tag{5.19}$$

The probability of error can now be expressed in terms of a trace of projection and density operators:

$$P_e = 1 - \sum_{j=1}^{M} \zeta_j \sum_{n} p_{jn} \langle x_n | \rho_j | x_n \rangle = 1 - \sum_{j=1}^{M} \zeta_j \mathrm{Tr}(\rho_j \Pi_j). \tag{5.20}$$

The *necessary condition* for projection operators to minimize P_e is [1,3]

$$\sum_{j=1}^{M} \zeta_j \Pi_j \rho_j = \sum_{j=1}^{M} \zeta_j \rho_j \Pi_j \tag{5.21}$$

The *sufficient condition* for the projection operators to represent the optimum solution would be [1,3]

$$\left\{ \sum_{i=1}^{M} \zeta_i \Pi_i \rho_i - \zeta_j \rho_j \right\} \text{ are positive semidefinite } \forall\ j = 1, \cdots, M \tag{5.22}$$

When the projection operators Π_j have a common complete set of eigenkets, the Π_j can be used as field observables measured simultaneously at the receiver side. In this case, we decide in favor of H_j whenever the observed value of Π_j is 1. Only one projection operator is needed for the binary quantum detection problem because the projection operators satisfy the completeness relationship (5.19).

5.3 Binary quantum detection

In the simplest binary communication problem, the digit 1 is represented by the presence of the pulse of duration T, while the digit 0 is represented by the absence of the pulse. Under hypothesis H_0, the modes in the receiver (Rx) are excited by random noise and described by mixture density operator ρ_0. On the other hand, under hypothesis H_1, the modes in Rx are described by mixture density operator ρ_1. There are two projection (detection) operators that effectively partition the outcome space R into two disjoint regions R_0 and R_1 [2]:

$$\Pi_0 = \sum_{n:\, b_n \in R_0} |b_n\rangle\langle b_n|,\ \Pi_1 = \sum_{n:\, b_n \in R_1} |b_n\rangle\langle b_n|;\ R_0 \cup R_1 = R, \tag{5.23}$$

where $|b_n\rangle$ are the eigenkets of the measurement Hermitian operator M. The *detection probability* Q_D and *false-alarm probability* Q_{FA} can be expressed in terms of projection operators as follows:

$$\begin{aligned} Q_D &= \Pr(\text{choosing } H_1 | \rho_1) = \sum_{n:\, b_n \in R_1} \langle b_n | \rho_1 | b_n \rangle = \mathrm{Tr}(\rho_1 \Pi_1), \\ Q_{FA} &= \Pr(\text{choosing } H_1 | \rho_0) = \sum_{n:\, b_n \in R_1} \langle b_n | \rho_0 | b_n \rangle = \mathrm{Tr}(\rho_0 \Pi_1). \end{aligned} \tag{5.24}$$

The conditional probabilities of correct decision are given by

$$P(C|H_1)=Q_D=\mathrm{Tr}(\rho_1\Pi_1),\ \ P(C|H_0)=1-Q_{FA}=1-\mathrm{Tr}(\rho_0\Pi_1), \tag{5.25}$$

so the probability of a correct decision is determined by

$$P(C)=(1-\zeta_0)\mathrm{Tr}(\rho_1\Pi_1)+\zeta_0[1-\mathrm{Tr}(\rho_0\Pi_1)],\ \ \zeta_0=\Pr(H_0). \tag{5.26}$$

The average probability of error is now simply

$$\begin{aligned} P_e &= 1-P(C)=1-(1-\zeta_0)\mathrm{Tr}(\rho_1\Pi_1)-\zeta_0[1-\mathrm{Tr}(\rho_0\Pi_1)] \\ &= \zeta_0\mathrm{Tr}(\rho_0\Pi_1)+(1-\zeta_0)[1-\mathrm{Tr}(\rho_1\Pi_1)]. \end{aligned} \tag{5.27}$$

Given that Tr(A)+Tr(B) = Tr($A + B$), we can rearrange the previous equation to obtain

$$P_e=(1-\zeta_0)\left\{1-\underbrace{\mathrm{Tr}\left[(\rho_1-\Lambda_0\rho_0)\Pi_1\right]}_{\mathrm{D}}\right\},\ \ \Lambda_0=\zeta_0/(1-\zeta_0), \tag{5.28}$$

where $D=(\rho_1-\Lambda_0\rho_0)\Pi_1$ is the decision operator. Minimizing the error probability is equivalent to maximizing the trace of the decision operator D. Given that $Tr(X)=\sum_i\langle a^i|X|a^i\rangle$, from the eigenvalue equation,

$$(\rho_1-\Lambda_0\rho_0)|\eta_k\rangle=\eta_k|\eta_k\rangle, \tag{5.29}$$

we can determine the eigenvalues η_k and eigenkets $|\eta_k\rangle$ of $\rho_1-\Lambda_0\rho_0$ and use them to determine the trace of the decision operator by

$$\mathrm{Tr}(D)=\mathrm{Tr}[(\rho_1-\Lambda_0\rho_0)\Pi_1]=\sum_k\langle\eta_k|(\rho_1-\Lambda_0\rho_0)\Pi_1|\eta_k\rangle=\sum_k\eta_k\langle\eta_k|\Pi_1|\eta_k\rangle \tag{5.30}$$

The trace of the decision operator is at maximum for the projection operator Π_1 for which

$$\langle\eta_k|\Pi_1|\eta_k\rangle=\begin{cases}1, & \eta_k\geq 0\\ 0, & \eta_k<0\end{cases}=U(\eta_k). \tag{5.31}$$

Therefore, the optimum projection operator is given by

$$\Pi_1^*=\sum_k U(\eta_k)|\eta_k\rangle\langle\eta_k|. \tag{5.32}$$

The probability of error for the optimum projector is now

$$P_e^*=(1-\zeta_0)\left\{1-\sum_k\eta_k U(\eta_k)\right\}. \tag{5.33}$$

The optimum receiver is simplest when operators ρ_0 and ρ_1 commute. The eigenvectors $|\eta_k\rangle$ are identical to eigenvectors $|\phi_k\rangle$ common to both ρ_0 and ρ_1, so we can write

$$\rho_j=\sum_k p_k^{(j)}|\phi_k\rangle\langle\phi_k|;\ j=0,1;\ \sum_k p_k^{(j)}=1. \tag{5.34}$$

In this case, the eigenvalues of $\rho_1 - \Lambda_0\rho_0$ are simply

$$\eta_k = p_k^{(1)} - \Lambda_0 p_k^{(0)}, \tag{5.35}$$

and when the system is in a state $|\phi_k\rangle$ for which $\eta_k > 0$, the decision rule is

$$p_k^{(1)}/p_k^{(0)} \underset{H_0}{\overset{H_1}{\gtrless}} \Lambda_0, \tag{5.36}$$

which is the same optimum decision rule as in the classical case.

5.3.1 Quantum binary decision for pure states

For pure states represented by $|\psi_0\rangle$ and $|\psi_1\rangle$, the corresponding density operators are $\rho_j = |\psi_j\rangle\langle\psi_j|$ $(j = 0,1)$, and the eigenvalues of $\rho_1 - \Lambda_0\,\rho_1$ are now linear combinations of pure states; we can write

$$|\eta_j\rangle = a_{j0}|\psi_0\rangle + a_{j1}|\psi_1\rangle. \tag{5.37}$$

Only two eigenvalues that are different from zero can be determined by solving the eigenvalue equation:

$$(\rho_1 - \Lambda_0\rho_0)|\eta_j\rangle = \eta_j|\eta_j\rangle. \tag{5.38}$$

To simplify the derivation, let us assume that $\zeta = 1/2$. Based on the previous section and Eq. (5.32), we conclude that the optimum projection operator would be $\Pi_1^* = |\eta_1\rangle\langle\eta_1|$. The corresponding detection probability Q_D and false-alarm probability Q_{FA} are given by

$$\begin{aligned} Q_D &= \mathrm{Tr}(\rho_1\Pi_1^*) = \langle\eta_1|\rho_1|\eta_1\rangle = \langle\eta_1|\psi_1\rangle\langle\psi_1|\eta_1\rangle = |\langle\psi_1|\eta_1\rangle|^2 = |a_{10}\langle\psi_1|\psi_0\rangle + a_{11}|^2 \\ &= \frac{1}{2}\left[1 + \sqrt{1 - |\langle\psi_1|\psi_0\rangle|^2}\right], \\ Q_{FA} &= \Pr(\text{choosing} H_1|\rho_0) = \mathrm{Tr}(\rho_0\Pi_1^*) = \frac{1}{2}\left[1 - \sqrt{1 - |\langle\psi_1|\psi_0\rangle|^2}\right], \end{aligned} \tag{5.39}$$

where the coefficient a_{10} is determined in such a way as to maximize the detection probability. For derivation, the interested reader is referred to Ref. [1] [see chapter 4, sect. 2(c)]. Below, we instead provide the geometric approach to maximizing the correct detection probability. The corresponding probability of correct detection is now

$$P^*(C) = \frac{1}{2}(Q_D + 1 - Q_{FA}) = \frac{1}{2}\left[1 + \sqrt{1 - |\langle\psi_1|\psi_0\rangle|^2}\right], \tag{5.40}$$

while the probability of error corresponding to the optimum detector is

$$P_e^* = 1 - P^*(C) = \frac{1}{2}\left[1 - \sqrt{1 - |\langle\psi_1|\psi_0\rangle|^2}\right]. \tag{5.41}$$

The optimization problem can be geometrically interpreted, as shown in Fig. 5.1. The eigenkets $|\eta_j\rangle$ are orthonormal and related to the signal states $|\psi_j\rangle$ by Eq. (5.37). The angle between signal states

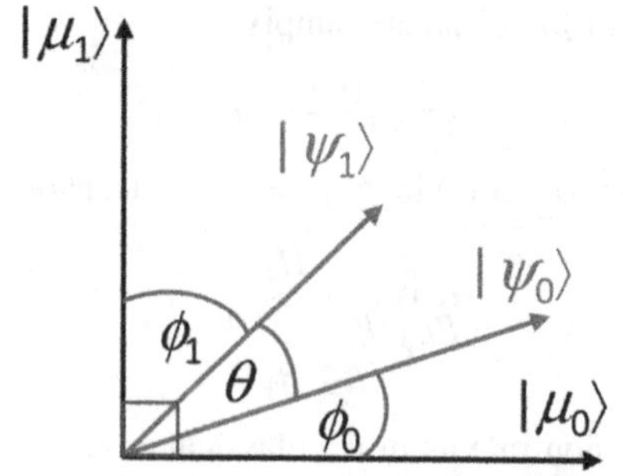

FIGURE 5.1

Geometric representation of signal pure states and measurement states related to the binary detection problem.

denoted by θ can be determined from the dot-product of signal states $\langle\psi_0|\psi_1\rangle = \cos\theta$. By denoting the angles between the signal states and eigenkets by ϕ_i as shown in Fig. 5.1, we can express the conditional probabilities for correct detection as follows:

$$P(C|H_0) = \cos^2(\phi_0) = |\langle\eta_0|\psi_0\rangle|^2, \quad P(C|H_1) = \cos^2(\phi_1) = |\langle\eta_1|\psi_1\rangle|^2. \tag{5.42}$$

From Fig. 5.1, we conclude that $\phi_0 = \pi/2 - \phi_1 - \theta$. The average probability of correct detection can now be expressed by

$$P(C) = \frac{1}{2}P(C|H_0) + \frac{1}{2}P(C|H_1) = \frac{1}{2}\cos^2\left(\frac{\pi}{2} - \phi_1 - \theta\right) + \frac{1}{2}\cos^2(\phi_1). \tag{5.43}$$

By taking the first derivative of $P(C)$ with respect to ϕ_1 and equating this derivative to zero, we obtain the optimum value for angle $\phi_1^* = (\pi/2 - \theta)/2$, which after substituting into Eq. (5.43) yields

$$P^*(C) = \frac{1}{2}\cos^2\left(\frac{\frac{\pi}{2} - \theta}{2}\right) + \frac{1}{2}\cos^2\left(\frac{\frac{\pi}{2} - \theta}{2}\right) = \frac{1}{2}\left[1 + \cos\left(\frac{\pi}{2} - \theta\right)\right] = \frac{1}{2}[1 + \sin(\theta)]$$

$$= \frac{1}{2}\left[1 + \sqrt{1 - \cos^2\theta}\right] \overset{\cos^2\theta = |\langle\psi_0|\psi_1\rangle|^2}{\widehat{=}} \frac{1}{2}\left[1 + \sqrt{1 - |\langle\psi_0|\psi_1\rangle|^2}\right], \tag{5.44}$$

thus proving Eq. (5.40).

When transmitted symbols are not equally likely, that is, $\zeta_0 = \Pr(H_0) \neq \zeta_1 = \Pr(H_1)$, the corresponding expression for the average error probability of optimum detector will be

$$P_e^* = 1 - P^*(C) = \frac{1}{2}\left[1 - \sqrt{1 - 4\zeta_0\zeta_1|\langle\psi_1|\psi_0\rangle|^2}\right]. \tag{5.45}$$

This expression is commonly referred to in the literature as the *Helstrom bound*.

Before continuing our discussion on quantum optical communication, it is convenient to introduce the coherent state representation.

5.4 Coherent states, quadrature operators, and uncertainty relations

Monochromatic plane wave can be represented as [8–10]

$$\boldsymbol{E}(\boldsymbol{r},t)=\boldsymbol{e}|\boldsymbol{E}|[I\cos(\boldsymbol{kr}-\omega t)-Q\sin(\boldsymbol{kr}-\omega t)] \tag{5.46}$$

wherein the first component represents the in-phase and the second the quadrature component, with (I, Q) representing the coordinates of the corresponding signal constellation defined in the signal space. The vector $\boldsymbol{e}$ denotes the polarization orientation, $\boldsymbol{k}$ is the wave vector, and $\boldsymbol{r}$ is the position vector.

In quantum optics [11–17], the quadrature components I and Q are replaced with corresponding operators $\widehat{I}$ and $\widehat{Q}$, equivalent to the position and momentum operators satisfying a similar commutation relationship:

$$[\widehat{I},\widehat{Q}]=2jN_0, \tag{5.47}$$

where N_0 represents the variance in vacuum fluctuation (shot noise). (Very often in quantum optics, different notation is used for the quadrature operators, namely, $\widehat{q}$ and $\widehat{p}$.) The eigenstates of $\widehat{I}$ can be denoted $|I\rangle$ since $\widehat{I}|I\rangle=I|I\rangle$. The annihilation a and creation $a^{\dagger}$ operators [4,6,18] are related to the quadrature operators by

$$a=\frac{\widehat{I}+j\widehat{Q}}{2\sqrt{N_0}},\quad a^{\dagger}=\frac{\widehat{I}-j\widehat{Q}}{2\sqrt{N_0}}. \tag{5.48}$$

In other words, we can express the quadrature operators in terms of the annihilation and creation operators as follows:

$$\widehat{I}=(a^{\dagger}+a)\sqrt{N_0},\quad \widehat{Q}=j(a^{\dagger}-a)\sqrt{N_0}. \tag{5.49}$$

The *photon number operator* is related to the creation/annihilation and quadrature operators by

$$N=a^{\dagger}a=\frac{\widehat{I}+j\widehat{Q}}{2\sqrt{N_0}}\frac{\widehat{I}-j\widehat{Q}}{2\sqrt{N_0}}=\frac{1}{4N_0}\left[(\widehat{I}^2+\widehat{Q}^2)-j\underbrace{(\widehat{I}\widehat{Q}-\widehat{Q}\widehat{I})}_{[\widehat{I},\widehat{Q}]=2N_0j}\right]=\frac{1}{4N_0}[(\widehat{I}^2+\widehat{Q}^2)+2N_0], \tag{5.50}$$

The Fock (photon number) states are eigenkets of the photon number operator:

$$N|n\rangle=n|n\rangle, \tag{5.51}$$

where n represents the number of photons.

The *vacuum state*, denoted $|0\rangle$, is an eigenket of the annihilation operator:

$$a|0\rangle=0\Rightarrow\langle 0|a^{\dagger}a|0\rangle=0. \tag{5.52}$$

The creation and annihilation operators can be respectively defined as follows [4,6,18]:

$$\begin{aligned}a^{\dagger}|n\rangle&=(n+1)^{1/2}|n+1\rangle,\quad n\geq 0\\ a|n\rangle&=n^{1/2}|n-1\rangle,\quad n\geq 1,\quad a|0\rangle=0.\end{aligned} \tag{5.53}$$

The creation operator raises the number of photons by one, while the annihilation operator reduces the number of photons by one.

The right eigenvector of the annihilation operator is commonly referred to as the *coherent state* and is defined as

$$a|\alpha\rangle = \alpha|\alpha\rangle, \quad \alpha = \alpha_I + j\alpha_Q, \tag{5.54}$$

where $\alpha = \alpha_I + j\alpha_Q$ is a complex number. On the other hand, the dual state $\langle\alpha|$ is the left eigenket of the creation operator $a^\dagger$; that is, $\langle\alpha|a^\dagger = \alpha^*\langle\alpha|$.

Let us now represent the coherent state in terms of number states. Using the completeness relationship, we can write

$$|\alpha\rangle = \sum_n (|n\rangle\langle n|)|\alpha\rangle = \sum_n |n\rangle\langle n|\alpha\rangle. \tag{5.55}$$

Since the number state $|n\rangle$ can be represented in terms of the ground state by Refs. [4,18]:

$$|n\rangle = (a^\dagger)^n|0\rangle/\sqrt{n!}, \tag{5.56}$$

the projection coefficient can be determined as

$$\langle n|\alpha\rangle = \langle 0|\underbrace{a^n|\alpha\rangle}_{\alpha^n|\alpha\rangle} / \sqrt{n!} = \langle 0|\alpha\rangle\alpha^n / \sqrt{n!}, \tag{5.57}$$

and the coherent state after substituting Eq. (5.57) into Eq. (5.55) becomes

$$|\alpha\rangle = \langle 0|\alpha\rangle \sum_n \frac{\alpha^n}{\sqrt{n!}}|n\rangle. \tag{5.58}$$

The projection along the ground state $\langle 0|\alpha\rangle$ can be determined from the normalization condition:

$$1 = \sum_n |\langle n|\alpha\rangle|^2 = \sum_n \langle 0|\alpha\rangle\frac{\alpha^n}{\sqrt{n!}}\langle\alpha|0\rangle\frac{\alpha^{*n}}{\sqrt{n!}} = |\langle 0|\alpha\rangle|^2 \underbrace{\sum_n \frac{\left(|\alpha|^2\right)^n}{n!}}_{e^{|\alpha|^2}} = |\langle 0|\alpha\rangle|^2 e^{|\alpha|^2}, \tag{5.59}$$

and by solving for $\langle 0|\alpha\rangle$, we obtain

$$\langle 0|\alpha\rangle = e^{-|\alpha|^2/2}, \tag{5.60}$$

and after substituting Eq. (5.60) into Eq. (5.58), the coherent state is represented by

$$|\alpha\rangle = e^{-|\alpha|^2/2}\sum_n \frac{\alpha^n}{\sqrt{n!}}|n\rangle. \tag{5.61}$$

Another convenient form can be obtained by expressing the number operator in terms of the ground state based on Eq. (5.56) to obtain

$$|\alpha\rangle = e^{-|\alpha|^2/2}\sum_n \frac{\alpha^n}{\sqrt{n!}}(a^\dagger)^n|0\rangle / \sqrt{n!} = e^{-|\alpha|^2/2}\underbrace{\sum_n \frac{(\alpha a^\dagger)^n}{n!}}_{\exp(\alpha a^\dagger)}|0\rangle = e^{-\frac{|\alpha|^2}{2}+\alpha a^\dagger}|0\rangle, \tag{5.62}$$

Given that the action of a^n $(n > 0)$ on the vacuum state is $a^n|0\rangle = 0$, we conclude that

$$\sum_n \frac{(-\alpha^* a)^n}{n!}|0\rangle = \exp(-\alpha^* a)|0\rangle = |0\rangle, \tag{5.63}$$

and we can rewrite the previous equation as

$$|\alpha\rangle = e^{-\frac{|\alpha|^2}{2}+\alpha a^\dagger}|0\rangle = e^{-\frac{|\alpha|^2}{2}+\alpha a^\dagger}e^{-\alpha^* a}|0\rangle. \tag{5.64}$$

Given that operators $A = \alpha a^\dagger$, $B = -\alpha^* a$ do not commute since $[A,B] = -|\alpha|^2 a^\dagger a + |\alpha|^2 a a^\dagger = |\alpha|^2 \overbrace{[a, a^\dagger]}^{\hat{1}} = |\alpha|^2$, while $[A,[A,B]] = [B,[A,B]] = 0$, so we can apply the disentangling theorem, which is the special case of the Baker–Campbell–Hausdorff theorem [4,18], namely,

$$e^A e^B e^{-\frac{1}{2}[A,B]} = e^{A+B} = e^B e^A e^{\frac{1}{2}[A,B]}, \tag{5.65}$$

and rewrite Eq. (5.64) to obtain

$$|\alpha\rangle = e^{-\frac{1}{2}\overbrace{|\alpha|^2}^{[A,B]} + \overbrace{\alpha a^\dagger}^{A}} e^{\overbrace{-\alpha^* a}^{B}}|0\rangle = \overbrace{e^{\alpha a^\dagger - \alpha^* a}}^{D(\alpha)}|0\rangle = D(\alpha)|0\rangle, \tag{5.66}$$

where $D(\alpha) = e^{\alpha a^\dagger - \alpha^* a}$ is the *displacement operator*. The displacement operator describes the coherent state as the displacement of another coherent state in the I-Q space. Therefore, we can also define the coherent state by applying the displacement operator on the vacuum state; that is,

$$|\alpha\rangle = D(\alpha)|0\rangle, \quad D(\alpha) = e^{\alpha a^\dagger - \alpha^* a}. \tag{5.67}$$

Because $D^\dagger(\alpha) = D(-\alpha)$, the displacement operator is unitary; namely, we can write

$$D(\alpha)D^\dagger(\alpha) = D(\alpha)D(-\alpha) = e^{-\frac{|\alpha|^2}{2}}e^{\alpha a^\dagger}\overbrace{e^{-\alpha^* a}e^{\alpha^* a}}^{\hat{1}}e^{-\alpha a^\dagger}e^{\frac{|\alpha|^2}{2}} = e^{\alpha a^\dagger}e^{-\alpha a^\dagger} = \hat{1}. \tag{5.68}$$

The product of the two displacement operators $D(\alpha_1)$ and $D(\alpha_2)$ is another displacement operator, $D(\alpha_1+\alpha_2)$, except for the global phase shift; namely, we can write

$$D(\alpha_1 + \alpha_2) = e^{(\alpha_1+\alpha_2)a^\dagger - (\alpha_1^*+\alpha_2^*)a} = e^{\overbrace{\alpha_1 a^\dagger - \alpha_1^* a}^{A} + \overbrace{\alpha_2 a^\dagger - \alpha_2^* a}^{B}}$$

$$= \overbrace{e^A}^{D(\alpha_1)} \overbrace{e^B}^{D(\alpha_2)} e^{-\frac{1}{2}\overbrace{\alpha_1\alpha_2^* - \alpha_1^*\alpha_2}^{[A,B]}} = e^{-j\mathrm{Im}\{\alpha_1\alpha_2^* - \alpha_1^*\alpha_2\}}D(\alpha_1)D(\alpha_2). \tag{5.69}$$

and thus, the displacement operator represents the semigroup operation.

The expected value of the number operator is

$$\overline{N} = \langle\alpha|N|\alpha\rangle = \underbrace{\langle\alpha|a^\dagger}_{\langle\alpha|\alpha^*} \underbrace{a|\alpha\rangle}_{\alpha|\alpha\rangle} = |\alpha|^2. \tag{5.70}$$

On the other hand, the expected value of the square of the number operator is

$$\langle N^2\rangle = \langle\alpha|N^2|\alpha\rangle = \langle\alpha|NN|\alpha\rangle = \underbrace{\langle\alpha|a^\dagger}_{\langle\alpha|\alpha^*}\, aa^\dagger\, \underbrace{a|\alpha\rangle}_{\alpha|\alpha\rangle} = |\alpha|^2\langle\alpha|\underbrace{aa^\dagger}_{\widehat{a^\dagger a+1}}|\alpha\rangle = |\alpha|^4 + |\alpha|^2 = \overline{N}^2 + \overline{N}. \quad (5.71)$$

The corresponding variance is

$$\sigma^2(N) = \mathrm{Var}(N) = \langle N^2\rangle - \langle N\rangle^2 = \overline{N}^2 + \overline{N} - \overline{N}^2 = \overline{N}, \quad (5.72)$$

and the fractional uncertainty in the photon number would be

$$\frac{\sigma(N)}{\overline{N}} = \frac{\sqrt{\overline{N}}}{\overline{N}} = \frac{1}{\sqrt{\overline{N}}} \xrightarrow{N\to\infty} 0. \quad (5.73)$$

The probability of detecting n photons is given by

$$P_n = |\langle n|\alpha\rangle|^2 = e^{-|\alpha|^2}\frac{|\alpha|^{2n}}{n!} = e^{-\overline{N}}\frac{\overline{N}^n}{n!}, \quad (5.74)$$

which represents the *Poisson distribution* with a mean value of $|\alpha|^2$. Fig. 5.2 illustrates the coherent state photon number distribution for two mean values.

The *density operator* of a coherent state $|\alpha\rangle$ is determined by

$$\rho = |\alpha\rangle\langle\alpha| = e^{-|\alpha|^2}\sum_{n=0}^{+\infty}\sum_{m=0}^{+\infty}(n!)^{-1/2}(m!)^{-1/2}\alpha^n(\alpha^*)^m|n\rangle\langle m|. \quad (5.75)$$

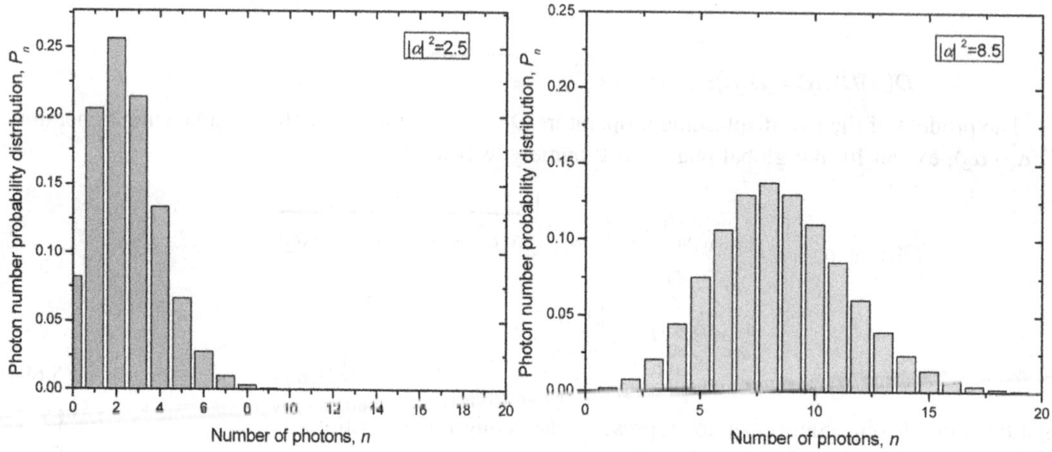

FIGURE 5.2

Photon number probability distributions for coherent states with (left) $|\alpha|^2 = 2.5$ and (right) $|\alpha|^2 = 8.5$.

The diagonal terms of ρ,

$$\langle n|\rho|n\rangle = e^{-|\alpha|^2}\frac{|\alpha|^{2n}}{n!} \tag{5.76}$$

follow the Poisson distribution as expected. On the other hand, the off-diagonal terms of ρ are given by

$$\langle m|\rho|n\rangle = e^{-|\alpha|^2}\frac{\alpha^n(\alpha^*)^m}{\sqrt{n!m!}}. \tag{5.77}$$

The coherent states are not orthogonal because

$$\begin{aligned} |\langle\alpha|\beta\rangle|^2 &= \left|e^{-(|\alpha|^2+|\beta|^2)/2}\sum_{n=0}^{+\infty}\sum_{m=0}^{+\infty}(n!)^{-1/2}(m!)^{-1/2}\alpha^n(\beta^*)^m\langle n|m\rangle\right|^2 \\ &= \left|e^{-(|\alpha|^2+|\beta|^2)/2}\sum_{n=0}^{+\infty}(n!)^{-1}\alpha^n(\beta^*)^n\right|^2 = \left|e^{-(|\alpha|^2+|\beta|^2-2\alpha\beta^*)/2}\right|^2 \\ &= e^{-|\alpha-\beta|^2}. \end{aligned} \tag{5.78}$$

and when $|\alpha-\beta|^2$ is large enough, they become orthogonal.

Very often, quadrature operators are expressed in terms of *shot noise units* (SNU), where N_0 is normalized to 1, that is, $a^\dagger = (\hat{I}-j\hat{Q})/2,\ a = (\hat{I}+j\hat{Q})/2$. In other words, the quadrature operators representing the in-phase and quadrature components of the electromagnetic field are related to the creation and annihilation operators (in SNU) by $\hat{I} = a^\dagger + a,\ \hat{Q} = j(a^\dagger - a)$. The quadrature operators satisfy the commutation relation:

$$\begin{aligned} [\hat{I},\hat{Q}] &= [a^\dagger + a, j(a^\dagger - a)] = [a^\dagger, j(a^\dagger - a)] + [a, j(a^\dagger - a)] \\ &= \underbrace{j[a^\dagger,a^\dagger]}_{=0}\ \underbrace{-j[a^\dagger,a]}_{-1}\ \underbrace{+j[a,a^\dagger]}_{1}\ \underbrace{-j[a,a]}_{0} \\ &= 2j, \end{aligned} \tag{5.79}$$

where in the first line, we employ the property of the commutators [1] $[A+B,C]=[A,C]+[B,C]$, while in the second line, the two commutation relations of the creation/annihilation operators of the harmonic oscillator [18] are $\left[a_j,a_k^\dagger\right] = \delta_{jk}\hat{1},\ \left[a_j^\dagger,a_k^\dagger\right] = [a_j,a_k] = 0$, thus proving Eq. (5.47).

The expected values of quadrature operators with respect to the coherent state $|\alpha\rangle$ are given by

$$\begin{aligned} \langle\hat{I}\rangle_\alpha &= \langle\alpha|\hat{I}|\alpha\rangle = \langle\alpha|a+a^\dagger|\alpha\rangle\sqrt{N_0} = \langle\alpha|\alpha+\alpha^*|\alpha\rangle\sqrt{N_0} = 2\mathrm{Re}\{\alpha\}\sqrt{N_0}, \\ \langle\hat{Q}\rangle_\alpha &= \langle\alpha|\hat{Q}|\alpha\rangle = j\langle\alpha|a^\dagger - a|\alpha\rangle\sqrt{N_0} = j\langle\alpha|\alpha^* - \alpha|\alpha\rangle\sqrt{N_0} = 2\mathrm{Im}\{\alpha\}\sqrt{N_0}. \end{aligned} \tag{5.80}$$

If we then apply the scaling factor $1/(2\sqrt{N_0})$ on α, the expected values of in-phase and quadrature operators will represent the real and imaginary parts of α. For the vacuum state, the expected values of the quadrature operators equal zero, and therefore, Eq. (5.80) defines the displacement. The expected values of $\hat{I}^2$ and $\hat{Q}^2$ are given by

$$\langle \hat{I}^2 \rangle_\alpha = \langle \alpha | \hat{I}^2 | \alpha \rangle = \langle \alpha | (a + a^\dagger)(a + a^\dagger) | \alpha \rangle N_0 = \langle \alpha | a^2 + \underbrace{aa^\dagger}_{a^\dagger a + \hat{1}} + a^\dagger a + (a^\dagger)^2 | \alpha \rangle N_0$$

$$= \langle \alpha | \alpha^2 + 2\alpha^* \alpha + \alpha^{*2} + 1 | \alpha \rangle = \left[(\alpha + \alpha^*)^2 + 1 \right] N_0 = \left(4\mathrm{Re}^2\{\alpha\} + 1 \right) N_0 = \left(4\alpha_I^2 + 1 \right) N_0,$$

$$\langle \hat{Q}^2 \rangle_\alpha = \langle \alpha | \hat{Q}^2 | \alpha \rangle = j^2 \langle \alpha | (a^\dagger - a)(a^\dagger - a) | \alpha \rangle N_0 = \left[1 - (\alpha^* - \alpha)^2 \right] N_0. \tag{5.81}$$

The corresponding variances for the quadrature operators will then be

$$\begin{aligned} \mathrm{Var}(\hat{I}) &= V(\hat{I}) = \langle \hat{I}^2 \rangle_\alpha - \left[\langle \hat{I} \rangle_\alpha \right]^2 = \left[(\alpha + \alpha^*)^2 + 1 \right] N_0 - (\alpha + \alpha^*)^2 N_0 = N_0, \\ \mathrm{Var}(\hat{Q}) &= V(\hat{Q}) = \langle \hat{Q}^2 \rangle_\alpha - \left[\langle \hat{Q} \rangle_\alpha \right]^2 = \left[1 - (\alpha^* - \alpha)^2 \right] N_0 + (\alpha^* - \alpha)^2 N_0 = N_0, \end{aligned} \tag{5.82}$$

indicating that the variances in the fluctuations for both quadratures are the same. By setting $\alpha = 0$, the same conclusion applies to the vacuum state. In the (I,Q) plane, also known as the *phase-space* plane, the coherent state can be represented by a circle with a radius related to the uncertainty, as illustrated in Fig. 5.3, where we denoted the uncertainties in the I- and Q-directions as ΔI and ΔQ. The center of the coherent state representation is determined by

$$\langle \hat{I} + j\hat{Q} \rangle_\alpha = \langle \hat{I} \rangle_\alpha + j \langle \hat{Q} \rangle_\alpha = \alpha + \alpha^* + j \cdot j(\alpha^* - \alpha) = 2\alpha = 2|\alpha| e^{j\mathrm{Arg}\{\alpha\}}.$$

Interestingly enough, the product of the variances of the quadratures is given by

$$V(\hat{I}) V(\hat{Q}) = N_0^2, \tag{5.83a}$$

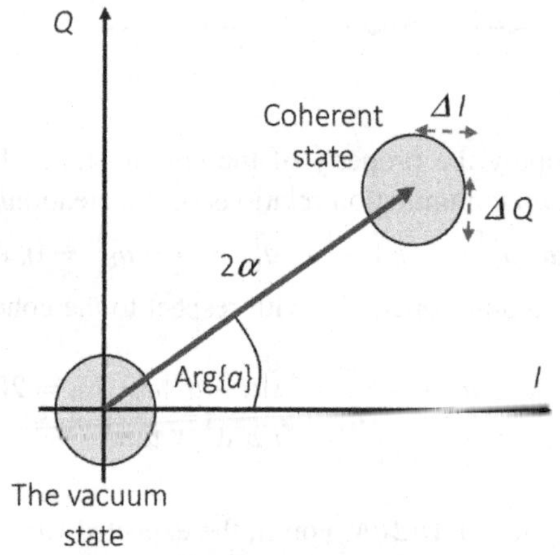

FIGURE 5.3

The coherent states.

indicating that the Heisenberg uncertainty principle is satisfied with the equality sign. In other words, the coherent state satisfies the minimum uncertainty relation, and this observation can be used as another definition for the coherent state. Very often, the uncertainty principle is expressed in terms of SNU as follows:

$$V(\hat{I})V(\hat{Q}) = 1, \ N_0 = 1 \tag{5.83b}$$

Now we have built sufficient background knowledge to continue our discussion on quantum optical communication.

5.5 Binary quantum optical communication in the absence of background radiation

We start our discussion with the classical photon-counting receiver.

5.5.1 Classical photon-counting receiver

In the OOK modulation scheme, the hypotheses H_1 and H_0 represent the presence and absence of the signal. In the absence of background radiation, either no photons or an average of $|\alpha|^2 = \lambda$ photons per signal is received, so the corresponding distributions are

$$P(n|H_1) = e^{-\lambda}\frac{\lambda^n}{n!}, \ P(n|H_0) = \begin{cases} 1, & n = 0 \\ 0, & n \geq 1 \end{cases}. \tag{5.84}$$

In the absence of background noise, hypothesis H_0 will always be properly detected. When the signal is sent but no photon is detected, an error will occur, with the average probability of error determined by

$$P_e = \frac{1}{2}P(n=0|H_1) = \frac{1}{2}e^{-\lambda} = \frac{1}{2}e^{-|\alpha|^2}. \tag{5.85}$$

The average number of photons transmitted, assuming equally probable transmission, will be $N_s = 0.5 \cdot 0 + 0.5 \cdot |\alpha|^2 = 0.5 \cdot |\alpha|^2$. The probability of error expressed in terms of the average number of photons per symbol N_s will then be

$$P_e = \frac{1}{2}e^{-|\alpha|^2} = \frac{1}{2}e^{-2N_s}. \tag{5.86}$$

5.5.2 Quantum photon-counting receiver

The transmitted symbols 0 and 1 are represented by the vacuum state and coherent state, respectively:

$$\begin{aligned} &H_0: |\psi_0\rangle = |0\rangle, \\ &H_1: |\psi_1\rangle = e^{-|\alpha|^2/2}\sum_{n=0}^{\infty}\frac{\alpha^n}{(n!)^{1/2}}|n\rangle. \end{aligned} \tag{5.87}$$

The corresponding density operators are given by

$$\rho_0 = |\psi_0\rangle\langle\psi_0| = |0\rangle\langle 0|,$$
$$\rho_1 = |\psi_1\rangle\langle\psi_1| = e^{-|\alpha|^2}\sum_{n=0}^{+\infty}\sum_{m=0}^{+\infty}(n!)^{-1/2}(m!)^{-1/2}\alpha^n(\alpha^*)^m|n\rangle\langle m|. \tag{5.88}$$

The projection operators corresponding to the hypotheses H_i $(i=0,\,1)$ need to determine whether the received state is in the ground state and can be defined by

$$\Pi_0 = |0\rangle\langle 0| \text{ and } \Pi_1 = \hat{1} - |0\rangle\langle 0| = \sum_{n=1}^{\infty}|n\rangle\langle n|. \tag{5.89}$$

Let us look into the action of the projection operator on received states:

$$\Pi_0|\psi_0\rangle = |0\rangle\langle 0|0\rangle \doteq |w_0\rangle,$$
$$\Pi_1|\psi_1\rangle = \sum_{n=1}^{\infty}|n\rangle\langle n|\left(e^{-|\alpha|^2/2}\sum_{m=0}^{\infty}\frac{\alpha^m}{(m!)^{1/2}}|m\rangle\right) = e^{-|\alpha|^2/2}\sum_{n=1}^{\infty}\sum_{m=0}^{\infty}\frac{\alpha^m}{(m!)^{1/2}}|n\rangle\underbrace{\langle n|m\rangle}_{\delta_{mn}}$$
$$= e^{-|\alpha|^2/2}\sum_{n=1}^{\infty}\frac{\alpha^n}{(n!)^{1/2}}|n\rangle \doteq |W_1\rangle \Rightarrow |w_1\rangle = \frac{|W_1\rangle}{\||W_1\rangle\|}. \tag{5.90}$$

The action of the projection operator Π_0 on the ground state is to recover the ground state, and the measurement state is $|w_0\rangle = |0\rangle$. On the other hand, the action of the projection operator Π_1 on the coherent state is to achieve the measurement state $|w_1\rangle$, which is the "complement" of the ground state (it contains only nonzero photon number terms). Those two measurement states are orthogonal since $\langle w_0|W_1\rangle = {}^{-|\alpha|^2/2}\sum_{n=1}^{\infty}\frac{\alpha^n}{(n!)^{1/2}}\underbrace{\langle 0|n\rangle}_{0} = 0.$

The conditional probabilities for correct decisions, corresponding to H_i $(i=0,\,1)$ are given, respectively, by

$$\Pr(n=0|H_0) = \mathrm{Tr}(\rho_0\Pi_0) = \langle 0|\Pi_0|0\rangle = |\langle w_0|\psi_0\rangle|^2 = 1 \text{ and}$$
$$\Pr(n\ge 1|H_1) = \mathrm{Tr}(\rho_1\Pi_1) = \langle\psi_1|\Pi_1|\psi_1\rangle = \underbrace{e^{-|\alpha|^2/2}\sum_{k=0}^{\infty}\frac{(\alpha^*)^k}{(k!)^{1/2}}\langle k|}_{\langle\psi_1|}\underbrace{\sum_{n=1}^{\infty}|n\rangle\langle n|}_{\Pi_1}\underbrace{e^{-|\alpha|^2/2}\sum_{m=0}^{\infty}\frac{\alpha^m}{(m!)^{1/2}}|m\rangle}_{|\psi_1\rangle}$$
$$= e^{-|\alpha|^2}\sum_{k=0}^{\infty}\sum_{n=1}^{\infty}\sum_{m=0}^{\infty}\frac{(\alpha^*)^k}{(k!)^{1/2}}\frac{\alpha^m}{(m!)^{1/2}}\underbrace{\langle k|n\rangle}_{\delta_{kn}}\underbrace{\langle n|m\rangle}_{\delta_{nm}} = e^{-|\alpha|^2}\sum_{n=1}^{\infty}\frac{|\alpha|^{2n}}{n!} = e^{-|\alpha|^2}\left(e^{|\alpha|^2}-1\right)$$
$$= 1 - e^{-|\alpha|^2}. \tag{5.91}$$

The average probability of a correct decision, assuming equally probable transmission, is obtained by averaging the conditional probabilities by

$$P_c = \underbrace{P(C|H_0)}_{1}\,\underbrace{P(H_0)}_{1/2} + \underbrace{P(C|H_1)}_{1-e^{-|\alpha|^2}}\,\underbrace{P(H_1)}_{1/2} = 1 - e^{-|\alpha|^2}/2. \tag{5.92}$$

The average probability of error is 1 minus the probability of a correct decision; that is,

$$P_e = 1 - P(C) = \frac{1}{2}e^{-|\alpha|^2} = \frac{1}{2}e^{-2N_s}, \tag{5.93}$$

which is the same as for the classical photon-counting receiver.

5.5.3 Optimum quantum detection for on–off keying

Similarly to the case of the quantum photon-counting receiver, the transmitted symbols 0 and 1 for OOK signaling for optimum quantum detection are represented by the vacuum state and coherent state, respectively:

$$H_0\colon |\psi_0\rangle = |0\rangle \text{ and } H_1\colon |\psi_1\rangle = e^{-|\alpha|^2/2}\sum_{n=0}^{\infty}\frac{\alpha^n}{(n!)^{1/2}}|n\rangle. \tag{5.94}$$

The error probability for a binary optimum quantum receiver when the symbols are represented by the pure states was derived in Section 5.3.1 (see Eq. 5.41). This equation requires the determination of the magnitude squared of the dot product between two states. Given that the dot product magnitude square of two coherent states $|\alpha\rangle$ and $|\beta\rangle$ is $|\langle\alpha|\beta\rangle|^2 = e^{-|\alpha-\beta|^2}$ (see Eq. 5.78), the corresponding norm of the dot product for OOK will be $|\langle 0|\alpha\rangle|^2 = e^{-|\alpha-0|^2} = e^{-|\alpha|^2}$, and after substituting into Eq. (5.41), we obtain the following expression for the average error probability:

$$P_e^* = \frac{1}{2}\left[1 - \sqrt{1 - |\langle\psi_1|\psi_0\rangle|^2}\right] = \frac{1}{2}\left[1 - \sqrt{1 - e^{-|\alpha|^2}}\right]. \tag{5.95a}$$

Because the *average number of photons/symbol* is $N_s = 0.5(0) + 0.5|\alpha|^2 = 0.5|\alpha|^2$, we can express the minimum error probability in terms of N_s by

$$P_e^* = \frac{1}{2}\left[1 - \sqrt{1 - e^{-2N_s}}\right]. \tag{5.95b}$$

When transmitted symbols are not equally likely, that is, $\zeta_0 = \Pr(H_0) \neq \zeta_1 = \Pr(H_1)$, the corresponding expression for the average error probability of the optimum OOK detector will be

$$P_{e,\text{OOK}}^* = \frac{1}{2}\left[1 - \sqrt{1 - 4\zeta_0\zeta_1 e^{-2N_s}}\right]. \tag{5.96}$$

5.5.4 Optimum quantum detection for binary phase-shift keying

The transmitted symbols 0 and 1 for BPSK signaling for optimum quantum detection are represented by coherent states with phase 0 and π, respectively:

$$H_0\colon |\psi_0\rangle = |-\alpha\rangle,\ \ H_1\colon |\psi_1\rangle = |\alpha\rangle = e^{-|\alpha|^2/2}\sum_{n=0}^{\infty}\frac{\alpha^n}{(n!)^{1/2}}|n\rangle. \tag{5.97}$$

The magnitude squared of the dot product between two signaling states is given by $|\langle -\alpha|\alpha\rangle|^2 = e^{-|\alpha-(-\alpha)|^2} = e^{-4|\alpha|^2}$. Given that the average number of photons/symbol is $N_s = 0.5(|\alpha|^2) + 0.5|-\alpha|^2 = |\alpha|^2$, we can express the minimum error probability for BPSK by

$$P^*_{e,\text{BPSK}} = \frac{1}{2}\left[1 - \sqrt{1 - |\langle\psi_1|\psi_0\rangle|^2}\right] = \frac{1}{2}\left[1 - \sqrt{1 - e^{-4|\alpha|^2}}\right] = \frac{1}{2}\left[1 - \sqrt{1 - e^{-4N_s}}\right]. \tag{5.98}$$

5.5.5 Near-optimum quantum detection for binary phase-shift keying (Kennedy receiver)

Near-optimum performance for BPSK can be achieved by adding a local laser field of the same amplitude in phase with the received signal followed by photon counting, as illustrated in Fig. 5.4. As this receiver was proposed by Kennedy [19], it is known as the Kennedy receiver. The key idea is to apply the displacement operator $D(\alpha)$ to the incoming coherent state $|\beta\rangle$, $\beta \in \{\alpha, -\alpha\}$ to obtain

$$D(\alpha)|\beta\rangle = |\beta + \alpha\rangle = \begin{cases} |2\alpha\rangle, & \beta = \alpha \\ |0\rangle, & \beta = -\alpha \end{cases} \tag{5.99}$$

The displacement operator effectively converts BPSK to OOK, so the single-photon detector or photon-counting receiver can be used to distinguish the received quantum states. The error probability of the Kennedy receiver will be

$$P_{e,\text{Kennedy}} = \frac{1}{2}e^{-|2\alpha|^2} = \frac{1}{2}e^{-4|\alpha|^2} = \frac{1}{2}e^{-4N_s}. \tag{5.100}$$

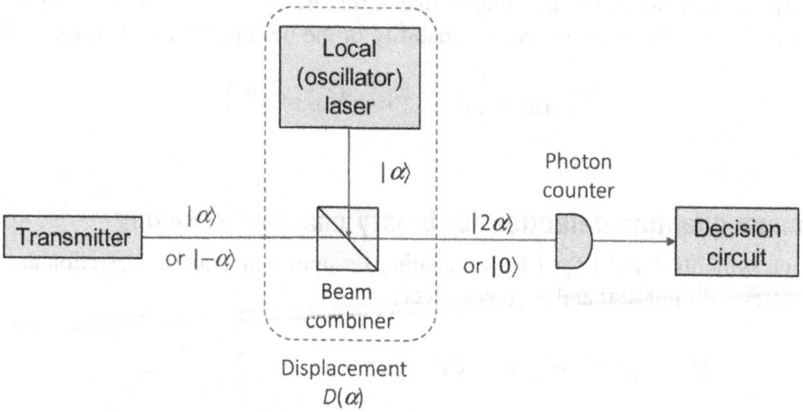

FIGURE 5.4

Kennedy's near-optimum binary phase-shift keying receiver.

5.5.6 Dolinar receiver

The Dolinar receiver [20] represents an extension of the Kennedy receiver that employs a photon counter and feedback control. It represents an adaptive displacement scheme that requires multiple copies of the transmitted states and adjusts the displacement as illustrated in Fig. 5.5. Whenever the photon counter detects the photon, the corresponding preliminary decision d_t becomes complemented, the phase modulator within the polar modulator introduces the phase shift of $d_t\pi$, and the amplitude modulator within the polar modulator pulse shapes the local oscillator (LO) laser signal to s_{d_t}. The key problem is determining the pulse shapes $s_i(t)$ so that $\Pr(d_t = d)$, where d is the transmitted bit at time instance t. In Ref. [21], this problem was solved by dynamic programming, whereas in Ref. [22], the authors provided an analytical approach that is briefly described here.

Without loss of generality, let us assume that the symmetric solution $s_0(t) = -s_1(t)$ exists. Assume further that process d_t can be modeled as a telegraphic process described by a nonhomogeneous Poisson process with the corresponding rates being [22]

$$\lambda_-(t) = |\alpha - s_0(t)|^2 \text{ and } \lambda_+(t) = |\alpha + s_0(t)|^2. \tag{5.101}$$

Let the conditional probability $P_0(t)$ denote $\Pr(d_t = 0|d = 0)$, and let the number of arrivals during interval $[t,t+\Delta t]$ be denoted $N_{[t,t+\Delta t]}$. The conditional probability $P_0(t+\Delta t)$ can then be estimated by Ref. [22]:

$$\begin{aligned} P_0(t+\Delta t) &\simeq \Pr\left[d_t = 0, N_{[t,t+\Delta t]} = 0\middle|d = 0\right] + \Pr\left[d_t = 1, N_{[t,t+\Delta t]} = 1\middle|d = 0\right] \\ &= \Pr\left[d_t = 0\middle|N_{[t,t+\Delta t]} = 0\right]P_0(t) + \Pr\left[d_t = 1\middle|N_{[t,t+\Delta t]} = 1\right][1 - P_0(t)] \\ &= [1 - \lambda_-(t)\Delta t]P_0(t) + \lambda_+(t)\Delta t[1 - P_0(t)]. \end{aligned} \tag{5.102}$$

As $\Delta t \to 0$, the previous equation becomes the differential equation

$$P_0'(t) = \lambda_-(t) - [\lambda_-(t) + \lambda_+(t)]P_0(t). \tag{5.103}$$

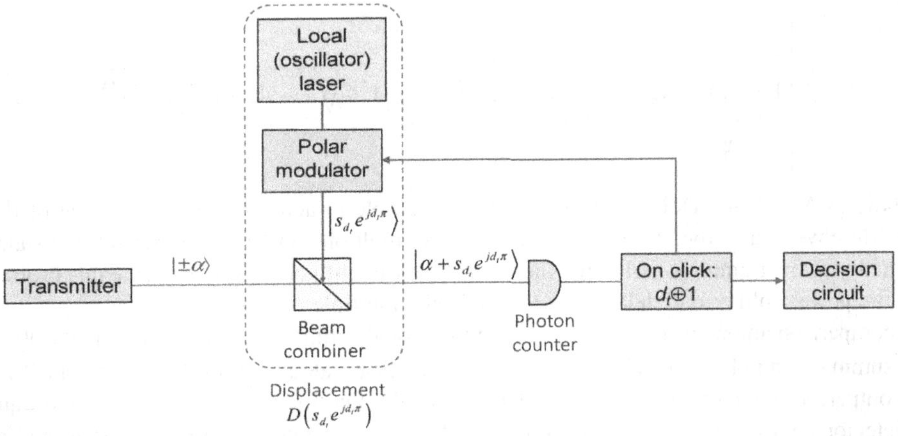

FIGURE 5.5

Dolinar receiver.

A similar differential equation holds for $P_1(t)$. The probability of a correct decision satisfies the following differential equation [22]:

$$P_c'(t) = \zeta_0 P_0'(t) + \zeta_1 P_0'(t) = \lambda_-(t) - [\lambda_-(t) + \lambda_+(t)]P_c(t). \tag{5.104}$$

Given that the goal is to achieve the Helstrom bound, we formulate the correct decision probability to be

$$P_c(t) = \frac{1}{2}\left[1 + \sqrt{1 - 4\zeta_0\zeta_1 e^{-4\alpha^2 t}}\right], \tag{5.105}$$

and from Eq. (5.104), we obtain the following solution for $s_0(t)$ [22]:

$$s_0(t) = \frac{\alpha}{\sqrt{1 - 4\zeta_0\zeta_1 e^{-4\alpha^2 t}}}, \quad 0 < t \le T \tag{5.106}$$

with T being the bit duration. As before, $\zeta_i = \Pr(H_i)$ $(i = 0,1)$.

The Dolinar receiver has been experimentally demonstrated in Ref. [23] and recently applied to the QPSK detection problem in Ref. [24]. Although the Dolinar receiver is much more complex than the Kennedy receiver, it is more robust against detector imperfections and nonzero dark current. The improved Kennedy receiver may be used as an alternative solution [25].

The main drawbacks of the Dolinar receiver are the requirement for precise optical-electrical feedback and the need for multiple copies of the transmitted quantum state. Using multiple copies of the same state can do a much better job than the Dolinar receiver. Namely, let us consider the discrimination problem of distinguishing between the following two product states, each containing multiple copies of the coherent states:

$$|\alpha_0\rangle = \overbrace{|-\alpha\rangle\cdots|-\alpha\rangle}^{N\text{ copies}} \text{ and } |\alpha_1\rangle = \overbrace{|\alpha\rangle\cdots|\alpha\rangle}^{N\text{ copies}}. \tag{5.107}$$

The optimum quantum receiver will have the following average error probability:

$$P_e^* = 1 - \frac{1}{2}\left[1 + \sqrt{1 - 4\zeta_0\zeta_1 \underbrace{|\langle\alpha_0|\alpha_1\rangle|}_{|\langle-\alpha|\alpha\rangle|^N}{}^2}\right] = 1 - \frac{1}{2}\left[1 + \sqrt{1 - 4\zeta_0\zeta_1 |\langle-\alpha|\alpha\rangle|^{2N}}\right], \tag{5.108}$$

and as long as $N > 1$, the Dolinar receiver will not be able to achieve the performance of this new scheme. Moreover, since the Dolinar receiver employs multiple copies of the transmitted state while the Kennedy receiver employs only one, the comparison might not be fair. To solve this problem, we can use the power splitter and delay lines to obtain the multiple copies we need.

The comparison among different binary quantum optical communication schemes in the absence of noise is summarized in Fig. 5.6. The classical coherent BPSK calculated by $P_e = 0.5\text{erfc}(\sqrt{2N_s})$ [8,9] slightly outperforms the optimum quantum OOK detector. On the other hand, the optimum quantum BPSK detector significantly outperforms its classical counterpart but slightly outperforms the Kennedy receiver.

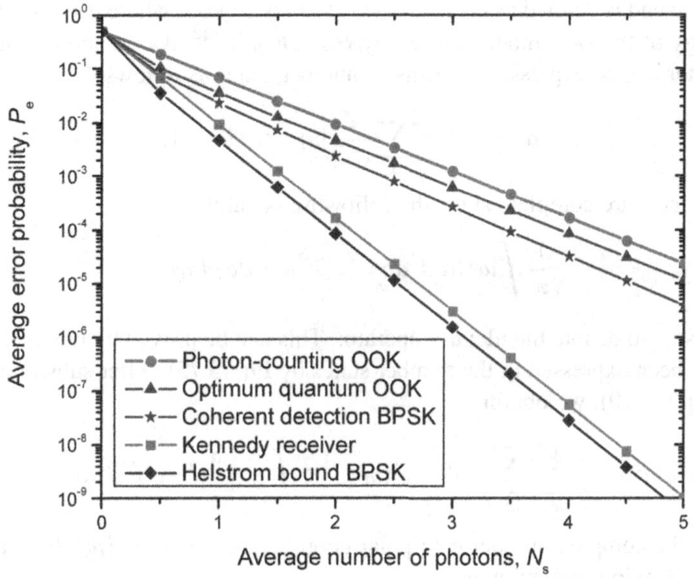

FIGURE 5.6

Comparison of different binary detection schemes in the absence of noise.

5.6 Field coherent states, P-representation, and noise representation

So far, we have considered single-mode coherent states in the absence of noise. This section extends our study to field coherent states and discusses both the noise representation and signal plus noise representation [1–3,6,7,15–17].

5.6.1 The field coherent states and P-representation

The CV quantum system lives in an infinite-dimensional Hilbert space, which is characterized by observables with continuous spectra [1,6,7]. The CV quantum system can be represented by M quantized radiation modes of the electromagnetic field (M bosonic modes). The quantum theory of radiation treats each radiation (bosonic) mode as a harmonic oscillator [1,2]. The bosonic modes live in a tensor-product Hilbert space $H^{\otimes M} = \otimes_{m=1}^{M} H_m$. The *radiation (bosonic) modes* are associated with M pairs of bosonic field operators $\left\{a_m^\dagger, a_m\right\}_{m=1}^{M}$, known as the creation and annihilation operators, respectively. The Hilbert space is infinitely dimensional but separable. Namely, a single-mode Hilbert space is spanned by a countable basis of number (Fock) states $\{|n\rangle\}_{n=1}^{\infty}$, wherein the Fock state $|n\rangle$ is the eigenstate of the number operator $N = a^\dagger a$ satisfying the eigenvalue equation $\widehat{n}|n\rangle = n|n\rangle$. The creation and annihilation operators have already been introduced by Eq. (5.53). The creation operator raises the number of photons by one; conversely, the annihilation operator reduces the number of photons by one. The right eigenvector of the annihilation operator is commonly referred to as the *coherent state* and is defined as $a|\alpha\rangle = \alpha|\alpha\rangle, \ \alpha = \alpha_I + j\alpha_Q$, where $\alpha = \alpha_I + j\alpha_Q$ is a complex number. The energy of the m-th mode is given by $H_m = \hbar\omega_m|\alpha_m|^2$. As we showed in Section 5.4, the coherent state vector can be expressed in terms of number states as follows:

$$|\alpha\rangle = e^{-|\alpha|^2/2}\sum_n \frac{\alpha^n}{\sqrt{n!}}|n\rangle, \quad \langle\alpha|\alpha\rangle = 1. \tag{5.109}$$

The coherent states are complete since the following is valid:

$$\frac{1}{\pi}\int|\alpha\rangle\langle\alpha|d^2\alpha = \hat{1}, \quad d^2\alpha = d\alpha_I d\alpha_Q, \tag{5.110}$$

where we again use $\hat{1}$ to denote the identity operator. This can be proved by noticing that $\rho = |\alpha\rangle\langle\alpha|$, which has already been expressed in the number states by Eq. (5.75). After substituting Eq. (5.75) on the left side of Eq. (5.110), we obtain

$$\frac{1}{\pi}\int|\alpha\rangle\langle\alpha|d^2\alpha = \frac{1}{\pi}\sum_{n=0}^{+\infty}\sum_{m=0}^{+\infty}(n!)^{-1/2}(m!)^{-1/2}\int e^{-|\alpha|^2}\alpha^n(\alpha^*)^m d^2\alpha|n\rangle\langle m|. \tag{5.111}$$

By expressing the complex number α in polar coordinates $\alpha = r\exp(j\theta)$, because $d^2\alpha = rdrd\theta$, we can rewrite the previous equation as

$$\begin{aligned}\frac{1}{\pi}\int|\alpha\rangle\langle\alpha|d^2\alpha &= \frac{1}{\pi}\sum_{n=0}^{+\infty}\sum_{m=0}^{+\infty}(n!)^{-1/2}(m!)^{-1/2}\overbrace{\int e^{-r^2}r^{m+n}rdr}^{\frac{1}{2}\int e^{-u}u^{\frac{m+n}{2}}du}\overbrace{\int_0^{2\pi}e^{j(n-m)}d\theta}^{2\pi\delta_{nm}}|n\rangle\langle m| \\ &= \sum_{n=0}^{+\infty}(n!)^{-1}\underbrace{\int e^{-u}u^n du}_{n!}|n\rangle\langle n| = \sum_{n=0}^{+\infty}|n\rangle\langle n| = \hat{1},\end{aligned} \tag{5.112}$$

thus proving Eq. (5.110).

The *entire field is in coherent state* $|\{\alpha_m\}\rangle$ when all modes $|\alpha_m\rangle$ are in coherent states, and the state vector $|\{\alpha_m\}\rangle$ is then a simultaneous right eigenket of all annihilation operators, that is, $a_m|\{\alpha_m\}\rangle = \alpha_m|\{\alpha_m\}\rangle$. The state vector can be represented as the tensor product of the individual state vectors:

$$|\{\alpha_m\}\rangle = |\alpha_1, \cdots, \alpha_m, \cdots\rangle = \prod_m|\alpha_m\rangle. \tag{5.113}$$

It was shown by Glauber and Sudarshan [15–17,26] that a large class of density operators ρ can be written in so-called *P-representation*, also known as the *Glauber-Sudarshan P-function*, by the following:

$$\rho = \int P(\{\alpha_m\})\prod_{m=1}^{\infty}|\alpha_m\rangle\langle\alpha_m|d^2\alpha_m, \tag{5.114}$$

where the weight function $P(\{\alpha_m\})$ has many properties of the classical probability density function (PDF) except that it is not always positive. In particular, from $\mathrm{Tr}\rho = 1$, we have

$$\int P(\{\alpha_m\}) \prod_{m=1}^{\infty} d^2\alpha_m = 1. \tag{5.115}$$

When the state of the field is described by Eq. (5.114), the expected value of an operator A is given by

$$\langle A\rangle = \mathrm{Tr}(\rho A) = \int P(\{\alpha_m\})\langle\{\alpha_m\}|A|\{\alpha_m\}\rangle \prod_{m=1}^{\infty} d^2\alpha_m. \tag{5.116}$$

5.6.2 Noise representation

Here we consider a scenario in which the field is composed of *thermal radiation* in thermal equilibrium at temperature $\mathfrak{T}$. For the moment, let us observe the single mode within the cavity with walls at temperature $\mathfrak{T}$. Statistical mechanics tells us that the probability of a mode being in the n-th exciting state is given by Refs. [27,28]:

$$P_n = \frac{e^{-E_n/k_B\mathfrak{T}}}{\sum_m e^{-E_m/k_B\mathfrak{T}}}, \quad E_n = \hbar\omega(n+1/2), \tag{5.117}$$

where k_B is the Boltzmann constant. The corresponding density operator of the thermal field will be

$$\rho_{\text{thermal}} = \frac{e^{-\hat{H}/k_B\mathfrak{T}}}{\mathrm{Tr}\left(e^{-\hat{H}/k_B\mathfrak{T}}\right)}, \quad \hat{H} = \hbar\omega\left(a^\dagger a + \frac{1}{2}\right), \tag{5.118}$$

and the trace-operation in the denominator can be determined by

$$\begin{aligned}\mathrm{Tr}\left(e^{-\hat{H}/k_B\mathfrak{T}}\right) &= \sum\nolimits_m \overbrace{\langle m|e^{-\hat{H}/k_B\mathfrak{T}}|m\rangle}^{e^{-E_m/k_B\mathfrak{T}}} = \sum\nolimits_m e^{-\overbrace{E_m}^{\hbar\omega(n+1/2)}/k_B\mathfrak{T}} = Z\\ &= \sum\nolimits_m e^{-\hbar\omega(n+1/2)/k_B\mathfrak{T}} = \frac{e^{-\hbar\omega/2k_B\mathfrak{T}}}{1-e^{-\hbar\omega/k_B\mathfrak{T}}},\end{aligned} \tag{5.119}$$

with Z being the partition function. The average photon number of the thermal field can now be calculated by

$$\begin{aligned}\mathfrak{N} &= Tr(N\rho_{\text{thermal}}) = \sum_n \langle n|N\rho_{\text{thermal}}|n\rangle = \sum_n n \overbrace{P_n}^{e^{-\hbar\omega(n+1/2)/k_B\mathfrak{T}}/Z} = \frac{1}{Z}\sum_n ne^{-\hbar\omega(n+1/2)/k_B\mathfrak{T}}\\ &= \underbrace{\frac{e^{-\hbar\omega/2k_BT}}{Z}}_{\frac{e^{-\hbar\omega/2k_B\mathfrak{T}}}{1-e^{-\hbar\omega/k_B\mathfrak{T}}}} \underbrace{\sum_n ne^{-\hbar\omega n/k_B\mathfrak{T}}}_{\frac{e^{-\hbar\omega/k_B\mathfrak{T}}}{\left[1-e^{-\hbar\omega/k_B\mathfrak{T}}\right]^2}} = \frac{1}{e^{\frac{\hbar\omega}{k_B\mathfrak{T}}}-1}.\end{aligned} \tag{5.120}$$

The corresponding density operator describing the state of the m-th mode in the P-representation is given by

$$\rho_m = \frac{1}{\pi \mathfrak{N}_m} \int \exp\left(- |\alpha|^2 / \mathfrak{N}_m \right) |\alpha\rangle\langle\alpha| d^2\alpha, \quad \mathfrak{N}_m = Tr\left(\rho a_m^\dagger a_m\right) = \frac{1}{e^{\hbar\omega/k_B \mathfrak{T}} - 1}, \tag{5.121}$$

where $\mathfrak{N}_m$ denotes the average number of photons in the m-th mode with frequency ω_m as given by the previous equation. In the classical limit, when $k_B\ \mathfrak{T} >> \hbar\omega_m$, the weight function $P(\alpha_m)$ becomes the classical PDF, and we can write

$$P(\alpha_m) = \frac{1}{\pi \mathfrak{N}_m} \exp\left(- |\alpha_m|^2 / \mathfrak{N}_m \right) = \frac{1}{\pi \mathfrak{N}_m} \exp\left[- \left(\alpha_{mI}^2 + \alpha_{mQ}^2 \right) / \mathfrak{N}_m \right]. \tag{5.122}$$

The average number of photons in the classical limit is $k_B\ \mathfrak{T}/\hbar\omega_m$, and the energy of the m-th mode, $k_B\ \mathfrak{T}$, is independent of the frequency and corresponds to the additive white Gaussian noise. By substituting Eq. (5.109) into Eq. (5.121), we can express the density operator ρ_m in terms of the number states $|n_m\rangle$ as follows:

$$\rho_m = \sum_{n_m=0}^{\infty} (1 - v_m) v_m^{n_m} |n_m\rangle\langle n_m|, \quad v_m = \frac{\mathfrak{N}_m}{\mathfrak{N}_m + 1} = e^{-\hbar\omega/k_B \mathfrak{T}}. \tag{5.123}$$

The density operator ρ of the whole field, composed of the thermal radiation in an equilibrium of M modes, can be represented as the direct product of the modes:

$$\rho = \pi^{-M} \prod_m \int e^{-\frac{|\alpha_m|^2}{\mathfrak{N}_m}} |\alpha_m\rangle\langle\alpha_m| d^2\alpha_m / \mathfrak{N}_m. \tag{5.124}$$

When the received signal contains both the *thermal noise* and the *coherent signal* represented by the coherent state vector $|\mu_m\rangle$, the center of the Gaussian weight function in the P-representation is shifted from the origin by a phasor μ_m, that is, $P_m = (\pi \mathfrak{N}_m)^{-1} \exp\left(- |\alpha - \mu_m|^2 / \mathfrak{N}_m \right)$. The corresponding density operator ρ_m in the P-representation is given by

$$\rho_m = \frac{1}{\pi \mathfrak{N}_m} \int e^{-\frac{|\alpha - \mu_m|^2}{\mathfrak{N}_m}} |\alpha\rangle\langle\alpha| d^2\alpha. \tag{5.125}$$

By substituting Eqs. (5.109) and (5.125), we can express the elements of the density operator ρ_m in terms of the number states $|n_m\rangle$ as follows [1]:

$$\langle l|\rho_m|k\rangle = \begin{cases} (1 - v_m) v_m^k \left(\mu_m^* / \mathfrak{N}_m\right)^{k-l} (l!/k!) e^{-(1-v_m)|\mu_m|^2} L_l^{k-l}\left[- (1 - v_m)^2 |\mu_m|^2 / v_m \right], & l \geq k \\ \langle k|\rho_m|l\rangle^*, & \end{cases} \tag{5.126}$$

where $v_m = \mathfrak{N}_m / (\mathfrak{N}_m + 1)$ as before and L() denotes the associated Laguerre polynomials with subscript and superscript denoting the degree and order, respectively.

The density operator of the entire received field composed of the coherent signal and Gaussian thermal noise can be written in P-representation as follows:

$$\rho = \int \cdots \int P(\{\alpha_m\}) \prod_{m=1}^{M} |\alpha_m\rangle\langle\alpha_m| d^2\alpha_m,$$
$$P(\{\alpha_m\}) = \frac{1}{\pi^M |\det \boldsymbol{\Sigma}|} e^{-\sum_k \sum_l (\alpha_k^* - \mu_k^*)(\boldsymbol{\Sigma}^{-1})_{kl} (\alpha_l - \mu_l)}, \tag{5.127}$$

where μ_k is the complex amplitude of the k-th coherent signal, and $\boldsymbol{\Sigma}$ represents the *mode correlation matrix* whose elements are determined by

$$\Sigma_{lk} = \mathrm{Tr}\left(\rho a_k^\dagger a_l\right) - \mathrm{Tr}\left(\rho a_k^\dagger\right)\mathrm{Tr}(\rho a_l). \tag{5.128}$$

When the modes are statistically independent, the mode correlation matrix is diagonal, with the elements being

$$\Sigma_{lk} = \mathcal{N}_k \delta_{lk}, \tag{5.129}$$

where $\mathcal{N}_k$ is the average number of thermal photons in mode k.

The P-representation of the density operator ρ is not unique. Let $|\{\beta_m\}\rangle$ be coherent states of the operators B_m spanning the same vector space as spanned by the vectors $|\{\alpha_m\}\rangle$. The P-representation in terms of coherent states $|\{\beta_m\}\rangle$ will now be

$$\rho = \int \cdots \int P(\{\beta_m\}) \prod_m |\{\beta_m\}\rangle\langle\{\beta_m\}| d^2\beta_m, \tag{5.130}$$

where the weight function $P(\{\beta_m\})$ is related to the previous weight function $P(\{\alpha_m\})$ by the following substitution relation:

$$\alpha_m = \sum_{j=1}^{\infty} \beta_j U_{jm}^\dagger, \tag{5.131}$$

where U_{jm} are elements of unitary matrix $\boldsymbol{U}$. The unitary matrix $\boldsymbol{U}$ should be chosen so that the matrix $\boldsymbol{U\Sigma}^{-1}\boldsymbol{U}^\dagger$ is diagonal, and thus, the density operator ρ can be simplified to

$$\rho = \pi^{-\nu} \int e^{-\sum_m \frac{|\beta_m - \mu'_m|^2}{\mathcal{N}'_m}} \prod_m |\{\beta_m\}\rangle\langle\{\beta_m\}| d^2\beta_m / \mathcal{N}'_m, \tag{5.132}$$

where N'_m is the m-th diagonal element of $\boldsymbol{U\Sigma}^{-1}\boldsymbol{U}^\dagger$, while

$$\mu'_m = \sum_{i=1}^{\infty} U_{mi}\mu_i. \tag{5.133}$$

Clearly, for any coherent signal in Gaussian thermal noise, the receiver can be chosen such that the individual modes are uncorrelated.

5.7 Binary quantum detection in the presence of noise

As shown earlier, for hypothesis H_0 in OOK, when only thermal noise is present, the density operator is

$$\rho_0 = \sum_{n=0}^{\infty} (1-\nu)\nu^n |n\rangle\langle n|, \quad \nu = \mathcal{N}/(\mathcal{N}+1), \quad \mathcal{N} = \frac{1}{e^{\frac{\hbar\omega}{k_B \mathcal{T}}} - 1}. \tag{5.134}$$

On the other hand, for hypothesis H_1, when both signal and thermal noise are present, the elements of density operator are given by Ref. [1]:

$$\langle n|\rho_1|m\rangle = \begin{cases} (1-v)\sqrt{\frac{n!}{m!}}v^m(\mu^*/\mathfrak{N})^{m-n}e^{-(1-v)|\mu|^2}L_n^{m-n}\left[-(1-v)^2|\mu|^2/v\right], & m \geq n \\ \langle m|\rho_k|n\rangle^*, & m < n \end{cases} \tag{5.135}$$

where $|\mu\rangle$ is the coherent state used to represent the transmitted nonzero symbol.

Now we need to solve the eigenvalue equation:

$$(\rho_1 - \Lambda_0\rho_0)|\eta_k\rangle = \eta_k|\eta_k\rangle, \quad \Lambda_0 = \zeta_0/(1-\zeta_0), \tag{5.136}$$

which is simplified for equal probable transmission $\Lambda_0 = \zeta_0/(1-\zeta_0)|_{\zeta_0=1/2} = 1$. The easiest way is to determine the element of difference matrix first as $\langle n|\rho_1 - \rho_0|m\rangle$ and perform the diagonalization to obtain the elements on main diagonal by $\langle \eta_k|\rho_1 - \rho_0|\eta_k\rangle$. The average symbol error probability was derived earlier [see Eq. 5.33]:

$$P_e = (1-\zeta_0)\left\{1 - \sum_k \eta_k U(\eta_k)\right\}\Bigg|_{\zeta_0=1/2} = \frac{1}{2}\left\{1 - \sum_k \eta_k U(\eta_k)\right\}, \quad U(\eta_k) = \begin{cases} 1, & \eta_k \geq 0 \\ 0, & \eta_k < 0 \end{cases}. \tag{5.137}$$

To illustrate how much the performance of the quantum optimum OOK detector is affected by the background noise, we plot the error probability for various average numbers of noise photons ($\mathfrak{N}$) in Fig. 5.7. The performance of the optimum quantum OOK detector is very sensitive to background

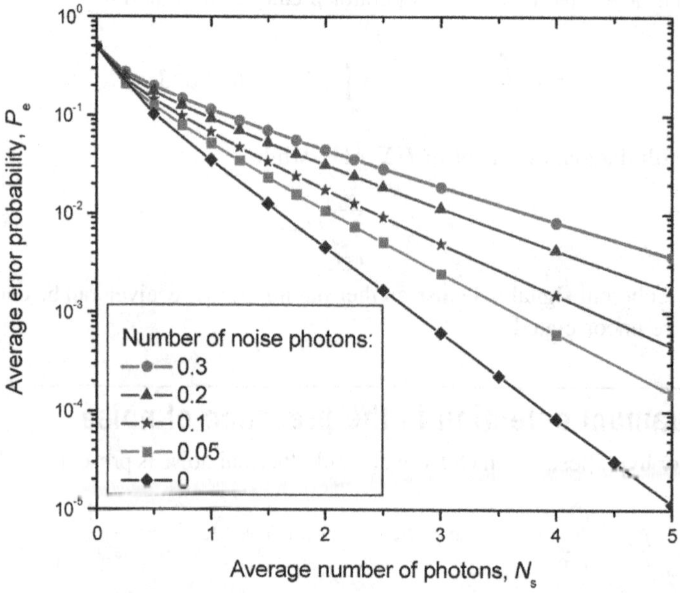

FIGURE 5.7

Performance of optimum quantum on–off keying detector in the presence of noise for different average noise numbers.

noise. We used 99 Fock (photon number) states to represent the density operators to evaluate the performance. To determine the eigenvalues of density matrix difference $\rho_1 - \rho_0$, we used the Jacobi rotation method [29].

For BPSK, the procedure is the same except that symbols 0 and 1 are represented by coherent states $|\mu_0 = -\alpha\rangle$ and $|\mu_1 = \alpha\rangle$, respectively, and we must use the matrix representation given by Eq. (5.135) by setting $\mu_0 = -\alpha$ for H_0 and $\mu_1 = \alpha$ for H_1.

5.8 Gaussian states, transformation, and channels, squeezed states, and Gaussian state detection

A CV system composed of M bosonic modes can also be described by quadrature operators $\{\widehat{I}_m, \widehat{Q}_m\}_{m=1}^{M}$, which can be arranged in vector form as

$$\widehat{\boldsymbol{x}} \doteq \left[\widehat{I}_1 \widehat{Q}_1 \cdots \widehat{I}_M \widehat{Q}_M\right]^T. \tag{5.138}$$

As a generalization of Eq. (5.47), these quadrature operators satisfy the commutation relations

$$[\widehat{x}_m, \widehat{x}_n] = 2j\Omega_{mn}, \tag{5.139}$$

where Ω_{mn} is the m-th row, n-th column element of a $2 \times 2M$ matrix:

$$\Omega \doteq \bigoplus_{m=1}^{M} \omega = \begin{bmatrix} \omega & & \\ & \ddots & \\ & & \omega \end{bmatrix}, \quad \omega = \begin{bmatrix} 0 & 1 \\ -1 & 0 \end{bmatrix}, \tag{5.140}$$

commonly referred to as the *symplectic form*.

For the quadrature operators, the first moment $\widehat{\boldsymbol{x}}$ represents the displacement and can be determined as

$$\overline{\boldsymbol{x}} = \langle \widehat{\boldsymbol{x}} \rangle = \text{Tr}(\widehat{\boldsymbol{x}}\widehat{\rho}), \tag{5.141}$$

where the density operator $\widehat{\rho}$ is the trace-one positive operator, which maps $\mathcal{H}^{\otimes M}$ to $\mathcal{H}^{\otimes M}$. The space of all density operators is known as the *state space*. When the density operator is the projector operator, that is, $\widehat{\rho}^2 = \widehat{\rho}$, we say that the density operator is *pure* and can be represented as $\widehat{\rho} = |\phi\rangle\langle\phi|$, $|\phi\rangle \in \mathcal{H}^{\otimes M}$. On the other hand, we can determine the elements of *covariance matrix* $\boldsymbol{\Sigma} = (\Sigma_{mn})_{2M \times 2M}$ as follows [6,7]:

$$\Sigma_{mn} = \frac{1}{2}\langle\{\Delta\widehat{x}_m, \Delta\widehat{x}_n\}\rangle = \frac{1}{2}(\langle\widehat{x}_m\widehat{x}_n\rangle + \langle\widehat{x}_n\widehat{x}_m\rangle) - \langle\widehat{x}_m\rangle\langle\widehat{x}_n\rangle, \quad \Delta\widehat{x}_m = \widehat{x}_m - \langle\widehat{x}_m\rangle, \tag{5.142}$$

where $\{\cdot,\cdot\}$ denotes the anticommutator. The diagonal elements of the covariance matrix are the variances of the quadratures:

$$V(\widehat{x}_i) = \Sigma_{ii} = \left\langle(\Delta\widehat{x}_i)^2\right\rangle = \left\langle(\widehat{x}_i)^2\right\rangle - \langle\widehat{x}_i\rangle^2. \tag{5.143}$$

The covariance matrix is a real, symmetric, and positive definite ($\mathbf{\Sigma} > 0$) matrix that satisfies the uncertainty principle, expressed as $\mathbf{\Sigma} + \mathrm{j}\,\mathbf{\Omega} \geq 0$. When we observe only diagonal terms, we arrive at the usual uncertainty principle of quadratures, that is, $V(\widehat{I}_m)V(\widehat{Q}_m) \geq 1$.

5.8.1 Gaussian and squeezed states

When the first two moments are sufficient to completely characterize the density operator, that is, $\widehat{\rho} = \widehat{\rho}(\overline{x}, \mathbf{\Sigma})$, we say they represent *Gaussian states*. Gaussian states are bosonic states for which the quasi-probability distribution, commonly referred to as the *Wigner function*, is a multivariate Gaussian distribution:

$$W(\boldsymbol{x}) = \frac{1}{(2\pi)^M (\det \mathbf{\Sigma})^{1/2}} \exp\left[-(\boldsymbol{x}-\overline{\boldsymbol{x}})^T \mathbf{\Sigma}^{-1} (\boldsymbol{x}-\overline{\boldsymbol{x}})/2\right], \tag{5.144}$$

where $\boldsymbol{x} \in \Re^{2M}$ ($\Re$—the set of real numbers) are eigenvalues of the quadrature operators $\widehat{\boldsymbol{x}}$. The Wigner function of the thermal state, given by Eq. (5.144), is a zero-mean Gaussian with a covariance matrix $\sum = (2\mathfrak{N} + 1)\mathbf{1}$, where $\mathfrak{N}$ is the average number of photons, and $\mathbf{1}$ denotes the identity matrix. Gaussian CVs span a real symplectic space $(\Re^{2M}, \Omega)$ commonly referred to as the *phase space*, as mentioned earlier. The corresponding *characteristic function* is given by

$$\chi(\zeta) = \exp\left[-j(\Omega\,\overline{x})^T \zeta - \frac{1}{2}\zeta^T (\Omega \Sigma \Omega^T)\zeta\right], \quad \zeta \in \Re^{2M}. \tag{5.145}$$

Not surprisingly, the Wigner function and the characteristic function are related by the Fourier transform:

$$W(\boldsymbol{x}) = \frac{1}{(2\pi)^{2M}} \int_{\Re^{2M}} \chi(\zeta) e^{-j\boldsymbol{x}^T \Omega \zeta} d^{2M}\zeta. \tag{5.146}$$

Let us introduce the *Weyl operator* as follows:

$$D(\zeta) = e^{j\widehat{x}^T \Omega \zeta}. \tag{5.147}$$

The displacement operator $D(\alpha)$ is just a complex version of the Weyl operator. The characteristic function is related to the Weyl operator by

$$\chi(\zeta) = \mathrm{Tr}[\widehat{\rho} D(\zeta)], \tag{5.148}$$

and represents the expected value of the Weyl operator. It can be applied to both Gaussian and non-Gaussian states. The Wigner function can be determined as the Fourier transform of the characteristic function, given by Eq. (5.146). Therefore, the Wigner function is an equivalent representation of an arbitrary quantum state defined over a $2M$-dimensional phase space. When the Wigner function is nonnegative, the pure state will be Gaussian.

A *squeezed state* is a Gaussian state with unequal fluctuations in quadratures:

$$V(\widehat{I}) = e^{2s}, \quad V(\widehat{Q}) = e^{-2s}, \tag{5.149}$$

where s is the squeezing parameter. (The fluctuations above are expressed in SNU.)

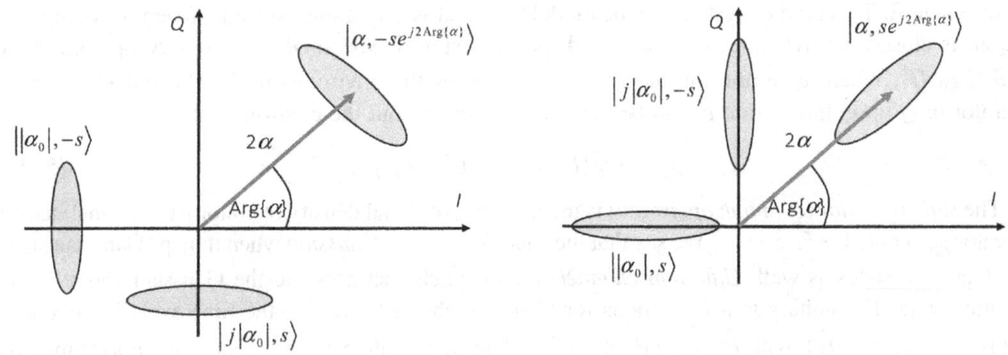

FIGURE 5.8

Squeezed-displaced states: (left) amplitude-squeezed states and (right) phase-squeezed states.

For a single mode, the PDF of the in-phase component, denoted $f(I)$, can be determined by averaging the Wigner function over the quadrature component, that is, $f(I) = \int_Q W(I,Q)dQ$. On the other hand, if the quadrature component is measured, the result will follow the PDF $f(Q) = \int_I W(I,Q)dI$. Therefore, we can interpret the Wigner function of the coherent state $|\alpha\rangle$ as the 2-D Gaussian distribution centered at (Re{α},Im{α})=(α_I,α_Q) with a variance of 1 SNU (with zero covariance). Therefore, in the (I,Q)-plane, the coherent state is represented by a circle with its radius related to the standard deviation (uncertainty), as already illustrated in Fig. 5.3. For the displaced-squeezed states in Fig. 5.8, we have identified the amplitude-squeezing states (on the left) and phase-squeezing states (on the right).

The squeezed-displaced coherent state $|\alpha,s\rangle$ can be represented in terms of the ground state $|0\rangle$ as follows:

$$\left|\alpha, \underline{s}\right\rangle = D(\alpha)S\left(\underline{s}\right)|0\rangle,\quad S\left(\underline{s}\right) = e^{\frac{1}{2}\left(\underline{s}a^{\dagger 2} - \underline{s}^* a^2\right)}, \tag{5.150}$$

where $S(\underline{s})$ is the *squeezing operator* and $\underline{s}$ is the squeezing factor, which is generally a complex number $\underline{s} = s\,\exp(j\theta)$. The squeezing factor $\underline{s}$ in quantum optics is often a real number, called the squeezing parameter, so the squeezing operator simplifies to $S(s) = \exp\left[s\left(a^{\dagger 2} - a^2\right)/2\right]$ Applying this squeezing operator to the vacuum state, we obtain the *squeezed vacuum state*:

$$|0, s\rangle = S(s)|0\rangle = (\cosh s)^{-1/2} \sum_{k=0}^{\infty} \frac{\sqrt{(2k)!}}{2^n k!} (\tanh s)^k |2k\rangle, \tag{5.151}$$

with the corresponding derivation provided in the next section.

5.8.2 Gaussian transformations, Gaussian channels, and squeezed states

As discussed in Chapter 4, a given quantum state undergoes a transformation described by the quantum operation (superoperator) [4]. Let the composite system C comprise quantum register Q and environment E. This kind of system can be modeled as a closed quantum system. When the composite system is closed, its dynamic is unitary, and the final state is specified by a unitary operator U as $U(\hat{\rho}\otimes\varepsilon_0)U^\dagger$, where ε_0 is the initial density operator of the environment E. The reduced density operator of Q upon interaction ρ_f can be obtained by tracing out the environment:

$$\hat{\rho}_f=\mathrm{Tr}_E\left[U(\hat{\rho}\otimes\varepsilon_0)U^\dagger\right]\doteq\xi(\hat{\rho}). \tag{5.152}$$

The *superoperator* (*quantum operation*) is the mapping of initial density operator ρ to the final density operator ρ_f, denoted as ξ: $\hat{\rho}\rightarrow\hat{\rho}_f$. We say that the *superoperator is Gaussian* when it maps Gaussian states into Gaussian states as well. *Gaussian channels* are channels that preserve the Gaussian nature of the quantum state. The unitary transformations for Gaussian channels are like the time-evolution operator, that is, $U=\exp(-j\hat{H})$, with $\hat{H}$ as the Hamiltonian. The Hamiltonian is the second-order polynomials of the creation and annihilation operators' vectors, defined as $\boldsymbol{a}^\dagger=\left[a_1^\dagger\cdots a_M^\dagger\right]$, $\boldsymbol{a}=[a_1\cdots a_M]$, and can be represented as follows [6,7]:

$$\hat{H}=j\left(\boldsymbol{a}^\dagger\boldsymbol{\alpha}+\boldsymbol{a}^\dagger\boldsymbol{F}_1\boldsymbol{a}+\boldsymbol{a}^\dagger\boldsymbol{F}_2\boldsymbol{a}\right)+H.c., \tag{5.153}$$

where $\alpha\in C^M$ (C—the set of complex numbers), while $\boldsymbol{F}_1$ and $\boldsymbol{F}_2$ are $M\times M$ complex matrices (and H.c. stands for the Hermitian conjugate).

In the Heisenberg picture, these kinds of transformations correspond to a unitary *Bogoliubov transformation* [6,7,30,31]:

$$\boldsymbol{a}\rightarrow\boldsymbol{a}_f=U^\dagger\boldsymbol{a}U=\boldsymbol{A}\boldsymbol{a}+\boldsymbol{B}\boldsymbol{a}^\dagger+\boldsymbol{\alpha}, \tag{5.154}$$

where $\boldsymbol{A}$ and $\boldsymbol{B}$ are $M\times M$ complex matrices satisfying $\boldsymbol{AB}^T=\boldsymbol{BA}^T$, $\boldsymbol{AA}^\dagger=\boldsymbol{BB}^\dagger+\boldsymbol{1}$, with $\boldsymbol{1}$ as the identity matrix. Bogoliubov transformation induces autoequivalence to the respective representations. As an illustration, let us consider creation and annihilation operators satisfying the commutation relation $\left[a,a^\dagger\right]=\hat{1}$. Let us define new pairs of operators $\hat{b}=ua+va^\dagger$, $\hat{b}^\dagger=u^*a^\dagger+v^*a$, where u and v are complex constants. Given that Bogoliubov transformation preserves commutativity relations, the new commutator $\left[\hat{b},\hat{b}^\dagger\right]$ must be $\hat{1}$, and consequently, $|u|^2-|v|^2$ must be 1. Since the hyperbolic identity $\cosh^2 x-\sinh^2 x=1$ has the same form, we can parametrize u and v by $u=e^{j\theta_u}\cosh s$, $u=e^{j\theta_v}\sinh s$. This transformation is, therefore, a *linear symplectic transformation* of the phase space. Two angles correspond to the orthogonal symplectic transformations (such as rotations), and the squeezing factor s corresponds to the diagonal transformation. For the quadrature operators, Gaussian transformation is described by an affine map $U_{S,c}$:

$$U_{S,c}:\boldsymbol{x}\rightarrow\boldsymbol{x}_f=\hat{\boldsymbol{S}}\boldsymbol{x}+\boldsymbol{c}, \tag{5.155}$$

where $\boldsymbol{S}$ is a $2M\times 2M$ real matrix and $\boldsymbol{c}$ is a real vector of length $2M$. This transformation preserves the commutation relations given by Eq. (5.139) when matrix $\boldsymbol{S}$ is a symplectic matrix, which is when it satisfies the following property:

$$\boldsymbol{S}\Omega\boldsymbol{S}^T=\Omega. \tag{5.156}$$

The affine mapping can be decomposed as

$$U_{S,c}=D(\boldsymbol{c})U_S,\ D(\boldsymbol{c}):\boldsymbol{x}\rightarrow\boldsymbol{x}+\boldsymbol{c},\ U_S:\ \boldsymbol{x}\rightarrow\boldsymbol{S}\boldsymbol{x}, \tag{5.157}$$

where $D(\boldsymbol{c})$ is the Weyl operator representing the translation in the phase-space, and U_S is a linear symplectic mapping. The action of Gaussian transformation on the first two moments of the quadrature field operators will simply be

$$\overline{\boldsymbol{x}} \rightarrow \boldsymbol{S}\overline{\mathbf{x}} + \boldsymbol{c}, \ \ \boldsymbol{\Sigma} \rightarrow \boldsymbol{S}\boldsymbol{\Sigma}\boldsymbol{S}^T. \tag{5.158}$$

Thus, Eq. (5.158) completely characterizes the action on the Gaussian state.

To illustrate, let us consider the *BS transformation*. The *BS* is a partially silvered piece of glass with a reflection coefficient parameterized by $\cos^2\theta$. BS transmittivity is given by $T = \cos^2\theta \in [0,1]$. It could be a piece of dielectric material as well. Finally, it can be implemented in fiber optics as a directional coupler. The BS acts on two modes that can be described by the creation (annihilation) operators a ($a^\dagger$) and b ($b^\dagger$). In the Heisenberg picture, the annihilation operators of the modes are transformed by the Bogoliubov transformation of the type

$$\begin{bmatrix} a \\ b \end{bmatrix} \rightarrow \begin{bmatrix} a_f \\ b_f \end{bmatrix} = \begin{bmatrix} \sqrt{T} & \sqrt{1-T} \\ -\sqrt{1-T} & \sqrt{T} \end{bmatrix} \begin{bmatrix} a \\ b \end{bmatrix} = \begin{bmatrix} \cos\theta & \sin\theta \\ -\sin\theta & \cos\theta \end{bmatrix} \begin{bmatrix} a \\ b \end{bmatrix}. \tag{5.159}$$

The action of the BS can be described by the following transformation:

$$BS(\theta) = e^{-j\hat{H}} = e^{\theta\left(a^\dagger b - ab^\dagger\right)} \tag{5.160}$$

The quadrature operators of the modes $\widehat{\boldsymbol{x}} = (\widehat{x}_1\ \widehat{x}_2\ \widehat{x}_3\ \widehat{x}_4)^T = \left(\widehat{I}_a\ \widehat{Q}_a\ \widehat{I}_b\ \widehat{Q}_b\right)^T$ are transformed by the following symplectic transformation:

$$\widehat{\boldsymbol{x}} \rightarrow \widehat{\boldsymbol{x}}_f = \begin{bmatrix} \sqrt{T}\mathbf{1} & \sqrt{1-T}\mathbf{1} \\ -\sqrt{1-T}\mathbf{1} & \sqrt{T}\mathbf{1} \end{bmatrix} \widehat{\boldsymbol{x}}, \tag{5.161}$$

where $\mathbf{1}$ is now the 2×2 identity matrix.

As another example, let us consider the *phase rotation* operator. The phase rotation can be described by the free-propagation Hamiltonian $\widehat{H} = \theta a^\dagger a$ as $\widehat{R}(\theta) = e^{-j\widehat{H}} = e^{-j\theta a^\dagger a}$. For the annihilation operator, the corresponding transformation is simply $a \rightarrow a_f = ae^{-j\theta}$. For the quadrature operators, the corresponding transformation is the symplectic mapping:

$$\widehat{\boldsymbol{x}} \rightarrow \widehat{\boldsymbol{x}}_f = \widehat{R}(\theta)\widehat{\boldsymbol{x}}, \ \ \widehat{R}(\theta) = \begin{bmatrix} \cos\theta & \sin\theta \\ -\sin\theta & \cos\theta \end{bmatrix}. \tag{5.162}$$

As a final example, let us revisit the *squeezing operator* $\widehat{S}(s) = \exp\left[s\left(a^{\dagger 2} - a^2\right)/2\right]$. Based on the definition of squeezing states [see Eq. (5.149)], we conclude that the squeezing operator, which is a unitary operator, must have the following action on quadratures:

$$S^\dagger(s)\widehat{I}S(s) = e^s\widehat{I} \text{ and } S^\dagger(s)\widehat{Q}S(s) = e^{-s}\widehat{Q}, \tag{5.163}$$

and the corresponding matrix representation of this transformation will simply be

$$\widehat{\boldsymbol{x}} \rightarrow \widehat{\boldsymbol{x}}_f = \underbrace{\begin{bmatrix} e^s & 0 \\ 0 & e^{-s} \end{bmatrix}}_{\boldsymbol{S}(s)} \widehat{\boldsymbol{x}} = \boldsymbol{S}(s)\widehat{\boldsymbol{x}}, \ \ \boldsymbol{S}(s) = \begin{bmatrix} e^s & 0 \\ 0 & e^{-s} \end{bmatrix}. \tag{5.164}$$

Given that the annihilation and creation operators can be expressed in terms of quadrature operators by $a^\dagger = (\hat{I} - j\hat{Q})/2$, $a = (\hat{I} + j\hat{Q})/2$, the corresponding transformation of these operators becomes

$$\begin{aligned} S^\dagger(s)aS(s) &= \frac{1}{2}\left[S^\dagger(s)\hat{I}S(s) + jS^\dagger(s)\hat{Q}S(s)\right] = e^s \overbrace{\hat{I}}^{a+a^\dagger} + je^{-s}\frac{\overbrace{\hat{Q}}^{j(a^\dagger - a)}}{2} \\ &= a\cosh s + a^\dagger \sinh s, \\ S^\dagger(s)a^\dagger S(s) &= \frac{1}{2}\left[\left[S^\dagger(s)\hat{I}S(s) - jS^\dagger(s)\hat{Q}S(s)\right]\right] = e^s \overbrace{\hat{I}}^{a+a^\dagger} - je^{-s}\frac{\overbrace{\hat{Q}}^{j(a^\dagger - a)}}{2} \\ &= a\sinh s + a^\dagger \cosh s. \end{aligned} \tag{5.165}$$

When s is the complex number given by $\underline{s} = s\exp(j\theta)$, by introducing the substitution $a' = ae^{-j\theta/2}$, the squeezing operator given in Eq. (5.150) becomes

$$S(\underline{s}) = e^{\frac{1}{2}\left(\underline{s}a^{\dagger 2} - \underline{s}^* a^2\right)} = e^{\frac{1}{2}\left(se^{j\theta}a^{\dagger 2} - se^{-j\theta}a^2\right)} = e^{\frac{1}{2}\left(s\left(a^\dagger e^{j\theta/2}\right)^2 - s\left(ae^{-j\theta/2}\right)^2\right)} = e^{\frac{1}{2}\left(sa'^{\dagger 2} - sa'^2\right)}, \tag{5.166}$$

which is the same as the squeezing operator for real parameter s, and thus, the transformations of Eq. (5.165) are directly applicable when we replace a with a' to obtain

$$\begin{aligned} S^\dagger(\underline{s})ae^{-j\theta/2}S(\underline{s}) &= ae^{-j\theta/2}\cosh s + a^\dagger e^{j\theta/2}\sinh s, \\ S^\dagger(\underline{s})a^\dagger e^{j\theta/2}S(\underline{s}) &= ae^{-j\theta/2}\sinh s + a^\dagger e^{j\theta/2}\cosh s. \end{aligned} \tag{5.167}$$

Now by multiplying the first line with exp(jθ/2) and the second line with exp(−jθ/2), we obtain

$$\begin{aligned} S^\dagger(\underline{s})aS(\underline{s}) &= a\cosh s + a^\dagger e^{j\theta}\sinh s, \\ S^\dagger(\underline{s})a^\dagger S(\underline{s}) &= ae^{-j\theta}\sinh s + a^\dagger \cosh s, \end{aligned} \tag{5.168}$$

which can be used to describe the transformation of a Gaussian state by the squeezing operator for a complex squeezing factor $\underline{s} = s\exp(j\theta)$.

We have now built sufficient knowledge to determine the *squeezed vacuum state representation* in terms of number states. We know that for the vacuum state $a|0\rangle = 0$, and by multiplying by $S(s)$ from the left, we obtain

$$S(\underline{s})a|0\rangle = 0. \tag{5.169}$$

Given that the squeezing operator is unitary, we can insert the identity operator to obtain

$$S(\underline{s})a\underbrace{S^\dagger(\underline{s})S(\underline{s})}_{\hat{1}}|0\rangle = S(\underline{s})aS^\dagger(\underline{s})\underbrace{S(\underline{s})|0\rangle}_{|\underline{s}\rangle} = S(\underline{s})aS^\dagger(\underline{s})|\underline{s}\rangle = 0. \tag{5.170}$$

Now by applying the first line in Eq. (5.168) and considering that $S^{\dagger}(s) = S(-s)$, we can rewrite the previous equation as

$$\underbrace{\left(a\cosh s - a^{\dagger}e^{j\theta}\sinh s\right)}_{S(\underline{s})aS^{\dagger}(\underline{s})}\left|\underline{s}\right\rangle = 0, \tag{5.171}$$

from which we conclude that the squeezed vacuum state is an eigenket of the operator $a\cosh s - a^{\dagger}e^{j\theta}\sinh s$ with a zero eigenvalue.

The squeezed vacuum state can be represented in terms of number states by

$$\left|\underline{s}\right\rangle = \sum_{n=0}^{\infty} c_n |n\rangle, \tag{5.172}$$

and after substituting Eq. (5.172) into Eq. (5.171), we determine that the coefficient related to the n-th number state must satisfy

$$\left(c_{n+1}\cosh s\sqrt{n+1} - c_{n-1}e^{j\theta}\sinh s\sqrt{n}\right)|n\rangle = 0, \tag{5.173}$$

and thus, the following recursion relation for the expansion coefficients can be established:

$$c_{n+1} = \frac{e^{j\theta}\sinh s}{\cosh s}\sqrt{\frac{n}{n+1}}c_{n-1}. \tag{5.174}$$

We also conclude that all expansion coefficients with an odd index are zero. From Eq. (5.174), we determine the following solution of the recursion relation:

$$c_{2k} = \left(\frac{e^{j\theta}\sinh s}{\cosh s}\right)^k\sqrt{\frac{(2k-1)!!}{(2m)!!}}c_0 = \left(e^{j\theta}\tanh s\right)^k\sqrt{\frac{(2k-1)!!}{(2k)!!}}c_0, \tag{5.175}$$

and c_0 can be determined from the normalization condition

$$\sum_{k=0}^{\infty}|c_{2k}|^2 = \left[1 + \sum_{k=1}^{\infty}\tanh^{2k}s\frac{(2k-1)!!}{(2k)!!}\right]|c_0|^2 = 1, \tag{5.176}$$

as

$$|c_0|^2 = \frac{\overbrace{1}^{\sqrt{1-\tanh^2 s}}}{1 + \sum_{k=1}^{\infty}\left(\tanh^2 s\right)^k\frac{(2k-1)!!}{(2k)!!}} \overset{\sqrt{1-\tanh^2 s} = 1/\cosh s}{=} \frac{1}{\cosh s}. \tag{5.177}$$

Therefore, $c_0 = (\cosh s)^{-1/2}$. Finally, by substituting Eq. (5.175) into Eq. (5.172), we obtain

$$\left|\underline{s}\right\rangle = \sum_{k=0}^{\infty}c_{2k}|2k\rangle = (\cosh s)^{-1/2}\sum_{k=0}^{\infty}\left(e^{j\theta}\tanh s\right)^k\sqrt{\frac{(2k-1)!!}{(2k)!!}}|2k\rangle. \tag{5.178}$$

Using the relationships $(2k)!! = 2^k k!$, $(2k-1)!! = 2^{-k}(2k)!/k!$, we can rewrite the previous equation as

$$\left|\underline{s}\right\rangle = (\cosh s)^{-1/2} \sum_{k=0}^{\infty} \left(e^{j\theta} \tanh s\right)^k \frac{\sqrt{(2k)!}}{2^k k!} |2k\rangle. \tag{5.179}$$

Clearly, for $\theta = 0$, we obtain Eq. (5.151).

The probability of detecting $2k$ photons in the field is simply

$$P_{2k} = |c_{2k}|^2 = \cosh^{-1} s (\tanh s)^{2k} \frac{(2k)!}{2^{2k}(k!)^2}. \tag{5.180}$$

In the rest of this section, we describe how to represent the *displaced-squeezed states* defined by Eq. (5.150) in terms of number states. Our starting point will be Eq. (5.170), which after multiplication with the displacement operator $D(\alpha)$ becomes

$$D(\alpha) \underbrace{S\left(\underline{s}\right) a S^{\dagger}\left(\underline{s}\right)}_{a \cosh s - a^{\dagger} e^{j\theta} \sinh s} D^{\dagger}(\alpha) \underbrace{D(\alpha)\left|\underline{s}\right\rangle}_{\left|\alpha, \underline{s}\right\rangle} = 0. \tag{5.181}$$

Since $D(\alpha) a D^{\dagger}(\alpha) = a - \alpha$, from the previous equation, we obtain

$$D(\alpha)\left(a \cosh s - a^{\dagger} e^{j\theta} \sinh s\right) D^{\dagger}(\alpha)\left|\alpha, \underline{s}\right\rangle = \left[(a-\alpha)\cosh s - \left(a^{\dagger} - \alpha^*\right) e^{j\theta} \sinh s\right]\left|\alpha, \underline{s}\right\rangle = 0, \tag{5.182}$$

or equivalently, we have just come up with the following eigenvalue equation:

$$\left(a \cosh s - a^{\dagger} e^{j\theta} \sinh s\right)\left|\alpha, \underline{s}\right\rangle = \underbrace{\left(\alpha \cosh s - \alpha^* e^{j\theta} \sinh s\right)}_{\beta}\left|\alpha, \underline{s}\right\rangle. \tag{5.183}$$

This eigenvalue equation is more general than Eq. (5.171) for $\alpha = 0$, and we can make the ansatz

$$C_n = K\left(\frac{1}{2} e^{j\theta} \tanh s\right)^{n/2} h_n(z), \quad z \in C \tag{5.184}$$

where K is the normalization constant, and $h_n(z)$ (with z being the complex number) is the function to be determined. The expansion $\left|\alpha, \underline{s}\right\rangle = \sum_n C_n |n\rangle$ must satisfy the eigenvalue Eq. (5.183), and by equating the corresponding n-th terms on both sides, we obtain the following recursion relation for $h_n(z)$:

$$\sqrt{n+1} h_{n+1}(z) - 2z h_n(z) - 2\sqrt{n} h_{n-1}(z) = 0, \quad z = \beta\left[e^{j\theta} \sinh(2s)\right]^{-1/2}, \tag{5.185}$$

which can be rewritten as

$$h_{n+1}(z) = \frac{2z}{\sqrt{n+1}} h_n(z) + 2\sqrt{\frac{n}{n+1}} h_{n-1}(z). \tag{5.186}$$

For $n = 0$, we can express $h_0(z) = C_0/K$. On the other hand, for $n = 1$, we obtain $h_1(z) = 2zh_0(z)$. Now we can use the recursive relation of Eq. (5.186) to determine the higher-order terms. The C_0 coefficient in expansion can be determined by

$$
\begin{aligned}
C_0 &= \left\langle 0 \middle| \alpha, \underline{s} \right\rangle = \underbrace{\langle 0| \underbrace{D(\alpha)}_{D^\dagger(-\alpha)}}_{\langle -\alpha|} \underbrace{S\left(\underline{s}\right)|0\rangle}_{\left|\underline{s}\right\rangle} \\
&= \left\langle -\alpha \middle| \underline{s} \right\rangle = e^{-\frac{1}{2}|\alpha|^2} \sum_{k=0}^{\infty} \frac{(\alpha^*)^{2k}}{\sqrt{(2k)!}} \underbrace{c_{2k}}_{\sqrt{\cosh s}\left(e^{j\theta} \tanh s\right)^k \frac{\sqrt{(2k)!}}{2^k k!}} = e^{-\frac{1}{2}|\alpha|^2} (\cosh s)^{-1/2} \underbrace{\sum_{k=0}^{\infty} \frac{\left(\frac{1}{2}\alpha^{*2} e^{j\theta} \tanh s\right)^k}{k!}}_{\exp\left(\frac{1}{2}\alpha^{*2} e^{j\theta} \tanh s\right)} \\
&= (\cosh s)^{-1/2} \exp\left(-\frac{1}{2}|\alpha|^2 + \frac{1}{2}\alpha^{*2} e^{j\theta} \tanh s \right).
\end{aligned}
\tag{5.187}
$$

By setting the $h_0(z)$ to be one, the constant K is simply equal to C_0.

The displaced-squeezed state $|\alpha,s\rangle$ can now be decomposed in terms of Fock states by

$$
\begin{aligned}
\left|\alpha, \underline{s}\right\rangle &= C_0 \sum_{n=0}^{\infty} \left(\frac{1}{2} e^{j\theta} \tanh s\right)^{n/2} h_n\left(\frac{\beta}{\sqrt{e^{j\theta} \sinh(2s)}}\right) |n\rangle \\
&= (\cosh s)^{-1/2} e^{-\frac{1}{2}|\alpha|^2 + \frac{1}{2}\alpha^{*2} e^{j\theta} \tanh s} \sum_{n=0}^{\infty} \left(\frac{1}{2} e^{j\theta} \tanh s\right)^{n/2} h_n\left(\frac{\beta}{\sqrt{e^{j\theta} \sinh(2s)}}\right) |n\rangle.
\end{aligned}
\tag{5.188}
$$

Someone may want to express the expansion in terms of Hermite polynomials [13–15], but this is possible only for $|-\alpha,s\rangle$. Unfortunately, Hermite polynomials grow quickly with increases in order and argument, which introduces numerical problems. Alternatively, someone may use a different definition of the squeezing parameter $\widehat{S}(s) = \exp\left[-s\left(a^{\dagger 2} - a^2\right)/2\right]$, such as in Refs. [5,15].

As an illustration, Fig. 5.9 provides the photon probability distributions for two different displaced-squeezed states. Clearly, the probability distributions are quite different from the Poisson distribution.

5.8.3 Thermal decomposition of Gaussian states and von Neumann entropy

The *thermal state* is a bosonic state that maximizes the von Neumann entropy:

$$
S(\widehat{\rho}) = -\operatorname{Tr}(\widehat{\rho}\log\widehat{\rho}) = -\sum_i \eta_i \log \eta_i, \tag{5.189}
$$

where η_i are eigenvalues of the density operator. The thermal state, as shown by Eq. (5.123), can be represented in terms of number states:

$$
\widehat{\rho}_{\text{thermal}} = \sum_{n=0}^{\infty} (1-v)v^n |n\rangle\langle n| = \frac{1}{\mathfrak{N}+1} \sum_{n=0}^{\infty} \left(\frac{\mathfrak{N}}{\mathfrak{N}+1}\right)^n |n\rangle\langle n|, \quad v = \frac{\mathfrak{N}}{\mathfrak{N}+1}. \tag{5.190}
$$

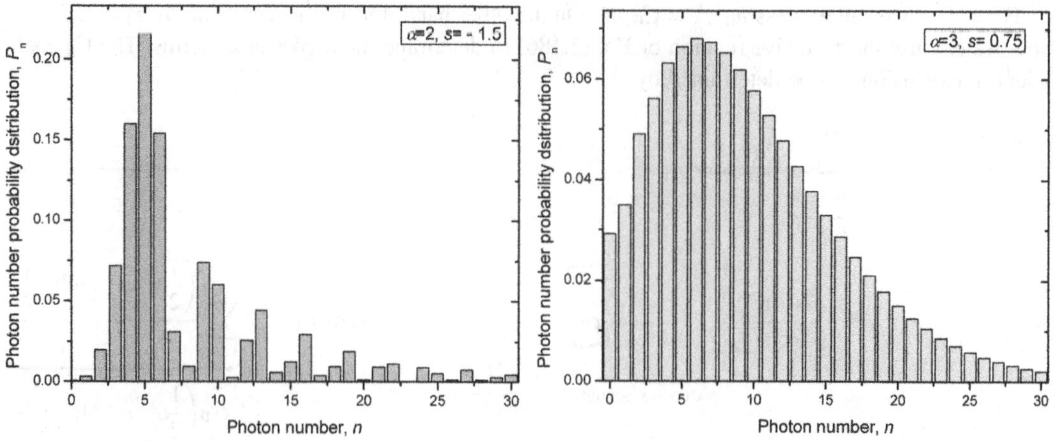

FIGURE 5.9

Photon number probability distributions for displaced-squeezed states: (left) $\alpha = 2$, $s = -1.5$ and (right) $\alpha = 3$, $s = 0.75$.

So the entropy of the thermal state is given by Ref. [32]:

$$S(\widehat{\rho}_{\text{thermal}}) = -\sum\nolimits_i \eta_i \log \eta_i = \mathfrak{N} \log\left(\frac{\mathfrak{N}+1}{\mathfrak{N}}\right) + \log(\mathfrak{N}+1)$$
$$= g(\mathfrak{N}), \quad g(x) = (x+1)\log(x+1) - x\log x, \quad x > 0, \tag{5.191}$$

where the g-function is a monotonically increasing concave function.

The density operator of the whole field, composed of M thermal states, can be represented as the direct product of modes, as given by Eq. (5.124), which can also be written as

$$\widehat{\rho}_{\text{total}} = \bigotimes_{m=1}^{M} \widehat{\rho}_{\text{thermal}}(\mathfrak{N}_m). \tag{5.192}$$

Given that thermal states are independent, the von Neumann entropy is additive, and we can write

$$S(\widehat{\rho}_{\text{total}}) = S\left[\bigotimes_{m=1}^{M} \widehat{\rho}_{\text{thermal}}(\mathfrak{N}_m)\right] = \sum_{m=1}^{M} S(\widehat{\rho}_{\text{thermal}}(\mathfrak{N}_m)) = \sum_{m=1}^{M} g(\mathfrak{N}_m), \tag{5.193}$$

where N_m is the average number of photons in the m-th thermal state.

By applying *Williamson's theorem* [33] on the M-mode covariance matrix $\mathbf{\Sigma}$, claiming that every positive-definite real matrix can be placed in a diagonal form by a suitable symplectic transformation, we can represent the covariance matrix $\mathbf{\Sigma}$, with the help of a symplectic matrix $\boldsymbol{S}$ [6,7,32]:

$$\mathbf{\Sigma} = \boldsymbol{S}\mathbf{\Sigma}^{\oplus}\boldsymbol{S}^T, \quad \mathbf{\Sigma}^{\oplus} = \bigoplus_{m=1}^{M} \nu_m \mathbf{1}, \tag{5.194}$$

where $\mathbf{\Sigma}^{\oplus}$ is the diagonal matrix known as the Williamson's form, and ν_m $(m = 1,\ldots,M)$ are symplectic eigenvalues of the covariance matrix determined as the modulus of ordinary eigenvalues of the matrix $j\mathbf{\Omega\Sigma}$. Given that only the modulus of the eigenvalues is relevant, the $2M \times 2M$ covariance matrix

$\mathbf{\Sigma}$ will have exactly M symplectic eigenvalues. The representation (5.194) is, in fact, the thermal decomposition of a zero-mean Gaussian state, described by the covariance matrix $\mathbf{\Sigma}$. The uncertainty principle can now be expressed as $\mathbf{\Sigma}^{\oplus} \geq 1,\ \mathbf{\Sigma} > 0$. Given that the covariance matrix is positive definite, its symplectic eigenvalues are larger or equal to 1, and the entropy can be calculated using Eq. (5.193):

$$S(\hat{\rho}(\mathbf{0}, \mathbf{\Sigma}^{\oplus})) = S\left[\bigotimes_{m=1}^{M} \hat{\rho}_{\text{thermal}}\left(\frac{\nu_m - 1}{2}\right)\right] = \sum_{m=1}^{M} S\left(\hat{\rho}_{\text{thermal}}\left(\frac{\nu_m - 1}{2}\right)\right) = \sum_{m=1}^{M} g\left(\frac{\nu_m - 1}{2}\right). \tag{5.195}$$

The *thermal decomposition* of an arbitrary Gaussian state can be represented as

$$\hat{\rho}(\overline{\mathbf{x}}, \mathbf{\Sigma}^{\oplus}) = D(\overline{\mathbf{x}}) U_S\left[\hat{\rho}(\mathbf{0}, \mathbf{\Sigma}^{\oplus})\right] U_S^{\dagger} D(\overline{\mathbf{x}})^{\dagger}, \tag{5.196}$$

where $D(\overline{\mathbf{x}})$ represents the phase-space translation, while U_S corresponds to the symplectic mapping described above.

5.8.4 Covariance matrices of two-mode Gaussian states

Two-mode Gaussian states $\hat{\rho} = \hat{\rho}(\overline{\mathbf{x}}, \mathbf{\Sigma})$ have been intensively studied [6,7,34–38], and simple analytical formulas have been derived. The covariance matrix for two-mode Gaussian states can be represented as [6,7,36,38]

$$\mathbf{\Sigma} = \begin{bmatrix} \mathbf{A} & \mathbf{C} \\ \mathbf{C}^T & \mathbf{B} \end{bmatrix}, \tag{5.197a}$$

where $\mathbf{A}$, $\mathbf{B}$, and $\mathbf{C}$ are 2×2 real matrices, while $\mathbf{A}$ and $\mathbf{B}$ are also orthogonal matrices. The symplectic eigenvalues' spectrum $\{\lambda_1, \lambda_2\} = \{\lambda_-, \lambda_+\}$ can be computed by Refs. [6,7,36,38]:

$$\lambda_{1,2} = \sqrt{\frac{\Delta(\mathbf{\Sigma}) \mp \sqrt{\Delta^2(\mathbf{\Sigma}) - 4 \det \mathbf{\Sigma}}}{2}}, \tag{5.197b}$$

$$\det \mathbf{\Sigma} = \lambda_1^2 \lambda_2^2,\ \Delta(\mathbf{\Sigma}) = \det \mathbf{A} + \det \mathbf{B} + 2 \det \mathbf{C} = \lambda_1^2 + \lambda_2^2.$$

The uncertainty principle is equivalent to the following conditions [38]:

$$\mathbf{\Sigma} > 0,\ \det \mathbf{\Sigma} \geq 1,\ \Delta(\mathbf{\Sigma}) \leq 1 + \det \mathbf{\Sigma}. \tag{5.198}$$

The Einstein–Podolsky–Rosen (EPR) states are an important class of two-mode Gaussian states and are defined by Refs. [6,7]:

$$|\Psi\rangle = \sqrt{1 - \lambda^2} \sum_{n=0}^{\infty} \lambda^n |n\rangle_a |n\rangle_b = \sqrt{\frac{2}{\nu + 1}} \sum_{n=0}^{\infty} \left(\frac{\nu - 1}{\nu + 1}\right)^{n/2} |n\rangle_a |n\rangle_b,\ \lambda^2 = \frac{\nu - 1}{\nu + 1}, \tag{5.199}$$

with $\lambda = \tanh(s)$ and $\nu = \cosh(2s)$ representing the variances of the quadratures. The corresponding density operator is given by $\rho_{\text{EPR}} = |\Psi\rangle\langle\Psi|$, while the covariance matrix can be represented by the *standard form* [34,35]:

$$\mathbf{\Sigma} = \begin{bmatrix} a\mathbf{1} & \mathbf{C} \\ \mathbf{C} & b\mathbf{1} \end{bmatrix},\ \mathbf{C} = \begin{bmatrix} c_1 & 0 \\ 0 & c_2 \end{bmatrix};\ a, b, c \in \Re. \tag{5.200}$$

In particular, when $c_1 = -c_2 = c \geq 0$, the symplectic eigenvalues are simplified to

$$\lambda_{1,2} = \left[\sqrt{(a+b)^2 - 4c^2} \mp (b-a)\right]/2. \tag{5.201}$$

The corresponding symplectic matrix S performing the symplectic decomposition $\mathbf{\Sigma} = \mathbf{S}\mathbf{\Sigma}^{\oplus}\mathbf{S}^T$ is given by Refs. [6,7]:

$$\mathbf{S} = \begin{bmatrix} d_+\mathbf{1} & d_-\mathbf{Z} \\ d_-\mathbf{Z} & d_+\mathbf{1} \end{bmatrix}, \quad d_{\pm} = \sqrt{\frac{a+b \pm \sqrt{(a+b)^2 - 4c^2}}{2\sqrt{(a+b)^2 - 4c^2}}}. \tag{5.202}$$

5.8.5 Gaussian state detection

A quantum measurement is specified by the set of measurement operators given by $\{M_m\}$, where index m stands for possible measurement result, satisfying the completeness relation $\sum_m M_m^\dagger M_m = \hat{1}$. The probability of finding the measurement result m, given the state $|\psi\rangle$, is given by Refs. [1,39–43]:

$$p_m = \Pr(m) = \operatorname{Tr}\left(\hat{\rho} M_m^\dagger M_m\right). \tag{5.203}$$

After the measurement, the system is left in the following state:

$$\hat{\rho}_f = M_m \hat{\rho} M_m^\dagger / p_m. \tag{5.204}$$

If we are only interested in the measurement outcome, we can define a set of positive operators $\left\{\Pi_m = M_m^\dagger M_m\right\}$ and describe the measurement as the *positive operator-valued measure* (POVM). The probability of getting the m-th measurement result will then be [39,42]

$$p_m = \Pr(m) = \operatorname{Tr}(\hat{\rho}\Pi_m). \tag{5.205}$$

In addition to being positive, the Π_m-operators satisfy completeness relation $\sum_m \Pi_m = \hat{1}$.

In CV systems, measurement results are real, that is, $m \in \Re$, so the probability of the outcome becomes the PDF. When measuring Gaussian states, the measurement distribution is Gaussian. If we perform Gaussian measurements on M of $M+N$ available modes, the distribution of measurements is Gaussian, while the remaining unmeasured N modes are in a Gaussian state.

5.8.5.1 Homodyne detection

In homodyne detection, the most common measurement in CV systems, we measure one of the two quadratures since we cannot measure both quadratures simultaneously with complete precision, according to the uncertainty principle. The measurement operators are projectors over the quadrature basis $|I\rangle\langle I|$ or $|Q\rangle\langle Q|$, and corresponding outcomes have the PDF $f(I)$ or $f(Q)$, obtained by the marginal integration of the Wigner function:

$$f(I) = \int_Q W(I,Q)dQ, \; f(Q) = \int_I W(I,Q)dI. \tag{5.206}$$

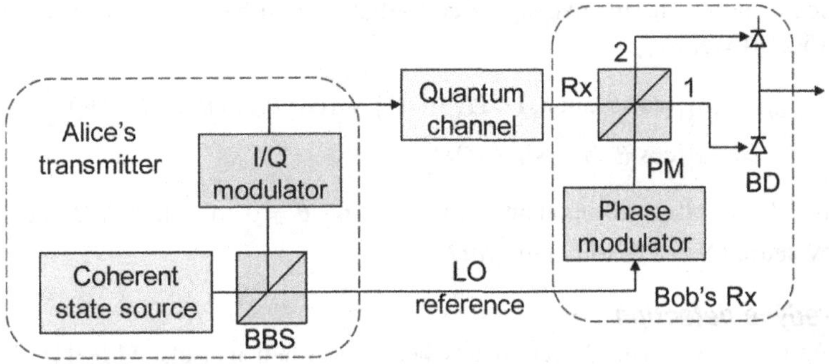

FIGURE 5.10

Homodyne detection. *BBS*, balanced beam splitter; *BD*, balanced detector.

When more than one bosonic mode is present in the optical beam, we must apply partial homodyning and perform integration over both quadratures of all other modes.

In the homodyne detection scheme [6,7,41–43], which is relevant for quantum communication and QKD and shown in Fig. 5.10, only one quadrature component is measured at a time. Alice sufficiently attenuates the laser beam, which serves as a weak coherent state source, and employs the balanced beam splitter (BBS) to split the signal into two parts. The upper output of the BBS is used as an input of the I/Q modulator to place the Gaussian state in the desired location in the I-Q plane to obtain $|\alpha\rangle = |I + jQ\rangle$ as well as control Alice's variance. The second output of the BBS is used to send the LO reference signal for Bob. Since the action of the BS is described by $BBS = 2^{-1/2}\begin{bmatrix} 1 & 1 \\ -1 & 1 \end{bmatrix}$ [see Eq. 5.159 for $T = 0.5$], the annihilation operators at the output of Bob's BBS will be

$$\begin{bmatrix} a_1 \\ a_2 \end{bmatrix} = \frac{1}{\sqrt{2}}\begin{bmatrix} 1 & 1 \\ -1 & 1 \end{bmatrix}\begin{bmatrix} a_{Rx} \\ a_{PM} \end{bmatrix} = \frac{1}{\sqrt{2}}\begin{bmatrix} a_{Rx} + a_{PM} \\ -a_{Rx} + a_{PM} \end{bmatrix}, \quad (5.207)$$

where the creation (annihilation) operators that receive BBS inputs Rx and PM are denoted $a_{Rx}^\dagger$ (a_{Rx}) and $a_{PM}^\dagger(a_{PM})$, respectively. The number operators that receive BBS outputs 1 and 2 can now be expressed as follows:

$$\widehat{n}_1 = a_1^\dagger a_1 = \frac{1}{2}\left(a_{PM}^\dagger + a_{Rx}^\dagger\right)(a_{PM} + a_{Rx}),\ \widehat{n}_2 = a_2^\dagger a_2 = \frac{1}{2}\left(a_{PM}^\dagger - a_{Rx}^\dagger\right)(a_{PM} - a_{Rx}), \quad (5.208)$$

The output of the balanced detector (BD), assuming a photodiode responsivity of $R_{\mathrm{ph}} = 1\mathrm{A/W}$, is given (in SNU) by

$$\widehat{i}_{BD} = 4(\widehat{n}_1 - \widehat{n}_2) = 4\left(a_{PM}^\dagger a_{Rx} + a_{Rx}^\dagger a_{PM}\right). \quad (5.209)$$

The phase modulation output can be represented in SNU by $a_{PM} = |\alpha|\exp(j\theta)\mathbf{1}/2$, and after substitution into Eq. (5.209), we obtain

$$\widehat{i}_{BD} = 4\left(\frac{|\alpha|e^{-j\theta}}{2}a_{Rx} + a_{Rx}^\dagger\frac{|\alpha|e^{j\theta}}{2}\right) = 2|\alpha|\left(e^{j\theta}a_{Rx}^\dagger + e^{-j\theta}a_{Rx}\right). \quad (5.210)$$

In the absence of noise, the receive signal constellation will be $\boldsymbol{a}_{Rx}=(I,Q)/2$, which after substitution into Eq. (5.210) yields

$$\begin{aligned}\hat{i}_{BD} &= |\alpha|\left[(\cos\theta,\sin\theta)\mathbf{1}\cdot\left(I\hat{I},-Q\hat{Q}\right)+(\cos\theta,\sin\theta)\mathbf{1}\cdot\left(I\hat{I},Q\hat{Q}\right)\right]\\ &= |\alpha|\left(\cos\theta\cdot I\hat{I}+\sin\theta\cdot Q\hat{Q}\right).\end{aligned} \tag{5.211}$$

Bob measures the in-phase component $\hat{I}$ by selecting $\theta=0$ to obtain $|\alpha|I$ and the quadrature component $\hat{Q}$ by setting $\theta=\pi/2$ to obtain $|\alpha|Q$.

5.8.5.2 Heterodyne detection

The theory of optical heterodyne detection has been described in Refs. [42–44]. The heterodyne detection corresponds to the projection onto coherent states, that is, $M(\alpha)\sim|\alpha\rangle\langle\alpha|$. This theory has been generalized to the POVM based on projections over pure Gaussian states [6,7,45]. The transmitter side is identical to the homodyne detection version for heterodyne detection. In a heterodyne detection receiver, as shown in Fig. 5.11, the quantum signal is split using the BBS—one output is used to measure the in-phase component (upper branch), while the other output (lower branch) is used as input to the second BD, to detect the quadrature component. The LO classical reference signal is also split by the BBS, with the upper output used as the input to the in-phase BD, and the other output, after phase modulation to introduce the $\pi/2$-phase shift, is used as the LO input to the quadrature BD to detect the quadrature component. In such a way, both quadratures can be measured simultaneously, and mutual information can be doubled. On the other hand, the BBS introduces 3 dB attenuation to the quantum signal.

5.8.5.3 Partial measurements

As indicated earlier, the measurement on one mode will affect the remaining modes of Gaussian state, depending on corresponding correlations [6,7]. Let B be the mode on which measurement is performed, and A denote remaining $M-1$ modes of the Gaussian state described by $2M\times 2M$ covariance matrix $\boldsymbol{\Sigma}$. Similar to Eq. (5.197a), we can decompose the covariance matrix as follows [6,7]:

$$\boldsymbol{\Sigma}=\begin{bmatrix}\boldsymbol{A} & \boldsymbol{C}\\ \boldsymbol{C}^T & \boldsymbol{B}\end{bmatrix}, \tag{5.212}$$

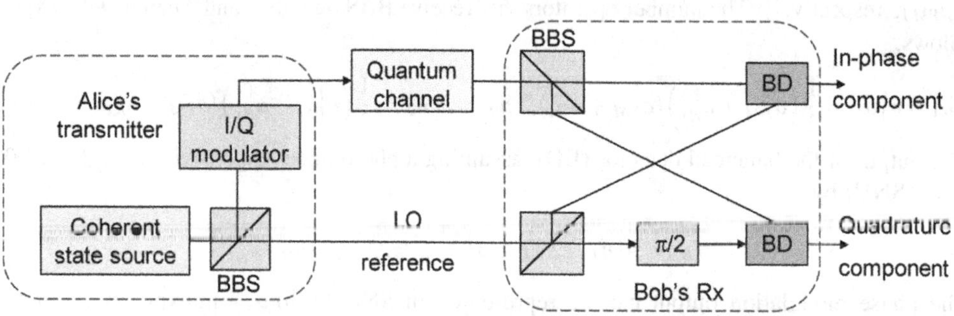

FIGURE 5.11

Heterodyne detection. *BBS*, balanced beam splitter; *BD*, balanced detector described in Fig. 5.10.

where $\boldsymbol{B}$ is the real matrix of size 2×2, $\boldsymbol{A}$ is the real matrix of size $2(M-1)\times 2(M-1)$, and $\boldsymbol{C}$ is the real matrix of size $2(M-1)\times 2$. For homodyne detection of in-phase I or quadrature Q, the partial measurement of $\boldsymbol{B}$ transforms the covariance matrix of $\boldsymbol{A}$ as follows [6,7]:

$$\boldsymbol{\Sigma}_{A|B}^{(\hat{I},\hat{Q})} = \boldsymbol{A} - \boldsymbol{C}\left(\boldsymbol{\Pi}_{I,Q}\boldsymbol{B}\,\boldsymbol{\Pi}_{I,Q}\right)^{-1}\boldsymbol{C}^T, \quad \boldsymbol{\Pi}_I = \text{diag}(1,0), \quad \boldsymbol{\Pi}_Q = \text{diag}(0,1), \tag{5.213}$$

where we use $(\bullet)^{-1}$ to denote the pseudoinverse operation. Given that $\boldsymbol{\Pi}_I\boldsymbol{B}\,\boldsymbol{\Pi}_I = \begin{bmatrix} B_{11} & 0 \\ 0 & 0 \end{bmatrix} = \begin{bmatrix} V(\hat{I}) & 0 \\ 0 & 0 \end{bmatrix}$ and $\boldsymbol{\Pi}_Q\boldsymbol{B}\,\boldsymbol{\Pi}_Q = \begin{bmatrix} 0 & 0 \\ 0 & B_{22} \end{bmatrix} = \begin{bmatrix} 0 & 0 \\ 0 & V(\hat{Q}) \end{bmatrix}$ matrices are diagonal, the previous equation simplifies to

$$\boldsymbol{\Sigma}_{A|B}^{(\hat{I},\hat{Q})} = \boldsymbol{A} - \frac{1}{V(\hat{I},\hat{Q})}\boldsymbol{C}\boldsymbol{\Pi}_{I,Q}\boldsymbol{C}^T. \tag{5.214}$$

In heterodyne detection, we measure both quadratures, and given that an additional BBS is used, with one of the inputs being in a vacuum state, we need to account for fluctuations due to the vacuum state and modify the partial covariance matrix [6,7]:

$$\boldsymbol{\Sigma}_{A|B} = \boldsymbol{A} - \boldsymbol{C}(\boldsymbol{B}+\boldsymbol{1})^{-1}\boldsymbol{C}^T. \tag{5.215}$$

5.8.6 The covariance matrices of multimode Gaussian systems and lossy transmission channel

For a bipartite system composed of two separable subsystems A and B described by corresponding density matrices ρ_A and ρ_B, the density operator of the bipartite system can be represented by $\rho = \rho_A \otimes \rho_B$. For Gaussian bipartite system, composed of two separable subsystems A and B, with A subsystem being a Gaussian state with M modes represented by quadrature operators $\hat{\boldsymbol{x}}_A$, while B subsystem being a Gaussian state with N modes represented by quadrature operators $\hat{\boldsymbol{x}}_B$; the Gaussian state of a bipartite system will contain $M+N$ modes and can be described by the quadrature operators $\hat{\boldsymbol{x}} = [\hat{\boldsymbol{x}}_A \ \hat{\boldsymbol{x}}_B]^T$, with corresponding covariance matrix of size $2(M+N)\times 2(M+N)$ being represented as

$$\boldsymbol{\Sigma} = \boldsymbol{\Sigma}_A \oplus \boldsymbol{\Sigma}_B = \begin{bmatrix} \boldsymbol{\Sigma}_A & \boldsymbol{0} \\ \boldsymbol{0} & \boldsymbol{\Sigma}_B \end{bmatrix}, \tag{5.216}$$

where $\boldsymbol{0}$ denotes the all-zeros matrix.

To model the *lossy transmission channel,* we can use a BS with attenuation $0 \le T \le 1$, as shown in Fig. 5.12. The output state can be represented in terms of input state $\boldsymbol{x}_A$ and ground state $\boldsymbol{x}_0$:

$$\boldsymbol{x}_B = \sqrt{T}\boldsymbol{x}_A + \sqrt{1-T}\boldsymbol{x}_0 = \sqrt{T}\left(\boldsymbol{x}_A + \sqrt{\chi_{\text{line}}}\,\boldsymbol{x}_0\right), \quad \chi_{\text{line}} = (1-T)/T = 1/T - 1 \tag{5.217}$$

For lossy transmission when Gaussian modulation of coherent states is employed instead of vacuum states, the previous expression must be modified as follows:

$$\boldsymbol{x}_B = \sqrt{T}\left(\boldsymbol{x}_A + \sqrt{\chi_{\text{line}}}\,\boldsymbol{x}_0\right), \quad \chi_{\text{line}} = (1-T)/T + \varepsilon = 1/T - 1 + \varepsilon, \varepsilon \ge 0, \tag{5.218}$$

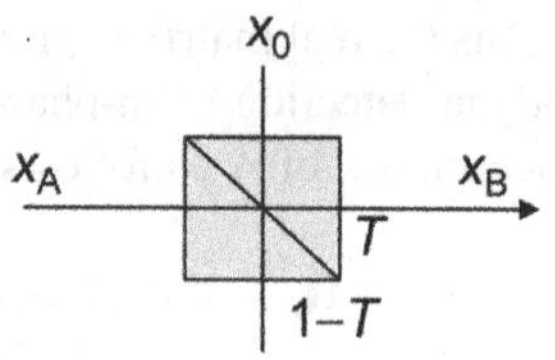

FIGURE 5.12

Lossy transmission channel model.

The first term in χ_{line} is contributed to the channel attenuation, while the second term ε, commonly referred to as the excess noise, is due to other noisy factors such as the laser phase noise, imperfect orthogonality of quadratures, and so on.

Let us assume that the EPR state, defined by Eq. (5.199), is used in the CV system as the source, with the covariance matrix given by

$$\boldsymbol{\Sigma}_{\mathrm{EPR}} = \begin{bmatrix} v\mathbf{1} & \sqrt{v^2-1}\mathbf{Z} \\ \sqrt{v^2-1}\mathbf{Z} & v\mathbf{1} \end{bmatrix}, \quad v = \cosh 2s, \tag{5.219}$$

wherein the covariance matrix elements are determined by applying the definitions of Eqs. (5.142) and (5.143) to the EPR state. The identity matrix can describe the thermal state at the input of the BS. Based on the previous equation, the corresponding bipartite system will have the following covariance matrix:

$$\boldsymbol{\Sigma} = \begin{bmatrix} \boldsymbol{\Sigma}_{\mathrm{EPR}} & \mathbf{0} \\ \mathbf{0} & \mathbf{1} \end{bmatrix} = \begin{bmatrix} v\mathbf{1} & \sqrt{v^2-1}\mathbf{Z} & \mathbf{0} \\ \sqrt{v^2-1}\mathbf{Z} & v\mathbf{1} & \mathbf{0} \\ \mathbf{0} & \mathbf{0} & \mathbf{1} \end{bmatrix}. \tag{5.220}$$

The BS acts on Bob's qubit and thermal state while leaving Alice's qubit unaffected; therefore, its action can be described as

$$BS_{\mathrm{combined}}(T) = \mathbf{1} \oplus \begin{bmatrix} \sqrt{T}\mathbf{1} & \sqrt{1-T}\mathbf{1} \\ -\sqrt{1-T}\mathbf{1} & \sqrt{T}\mathbf{1} \end{bmatrix} = \begin{bmatrix} \mathbf{1} & \mathbf{0} & \mathbf{0} \\ \mathbf{0} & \sqrt{T}\mathbf{1} & \sqrt{1-T}\mathbf{1} \\ \mathbf{0} & -\sqrt{1-T}\mathbf{1} & \sqrt{T}\mathbf{1} \end{bmatrix}. \tag{5.221}$$

To determine the covariance matrix after the BS, we apply the symplectic operation by BS_{combined} on input covariance matrix to obtain

$$\boldsymbol{\Sigma}' = BS_{\mathrm{combined}}(T)\boldsymbol{\Sigma}[BS_{\mathrm{combined}}(T)]^T$$

$$= \begin{bmatrix} \mathbf{1} & \mathbf{0} & \mathbf{0} \\ \mathbf{0} & \sqrt{T}\mathbf{1} & \sqrt{1-T}\mathbf{1} \\ \mathbf{0} & -\sqrt{1-T}\mathbf{1} & \sqrt{T}\mathbf{1} \end{bmatrix} \begin{bmatrix} v\mathbf{1} & \sqrt{v^2-1}\mathbf{Z} & \mathbf{0} \\ \sqrt{v^2-1}\mathbf{Z} & v\mathbf{1} & \mathbf{0} \\ \mathbf{0} & \mathbf{0} & \mathbf{1} \end{bmatrix} \begin{bmatrix} \mathbf{1} & \mathbf{0} & \mathbf{0} \\ \mathbf{0} & \sqrt{T}\mathbf{1} & \sqrt{1-T}\mathbf{1} \\ \mathbf{0} & -\sqrt{1-T}\mathbf{1} & \sqrt{T}\mathbf{1} \end{bmatrix}^T. \tag{5.222}$$

Now by keeping Alice and Bob submatrices, we obtain

$$\Sigma'_{AB} = \begin{bmatrix} v\mathbf{1} & \sqrt{T(v^2-1)}\mathbf{Z} \\ \sqrt{T(v^2-1)}\mathbf{Z} & (vT+1-T)\mathbf{1} \end{bmatrix}. \tag{5.223}$$

In the presence of excess noise, we need to replace the $(\Sigma'_{AB})_{22}$-element by $vT+1+\varepsilon'-T$ to obtain the following *covariance matrix for homodyne detection*:

$$\Sigma_{AB}^{(\text{homodyne})} = \begin{bmatrix} v\mathbf{1} & \sqrt{T(v^2-1)}\mathbf{Z} \\ \sqrt{T(v^2-1)}\mathbf{Z} & (vT+1+\varepsilon'-T)\mathbf{1} \end{bmatrix}. \tag{5.224}$$

In *heterodyne detection*, Bob employs the BBS to split the quantum signal, while the second input to the BBS is the vacuum state, and the operation of the BBS can be represented by Eq. (5.159) by setting $T=1/2$. The corresponding covariance matrix at the input of the BBS is

$$\Sigma_{\text{in}}^{(\text{BBS})} = \begin{bmatrix} \Sigma_{\text{AB}}^{(\text{homodyne})} & \mathbf{0} \\ \mathbf{0} & \mathbf{1} \end{bmatrix} = \begin{bmatrix} v\mathbf{1} & \sqrt{T(v^2-1)}\mathbf{Z} & \mathbf{0} \\ \sqrt{T(v^2-1)}\mathbf{Z} & (vT+1+\varepsilon'-T)\mathbf{1} & \mathbf{0} \\ \mathbf{0} & \mathbf{0} & \mathbf{1} \end{bmatrix}. \tag{5.225}$$

The covariance matrix and BBS output can be determined by applying the symplectic operation by $BS_{\text{combined}}(1/2)$ on the input covariance matrix to obtain

$$\begin{aligned}
\Sigma_{\text{AB}}^{(\text{het})} &= BS_{\text{combined}}(1/2)\Sigma_{\text{in}}^{(\text{BBS})}[BS_{\text{combined}}(1/2)]^T \\
&= \begin{bmatrix} \mathbf{1} & \mathbf{0} & \mathbf{0} \\ \mathbf{0} & \frac{1}{\sqrt{2}}\mathbf{1} & \frac{1}{\sqrt{2}}\mathbf{1} \\ \mathbf{0} & -\frac{1}{\sqrt{2}}\mathbf{1} & \frac{1}{\sqrt{2}}\mathbf{1} \end{bmatrix} \begin{bmatrix} v\mathbf{1} & \sqrt{T(v^2-1)}\mathbf{Z} & \mathbf{0} \\ \sqrt{T(v^2-1)}\mathbf{Z} & (vT+1+\varepsilon'-T)\mathbf{1} & \mathbf{0} \\ \mathbf{0} & \mathbf{0} & \mathbf{1} \end{bmatrix} \begin{bmatrix} \mathbf{1} & \mathbf{0} & \mathbf{0} \\ \mathbf{0} & \frac{1}{\sqrt{2}}\mathbf{1} & \frac{1}{\sqrt{2}}\mathbf{1} \\ \mathbf{0} & -\frac{1}{\sqrt{2}}\mathbf{1} & \frac{1}{\sqrt{2}}\mathbf{1} \end{bmatrix}^T \\
&= \begin{bmatrix} v\mathbf{1} & \sqrt{\frac{T}{2}(v^2-1)}\mathbf{Z} & -\sqrt{\frac{T}{2}(v^2-1)}\mathbf{Z} \\ \sqrt{\frac{T}{2}(v^2-1)}\mathbf{Z} & \frac{vT+2+\varepsilon'-T}{2}\mathbf{1} & -\frac{vT+\varepsilon'-T}{2}\mathbf{1} \\ -\sqrt{\frac{T}{2}(v^2-1)}\mathbf{Z} & -\frac{vT+\varepsilon'-T}{2}\mathbf{1} & \frac{vT+2+\varepsilon'-T}{2}\mathbf{1} \end{bmatrix}.
\end{aligned} \tag{5.226}$$

5.9 Generation of quantum states

Most optical media are dielectric [8,9,13], so the magnetic flux density $\boldsymbol{B}$ and magnetic field $\boldsymbol{H}$ are related by $\boldsymbol{B} = \mu\boldsymbol{H}$ (with μ being the permittivity of the optical medium), while the electric flux density $\boldsymbol{D}$, electric field $\boldsymbol{E}$, and induced electric density (polarization) $\boldsymbol{P}$ are related by the constitutive relation $\boldsymbol{D} = \varepsilon\boldsymbol{E} + \boldsymbol{P}$ (with ε being the permeability). For the NL optical medium, the electric polarization is related to the electric field by

$$\boldsymbol{P} = \overrightarrow{\chi}^{(1)}: \boldsymbol{E} + \underbrace{\overrightarrow{\chi}^{(2)}: \boldsymbol{EE} + \overrightarrow{\chi}^{(3)}: \boldsymbol{EEE} + \cdots}_{\boldsymbol{P}_{NL}} \tag{5.227}$$

where $\overrightarrow{\chi}^{(i)}$ $(i = 1, 2, 3, \cdots)$ are tensors and the operation (:) denotes the tensor product. The first term is the linear term and is applicable in linear optics. The second and higher terms are responsible for different phenomena in NL optics, as illustrated in Fig. 5.13. The second-order term dominates in bulk optics materials such as various crystals, whereas the third-order term dominates in optical fibers and gaseous atomic/molecular media. When due to the symmetry of randomly oriented particles (atoms and molecules), the second term becomes zero. NL devices introduce interaction among field modes, with interaction Hamiltonian being represented as

$$\widehat{H}_{\text{interaction}} = \frac{1}{2}\int \widehat{\boldsymbol{P}}_{NL}(\boldsymbol{r}, t)\cdot\widehat{\boldsymbol{E}}(\boldsymbol{r}, t),\ \ \widehat{\boldsymbol{P}}_{NL} = \overrightarrow{\chi}^{(2)}: \widehat{\boldsymbol{E}}\widehat{\boldsymbol{E}} + \overrightarrow{\chi}^{(3)}: \widehat{\boldsymbol{E}}\widehat{\boldsymbol{E}}\widehat{\boldsymbol{E}} + \cdots. \tag{5.228}$$

The *evolution of the quantum state* in the interaction picture is described by the following unitary operator:

$$\widehat{U} = \exp\left(-j\int \widehat{H}_{\text{interaction}}dt\right), \tag{5.229}$$

which gives rise to momentum conservation. The total electromagnetic energy localized in the medium of volume V is given by Ref. [10]:

$$W = \int_V [\boldsymbol{E}(\boldsymbol{r}, t)\cdot\boldsymbol{D}(\boldsymbol{r}, t) + \boldsymbol{H}(\boldsymbol{r}, t)\cdot\boldsymbol{B}(\boldsymbol{r}, t)]d^3r, \tag{5.230}$$

and gives rise to energy conservation.

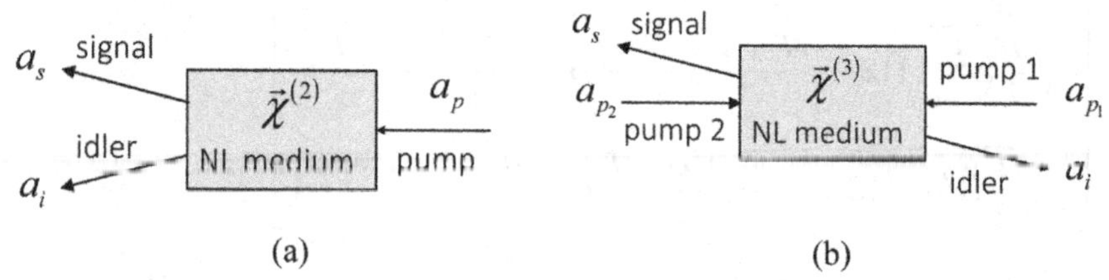

FIGURE 5.13

Nonlinear optics phenomena: (A) three-wave mixing and (B) four-wave mixing.

In second-order-dominating devices, as illustrated in Fig. 5.13A, the strong pump photon generates signal and idler photons. This interaction is *TWM* and described by the following interaction Hamiltonian:

$$\hat{H}_{TWM} = j\eta a_s^\dagger a_i^\dagger a_p + H.c., \tag{5.231}$$

where a_p denotes the pump photon annihilation operators, while $a_s^\dagger$ and $a_i^\dagger$ denote the signal and idler photon creation operators. Parameter η is related to conversion efficiency. On the other hand, in the $\chi^{(3)}$-device, see Fig. 5.13B, two bright pump photons interact and generate the signal and idler photons. This interaction is *FWM* and can be described by the following interaction Hamiltonian:

$$\hat{H}_{FWM} = j\eta a_s^\dagger a_i^\dagger a_{p_1} a_{p_2} + H.c., \tag{5.232}$$

The second- and third-order NL effects are typically weak, and we need to use the strong pump beams to generate any observable quantum effect. When the pump beams are very strong, they become classical and can be represented by the corresponding complex variables; for instance, we can use the complex variable A_p instead of a_p. The corresponding two-photon processes are known as the *parametric process*. By substituting ηA_p in TWM Hamiltonian with ζ, and $\eta A_{p1} A_{p2}$ in FWM Hamiltonian with as well, the parametric processes Hamiltonian can be written as

$$\hat{H}_P = j\zeta a_s^\dagger a_i^\dagger + H.c., \tag{5.233}$$

The corresponding evolution operator becomes

$$\hat{U}_P = \exp\left(-j\int \hat{H}_P dt\right) = \exp\left(\underbrace{\zeta t}_{-\xi}\ a_s^\dagger a_i^\dagger - \zeta^* t a_s a_i\right) = \exp\left(-\xi a_s^\dagger a_i^\dagger + \xi^* a_s a_i\right), \xi = -\zeta t. \tag{5.234}$$

When both signal and idler photons are in vacuum states, the state at the output of the NL device is the TMS state:

$$|\Psi\rangle = \exp\left(-\xi a_s^\dagger a_i^\dagger + \xi^* a_s a_i\right)|0\rangle|0\rangle = S_2(\xi)|0\rangle|0\rangle, \ \ S_2(\xi) = e^{-\xi a_s^\dagger a_i^\dagger + \xi^* a_s a_i}. \tag{5.235}$$

In this case, the parameter ξ is a real number and can be denoted s to specify the squeezing level. The corresponding Hamiltonian in this case is commonly referred to as the *two-mode squeezing operator*:

$$S_2(s) = \exp\left(-s\left(a_s^\dagger a_i^\dagger - a_s a_i\right)\right). \tag{5.236}$$

In the Heisenberg picture, the quadrature operators $\hat{x} = (\hat{x}_1 \hat{x}_2 \hat{x}_3 \hat{x}_4)^T = \left(\hat{I}_a \hat{Q}_a \hat{I}_b \hat{Q}_b\right)^T$ are transformed to Ref. [4]

$$\hat{x} \to \hat{x}_f = \hat{S}_2(s)\hat{x}, \ \ \hat{S}_2(s) = \begin{bmatrix} \cosh s\, \mathbf{1} & \sinh s\, \mathbf{Z} \\ \sinh s\, \mathbf{Z} & \cosh s\, \mathbf{1} \end{bmatrix}, \tag{5.237}$$

where $\mathbf{Z} = \text{diag}(1,-1)$ is the Pauli Z-matrix. Now by applying $S_2(s)$ on two vacuum states, as described by Eq. (5.233), we can represent the TMS state, also known as the EPR state, introduced by Eq. (5.199).

When the signal and idler quantum fields are identical, that is, $a_s = a_i = a$, the interaction Hamiltonian (5.233) simplifies to $\widehat{H}_P = j(\zeta a^{\dagger 2} - \zeta^* a^2)$, and the corresponding evolution operator becomes $\widehat{U}_P = \exp(-j\widehat{H}_P t) = \exp[(\xi a^{\dagger 2} - \xi^* a^2)/2]$, $\xi = 2\zeta t$. The squeezing parameter ξ in quantum optics is typically a real number, say s, so the *squeezing operator* becomes $S(s) = \exp[s(a^{\dagger 2} - a^2)/2]$, which is consistent with Section 5.8.2.

For $|\xi| << 1$, by Taylor expansion in the evaluation operator of Eq. (5.235), we obtain the following approximation for the TMS state:

$$|\Psi\rangle \cong \left(\mathbf{1} - \xi a_s^\dagger a_i^\dagger + \xi^{*2} a_s^{\dagger 2} a_i^{\dagger 2} + \cdots\right)|0\rangle|0\rangle = |0\rangle_s|0\rangle_i - \xi|1\rangle_s|1\rangle_i + \xi^2|2\rangle_s|2\rangle_i + \cdots \tag{5.238}$$

Clearly, for a sufficiently small ξ, the third and higher terms are negligible, and the two-photon state term dominates; these are widely used in two-photon interference studies. On the other hand, the first two terms are irrelevant in four-photon coincidence measurements. Now by employing the BS on two-photon and four-photon states, various entangled states can be generated.

Because S_2-operator cannot be decomposed as the product of two single-mode operators, it describes an entanglement operation in which the squeezing is not in individual modes but rather in superposition modes. So we must define the superposition quadrature operators:

$$\widehat{I} = 2^{-3/2}(a^\dagger + a + b^\dagger + b), \quad \widehat{Q} = 2^{-3/2} j(a^\dagger - a + b^\dagger - b). \tag{5.239}$$

Now by applying a procedure similar to that of Section 5.8.2—see Eqs. (5.163)–(5.168)—we obtain

$$\begin{aligned} S^\dagger\left(\underline{s}\right) a S\left(\underline{s}\right) &= a \cosh s - b^\dagger e^{j\theta} \sinh s, \\ S^\dagger\left(\underline{s}\right) b S\left(\underline{s}\right) &= b \cosh s - a^\dagger e^{j\theta} \sinh s. \end{aligned} \tag{5.240}$$

To represent the TMS state in terms of two-mode number states $|n,m\rangle = |n\rangle|m\rangle$, we employ a procedure similar to the one used in Section 5.8.2. The starting point is $S_2(\xi)a|0,0\rangle = 0$, and after inserting the unitary operator, we obtain

$$\underbrace{S_2(\xi) a S_2^\dagger(\xi)}_{a \cosh s + b^\dagger e^{j\theta} \sinh s} \quad S_2(\xi)|0,0\rangle = \left(a \cosh s + b^\dagger e^{j\theta} \sinh s\right)|\Psi\rangle = 0. \tag{5.241}$$

The expansion in terms of two-modes product states,

$$|\Psi\rangle = \sum_{n,m=0}^{\infty} C_{n,m}|n\rangle|m\rangle, \tag{5.242}$$

must satisfy the eigenvalue Eq. (5.241), and after substituting Eq. (5.242) into Eq. (5.241), we obtain the following recursion relation for the expansion coefficients:

$$\sum_{n,m=0}^{\infty} C_{n,m}\left[\cosh s \cdot \sqrt{n}\,|n-1\rangle|m\rangle + e^{j\theta} \cosh s \cdot \sqrt{m+1}\,|n\rangle|m+1\rangle\right] = 0, \tag{5.243}$$

and out of many possible solutions, we choose one that contains a two-mode vacuum state so that

$$C_{n,m}=\left(-e^{j\theta}\tanh s\right)^{n}\delta_{n,m}C_{0,0},\quad C_{0,0}=(\cosh s)^{-1}, \tag{5.244}$$

and after substitution into Eq. (5.242), we obtain

$$|\Psi\rangle=\frac{1}{\cosh s}\sum_{n=0}^{\infty}\left(-e^{j\theta}\tanh s\right)^{n}|n\rangle|n\rangle. \tag{5.245}$$

The probability of getting n_1 photons in mode a and n_2 photons in mode b can be obtained by

$$P_{n_1,n_2}=|\langle n_1,n_2|\Psi\rangle|^2=\frac{(\tanh s)^{2n}}{\cosh^2 s}\delta_{n_1,n}\delta_{n_2,n}, \tag{5.246}$$

and the corresponding joint probability will be diagonal with a monotonically decreasing envelope. Given that only paired states $|n\rangle|n\rangle$ occur in the expansion of Eq. (5.245), these states are often called *twin beams* or *twin waves*. Twin waves are relevant in quantum and classical optical communication to deal with Kerr NL effects [46].

Let us now consider the case in which the a_s field in TWM and the a_s and a_{p2} fields in FWM are strong enough to be considered classical, in which case the interaction Hamiltonian can be represented by

$$\hat{H}_{FC}=j\zeta a_i^{\dagger}a_p-j\zeta^{*}a_i a_p^{\dagger},\quad \zeta=\begin{cases}\eta A_s^{*}, & \text{TWM}\\ \eta A_s^{*}A_{p2}, & \text{FWM}\end{cases} \tag{5.247}$$

and describes a *one-photon process* in which one photon is annihilated or created at a given instance in time. This process annihilates the photon in mode a_p and creates one photon in mode a_i. When these two photons have different frequencies, the *photon (wavelength) conversion process* takes place. The corresponding evolution operator will be

$$\hat{U}_{FC}=\exp\left(\xi a_i^{\dagger}a_p-\xi^{*}a_i a_p^{\dagger}\right),\xi=\zeta t, \tag{5.248}$$

which is for real ξ equivalent to the BS evolution operator.

5.10 Multilevel quantum optical communication

To analyze the M-ary quantum communication systems in the absence of noise, we can employ the theory from Sections 5.2 and 5.3. On the other hand, to analyze the M-ary quantum systems in the presence of noise, we should use the material from Section 5.6. Assuming that the transmitted states are represented by the density operators ρ_j ($j=1,\ldots,M$), we must determine the detection operators Π_j ($j=1\ldots,M$) to minimize the probability of error given by Eq. (5.20) subject to the constraints of Eqs (5.21) and (5.22), or equivalently, to maximize the average probability of a correct decision:

$$P_c=\sum_{j=1}^{M}\zeta_j\mathrm{Tr}\left(\rho_j\Pi_j\right),\quad \zeta_j=\Pr\left(H_j\right);\ j=1,\cdots,M \tag{5.249}$$

So the optimization problem can be formulated by

$$\Pi_j^{(\mathrm{opt})} = \underset{\{\Pi_j\}}{\operatorname{argmax}} \sum_{j=1}^{M} \zeta_j \mathrm{Tr}(\rho_j \Pi_j), \tag{5.250}$$

subject to the constraints of Eqs. (5.21) and (5.22).

In particular, when the transmitted states are pure coherent states $\{|\alpha_i\rangle\}$ $(i = 1,\ldots,M)$ and linearly independent in the absence of background noise, the optimum decision strategy is expressed in terms of M detection (projection) operators [1,19,45],

$$\Pi_j^{(\mathrm{opt})} = |w_j\rangle\langle w_j|. \tag{5.251}$$

onto the orthogonal axes (measurement states) $|w_j\rangle$, wherein the angles between measurement states and signal states are chosen to maximize the correct probability:

$$\left|w_j^{(\mathrm{opt})}\right\rangle = \underset{\{|w_j\rangle\}}{\operatorname{argmax}} \sum_{j=1}^{M} \zeta_j \mathrm{Tr}(\underbrace{|\alpha_j\rangle\langle\alpha_j|}_{\rho_j}\,\underbrace{|w_j\rangle\langle w_j|}_{\Pi_j}) = \underset{\{|w_j\rangle\}}{\operatorname{argmax}} \sum_{j=1}^{M} \zeta_j |\langle\alpha_j|w_j\rangle|^2. \tag{5.252}$$

The probability of correct decision for this optimum receiver will then be

$$P_c^{(\mathrm{opt})} = \sum_{j=1}^{M} \zeta_j \left|\left\langle \alpha_j | w_j^{(\mathrm{opt})} \right\rangle\right|^2. \tag{5.253}$$

Given that $\langle\alpha_j|w_j\rangle = \cos\theta_j$ where θ_j is angle between kets $|\alpha_j\rangle$ and $|w_j\rangle$, the optimization problem becomes

$$\left|\theta_j^{(\mathrm{opt})}\right\rangle = \underset{\{\theta_j\}}{\operatorname{argmax}} \sum_{j=1}^{M} \zeta_j \cos^2\theta_j, \tag{5.254}$$

and we can use the geometric approach in optimization, as illustrated in Fig. 5.14 for two vectors lying in the plane of two axes. So we need to rotate the constellation vectors $|\alpha_j\rangle$ around the corresponding axes $|w_i\rangle$ to maximize the correct probability.

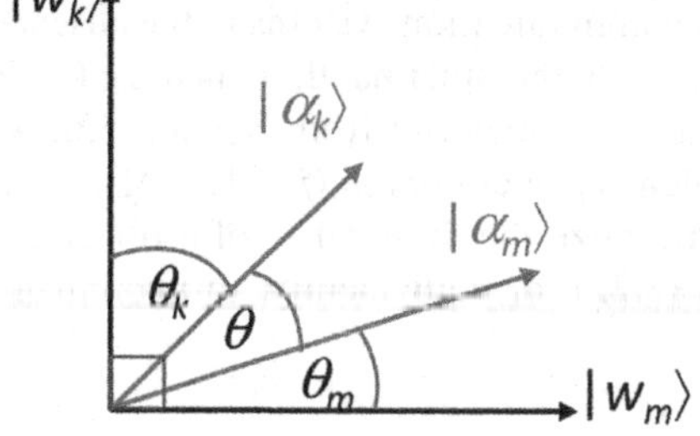

FIGURE 5.14

Optimization procedure.

We rewrite Eq. (5.253) as follows:

$$P_c^{(\mathrm{opt})} = \sum_{k=1}^{M} \zeta_k \langle w_k^{(\mathrm{opt})} | \alpha_k \rangle \langle \alpha_k | w_k^{(\mathrm{opt})} \rangle \overset{\sum_m |w_m^{(\mathrm{opt})}\rangle\langle w_m^{(\mathrm{opt})}| = \hat{1}}{=} \sum_{k=1}^{M}\sum_{m=1}^{M} \zeta_k \langle w_k^{(\mathrm{opt})} | \alpha_k \rangle \langle \alpha_k | \left(|w_m^{(\mathrm{opt})}\rangle\langle w_m^{(\mathrm{opt})}| \right) | w_k^{(\mathrm{opt})} \rangle$$

$$= \sum_{k=1}^{M}\sum_{m=1}^{M} \zeta_k \langle w_k^{(\mathrm{opt})} | \alpha_k \rangle \langle \alpha_k | w_m^{(\mathrm{opt})} \rangle \underbrace{\langle w_m^{(\mathrm{opt})} | w_k^{(\mathrm{opt})} \rangle}_{\delta_{mk}}$$

$$= \sum_{k=1}^{M}\sum_{m=1}^{M} \zeta_m \langle w_m^{(\mathrm{opt})} | \alpha_m \rangle \langle \alpha_m | w_k^{(\mathrm{opt})} \rangle \delta_{km}, \tag{5.255}$$

and conclude that the $0.5M(M-1)$ equations below must be satisfied:

$$\zeta_k \langle w_k^{(\mathrm{opt})} | \alpha_k \rangle \langle \alpha_k | w_m^{(\mathrm{opt})} \rangle = \zeta_m \langle w_m^{(\mathrm{opt})} | \alpha_m \rangle \langle \alpha_m | w_k^{(\mathrm{opt})} \rangle. \tag{5.256}$$

Given that the scalar products of signaling states can be easily determined, we have

$$\gamma_{km} = \langle \alpha_k | \alpha_m \rangle = \left\langle \alpha_k \left| \underbrace{\left(\sum_j |w_j^{(\mathrm{opt})}\rangle\langle w_j^{(\mathrm{opt})}| \right)}_{\hat{1}} \right| \alpha_m \right\rangle = \sum_{j=1}^{M} \left\langle \alpha_k \middle| w_j^{(\mathrm{opt})} \right\rangle \left\langle w_j^{(\mathrm{opt})} \middle| \alpha_m \right\rangle, \tag{5.257}$$

representing the additional $0.5M(M+1)$ equations. In total, we have M^2 equations to determine M^2 unknown coefficients that represent the measurement vectors in terms of signaling vectors:

$$|w_m^{(\mathrm{opt})}\rangle = \sum_{j=1}^{M} \langle \alpha_j | w_m^{(\mathrm{opt})} \rangle |\alpha_j\rangle. \tag{5.258}$$

The calculation of coefficients of the measurement vectors can be computationally intensive. Alternatively, we can use least-squares measurements, also known as SRMs. Interestingly enough, this method for the pure states and the signal constellation with geometrically uniform symmetry (GUS) has the same performance as the optimum method described above.

5.10.1 The square root measurement-based quantum decision

The signaling pure states $|\alpha_m\rangle$ $(m = 1,\ldots,M)$ can be organized in terms of the *state matrix*:

$$\boldsymbol{\Gamma} = [|\alpha_1\rangle, \ |\alpha_2\rangle, \cdots, |\alpha_M\rangle], \tag{5.259}$$

which is clearly of size $n \times M$, where n is the dimensionality of the signaling Hilbert space. The singular value decomposition (SVD) [8] of the state matrix is given by

$$\boldsymbol{\Gamma} = [|\alpha_1\rangle, \ |\alpha_2\rangle, \cdots, |\alpha_M\rangle] = \boldsymbol{U\Sigma V}^\dagger = \sum_{m=1}^{r} \sigma_m |u_m\rangle\langle v_m| = \boldsymbol{U}_r\boldsymbol{\Sigma}_r\boldsymbol{V}_r^\dagger, \ \sigma_m = \sqrt{\lambda_m}, \tag{5.260}$$

where σ_m is the m-the singular value (the square root of eigenvalue λ_m), $\boldsymbol{\Sigma}_r = \mathrm{diag}(\sigma_1, \cdots, \sigma_r)$ with r being the rank of the state matrix, and $\boldsymbol{U}$ and $\boldsymbol{V}$ are unitary matrices.

We can define the *measurement matrix* by listing the measurement vectors $|w_m\rangle$ as corresponding columns:

$$\boldsymbol{W} = [|w_1\rangle, \ |w_2\rangle, \cdots, |w_M\rangle]. \tag{5.261}$$

We define the *Gram matrix* **G** and *Gram operator* ***T***, respectively, as

$$\begin{aligned} \boldsymbol{G} &= \boldsymbol{\Gamma}^\dagger \boldsymbol{\Gamma}, \\ \boldsymbol{T} &= \boldsymbol{\Gamma}\boldsymbol{\Gamma}^\dagger = \sum_{m=1}^{M} |\alpha_m\rangle\langle\alpha_m|. \end{aligned} \tag{5.262}$$

The *spectral decomposition*, also known as the eigenvalue decomposition (EID), is given for the Gram matrix and operator, respectively, as

$$\boldsymbol{G} = \boldsymbol{V}\boldsymbol{\Sigma}_G\boldsymbol{V}^\dagger = \sum_{m=1}^{r} \overbrace{\lambda_m}^{\sigma_m^2} |v_m\rangle\langle v_m| = \sum_{m=1}^{r} \sigma_m^2 |v_m\rangle\langle v_m| = \boldsymbol{V}_r\boldsymbol{\Sigma}_r^2\boldsymbol{V}_r^\dagger,$$

$$\boldsymbol{T} = \boldsymbol{U}\boldsymbol{\Sigma}_T\boldsymbol{U}^\dagger = \sum_{m=1}^{r} \sigma_m^2 |u_m\rangle\langle u_m| = \boldsymbol{U}_r\boldsymbol{\Sigma}_r^2\boldsymbol{U}_r^\dagger, \ \ \boldsymbol{\Sigma}_r^2 = \mathrm{diag}(\sigma_1^2, \cdots, \sigma_r^2). \tag{5.263}$$

with $\boldsymbol{U}$ and $\boldsymbol{V}$ as unitary matrices, λ_m as eigenvalues of $\boldsymbol{G}$ and $\boldsymbol{T}$, and $|u_m\rangle$, and $|v_m\rangle$ as the corresponding eigenkets. With r, we denote the rank of $\boldsymbol{G}$.

The key idea behind SRM is to make the error vectors, defined as the difference between the signal states and measurement states, as small as possible, as illustrated in Fig. 5.15.

More specifically, we want to determine the measurement kets $|w_m\rangle$ and relate them to the signaling states $|\alpha_m\rangle$ such that the quadratic error is minimized:

$$\varepsilon = \sum_{m=1}^{M} \langle e_m|e_m\rangle = \sum_{m=1}^{M} (\langle\alpha_m| - \langle w_m|)(|\alpha_m\rangle - |w_m\rangle), \tag{5.264}$$

subject to the completeness relation of the measurement operator W; that is,

$$\boldsymbol{W}\boldsymbol{W}^\dagger = \sum_{m=1}^{M} |w_m\rangle\langle w_m| = \mathbf{1}. \tag{5.265}$$

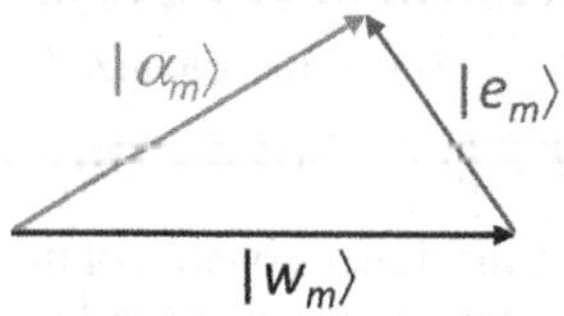

FIGURE 5.15

Square root measurement method.

By defining the error matrix as the difference between the state and the measurement matrices,

$$\boldsymbol{E} = [|e_1\rangle, \cdots, |e_m\rangle, \cdots |e_M\rangle] = \boldsymbol{\Gamma} - \boldsymbol{W}, \tag{5.266}$$

we can alternatively express the quadratic error by

$$\begin{aligned}\varepsilon = \mathrm{Tr}\left(\boldsymbol{E}\boldsymbol{E}^\dagger\right) = \mathrm{Tr}\left(\boldsymbol{E}^\dagger\boldsymbol{E}\right) &= \sum\nolimits_m \overbrace{\langle v_m|\boldsymbol{E}^\dagger}^{\langle v_m|(\boldsymbol{\Gamma}-\boldsymbol{W})^\dagger} \overbrace{\boldsymbol{E}|v_m\rangle}^{(\boldsymbol{\Gamma}-\boldsymbol{W})|v_m\rangle} = \sum\nolimits_m \overbrace{\langle v_m|(\boldsymbol{\Gamma}-\boldsymbol{W})^\dagger}^{\langle u_m|\sigma_m - \langle\eta_m|} \overbrace{(\boldsymbol{\Gamma}-\boldsymbol{W})|v_m\rangle}^{\sigma_m|u_m\rangle - |\eta_m\rangle} \\ &= \sum\nolimits_m (\langle u_m|\sigma_m - \langle\eta_m|)(\sigma_m|u_m\rangle - |\eta_m\rangle) = \sum\nolimits_m \left(\sigma_m^2 - \sigma_m\langle u_m|\eta_m\rangle - \sigma_m\langle\eta_m|u_m\rangle + 1\right),\end{aligned} \tag{5.267}$$

where σ_m is the m-the singular value (the square root of eigenvalue λ_m). Given that $|\langle u_m|\eta_m\rangle| \le 1$, the quadratic error is minimized when $|\eta_m\rangle = |u_m\rangle$. So the minimum quadratic error is

$$\varepsilon_{\min} = \sum_m \left(\sigma_m^2 - 2\sigma_m + 1\right) = \sum_m (\sigma_m - 1)^2. \tag{5.268}$$

From Eq. (5.267), we conclude that $\boldsymbol{W}|v_m\rangle = |u_m\rangle$, and the optimum measurement matrix $\boldsymbol{W}$ becomes

$$\boldsymbol{W}^{(\mathrm{opt})} = \sum_{m=1}^{r} |u_m\rangle\langle w_m| = \boldsymbol{U}_r\boldsymbol{V}_r^\dagger. \tag{5.269}$$

The optimum measurement matrix can be expressed in terms of the Gram matrix by

$$\boldsymbol{W}^{(\mathrm{opt})} = \boldsymbol{\Gamma}\boldsymbol{G}^{-1/2}, \tag{5.270}$$

and in terms of the Gram operator by

$$\boldsymbol{W}^{(\mathrm{opt})} = \boldsymbol{T}^{-1/2}\boldsymbol{\Gamma}. \tag{5.271}$$

Using the EID of $\boldsymbol{G}$ and the SVD of $\boldsymbol{\Gamma}$, we can easily prove Eq. (5.270):

$$\boldsymbol{\Gamma}\boldsymbol{G}^{-1/2} = \boldsymbol{U}_r\boldsymbol{\Sigma}_r\boldsymbol{V}_r^\dagger\left(\boldsymbol{V}_r\boldsymbol{\Sigma}_r^2\boldsymbol{V}_r^\dagger\right)^{-1/2} = \boldsymbol{U}_r\boldsymbol{\Sigma}_r\overbrace{\boldsymbol{V}_r^\dagger\boldsymbol{V}_r}^{\mathbf{1}}\boldsymbol{\Sigma}_r^{-1}\boldsymbol{V}_r^\dagger = \boldsymbol{U}_r\overbrace{\boldsymbol{\Sigma}_r\boldsymbol{\Sigma}_r^{-1}}^{\mathbf{1}}\boldsymbol{V}_r^\dagger = \boldsymbol{U}_r\boldsymbol{V}_r^\dagger = \boldsymbol{W}^{(\mathrm{opt})}. \tag{5.272}$$

In a similar fashion, using the EID of $\boldsymbol{T}$ and SVD of $\boldsymbol{\Gamma}$, we can easily prove Eq. (5.271) by

$$\boldsymbol{T}^{-1/2}\boldsymbol{\Gamma} = \left(\boldsymbol{U}_r\boldsymbol{\Sigma}_r^2\boldsymbol{U}_r^\dagger\right)^{-1/2}\boldsymbol{U}_r\boldsymbol{\Sigma}_r\boldsymbol{V}_r^\dagger = \boldsymbol{U}_r\boldsymbol{\Sigma}_r^{-1}\overbrace{\boldsymbol{U}_r^\dagger\boldsymbol{U}_r}^{\mathbf{1}}\boldsymbol{\Sigma}_r\boldsymbol{V}_r^\dagger = \boldsymbol{U}_r\overbrace{\boldsymbol{\Sigma}_r^{-1}\boldsymbol{\Sigma}_r}^{\mathbf{1}}\boldsymbol{V}_r^\dagger = \boldsymbol{U}_r\boldsymbol{V}_r^\dagger = \boldsymbol{W}^{(\mathrm{opt})}. \tag{5.273}$$

From Eq. (5.270), we conclude that we can express the measurement vectors as the following linear combination of the signal states:

$$|w_k\rangle = \sum_{m=1}^{M} \left(\boldsymbol{G}^{-1/2}\right)_{km}|\alpha_m\rangle. \tag{5.274}$$

The transition probabilities can now be determined as

$$P(k|m) = \mathrm{Tr}(\rho_m W_k) = |\langle w_k|\alpha_m\rangle|^2 = \left|\left(\boldsymbol{G}^{1/2}\right)_{mk}\right|^2. \tag{5.275}$$

The average error probability for equal probable transmission will then be

$$P_e = 1 - \frac{1}{M}\sum_{m=1}^{M}\left|\left(G^{1/2}\right)_{mm}\right|^2. \tag{5.276}$$

5.10.2 Geometrically uniform symmetry constellations and M-ary phase-shift keying

For signal constellations possessing GUS, the m-th signaling state can be represented by

$$|\alpha_m\rangle = S^m|\alpha_0\rangle;\ m = 0, 1, \cdots, M-1 \tag{5.277}$$

with S being the unitary operator called the *symmetry operator*, which is the M-th root of the identity operator, that is, $S^M = \widehat{1}$. State $|\alpha_0\rangle$ is the referent state.

As an illustration, the M-ary phase-shift keying (PSK) constellation can be expressed in terms of rotation operator $R(\theta) = \exp(jN\theta)$ by

$$S = R\left(\frac{2\pi}{M}\right) = e^{j\frac{2\pi}{M}N} = W_M^N,\ W_M = e^{j\frac{2\pi}{M}}. \tag{5.278}$$

The Gram matrix for GUS states is circulant and can be represented by

$$\boldsymbol{G} = \boldsymbol{DFT}_{[M]}\boldsymbol{\Sigma}_G\boldsymbol{DFT}^{\dagger}_{[M]} = \sum_{m=0}^{M-1}\sigma_m^2|\omega_m\rangle\langle\omega_m|, \tag{5.279}$$

where $|\omega_m\rangle$ is the m-th column of the discrete Fourier transform matrix $\mathrm{DFT}_{[M]}$, given by

$$|\omega_m\rangle = M^{-1/2}\left[W_M^{-m}, W_M^{-2m}, \cdots, W_M^{-m(M-1)}\right]^{\mathrm{T}};\ m = 0, 1, \cdots, M-1 \tag{5.280}$$

The eigenvalues are determined by the DFT of the first row of the Gram matrix [1]:

$$\lambda_m = \sigma_m^2 = \frac{1}{M}\sum_{m=0}^{M-1}G_{0m}W_M^{-mk},\ G_{0m} = \langle\alpha_0|\alpha_m\rangle. \tag{5.281}$$

The elements of the square root of the Gram matrix are given by

$$\left(G^{\pm 1/2}\right)_{mk} = \frac{1}{M}\sum_{l=0}^{M-1}\lambda_l^{\pm 1/2}W_M^{-l(m-k)}. \tag{5.282}$$

Clearly, the diagonal elements are equal.

As before, the measurement kets can be determined as a linear combination of the signaling states by

$$|w_k\rangle = \sum_{m=1}^{M}\left(G^{-1/2}\right)_{km}|\alpha_m\rangle. \tag{5.283}$$

Finally, the average probability of error is simplified:

$$P_e = 1 - \frac{1}{M}\sum_{m=1}^{M}\left|\left(\mathbf{G}^{1/2}\right)_{mm}\right|^2 = 1 - \frac{1}{M}\sum_{m=1}^{M}\left|\left(\mathbf{G}^{1/2}\right)_{00}\right|^2$$

$$= 1 - \left|\left(\mathbf{G}^{1/2}\right)_{00}\right|^2 = 1 - \frac{1}{M^2}\left(\sum_{l=0}^{M-1}\lambda_l^{1/2}\right)^2. \tag{5.284}$$

Fig. 5.16 provides the average symbol error probability for various constellation sizes obtained by employing Eq. (5.284).

BPSK and QPSK have comparable performance, where the performance of 8-PSK is worse. The performance of 16-PSK is significantly worse than that of QPSK.

The methodology to calculate the error probability described above is applicable to squeezed-displaced states as well, as given by Eq. (5.188) and illustrated in Fig. 5.17 for QPSK. The signal constellation points are represented by

$$\left|\alpha_m, \underline{s}_m\right\rangle = \left|\alpha e^{j2\pi m/M}, s e^{j\theta} e^{j2\pi 2m/M}\right\rangle; \quad m = 0, 1, \cdots, M-1 \tag{5.285}$$

The average number of photons per symbol N_s is related to the squeezing parameter s by $N_s = |\alpha|^2 + \sinh^2 s$. The number of signal photons must be sufficient to enable both squeezing and

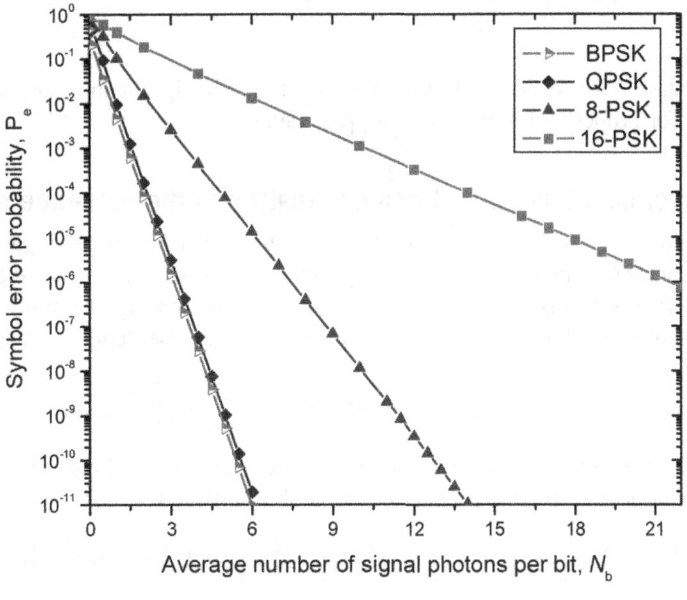

FIGURE 5.16

Symbol error probability versus average number of signal photons per bit.

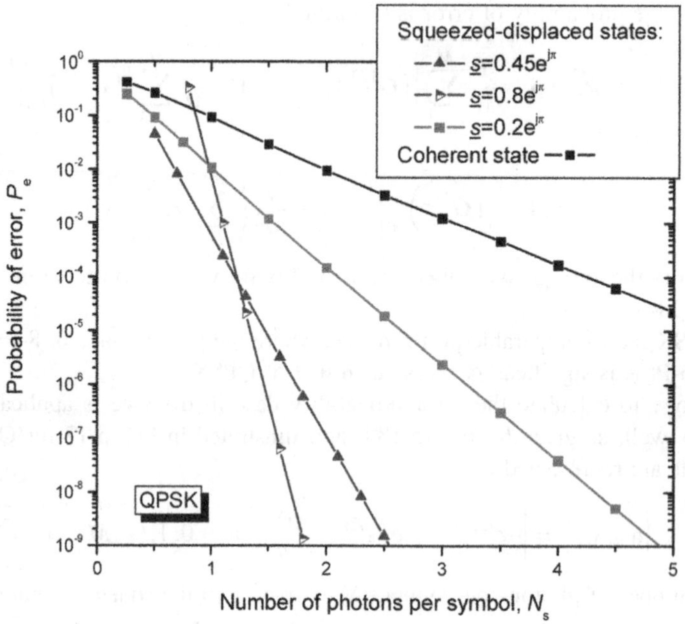

FIGURE 5.17

Symbol error probability versus average number of signal photons per symbol for squeezed-displaced state-based quantum phase-shift keying (QPSK).

displacement operators. Squeezed-displaced state-based QPSK significantly outperforms coherent state-based QPSK for properly chosen squeezing parameters.

5.10.3 Multilevel quantum optical communication in the presence of noise

We can use the theory derived in Section 5.6 to analyze the performance of single-mode multilevel quantum optical communication systems in the presence of noise. Given that the received signal contains both the *thermal noise* and the *coherent signal* representing the transmitted state $|\mu_m\rangle$ ($m = 1,\ldots,M$), the corresponding density operator ρ_m in the P-representation is given by

$$\rho_m = \frac{1}{\pi \mathfrak{N}} \int e^{-\frac{|\alpha-\mu_m|^2}{\mathfrak{N}}} |\alpha\rangle\langle\alpha| d^2\alpha, \quad \mathfrak{N} = \frac{1}{e^{\hbar\omega/k_B \mathfrak{T}} - 1}, \tag{5.286}$$

with k_B being the Boltzmann constant, $\mathfrak{T}$ is the absolute temperature, and $\hbar\omega$ is a photon energy.

The density operator ρ_m can be expressed in terms of number states $|n\rangle$:

$$\langle l|\rho_m|k\rangle = \begin{cases} (1-v)v^k \left(\mu_m^*/\mathfrak{N}\right)^{k-l} (l!/k!) e^{-(1-v)|\mu_m|^2} L_l^{k-l}\left[-(1-v)^2|\mu_m|^2/v\right], & l \geq k \\ \langle k|\rho_m|l\rangle^*, & \end{cases} \tag{5.287}$$

where $v = \mathfrak{N}/(\mathfrak{N}+1)$. We must approximate the density matrix with fixed number of terms of n:

$$\widetilde{\rho}_m \cong \sum_{k=0}^{n-1}\sum_{l=0}^{n-1}(\rho_m)_{kl}|k\rangle\langle l|, \quad (\rho_m)_{kl} = \langle l|\rho_m|k\rangle. \tag{5.288}$$

We can then perform the spectral decomposition (EID) [5]:

$$\widetilde{\rho}_m = \mathbf{Z}\mathbf{\Sigma}_{\rho_m}\mathbf{Z}^\dagger = \sum_{l=1}^{r}\delta_l^2|z_l\rangle\langle z_l| = \mathbf{Z}_r\mathbf{D}_r\mathbf{Z}_r^\dagger, \quad \mathbf{\Sigma}_{\rho_m} = \mathrm{diag}\left(\delta_1^2, \cdots, \delta_r^2, \underbrace{0, \cdots, 0}_{n-r \text{ times}}\right), \tag{5.289}$$

where $r \le n$ is the rank of the density matrix, δ_l^2 are eigenvalues, $\mathbf{Z}$ is the unitary matrix with columns being the orthonormal basis kets $\{|z_l\rangle\}$, $\mathbf{Z}_r$ is $r\times r$ unitary composed of the first r columns of $\mathbf{Z}$, and $\mathbf{D}_r = \mathrm{diag}(\delta_1^2, \cdots, \delta_r^2)$. Now by introducing $\boldsymbol{\beta} = \mathbf{Z}_r\mathbf{D}_r^{1/2}$, we can represent the density matrix by

$$\widetilde{\rho}_m = \boldsymbol{\beta}_m\boldsymbol{\beta}_m^\dagger, \quad \boldsymbol{\beta}_m = \mathbf{Z}_r(\widetilde{\rho}_m)\mathbf{D}_r^{1/2}(\widetilde{\rho}_m), \tag{5.290}$$

and thus reduce the complexity of representation.

We create the *state matrix* by

$$\mathbf{\Gamma} = [\boldsymbol{\beta}_0, \boldsymbol{\beta}_1, \cdots, \boldsymbol{\beta}_{M-1}], \tag{5.291}$$

with the number of rows being n and the number of columns given by $r_0 + r_1 + \ldots + r_{M-1}$, $r_m = \mathrm{rank}(\boldsymbol{\beta}_m);\ m = 0, 1, \cdots, M-1$.

The *Gram matrix* **G** and *Gram operator* ***T*** can now be created by

$$\mathbf{G} = \mathbf{\Gamma}^\dagger\mathbf{\Gamma}, \quad \boldsymbol{T} = \mathbf{\Gamma}\mathbf{\Gamma}^\dagger = \sum_{m=0}^{M-1}\boldsymbol{\beta}_m\boldsymbol{\beta}_m^\dagger. \tag{5.292}$$

The *measurement matrix* **W** can now be determined as

$$\mathbf{W} = \mathbf{\Gamma}\mathbf{G}^{-1/2} = \boldsymbol{T}^{-1/2}\mathbf{\Gamma}. \tag{5.293}$$

The m-th measurement operator $\mathbf{W}_m$ can be represented by

$$\mathbf{W}_m = \mathbf{w}_m\mathbf{w}_m^\dagger, \quad \mathbf{w}_m = \boldsymbol{T}^{-1/2}\boldsymbol{\beta}_m. \tag{5.294}$$

The transition probability $P(k|m)$ can be determined as

$$P(k|m) = \mathrm{Tr}(\widetilde{\rho}_m\mathbf{W}_k) = \mathrm{Tr}\left(\boldsymbol{\beta}_m\boldsymbol{\beta}_m^\dagger\mathbf{w}_k\mathbf{w}_k^\dagger\right). \tag{5.295}$$

Finally, the average probability of error can be determined by

$$P_e = 1 - \frac{1}{M}\sum_{m=0}^{M-1}\mathrm{Tr}\left(\boldsymbol{\beta}_m\boldsymbol{\beta}_m^\dagger\mathbf{w}_m\mathbf{w}_m^\dagger\right) \tag{5.296}$$

5.11 Summary

This chapter was devoted to quantum detection theory, quantum communication, and Gaussian quantum information theory. The density operators were briefly reviewed in Section 5.1 before describing quantum detection theory concepts in Section 5.2. This theory was applied to the binary detection problem in Section 5.3. Coherent states were described in Section 5.4, followed by an introduction to quadrature operators and derivation of the uncertainty principle for CV systems. In Section 5.5, binary quantum optical communication in the absence of noise was discussed, including the photon-counting (Subsections 5.5.1 and 5.5.2), OOK (Subsection 5.5.3), BPSK (Subsection 5.5.4), Kennedy (Subsection 5.5.5), and Dolinar receivers (Subsection 5.5.6).

The *P*-representation was introduced in Section 5.6 and applied to represent thermal noise and the coherent state in the presence of thermal noise. Binary optical communication in the presence of noise was discussed in Section 5.7, focusing on OOK and BPSK. Gaussian and squeezed states were introduced in Section 5.8, followed by definitions of the Wigner function and covariance matrices in Subsection 5.8.1. Gaussian transformation and Gaussian channels were covered in Subsection 5.8.2, with BS operation, phase rotation operation, and the squeezing operator as examples. The thermal decomposition of Gaussian states was discussed in Subsection 5.8.3, and the von Neumann entropy for thermal states was derived. The covariance matrices for two-mode Gaussian states were discussed in Subsection 5.8.4, along with the calculation of symplectic eigenvalues relevant in the von Neumann entropy calculation. Gaussian state measurement and detection were discussed in Subsection 5.8.5, emphasizing homodyne detection, heterodyne detection, and partial measurements. In Section 5.9, the focus moved to NL quantum optics fundamentals, particularly TWM and FWM. In the same section, the generation of Gaussian states was discussed, particularly TMS generation. Section 5.10 was devoted to multilevel quantum optical communication, describing scenarios in the absence and presence of noise. Analyzing the performance of these schemes through SRM was described in detail in Subsection 5.10.1. This method was applied to signal constellations with GUS, such as M-ary PSK in Subsection 5.10.2. Finally, Subsection 5.10.3 was devoted to M-ary quantum communication in the presence of noise.

References

[1] C.W. Helstrom, Quantum Detection and Estimation Theory, Academic Press, New York, 1976.

[2] V. Vilnrotter, C.-W. Lau, Quantum Detection Theory for the Free-Space Channel, August 15, 2001, pp. 1–34. IPN Progress Report 42-146.

[3] C.W. Helstrom, J.W.S. Liu, J.P. Gordon, Quantum-mechanical communication theory, Proc. IEEE 58 (10) (1970) 1578–1598.

[4] I.B. Djordjevic, Quantum Information Processing and Quantum Error Correction: An Engineering Approach, Elsevier/Academic Press, Amsterdam-Boston, 2012.

[5] G. Cariolaro, Quantum Communications, Springer International Publishing Switzerland, Cham: Switzerland, 2015.

[6] I.B. Djordjevic, Physical-Layer Security and Quantum Key Distribution, Springer Nature Switzerland AG, Cham, Switzerland, 2019.

[7] C. Weedbrook, S. Pirandola, R. García-Patrón, N.J. Cerf, T.C. Ralph, J.H. Shapiro, S. Lloyd, Gaussian quantum information, Rev. Mod. Phys. 84 (2) (2012) 621.

[8] I.B. Djordjevic, Advanced Optical and Wireless Communications Systems, Springer International Publishing, Switzerland, 2017.
[9] M. Cvijetic, I.B. Djordjevic, Advanced Optical Communications and Networks, Artech House, 2013.
[10] D.M. Velickovic, Electromagnetics – Vol. I, second ed., Faculty of Electronic Engineering, University of Nis, Serbia, 1999.
[11] L. Mandel, E. Wolf, Optical Coherence and Quantum Optics, Cambridge University Press, Cambridge-New York-Melbourne, 1995.
[12] M.O. Scully, M.S. Zubairy, Quantum Optics, Cambridge University Press, Cambridge-New York-Melbourne, 1997.
[13] Z.-Y.J. Ou, Quantum Optics for Experimentalists, World Scientific, Hackensack-London, 2017.
[14] H.P. Yuen, Two-photon coherent states of the radiation field, Phys. Rev. A 13 (1976) 2226.
[15] C.C. Gerry, P.L. Knight, Introductory Quantum Optics, Cambridge University Press, Cambridge, UK, 2008.
[16] R.J. Glauber, Coherent and incoherent states of the radiation field, Phys. Rev. 131 (1963) 2766.
[17] R.J. Glauber, Quantum Theory of Optical Coherence, Selected Papers and Lectures, Wiley-VCH, Weinheim, 2007.
[18] J.J. Sakurai, Modern Quantum Mechanics, Addison-Wisley, 1994.
[19] R.S. Kennedy, Near-optimum Receiver for the Binary Coherent State Quantum Channel, 1973, pp. 219–225. MIT Research Laboratory of Electronics Quarterly Progress Report, no. 108.
[20] S.J. Dolinar, An Optimum Receiver for the Binary Coherent State Quantum Channel, 1973, pp. 115–120. MIT Research Laboratory of Electronics Quarterly Progress Report, no. 111.
[21] J. Geremia, Distinguishing between optical coherent states with imperfect detection, Phys. Rev. A 70 (2004) 062303.
[22] A. Assalini, N. Dolla Pozza, G. Pierobon, Revisiting the Dolinar receiver through multiple-copy state discrimination theory, Phys. Rev. A 84 (2011) 022342.
[23] R.L. Cook, P.J. Martin, J.M. Geremia, Optical coherent state discrimination using a closed-loop quantum measurement, Nature 446 (7137) (2007) 774–777.
[24] F.E. Becerra, J. Fan, G. Baumgartner, J. Goldhar, J.T. Kosloski, A. Migdall, Experimental demonstration of a receiver beating the standard quantum limit for multiple nonorthogonal state discrimination, Nat. Photonics 7 (2013) 147–152.
[25] M. Takeoka, M. Sasaki, Discrimination of the binary coherent signal: Gaussian-operation limit and simple non-Gaussian near-optimal receivers, Phys. Rev. A 78 (2008) 022320.
[26] E.C.G. Sudarshan, Equivalence of semiclassical and quantum mechanical descriptions of statistical light beams, Phys. Rev. Lett. 10 (7) (1963) 277.
[27] H.H. Wannier, Statistical Physics, Dover, New York, 1987.
[28] I.B. Djordjevic, Quantum Biological Information Theory, Springer International Publishing Switzerland, Cham, 2016.
[29] N. Le Bihan, S.J. Sangwine, Jacobi method for quaternion matrix singular value decomposition, Appl. Math. Comput. 187 (2) (2007) 1265–1271.
[30] N.N. Bogoliubov, On the theory of superfluidity, J. Phys. 11 (1947) 23 (Izv. Akad. Nauk Ser. Fiz. 11: 77 (1947)).
[31] N.N. Bogoljubov, On a new method in the theory of superconductivity, Il Nuovo Cimento 7 (6) (1958) 794–805.
[32] A.S. Holevo, M. Sohma, O. Hirota, Capacity of quantum Gaussian channels, Phys. Rev. A 59 (3) (1999) 1820–1828.
[33] J. Williamson, On the algebraic problem concerning the normal forms of linear dynamical systems, Am. J. Math. 58 (1) (1936) 141–163.

[34] R. Simon, Peres-horodecki separability criterion for continuous variable systems, Phys. Rev. Lett. 84 (2000) 2726.
[35] L.-M. Duan, G. Giedke, J.I. Cirac, P. Zoller, Inseparability criterion for continuous variable systems, Phys. Rev. Lett. 84 (2000) 2722.
[36] A. Serafini, F. Illuminati, S. De Siena, Symplectic invariants, entropic measures and correlations of Gaussian states, J. Phys. B 37 (2004) L21–L28.
[37] A. Serafini, Multimode uncertainty relations and separability of continuous variable states, Phys. Rev. Lett. 96 (2006) 110402.
[38] S. Pirandola, A. Serafini, S. Lloyd, Correlation matrices of two-mode bosonic systems, Phys. Rev. A 79 (2009) 052327.
[39] T.C. Ralph, Continuous variable quantum cryptography, Phys. Rev. A 61 (1999) 010303(R).
[40] F. Grosshans, P. Grangier, Continuous variable quantum cryptography using coherent states, Phys. Rev. Lett. 88 (2002) 057902.
[41] F. Grosshans, Collective attacks and unconditional security in continuous variable quantum key distribution, Phys. Rev. Lett. 94 (2005) 020504.
[42] H.P. Yuen, J.H. Shapiro, Optical communication with two-photon coherent states—part III: quantum measurements realizable with photoemissive detectors, IEEE Trans. Inf. Theor. 26 (1) (1980) 78–92.
[43] G. Giedke, J.L. Cirac, Characterization of Gaussian operations and distillation of Gaussian states, Phys. Rev. A 66 (2002) 032316.
[44] J. Eisert, S. Scheel, M.B. Plenio, Distilling Gaussian states with Gaussian operations is impossible, Phys. Rev. Lett. 89 (2002) 137903.
[45] I.B. Djordjevic, LDPC-coded *M*-ary PSK optical coherent state quantum communication, IEEE/OSA J. Lightw. Technol. 27 (5) (2009) 494–499.
[46] X. Liu, A.R. Chraplyvy, P.J. Winzer, R.W. Tkach, S. Chandrasekhar, Phase-conjugated twin waves for communication beyond the Kerr nonlinearity limit, Nat. Photonics 7 (2013) 560–568.

CHAPTER

Quantum key distribution

6

Chapter outline

Quantum Communication, Quantum Networks, and Quantum Sensing. https://doi.org/10.1016/B978-0-12-822942-2.00012-1

6.1 Cryptography basics

The basic key-based cryptographic system is shown in Fig. 6.1. The source emits the message (plaintext) $\boldsymbol{M}$ toward the encryption block, which generates the cryptogram with the help of key K obtained from the key source. On the receiver side, the cryptogram transmitted over an insecure channel is processed by the decryption algorithm together with key $\boldsymbol{K}$ obtained through the secure channel, which reconstructs the original plaintext to deliver to the authenticated user. The encryption process can be mathematically described as $\boldsymbol{E_K(M) = C}$, while the decryption process is described by $\boldsymbol{D_K(C) = M}$. The composition of decryption and encryption functions yields to identity mapping $\boldsymbol{D_K(E_K(M)) = M}$. The key source typically generates the key randomly from the *keyspace* (the range of all possible key values).

The key-based algorithms can be categorized into two broad categories:

- In *symmetric algorithms*, the decryption key can be derived from the encryption key and vice versa. Alternatively, the same key can be used for both encryption and decryption stages. Symmetric algorithms are also known as one-key (single-key) or secret-key algorithms. A well-known system employing this type of algorithm is the digital encryption standard [1–4].
- In *asymmetric algorithms*, the encryption and decryption keys are different. Moreover, the decryption key cannot be determined from the encryption key, at least in any reasonable amount of time. For this reason, the encryption keys can even be made public because an eavesdropper will be unable to determine the decryption key. *Public-key systems* [5] are based on this concept. The encryption keys are made public in public-key systems, while the decryption key is known only to the intended user. The encryption key is then called the *public* key, while the decryption key is known as the *secret (private)* key. The keys can be applied in an arbitrary order to create the cryptogram from plaintext and reconstruct plaintext from the cryptogram.

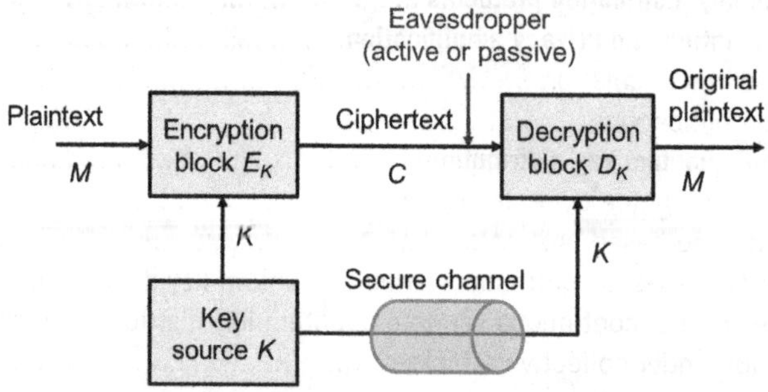

FIGURE 6.1

Basic key-based cryptographic scheme.

The simplest private key cryptosystem is the *Vernam cipher* (*one-time pad*). In a *one-time pad* [6], a completely random sequence of characters, with the sequence length equal to the message sequence length, is used as a key. Another random sequence is used as a key for each new message, and thus, the one-time pad scheme provides perfect security. Namely, the brute-force search approach would require verifying m^n possible keys, where m is the employed alphabet size, and n is the length of the intercepted cryptogram. In practice in digital and computer communications, we typically operate on binary alphabet GF(2) = {0,1}. To obtain the key, we need a special random generator, and to encrypt using a one-time pad scheme, we simply perform addition mod 2, i.e., an XOR operation, as illustrated in Fig. 6.2. Despite offering perfect security, the one-time pad scheme has several drawbacks [7–9]: it requires that keys are securely distributed, are at least as long as the message, have key bits that are not reused, are delivered in advance delivery, are stored securely until use, and are destroyed after use.

According to Shannon [10], *perfect security*, also known as *unconditional security*, is achieved when the messages $\boldsymbol{M}$ and cryptograms $\boldsymbol{C}$ are statistically independent so that the corresponding mutual information equals zero:

$$I(\boldsymbol{M},\boldsymbol{C}) = H(\boldsymbol{M}) - H(\boldsymbol{M}|\boldsymbol{C}) = 0 \Leftrightarrow H(\boldsymbol{M}|\boldsymbol{C}) = H(\boldsymbol{M}) \tag{6.1}$$

By employing the chain rule [2,11] given by $H(X,Y) = H(X) + H(Y|X)$, we can write

$$\begin{aligned} H(\boldsymbol{M}) &= H(\boldsymbol{M}|\boldsymbol{C}) \\ &\leq H(\boldsymbol{M},\boldsymbol{K}|\boldsymbol{C}) = \\ &= H(\boldsymbol{K}|\boldsymbol{C}) + \underbrace{H(\boldsymbol{M}|\boldsymbol{C},\boldsymbol{K})}_{=0} \\ &= H(\boldsymbol{K}|\boldsymbol{C}) = H(\boldsymbol{K}), \end{aligned} \tag{6.2a}$$

where we use the fact that when both the cryptogram and the key are known, the message has been completely specified—that is, $H(\boldsymbol{M}|\boldsymbol{K},\boldsymbol{C}) = 0$. The perfect secrecy condition can therefore be summarized as

$$H(\boldsymbol{M}) \leq H(\boldsymbol{K}). \tag{6.2b}$$

In other words, the entropy (uncertainty) of the key cannot be lower than the entropy of the message for an encryption scheme to be perfectly secure. Given that the length of the key in the Vernam cipher is at least equal to the message length, the one-time pad scheme appears to be perfectly secure.

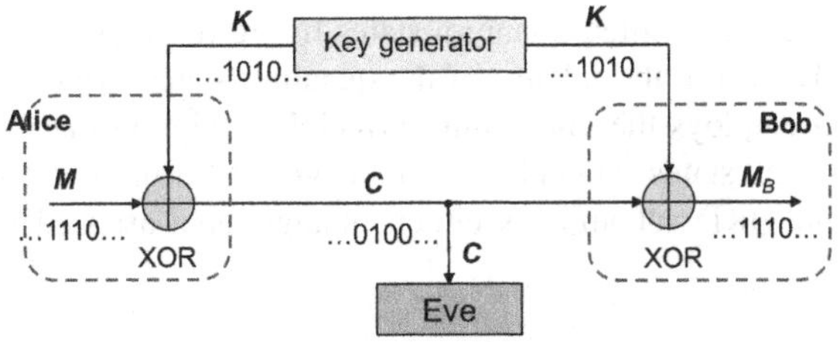

FIGURE 6.2

One-time pad encryption scheme.

However, because this condition is difficult to satisfy in conventional cryptography compared with information-theoretic security, *computational security* is used [1,4,8,12–14]. Computational security introduces two relaxations concerning information-theoretic security [4]:

- Security is guaranteed against an efficient eavesdropper running cryptanalytic attacks for a certain, limited amount of time. Of course, an eavesdropper with sufficient computational resources or time can break the security of the encryption scheme.
- Although eavesdroppers can successfully break security protocols, their probability of doing so is small.

A reader interested in learning more about computational security is referred to an excellent book by Katz and Lindell [4]. However, using quantum computing, any conventional cryptographic scheme can be broken in a reasonable amount of time by employing Shor's factorization algorithm [15–19].

6.2 Quantum key distribution basics

Significant achievements have recently been made in quantum computing [7–9]. Many companies are currently developing medium-scale quantum computers. Given that most cryptosystems depend on the computational hardness assumption, quantum computing represents a serious challenge to modern cybersecurity systems. As an illustration, to break the Rivest–Shamir–Adleman (aka RSA) protocol [20], one must determine the period r of the function $f(x) = m^x \bmod n = f(x + r)$ ($r = 0,1, \ldots,2^L-1$; m is an integer smaller than $n-1$). This period is determined in one step of Shor's factorization algorithm [7–9,17–19].

QKD with symmetric encryption can be interpreted as a physical-layer security scheme that provides provable security against quantum computer-initiated attacks [21]. The first QKD scheme was introduced by Bennett and Brassard, who proposed it in 1984 [15,16], and is now known as the BB84 protocol. The security of QKD is guaranteed by the laws of quantum mechanics. Various photon degree-of-freedom (DOFs), such as polarization, time, frequency, phase, and orbital angular momentum, can be employed to implement various QKD protocols. Generally speaking, there are two generic QKD schemes, discrete-variable QKD (DV-QKD) and continuous-variable QKD (CV-QKD), depending on the strategy applied on Bob's side. In DV-QKD schemes, a single-photon detector (SPD) is applied on Bob's side, while in CV-QKD, field quadratures are measured with the help of homodyne/heterodyne detection. The DV-QKD scheme achieves unconditional security by employing the no-cloning theorem and theorem on the indistinguishability of arbitrary quantum states. The no-cloning theorem claims that arbitrary quantum states cannot be cloned, indicating that Eve cannot duplicate nonorthogonal quantum states even with the help of a quantum computer. On the other hand, the second theorem claims that nonorthogonal states cannot be unambiguously distinguished. Namely, when Eve interacts with the transmitted quantum states trying to obtain information on transmitted bits, she will inadvertently disturb the fidelity of the quantum states, which Bob will detect. On the other hand, the CV-QKD employs the uncertainty principle, claiming that both in-phase and quadrature components of coherent states cannot be simultaneously measured with complete precision. We can also classify different QKD schemes as either entanglement-assisted or prepare-and-measure types.

The research on QKD is gaining momentum, particularly after the first satellite-to-ground QKD demonstration [22]. Recently, a QKD over 404 km of ultralow-loss optical fiber was demonstrated; however, it had an ultralow secure key rate ($3.2 \cdot 10^{-4}$ b/s). Given that quantum states cannot be amplified, fiber attenuation limits distance. On the other hand, the dead time (the time over which an SPD remains unresponsive to incoming photons due to long recovery time) of the SPDs, typically in the 10–100 ns range, limits the baud rate and therefore the secure key rate. CV-QKD schemes, since they employ homodyne/heterodyne detection, do not have a deadtime limitation; however, the typical distances are shorter.

By transmitting the nonorthogonal qubit states between them and checking for disturbance in the transmitted state caused by the channel or Eve's activity, Alice and Bob can establish an upper bound on noise/eavesdropping in their quantum communication channel [7]. The *threshold for maximum tolerable error rate* is dictated by the efficiency of the best information reconciliation and privacy amplification steps [7].

QKD protocols can be categorized into several general categories:

- **Device-dependent QKD**, in which the quantum source is typically placed on Alice's side, with the quantum detector at Bob's side. Popular classes include DV-QKD, CV-QKD, entanglement-assisted QKD (EA-QKD), distributed phase reference, etc. For EA-QKD, the entangled source can be placed in the middle of the channel to extend the transmission distance.
- **Source-device-independent QKD**, in which the quantum source is placed at Charlie's (Eve's) side, while the quantum detectors are at both Alice's and Bob's sides.
- **Measurement-device-independent QKD (MDI-QKD)**, in which the quantum detectors are placed at Charlie's (Eve's) side, while the quantum sources are placed at both Alice's and Bob's sides. The quantum states get prepared at both Alice's and Bob's sides and get transmitted toward the Charlie's detectors. Charlie performs partial Bell state measurements (BSMs) and announces when the desired partial Bell states are detected, with details to be provided later.

The *classical postprocessing steps* are summarized in Fig. 6.3 [2,7–9].

The raw key is imperfect, and we must perform the *information reconciliation* and *privacy amplification* to increase the correlation between sequences $\boldsymbol{X}$ (generated by Alice) and sequence $\boldsymbol{Y}$ (received by Bob) strings while reducing eavesdropper Eve's mutual information about the result to a desired level of security.

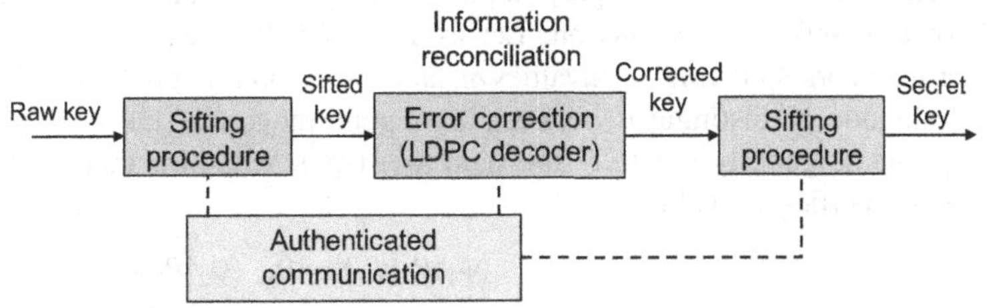

FIGURE 6.3

Classical postprocessing steps.

Information reconciliation is nothing more than error correction performed over a public channel, which reconciles errors between X and Y to obtain a shared bit string K while divulging the information to Eve as little as possible.

Privacy amplification [23, 24] is used between Alice and Bob to distill from K a smaller set of bits S whose correlation with Eve's string Z is below the desired threshold. One way to accomplish privacy amplification is through the use of the *universal hash functions* G, which map the set of n-bit strings A to the set of m-bit strings B such that for any distinct $a_1, a_2 \in A$, when g is chosen uniformly at random from G, the probability of having $g(a_1) = g(a_2)$ is at most $1/|B|$.

Additional details on information reconciliation and privacy amplification will be provided in Section 6.9 of this chapter.

6.3 No-cloning theorem and distinguishing quantum states

No-cloning theorem: No quantum copier exists that can clone an arbitrary quantum state.

Proof: If the input state is $|\psi\rangle$, then the output of the copier will be $|\psi\rangle|\psi\rangle$. For an arbitrary quantum state, such a copier raises a fundamental contradiction. Consider two arbitrary states $|\psi\rangle$ and $|\chi\rangle$ that are the input to the copier. When they are input individually, we expect to obtain $|\psi\rangle|\psi\rangle$ and $|\chi\rangle|\chi\rangle$. Now consider a superposition of these two states given by

$$|\varphi\rangle = \alpha|\psi\rangle + \beta|\chi\rangle \Rightarrow |\varphi\rangle|\varphi\rangle = (\alpha|\psi\rangle + \beta|\chi\rangle)(\alpha|\psi\rangle + \beta|\chi\rangle)$$

$$= \alpha^2|\psi\rangle|\psi\rangle + \alpha\beta|\psi\rangle|\chi\rangle + \alpha\beta|\chi\rangle|\psi\rangle + \beta^2|\chi\rangle|\chi\rangle \tag{6.3}$$

On the other hand, the linearity of quantum mechanics tells us that the quantum copier can be represented by a unitary operator that performs the cloning. If such a unitary operator were to act on the superposition state $|\varphi\rangle$, the output would be a superposition of $|\psi\rangle|\psi\rangle$ and $|\chi\rangle|\chi\rangle$, that is,

$$|\varphi'\rangle = \alpha|\psi\rangle|\psi\rangle + \beta|\chi\rangle|\chi\rangle \tag{6.4}$$

The difference between the previous two equations leads to the contradiction mentioned above. Consequently, there is no unitary operator that can clone $|\varphi\rangle$.

This result raises a related question: Do specific states exist for which cloning is possible? The answer to this question is (surprisingly) yes. Cloning is possible only for mutually orthogonal states.

Theorem: It is impossible to unambiguously distinguish nonorthogonal quantum states.

In other words, there is no measurement device we can create that can reliably distinguish nonorthogonal states. This fundamental result plays an important role in quantum cryptography.

Proof: Its proof is based on contradiction. Let assume that the measurement operator M is a Hermitian operator with corresponding eigenvalues m_i and corresponding projection operators P_i, that allows us to unambiguously distinguish between two nonorthogonal states $|\psi_1\rangle$ and $|\psi_2\rangle$. The eigenvalue m_1 (m_2) unambiguously identifies the state $|\psi_1\rangle$ ($|\psi_2\rangle$). We know that for projection operators, the following properties are valid:

$$\langle\psi_1|P_1|\psi_1\rangle = 1,\ \ \langle\psi_2|P_2|\psi_2\rangle = 1,\ \ \langle\psi_1|P_2|\psi_1\rangle = 0,\ \ \langle\psi_2|P_1|\psi_2\rangle = 0. \tag{6.5}$$

Given the foregoing, $|\psi_2\rangle$ can be represented in terms of $|\psi_1\rangle$ and another state, $|\chi\rangle$ orthogonal to $|\psi_1\rangle$, as follows:

$$|\psi_2\rangle = \alpha|\psi_1\rangle + \beta|\chi\rangle. \tag{6.6}$$

To satisfy Eq. (6.5), $|\psi_2\rangle$ must equal $|\chi\rangle$, thus representing the contradiction.

6.4 Discrete variable quantum key distribution protocols

In this section, we describe some relevant DV-QKD protocols including BB84, B92, time-phase encoding, E91, and Einstein–Podolsky–Rosen (EPR) protocols.

6.4.1 BB84 protocols

The BB84 protocol was named after Bennett and Brassard, who proposed it in 1984 [15,16]. The three key principles being employed are the no-cloning theorem, state collapse during measurement, and the irreversibility of measurements [7]. The BB84 protocol can be implemented using various DOFs, including the polarization DOF [25] or the phase of the photons [22]. Experimentally, the BB84 protocol has been demonstrated over both fiber-optic and free-space-optical (FSO) channels [25, 26]. The polarization-based BB84 protocol over fiber-optic channels is affected by polarization mode dispersion, polarization-dependent loss, and fiber loss that affect transmission distance. To extend transmission distance, phase encoding is employed as in Ref. [26]. Unfortunately, the secure key rate over 405 km of ultralow-loss fiber is extremely low at only 6.5 b/s. In an FSO channel, the polarization effects are minimized; however, the atmospheric turbulence can introduce wavefront distortion and random phase fluctuations. Previous QKD demonstrations include a satellite-to-ground FSO link demonstration and a demonstration over an FSO link between two locations in the Canary Islands [25].

Two bases used in the BB84 protocol are the computational basis CB $= \{|0\rangle, |1\rangle\}$ and diagonal basis:

$$DB = \{|+\rangle = (|0\rangle + |1\rangle)/\sqrt{2}, |-\rangle = (|0\rangle - |1\rangle)/\sqrt{2}\}. \tag{6.7}$$

These bases belong to the class of *mutually unbiased bases* (MUBs) [27–31]. Two orthonormal bases $\{|e_1\rangle, \ldots, |e_N\rangle\}$ and $\{|f_1\rangle, \ldots, |f_N\rangle\}$ in Hilbert space C^N are MUBs when the square of the magnitude of the inner product between any two basis states $\{|e_m\rangle\}$ and $\{|f_n\rangle\}$ equals the inverse of the dimension:

$$|\langle e_m | f_n\rangle|^2 = \frac{1}{N} \; \forall m, n \in \{1, \cdots, N\} \tag{6.8}$$

The keyword *unbiased* means that if a system is prepared in a state belonging to one of the bases, all outcomes of the measurement with respect to the other bases are equally likely.

Alice randomly selects the MUB, followed by random selection of the basis state. The logical 0 is represented by $|0\rangle$, $|+\rangle$and logical 1 by $|1\rangle$, $|-\rangle$. Bob measures each qubit by randomly selecting the basis, computational or diagonal. In a sifting procedure, Alice and Bob announce the bases being used for each qubit and only keep instances using the same basis.

Now we describe the *polarization-based BB84 QKD protocol* with four corresponding states employed, as provided in Fig. 6.4. The bulky optics implementation of BB84 protocol is illustrated in Fig. 6.5.

The laser output state is known as the *coherent state* [31–36], representing the right eigenket of the *annihilation operator a*; namely, $a\,|\alpha\rangle = \alpha\,|\alpha\rangle$, where α is the complex eigenvalue. The coherent state vector $|\alpha\rangle$ can be represented in terms of orthonormal eigenkets $|n\rangle$ (the number or Fock state) of the number operator $N = a^\dagger a$ as follows:

$$|\alpha\rangle = \exp\left[-|\alpha|^2/2\right] \sum_{n=0}^{+\infty} (n!)^{-1/2} \alpha^n |n\rangle \tag{6.9}$$

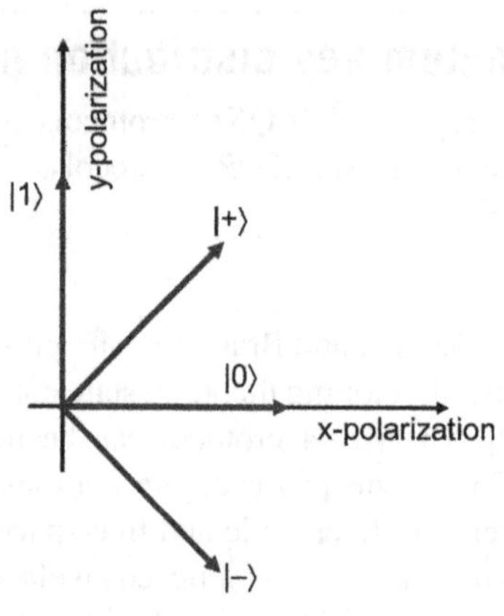

FIGURE 6.4

The four states employed in the BB84 protocol.

The coherent states are not orthogonal since the following is valid:

$$|\langle\alpha|\beta\rangle|^2 = \left| e^{-(|\alpha|^2+|\beta|^2)/2} \sum_{n=0}^{+\infty}\sum_{m=0}^{+\infty} (n!)^{-1/2}(m!)^{-1/2}\alpha^n(\beta^*)^m\langle n|m\rangle \right|^2$$

$$= \left| e^{-(|\alpha|^2+|\beta|^2)/2} \sum_{n=0}^{+\infty} (n!)^{-1}\alpha^n(\beta^*)^n \right|^2 = \left| e^{-(|\alpha|^2+|\beta|^2-2\alpha\beta^*)/2} \right|^2 = e^{-|\alpha-\beta|^2}. \quad (6.10)$$

By sufficiently attenuating the coherent state, we will obtain the weak coherent state (WCS), for which the probability of generating more than one photon is very low. For example, for $\alpha = 0.1$ we obtain $|0.1\rangle = \sqrt{0.90}\,|0\rangle + \sqrt{0.09}\,|1\rangle + \sqrt{0.002}\,|2\rangle + \ldots$. The number state $|n\rangle$ can be expressed in terms of the ground state ($|0\rangle$) with the help of the creation operator by

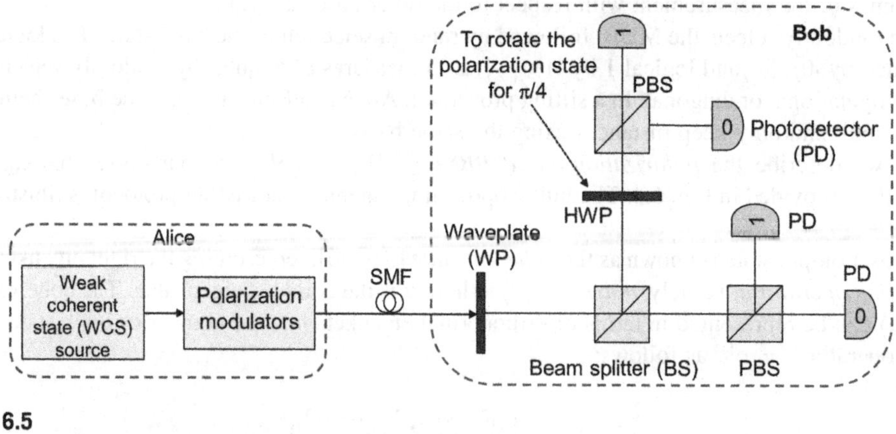

FIGURE 6.5

Polarization-based BB84 protocol.

$$|n\rangle = \frac{(a^\dagger)^n|0\rangle}{(n!)^{1/2}}. \tag{6.11}$$

The density operator of a coherent state is given by

$$\rho = |\alpha\rangle\langle\alpha| = e^{-|\alpha|^2}\sum_{n=0}^{+\infty}\sum_{m=0}^{+\infty}(n!)^{-1/2}(m!)^{-1/2}\alpha^n(\alpha^*)^m|n\rangle\langle m|. \tag{6.12}$$

The diagonal terms of ρ are determined by

$$\langle n|\rho|n\rangle = e^{-|\alpha|^2}\frac{|\alpha|^{2n}}{n!}, \tag{6.13}$$

which is clearly a Poisson distribution for the average number of photons $|\alpha|^2$. Therefore, the probability that n photons can be found in a coherent state $|\alpha\rangle$ follows the Poisson distribution:

$$P(n) = |\langle n|\alpha\rangle|^2 = e^{-\mu}\frac{\mu^n}{n!}, \tag{6.14}$$

where we use $\mu = |\alpha|^2$ to denote the average number of photons.

Given that the probability that multiple photons are transmitted is now nonzero, Eve can exploit this fact to gain information from the channel, known as a *photon number splitting (PNS) attack*. Assuming that she can detect the number of photons that reach her without performing a system measurement, when multiple photons are detected, Eve will take a single photon and place it into quantum memory until Alice and Bob perform time sifting and basis reconciliation. By learning Alice's and Bob's measurement bases from the discussion over the authenticated public channel, the photons stored in quantum memory will be measured on the correct basis, providing Eve with all the information contained in the photon.

To overcome a PNS attack, the use of *decoy-state*-based QKD was proposed in Ref. [37]. In decoy-state QKD, the average number of photons transmitted is increased during random timeslots, allowing Alice and Bob to detect if Eve is stealing photons when multiple photons are transmitted.

6.4.2 B92 protocol

The B92 protocol, introduced by Bennet in 1992 [38], employs just two nonorthogonal states, as illustrated in Fig. 6.6. Alice randomly generates a classical bit d, and depending on its value (0 or 1), she sends the following state to Bob [7]:

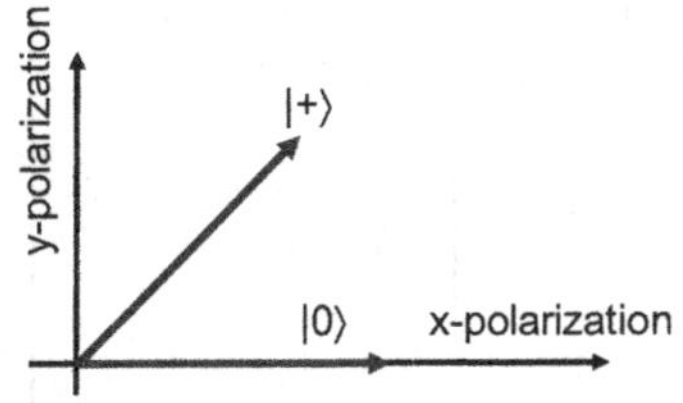

FIGURE 6.6

The two states employed in the B92 protocol.

$$|\psi\rangle = \begin{cases} |0\rangle, \ d = 0 \\ |+\rangle = (|0\rangle + |1\rangle)/\sqrt{2}, d = 1 \end{cases} \tag{6.15}$$

Bob randomly generates a classical bit d' and subsequently measures the received qubits in the CB $= \{|0\rangle, |1\rangle\}$ if $d' = 0$ or DB $= \{|+\rangle, |-\rangle\}$ if $d' = 1$; with the result of measurement being $r = 0$ or 1, corresponding to -1 and $+1$ eigenstates of Z and X observables, as illustrated in Table 6.1. Clearly, when the result of a measurement is $r = 0$, the possible states sent by Alice are $|0\rangle$ and $|+\rangle$, and Bob does not know which state Alice has sent. On the other hand, when the measurement result is $r = 1$, the possible states detected by Bob are $|1\rangle$ and $|-\rangle$, and Bob's detected bit is a complement of Alice's bit. Bob publicly announces r (and, of course, keeps d' secret). Alice and Bob keep only those pairs $\{d,d'\}$ for which the measurement result was $r = 1$. The *final key bit* is d for Alice and $1-d'$ for Bob. After that, the information reconciliation and privacy amplification stages take place.

6.4.3 Ekert (E91) and Einstein–Podolsky–Rosen protocols

The Ekert (E91) protocol was introduced by Ekert in 1991 [39] (see also [7]). Suppose Alice and Bob share n entangled pairs of qubits in the Bell state (EPR pair) $|B_{00}\rangle$:

$$|B_{00}\rangle = \frac{|00\rangle + |11\rangle}{\sqrt{2}}. \tag{6.16}$$

Alice generates a random classical bit b to determine the base and measures her half of the Bell state as either CB $= \{|0\rangle, |1\rangle\}$ basis for $b = 0$ or DB $= \{|+\rangle, |-\rangle\}$ basis for $b = 1$ to obtain the data bit d. On the other hand but in a similar fashion, Bob randomly generates the basis bit b' and measures either CB or DB to obtain d'. Alice and Bob compare the bases being used, namely b and b', and keep for their raw key only those instances of $\{d,d'\}$ for which the basis was the same $b = b'$. Clearly, the raw key generated is *truly random*; it is undetermined until Alice and Bob perform measurements on their Bell state half. For this reason, QKD is essentially a *secret key generation* method, as dictated by the entangled source. If instead the Bell state $|B_{01}\rangle = (|01\rangle + |10\rangle)/\sqrt{2}$ were used, Bob's raw key sequence would complement Alice's raw key sequence.

Table 6.1 Explanation of the B92 protocol.

Bits generated by Alice	$d = 0$		$d = 1$	
States sent by Alice	$\|0\rangle$		$\|+\rangle$	
Bits generated by Bob	$d' = 0$	$d' = 1$	$d' = 0$	$d' = 1$
Bob's measurement base	CB: $\{\|0\rangle,\|1\rangle\}$	DB: $\{\|+\rangle,\|-\rangle\}$	CB: $\{\|0\rangle,\|1\rangle\}$	DB: $\{\|+\rangle,\|-\rangle\}$
The possible resulting states to be detected by Bob	$\|0\rangle$ $\|1\rangle$	$\|+\rangle$ $\|-\rangle$	$\|0\rangle$ $\|1\rangle$	$\|+\rangle$ $\|-\rangle$
The probability of Bob measuring a given state	1 0	1/2 1/2	1/2 1/2	1 0
Result of Bob's measurement r	0 -	0 1	0 1	0 -

The *EPR protocol* is a three-state protocol that employs Bell's inequality to detect the presence of Eve as a hidden random variable, as described in Ref. [40]. Let $|\theta\rangle$ denote the polarization state of a photon linearly polarized at an angle θ. The three possible polarization states for the EPR pair are [40]

$$\begin{aligned} |B_0\rangle &= \frac{|0\rangle|3\pi/6\rangle + |3\pi/6\rangle|0\rangle}{\sqrt{2}}, \\ |B_1\rangle &= \frac{|\pi/6\rangle|4\pi/6\rangle + |4\pi/6\rangle|\pi/6\rangle}{\sqrt{2}}, \\ |B_2\rangle &= \frac{|2\pi/6\rangle|5\pi/6\rangle + |5\pi/6\rangle|2\pi/6\rangle}{\sqrt{2}}. \end{aligned} \tag{6.17}$$

For each EPR pair, we apply the following encoding rules [40]:

$$\begin{aligned} |\psi\rangle_1 &= \begin{cases} |0\rangle, & d=0 \\ |3\pi/6\rangle, & d=1 \end{cases} \\ |\psi\rangle_2 &= \begin{cases} |\pi/6\rangle, & d=0 \\ |4\pi/6\rangle, & d=1 \end{cases} \\ |\psi\rangle_3 &= \begin{cases} |2\pi/6\rangle, & d=0 \\ |5\pi/6\rangle, & d=1 \end{cases} \end{aligned} \tag{6.18}$$

The corresponding measurement operators will be [40]

$$M_0 = |0\rangle\langle 0|, M_1 = |\pi/6\rangle\langle\pi/6|, M_2 = |2\pi/6\rangle\langle 2\pi/6|. \tag{6.19}$$

The EPR protocol can be described as follows. The EPR state $|B_i\rangle$ ($i = 0,1,2$) is randomly selected. The first photon in the EPR pair is sent to Alice, and the second is sent to Bob. Alice and Bob select randomly, independently, and with equal probability one of the measurement operators M_0, M_1, or M_2. They subsequently measure their respective photon (qubit), and the measurement results determine the corresponding bit for the sifted key. Clearly, selecting EPR pairs as in Eq. (6.17) results in complementary raw keys. Alice and Bob then communicate over an authenticated public channel to determine the instances when they used the same measurement operator, and they keep those instances. The corresponding shared key represents the *sifted key*. The instances when they used different measurement operators represent the *rejected key*, and these are used to detect Eve's presence by employing the Bell's inequality on the rejected key. Alice and Bob then perform information reconciliation and privacy amplification steps.

Let us now describe how the rejected key can be used to detect Eve's presence [40]. Let $P(\neq|i,j)$ denote the probability that Alice's and Bob's bits in the rejected key are different given that Alice's and Bob's measurement operators are either M_i and M_j or M_j and M_i, respectively. Further, let $P(=|i,j) = 1- P(\neq|i,j)$. Let us denote the difference between these two probabilities by $\Delta P(i,j) = P(\neq|i,j) - P(=|i,j)$. The Bell's parameter can be defined by Ref. [40]:

$$\beta = 1 + \Delta P(1,2) - |\Delta P(0,1) - \Delta P(0,2)|, \tag{6.20}$$

which for the measurement operators above reduces to $\beta \geq 0$. However, the quantum mechanics prediction gives $\beta = -1/2$, clearly violating Bell's inequality.

6.4.4 Time-phase encoding

The BB84 protocol can also be implemented using various DOFs in addition to polarization states such as time-phase encoding [41–43] and orbital angular momentum [44, 45]. The time-phasing encoding states for the BB84 protocol are provided in Fig. 6.7. The time-basis corresponds to the computational basis, while the phase-basis to the diagonal basis. The pulse is localized within the time bin of duration $\tau = T/2$. Time-basis is like pulse–position modulation (PPM). The state in which the photon is placed in the first-time bin is denoted by $|t_0\rangle$, while the state in which the photon s placed in the second-time bin is denoted by $|t_1\rangle$. The phase MUB states are defined by

$$|f_0\rangle = \frac{1}{\sqrt{2}}(|t_0\rangle + |t_1\rangle), |f_1\rangle = \frac{1}{\sqrt{2}}(|t_0\rangle - |t_1\rangle). \tag{6.21}$$

Alice randomly selects either the time-basis or the phase-basis, followed by random selection of the basis state. The logical 0 is represented by $|t_0\rangle$, $|f_0\rangle$; while logical one by $|t_1\rangle$, $|f_1\rangle$. Bob measures each qubit by randomly selecting the basis, either time or phase. Alice and Bob announce the bases for each qubit during the sifting procedure and keep only instances that use the same basis. To implement the time-phase encoder, the electrooptical polar modulator, which is a concatenation of an amplitude modulator and phase modulator, or I/Q modulator can be used [46]. The amplitude modulator (Mach–Zehnder modulator) is used for pulse shaping, and the phase modulator introduces the desired phase shift, as illustrated in Fig. 6.8. The arbitrary waveform generator (AWG) is used to generate corresponding RF waveforms as needed. The variable optical attenuator (VOA) is used to attenuate the modulated beam down to the single-photon level.

On the receiver side, time-basis states can be detected with the help of a single-photon counter and properly designed electronics. On the other hand, to detect phase-basis states, a time-delay interferometer can be used as described in Ref. [47]; see Fig. 6.9. The difference in the path between two arms is [47] $\Delta L = \Delta L_0 + \delta L$, where ΔL_0 is the path difference equal to $c\tau$ (c is the speed of light), and $\delta L << \Delta L_0$ is small path difference used to adjust the phase since $\phi = k\delta L$ (k is the wave number).

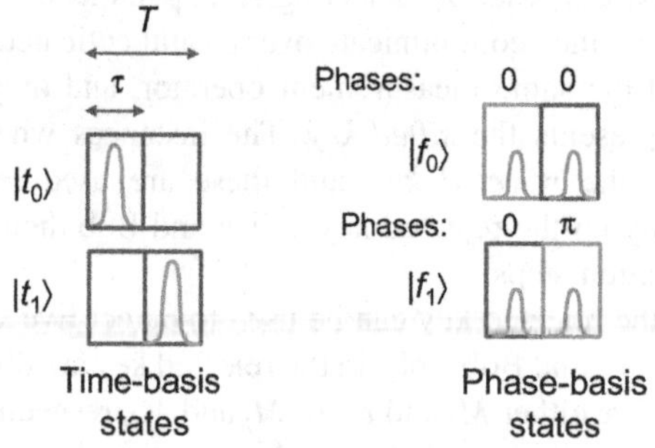

FIGURE 6.7

Time-phase encoding states used in the BB84 protocol.

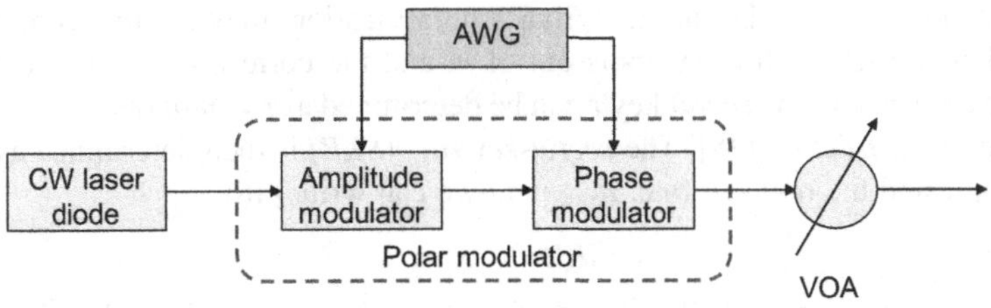

FIGURE 6.8

Time-phase encoder BB84 quantum key distribution protocol. *AWG*, arbitrary waveform generator; *VOA*, variable optical attenuator.

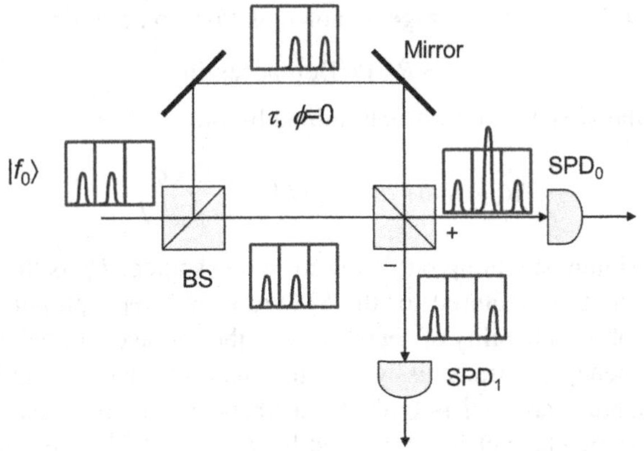

FIGURE 6.9

Time-phase decoder for BB84 quantum key distribution protocol. *BS*, 50:50 beam splitter; *SPD*, single-photon detector.

When the phase-state $|f_0\rangle$ is incident to the time-phase decoder, the outputs of the second 50:50 beam splitter (BS) occupy three time slots, and the + output denotes the case when the interferometer outputs interfere constructively, while the – output denotes that corresponding interferometer outputs interfere destructively. For constructive interference, the middle pulse doubles, whereas for destructive interference, the middle pulses cancel each other. Therefore, the click of the SPD in the middle slot at the + output identifies the $|f_0\rangle$-state with a corresponding click at the – output identifies the $|f_1\rangle$-state.

6.5 Quantum key distribution security

As discussed in the section on the BB84 protocol, Alice and Bob perform a sifting procedure and parameter estimation after the raw key transmission phase is completed. They are left with n symbols out of N transmitted, and the corresponding key is known as the *sifted key*. After that, they perform the classical postprocessing such as information reconciliation and privacy amplification. In the information reconciliation stage, they employ error correction to correct for the error introduced by both quantum channel and Eve, and the corresponding key after this stage is completed is commonly referred to as the *corrected key.* During the privacy amplification, they remove correlation remained with Eve, and they are left with m symbols out of n, and the corresponding key is the *secure key.* Clearly, the *fractional rate* of the secret key r can be determined as $r = m/n$ (as $N \rightarrow \infty$), and it is often referred to as the *secret fraction* [48]. The *secret-key rate* (*SKR*) is then determined as the product of the secret fraction r and the raw key rate, R_{raw}, and we can write [48]

$$SKR = r \cdot R_{\text{raw}}. \tag{6.22}$$

Interestingly, in many papers, *SKR* refers to the secret fraction r. Given that the raw key rate is dictated by the devices employed (particularly the SPD) and the quantum channel over which transmission takes place and not by the laws of quantum physics, the fraction rate can also be called the normalized SKR, given that $r = SKR/R_{\text{raw}}$.

The raw key rate can be determined as the product of signaling rate R_s and the probability of Bob accepting the transmitted symbol on average, denoted as Pr(Bob accepting), so we can write

$$R_{\text{raw}} = R_s \cdot \Pr(\text{Bob accepting}), \tag{6.23}$$

On the other hand, the signaling rate is determined by [48]

$$R_s = \min\left(R_{s,\max}, 1/T_{dc}, \frac{\mu T T_B \eta}{\tau_{\text{dead}}}\right), \tag{6.24}$$

where $R_{s,\max}$ is the maximum signaling rate allowed by the source, T_{dc} is the duty cycle, τ_{dead} is the dead time of the SPD (the time required for the SPD to reset after a photon is detected). The term $\mu T T_B \eta$ corresponds to Bob's probability of detection; μ is the average number of photons generated by the source, η is SPD efficiency, T is transmissivity of the quantum channel, and T_B denotes the losses in Bob's receiver. The quantum channel is typically a fiber-optic or free-space optical channel. The attenuation in the fiber-optic channel is determined by $T = 10^{-\alpha L/10}$, where α is the attenuation coefficient in dB/km, while L is the transmission distance in km. The typical attenuation coefficient in standard single-mode fiber (SSMF) is 0.2 dB/km at 1550 nm. Regarding, the FSO link, if we ignore the atmospheric turbulence effects, for the line-of-site link of transmission distance L, we can estimate the FSO link loss by Ref. [49] $T = [d_{\text{Rx}}/(d_{\text{Tx}} + DL)]^2 \cdot 10^{-\alpha L/10}$, where d_{Tx} and d_{Rx} represent diameters of transmit and receive telescopes' apertures, while D is the divergence. Therefore, the first term represents the geometric losses, while the second term is the average attenuation due to scattering, with scattering attenuation coefficient being <0.1 dB/km in clear sky condition at 1520–1600 nm.

In BB84 protocol, the probability that Bob accepts the transmitted symbol is related to the probability he used the same basis, denoted as p_{sift}, so that the signaling rate becomes $R_s p_{\text{sift}}$, and the raw key rate when Alice sends n photons to Bob will be [48]

$$R_{\text{raw},n} = R_s p_{\text{sift}} p_{A,n} p_n, \tag{6.25}$$

where $p_{A,n}$ is the probability that the pulse sent by Alice contains n photons (see section below), and p_n is the probability that Eve sends Bob a pulse with n photons. The overall total raw key rate will be $R_{\text{raw}} = \sum_n R_{\text{raw},n}$.

What remains to be determined is the fraction rate. The QKD can achieve the unconditional security, which means that its security can be verified without imposing any restrictions on either Eve's computational power or eavesdropping strategy. The bounds on the fraction rate are dependent on the classical postprocessing steps. The most common is the one-way postprocessing, in which either Alice or Bob holds the reference key and sends the classical information to the other party through the public channel, while the other party performs certain procedure on data without providing the feedback. The most common one-way processing consists of two steps, the information reconciliation and privacy amplification. The expression for *secret fraction* obtained by *one-way postprocessing* is given by Ref. [48]:

$$r = I(A;B) - \min_{\text{Eve's strategies}} (I_{EA}, I_{EB}), \tag{6.26}$$

where $I(A;B)$ is the mutual information between Alice and Bob, while the second term corresponds to Eve's information I_E about Alice's or Bob's raw key, where minimization is performed over all possible eavesdropping strategies. Alice and Bob will decide to employ either direct or reverse reconciliation so that they can minimize Eve's information.

We now describe different eavesdropping strategies that Eve may employ, which determine Eve's information I_E.

6.5.1 Independent (individual) or incoherent attacks

This is the most constrained family of attacks, in which Eve attacks each qubit independently, and interacts with each qubit by applying the same strategy. Moreover, she measures the quantum states before the classical postprocessing takes place. The security bound for incoherent attacks is the same as that for classical physical-layer security (PLS) [2], and it is determined by Csiszár-Körner bound [50], given by Eq. (6.26), wherein the mutual information between Alice and Eve is given by

$$I_{EA} = \max_{\text{Eve's strategies}} I(A;E), \tag{6.27}$$

where the maximization is performed over all possible incoherent eavesdropping strategies. A similar definition holds for I_{BE}.

An important family of incoherent attacks are the *intercept-resend (IR)* attacks; in which Eve intercepts the quantum signal send by Alice, performs the measurement on it, and based on the measurement result she prepares new quantum signal (in the same MUB as the measured quantum state) and sends such prepared quantum signal to Bob. In BB84 protocol, Eve's basis choice will match Alice's basis 50% of the time. When she uses the wrong basis there is still 50% chance to guess the correct bit. When Bob does the measurement there is 50% chance that Bob will use the same basis as Eve. However, these bits will correlate with Eve's sequence not Alice one, and overall probability of error will be 1/4, while the mutual information between Alice and Eve will be $I(A;E) = 1/2$. Given high probability of error introduced by the IR attack Alice and Bob can easily identify Eve's activity. However, Eve might choose to apply the IR attack with probability of p_{IR}. In that case the probability of error will be $q = p_{\text{IR}}/4$. The mutual information between Alice and Eve will be then $I(A;E) = h(p_{\text{IR}}/4)$, where $h(\bullet)$ is the binary entropy function $h(q) = -q\log_2 q - (1-q)\log_2(1-q)$. If all errors get contributed to Eve, the secret fraction can be estimated as

$$r = \underbrace{1 - h(q)}_{I(A;B)} - h(q) = 1 - 2h(q). \tag{6.28}$$

Another important family of incoherent attacks are PNS attacks [51], in which Eve acts on multiphoton quantum states. In a WCS-based QKD system, the quantum states are generated by modulating the beam from the coherent light source, which is then attenuated so that on average, one photon per pulse is transmitted by Alice. However, as discussed above, the coherent source emits the photons based on Poisson distribution so that the probability of emitting n photons in a state-generated mean photon number μ is determined by the following equation:

$$p_{A,n} = p(n|\mu) = e^{-\mu}\frac{\mu^n}{n!}. \tag{6.29}$$

As the parameter μ increases, the probability of getting more than one photon increases as well. To perform the PNS attack, Eve can employ the beam splitter to take one of the photons from the multiphoton state and pass the rest to Bob, as illustrated in Fig. 6.10. She can measure the photon by randomly selecting the basis. Eve can even replace the quantum link with ultralow-loss fiber so that Alice and Bob cannot determine that the transmitted signal has been attenuated. To solve for the PNS attack, Alice and Bob can employ a decoy-state-based protocol [37], in which Alice transmits the quantum states with different mean photon numbers representing signal and decoy states. Eve cannot distinguish between the decoy and signal states, and given that Alice and Bob know the decoy signal level, they can identify the PNS attack.

6.5.2 Collective attacks

The collective attacks represent generalization of the incoherent attacks, given that Eve's interaction with each qubit is also independent and identically distributed (i.i.d.). However, in these attacks, Eve can store her ancilla qubits in a quantum memory until the end of classical postprocessing steps. For instance given that information reconciliation requires the exchange of the parity bits over an authenticated classical channel, Eve can apply the best-known classical attacks to learn the content of the parity bits. Based on all information available to her, Eve can perform the best measurement strategy on her ancilla qubits (stored in the quantum memory).

The PNS attack is stronger when Eve applies the quantum memory. Namely, from the sifting procedure, Eve can learn which basis was used by Alice and can apply the correct basis on her photon stored in the quantum memory and thus double the number of identified bits compared with the case when quantum memory is not used.

The security bound for collective attacks, assuming one-way postprocessing, is given by Eq. (6.26), wherein Eve's information about Alice sequence is determined from Holevo information as follows [52] (see also [48]):

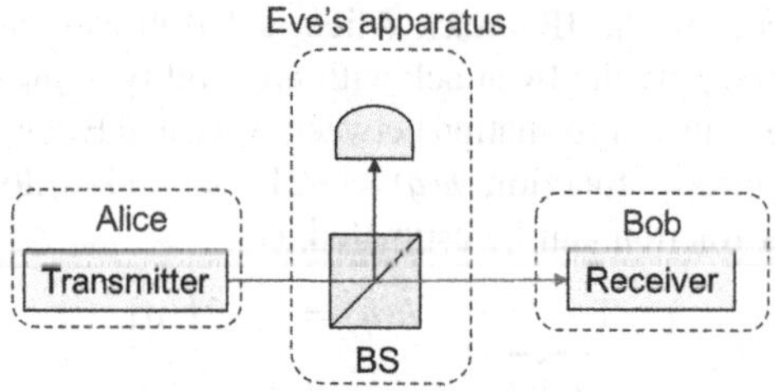

FIGURE 6.10

Illustration of Eve's realization of a photon number splitting attack. *BS*, beam splitter.

$$I_{EA} = \max_{\text{Eve's strategies}} \chi(A;E), \tag{6.30}$$

where maximization is performed over all possible collective eavesdropping strategies. The similar definition holds for I_{BE}. This bound is also known as the Devetak–Winter bound. The Holevo information, introduced in Ref. [53], is defined here as

$$\chi(A;E) = S(\rho_E) - \sum\nolimits_a S\left(\rho_{E|a}\right)p(a), \tag{6.31}$$

where $S(\rho)$ is the von Neumann entropy defined as $S(\rho) = -\mathrm{Tr}(\log(\rho)) = -\sum_i \lambda_i \log \lambda_i$, with λ_i being the eigenvalues of the density operator (state) ρ. In (6.31), $p(a)$ represents the probability of occurrence of symbol a from Alice's classical alphabet, while $\rho_{E|a}$ is the corresponding density operator of Eve's ancilla. Finally, ρ_E is Eve's partial density state defined by $\rho_E = \sum_a p(a)\rho_{E|a}$. In other words, the Holevo information corresponds to the average reduction in the von Neumann entropy given that we know how ρ_E get prepared.

6.5.3 Quantum hacking attacks and side-channel attacks

Quantum hacking attacks exploit the weakness of the particular QKD system implementation. Many of hacking attacks are feasible with current technology. In a *Trojan horse attack* [54] Eve sends the bright laser beam toward Alice's encoder and measures the reflected photons to gain the information about the secret. This attack can be avoided using an optical isolator. It was also noticed that silicon-based photon counters employing the avalanche photodiodes (APDs) emit some light at different wavelengths when they detect a single-photon, which can be exploited by Eve [55]. In time-shifting attack [56], Eve exploits the efficiency mismatch of Bob's SPDs to estimate Bob's basis selection. In a *blinding attack* [57], Eve exploits APD-based single-photon counter properties to force Bob to pick up the same basis as Eve does. The APD for photon counting often operate in gated mode so that the APD is active only when photon arrival is expected. When in off mode the APD operates in a linear regime, meaning that the photocurrent is proportional the receive optical power, and therefore, it behaves as the classical photodetector. To exploit this, Eve can send the weak classical light beam, sufficiently strong to generate the click when Bob employs the same basis, and not to register the click otherwise. In such scenario 50% of events will be recorded as inconclusive. Fortunately, the common property of quantum hacking attacks is that they are preventable once the attack is known. However, the threat from an unknown quantum hacking attack is still present.

Many *side-channel attacks* are simultaneously time zero-error attacks [48]. Namely, Eve can exploit the quantum channel loss to hide her side-channel attack. The beam-splitting attack exploits the losses of the quantum channel, and Eve can place the beam splitter just after Alice's transmitter, take the portion of the beam and pass the rest to Bob. Since Alice does not modify the optical mode, this attack will not be detected by Bob. Interestingly, the Poisson distribution is preserved during the PNS attack [58]. In realistic implementations, Eve can perform *unambiguous state discrimination* followed by signal resending, when Alice employs the linearly independent signal states, and achieve better than beam-slitting efficiency as shown in Ref. [59]. Fortunately, the existence of side-channel attacks does not compromise the security as long as the corresponding side-channel attacks are taken into account during the privacy amplification stage [48].

6.5.4 Security of BB84 protocol

The secret fraction that can be achieved with BB84 protocol is lower bounded by

$$r = q_1^{(Z)}\left[1 - h_2\left(e_1^{(X)}\right)\right] - q^{(Z)} f_e h_2\left(e^{(Z)}\right), \tag{6.32}$$

where $q_1^{(Z)}$ denotes the probability of declaring a successful result ("the gain") when Alice sent a single-photon and Bob detected it in the Z-basis, f_e denotes the error correction inefficiency ($f_e \geq 1$), $e^{(X)}$ [$e^{(Z)}$] denotes the quantum bit error rate (QBER) in the X-basis (Z-basis), and $h_2(x)$ is the binary entropy function, defined as

$$h_2(x) = -x\log_2(x) - (1-x)\log_2(1-x). \tag{6.33}$$

The second term $q^{(Z)}h_2[e^{(X)}]$ denotes the amount of information Eve was able to learn during the raw key transmission, and this information is typically removed from the final key during the privacy amplification stage. The last term $q^{(Z)}f_e\, h_2[e^{(Z)}]$ denotes the amount of information revealed during the information reconciliation (error correction) stage, typically related to the parity-bits exchanged over a noiseless public channel (when systematic error-correction is used).

6.6 Decoy-state protocols

When a WCS source is used instead of a single-photon source, this scheme is sensitive to the PNS attack. Given that the probability that multiple photons are transmitted from WCS is now nonzero, Eve can exploit this fact to gain information from the channel. To overcome the PNS attack, the use of *decoy-state* quantum key distribution systems are advocated in Refs. [37, 61–70]. In decoy-state QKD the average number of photons transmitted is increased during random timeslots, allowing Alice and Bob to detect if Eve is stealing photons when multiple photons are transmitted. There exist different versions of decoy-state-based protocols. In one-decoy-state-based protocol, in addition to signal state with mean photon number μ, the decoy state with mean photon number $\nu<\mu$ is employed. Alice first decides on signal and decoy mean photon number levels, and then determines the optimal probabilities for these two levels to be used, based on corresponding SKR expression. In weak-plus-vacuum decoy-state-based protocol, in addition to μ and ν levels, the vacuum state is also used as the decoy state. Both protocols have been evaluated in Ref. [67], both theoretically and experimentally. It was concluded that the use of one signal state and two decoy states is enough, which agrees with findings in Refs. [68, 69].

The probability that Alice can successfully detect the photon when Alice employs the WCS with mean photon number μ, denoted as q_μ, also known as the *gain*, can be determined by

$$q_\mu = \sum_{n=0}^{\infty} y_n e^{-\mu}\frac{\mu^n}{n!}, \tag{6.34}$$

where y_n, also known as the *yield*, is the probability of Bob's successful detection when Alice has sent n photons. The similar expression holds for decoy states with the mean photon number σ_i; that is,

$$q_{\sigma_i} = \sum_{n=0}^{\infty} y_n e^{-\sigma_i}\frac{\sigma_i^n}{n!}. \tag{6.35}$$

Let e_n denote the QBER corresponding to the case when Alice has sent n-photons, the average QBER for the coherent state $|\mu\exp(j\phi)\rangle$ (where ϕ is the arbitrary phase) can then be estimated by

$$\overline{e}_\mu = \frac{1}{q_\mu}\sum_{n=0}^{\infty} y_n e_n e^{-\mu}\frac{\mu^n}{n!}. \tag{6.36}$$

If T is the overall transmission efficiency composed of the channel transmissivity T_{ch}, photodetector efficiency η, and Bob's optics transmissivity T_{Bob}, the transmission efficiency of the n-photon pulse will be given by

$$T_n = 1 - (1 - T)^n. \tag{6.37}$$

In the absence of Eve, for $n = 0$, the yield y_0 is affected by the background detection rate, denoted as p_d, so we can write $y_0 = p_d$. On the other hand, the yield for $n \geq 1$ is contributed by both the signal photons and the background rate as follows:

$$y_n = T_n + y_0 - T_n y_0 = T_n + y_0(1 - T_n) = 1 - (1 - y_0)(1 - T)^n. \tag{6.38}$$

The error rate of n-photon states e_n can be estimated by Refs. [37, 70]:

$$e_n = \frac{1}{y_n}\left(e_0 y_0 + e_d \underbrace{T_n}_{\cong y_n - y_0} \right) \cong e_d + \frac{y_0}{y_n}(e_0 - e_d), \tag{6.39}$$

where e_d is the probability of a photon hitting an erroneous detector. For time-phase encoding, this represents the intrinsic misalignment error rate. In the absence of signal photons, assuming that two detectors have equal background rates, we can write $e_0 = 1/2$. Similarly, the average QBER can be estimated by

$$\overline{e}_\mu = e_d + \frac{y_0}{q_\mu}(e_0 - e_d). \tag{6.40}$$

The secrecy fraction can now be lower bounded by modifying the corresponding expression for BB84 protocols as follows:

$$r = q_1^{(Z)}\left[1 - h_2\left(e_1^{(X)}\right)\right] - q_\mu^{(Z)} f_e h_2\left(\overline{e}_\mu^{(Z)}\right), \tag{6.41}$$

where a subscripted 1 denotes single-photon pulses and μ denotes the pulse with the mean photon number μ.

As an illustration, Fig. 6.11 provides secrecy fraction results for the decoy-state protocol and the WCS-based, prepare-and-measure, BB84 protocol, assuming that the standard SMF with attenuation coefficient 0.21 dB/km is used.

Other parameters, used in simulations are set as follows [37, 71–73]: the detector efficiency $\eta = 0.045$, the dark count rate $p_d = 1.7 \cdot 10^{-6}$, the probability of a photon hitting erroneous detector $e_d = 0.033$, and reconciliation inefficiency $f_e = 1.22$. For decoy-state-based BB84 protocol to calculate the secrecy fraction we use Eq. (6.41) for the optimum value of mean photon number per pulse μ, wherein q_μ is determined by $q_\mu = 1-(1-y_0)\exp(-\mu T)$, $\overline{e}_\mu$ is determined by Eq. (6.40), e_1 is

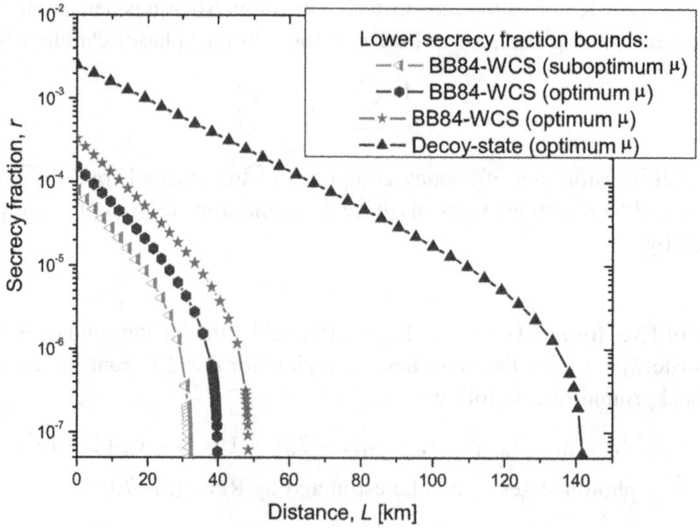

FIGURE 6.11

Secrecy fraction versus transmission distance for decoy-state-based protocol and weak-coherent-state-based BB84 protocol.

determined by Eq. (6.39), and q_1 as the argument of (6.34) for $n=1$, that is, by $q_1 = y_1\mu \exp(-\mu)$. For WCS-based BB84 protocol, without the decoy state, we use Eq. (6.41), assuming that Eve controls the channel, performs intercept-resent attacks, and she is transparent to multiphoton states. In this lower-bound scenario, the error from single-photon events dominates. Now by calculating $q^{(Z)}$ by $q_\mu = 1-(1-y_0)\exp(-\mu T)$, $e^{(Z)}$ by $E_\mu = e_d+(e_0-e_d)y_0/q_\mu$, and assuming suboptimum μ proportional to T, we obtained the yellow curve in Fig. 6.11, which is consistent with ref. [37]. When using the optimum μ instead (maximizing the secrecy fraction), we obtain the violet curve, which is consistent with ref. [71]. For these two curves, we estimate the $q_1^{(Z)}$ from (6.34), by setting $y_n = 1$ for $n \geq 2$ (see Ref. [70]) as follows:

$$q_1^{(X)} \geq q_\mu - \sum_{n=2}^{\infty} e^{-\mu} \frac{\mu^n}{n!}. \tag{6.42}$$

Given that multiphoton states do not introduce errors—that is, $e_n = 0$ for $n \geq 2$, we can bound e_1 as follows:

$$e_1^{(X)} \leq q_\mu E_\mu / q_1^{(X)}. \tag{6.43}$$

More accurate expression for q_μ, for WCS-based BB84 protocol, will be the following:

$$q_\mu = y_0 e^{-\mu} + y_1 e^{-\mu}\mu + \sum_{n=2}^{\infty} e^{-\mu}\frac{\mu^n}{n!}, \tag{6.44}$$

which is obtained from Eq. (6.34) by setting $y_n = 1$ for $n \geq 2$. The corresponding curve is denoted in pink, indicating that the distance $L = 48.62$ km is achievable when the WCS-based BB84 protocol is used. In contrast, for decoy-state-based BB84 protocol and optimum μ, it is possible to achieve the distance close to 141.8 km (see blue curve), which agrees with ref. [71].

6.7 Measurement-device-independent quantum key distribution protocols

Any discrepancy in parameters of devices from assumptions made in QKD protocols, can be exploited by adversaries through side-channel and quantum hacking attacks [54–58] to compromise security of the corresponding protocols. One way to solve for this problem is to invent different ways to overcome the known side-channel/quantum hacking attacks. Unfortunately, the eavesdropper can always come up with a more sophisticated attack. Another, more effective strategy is to invent new protocols in which minimal assumptions have been made on devices being employed in the protocol. One such protocol is known as the device-independent QKD (DI-QKD) protocol [74–77] in which the statistics of detection has been determined to secure the key without any assumption being made on the operation of devices. In MDI-QKD protocols, Charlie performs BSMs and announces the results; this is why describe the BSMs first.

6.7.1 Photonic bell state measurements

BSMs play a key role in photonic quantum computation and communication including quantum teleportation [78], superdense coding [79], quantum swapping [80], quantum repeaters [81], QKD [82–84], to mention few. A *complete BSM* represents a projection of any two-photon state to maximally entangled Bell states, defined by

$$|\psi^{\pm}\rangle = \frac{1}{\sqrt{2}}(|01\rangle \pm |10\rangle), |\phi^{\pm}\rangle = \frac{1}{\sqrt{2}}(|00\rangle \pm |11\rangle). \tag{6.45}$$

Unfortunately, the complete BSM is impossible to achieve using only linear optics without auxiliary photons. It has been shown in Ref. [85] that the probability of successful BSM for two photons being in complete mixed input states is limited to 50%. The conventional approach to the Bell state analysis is to employ 50:50 BS, followed by SPDs to allow discrimination of orthogonal states. As an illustration, the setup to perform the *BSM on polarization states* is provided in Fig. 6.12. The states emitted by Alice and Bob are denoted by density operators ρ_A and ρ_B, respectively. When photons at BS inputs one and two are of the same polarization, that is $|\mathrm{HH}\rangle = |00\rangle$ or $|\mathrm{VV}\rangle = |11\rangle$, according to the Hong-Ou-Mandel (HOM) effect [2,86,87], they will appear simultaneously at either output port 3 or port 4, and the Bell states $|\phi^{\pm}\rangle$ transform to

$$|\phi^{\pm}\rangle = \frac{1}{\sqrt{2}}(|00\rangle_{12} \pm |11\rangle_{12}) \overset{BS}{\rightarrow} |\phi^{\pm}_{\mathrm{out}}\rangle = \frac{j}{2\sqrt{2}}(|00\rangle_{33} + |00\rangle_{44} \pm |11\rangle_{33} \pm |11\rangle_{44}), \tag{6.46}$$

and clearly cannot be distinguished by the corresponding polarizing beam splitters (PBSs).

On the other hand, when photons at the BS input ports one and two are of different polarizations, they can exit either the same output port or different ports. When both photons exit the output port 3,

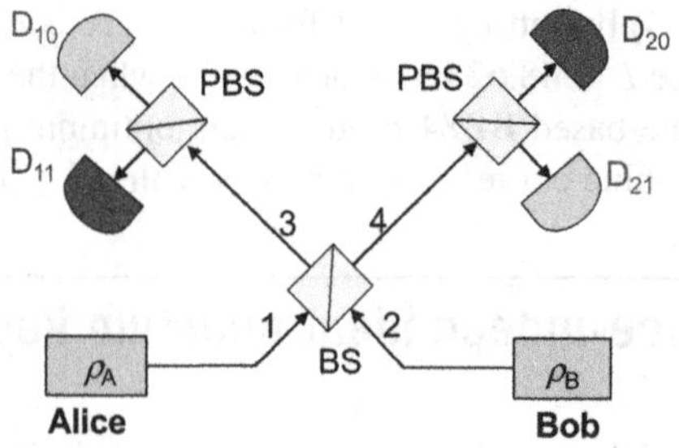

FIGURE 6.12

Illustration of setup for performing Bell state measurements on polarization states.

after the PBS, the SPDs D_{10} and D_{11} will simultaneously click. When both photons exit the output port 4, after the PBS the SPDs D_{20} and D_{21} will simultaneously click. When input photons of different polarizations exit different ports there exist two options. When horizontal photon exits port three and vertical photon port 4, the SPDs D_{11} and D_{20} will simultaneously click. Finally, when horizontal photon exits port four and vertical photon port 3, the SPDs D_{10} and D_{21} will simultaneously click. Given that the Bell states $|\psi^{\pm}\rangle$ are transformed (by the BS) to

$$\begin{aligned} |\psi^+\rangle &= \frac{1}{\sqrt{2}}(|01\rangle_{12}+|10\rangle_{12}) \overset{BS}{\rightarrow} |\psi^+_{\text{out}}\rangle = j(|01\rangle_{33}+|10\rangle_{44}), \\ |\psi^-\rangle &= \frac{1}{\sqrt{2}}(|01\rangle_{12}-|10\rangle_{12}) \overset{BS}{\rightarrow} |\psi^-_{\text{out}}\rangle = |01\rangle_{34}-|10\rangle_{34}, \end{aligned} \tag{6.47}$$

they can be distinguished by PBSs and SPDs. Clearly, based on discussion above we conclude that the projection on $|\psi^+\rangle$ occurs when both SPDs after any of two PBSs simultaneously click. In other words, when either D_{10} and D_{11} or D_{20} and D_{21} simultaneously click, the projection on $|\psi^+\rangle$ is identified. Similarly, when either D_{10} and D_{21} or D_{20} and D_{11} simultaneously click, the projection on $|\psi^-\rangle$-state is identified. In conclusion, when ideal SPDs are used the BSM efficiency is ½. If the SPDs have nonideal detection efficiency η, the BSM efficiency will be $0.5\eta^2$.

Another widely used DOF to encode the quantum states is *time-bin DOF*. For 2-D quantum communications, the photon can be encoded as the superposition of early-state, denoted as $|0\rangle$, and late-state, denoted as $|1\rangle$, as follows $|\xi\rangle = \alpha|0\rangle + \beta|1\rangle$, $|\alpha|^2 + |\beta|^2 = 1$. To perform the BSM we can use the setup provided in Fig. 6.13, which requires only two SPDs in addition to the BS. Similarly to polarization, the Bell states are defined by Eq. (6.45), but with different basis-kets. Again, when two photons arrive at the same time at BS input ports, they will exit the same output port according to the HOM effect, and we cannot distinguish the $|\phi^{\pm}\rangle$ states. On the other hand, the projection on $|\psi^+\rangle$ is identified when either detector registers the click in early and late time-bins. Similarly, the projection on $|\psi^-\rangle$ occurs when either detectors D_1 and D_2 or D_2 and D_1 click in consecutive (early–late) time-bins.

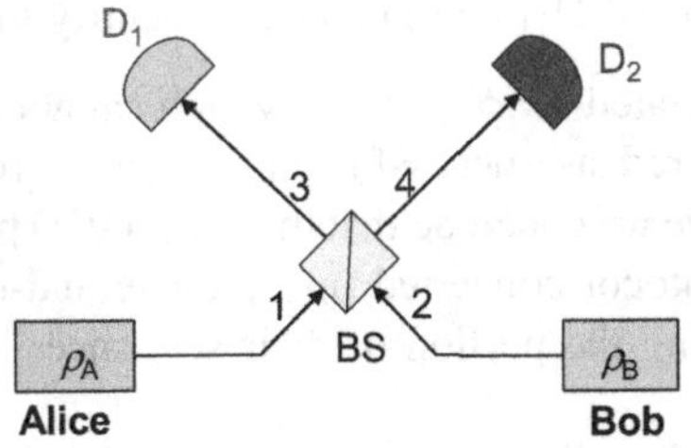

FIGURE 6.13

Illustration of the setup for performing Bell state measurements on time-bin states.

6.7.2 Description of measurement-device-independent quantum key distribution protocol

In MDI-QKD protocols [82,88–91], Alice and Bob are connected to a third party-Charlie, through a quantum channel, such as a free-space optical (FSO) or a fiber-optic link. The SPDs are located at Charlie's side, as illustrated in Fig. 6.14, for polarization-based MDI-QKD protocol.

Alice and Bob can employ either single-photon sources or WCS sources, and randomly select one of four states $|0\rangle$, $|1\rangle$, $|+\rangle$, $|-\rangle$ to be sent to Charlie, with the help of polarization modulator (PolM), which is very similar to that used in BB84 protocol. The intensity modulator is employed to impose different decoy states. Charlie can essentially be Eve as well. Charlie performs the partial BSM with the help of one BS, two PBSs, and four SPDs as described in Fig. 6.14, and announces the events for each the measurement resulted into either $|\psi^+\rangle = (|01\rangle + |10\rangle)/\sqrt{2}$ state or $|\psi^-\rangle = (|01\rangle - |10\rangle)/\sqrt{2}$ state. Charlie also discloses the results of the BSM. Given that Charlie's BSMs are used only to check the parity of Alice's and Bob's bits, these measurements will not disclose any information related to the individual bits themselves [88].

A successful BSM corresponds to the observation when precisely two detectors (with orthogonal polarizations) are triggered as follows:

- Detectors' D_{10} and D_{21} clicks identify the Bell state $|\psi^-\rangle$,

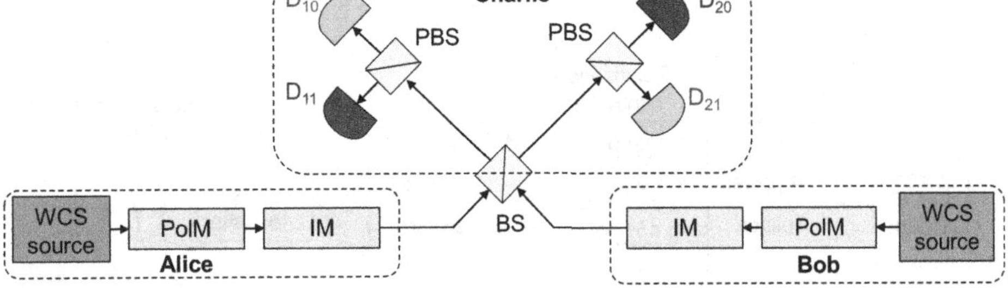

FIGURE 6.14

Polarization state-based measurement-device-independent quantum key distribution system configuration.

- On the other hand, then detectors' D_{11} and D_{20} clicks identify the Bell state $|\psi^+\rangle$.

Other detection patterns are related to $|\phi^{\pm}\rangle$-states, which cannot be distinguished with SPDs, and therefore, cannot not be considered as successful and as such are discarded from consideration. Clearly, based on description above we conclude that the MDI-QKD protocol represents generalization of both the EPR-based QKD protocol converted into prepare-and-measure scheme and the decoy-state-based protocol. By comparing the portion of their sequences, Alice and Bob can verify Charlie's honesty.

Let us now describe one possible implementation of the PolM, that is decoy-state-based BB84 encoder configuration, which is provided in Fig. 6.15. The laser beam signal is split by 1:4 coupler into four branches, each containing the polarization controller to impose the following polarization states $|0\rangle, |1\rangle, |+\rangle, |-\rangle$. With the help of 4:1 optical space switch, we randomly select the polarization state to be sent to Charlie. For this purpose the use of switches with electrooptic switching is desirable, such as those reported in Refs. [92, 93], since in these switches the switching state can be changed within few ns. The phase modulator at the output is used to perform the phase randomization of the coherent state $|\alpha\rangle = \exp(-|\alpha|^2/2)\sum_n \alpha^n |n\rangle/\sqrt{n!}$ as follows:

$$\frac{1}{2\pi}\int_0^{2\pi} \left||\alpha|e^{j\vartheta}\right\rangle\left\langle|\alpha|e^{-j\vartheta}\right| d\vartheta = \frac{1}{2\pi}\sum_{n=0}^{\infty}\sum_{m=0}^{\infty}\frac{|\alpha|^{m+n}}{\sqrt{m!n!}}e^{-|\alpha|^2}|n\rangle\langle m|\underbrace{\int_0^{2\pi} e^{j(n-m)\vartheta}d\vartheta}_{\delta_{nm}}$$

$$= \sum_{n=0}^{\infty}\frac{\left(|\alpha|^2\right)^n}{n!}e^{-|\alpha|^2}|n\rangle\langle n| = \sum_{n=0}^{\infty}\frac{\mu^n}{n!}e^{-\mu}|n\rangle\langle n|, \mu = |\alpha|^2. \tag{6.48}$$

where $\mu = |\alpha|^2$ is mean photon number per pulse. Variable optical attenuator is then used to reduce the mean photon number down to the quantum level. Other PolM configurations can be used, such as one described in Ref. [88], employing four laser diodes. Alice and Bob then exchange the information, over an authenticated classical, noiseless, channel about the bases being used: Z-basis spanned by $|0\rangle$ and $|1\rangle$ [computational basis (CB)] states, or X-basis spanned by $|+\rangle$ and $|-\rangle$ (diagonal basis) states. Alice and Bob than keep only events in which they used the same basis, while at the same time Charlie announced the BSM states $|\psi^-\rangle$ or $|\psi^+\rangle$. One of them (say Bob) flips the sent bits except when both used diagonal basis and Charlie announces the state $|\psi^+\rangle$. The remaining events are discarded.

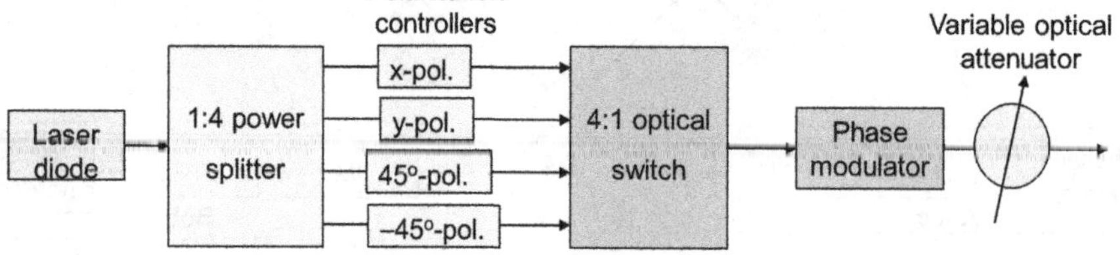

FIGURE 6.15

Decoy-state-based BB84 encoder (PolM) configuration.

Further, the X-key is formed out of those bits when Alice and Bob prepared their photons in X-basis (diagonal basis). The error rate on these bits is used to bound the information obtained by Eve. Next, Alice and Bob form the Z-key out of those bits for which they both used the Z-basis (computational basis). Finally, they perform information reconciliation and privacy amplification on the Z-key to get the secret key.

In addition to being insensitive to the SPDs used by Charlie, the MDI-QKD protocol allows to connect multiple participants in a star-like QKD network, in which Charlie is located in the center (serving as a hub) [94, 95]. In this QKD network, illustrated in Fig. 6.16, the most expensive equipment (SPDs) is shared among multiple users, and the QKD network can easily be upgraded without disruption to other users. With the help of *N*:2 optical switch, any two users can participate in MDI-QKD protocol to exchange the secure keys.

6.7.3 Time-phase-encoding-based measurement-device-independent quantum key distribution protocol

The time-phase encoding basis states for BB84 protocol ($N = 2$) are provided in Fig. 6.17.

The time-basis corresponds to the computational basis $\{|0\rangle, |1\rangle\}$, while the phase-basis to the diagonal basis $\{|+\rangle, |-\rangle\}$. The pulse is localized within the time bin of duration $\tau = T/2$. The time-basis is similar to PPM. The state in which the photon is placed in the first-time bin (early state) is denoted by $|0\rangle = |e\rangle$, while the state in which the photon is placed in the second-time bin (late state) is denoted by $|1\rangle = |l\rangle$. The time-phase states are defined by $|+\rangle = (|0\rangle + |1\rangle)/\sqrt{2}, |-\rangle = (|0\rangle - |1\rangle)/\sqrt{2}$. Alice randomly selects either the time-basis or the phase-basis, followed by random selection of the basis state. The logical 0 is represented by $|0\rangle$, $|+\rangle$; while logical one by $|1\rangle$, $|-\rangle$. Alice and Bob randomly select one of four states $|0\rangle$, $|1\rangle$, $|+\rangle$, $|-\rangle$ to be sent to Charlie. Charlie performs the partial BSM with the help of one 50:50 BS, and two SPDs as shown in Fig. 6.18, and announces the events for each the measurement resulted into either $|\psi^+\rangle = (|01\rangle + |10\rangle)/\sqrt{2}$ state or $|\psi^-\rangle = (|01\rangle - |10\rangle)/\sqrt{2}$ state.

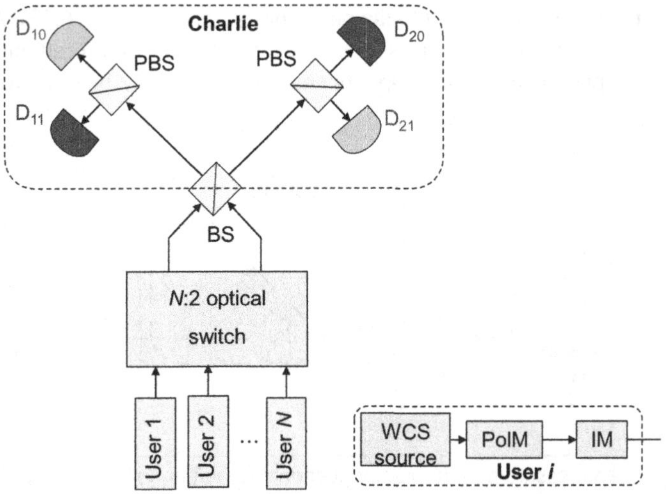

FIGURE 6.16

Measurement-device-independent quantum key distribution-based network.

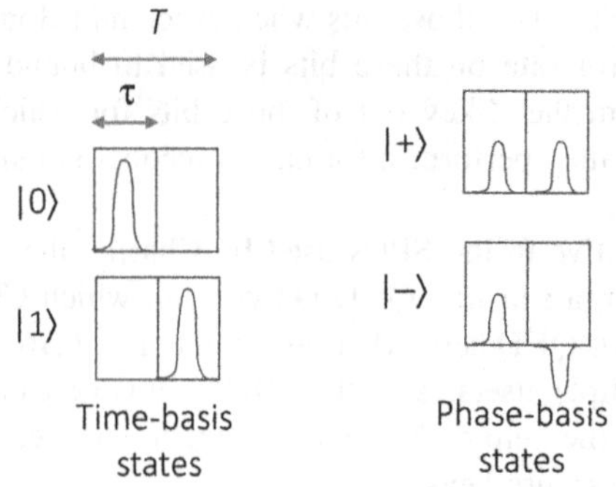

FIGURE 6.17

Time-basis and phase-basis states used in time-phase encoding for BB84 protocol.

Charlie also discloses the result of the BSM. A successful BSM corresponds to the observation when the SPDs are triggered as follows:

- Two consecutive (both early and late time-bins) clicks on one of SPDs identifies the Bell state $|\psi^{-}\rangle$.
- On the other hand, when one SPD detects the presence of photon in early-time-bin and the second one in the late-time-bin we identify the Bell state $|\psi^{+}\rangle$.
- Other detection patterns are related to $|\phi^{\pm}\rangle$-states and are discarded from consideration.

The rest of the protocol is similar to the polarization state-based MDI-QKD protocol. To implement the time-phase encoder, Alice and Bob can employ either the electrooptical polar modulator, composed of the concatenation of an amplitude modulator and phase modulator, as illustrated in Fig. 6.18; or alternatively, an electrooptical I/Q modulator can be used [46, 96]. The amplitude modulator (Mach-Zehnder modulator) is used for pulse shaping, and the phase modulator is used to introduce the desired phase shift according to the phase-basis states (see Fig. 6.17). AWGs can be used to generate the corresponding RF waveforms as needed. VOAs are used to attenuate the modulated beam down to the single-photon level. For proper operation, AWGs must be synchronized. Further, the

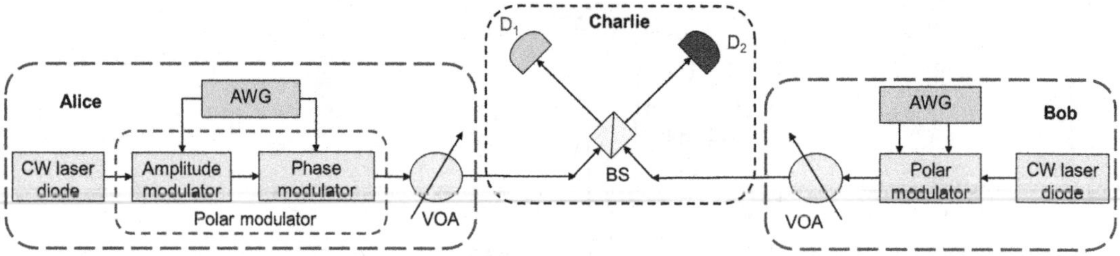

FIGURE 6.18

Time-phase encoding-based measurement-device-independent quantum key distribution system configuration.

continuous-wave (CW) laser diodes used by Alice and Bob should be frequency locked. Finally, incoming photons (from Alice and Bob) to the BSM must have the same polarization.

6.7.4 The secrecy fraction of measurement-device-independent quantum key distribution protocols

The secret fraction of the MDI-QKD protocol, when both Alice and Bob employ single-photon sources, is similar to that of the BB84 protocol:

$$r = q^{(Z)}\left[1 - h_2\left(e^{(X)}\right)\right] - q^{(Z)} f_e h_2\left(e^{(Z)}\right), \tag{6.49}$$

where $q^{(Z)}$ now denotes the probability of successful BSM result ("the gain"). On the other hand, when both Alice and Bob employ WCS sources, the decoy-state method must be employed so that the secret fraction can be estimated by Ref. [89],

$$r = q_{11}^{(Z)}\left[1 - h_2\left(e_{11}^{(X)}\right)\right] - q_{\mu\sigma}^{(Z)} f_e h_2\left(e_{\mu\sigma}^{(Z)}\right), \tag{6.50}$$

where we introduced the following notation:

- $q_{11}^{(Z)}$ denotes the probability that Charlie declares a successful result ("the gain") when both Alice and Bob send a single photon each in the Z-basis,
- $e_{11}^{(X)}$ denotes the phase-error rate of the single-photon signals, both sent in the Z-basis,
- $q_{\mu\sigma}^{(Z)}$ denotes the gain in the Z-basis when Alice sends the WCS of intensity μ to Charlie, while Bob simultaneously sends the WCS of intensity σ.

In the equation above, the information removed from the final key in the privacy amplification step is described by the following term: $q_{11}^{(Z)} h_2\left(e_{11}^{(X)}\right)$. The information revealed by Alice in the information reconciliation's step is described by the following term:

$$q_{\mu\sigma}^{(Z)} f_e h_2\left(e_{\mu\sigma}^{(Z)}\right),$$

with f_e being the error correction inefficiency ($f_e \geq 1$), as before. The error rates $q_{11}^{(Z)}$ and $e_{11}^{(X)}$ cannot be measured directly but instead are bounded by employing the decoy-state approach in a fashion similar to that described in the previous section.

6.8 Twin-field quantum key distribution protocols

The quantum channel can be characterized by the transmissivity T, which can also be interpreted as the probability of a photon being successfully transmitted by the channel. The secret-key capacity of the quantum channel can be used as an upper bound on the maximum possible secret key rate. Very often, the Pirandola–Laurenza–Ottaviani–Banchi (PLOB) bound on a linear key rate is used given by Ref. [97] $r_{\text{PLOB}} = -\log_2(1-T)$. To overcome the rate-distance limit of DV-QKD protocols, two approaches have been pursued until recently: (i) developing quantum relays [98] and (ii) employing trusted relays [99]. Unfortunately, the quantum relays require the use of long-time quantum memories and high-fidelity entanglement distillation, which are still out of reach with current technology. On the other hand, the trusted-relay methodology assumes that the relay between two users can be trusted, and this assumption is difficult to verify in practice. The MDI-QKD approach, described in the previous section, was able to close the detection loopholes and extend the transmission distance; however, its SKR is still bounded by the $O(T)$-dependence of the upper limit.

Recently, twin-field QKD (TF-QKD) has been proposed to overcome the rate-distance limit [100]. The authors in Ref. [100] have shown that the TF-QKD upper limit scales with the square root of transmittance, that is, $r \sim O(\sqrt{T})$. In TF-QKD, Alice and Bob generate a pair of optical fields with random phases and phase-encoding in X and Y bases at remote destinations and send them to untrusted Charlie (or Eve), who then performs single-photon detection. This scheme retains the advantages of the MDI-QKD scheme, such as immunity to detector attacks, SPD multiplexing, and star network architecture while concurrently overcoming the PLOB bound. The unconditional security proof was missing in Ref. [100], and in a series of papers, it was proved that this scheme is unconditionally secure [101−104]. The TF-QKD scheme can be interpreted as the generalization of both (1) the time-phase encoding-based MDI-QKD scheme shown in Fig. 6.18 and (2) phase-encoding-based BB84 (see Section 6.4.4).

The authors have shown in Ref. [104] that the TF-QKD scheme can be interpreted as a particular MDI-QKD scheme implemented in a two-dimensional subspace of vacuum state $|0\rangle$ and one-photon state $|1\rangle$, as illustrated in Fig. 6.19, representing an equivalent model for TF-QKD with single-photon sources and laser diodes. Given that the vacuum state is insensitive to the channel loss, this state can always be detected, namely, "no click" means successful detection of the vacuum state. This improvement in detection probability of the vacuum state practically yields a higher SKR than traditional MDI-QKD and beating-up the linear key rate bound. The computational basis or Z basis is defined by $\{|0\rangle, |1\rangle\}$ (with $|0\rangle$ and $|1\rangle$, representing the vacuum and one-photon state, respectively, as mentioned above). The X-basis (diagonal basis) is defined by $\{|+\rangle=(|0\rangle+|1\rangle)/\sqrt{2}, |-\rangle=(|0\rangle-|1\rangle)/\sqrt{2}\}$, while the Y-basis is defined by $\{|+j\rangle=(|0\rangle+j|1\rangle)/\sqrt{2}, |-j\rangle=(|0\rangle-j|1\rangle)/\sqrt{2}\}$. The Bell states are defined in a manner similar to before; that is, $|\psi^{\pm}\rangle = (|01\rangle \pm |10\rangle)/\sqrt{2}$.

Alice (Bob) with the help of second optical switch randomly selects whether to use Z-basis or X/Y-basis. When Z-basis is selected, Alice (Bob) employs the first optical switch to randomly select either the vacuum state $|0\rangle$ or single-photon state, and selected state is sent over an optical channel (SMF or FSO link) toward Charlie. When X/Y basis branch is selected, CW laser generates the coherent state $|\alpha\rangle$,

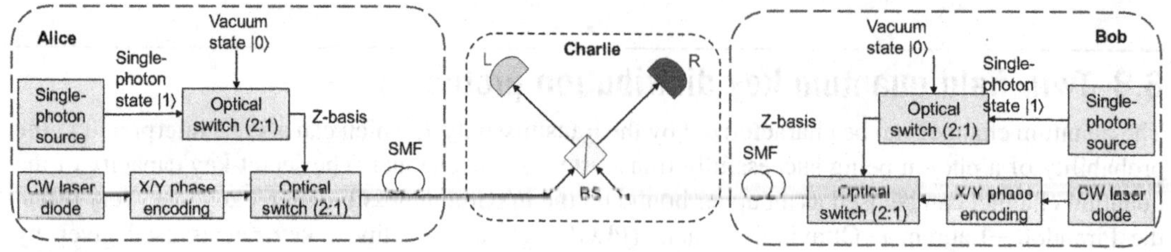

FIGURE 6.19

The conceptual twin-field quantum key distribution system with single-photon and continuous wave laser sources.

which is attenuated by the VOA, so that superposition state of vacuum and single-photon state is obtained, that is $|+\rangle=(|0\rangle+|1\rangle)/\sqrt{2}$. Alice (Bob) with the help of a phase modulator randomly selects one of four phases form the set $\{0,\pi,\pi/2,3\pi/2\}$. On such a way, both states for X-basis and Y-basis are selected at random. (Alternatively, instead of two 2:1 optical switches, one 4:1 optical can be used.)

Charlie performs the BSM measurements. The coincidence detection with SPD R click and no click on SPD L indicates the Bell state $|\psi^-\rangle$. On the other hand, the coincidence detection with SPD L clicks and no clicks on SPD R indicates the Bell state $|\psi^+\rangle$. Alice and Bob then exchange the basis being used through an authenticated public channel. Bob will flip his sequence when the Z-basis is used and Charlie identifies the Bell states $|\psi^\pm\rangle$. Bob will also flip his sequence when either the X- or Y-basis is used and Charlie identifies the Bell state $|\psi^-\rangle$. After the sifting procedure is completed, Alice and Bob will use the sequences corresponding to the Z-basis for the key, while the sequences corresponding to the X-basis/Y-basis will be used for quantum bit error rate estimation. Finally, Alice and Bob will perform classical postprocessing steps, information reconciliation, and privacy amplification to obtain the common secure key.

We now direct our attention to the more realistic, *practical TF-QKD* scheme shown in Fig. 6.20, which does not require the use of single-photon sources. To explain this scheme, we use an interpretation similar to Ref. [104]. Both Alice and Bob employ stabilized CW lasers of low linewidth to generate global phase-stabilized optical pulses with the help of an amplitude modulator. With the help of phase modulators, Alice and Bob choose the random phases $\phi_A\in[0,2\pi]$ and $\phi_B\in[0,2\pi]$, respectively. The random phase differences between Alice and Bob are discretized so that

$$\phi_{A,B}\in\Delta\phi_{k_{A,B}}=\left[\frac{2\pi}{M}k_{A,B},\frac{2\pi}{M}(k_{A,B}+1)\right)\Bigg), \tag{6.51}$$

where the phase-bin $\Delta\phi_{k_{A,B}}$ is discretized by $k_{A,B}\in\{0,1,\ldots,M-1\}$. Alice (Bob) then randomly selects whether to use the Z-basis or X/Y-basis. When the Z-basis is selected, the phase-randomized coherent state is sent with an intensity of either μ or 0, representing logic bits 1 and 0, with probability t and $1-t$, respectively. For the lossy channel, the twin-field (TF) state $|1\rangle_A|1\rangle_B$ can create Bell state detection with errors, so logic bits 0 and 1 are not sent with the same probability but rather with optimized probabilities. When X(Y) basis encoding is selected, Alice and Bob employ corresponding phase and amplitude modulators to randomly select 0 ($\pi/2$) and π ($3\pi/2$), representing logic bits 0 and 1, and such phase-encoded pulses are sent with randomly selected intensities $\{v/2,w/2,0\}$. The corresponding

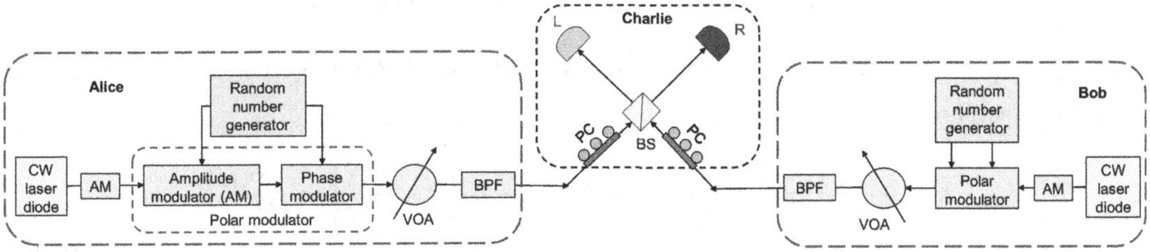

FIGURE 6.20

Practical twin-field quantum key distribution system configuration. *BPF*, optical bandpass filter; *PC*, polarization controller.

quantum states generated by Alice and Bob are sent toward Charlie over the optical (SMF or FSO) link. The polarization controllers ensure that Alice's and Bob's pulses have the same polarization. Charlie performs the BSM and announces the results in the same fashion as described above. Alice and Bob then disclose their phase information—that is, $k_{A,B}$ and intensities—which are used for parameter estimation. They keep the information related to Z-basis confidential to Charlie, and these data are used for the raw key. Alice and Bob then perform classical postprocessing steps, information reconciliation and privacy amplification, to obtain the common secure key.

For the TF-based BB84 protocol, Alice and Bob retain raw data for phase-encoding only if $|k_A - k_B| = 0$ or $M/2$, when both used the X-basis, and other raw data results are discarded. For $|k_A - k_B| = 0$ Bob flips his key bit when Charlie detects the Bell state $|\psi^-\rangle$. Similarly, for $|k_A - k_B| = M/2$, Bob flips his key bit when Charlie detects the Bell state $|\psi^+\rangle$. The secret fraction for TF-based BB84 protocol can be estimated by Ref. [104]:

$$r_{\text{TF-BB84}} = 2t(1-t)\mu e^{-\mu} y_{1ZZ}^{(TF)} \left[1 - h_2\left(e_{XX}^{(b_1)}\right)\right] - Q_{ZZ} f\, h_2(E_{ZZ}), \tag{6.52}$$

where Q_{ZZ} is the gain when both Alice and Bob used Z-basis and can be determined experimentally. The parameter f denotes error reconciliation inefficiency ($f \geq 1$). [$h_2(\cdot)$ denotes the binary entropy function, defined above.]

The gain Q_{ZZ} and the QBER for Z-basis can be determined by Ref. [104]:

$$\begin{aligned} Q_{ZZ} &= 2p_d(1-p_d)(1-t)^2 + 4(1-p_d)e^{-\frac{\mu}{2}T_{tot}}\left[1-(1-p_d)e^{-\frac{\mu}{2}T_{tot}}\right]t(1-t) \\ &\quad + 2(1-p_d)e^{-\mu T_{tot}}\left[I_0(\mu T_{tot}) - (1-p_d)e^{-\mu T_{tot}}\right]t^2, \\ E_{ZZ} &= \frac{2p_d(1-p_d)(1-t)^2 + 2(1-p_d)e^{-\mu T_{tot}}\left[I_0(\mu T_{tot}) - (1-p_d)e^{-\mu T_{tot}}\right]t^2}{Q_{ZZ}}, \end{aligned} \tag{6.53}$$

where $T_{\text{tot}} = T\eta_d$, with T being transmissivity and η_d being detector efficiency. Parameter p_d denotes the dark current rate of the threshold detector. Based on Eqs. (6.34)–(6.35), by setting $\mu = v/2$, $\sigma_1 = w/2$, and $\sigma_2 = 0$; the yield $y_{1ZZ}^{(TF)}$ becomes lower bounded by Ref. [2]:

$$y_{1ZZ}^{(TF)} \geq \frac{2v}{vw - w^2}\left[q_{w/2}e^{w/2} - \frac{w^2}{v^2}q_{v/2}e^{v/2} - q_0\left(1 - \frac{w^2}{v^2}\right)\right], \tag{6.54}$$

where $q_0 = 2p_d(1-p_d)$. By employing the decoy-state method [37–69, 103, 104], the yield $y_{1XX}^{(TF)}$ and QBER $e_{XX}^{(b_1)}$ are bounded as follows:

$$\begin{aligned} y_{1XX}^{(TF)} &\geq \frac{v}{vw - w^2}\left[q_w^{(XX)}e^w - \frac{w^2}{v^2}q_v^{(XX)}e^v - q_0\left(1 - \frac{w^2}{v^2}\right)\right] = y_{1XX}^{(TF,LB)}, \\ e_{XX}^{(b_1)} &\leq \frac{q_w^{(XX)}e^w Q_w^{(XX)} - q_0 e^{b_0}}{w y_{1XX}^{(TF,LB)}}, \end{aligned} \tag{6.55}$$

where for *TF* vacuum state, $e^{b_0} = 1/2$, and $Q_w^{(XX)}$ should be determined experimentally.

As an illustration, in Fig. 6.21 we provide comparisons of phase-matching (PM) TF-QKD protocol introduced in Ref. [101] with the MDI-QKD protocol [83] and decoy-state-based BB84 protocol [37]. The system parameters are selected as follows: the detector efficiency $\eta_d = 0.25$, reconciliation inefficiency $f_e = 1.15$, the dark count rate $pd = 8 \times 10^{-8}$, misalignment error $e_d = 1.5\%$, and the number of phase slices for PM TF-QKD is set to $M = 16$. Regarding the transmission medium, it is assumed that recently reported ultra-low-loss fiber of attenuation 0.1419 dB/km (at 1560 nm) is used [105]. In the same figure, the PLOB bound on a linear key rate is provided as well. Clearly, the PM TF-QKD scheme outperforms the decoy-state BB84 protocol for distances larger than 162 km. It outperforms MDI-QKD protocol for all distances, and exceeds the PLOB bound at distance 322 km. Finally, the PM TF-QKD protocol can achieve the distance of even 623 km. Unfortunately, the normalized SKR value is rather low at this distance. For calculations, the expressions provided in Appendix B of ref. [101] are used, with few obvious typos being corrected.

6.9 Information reconciliation and privacy amplification

As discussed earlier, the raw key is imperfect, and we must perform *information reconciliation* and *privacy amplification* to increase the correlation between Alice's and Bob's sequences, while at the same time reducing eavesdropper Eve's mutual information about the result, to a desired level of security.

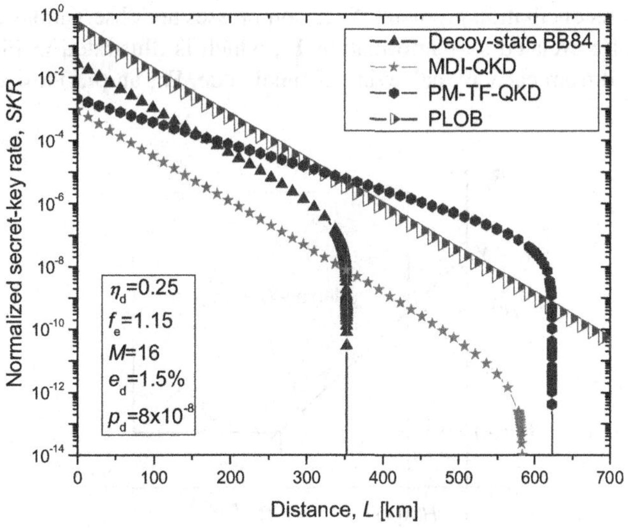

FIGURE 6.21

The prepare-and-measure twin-field quantum key distribution (QKD) protocol against measurement-device-independent QKD and decoy-state BB84 protocols in normalized secret-key rate versus transmission distance.

6.9.1 Information reconciliation

EA-QKD has certain similarities to the source-type secret-key generation (agreement) protocol, relevant in physical-layer security [2]. In this model, Alice, Bob, and Eve have access to the respective output ports X, Y, Z of the source characterized by the joint PDF f_{XYZ}. Each of them sees the i.i.d. realizations of the source denoted respectively as X^n, Y^n, and Z^n. Given that Alice's and Bob's realizations are not perfectly correlated they need to correct the discrepancies' errors. This step in secret-key generation and QKD is commonly referred to as the *information reconciliation* (or just reconciliation). The reconciliation process can be considered as a special case of the *source coding with the side information* [11,106]. To encode the source X we know from Shannon's source coding theorem that the rate R_x must be as small as possible but not lower than the entropy of the source, in other words $R_x > H(X)$. To jointly encode the source (XY, f_{XY}) we need the rate $R > H(X,Y)$. If we separately encode the sources, the required rate would be $R > H(X)+H(Y)$. However, Slepian and Wolf have shown in Ref. [106] that the rate $R = H(X,Y)$ is still sufficient even for separate encoding of correlated sources. This claim is commonly referred to as the *Slepian-Wolf theorem*, and can be formulated as follows [11,106]: The *achievable rate region R* (for which the error probability tends to zero) for the separate encoding of the correlated source (XY, f_{XY}) is given by

$$R \doteq \left\{ (R_x, R_y): \begin{array}{c} R_x \geq H(X|Y) \\ R_y \geq H(Y|X) \\ R_x + R_y \geq H(X,Y) \end{array} \right\} \tag{6.56}$$

The typical shape of the Slepian-Wolf region is provided in Fig. 6.22. Now if we assume that (X, f_X) should be compressed when (Y, f_Y) is available as the side information, then $H(X|Y)$ is sufficient to describe X.

Going back to our reconciliation problem, Alice compresses her observations X^n and Bob decodes them with the help of correlated side information Y^n, which is illustrated in Fig. 6.23. The source encoder can be derived from capacity achieving channel code. For an (n,k) low-density parity-check

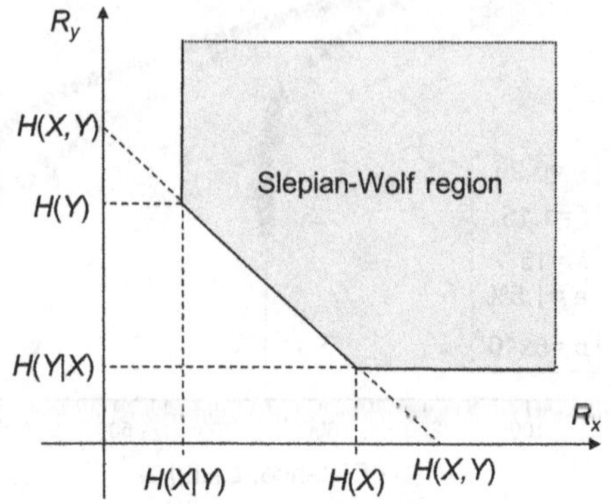

FIGURE 6.22

Typical Slepian–Wolf achievable rate region for a correlated source (XY, f_{XY}).

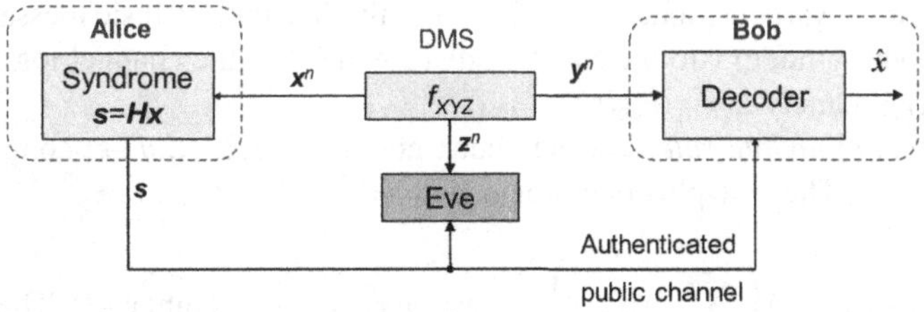

FIGURE 6.23

Reconciliation problem represented as source coding with syndrome-based side information.

(LDPC) code, a parity-check matrix $\boldsymbol{H}$ of size $(n-k)\times n$ can be used to generate the syndrome by $\boldsymbol{s}=\boldsymbol{Hx}$, where $\boldsymbol{x}$ is Alice's DMS observation vector of length n. Since $\boldsymbol{x}$ is not a codeword, the syndrome vector is different from all-zeros vector. The syndrome vector is of length $n-k$ and can be represented as $\boldsymbol{s}=[s_1\ s_2\ \ldots\ s_{n-k}]^{\mathrm{T}}$. Clearly, the syndrome vector is transmitted toward Bob over the (error-free) authenticated public channel.

Once Bob receives the syndrome vector $\boldsymbol{s}$, his decoder tries to determine the vector $\boldsymbol{x}$ based on $\boldsymbol{s}$ and Bob's observation vector $\boldsymbol{y}$, which maximizes the posterior probability:

$$\widehat{\boldsymbol{x}}=\max P(\boldsymbol{y}|\boldsymbol{x},\boldsymbol{s}), \tag{6.57}$$

and clearly, the decoding problem is similar to the maximum a posteriori probability (MAP) decoding. The compression rate is $(n-k)/n=1-k/n$, while the linear code rate is k/n. The number of syndrome bits required is

$$n-k\geq H(\boldsymbol{X}^n|\boldsymbol{Y}^n)=nH(X|Y). \tag{6.58}$$

In practice, the practical reconciliation algorithms introduce the overhead $OH>0$, so that the number of transmitted bits over the public channel is $nH(X|Y)\,(1+OH)$. At the end of the reconciliation step, Alice and Bob share the common sequence $\boldsymbol{X}^n$ with entropy $nH(X)$ with high probability, so the *reconciliation efficiency* β can be written as

$$\beta=\frac{nH(X)-nH(X|Y)(1+OH)}{nI(X,Y)}=1-OH\frac{H(X|Y)}{I(X,Y)}\leq 1. \tag{6.59}$$

MAP decoding complexity is prohibitively high when LDPC coding is used, instead the low-complexity sum-product algorithm (SPA) (also known as belief-propagation algorithm) should be used as described in Ref. [107]. Given that Alice observation $\boldsymbol{x}$ is not a codeword, the syndrome bits are nonzero. To account for this problem, we need to change the sign of log-likelihood to be sent from the c-th check-node to the v-th variable node when the corresponding syndrome bit s_c is 1. Based on ref. [108], we can summarize the *log-domain SPA* below. To improve the clarity of presentation we use the index $v=1,\ldots,n$ to denote variable nodes and index $c=1,\ldots,n-k$ to denote the check nodes. Additionally, we use the following notation: (i) $N(c)$ to denote $\{v$-nodes connected to c-node $c\}$, (ii) $N(c)\backslash\{v\}$ denoting $\{v$-nodes connected to c-node c except v-node $v\}$, (iii) $N(v)$ to denote $\{c$-nodes connected to v-node $v\}$, and (iv) $N(v)\backslash\{c\}$ representing $\{c$-nodes connected to v-node v except c-node $c\}$.

The formulation of the *log-domain SPA for source coding with the side information* is now provided:

(1) *Initialization*: For the variable nodes $v = 1,2, \ldots,n$; initialize the extrinsic messages L_{vc} to be sent from variable node (v-node) v to the check node (c-node) c to the channel log-likelihood ratios, denoted as $L_{\text{ch}}(v)$, namely $L_{vc} = L_{\text{ch}}(v)$.

(2) *Check-node (c-node) update rule*: For the check nodes $c = 1,2, \ldots,n-k$; compute $L_{cv} = (1-2s_c) \underset{N(c)/\{v\}}{\boxplus} L_{vc}$. The box-plus operator is defined by $L_1 \boxplus L_2 = \prod_{k=1}^{2} \text{sign}(L_k)\cdot\phi\left(\sum_{k=1}^{2}\phi(|L_k|)\right)$, where $\phi(x) = -\log\tanh(x/2)$. The box operator for $|N(c)\backslash\{v\}|$ components is obtained by recursively applying the two-component version defined above.

(3) *Variable node (v-node) update rule*: For $v = 1,2, \ldots,n$; set $L_{vc} = L_{\text{ch}}(v) + \sum_{N(v)\backslash\{c\}} L_{cv}$ for all c-nodes for which the c-th row/v-the column element of the parity-check matrix $\boldsymbol{H}$ is one, that is $h_{cv} = 1$.

(4) *Bit decisions*: Update L(v) ($v = 1, \ldots,n$) by $L(v) = L_{\text{ch}}(v) + \sum_{N(v)} L_{cv}$ and make decisions as follows: $\hat{x}_v = 1$ when $L(v) < 0$ (otherwise set $\hat{x}_v = 0$).

So the only difference with respect to conventional log-domain SPA is the introduction of the multiplication factor $(1-2s_c)$ in the c-node update rule. Moreover, the channel likelihoods $L_{\text{ch}}(v)$ should be calculated from the joint distribution f_{XY} by $L_{\text{ch}}(v) = \log[f_{XY}(x_v = 0|y_v)/f_{XY}(x_v = 1|y_v)]$ Because the c-node update rule involves log and tanh functions, it can be computationally intensive, and to lower the complexity the reduced complexity approximations should be used.

6.9.2 Privacy amplification

The purpose of the privacy amplification is to extract the secret key from the reconciled (corrected) sequence. The key idea is to apply a well-chosen compression function, Alice and Bob have agreed on, to the reconciled binary sequence of length n as follows g: $(0,1)^n \rightarrow (0,1)^k$, where $k < n$, so that Eve gets negligible information about the secret key [8,23,109–111]. Typically, g is selected at random from the set of compression functions G so that Eve does not know which g has been used. In practice, the set of functions G is based on the family of universal hash functions, introduced by Carter and Wegman [112, 113]. We say that the family of functions G: $(0,1)^n \rightarrow (0,1)^k$ is a *universal$_2$ family of hash functions* if for any two distinct $\boldsymbol{x}_1$ and $\boldsymbol{x}_2$ from $(0,1)^n$, the probability that $g(\boldsymbol{x}_1) = g(\boldsymbol{x}_2)$ when g is chosen uniformly at random from G is smaller than 2^{-k}. As an illustrative example, let us consider the multiplication in GF(2^n) [23]. Let a and x be two elements from GF(2^n). Let the g function be defined as the mapping $(0,1)^n \rightarrow (0,1)^k$ that assigns to argument x the first k bits of the product $ax \in$ GF(2^n). It can be shown that the set of such functions is a universal class of functions for $k \in [1,n]$.

The analysis of privacy amplification does not rely on Shannon entropy, but Rényi entropy [114 116]. For discrete random variable X, taking the values from the sample space $\{x_1, \ldots,x_n\}$ and distributed according to $P_X = \{p_1, \ldots,p_n\}$, the *Rényi entropy of X of order α* is defined by Ref. [115]:

$$R_\alpha(X) = \frac{1}{1-\alpha}\log_2\left\{\sum\nolimits_x [P_X(x)]^\alpha\right\}. \tag{6.60}$$

The Réniy entropy of order two is also known as the *collision entropy.* From Jensen's inequality [11], we conclude that the collision entropy is upper limited by the Shannon entropy, that is, $R_2(X) \leq H(X)$. In the limit when $\alpha \to \infty$, the Réniy entropy becomes the *min-entropy*:

$$\lim_{\alpha \to \infty} R_\alpha(X) = -\log_2 \max_{x \in X} P_X(x) = H_\infty(X). \tag{6.61}$$

Finally, by applying the l'Hôpital's rule as $\alpha \to 1$, the Réniy entropy becomes the Shannon entropy:

$$\lim_{\alpha \to 1} R_\alpha(X) = -\sum\nolimits_x P_X(x) \log_2 P_X(x) = H(X). \tag{6.62}$$

The properties of the Réniy entropy are summarized in Ref. [115]. Regarding conditional Réniy entropy, three different definitions are discussed in detail in Ref. [116]. We adopt the first of these. Let the joint distribution for (X, Y) be denoted as $P_{X,Y}$ and the distribution for X be denoted as P_X. Further, let the Rényi entropy of the conditional random variables $X|Y=y$ distributed according to $P_{X|Y=y}$ be denoted as $R_a(X|Y=y)$. Then the *conditional Rényi entropy* can be defined by

$$R_\alpha(X|Y) = \sum\nolimits_y P_Y(y) R_\alpha(X|Y=y) = \frac{1}{1-\alpha} \sum\nolimits_y P_Y(y) \log_2 \sum\nolimits_x P_{X|Y}(x|y)^\alpha. \tag{6.63}$$

The mutual Rényi information can then be defined as

$$I_\alpha(X, Y) = R_\alpha(X) - R_\alpha(X|Y), \tag{6.64}$$

and is not symmetric to $R_\alpha(Y) - R_\alpha(Y|X)$. Moreover, it can even be negative, as discussed in Refs. [115, 116].

Bennett et al. have proved the following *privacy amplification theorem* in Ref. [23] (see theorem 3): Let X be a random variable with distribution P_X and Rényi entropy of order two $R_2(X)$. Further, let G be the random variable representing uniform selection at random of a member of universal hash functions $(0,1)^n \to (0,1)^k$. Then the following inequalities are satisfied [23]:

$$\begin{aligned} H(G(X)|G) &\geq R_2(G(X)|G) \geq k - \log_2\left[1 + 2^{k-R_2(X)}\right] \\ &\geq k - \frac{2^{k-R_2(X)}}{\ln 2}. \end{aligned} \tag{6.65}$$

Bennett et al. have also proved the following *corollary of the privacy amplification theorem* [23] (see also [110]): Let $S \in \{0,1\}^n$ be a random variable representing the sequence shared between Alice and Bob, and let E be a random variable representing the knowledge Eve was able to get about the S, with a particular realization of E denoted by e. If the conditional Rényi entropy of order two $R_2(S|E=e)$ is known to be at least r_2, and Alice and Bob choose the secret key by $K = G(S)$, where G is a hash function chosen uniformly at random from the universal$_2$ family of hash functions $(0,1)^n \to (0,1)^k$, then the following is valid:

$$H(K|G, E=e) \geq k - \frac{2^{k-r_2}}{\ln 2}. \tag{6.66}$$

Cachin has proved the following Rényi entropy *lemma* in Ref. [117] (see also [118]): Let X and Q be two random variables and $s > 0$. Then with a probability of at least $1-2^{-s}$, the following is valid [117, 118]:

$$R_2(X) - R_2(X|Q=q) \leq \log|Q| + 2s + 2 \tag{6.67}$$

The total information available to Eve is composed of her observations $\mathbf{Z}^n$ from the source of common randomness and additional bits exchanged in the information reconciliation phase over the authenticated public channel, represented by random variable Q. Based on Cachin's lemma, we can write

$$R_2(S|\mathbf{Z}^n = z^n) - R_2(S|\mathbf{Z}^n = z^n, Q=q) \leq \log|Q| + 2s + 2 \text{ with probability } 1 - 2^{-s} \tag{6.68}$$

We know that $R_2(X) \leq H(X)$, so we conclude that $R_2(S|\mathbf{Z}^n = z^n) \leq H(S|\mathbf{Z}^n = z^n)$, and therefore, we can upper bound $R_2(S|\mathbf{Z}^n = z^n)$ as follows:

$$R_2(S|\mathbf{Z}^n = z^n) \leq nH(X|Z). \tag{6.69}$$

Based on (6.68), we can lower bound $R_2(S|\mathbf{Z}^n = z^n, Q = q)$ by

$$R_2(S|\mathbf{Z}^n = z^n, Q=q) \geq nH(X|Z) - \log|Q| - 2s - 2 \text{ with probability } 1 - 2^{-s} \tag{6.70}$$

Since the number of bits exchanged over the public channel $\log|Q|$ is approximately $nH(X|Y)(1+OH)$, for sufficiently large n, the previous inequality becomes

$$R_2(S|\mathbf{Z}^n = z^n, Q = q) \geq \underbrace{nH(X|Z) - nH(X|Y)(1+OH) - 2s - 2}_{r_2} \text{ with probability } 1 - 2^{-s} \tag{6.71}$$

From Eq. (6.66), we conclude that $k - r_2 = -k_2 < 0$, which guarantees that for $k_2 > 0$, Eve's uncertainty of the key is lower bounded by

$$H(K|E) \geq k - 2^{-k_2}/\ln 2 \text{ with probability } 1 - 2^{-s} \tag{6.72}$$

6.10 Continuous variable quantum key distribution

The readers not familiar with quantum optics should first read Sections 5.4 and 5.8 in Chapter 5. The CV-QKD can be implemented by employing either homodyne detection, where only one quadrature component is measured at a time (because of the uncertainty principle) as illustrated in Fig. 6.24, or with heterodyne detection (HD), where one BS and two balanced photodetectors are used to measure both quadrature components simultaneously as illustrated in Fig. 6.25. HD can double the mutual

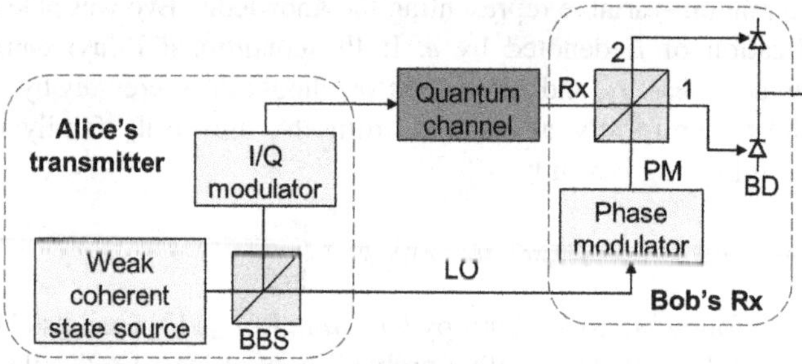

FIGURE 6.24

Gaussian modulation-based continuous-variable quantum key distribution protocols with homodyne detection. *BBS*, balanced beam splitter; *BD*, balanced detector.

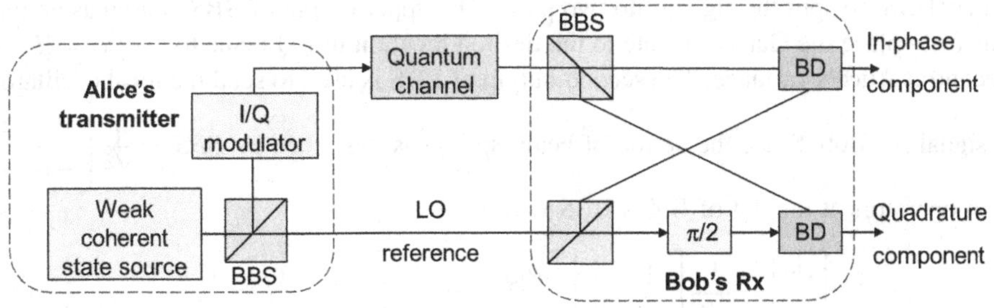

FIGURE 6.25

Gaussian modulation-based continuous-variable quantum key distribution protocols with heterodyne detection. *BBS*, balanced beam splitter; *BD*, balanced detector (shown in Fig. 6.24).

information between Alice and Bob compared with the homodyne detection scheme at the expense of an additional 3 dB loss of the BS. To reduce phase noise, quantum signals are typically copropagated with the time-domain multiplexed high-power pilot-tone to align Alice's and Bob's measurement bases. CV-QKD studies are gaining momentum, judging by the increasing number of papers related to CV-QKD [8,119–145], thanks to their compatibility with start-of-the-art optical communication technologies. The CV-QKD system experiences a 3 dB loss limitation in transmittance when direct reconciliation is used. To avoid this problem, reverse reconciliation [146] or postselection [127] methods are used. Various eavesdropping strategies discussed earlier are applicable to CV-QKD systems as well. It has been shown that for Gaussian modulation, Gaussian attack is the optimum attack for both individual attacks [133] and collective attacks [134, 135]. In this section of the chapter, we focus our attention to the CV-QKD with Gaussian modulation. To implement CV-QKD both squeezed states and coherent states can be employed.

6.10.1 Homodyne and heterodyne detection schemes

In *homodyne detection* scheme, provided in Fig. 6.24, which represents the most common measurement in CV systems, we measure one of the two quadratures since we cannot measure both quadratures simultaneously, according to the uncertainty principle. The measurement operators are projectors over the quadrature basis $|I\rangle\langle I|$ or $|Q\rangle\langle Q|$, and corresponding outcomes have PDF $f(I)$ or $f(Q)$, obtained by the marginal integration the *Wigner function*, introduced Eq. (5.144), which for the single mode becomes

$$f(I) = \int_Q W(I,Q)dQ, \; f(Q) = \int_I W(I,Q)dI.$$

When more than one bosonic modes are in the optical beam, we must apply partial homodyning and also perform integration over both quadratures of all other modes.

In Gaussian modulation-based CV-QKD with homodyne detection [2,8], shown in Fig. 6.24, only one quadrature component is measured at a time. This protocol is commonly referred to as the *GG02 CV-QKD protocol*, named after the authors Grosshans and Grangier who proposed it in 2002 [146]. Alice sufficiently attenuates the laser beam, which serves as a WCS source and employs the balanced beam splitter (BBS) to split the signal into two parts. The upper output of BBS is used as an input the I/Q modulator to place the Gaussian state to the desired location in I-Q plane to get $|\alpha\rangle = |I + jQ\rangle$ as well as to control Alice's variance. The second output of BBS is used to send the local oscillator (LO) reference signal for Bob. Since the action of beam splitter is described by $BBS = \frac{1}{\sqrt{2}}\begin{bmatrix} 1 & 1 \\ -1 & 1 \end{bmatrix}$, the annihilation operators at the out of Bob's BBS will be

$$\begin{bmatrix} a_1 \\ a_2 \end{bmatrix} = \frac{1}{\sqrt{2}}\begin{bmatrix} 1 & 1 \\ -1 & 1 \end{bmatrix}\begin{bmatrix} a_{Rx} \\ a_{PM} \end{bmatrix} = \frac{1}{\sqrt{2}}\begin{bmatrix} a_{Rx} + a_{PM} \\ -a_{Rx} + a_{PM} \end{bmatrix}, \tag{6.73}$$

where the creation (annihilation) operators at receive BBS inputs Rx and PM are denoted by $a_{Rx}^{\dagger}(a_{Rx})$ and $a_{PM}^{\dagger}(a_{PM})$, respectively. The number operators at receive BBS outputs one and two can now be expressed as follows:

$$\hat{n}_1 = a_1^{\dagger}a_1 = \frac{1}{2}\left(a_{PM}^{\dagger} + a_{Rx}^{\dagger}\right)(a_{PM} + a_{Rx}),\ \hat{n}_2 = a_2^{\dagger}a_2 = \frac{1}{2}\left(a_{PM}^{\dagger} - a_{Rx}^{\dagger}\right)(a_{PM} - a_{Rx}), \tag{6.74}$$

The output of a balanced detector (BD), assuming the photodiode responsivity of $R_{\text{ph}} = 1\text{A/W}$, is given by (in SNU)

$$\hat{i}_{BD} = 4(\hat{n}_1 - \hat{n}_2) = 4\left(a_{PM}^{\dagger}a_{Rx} + a_{Rx}^{\dagger}a_{PM}\right). \tag{6.75}$$

The phase modulation output can be represented in SNU by $a_{PM} = |\alpha|\exp(j\theta)\mathbf{1}/2$, and after substitution into (6.75) we obtain

$$\hat{i}_{BD} = 4\left(\frac{|\alpha|e^{-j\theta}}{2}a_{Rx} + a_{Rx}^{\dagger}\frac{|\alpha|e^{j\theta}}{2}\right) = 2|\alpha|\left(e^{j\theta}a_{Rx}^{\dagger} + e^{-j\theta}a_{Rx}\right). \tag{6.76}$$

In the absence of Eve, the receive signal constellation will be $\boldsymbol{a}_{Rx} = (I,Q)/2$, which after substitution into the previous equation, we obtain

$$\begin{aligned} \hat{i}_{BD} &= |\alpha|\left[(\cos\theta, \sin\theta)\mathbf{1}\cdot\left(I\hat{I}, -Q\hat{Q}\right) + (\cos\theta, \sin\theta)\mathbf{1}\cdot\left(I\hat{I}, Q\hat{Q}\right)\right] \\ &= |\alpha|\left(\cos\theta\cdot I\hat{I} + \sin\theta\cdot Q\hat{Q}\right). \end{aligned} \tag{6.77}$$

Bob measures in-phase component by selecting $\theta = 0$ to get $|\alpha|I$ and the quadrature component by setting $\theta = \pi/2$ to get $|\alpha|Q$.

The theory of optical HD has been developed in Ref. [147]. The HD corresponds to the projection onto coherent states, that is, $M(\alpha) \sim |\alpha\rangle\langle\alpha|$. This theory has been generalized to the to POVM based on projections over pure Gaussian states [148, 149]. For the HD-based Gaussian modulation CV-QKD, the transmitter side is identical to coherent detection version. In the HD receiver shown in Fig. 6.25, the quantum signal is split using the BBS, one output is used to measure the in-phase component (upper branch), while the other output (lower branch) is used as input to the second BD, to detect the quadrature component. The LO classical reference signal is also split by BBS, with upper output being used as input to the in-phase BD, and the other output, after the phase modulation to introduce the $\pi/2$-phase shift, is used as the LO input to the quadrature BD, to detect the quadrature component. On such a way both quadratures can be measured simultaneously, and mutual information can be doubled. On the other hand, the BBS introduced 3 dB attenuation of the quantum signal.

6.10.2 Squeezed state-based protocols

Squeezed state-based protocols employ the Heisenberg uncertainty principle, claiming that it is impossible to measure both quadratures with complete precision. To impose the information, Alice randomly selects to use either in-phase or quadrature degree-of-freedom (DOF), as illustrated in Fig. 6.26A. In Alice encoding rule I (when in-phase DOF is used), the squeezed state is imposed on the in-phase component (with squeezed parameter $s_I < 1$) by (in SNU)

$$X_{A,I} \to |X_{A,I}, s_I\rangle, s_I < 1, X_{A,I} \sim N\left(0, \sqrt{v_{A,I}}\right) \tag{6.78}$$

The amplitude is selected from a zero-mean Gaussian source of variance $v_{A,I}$. On the other hand, in Alice encoding rule Q (when the quadrature is used), the squeezed state is imposed on the quadrature (with squeezed parameter $s_Q > 1$) and we can write (in SNU)

$$X_{A,Q} \to |jX_{A,Q}, s_Q\rangle, s_Q > 1, X_{A,Q} \sim N\left(0, \sqrt{v_{A,Q}}\right) \tag{6.79}$$

The variances of squeezed states I and Q will be $V(\hat{I}) = \sigma_I^2 = s_I$ and $V(\hat{Q}) = \sigma_Q^2 = 1/s_Q$, respectively (as described above). On the receiver side Bob randomly selects whether to measure either in-phase or quadrature component. Alice and Bob exchange the encoding rules being used by them to measure the quadrature for every squeezed state and keep only instances when they measured the same quadrature in the sifting procedure. Therefore, this protocol is very similar to the BB84 protocol. After that the information reconciliation takes place, followed by the privacy amplification.

From Eve's point of view, the corresponding mixed states when Alice employed either I or Q DOF will be [8]

$$\begin{aligned} \rho_I &= \int f_N\left(0, \sqrt{v_{A,I}}\right)|x, s_I\rangle\langle x, s_I|dx, \\ \rho_Q &= \int f_N\left(0, \sqrt{v_{A,Q}}\right)|jQ, s_Q\rangle\langle jQ, s_Q|dx \end{aligned} \tag{6.80}$$

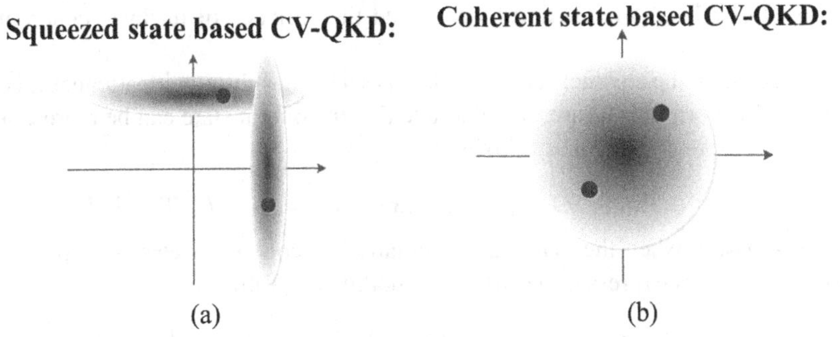

FIGURE 6.26

Illustration of operating principles of squeezed (A) and coherent (B) state-based protocols.

where $f_N(0,\sigma)$ is zero-mean Gaussian distribution with variance σ^2. When the condition $\rho_I = \rho_Q$ is satisfied, Eve cannot distinguish whether Alice was imposing the information on the squeezed state I or Q. Given that squeezed states are much more difficult to generate than the coherent states, we focus our attention to the coherent state-based CV-QKD protocols.

6.10.3 Coherent state-based continuous-variable quantum key distribution protocols

In coherent state-based protocols, with operation principle provided in Fig. 6.26B, there is only one encoding rule for Alice:

$$(X_{A,I}, X_{A,Q}) \to |X_{A,I} + jX_{A,Q}\rangle; X_{A,I} \sim N(0, \sqrt{v_A}), X_{A,Q} \sim N(0, \sqrt{v_A}). \tag{6.81}$$

In other words, Alice randomly selects a point in 2-D (I,Q) space from a zero-mean circular symmetric Gaussian distribution, with the help of apparatus provided in Fig. 6.25. Clearly, both quadratures have the same uncertainty. Here we again employ the Heisenberg uncertainty, claiming that it is impossible to measure both quadratures with complete precision. For homodyne receiver, Bob performs the random measurement on either $\hat{I}$ or $\hat{Q}$, and let the result of his measurement be denoted by Y_B. When Bob measures $\hat{I}$, his measurement Y_B is correlated $X_{A,I}$. On the other hand, when Bob measures $\hat{Q}$, his measurement result Y_B is correlated $X_{A,Q}$. Clearly, Bob is able to measure a single coordinate of the signal constellation point sent by Alice using Gaussian coherent state. Bob then announces which quadrature he measured in each signaling interval, and Alice selects the coordinate that agrees with Bob's measurement quadrature. The rest of the protocol is the same as for the squeezed state-based protocol.

From Eve's point of view, the mixed state will be

$$\rho = \int f_N(0, \sqrt{v_A}) f_N(0, \sqrt{v_A}) |I + jQ\rangle\langle I + jQ| dIdQ, \tag{6.82}$$

v_A is the Alice modulation variance in SNU. For lossless transmission, the total variance of Bob's random variable in SNU will be

$$V(Y_B) = \text{Var}(Y_B) = v_A + 1, \tag{6.83}$$

which is composed of Alice's modulation term v_A and the variance of intrinsic coherent state fluctuations is 1.

For lossy transmission, the channel can be modeled as a beam splitter with attenuation $0 \leq T \leq 1$, as illustrated in Fig. 6.27. Form this figure, we conclude that the output state can be represented in terms of input state $\boldsymbol{x}_A$ and ground state $\boldsymbol{x}_0$ as follows:

$$\boldsymbol{x}_B = \sqrt{T}\boldsymbol{x}_A + \sqrt{1-T}\boldsymbol{x}_0 = \sqrt{T}(\boldsymbol{x}_A + \sqrt{\chi_{\text{line}}}\boldsymbol{x}_0), \chi_{\text{line}} = (1-T)/T = 1/T - 1 \tag{6.84}$$

For lossy transmission when the Gaussian modulation of coherent states is employed instead of vacuum states, the previous expression needs to be modified as follows:

$$\boldsymbol{x}_B = \sqrt{T}(\boldsymbol{x}_A + \sqrt{\chi_{\text{line}}}\boldsymbol{x}_0), \chi_{\text{line}} = (1-T)/T + \varepsilon = 1/T - 1 + \varepsilon, \varepsilon \geq 0, \tag{6.85}$$

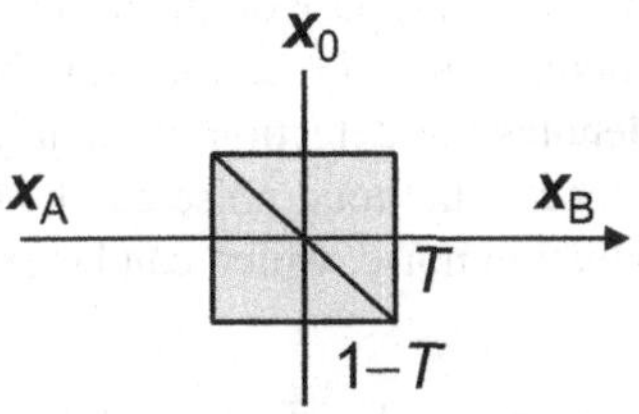

FIGURE 6.27

Illustration of lossy transmission channel.

The first term in χ_{line} is contributed to the channel attenuation, while the second term ε, commonly referred to as the *excess noise*, is due to other noisy factors such as the laser phase noise, imperfect orthogonality of quadratures, and so on. The total variance of Bob's random variable will now comprise three terms:

$$V(Y_B) = \text{Var}(Y_B) = T\left(\underbrace{v_A + 1}_{v} + \chi_{\text{line}}\right) = T(v + \chi_{\text{line}}), \tag{6.86}$$

where $v = v_A + 1$.

Mutual information between Alice and Bob, based on Shannon's formula, can be calculated by

$$I(A;B) \doteq I(X_A;Y_B) = \frac{1}{2}\log\left(1 + \frac{Tv_A}{T(1+\chi_{\text{line}})}\right)$$

$$= \frac{1}{2}\log\left(1 + \frac{v_A}{1+\chi_{\text{line}}}\right). \tag{6.87}$$

Eve's information in CV-QKD schemes is determined by reconciliation strategy.

Assuming that all source to excess-noise are statistically independent and additive, the total variance (in SNU) can be represented as a summation of individual contributions:

$$\varepsilon = \varepsilon_{\text{modulator}} + \varepsilon_{PN} + \varepsilon_{RIN} + \cdots \tag{6.88}$$

where the first term denotes the modulation imperfections, the second term corresponds the phase noise, while the third term corresponds to the relative intensity noise of the LO reference signal.

Referring now to Bob's input (or equivalently to the *channel output*), we can represent Bob's equivalent detector noise variance by

$$\chi_{\text{det}} = \frac{d(1 + v_{\text{el}}) - \eta}{\eta}, \tag{6.89}$$

where $d = 1$ for homodyne detection (either in-phase or quadrature can be measured at given signaling interval in homodyne detection scheme), and $d = 2$ for HD (both quadratures can be measured simultaneously). The parameter η denotes the detection efficiency, while v_{el} denotes the variance of equivalent electrical noise. The variance of the total noise can be determined by the summing up the variance of the channel noise and detection noise, which can be expressed by referring to the *channel input* as follows:

$$\chi_{\text{total}} = \chi_{\text{line}} + \frac{\chi_{\text{det}}}{T} = \underbrace{\frac{1-T}{T} + \varepsilon}_{\chi_{\text{line}}} + \frac{d(1+v_{el}) - \eta}{T\eta}. \tag{6.90}$$

When referring to the channel output, variance of the total noise at Bob's side can be determined by

$$\begin{aligned}\chi_{\text{total, Bob}} &= T\eta\left(\chi_{\text{line}} + \frac{\chi_{\text{det}}}{T}\right) = T\eta\frac{1-T}{T} + T\eta\varepsilon + T\eta\frac{d(1+v_{el}) - \eta}{T\eta} \\ &= \eta(1-T) + T\eta\varepsilon + d(1+v_{el}) - \eta.\end{aligned} \tag{6.91}$$

For the bipartite system composed of two separable subsystems A and B each described by corresponding density matrices ρ_A and ρ_B, the density operator of the bipartite system can be represented by $\rho = \rho_A \otimes \rho_B$. For Gaussian bipartite system, composed of two separable subsystems A and B, with A subsystem being a Gaussian state with M modes represented by quadrature operators $\hat{\boldsymbol{x}}_A$, while B subsystem being a Gaussian state with N modes represented by quadrature operators $\hat{\boldsymbol{x}}_B$; the Gaussian state of bipartite system will contain $M + N$ modes and can be described by the quadrature operators $\hat{\boldsymbol{x}} = [\hat{\boldsymbol{x}}_A \ \hat{\boldsymbol{x}}_B]^T$, with corresponding covariance matrix of size $2(M+N) \times 2(M+N)$ being represented as

$$\boldsymbol{\Sigma} = \boldsymbol{\Sigma}_A \oplus \boldsymbol{\Sigma}_B = \begin{bmatrix} \boldsymbol{\Sigma}_A & \mathbf{0} \\ \mathbf{0} & \boldsymbol{\Sigma}_B \end{bmatrix}, \tag{6.92}$$

where $\mathbf{0}$ denotes the all-zero matrix.

Let us assume that the EPR state, defined by Eq. (5.199), is used in CV-QKD system, with covariance matrix defined by Eq. (5.219). The identity matrix can be used to describe the thermal state at the input of beam splitter, see Fig. 6.27. Based on previous equation, corresponding bipartite system will have the following correlation matrix:

$$\boldsymbol{\Sigma} = \begin{bmatrix} \boldsymbol{\Sigma}_{EPR} & \mathbf{0} \\ \mathbf{0} & \mathbf{1} \end{bmatrix} = \begin{bmatrix} v\mathbf{1} & \sqrt{v^2 - 1}\mathbf{Z} & \mathbf{0} \\ \sqrt{v^2 - 1}\mathbf{Z} & v\mathbf{1} & \mathbf{0} \\ \mathbf{0} & \mathbf{0} & \mathbf{1} \end{bmatrix}. \tag{6.93}$$

The beam splitter acts on Bob's qubit and thermal state, while leaves Alice qubit unaffected; therefore, its action can be described as

$$BS_{\text{combined}}(T) = \mathbf{1} \oplus \begin{bmatrix} \sqrt{T}\mathbf{1} & \sqrt{1-T}\mathbf{1} \\ -\sqrt{1-T}\mathbf{1} & \sqrt{T}\mathbf{1} \end{bmatrix} = \begin{bmatrix} \mathbf{1} & \mathbf{0} & \mathbf{0} \\ \mathbf{0} & \sqrt{T}\mathbf{1} & \sqrt{1-T}\mathbf{1} \\ \mathbf{0} & -\sqrt{1-T}\mathbf{1} & \sqrt{T}\mathbf{1} \end{bmatrix}. \tag{6.94}$$

To determine the covariance matrix after the beam splitter, we apply the symplectic operation by BS_{combined} on input covariance matrix to obtain

$$\boldsymbol{\Sigma}' = BS_{\text{combined}}(T)\boldsymbol{\Sigma}[BS_{\text{combined}}(T)]^T$$

$$= \begin{bmatrix} \mathbf{1} & \mathbf{0} & \mathbf{0} \\ \mathbf{0} & \sqrt{T}\mathbf{1} & \sqrt{1-T}\mathbf{1} \\ \mathbf{0} & -\sqrt{1-T}\mathbf{1} & \sqrt{T}\mathbf{1} \end{bmatrix} \begin{bmatrix} v\mathbf{1} & \sqrt{v^2-1}\mathbf{Z} & \mathbf{0} \\ \sqrt{v^2-1}\mathbf{Z} & v\mathbf{1} & \mathbf{0} \\ \mathbf{0} & \mathbf{0} & \mathbf{1} \end{bmatrix} \begin{bmatrix} \mathbf{1} & \mathbf{0} & \mathbf{0} \\ \mathbf{0} & \sqrt{T}\mathbf{1} & \sqrt{1-T}\mathbf{1} \\ \mathbf{0} & -\sqrt{1-T}\mathbf{1} & \sqrt{T}\mathbf{1} \end{bmatrix}^T. \tag{6.95}$$

Now by keeping only Alice and Bob submatrices, we obtain

$$\boldsymbol{\Sigma}'_{AB} = \begin{bmatrix} v\mathbf{1} & \sqrt{T(v^2-1)}\mathbf{Z} \\ \sqrt{T(v^2-1)}\mathbf{Z} & (vT+1-T)\mathbf{1} \end{bmatrix}. \tag{6.96}$$

In the presence of excess noise, we need to replace the $(\Sigma'_{AB})_{22}$-element by $vT+1+\varepsilon'-T$, to obtain the following *covariance matrix for homodyne detection*:

$$\boldsymbol{\Sigma}_{AB}^{(\text{homodyne})} = \begin{bmatrix} v\mathbf{1} & \sqrt{T(v^2-1)}\mathbf{Z} \\ \sqrt{T(v^2-1)}\mathbf{Z} & (vT+1+\varepsilon'-T)\mathbf{1} \end{bmatrix}. \tag{6.97}$$

In *HD*, Bob employs the BBS to split the quantum signal, while the second input to the BBS is the vacuum state, and the operation of the BBS can be represented by Eq. (5.159) by setting $T = 1/2$. The corresponding covariance matrix at the input of BBS will be

$$\boldsymbol{\Sigma}_{\text{in}}^{(\text{BBS})} = \begin{bmatrix} \boldsymbol{\Sigma}_{\text{AB}}^{(\text{homodyne})} & \mathbf{0} \\ \mathbf{0} & \mathbf{1} \end{bmatrix} = \begin{bmatrix} v\mathbf{1} & \sqrt{T(v^2-1)}\mathbf{Z} & \mathbf{0} \\ \sqrt{T(v^2-1)}\mathbf{Z} & (vT+1+\varepsilon'-T)\mathbf{1} & \mathbf{0} \\ \mathbf{0} & \mathbf{0} & \mathbf{1} \end{bmatrix}. \tag{6.98}$$

The covariance matrix and the BBS output can be determined by applying the symplectic operation by $BS_{\text{combined}}(1/2)$ on input covariance matrix to obtain

$$\boldsymbol{\Sigma}_{AB}^{(\text{het})} = BS_{\text{combined}}(1/2)\boldsymbol{\Sigma}_{\text{in}}^{(\text{BBS})}[BS_{\text{combined}}(1/2)]^T = \begin{bmatrix} \mathbf{1} & \mathbf{0} & \mathbf{0} \\ \mathbf{0} & \frac{1}{\sqrt{2}}\mathbf{1} & \frac{1}{\sqrt{2}}\mathbf{1} \\ \mathbf{0} & -\frac{1}{\sqrt{2}}\mathbf{1} & \frac{1}{\sqrt{2}}\mathbf{1} \end{bmatrix}$$

$$\begin{bmatrix} v\mathbf{1} & \sqrt{T(v^2-1)}\mathbf{Z} & \mathbf{0} \\ \sqrt{T(v^2-1)}\mathbf{Z} & (vT+1+\varepsilon'-T)\mathbf{1} & \mathbf{0} \\ \mathbf{0} & \mathbf{0} & \mathbf{1} \end{bmatrix} \begin{bmatrix} \mathbf{1} & \mathbf{0} & \mathbf{0} \\ \mathbf{0} & \frac{1}{\sqrt{2}}\mathbf{1} & \frac{1}{\sqrt{2}}\mathbf{1} \\ \mathbf{0} & -\frac{1}{\sqrt{2}}\mathbf{1} & \frac{1}{\sqrt{2}}\mathbf{1} \end{bmatrix}^T$$

$$= \begin{bmatrix} v\mathbf{1} & \sqrt{\frac{T}{2}(v^2-1)}\mathbf{Z} & -\sqrt{\frac{T}{2}(v^2-1)}\mathbf{Z} \\ \sqrt{\frac{T}{2}(v^2-1)}\mathbf{Z} & \frac{vT+2+\varepsilon'-T}{2}\mathbf{1} & \frac{vT+\varepsilon'-T}{2}\mathbf{1} \\ -\sqrt{\frac{T}{2}(v^2-1)}\mathbf{Z} & -\frac{vT+\varepsilon'-T}{2}\mathbf{1} & \frac{vT+2+\varepsilon'-T}{2}\mathbf{1} \end{bmatrix}. \tag{6.99}$$

As discussed in previous sections, instead of employing EPR state, Alice employs the prepares-and-measure scheme, in which she employs a WCS source, and with the help of an I/Q modulator, generates the Gaussian state with variance v_A. Given that the minimum uncertainty of the coherent source is one SNU, Bob's variance will be v_A+1, and we can represent Bob's in-phase operator by

$$\hat{I}_B \sim \hat{I}_A + N(0,1), \tag{6.100}$$

where $N(0,1)$ is a zero-mean Gaussian source of variance 1 SNU. Clearly, the variance of Bob's in-phase operator is $V(\hat{I}_B) = V(\hat{I}_A) + 1 = v_A + 1$, as expected. On the other hand, the covariance of Alice's and Bob's in-phase operators will be

$$\text{Cov}(\hat{I}_A, \hat{I}_B) = \left\langle \left(\hat{I}_A - \underbrace{\langle \hat{I}_A \rangle}_{=0} \right) \left(\hat{I}_B - \underbrace{\langle \hat{I}_B \rangle}_{0} \right) \right\rangle = \left\langle \hat{I}_A \underbrace{\hat{I}_B}_{\hat{I}_A + N(0,1)} \right\rangle = \langle \hat{I}_A^2 \rangle + \underbrace{\langle \hat{I}_A N(0,1) \rangle}_{=0}$$

$$= v_A = \text{Cov}(\hat{I}_B, \hat{I}_A). \tag{6.101}$$

Therefore, the *covariance matrix for the prepare-and-measure protocol* is as follows:

$$\Sigma_{AB}^{(PM\ protocol)} = \begin{bmatrix} \overbrace{V(\hat{I}_A)\mathbf{1}}^{v_A} & \overbrace{\mathrm{Cov}(\hat{I}_A, \hat{I}_B)\mathbf{1}}^{v_A} \\ \underbrace{\mathrm{Cov}(\hat{I}_B, \hat{I}_A)\mathbf{1}}_{v_A} & \underbrace{V(\hat{I}_B)\mathbf{1}}_{v_B = v_A + 1} \end{bmatrix}$$

$$= \begin{bmatrix} v_A\mathbf{1} & v_A\mathbf{1} \\ v_A\mathbf{1} & \underbrace{(v_A + 1)\mathbf{1}}_{v} \end{bmatrix} = \begin{bmatrix} (v-1)\mathbf{1} & (v-1)\mathbf{1} \\ (v-1)\mathbf{1} & v\mathbf{1} \end{bmatrix}. \tag{6.102}$$

6.10.4 Secret key rate of continuous-variable quantum key distribution with Gaussian modulation under collective attacks

In this subsection, we describe how to calculate the secret fraction for coherent state homodyne detection-based CV-QKD scheme when Eve employs the Gaussian collective attack [131–135], which is known to be optimum for Gaussian modulation. This attack is completely characterized by estimating the covariance matrix by Alice and Bob Σ_{AB}, which is based on Eq. (6.97) for homodyne detection and Eq. (6.99) for HD. Let us re-write the covariance matrix for homodyne detection as follows:

$$\Sigma_{AB}^{(homodyne)} = \begin{bmatrix} v\mathbf{1} & \sqrt{T(v^2-1)}\mathbf{Z} \\ \sqrt{T(v^2-1)}\mathbf{Z} & T\left(v + 1/T - 1 + \underbrace{\varepsilon'/T}_{\varepsilon}\right)\mathbf{1} \end{bmatrix}$$

$$= \begin{bmatrix} v\mathbf{1} & \sqrt{T(v^2-1)}\mathbf{Z} \\ \sqrt{T(v^2-1)}\mathbf{Z} & T\left(v + \underbrace{1/T - 1 + \varepsilon}_{\chi_{line}}\right)\mathbf{1} \end{bmatrix} = \begin{bmatrix} v\mathbf{1} & \sqrt{T(v^2-1)}\mathbf{Z} \\ \sqrt{T(v^2-1)}\mathbf{Z} & T(v + \chi_{line})\mathbf{1} \end{bmatrix}. \tag{6.103}$$

where $\varepsilon = \varepsilon'/T$ is the access noise and $1/T-1$ is the channel loss term, both observed from Alice's side (at the channel input). The total variance of both terms is denoted by $\chi_{line} = 1/T - \varepsilon$. Now by adding the detector noise term, the covariance matrix for homodyne detection becomes

$$\boldsymbol{\Sigma}_{AB}^{(\text{hom.,complete})} = \begin{bmatrix} v\mathbf{1} & \sqrt{T\eta(v^2-1)}\mathbf{Z} \\ \sqrt{T\eta(v^2-1)}\mathbf{Z} & T\eta(v+\chi_{\text{total}})\mathbf{1} \end{bmatrix}, v = v_A + 1, \tag{6.104}$$

whereη detection efficiency, and χ_{total} is determined earlier by Eq. (6.90), that is, $\chi_{\text{total}} = \chi_{\text{line}} + \chi_{\text{det}}/T = \chi_{\text{line}} + [d(1+v_{el}) - \eta]/(T\eta)$, where v_{el} is the variance of the equivalent electrical noise.

In CV-QKD for direct reconciliation with nonzero SKR, the channel loss is lower limited by $T \geq 1/2$ [2,8], and for higher channel losses, we need to use reverse reconciliation. In the reverse reconciliation, the *secret fraction* is given by

$$r = \beta I(A;B) - \chi(B;E), \tag{6.105}$$

where β is the reconciliation efficiency and $I(A;B)$ is the mutual information between Alice and Bob that is identical for both individual and collective attacks [2,8]:

$$I(A;B) = \frac{d}{2}\log_2\left(\frac{v+\chi_{\text{total}}}{1+\chi_{\text{total}}}\right), \ d = \begin{cases} 1, \text{homodyne detection} \\ 2, \text{heterodyne detection} \end{cases}, \tag{6.106}$$

which is nothing else by the definition formula for channel capacity of Gaussian channel. What remains to determine is the Holevo information between Bob and Eve, denoted as $\chi(B;E)$, which is much more challenging to derive, and it will be described below. After that, the SKR can be determined as the product of the secret fraction and signaling rate. Given that homodyne/heterodyne detection is not limited by the dead time, typical SKRs for CV-QKD schemes are significantly higher than that of DV-QKD schemes.

The *Holevo information* between Eve and Bob can be calculated by

$$\chi(E;B) = S_E - S_{E|B}, \tag{6.107}$$

where S_E is the von Neumann entropy of Eve's density operator $\hat{\rho}_E$, defined by

$$S(\hat{\rho}_E) = -\text{Tr}(\hat{\rho}_E \log \hat{\rho}_E); \tag{6.108}$$

while $S_{E|B}$ is the von Neumann entropy of Eve's state when Bob performs the projective measurement, be homodyne or heterodyne. In principle, the symplectic eigenvalues of corresponding correlation matrices can be determined first to calculate the von Neumann entropies, as described in Section 5.8.3. To simplify the analysis, typically we assume that Eve possesses the purification [7,9] of Alice and Bob's joint state $\hat{\rho}_{AB}$ so that the resulting state is a pure state $|\psi\rangle$, and the corresponding density operator will be

$$\hat{\rho}_{ABE} = |\psi\rangle\langle\psi|. \tag{6.109}$$

By tracing out Eve's subspace, we obtain Alice-Bob mixed state, that is $\hat{\rho}_{AB} = \text{Tr}_E\hat{\rho}_{ABE}$. Similarly, by tracing out the Alice–Bob subspace, we get Eve's mixed state $\hat{\rho}_E = \text{Tr}_{AB}\hat{\rho}_{ABE}$. By applying the *Schmidt decomposition theorem* [2,7,9], we can represent the bipartite state $|\psi\rangle \in H_{AB} \otimes H_E$ in terms of *Schmidt bases* for subsystem AB, denoted as $\{|i_{AB}\rangle\}$, and subsystem E, denoted as $\{|i_E\rangle\}$, as follows:

$$|\psi\rangle = \sum_i \sqrt{p_i}|i_{AB}\rangle|i_E\rangle, \tag{6.110}$$

where $\sqrt{p_i}$ are Schmidt coefficients satisfying relationship $\sum_i p_i = 1$. By tracing out Eve's subsystem, we obtain the following mixed density operator for Alice-Bob subsystem:

$$\widehat{\rho}_{AB} = \mathrm{Tr}_E(|\psi\rangle\langle\psi|) = \sum_i p_i |i_{AB}\rangle\langle i_{AB}|. \tag{6.111}$$

On the other hand, by tracing out Alice-Bob subsystem, we obtained the following mixed state for Eve:

$$\widehat{\rho}_E = \mathrm{Tr}_{AB}(|\psi\rangle\langle\psi|) = \sum_i p_i |i_E\rangle\langle i_E|. \tag{6.112}$$

Clearly, both Alice-Bob and Eve's density operators have the same spectra (of eigenvalues), and therefore, the same von Neumann entropy:

$$S_{AB} = S_E = -\sum_i p_i \log p_i. \tag{6.113}$$

The calculation of Holevo information is now greatly simplified since the following is valid:

$$\chi(E;B) = S_E - S_{E|B} = S_{AB} - S_{A|B}. \tag{6.114}$$

To determine the von Neuman entropy for Alice-Bob subsystem, which is independent on Bob's measurement, we use the following covariance matrix from Eq. (6.104), observed at the channel input:

$$\mathbf{\Sigma}_{AB} = \begin{bmatrix} v\mathbf{1} & \sqrt{T(v^2-1)}\mathbf{Z} \\ \sqrt{T(v^2-1)}\mathbf{Z} & T(v+\chi_{\text{line}})\mathbf{1} \end{bmatrix}. \tag{6.115}$$

To determine the symplectic eigenvalues, we employ the theory from Sections 5.8.4 and 5.8.6. Clearly, the covariance matrix has the same form as Eq. (5.200), with $a = v$, $b = T(v+\chi_{\text{line}})$, and $c = \sqrt{T(v^2-1)}$. We can directly apply Eq. (5.201) to determine the symplectic eigenvalues. Alternatively, given that symplectic eigenvalues satisfy the following:

$$\lambda_1^2 + \lambda_2^2 = a^2 + b^2 - 2c^2,\ \lambda_1^2\lambda_2^2 = \left(ab - c^2\right)^2, \tag{6.116}$$

by introducing

$$\begin{aligned} A &= a^2 + b^2 - 2c^2 = v^2(1-2T) + 2T + T^2\left(v + \chi_{\text{line}}\right)^2, \\ B &= \left(ab - c^2\right)^2 = T^2\left(1 + v\chi_{\text{line}}\right)^2, \end{aligned} \tag{6.117}$$

the corresponding symplectic eigenvalues for Alice-Bob covariance matrix can be expressed as

$$\lambda_{1,2} = \sqrt{\frac{1}{2}\left(A \pm \sqrt{A^2 - 4B}\right)}, \tag{6.118}$$

which represents a form very often found in open literature [129–131, 150, 151]. The von Neumann entropy of, based on thermal state decomposition theorem and Eq. (5.195), is given by

$$S_{AB} = S(\widehat{\rho}_{AB}) = g\left(\frac{\lambda_1 - 1}{2}\right) + g\left(\frac{\lambda_2 - 1}{2}\right). \tag{6.119}$$

The calculation of $S_{A|B}$ is dependent on Bob's measurement strategy, homodyne or heterodyne, and we can apply the theory of partial measurements from Section 5.8.5. For homodyne detection, based on Eq. (5.214), we conclude that $\boldsymbol{A}=v\mathbf{1}$, $\boldsymbol{B}=T(v+\chi_{\text{line}})\mathbf{1}$, and $\boldsymbol{C}=\sqrt{T(v^2-1)}\boldsymbol{Z}$ so that Bob's homodyne measurement transforms the covariance matrix above to

$$\boldsymbol{\Sigma}_{A|B}^{(\hat{I})}=\boldsymbol{A}-\frac{1}{V(\hat{I})}\boldsymbol{C}\boldsymbol{\Pi}_I\boldsymbol{C}^T=\begin{bmatrix} v-\dfrac{v^2-1}{v+\chi_{\text{line}}} & 0 \\ 0 & v \end{bmatrix}. \tag{6.120}$$

The symplectic eigenvalue is determined now by solving for

$$\det\left(j\Omega\boldsymbol{\Sigma}_{A|B}-\lambda\mathbf{1}\right)=\lambda^2+v\left(\frac{v^2-1}{v+\chi_{\text{line}}}-v\right)=0, \tag{6.121}$$

and by solving for λ, we obtain

$$\lambda_3=\sqrt{v\left(v-\frac{v^2-1}{v+\chi_{\text{line}}}\right)}. \tag{6.122}$$

so $S_{A|B}$ becomes $g(\lambda_3)$. This expression is consistent with that provided in Ref. [150], except that a slightly different correlation matrix was used. Finally, for homodyne detection, the Holevo information is given by

$$\chi(E;B)=S_{AB}-S_{A|B}=g\left(\frac{\lambda_1-1}{2}\right)+g\left(\frac{\lambda_2-1}{2}\right)-g\left(\frac{\lambda_3-1}{2}\right). \tag{6.123}$$

In *HD*, Bob measures both quadratures, and given that an additional BBS is used with one input in a vacuum state, we must account for fluctuations due to the vacuum state and modify the partial correlation matrix as follows [2,147]:

$$\boldsymbol{\Sigma}_{A|B}=\boldsymbol{A}-\boldsymbol{C}(\boldsymbol{B}+\mathbf{1})^{-1}\boldsymbol{C}^T=K\mathbf{1},\, K=v-\frac{T(v^2-1)}{T(v+\chi_{\text{line}})+1}. \tag{6.124}$$

The symplectic eigenvalue can be determined from the following characteristic equation:

$$\det\left(j\Omega\boldsymbol{\Sigma}_{A|B}^{(\text{heterodyne})}-\lambda\mathbf{1}\right)=\lambda^2-K^2=0, \tag{6.125}$$

to obtain

$$\lambda_3^{(\text{heterodyne})}=K=v-\frac{T(v^2-1)}{T(v+\chi_{\text{line}})+1}. \tag{6.126}$$

Finally, for HD, the Holevo information is given by

$$\chi(E;B)^{(\text{heterodyne})}=g\left(\frac{\lambda_1-1}{2}\right)+g\left(\frac{\lambda_2-1}{2}\right)-g\left(\frac{\lambda_3^{(\text{heterodyne})}-1}{2}\right). \tag{6.127}$$

6.10.5 Reverse reconciliation results for Gaussian modulation-based continuous-variable quantum key distribution

This section provides illustrative secrecy fractions (normalized SKRs) for different scenarios by employing the derivations from the previous subsection. We consider reverse reconciliation cases only. In Fig. 6.28, we provide normalized SKRs versus channel loss when electrical noise variance in SNU is used as a parameter. The excess noise variance is set to $\varepsilon = 10^{-3}$, detector efficiency is $\eta = 0.85$, and reconciliation efficiency is set to $\beta = 0.85$. Interestingly, when electrical noise variance $v_{el} \leq 0.1$, there is not much degradation in SKRs, which indicates that commercially available p.i.n. photodetectors can be used in CV-QKD.

In Fig. 6.29, we show normalized SKRs versus channel loss when excess noise variance (expressed in SNU) is used as a parameter. The electrical noise variance is set to $v_{el} = 10^{-2}$, detector efficiency is $\eta = 0.85$, and reconciliation efficiency is set to $\beta = 0.85$. Evidently, the excess noise variance has high impact on SKR performance.

In Fig. 6.30, we provide normalized SKRs versus channel loss when detector efficiency η is used as a parameter. The excess noise variance is set to $\varepsilon = 10^{-3}$, the electrical noise variance is set to $v_{el} = 10^{-2}$, and reconciliation efficiency is set to $\beta = 0.85$. Clearly, the detection efficiency does not have high impact on SKRs.

In Fig. 6.31, we show normalized SKRs versus channel loss when reconciliation efficiency β is used as a parameter. The excess noise variance is set to $\varepsilon = 10^{-3}$, the electrical noise variance is set to $v_{el} = 10^{-2}$, and detector efficiency is set to $\eta = 0.85$. Evidently, the reconciliation efficiency does have a relevant impact on SKRs.

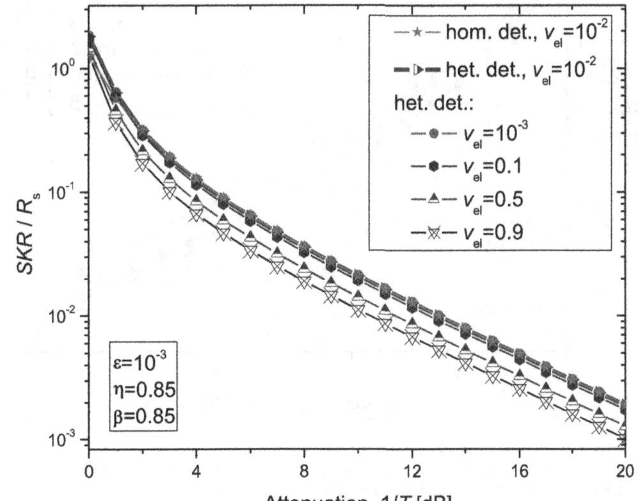

FIGURE 6.28

The secrecy fraction for continuous-variable quantum key distribution scheme with Gaussian modulation versus channel loss when electrical noise variance is a parameter.

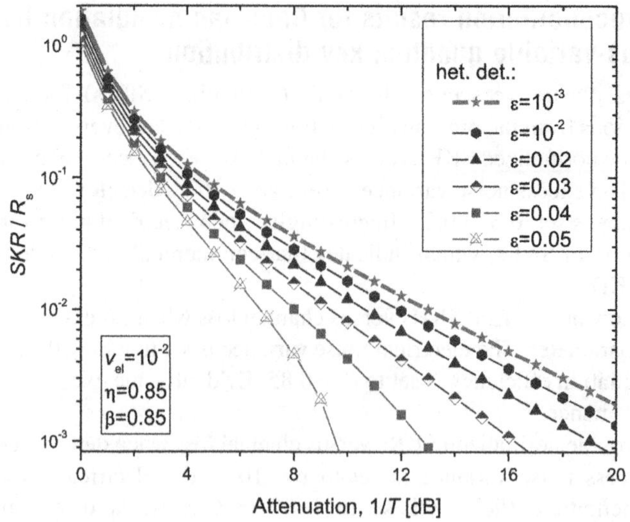

FIGURE 6.29

The secrecy fraction for continuous-variable quantum key distribution scheme with Gaussian modulation versus channel loss when excess noise variance is a parameter.

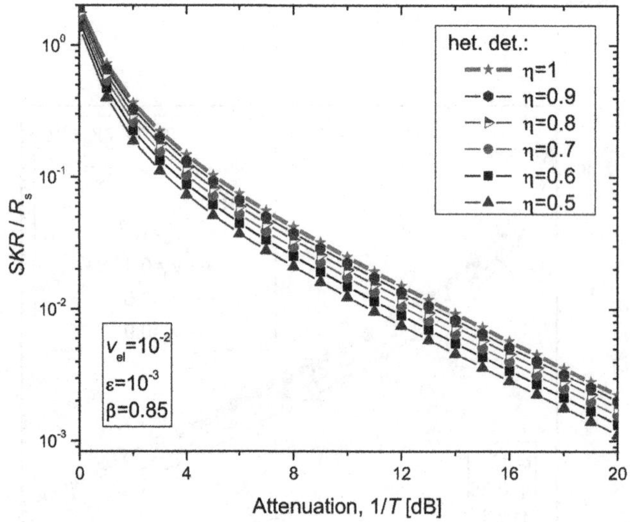

FIGURE 6.30

Secrecy fraction for continuous-variable quantum key distribution scheme with Gaussian modulation versus channel loss when detector efficiency is a parameter.

6.11 Summary

This chapter has been devoted to the fundamental concepts of QKD. The main differences between conventional cryptography and QKD were described in Section 6.1. In the section on QKD basics (Section 6.2), after a historical overview, various QKD types were briefly introduced. Section 6.3

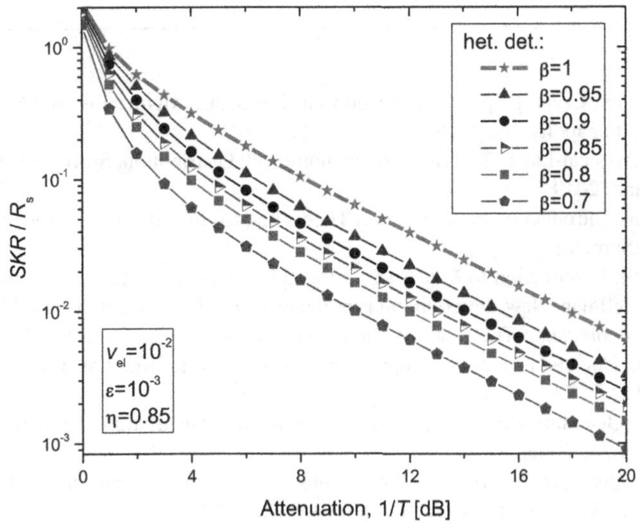

FIGURE 6.31

Secrecy fraction for continuous-variable quantum key distribution scheme with Gaussian modulation versus channel loss when reconciliation efficiency is a parameter.

described the two fundamental theorems on which QKD relies, namely the no-cloning theorem and the theorem of inability to unambiguously distinguish nonorthogonal quantum states. In Section 6.4, related to DV-QKD systems, the BB84, B92, Ekert (E91), EPR, and time-phase encoding protocols were described in their corresponding Subsections 6.4.1–6.4.4. Section 6.5 was devoted to QKD security, wherein the SKR is represented as the product of raw key and fractional rates. Moreover, the generic expression for the fractional rate was provided, followed by a description of various eavesdropping strategies, including individual (Section 6.5.1), collective (Section 6.5.2), and quantum hacking/side-channel attacks (Section 6.5.3). For individual and coherent attacks, the corresponding secret fraction expressions were provided, including the secret fraction for the BB84 protocol in Section 6.5.4. In Section 6.6, the decoy-state protocol was described with the corresponding SKR calculation.

The key MDI-QKD protocol concepts were introduced in Section 6.7, including photonic BSMs (Section 6.7.1), the polarization-based MDI-QKD protocol (Section 6.7.2), and the time-phase encoding-based MDI-QKD protocol (Section 6.7.3), while the secrecy fraction calculation was provided in Section 6.7.4. TF-QKD protocols were further described in Section 6.8, and their performance was evaluated against decoy-state and MDI-QKD protocols. In Section 6.9, the discussion moved to the classical processing steps, including information reconciliation (Section 6.9.1) and privacy amplification (Section 6.9.2) steps.

Section 6.10 was devoted to CV-QKD protocols, with homodyne and heterodyne detection schemes described in Section 6.10.1 after a brief description of squeezed state-based protocols in Section 6.10.2. Given that coherent states are much easier to generate and manipulate, coherent state-based protocols were described in detail in Section 6.10.3. For a lossy transmission channel, the corresponding covariance matrices were derived for homodyne and heterodyne coherent detection schemes, followed by SKR derivation for prepare-and-measure Gaussian modulation-based CV-QKD in Section 6.10.4. Some illustrative reconciliation results were provided in Section 6.10.5 related to Gaussian modulation-based CV-QKD schemes.

References

[1] B. Schneier, Applied Cryptography, Second Edition: Protocols, Algorithms, and Source Code in C, John Wiley & Sons, Indianapolis, IN, 2015.
[2] I.B. Djordjevic, Physical-Layer Security and Quantum Key Distribution, Springer Nature Switzerland AG, Cham: Switzerland, 2019.
[3] D. Drajic, P. Ivanis, Introduction to Information Theory and Coding, third ed., Akademska Misao, Belgrade, Serbia, 2009 (in Serbian).
[4] J. Katz, Y. Lindell, Introduction to Modern Cryptography, second ed., CRC Press, Boca Raton, FL, 2015.
[5] W. Diffie, M.E. Hellman, New direction in cryptography, IEEE Trans. Inf. Theor. IT-22 (1976) 644−654.
[6] D. Kahn, The Codebreakers: The Story of Secret Writing, Macmillan Publishing Co., New York, 1967.
[7] M.A. Neilsen, I.L. Chuang, Quantum Computation and Quantum Information, Cambridge University Press, Cambridge, 2000.
[8] G. van Assche, Quantum Cryptography and Secrete-Key Distillation, Cambridge University Press, Cambridge-New York, 2006.
[9] I.B. Djordjevic, Quantum Information Processing and Quantum Error Correction: An Engineering Approach, Elsevier/Academic Press, Amsterdam-Boston, 2012.
[10] C.E. Shannon, Communication theory of secrecy systems, Bell Syst. Tech. J. 28 (1949) 656−715.
[11] T.M. Cover, J.A. Thomas, Elements of Information Theory, John Wiley and Sons, New York, 1991.
[12] J.-P. Aumasson, Serious Cryptography: A Practical Introduction to Modern Encryption, No Starch Press, San Francisco, CA, 2018.
[13] J. Sebbery, J. Pieprzyk, Cryptography: An Introduction to Computer Security, Prentice Hall, New York, 1989.
[14] H. Delfs, H. Knebl, Introduction to Cryptography: Principles and Applications (Information Security and Cryptography), third ed., Springer, Heidelberg •New York, 2015.
[15] C.H. Bennet, G. Brassard, Quantum Cryptography: public key distribution and coin tossing, in: Proc. IEEE International Conference on Computers, Systems, and Signal Processing, 1984, pp. 175−179. Bangalore, India.
[16] C.H. Bennett, Quantum cryptography: uncertainty in the service of privacy, Science 257 (1992) 752−753.
[17] M. Le Bellac, An Introduction to Quantum Information and Quantum Computation, Cambridge University Press, 2006.
[18] P.W. Shor, Polynomial-time algorithms for prime number factorization and discrete logarithms on a quantum computer, SIAM J. Comput. 26 (5) (1997) 1484−1509.
[19] A. Ekert, R. Josza, Quantum computation and Shor's factoring algorithm, Rev. Mod. Phys. 68 (3) (1996) 733−753.
[20] R.L. Rivest, A. Shamir, L. Adleman, A method for obtaining digital signatures and public-key cryptosystems, Commun. ACM 21 (2) (1978) 120−126.
[21] S.M. Barnett, Quantum Information, Oxford University Press, Oxford, 2009.
[22] S.-K. Liao, et al., Satellite-to-ground quantum key distribution, Nature 549 (2017) 43−47.
[23] C.H. Bennett, G. Brassard, C. Crepeau, U. Maurer, Generalized privacy amplification, IEEE Inform. Theory 41 (6) (1995) 1915−1923.
[24] A. Chorti, et al., Physical layer security: a paradigm shift in data confidentiality, ser. Lecture Notes in Electrical Engineering 358, in: Physical and Data-Link Security Techniques for Future Communications Systems, Springer, 2016, pp. 1−15.
[25] R. Ursin, et al., Entanglement-based quantum communication over 144 km, Nat. Phys. 3 (7) (2007) 481−486.

[26] A. Boaron, et al., Secure quantum key distribution over 421 km of optical fiber, Phys. Rev. Lett. 121 (2018) 190502.
[27] J. Schwinger, Unitary operator bases, Proc. Natl. Acad. Sci. U.S.A. 46 (1960) 570–579.
[28] W.K. Wootters, B.D. Fields, Optimal state-determination by mutually unbiased measurements, Ann. Phys. 191 (1989) 363–381.
[29] I.D. Ivanovic, Geometrical description of quantal state determination, J. Physiol. Anthropol. 14 (1981) 3241–3245.
[30] I. Bengtsson, Three ways to look at mutually unbiased bases, AIP Conf. Proc. 889 (1) (2007) 40.
[31] I.B. Djordjevic, FBG-based weak coherent state and entanglement assisted multidimensional QKD, IEEE Photon. J. 10 (4) (2018) 7600512.
[32] C.W. Helstrom, Quantum Detection and Estimation Theory, Academic Press, New York, 1976.
[33] C.W. Helstrom, J.W.S. Liu, J.P. Gordon, Quantum-mechanical communication theory, Proc. IEEE 58 (10) (1970) 1578–1598.
[34] C. Vilnrotter, C.-W. Lau, Quantum detection theory for the free-space channel, IPN Prog. Rep. 42–146 (2001) 1–34.
[35] I.B. Djordjevic, LDPC-coded optical coherent state quantum communications, IEEE Photon. Technol. Lett. 19 (24) (2007) 2006–2008.
[36] I.B. Djordjevic, LDPC-coded *M*-ary PSK optical coherent state quantum communication, IEEE/OSA J. Lightw. Technol. 27 (5) (2009) 494–499.
[37] H.-K. Lo, X. Ma, K. Chen, Decoy state quantum key distribution, Phys. Rev. Lett. 94 (2005) 230504.
[38] C.H. Bennett, Quantum cryptography using any two nonorthogonal states, Phys. Rev. Lett. 68 (21) (1992) 3121–3124.
[39] A.K. Ekert, Quantum cryptography based on Bell's theorem, Phys. Rev. Lett. 67 (6) (1991) 661–663.
[40] S.J. Lomonaco, A quick glance at quantum cryptography, Cryptologia 23 (1) (1999) 1–41.
[41] P.D. Townsend, J.G. Rarity, P.R. Tapster, Single photon interference in 10 km long optical fibre interferometer, Electron. Lett. 29 (7) (1993) 634–635.
[42] C. Marand, P.D. Townsend, Quantum key distribution over distances as long as 30 km, Opt. Lett. 20 (16) (1995) 1695–1697.
[43] N.T. Islam, C.C.W. Lim, C. Cahall, J. Kim, D.J. Gauthier, Provably secure and high-rate quantum key distribution with time-bin qudits, Sci. Adv. 3 (11) (2017) e1701491.
[44] F.M. Spedalieri, Quantum key distribution without reference frame alignment: exploiting photon orbital angular momentum, Opt Commun. 260 (1) (2006) 340–346.
[45] I.B. Djordjevic, Multidimensional QKD based on combined orbital and spin angular momenta of photon, IEEE Photon. J. 5 (6) (2013) 7600112.
[46] I.B. Djordjevic, Advanced Optical and Wireless Communications Systems, Springer International Publishing, Switzerland, 2017.
[47] N.T. Islam, High-Rate, High-Dimensional Quantum Key Distribution Systems, PhD Dissertation, Duke University, 2018.
[48] V. Scarani, H. Bechmann-Pasquinucci, N.J. Cerf, M. Dušek, N. Lütkenhaus, M. Peev, The security of practical quantum key distribution, Rev. Mod. Phys. 81 (2009) 1301.
[49] S. Bloom, E. Korevaar, J. Schuster, H. Willebrand, Understanding the performance of free-space optics [Invited], J. Opt. Netw. 2 (6) (2003) 178–200.
[50] I. Csiszár, J. Körner, Broadcast channels with confidential messages, IEEE Trans. Inf. Theory IT- 24 (3) (1978) 339–348.
[51] G. Brassard, N. Lütkenhaus, T. Mor, B.C. Sanders, Limitations on practical quantum cryptography, Phys. Rev. Lett. 85 (2000) 1330.

[52] I. Devetak, A. Winter, Distillation of secret key and entanglement from quantum states, Proc. R. Soc. London Ser. A 461 (2053) (2005) 207–235.
[53] A.S. Holevo, Bounds for the quantity of information transmitted by a quantum communication channel, Probl. Inf. Transm. 9 (3) (1973) 177–183.
[54] M. Lucamarini, I. Choi, M.B. Ward, J.F. Dynes, Z.L. Yuan, A.J. Shields, Practical security bounds against the Trojan-horse attack in quantum key distribution, Phys. Rev. X 5 (2015) 031030.
[55] C. Kurtsiefer, P. Zarda, S. Mayer, H. Weinfurter, The breakdown flash of silicon avalanche photodiodes - back door for eavesdropper attacks? J. Mod. Opt. 48 (13) (2001) 2039–2047.
[56] B. Qi, C.-H.F. Fung, H.-K. Lo, X. Ma, Time-shift attack in practical quantum cryptosystems, Quant. Inf. Comput. 7 (1) (2006) 73–82. Available at: _HYPERLINK, https://arxiv.org/abs/quant-ph/0512080, https://arxiv.org/abs/quant-ph/0512080_.
[57] L. Lydersen, C. Wiechers, S. Wittmann, D. Elser, J. Skaar, V. Makarov, Hacking commercial quantum cryptography systems by tailored bright illumination, Nat. Photonics 4 (10) (2010) 686.
[58] N. Lütkenhaus, M. Jahma, Quantum key distribution with realistic states: photon-number statistics in the photon-number splitting attack, New J. Phys. 4 (2002) 44.
[59] M. Dušek, M. Jahma, N. Lütkenhaus, Unambiguous state discrimination in quantum cryptography with weak coherent states, Phys. Rev. 62 (2) (2000) 022306.
[60] W.-Y. Hwang, Quantum key distribution with high loss: toward global secure communication, Phys. Rev. Lett. 91 (2003) 057901.
[61] X. Ma, C.-H.F. Fung, F. Dupuis, K. Chen, K. Tamaki, H.-K. Lo, Decoy-state quantum key distribution with two-way classical postprocessing, Phys. Rev. 74 (2006) 032330.
[62] Y. Zhao, B. Qi, X. Ma, H.-K. Lo, L. Qian, Experimental quantum key distribution with decoy states, Phys. Rev. Lett. 96 (2006) 070502.
[63] D. Rosenberg, J.W. Harrington, P.R. Rice, P.A. Hiskett, C.G. Peterson, R.J. Hughes, A.E. Lita, S.W. Nam, J.E. Nordholt, Long-distance decoy-state quantum key distribution in optical fiber, Phys. Rev. Lett. 98 (1) (2007) 010503.
[64] Z.L. Yuan, A.W. Sharpe, A.J. Shields, Unconditionally secure one-way quantum key distribution using decoy pulses, Appl. Phys. Lett. 90 (2007) 011118.
[65] J. Hasegawa, M. Hayashi, T. Hiroshima, A. Tanaka, A. Tomita, Experimental decoy state quantum key distribution with unconditional security incorporating finite statistics, Available at: arXiv (2007), 0705.3081.
[66] T. Tsurumaru, A. Soujaeff, S. Takeuchi, Exact minimum and maximum of yield with a finite number of decoy light intensities, Phys. Rev. 77 (2008) 022319.
[67] M. Hayashi, General theory for decoy-state quantum key distribution with an arbitrary number of intensities, New J. Phys. 9 (2007) 284.
[68] Y. Zhao, B. Qi, X. Ma, H. Lo, L. Qian, Simulation and implementation of decoy state quantum key distribution over 60km Telecom fiber, in: Proc. 2006 IEEE International Symposium on Information Theory, 2006, pp. 2094–2098. Seattle, WA.
[69] X.-B. Wang, Three-intensity decoy-state method for device-independent quantum key distribution with basis-dependent errors, Phys. Rev. 87 (1) (2013) 012320.
[70] X. Sun, I.B. Djordjevic, M.A. Neifeld, Secret key rates and optimization of BB84 and decoy state protocols over time-varying free-space optical channels, IEEE Photon. J. 8 (3) (2016) 7904713.
[71] X. Ma, Quantum Cryptography: From Theory to Practice, Ph.D. Dissertation, University of Toronto, 2008.
[72] C.-H.F. Fung, K. Tamaki, H.-K. Lo, Performance of two quantum-key-distribution protocols, Phys. Rev. 73 (2006) 012337.

[73] M. Ben-Or, M. Horodecki, D.W. Leung, D. Mayers, J. Oppenheim, The universal composable security of quantum key distribution, in: Theory of Cryptography: Second Theory of Cryptography Conference, TCC 2005, Lecture Notes in Computer Science 3378, Springer Verlag, Berlin, 2005, pp. 386−406.

[74] L. Masanes, S. Pironio, A. Acín, Secure device-independent quantum key distribution with causally independent measurement devices, Nat. Commun. 2 (2011) 238.

[75] U. Vazirani, T. Vidick, Fully device-independent quantum key distribution, Phys. Rev. Lett. 113 (14) (2014) 140501.

[76] S. Pironio, A. Acin, N. Brunner, N. Nicolas Gisin, S. Massar, V. Scarani, Device-independent quantum key distribution secure against collective attacks, New J. Phys. 11 (4) (2009) 045021.

[77] C.C.W. Lim, C. Portmann, M. Tomamichel, R. Renner, N. Gisin, Device-independent quantum key distribution with local Bell test, Phys. Rev. X 3 (3) (2013) 031006.

[78] C.H. Bennett, G. Brassard, C. Crépeau, R. Jozsa, A. Peres, W.K. Wootters, Teleporting an unknown quantum state via dual classical and Einstein-Podolsky-Rosen channels, Phys. Rev. Lett. 70 (13) (1993) 1895.

[79] K. Mattle, H. Weinfurter, P.G. Kwiat, A. Zeilinger, Dense coding in experimental quantum communication, Phys. Rev. Lett. 76 (25) (1996) 4656.

[80] M. Źukowski, A. Zeilinger, M.A. Horne, A.K. Ekert, "Event-ready-detectors" Bell experiment via entanglement swapping, Phys. Rev. Lett. 71 (26) (1993) 4287.

[81] N. Sangouard, C. Simon, H. De Riedmatten, N. Gisin, Quantum repeaters based on atomic ensembles and linear optics, Rev. Mod. Phys. 83 (1) (2011) 33.

[82] H.-L. Yin, et al., Measurement-device-independent quantum key distribution over a 404 km optical fiber, Phys. Rev. Lett. 117 (2016) 190501.

[83] H.-K. Lo, M. Curty, B. Qi, Measurement-device-independent quantum key distribution, Phys. Rev. Lett. 108 (13) (2012) 130503.

[84] R. Valivarthi, I. Lucio-Martinez, A. Rubenok, P. Chan, F. Marsili, V.B. Verma, M.D. Shaw, J.A. Stern, J.A. Slater, D. Oblak, S.W. Nam, W. Tittel, Efficient Bell state analyzer for time-bin qubits with fast-recovery WSi superconducting single photon detectors, Opt Express 22 (2014) 24497−24506.

[85] N. Lütkenhaus, J. Calsamiglia, K.-A. Suominen, Bell measurements for teleportation, Phys. Rev. 59 (1999) 3295−3300.

[86] C.K. Hong, Z.Y. Ou, L. Mandel, Measurement of subpicosecond time intervals between two photons by interference, Phys. Rev. Lett. 59 (1987) 2044.

[87] Z.-Y.J. Ou, Quantum Optics for Experimentalists, World Scientific, New Jersey-London-Singapore, 2017.

[88] F. Xu, M. Curty, B. Qi, H.-K. Lo, Measurement-device-independent quantum cryptography, IEEE J. Sel. Top. Quant. Electron. 21 (3) (2015) 148−158.

[89] P. Chan, J.A. Slater, I. Lucio-Martinez, A. Rubenok, W. Tittel, Modeling a measurement-device-independent quantum key distribution system, Opt Exp. 22 (11) (2014) 12716−12736.

[90] M. Curty, F. Xu, C.C.W. Lim, K. Tamaki, H.-K. Lo, Finite-key analysis for measurement-device-independent quantum key distribution, Nat. Commun. 5 (2014) 3732.

[91] R. Valivarthi, et al., Measurement-device-independent quantum key distribution: from idea towards application, J. Mod. Opt. 62 (14) (2015) 1141−1150.

[92] I.B. Djordjevic, R. Varrazza, M. Hill, S. Yu, Packet switching performance at 10Gb/s across a 4x4 optical crosspoint switch matrix, IEEE Photon. Technol. Lett. 16 (2004) 102−104.

[93] R. Varrazza, I.B. Djordjevic, S. Yu, Active vertical-coupler based optical crosspoint switch matrix for optical packet-switching applications, IEEE/OSA J. Lightw. Technol. 22 (2004) 2034−2042.

[94] Y.-L. Tang, et al., Measurement-device independent quantum key distribution over untrustful metropolitan network, Phys. Rev. X 6 (1) (2016) 011024.

[95] R. Valivarthi, et al., Quantum teleportation across a metropolitan fibre network, Nat. Photonics 10 (2016) 676−680.

[96] M. Cvijetic, I.B. Djordjevic, Advanced Optical Communications and Networks, Artech House, 2013.

[97] S. Pirandola, R. Laurenza, C. Ottaviani, L. Banchi, Fundamental limits of repeaterless quantum communications, Nat. Commun. 8 (2017) 15043.
[98] L.-M. Duan, M. Lukin, J.I. Cirac, P. Zoller, Long-distance quantum communication with atomic ensembles and linear optics, Nature 414 (2001) 413–418.
[99] J. Qiu, et al., Quantum communications leap out of the lab, Nature 508 (2014) 441–442.
[100] M. Lucamarini, Z.L. Yuan, J.F. Dynes, A.J. Shields, Overcoming the rate–distance limit of quantum key distribution without quantum repeaters, Nature 557 (2018) 400–403.
[101] X. Ma, P. Zeng, H. Zhou, Phase-matching quantum key distribution, Phys. Rev. X 8 (2018) 031043.
[102] K. Tamaki, H.-K. Lo, W. Wang, M. Lucamarini, Information theoretic security of quantum key distribution overcoming the repeaterless secret key capacity bound, Avalable at: arXiv (2018), 1805.05511 [quant-ph].
[103] J. Lin, N. Lütkenhaus, Simple security analysis of phase-matching measurement-device-independent quantum key distribution, Phys. Rev. 98 (2018) 042332.
[104] H.-L. Yin, Y. Fu, Measurement-device-independent twin-field quantum key distribution, Sci. Rep. 9 (2019) 3045.
[105] Y. Tamura, et al., The First 0.14-dB/km loss optical fiber and its impact on submarine transmission, J. Lightwave Technol. 36 (2018) 44–49.
[106] D. Slepian, J.K. Wolf, Noiseless coding of correlated information sources, IEEE Trans. Inf. Theor. 19 (4) (1973) 471–480.
[107] A.D. Liveris, Z. Xiong, C.N. Georghiades, Compression of binary sources with side information using low-density parity-check codes, in: Proc. IEEE Global Telecommunications Conference 2002 (GLOBECOM '02), vol. 2, 2002, pp. 1300–1304. Taipei, Taiwan.
[108] I.B. Djordjevic, On advanced FEC and coded modulation for ultra-high-speed optical transmission, IEEE Commun. Surv. Tutor. 18 (3) (2016) 1920–1951, https://doi.org/10.1109/COMST.2016.2536726.
[109] M. Bloch, Physical-layer Security, PhD dissertation, School of Electrical and Computer Engineering, Georgia Institute of Technology, 2008.
[110] M. Bloch, J. Barros, Physical-layer Security: From Information Theory to Security Engineering, Cambridge University Press, Cambridge, 2011.
[111] C. Cachin, U.M. Maurer, Linking information reconciliation and privacy amplification, J. Cryptol. 10 (1997) 97–110.
[112] J.L. Carter, M.N. Wegman, Universal classes of hash functions, J. Comput. Syst. Sci. 18 (2) (1979) 143–154.
[113] M.N. Wegman, J. Carter, New hash functions and their use in authentication and set equality, J. Comput. Sci. Syst. 22 (1981) 265–279.
[114] A. Rényi, On measures of entropy and information, in: Proc. 4th Berkeley Symp. On Mathematical Statistics and Probability, vol. 1, Univ. Claifornia Press, 1961, pp. 547–561.
[115] I. Ilic, I.B. Djordjevic, M. Stankovic, On a general definition of conditional Rényi entropies, in: Proc. 4th International Electronic Conference on Entropy and its Applications, 21 November–1 December 2017, vol. 4, Sciforum Electronic Conference Series, 2017, https://doi.org/10.3390/ecea-4-05030.
[116] A. Teixeira, A. Matos, L. Antunes, Conditional Rényi entropies, IEEE Trans. Inf. Theor. 58 (7) (2012) 4273–4277.
[117] C. Cachin, Entropy Measures and Unconditional Security in Cryptography. PhD Dissertation, ETH Zurich, Hartung-Gorre Verlag, Konstanz, 1997.
[118] U. Maurer, S. Wolf, Information-theoretic key agreement: from weak to strong secrecy for free, in: B. Preneel (Ed.), Advances in Cryptology — EUROCRYPT 2000. EUROCRYPT 2000. Lecture Notes in Computer Science, vol. 1807, Springer, Berlin, Heidelberg, 2000.
[119] T.C. Ralph, Continuous variable quantum cryptography, Phys. Rev. 61 (1999) 010303(R).
[120] F. Grosshans, P. Grangier, Continuous variable quantum cryptography using coherent states, Phys. Rev. Lett. 88 (2002) 057902.
[121] F. Grosshans, Collective attacks and unconditional security in continuous variable quantum key distribution, Phys. Rev. Lett. 94 (2005) 020504.

[122] A. Leverrier, P. Grangier, Unconditional security proof of long-distance continuous-variable quantum key distribution with discrete modulation, Phys. Rev. Lett. 102 (2009) 180504.

[123] A. Becir, F.A.A. El-Orany, M.R.B. Wahiddin, Continuous-variable quantum key distribution protocols with eight-state discrete modulation, Int. J. Quant. Inf. 10 (2012) 1250004.

[124] Q. Xuan, Z. Zhang, P.L. Voss, A 24 km fiber-based discretely signaled continuous variable quantum key distribution system, Opt Express 17 (26) (2009) 24244−24249.

[125] D. Huang, P. Huang, D. Lin, G. Zeng, Long-distance continuous-variable quantum key distribution by controlling excess noise, Sci. Rep. 6 (2016) 19201.

[126] Z. Qu, I.B. Djordjevic, High-speed free-space optical continuous-variable quantum key distribution enabled by three-dimensional multiplexing, Opt Express 25 (7) (2017) 7919−7928.

[127] C. Silberhorn, T.C. Ralph, N. Lütkenhaus, G. Leuchs, Continuous variable quantum cryptography: beating the 3 dB Loss Limit, Phys. Rev. 89 (2002) 167901.

[128] K.A. Patel, J.F. Dynes, I. Choi, A.W. Sharpe, A.R. Dixon, Z.L. Yuan, R.V. Penty, J. Shields, Coexistence of high-bit-rate quantum key distribution and data on optical fiber, Phys. Rev. X 2 (2012) 041010.

[129] Z. Qu, I.B. Djordjevic, Four-dimensionally multiplexed eight-state continuous-variable quantum key distribution over turbulent channels, IEEE Photon. J. 9 (6) (2017) 7600408.

[130] Z. Qu, I.B. Djordjevic, Approaching Gb/s secret key rates in a free-space optical CV-QKD system Affected by atmospheric turbulence, in: Proc. ECOC 2017, P2.SC6.32, (Gothenburg, Sweden), 2017.

[131] Z. Qu, I.B. Djordjevic, High-speed free-space optical continuous variable-quantum key distribution based on Kramers-Kronig scheme, IEEE Photon. J. 10 (6) (2018) 7600807.

[132] M. Heid, N. Lütkenhaus, Efficiency of coherent-state quantum cryptography in the presence of loss: influence of realistic error correction, Phys. Rev. 73 (2006) 052316.

[133] F. Grosshans, N.J. Cerf, Continuous-variable quantum cryptography is secure against non-Gaussian attacks, Phys. Rev. Lett. 92 (2004) 047905.

[134] R. García-Patrón, N.J. Cerf, Unconditional optimality of Gaussian attacks against continuous-variable quantum key distribution, Phys. Rev. Lett. 97 (2006) 190503.

[135] M. Navascués, F. Grosshans, A. Acín, Optimality of Gaussian attacks in continuous-variable quantum cryptography, Phys. Rev. Lett. 97 (2006) 190502.

[136] F. Grosshans, G. Van Assche, J. Wenger, R. Tualle-Brouri, N.J. Cerf, P. Grangier, Quantum key distribution using Gaussian-modulated coherent states, Nature 421 (2003) 238.

[137] F. Grosshans, Communication et Cryptographie Quantiques avec des Variables Continues, PhD Dissertation, Université Paris XI, 2002.

[138] J. Wegner, Dispositifs Imulsionnels pour la Communication Quantique à Variables Contines, PhD Dissertation, Université Paris XI, 2004.

[139] Z. Qu, Secure High-Speed Optical Communication Systems, PhD Dissertation, University of Arizona, 2018.

[140] I.B. Djordjevic, Hybrid QKD protocol outperforming both DV- and CV-QKD protocols, IEEE Photon. J. 12 (1) (Feb. 2020) 7600108 (published online: 11 October 2019).

[141] I.B. Djordjevic, On the photon subtraction-based measurement-device-independent CV-QKD protocols, IEEE Access 7 (October 11, 2019) 147399−147405.

[142] T.-L. Wang, I.B. Djordjevic, J. Nagel, Laser beam propagation effects on secure key rates for satellite-to-ground discrete modulation CV-QKD, Appl. Opt. 58 (29) (October 2019) 8061−8068.

[143] I.B. Djordjevic, Proposal for slepian-states-based DV- and CV-QKD schemes suitable for implementation in integrated photonics platforms, IEEE Photon. J. 11 (4) (August 2019) 7600312.

[144] I.B. Djordjevic, Optimized-Eight-state CV-QKD protocol outperforming Gaussian modulation based protocols, IEEE Photon. J. 11 (4) (August 2019) 4500610.

[145] I.B. Djordjevic, On the discretized Gaussian modulation (DGM)-based continuous variable-QKD, IEEE Access 7 (May 17, 2019) 65342−65346.

[146] F. Grosshans, P. Grangier, Reverse reconciliation protocols for quantum cryptography with continuous variables, available at: arXiv (2002). quant-ph/0204127.

[147] H.P. Yuen, J.H. Shapiro, Optical communication with two-photon coherent states–Part III: quantum measurements realizable with photoemissive detectors, IEEE Trans. Inf. Theor. 26 (1) (1980) 78–92.
[148] C. Weedbrook, S. Pirandola, R. García-Patrón, N.J. Cerf, T.C. Ralph, J.H. Shapiro, L. Seth Lloyd, Gaussian quantum information, Rev. Mod. Phys. 84 (2012) 621.
[149] G. Giedke, J.L. Cirac, Characterization of Gaussian operations and distillation of Gaussian states, Phys. Rev. 66 (2002) 032316.
[150] S. Fossier, E. Diamanti, T. Debuisschert, R. Tualle-Brouri, P. Grangier, Improvement of continuous-variable quantum key distribution systems by using optical preamplifiers, J. Physiol. Biochem. 42 (2009) 114014.
[151] R. Garcia-Patron, Quantum Information with Optical Continuous Variables: From Bell Tests to Key Distribution, Ph.D. thesis, Université Libre de Bruxelles, 2007.

Further reading

[1] Z. Qu, I.B. Djordjevic, RF-assisted coherent detection based continuous variable (CV) QKD with high secure key rates over atmospheric turbulence channels, paper Tu.D2.2, Girona, Spain, in: Proc. ICTON 2017, 2017 (Invited Paper.).
[2] R. König, R. Renner, A. Bariska, U. Maurer, Small accessible quantum information does not imply security, Phys. Rev. Lett. 98 (2007) 140502.
[3] L. Mandel, E. Wolf, Optical Coherence and Quantum Optics, Cambridge University Press, Cambridge-New York-Melbourne, 1995.
[4] M.O. Scully, M.S. Zubairy, Quantum Optics, Cambridge University Press, Cambridge-New York-Melbourne, 1997.
[5] N.N. Bogoliubov, On the theory of superfluidity, J. Phys. (USSR) 11: 23, (Izv. Akad. Nauk Ser. Fiz. 11 (1947) 77.
[6] N.N. Bogoljubov, On a new method in the theory of superconductivity, Il Nuovo Cimento 7 (6) (1958) 794–805.
[7] A.S. Holevo, M. Sohma, O. Hirota, Capacity of quantum Gaussian channels, Phys. Rev. 59 (3) (1999) 1820–1828.
[8] J. Williamson, On the algebraic problem concerning the normal forms of linear dynamical systems, Am. J. Math. 58 (1) (1936) 141–163.
[9] R. Simon, Peres-horodecki separability criterion for continuous variable systems, Phys. Rev. Lett. 84 (2000) 2726.
[10] L.-M. Duan, G. Giedke, J.I. Cirac, P. Zoller, Inseparability criterion for continuous variable systems, Phys. Rev. Lett. 84 (2000) 2722.
[11] A. Serafini, F. Illuminati, S. De Siena, Symplectic invariants, entropic measures and correlations of Gaussian states, J. Physiol. Biochem. 37 (2004) L21–L28.
[12] A. Serafini, Multimode uncertainty relations and separability of continuous variable states, Phys. Rev. Lett. 96 (2006) 110402.
[13] S. Pirandola, A. Serafini, S. Lloyd, Correlation matrices of two-mode bosonic systems, Phys. Rev. 79 (2009) 052327.
[14] I.B. Djordjevic, Quantum Biological Information Theory, Springer International Publishing Switzerland, Cham-Heidelberg-New York, 2016.
[15] D. McMahon, Quantum Computing Explained, John Wiley & Sons, Hoboken, NJ, 2008.

CHAPTER

Quantum error correction fundamentals

7

Chapter outline

Quantum Communication, Quantum Networks, and Quantum Sensing. https://doi.org/10.1016/B978-0-12-822942-2.00003-0

7.1 Pauli operators (revisited)

Pauli operators $G = \{I,X,Y,Z\}$ [1–26], as already introduced in Chapter 3, can be represented in matrix form as follows:

$$X = \sigma_x \doteq \begin{bmatrix} 0 & 1 \\ 1 & 0 \end{bmatrix}, \quad Y = \sigma_y \doteq \begin{bmatrix} 0 & -j \\ j & 0 \end{bmatrix}, \quad Z = \sigma_z \doteq \begin{bmatrix} 1 & 0 \\ 0 & -1 \end{bmatrix}. \tag{7.1}$$

Their action on qubit $|\psi\rangle = a|0\rangle + b|1\rangle$ can be described as follows:

$$X(a|0\rangle + b|1\rangle) = a|1\rangle + b|0\rangle \quad Y(a|0\rangle + b|1\rangle) = j(a|1\rangle - b|0\rangle) \quad Z(a|0\rangle + b|1\rangle) = a|0\rangle - b|1\rangle \tag{7.2}$$

Therefore, the action of the X-operator is to introduce the bit-flip, the action of the Z-operator is to introduce the phase-flip, and the action of the Y-operator is to simultaneously introduce the bit- and phase-flips. The properties of Pauli operators can be summarized as

$$\begin{aligned} &X^2 = I, \ Y^2 = I, \ Z^2 = I \\ &XY = jZ \quad YX = -jZ \quad YZ = jX \quad ZY = -jX \quad ZX = jY \quad XZ = -jY \end{aligned} \tag{7.3}$$

The properties of Pauli operators given by Eq. (7.3) are summarized in the following multiplication table:

×	I	X	Y	Z
I	I	X	Y	Z
X	X	I	jZ	$-jY$
Y	Y	$-jZ$	Y	jX
Z	Z	jY	$-jX$	I

It is interesting to note that two Pauli operators commute only when they are identical or one of them is an identity operator; otherwise, they anticommute. Because set G is not closed under multiplication, it is not multiplicative. However, from Chapter 3 we know that two states $|\psi\rangle$ and $e^{j\theta}|\psi\rangle$ are equal up to the global phase shift because the measurement results are the same. Therefore, in quantum error correction, we often can simply omit the imaginary unit j. The corresponding multiplication table in which the imaginary unit is ignored is as follows:

×	I	X	Y	Z
I	I	X	Y	Z
X	X	I	Z	Y
Y	Y	Z	Y	X
Z	Z	Y	X	I

Clearly, all operators now commute, and the corresponding set G' with such defined multiplication is a commutative (Abelian) group. Such a group can be called a "projective" Pauli group [16] and is isomorphic to the group of binary 2-tuples $(Z_2)^2 = \{00,01,10,11\}$, with the addition table given by

+	00	01	11	10
00	00	01	11	10
01	01	00	10	11
11	11	10	00	01
10	10	11	01	00

The projective Pauli group is also isomorphic to the quarternary group $F_4 = \{0, 1, \omega, \overline{\omega}\}$, $\overline{\omega} = 1 + \omega$, with the corresponding addition table given by

+	0	$\overline{\omega}$	1	ω
0	0	$\overline{\omega}$	1	ω
$\overline{\omega}$	$\overline{\omega}$	0	ω	1
1	1	ω	0	$\overline{\omega}$
ω	ω	1	$\overline{\omega}$	0

Based on these tables, we can establish the following correspondence among the projective Pauli groups G', $(Z_2)^2$, and F_4:

G'	$(Z_2)^2$	F_4
I	00	0
X	01	$\overline{\omega}$
Y	11	1
Z	10	ω

This correspondence can relate the quantum codes to classical codes over $(Z_2)^2$ and F_4.

In addition to Pauli operators, the Hadamard operator (gate) will be used a great deal in this chapter, and its action can be described as

$$H(a|0\rangle + b|1\rangle) = a\frac{1}{\sqrt{2}}(|0\rangle + |1\rangle) + b\frac{1}{\sqrt{2}}(|0\rangle - |1\rangle) = \frac{1}{\sqrt{2}}[(a+b)|0\rangle + (a-b)|1\rangle] \tag{7.4}$$

The *quantum error correction code* (QECC) can be defined as a mapping from the K-qubit space to the N-qubit space. To facilitate its definition, we introduce the concept of Pauli operators on multiple qubits. A *Pauli operator on N qubits* has the form $cO_1O_2 \ldots O_N$, where each operator $O_i \in \{I, X, Y, Z\}$, and $c = j^l$ $(l = 1,2,3,4)$. This operator takes $|i_1 i_2 \ldots i_N\rangle$ to $cO_1|i_1\rangle \otimes O_2|i_2\rangle \ldots \otimes O_N|i_N\rangle$. For example, the action of $IXZ(|000\rangle + |111\rangle) = |010\rangle - |101\rangle$ is to bit-flip the second qubit and phase-flip the third qubit if it was $|1\rangle$. For convenience, we often use a shorthand notation to represent Pauli operators in which only the nonidentity operators O_i are written, while the identity operators are assumed. For example, *IXIZI* can be denoted X_2Z_4, meaning that operator X acts on the second qubit and operator Z on the fourth qubit (the action of identity operators I to other qubits are simply omitted since they do not cause any change). Two Pauli operators on multiple qubits *commute* if and only if there is an even number of places where they have different Pauli matrices, neither of which is identity I. For example,

XXI and IYZ do not commute, whereas XXI and ZYX do commute. If two Pauli operators do not commute, they anticommute since their individual Pauli matrices either commute or anticommute.

The set of Pauli operators on *N*-qubits forms the *multiplicative Pauli group* G_N. For the multiplicative group we can define the *Clifford operator* [8] U as the operator that preserves the elements of Pauli group under conjugation, namely $\forall\ O \in G_N : UOU^{\dagger} \in G_N$. The encoded operator for quantum error correction typically belongs to the Clifford group. To implement any unitary operator from the Clifford group, the use of CNOT U_{CNOT}, Hadamard H, and phase gate P is sufficient. The operation of H-gate is already given by (7.4), the action of P-gate can be described as $P(a|0\rangle + b|1\rangle) = a|0\rangle + jb|1\rangle$, and the action of CNOT-gate can be described as $U_{CNOT}|i\rangle|j\rangle = |i\rangle|i \oplus j\rangle$.

Every quantum channel error E (discrete or continuous) can be described as a linear combination of elements from the following discrete set $\{I,X,Y,Z\}$, given by $E = e_1I + e_2X + e_3Y + e_4Z$. For example, the following error $E = \begin{bmatrix} 1 & 0 \\ 0 & 0 \end{bmatrix}$ can be represented in terms of Pauli operators as $E = (I + Z)/2$. An error operator that affects several qubits can be written as a weighted sum of Pauli operators $\sum c_iP_i$ acting on ith qubits. An error may act not only on the code qubits but also on the environment. Given an initial state $|\psi\rangle\ |\phi\rangle^e$, which is a tensor product of code qubits $|\psi\rangle$ and environmental states $|\phi\rangle^e$, any error acting on both the code and the environment can be written as a weighted sum $\sum c_{i,j} P_iP_j^e$ of Pauli operators that act on both code and environment qubits. If S_i are syndrome operators that identify the error term $P_{\alpha'}|\psi\rangle$, then the operators S_iI^e will pick up terms of the form $\sum_{\beta'} c_{\alpha',\beta'}\, P_{\alpha'}|\psi\rangle\, P^e_{\beta'}\ |\phi\rangle^e$ from the quantum noise affected state, and these terms can be written as $P_{\alpha'}|\psi\rangle\ |\mu\rangle^e$ for some new environmental state $|\mu\rangle^e$. Therefore, measuring the syndrome restores a tensor product state of qubits and the environment, suggesting that the code and the environment evolve independently.

7.2 Quantum error correction concepts

Quantum error correction is essentially more complicated than classical error correction. The difficulties for quantum error correction can be summarized as follows. (1) The no-cloning theorem indicates that it is impossible to make a copy of an arbitrary quantum state $|\psi\rangle = \alpha|0\rangle + \beta|0\rangle$. If this is possible, then $|\psi\rangle|\psi\rangle = \alpha^2|00\rangle + \alpha\beta|01\rangle + \alpha\beta|10\rangle + \beta^2|11\rangle$ must be $\alpha\beta = 0$, so state $|\psi\rangle$ is no longer arbitrary. (2) Quantum errors are continuous, and a qubit can be in any superposition of the two bases states. (3) The measurements destroy the quantum information. A quantum error correction consists of four major steps: encoding, error detection, error recovery, and decoding. The elements of QECCs are shown in Fig. 7.1. Based on the discussion on quantum errors and Pauli operators on *N*-qubits, the

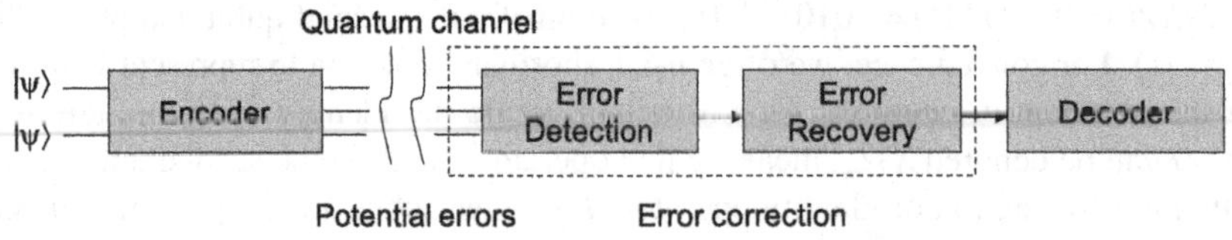

FIGURE 7.1

Quantum error correction principle.

QECC can be defined as follows. The [N,K] *QECC* performs encoding of the quantum state of K qubits specified by 2^K complex coefficients α_s ($s = 0,1, \ldots,2^K-1$), into a quantum state of N qubits in such a way that errors can be detected and corrected, and all 2^K complex coefficients can be perfectly restored up to the global phase shift. The sender (Alice) encodes quantum information in state $|\psi\rangle$ with the help of local ancilla qubits $|0\rangle$ and then sends the encoded qubits over a noisy quantum channel (say, free-space optical or optical fiber). The receiver (Bob) performs multiqubit measurement on all qubits to diagnose the channel error and performs a recovery unitary operation R to reverse the action of the channel. The quantum error correction principles will be more evident after several simple quantum codes, as provided below.

7.2.1 Three-qubit flip code

Assume we want to send a single qubit $|\Psi\rangle = \alpha|0\rangle + \beta|1\rangle$ through the quantum channel in which, during transmission, the transmitted qubit can be flipped to $X\,|\Psi\rangle = \beta\,|0\rangle + \alpha\,|1\rangle$ with probability p. Such a quantum channel is called a *bit-flip channel*, as shown in Fig. 7.2A.

The three-qubit flip code sends the same qubit three times and therefore represents the repetition code equivalent. The corresponding codewords in this code are $|\bar{0}\rangle = |000\rangle$ and $|\bar{1}\rangle = |111\rangle$. The three-qubit flip code encoder is shown in Fig. 7.2B. One input qubit and two ancillae are used at the encoder inputs and can be represented by $|\psi_{123}\rangle = \alpha|000\rangle + \beta|100\rangle$. The first ancilla qubit (the second qubit at the encoder input) is controlled by the information qubit (the first qubit at the encoder input), so its output is represented by $CNOT_{12}(\alpha|000\rangle + \beta|100\rangle) = \alpha|000\rangle + \beta|110\rangle$ (if the control qubit is $|1\rangle$, the target qubit gets flipped; otherwise, it stays unchanged). The output of the first CNOT gate is used as input to the second CNOT gate in which the second ancilla qubit (the third qubit) is controlled by the information qubit (the first qubit), so the corresponding encoder output is obtained as $CNOT_{13}(\alpha|000\rangle + \beta|110\rangle) = \alpha|000\rangle + \beta|111\rangle$, which indicates that the basis codewords are indeed $|\bar{0}\rangle$ and $|\bar{1}\rangle$.

With this code, we can correct a single qubit-flip, which occurs with probability $(1-p)^3 + 3p(1-p)^2 = 1 - 3p^2 + 2\,p^3$. Therefore, the probability of an error remaining uncorrected or wrongly corrected with this code is $3p^2 - 2\,p^3$. It is clear from Fig. 7.2B that the three-qubit bit-flip encoder is a *systematic encoder* in which the information qubit is unchanged, and the ancilla qubits are used to impose the encoding operation and create the parity qubits (output qubits 2 and 3).

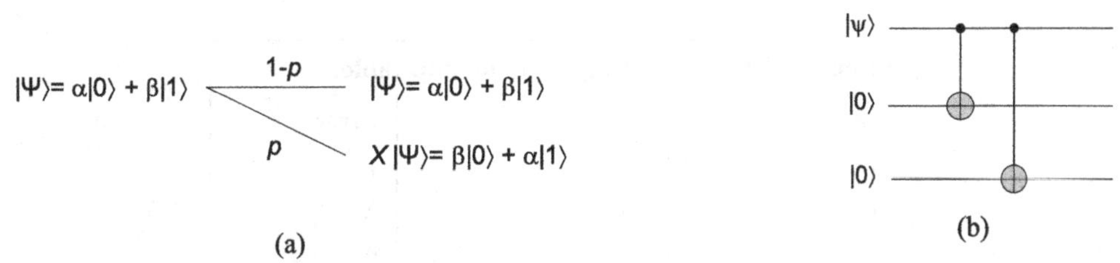

FIGURE 7.2

(A) Bit-flipping channel model. (B) Three-qubit flip code encoder.

Example. Let us assume that a qubit flip occurred on the first qubit leading to received quantum word $|\Psi_r\rangle = \alpha|100\rangle + \beta|011\rangle$. The error correction, as indicated above, consists of two steps. The first step is the error detection, in which the measurement is performed with projection operators P_i $(i = 0,1,2,3)$, defined as

$$P_0 = |000\rangle\langle 000| + |111\rangle\langle 111| \quad P_1 = |100\rangle\langle 100| + |011\rangle\langle 011|$$
$$P_2 = |010\rangle\langle 010| + |101\rangle\langle 101| \quad P_3 = |001\rangle\langle 001| + |110\rangle\langle 110|$$

The projection operator P_i determines whether bit-flip error occurred at the ith qubit location, and P_0 means there is no bit-flip error at all. The syndrome measurements give the following result:

$$\langle\psi_r|P_0|\psi_r\rangle = 0 \;\; \langle\psi_r|P_1|\psi_r\rangle = 1 \;\; \langle\psi_r|P_2|\psi_r\rangle = 0 \;\; \langle\psi_r|P_3|\psi_r\rangle = 0,$$

which can be represented as syndrome vector $S = [0\ 1\ 0\ 0]$, indicating that the error occurred in the first qubit. The second step is error recovery, in which we flip the first qubit back to the original by applying the X_1 operator to it. Another approach is to perform the measurements on the observables Z_1Z_2 and Z_2Z_3. The measurement result is eigenvalue ± 1, and the corresponding eigenvectors are two valid codewords, namely $|000\rangle$ and $|111\rangle$. The observables can be represented as follows:

$$Z_1Z_2 = (|00\rangle\langle 11| + |11\rangle\langle 11|)\otimes I - (|01\rangle\langle 01| + |10\rangle\langle 10|)\otimes I$$
$$Z_2Z_3 = I\otimes(|00\rangle\langle 11| + |11\rangle\langle 11|) - I\otimes(|01\rangle\langle 01| + |10\rangle\langle 10|)$$

It can be shown that $\langle\psi_r|Z_1Z_2|\psi_r\rangle = -1$, $\langle\psi_r|Z_2Z_3|\psi_r\rangle = +1$, indicating that an error occurred on either the first or second qubit but not on the second or third qubit. The intersection reveals that the first qubit was in error. Using this approach, we can create the tree-qubit lookup table (LUT), given as Table 7.1.

Three-qubit flip code error detection and error correction circuit is shown in Fig. 7.3.

In Chapter 3, we learned how to measure an observable without completely destroying the quantum information with the help of an ancilla and two Hadamard gates to perform the measurement on the ancilla instead. Because we need two observables (Z_1Z_2 and Z_2Z_3) for error detection, we need two ancillae and four Hadamard gates (see Fig. 7.3A). The results of measurements on ancillae determine the error syndrome $[\pm 1 \pm 1]$, and based on the LUT given by Table 7.1, we identify the error event and apply the corresponding X_i gate on the ith qubit being in error, and the error is corrected since $X^2 = I$. The control logic operation is described in Table 7.1. For example, if both outputs at the measurements' circuits are -1, the operator X_2 is activated. The last step is to perform decoding, as shown in Fig. 7.3B, by simply reversing the order of elements in the corresponding encoder.

Table 7.1 Three-qubit flip-code lookup table.

$Z_1 Z_2$	$Z_2 Z_3$	Error
+1	+1	I
+1	−1	X_3
−1	+1	X_1
−1	−1	X_2

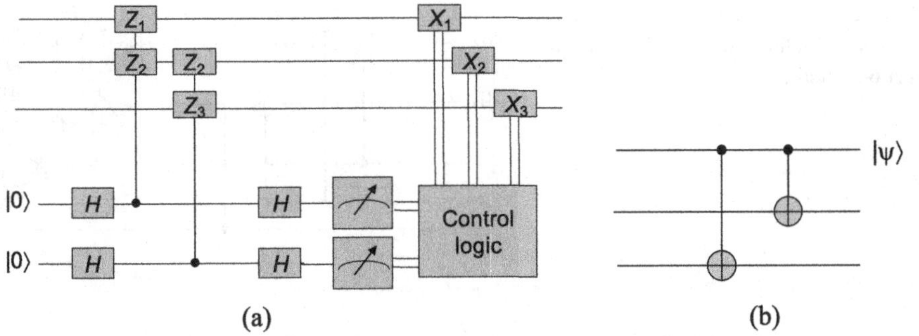

FIGURE 7.3

(A) Three-qubit flip code error detection and error correction circuit. (B) Decoder circuit configuration.

From the universal quantum gates section, we know that CNOT and Hadamard gates are sufficient for many QECCs. Therefore, we can use the equivalent circuits to measure the Z and X operators shown in Fig. 7.4A and B, respectively, and implement the three-qubit flip code error detection and correction circuits as shown in Fig. 7.5. The corresponding LUT for this representation is given as Table 7.2. The control logic operates as described in this table. For example, if $M_1 = M_2 = 1$, operator X_2 is activated.

7.2.2 Three-qubit phase flip code

Assume we want to send a single qubit $|\Psi\rangle = \alpha|0\rangle + \beta|1\rangle$ through a quantum channel for which, during transmission, the qubit can be phase-flipped as $Z|\Psi\rangle = \alpha|0\rangle - \beta|1\rangle$ with a certain probability p. Such a quantum channel is known as a quantum *phase-flip channel*, and the corresponding channel model is shown in Fig. 7.6.

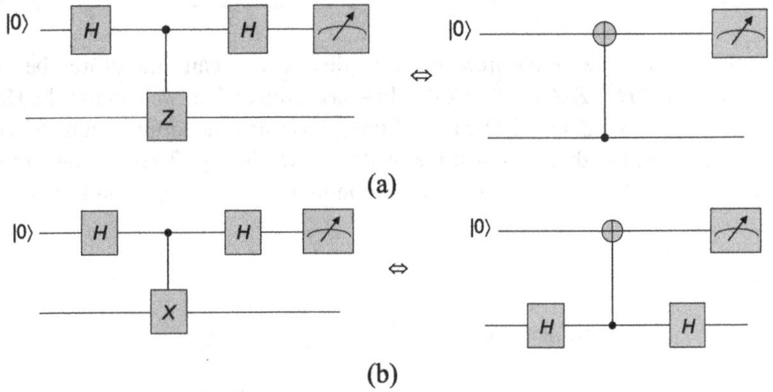

FIGURE 7.4

Equivalent circuits for measuring the (A) Z operator and (B) X operator.

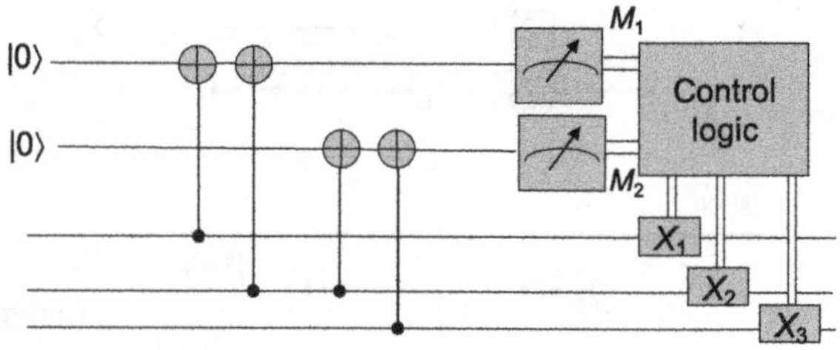

FIGURE 7.5

Equivalent three-qubit flip code error detection and error-correction circuit.

Table 7.2 Three-qubit code syndrome lookup table for the equivalent circuit shown in Fig. 7.6.

M_1	M_2	Error
0	0	I
1	0	X_1
1	1	X_2
0	1	X_3

To protect against phase-flip errors, we work with a different computational basis (CB) instead, known as the diagonal basis:

$$|+\rangle=\frac{|0\rangle+|1\rangle}{\sqrt{2}} \quad |-\rangle=\frac{|0\rangle-|1\rangle}{\sqrt{2}} \tag{7.5}$$

In this basis, the phase-flip operator Z acts as an ordinary qubit-flip operator because the action on new CB states is

$$Z|+\rangle=\frac{Z|0\rangle+Z|1\rangle}{\sqrt{2}}=\frac{|0\rangle-|1\rangle}{\sqrt{2}}=|-\rangle \quad Z|-\rangle=\frac{Z|0\rangle-Z|1\rangle}{\sqrt{2}}=\frac{|0\rangle+|1\rangle}{\sqrt{2}}=|+\rangle \tag{7.6}$$

The alternative syndrome measurement for this code can therefore be described as $H^{\otimes 3}Z_1Z_2H^{\otimes 3}=X_1X_2$ and $H^{\otimes 3}Z_2Z_3H^{\otimes 3}=X_2X_3$. For base conversion, we can use the Hadamard gate, whose action is described by Eq. (7.4). The three-qubit phase-flip encoding circuit shown in Fig. 7.7A can be implemented using the three-qubit flip encoder shown in Fig. 7.3B, followed by three Hadamard gates to perform the basis conversion. Corresponding error detection and recovery circuits are provided in Fig. 7.7B.

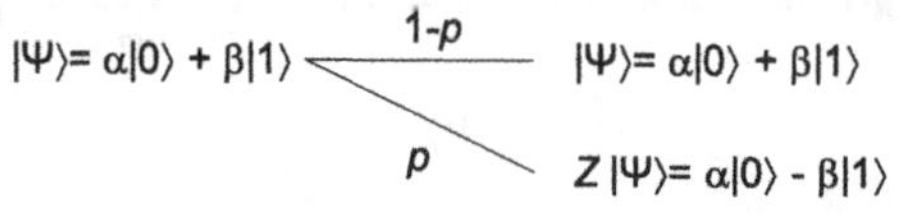

FIGURE 7.6

Phase-flipping channel model.

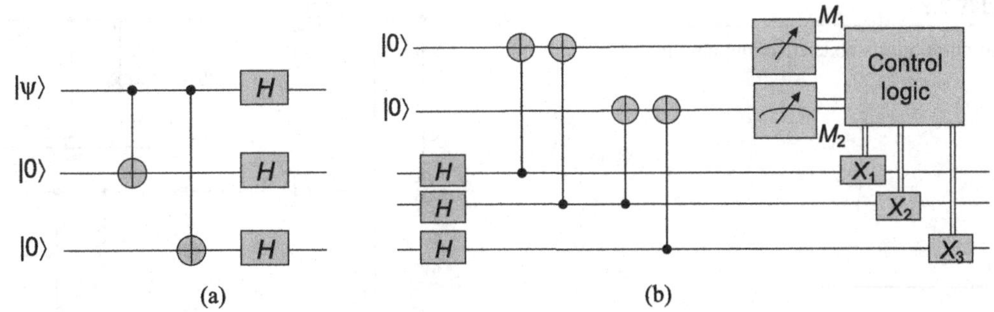

FIGURE 7.7

Three-qubit phase-flip: (A) Encoder circuit and (B) Error detection and recovery circuits.

7.2.3 Shor's nine-qubit code

The key idea of Shor's nine-qubit code is to first encode the qubit using the phase-flip encoder and then encode each of the three resulting qubits again with the bit-flip encoder, therefore providing protection against both qubit-flip and phase-flip errors. The two base codewords can be obtained, based on this description, as follows:

$$
\begin{aligned}
|\overline{0}\rangle &= \frac{1}{2\sqrt{2}}(|000\rangle + |111\rangle)(|000\rangle + |111\rangle)(|000\rangle + |111\rangle) \\
&= \frac{1}{2\sqrt{2}}(|000000000\rangle + |000000111\rangle + |000111000\rangle + |111000000\rangle + |000111111\rangle \\
&\quad + |111111000\rangle + |111111111\rangle) \\
|\overline{1}\rangle &= \frac{1}{2\sqrt{2}}(|000\rangle - |111\rangle)(|000\rangle - |111\rangle)(|000\rangle - |111\rangle) \\
&= \frac{1}{2\sqrt{2}}(|000000000\rangle - |000000111\rangle - |000111000\rangle - |111000000\rangle + |000111111\rangle \\
&\quad + |111111000\rangle - |111111111\rangle)
\end{aligned}
\tag{7.7}
$$

The Shor's encoder, shown in Fig. 7.8, is composed of two sections: (1) the phase-flip encoder and (2) every qubit from the phase-flip encoder, as further encoded using the qubit-flip encoder. Therefore, this code can be considered a *concatenated code*.

The Shor's nine-qubit error correction and decoder circuit, shown in Fig. 7.9, comprises two stages: (1) the first stage, composed of three three-qubit bit-flip detector, recovery, and decoder circuits followed by three Hadamard gates to perform CB conversion; and (2) an additional three-qubit bit-flip error detector, recovery, and decoder circuit.

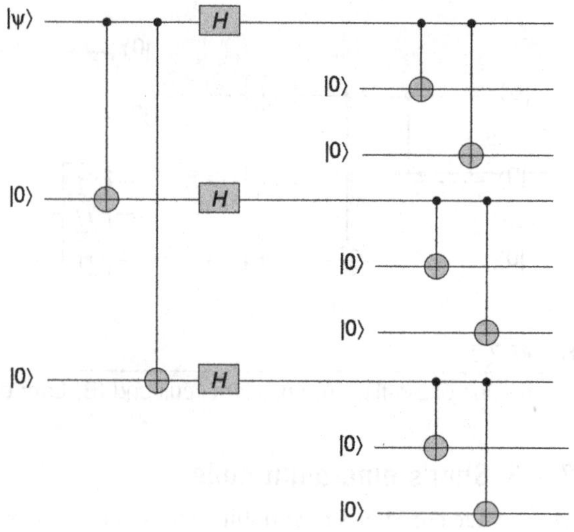

FIGURE 7.8

Nine-qubit Shor's code encoding circuit.

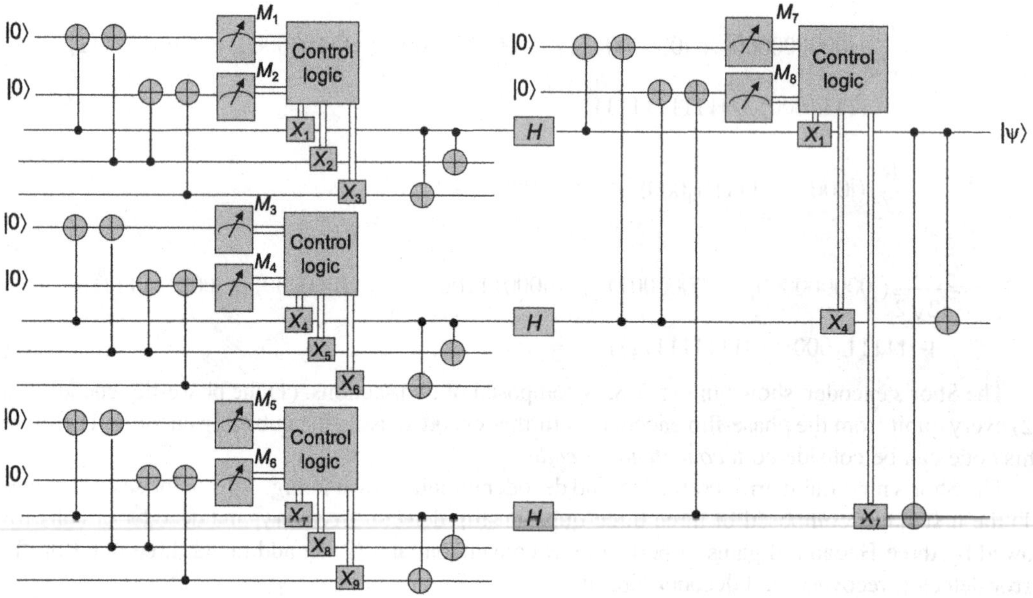

FIGURE 7.9

Error correction and decoding circuit for nine-qubit Shor's code.

Because the component codes in Shor's code are systematic, while the overall code is not systematic, the first, fourth, and seventh qubits are used after CB conversion as inputs to the second-stage three-qubit bit-flip code detector, recovery, and detector circuit.

In the next section, we describe an important class of QECCs, namely stabilizer codes. Stabilizer codes have been described in this section on a conceptual level, while in Section 7.3, they are described more formally. Because this is an important class of quantum codes, the next chapter is devoted entirely to stabilizer codes.

7.2.4 Stabilizer code concepts

To facilitate the description of stabilizer codes, we introduce several definitions. The Pauli group over one qubit is defined as $G_1 = \{\pm I, \pm jI, \pm X, \pm jX, \pm Y, \pm jY, \pm Z, \pm jZ\}$. The Pauli group G_N on N qubits consists of all Pauli operators of the form $cO_1O_2 \ldots O_N$, where $O_i \in \{I, X, Y, \text{or } Z\}$, and $c \in \{1, -1, j, -j\}$. Therefore, any Pauli vector on N qubits can be written uniquely as the product of X- and Z-containing operators together with the phase factor (± 1 or $\pm$ j). In a stabilizer framework, a codeword is defined to be a ket $|\psi\rangle$ that is a $+1$ eigenket of all stabilizers s_i, so $s_i\,|\psi\rangle = |\psi\rangle$ for all i. In other words, the set of operators s_i that "stabilizes" ("fixes") the codeword states forms the group S known as the stabilizer group. A *stabilizer group* S consists of a set of Pauli matrices (X, Y, Z together with multiplicative factors ± 1, $\pm$ j) with the property that any two operators in group S commute, so they can be simultaneously measured. We say that an operator fixes the state if the state is an eigenket with eigenvalue $+1$ for this operator.

Example: An alternative syndrome measurement basis for Shor's code is given by Z_1Z_2; Z_2Z_3; Z_3Z_4; Z_4Z_5; Z_5Z_6; Z_7Z_8; Z_8Z_9; $X_1X_2X_3X_4X_5X_6$; and $X_4X_5X_6X_7X_8X_9$. The Z-syndromes are used for bit-flip error detection and X-syndromes for phase-flip error detection. The key idea of stabilizer formalism is that quantum states can be more efficiently described by working on the operators that stabilize them than by working explicitly on the state. All operators that fix the base codewords of Shor's code $|\overline{0}\rangle$ and $|\overline{1}\rangle$, introduced earlier, can be written as the product of eight operators M_i, as shown in Table 7.3.

The stabilizer S is an Abelian subgroup of G_N. Let V_S denote the vector space stabilized by S, that is, the set of states on N qubits that are fixed by every element from S. The stabilizer S is described by its generators g_i. Any subgroup of G_N can be used as a stabilizer provided that the following two

Table 7.3 Stabilizer table for Shor's code.

	1	2	3	4	5	6	7	8	9	
M_1	Z	Z	I	I	I	I	I	I	I	Z_1Z_2
M_2	I	Z	Z	I	I	I	I	I	I	Z_2Z_3
M_3	I	I	I	Z	Z	I	I	I	I	Z_4Z_5
M_4	I	I	I	I	Z	Z	I	I	I	Z_5Z_6
M_5	I	I	I	I	I	I	Z	Z	I	Z_7Z_8
M_6	I	I	I	I	I	I	I	Z	Z	Z_8Z_9
M_7	X	X	X	X	X	X	I	I	I	$X_1X_2X_3X_4X_5X_6$
M_8	I	I	I	X	X	X	X	X	X	$X_4X_5X_6X_7X_8X_9$

conditions are satisfied: (1) the elements of S commute with each other, and (2) $-I$ is not an element of S.

An $[N,K]$ *stabilizer code* is defined as a vector subspace V_S stabilized by $S=<g_1,g_2,\ldots,g_{N-K}>$. Any element s from S can be written as a unique product of the powers of the generators, as described in Section 7.3. This interpretation is very similar to classical error correction, in which the basis vectors g_i in the code space have a role similar to that of stabilizer generators in quantum error correction.

Example: Let us observe the Bell (EPR) state, $|\psi\rangle = (|00\rangle + |11\rangle)/\sqrt{2}$. It can be easily shown that $X_1X_2|\psi\rangle = |\psi\rangle$ and $Z_1Z_2|\psi\rangle = |\psi\rangle$. Therefore, the EPR state is stabilized by X_1X_2 and Z_1Z_2. It can be shown that $|\psi\rangle$ is the unique state stabilized by the two operators.

Let us analyze what happens when a unitary gate U is applied to a vector subspace V_S stabilized by S:

$$s \in S:\ U|\psi\rangle = Us|\psi\rangle = UsU^+U|\psi\rangle = \left(UsU^+\right)U|\psi\rangle \tag{7.8}$$

Therefore, the state $U|\psi\rangle$ is stabilized by UsU^+. We say the set of U such that $UG_NU^+ = G_N$ is the *normalizer* of G_N, denoted $N(G_N)$. With this formalism, we can define the *distance* of a stabilizer code to be the minimum weight of an element of $N(S)-S$. More details of the stabilizer group and stabilizer codes can be found in Section 7.3. In the next section, we instead turn our attention to establishing the relationship between quantum and classical codes.

7.2.5 Relationship between quantum and classical codes

As indicated previously, we can write any Pauli operator on N qubits uniquely as a product of X-containing operators and Z-containing operators and a phase factor (± 1, $\pm$j). For instance, XIYZYI = – (XIXIXI) • (IIZZZI). We can now express the X operator as a binary string of length N, with "1" standing for X and "0" for I, and do the same for the Z operator. Thus, each stabilizer can be written as the X string followed by the Z string, giving a matrix of width $2N$. We mark the boundary between the two types of strings with vertical bars, so, for instance, the set of generators of Shor's code appears as the *quantum check matrix* A. The quantum check matrix of Shor's code (see Table 7.3) is given by

$$A = \begin{bmatrix} X \quad\quad Z \\ 111111000|000000000 \\ 000111111|000000000 \\ 000000000|110000000 \\ 000000000|011000000 \\ 000000000|000110000 \\ 000000000|000011000 \\ 000000000|000000110 \\ 000000000|000000011 \end{bmatrix}$$

The commutativity of stabilizers now appears as the *orthogonality of rows* with respect to a *twisted (symplectic) product*, formulated as follows—if the *k*th row is $r_k = (x_k; z_k)$, where x_k is the *X* binary string and z_k the *Z* string, the twisted product of rows *k* and *l* is defined by

$$r_k \odot r_l = x_k \cdot z_l + x_l \cdot z_k \text{mod} 2, \tag{7.9}$$

where $x_k \cdot z_l$ is the dot (scalar) product defined by $x_k \cdot z_l = \sum_j x_{kj} z_{lj}$. The twisted product is zero if and only if there is an even number of places where the operators corresponding to rows *k* and *l* differ (and are not the identity), i.e., if the operators commute. If we write the quantum-check **A** as $\mathbf{A} = (\mathbf{A}_1|\mathbf{A}_2)$, the condition that the twisted product is zero for all *k* and *l* can be written compactly as

$$\mathbf{A}_1\mathbf{A}_2^T + \mathbf{A}_2\mathbf{A}_1^T = \mathbf{0} \tag{7.10}$$

A *Pauli error operator E* can be interpreted as a binary string ***e*** of length 2*N* in which we reverse the order of *X* and Z strings. For example, $E = Z_1X_2Y_9$ can be written as $\mathbf{e} = [100000001|010000001]$. Thus, the quantum syndrome for the noise is exactly the classical syndrome $\mathbf{eA}^{\mathrm{T}}$, considering **A** as a parity-check matrix and ***e*** as a binary noise vector. In this formalism, we perform the ordinary dot product (mod 2) of error vector ***e*** with the corresponding row of the quantum-check matrix. If the result is zero, the error operator and the corresponding stabilizer of the row commute; otherwise, the result is 1, and they do not commute.

In conclusion, the properties of stabilizer codes can be inferred from those of a special class of classical codes. Given any binary matrix of size $M \times 2N$ that has the property that the twisted product of any two rows is zero, an equivalent quantum code can be constructed that encodes $N-M$ qubits in *N* qubits. Several examples are provided in the next three subsections.

7.2.6 Quantum cyclic codes

An $(N;K) = (5;\ 1)$ quantum code is generated by the following four stabilizers: XZZXI, IXZZX, XIXZZ and ZXIXZ. The corresponding quantum-check matrix is given by

$$\mathbf{A} = \begin{bmatrix} X \quad Z \\ 10010|01100 \\ 01001|00110 \\ 10100|00011 \\ 01010|10001 \end{bmatrix}$$

A correctable set of errors consists of all operators with one nonidentity term, e.g., XIIII, IIIYI, or IZIII. These correspond to binary strings such as 00000|10000 for XIIII, 00010|00010 for IIIYI, and so on. There are 15 of these, and each has a distinct syndrome, thereby using 2^4-1 possible nonzero syndromes. Because this code satisfies the quantum Hamming inequality (see Section 7.4) with the equality sign, it belongs to the class of *perfect quantum codes*.

7.2.7 Calderbank–Shor–Steane codes

An important class of codes invented by Calderbank, Shor and Steane, known as CSS [1,2], has the form

$$A = \begin{bmatrix} H & | & 0 \\ 0 & | & G \end{bmatrix}, \quad HG^T = 0 \tag{7.11}$$

where $\boldsymbol{H}$ and $\boldsymbol{G}$ are M x N matrices. The condition $\boldsymbol{HG}^{\mathrm{T}} = \boldsymbol{0}$ ensures that the twisted product condition is satisfied. As there are $2M$ stabilizer conditions applying to N qubit states, $N - 2M$ qubits are encoded in N qubits. The special case of CSS codes is dial-containing codes, also known as weakly self-dual codes, in which $\boldsymbol{H} = \boldsymbol{G}$, for which the quantum check matrix $\boldsymbol{A}$ has the following form:

$$A = \begin{bmatrix} H & | & 0 \\ 0 & | & H \end{bmatrix}, \quad HH^T = 0 \tag{7.12}$$

The $\boldsymbol{HH}^{\mathrm{T}} = 0$ condition is equivalent to $C^{\perp}(\boldsymbol{H}) \subset C(\boldsymbol{H})$, where $C(\boldsymbol{H})$ is the code having $\boldsymbol{H}$ as its parity check matrix and $C^{\perp}(\boldsymbol{H})$ as its dual code. An example of a dual-containing code is *Steane's* seven-qubit code, defined by the Hamming (7,4) code:

$$H = \begin{bmatrix} 0001111 \\ 0110011 \\ 1010101 \end{bmatrix}$$

The rows have an even number of 1's, and any two of them overlap by an even number of 1's, so $C^{\perp}(\boldsymbol{H}) \subset C(\boldsymbol{H})$. Here $M = 3$, $N = 7$, and thus, $N - 2M = 1$, so one qubit is encoded in seven qubits, representing the [1,7] code.

7.2.8 Quantum codes over GF(4)

Let the elements of GF(4) be 0, 1, w, and $w^2 = w' = 1 + w$. One can write a row $r_1 = (a_1a_2 \ldots a_n|b_1b_2 \ldots b_n)$ of the quantum-check matrix in the binary string representation as a vector over GF(4) as follows: $\rho_1 = (a_1 + b_1w;\ a_2 + b_2w;\ \ldots;\ a_n + b_nw)$. Given a second row $r_2 = (c_1c_2 \ldots c_n|d_1d_2 \ldots d_n)$, with $\rho_2 = (c_1 + d_1w;\ c_2 + d_2w;\ \ldots;\ c_n + d_nw)$, the *Hermitian inner product* is defined by

$$\rho_1 \cdot \rho_2 = \sum_i (a_i + b_iw')(c_i + d_iw) = \sum_i [(a_ic_i + b_id_i + b_ic_i) + (a_id_i + b_ic_i)w]. \tag{7.13}$$

Because $a_id_i + b_ic_i = r_1 \odot r_2$, the orthogonality in the Hermitian sense $(\rho_1 \cdot \rho_2 = 0)$ leads to orthogonality in the simplectic sense, too $(r_1 \odot r_2 = 0)$. The opposite is not true, because the term $a_ic_i + b_id_i + b_ic_i$ is not necessarily zero when $a_id_i + b_ic_i = 0$. Therefore, the codes over GF(4) satisfying the property that any two rows are orthogonal in the Hermitian inner product sense can be used as quantum codes.

7.3 Quantum error correction

The previous section introduced quantum error correction on a conceptual level. Here, we describe quantum codes more formally. The section starts with redundancy, followed by the stabilizer group, and then quantum syndrome decoding. We also formally establish the connection between classical and quantum codes. We then discuss the necessary and sufficient conditions for quantum error correction and provide a detailed example of quantum error correction. Further, the distance properties of quantum codes are discussed, with distance related to error correction capability. Finally, we describe quantum encoder and decoder implementations.

7.3.1 Redundancy and quantum error correction

A QECC that encodes K qubits into N qubits, denoted $[N,K]$, is defined by an encoding mapping U from the K-qubit Hilbert space H_2^K onto a 2^K-dimensional subspace C_q of the N-qubit Hilbert space H_2^N. The subspace C_q is called the *code space*; the states belonging to C_q are called the *codewords*; and the encoded CB kets are called the *basis codewords*. The single-qubit CB states are typically chosen to be the eigenkets of Z_j:

$$Z_j|b_j\rangle = (-1)^{b_j}|b_j\rangle; \quad b_j = 0, 1; \quad j = 1, 2, \ldots, K \tag{7.14}$$

The CB states of $H^k{}_2$ are given by

$$|\boldsymbol{b}\rangle \equiv |b_1 \ldots b_k\rangle = |b_1\rangle \otimes \ldots \otimes |b_k\rangle \tag{7.15}$$

The encoded CB states are obtained by applying the encoding mapping to the CB states as follows:

$$|\overline{\boldsymbol{b}}\rangle \equiv |\overline{b_1 \ldots b_k}\rangle = U|b_1 \ldots b_k\rangle, \tag{7.16}$$

where the action of encoding mapping is given by $Z_j \overset{U}{\rightarrow} \overline{Z}_j = UZ_jU^\dagger$. The encoded CB states are simultaneous eigenkets of $\{Z_j: j = 1, \ldots, K\}$ because

$$\overline{Z}_j|\overline{\boldsymbol{b}}\rangle = UZ_jU^\dagger|\overline{b_1 \ldots b_k}\rangle = UZ_jU^\dagger(U|b_1 \ldots b_k\rangle) = UZ_j|\boldsymbol{b}\rangle = U(-1)^{b_j}|\boldsymbol{b}\rangle = (-1)^{b_j}|\overline{\boldsymbol{b}}\rangle \tag{7.17}$$

Therefore, the encoding mapping preserves the eigenvalues.

The simplest QECCs are *canonical codes* [17,23], which can be introduced by the following trivial encoding mapping U_c:

$$U_c: |\psi\rangle \rightarrow |0\rangle|\psi\rangle. \tag{7.18}$$

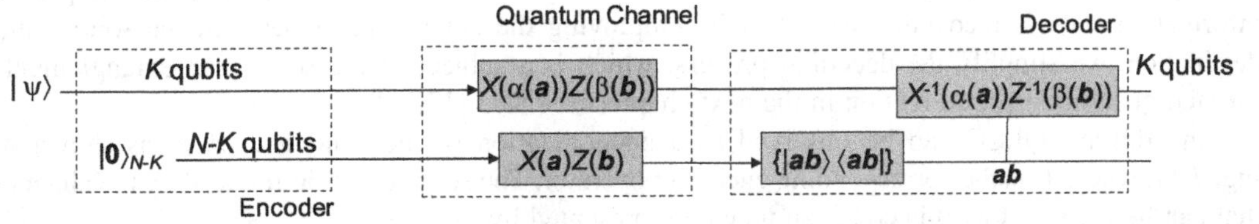

FIGURE 7.10

The canonical quantum error correction coding.

In canonical codes (see Fig. 7.10), the quantum register containing $N-K$ ancillae $|\mathbf{0}\rangle_{N-K} = \underbrace{|0\rangle \otimes \cdots \otimes |0\rangle}_{N-K}$ is appended to the information quantum register containing K-qubits. The basis for single-qubit errors is given by $\{I,X,Y,Z\}$. The basis for N-qubit errors is obtained by forming all possible direct products:

$$E = j^l O_1 \otimes \cdots \otimes O_N;\, O_i \in \{I, X, Y, Z\}, l = 0, 1, 2, 3. \tag{7.19}$$

The N-qubit error-basis can be transformed into the multiplicative *Pauli group* G_N if we allow E to be premultiplied by $j^{l'} = 1, j, -1, -j$ (for $l' = 0,1,2,3$). By noticing that $Y = -jXZ$, any error in the Pauli group of N-qubit errors can be represented by

$$\begin{aligned} &E = j^{l'} X(\boldsymbol{a})Z(\boldsymbol{b});\ \boldsymbol{a} = a_1 \cdots a_N;\ \boldsymbol{b} = b_1 \cdots b_N;\ a_i, b_i = 0, 1; l' = 0, 1, 2, 3 \\ &X(\boldsymbol{a}) \equiv X_1^{a_1} \otimes \cdots \otimes X_N^{a_N};\ Z(\boldsymbol{b}) \equiv Z_1^{b_1} \otimes \cdots \otimes Z_N^{b_N}. \end{aligned} \tag{7.20}$$

In Eq. (7.20), the subscript i ($i = 0,1, \ldots ,N$) denotes the location of the qubit to which operator X_i (Z_i) is applied; a_i (b_i) takes the value 1 if the ith X (Z) operator is to be included and the value 0 if the same operator is to be excluded. Therefore, to uniquely determine the error operator (up to the phase constant $j^{l'} = 1, j, -1, -j$ for $l' = 0,1,2,3$) it is sufficient to specify vectors $\boldsymbol{a}$ and $\boldsymbol{b}$. For example, the error $E = Z_1X_2Y_9 = -jZ_1X_2X_9Z_9$ can be identified (up to the phase constant) by specifying $\boldsymbol{a} = (010000001)$ and $\boldsymbol{b} = (100000001)$. Note that this representation is equivalent to the classical representation of error E as $\boldsymbol{e} = [100000001|010000001]$ by reversing the positions of vectors $\boldsymbol{a}$ and $\boldsymbol{b}$.

In Fig. 7.10, we employ this simplified notation and represent the quantum error as the product of X- and Z-containing operators. We observe separately possible errors introduced on information qubits, denoted $X(\alpha(\boldsymbol{a}))$ and $Z(\beta(\boldsymbol{b}))$, and errors introduced on ancillae, denoted $X(\boldsymbol{a})$ and $Z(\boldsymbol{b})$, where α and β are functions of $\boldsymbol{a}$ and $\boldsymbol{b}$, respectively. The action of a correctable quantum error E,

$$E \in \boldsymbol{E}_c = \left\{X(\boldsymbol{a})Z(\boldsymbol{b}) \otimes X(\alpha(\boldsymbol{a}))Z(\beta(\boldsymbol{b})) : \boldsymbol{a}, \boldsymbol{b} \in F_2^{N-K}; F_2 = \{0, 1\}\right\};\quad \alpha, \beta: F_2^{N-K} \to F_2^{N-K} \tag{7.21}$$

can be described as follows:

$$E(|\mathbf{0}\rangle|\psi\rangle) = X(\boldsymbol{a})Z(\boldsymbol{b})|0\rangle \otimes X(\alpha(\boldsymbol{a}))Z(\beta(\boldsymbol{b}))|\psi\rangle. \tag{7.22}$$

Since $X(\boldsymbol{a})Z(\boldsymbol{b})|\mathbf{0}\rangle = X(\boldsymbol{a})|\mathbf{0}\rangle = |\boldsymbol{a}\rangle$, we obtain

$$E(|\mathbf{0}\rangle|\psi\rangle) = |\boldsymbol{a}\rangle \otimes \overbrace{X(\alpha(\boldsymbol{a}))Z(\beta(\boldsymbol{b}))|\psi\rangle}^{|\psi'\rangle} = |\boldsymbol{a}\rangle|\psi'\rangle. \tag{7.23}$$

On the receiver side, we perform the measurements on ancillae to determine the syndrome $\boldsymbol{S} = (\boldsymbol{a},\boldsymbol{b})$ without affecting the information qubits. Once the syndromes are determined, we perform reverse recovery operator actions, denoted in Fig. 7.10 as $X^{-1}(\alpha(\boldsymbol{a}))$ and $Z^{-1}(\beta(\boldsymbol{b}))$, and the proper information state is recovered. Note that by employing the entanglement between the source and destination, we simplify the decoding process, which is a subject of the section on entanglement-assisted quantum error correction in the next chapter (see also [17,23]).

An arbitrary QECC can be observed as a generalization of the canonical code, as shown in Fig. 7.11, where U is the corresponding encoding operator. The set of errors introduced by the channel that can be corrected by this code can now be represented by

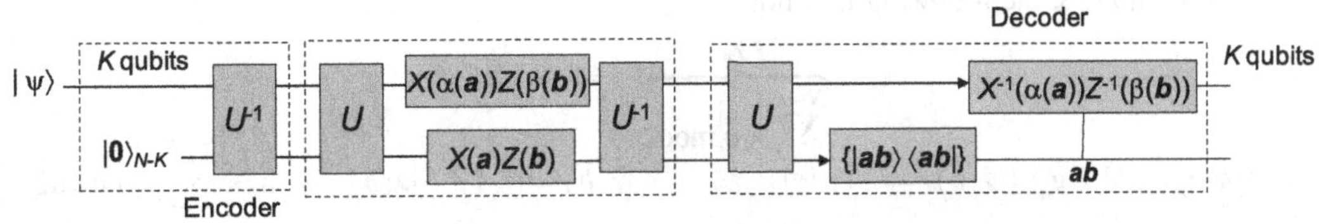

FIGURE 7.11

An arbitrary quantum error correcting code represented as a generalization of the canonical code.

$$E=\{U[X(\boldsymbol{a})Z(\boldsymbol{b})\otimes X(\alpha(\boldsymbol{a}))Z(\beta(\boldsymbol{b}))]U^{-1}:\boldsymbol{a},\boldsymbol{b}\in F_2^{N-K};F_2=\{0,1\}\};\quad \alpha,\beta:F_2^{N-K}\rightarrow F_2^{N-K} \tag{7.24}$$

7.3.2 Stabilizer group *S*

The quantum stabilizer code C_q with parameters $[N,K]$ can be defined as the unique subspace of H_2^N fixed by the elements from stabilizer S of C_q, as given by

$$\forall\ s\in S,\ |c\rangle\in C_q:\quad s|c\rangle=|c\rangle. \tag{7.25}$$

The stabilizer group S is constructed from a set of $N-K$ operators $g_1, \ldots, g_{N-K}$, also known as the *generators* of S. Generators have the following properties: (1) they commute among each other, (2) they are unitary and Hermitian, and (3) they have the order of 2 because $g^2{}_i=I$. Any element $s\in S$ can be written as a unique product of powers of the generators:

$$s=g_1^{a_1}\cdots g_{N-K}^{a_{N-K}},\ a_i\in\{0,1\};\ i=1,\cdots,N-K \tag{7.26}$$

When a_i is set to 1 (0), the ith generator is included (excluded) from the product of generators. Therefore, the elements from S can be labeled by simple binary strings of length $N-K$, namely $\boldsymbol{a}=a_1 \ldots a_{N-K}$. Here we can draw a parallel with classical error correction, in which any codeword in an LBC can be represented as a linear combination of basis codewords. The key difference is that the corresponding group in quantum error correction is multiplicative rather than additive. The eigenvalues of generators g_i, $\{\text{eig}(g_i)\}$, can be found starting from property 3:

$$|\text{eig}(g_i)\rangle=I|\text{eig}(g_i)\rangle=g_i^2|\text{eig}(g_i)\rangle=\text{eig}^2(g_i)|\text{eig}(g_i)\rangle \tag{7.27}$$

From Eq. (7.27), it is clear that $\text{eig}^2(g_i)=1$, meaning that the eigenvalues are $\text{eig}(g_i)=\pm1=(-1)^{\lambda_i};\lambda_i=0,1$. Because the "parent" space for C_q is a 2^N-dimensional space, we need to determine N commuting operators that specify a unique state $|\psi\rangle\in H_2^N$. The following 2^N simultaneous eigenstates of $\{g_1,g_2,\cdots,g_{N-K};\overline{Z}_1,\overline{Z}_2,\cdots,\overline{Z}_K\}$ can be used as the basis for H_2^N. Namely, we have chosen the subset $\{\overline{Z}_i\}$ in such a way that each element for this subset commutes with all generators $\{g_i\}$ and among all elements of the subset itself. The eigenkets can be labeled $\boldsymbol{\lambda}=\lambda_1 \ldots \lambda_{N-K}$, and the encoded CBs will be

$$g_i|\boldsymbol{\lambda};\overline{\boldsymbol{b}}\rangle=(-1)^{\lambda_i}|\boldsymbol{\lambda};\overline{\boldsymbol{b}}\rangle\quad \overline{Z}_j|\boldsymbol{\lambda};\overline{\boldsymbol{b}}\rangle=(-1)^{b_j}|\boldsymbol{\lambda};\overline{\boldsymbol{b}}\rangle;\ i=1,\ldots,N-K;j=1,\ldots,K;\lambda_i,b_j=0,1 \tag{7.28}$$

For every $s(\boldsymbol{a}) \in S$, the following is valid:

$$s(\boldsymbol{a})|\lambda;\bar{\boldsymbol{b}}\rangle = g_1^{a_1} \cdots g_{n-k}^{a_{n-k}}|\lambda;\bar{\boldsymbol{b}}\rangle = (-1)^{\overbrace{\sum_{i=1}^{N-K} \lambda_i a_i \bmod 2}^{\lambda \cdot a}} |\lambda;\bar{\boldsymbol{b}}\rangle = (-1)^{\lambda \cdot \boldsymbol{a}}|\lambda;\bar{\boldsymbol{b}}\rangle; \ \ \lambda \cdot \boldsymbol{a} = \sum_{i=1}^{N-K} \lambda_i a_i \bmod 2 \tag{7.29}$$

From Eq. (7.25), it is clear that

$$s(\boldsymbol{a})|\lambda;\bar{\boldsymbol{b}}\rangle = |\lambda;\bar{\boldsymbol{b}}\rangle \tag{7.30}$$

when $\lambda \cdot \boldsymbol{a} = 0 \bmod 2$. The simplest way to satisfy Eq. (7.30) is for $\lambda = 0 \ldots 0$, meaning that eigenkets $\{|\lambda = 0\ldots0;\bar{\boldsymbol{b}}\rangle\}$ can be used as encoded CB states.

The following interpretation of the same result is due to Gaitan [14]. The set $C(\lambda) = \{|\lambda;\bar{\boldsymbol{b}}\rangle : \bar{\boldsymbol{b}} \in F_2^K\}$ (where $F_2 = \{0,1\}$) is clearly the subspace of H_N^2, whose elements are simultaneous eigenvectors of the generators $g_1, \ldots, g_{N-K}$ with corresponding eigenvalues $(-1)^{\lambda_1}, \cdots, (-1)^{\lambda_{N-K}}$. It can be shown that the set of 2^{N-K} subspaces $\{C(\lambda)\text{: } \lambda \in F_2^{N-K}\}$ partitions H_2^N. Because the quantum stabilizer code C_q is fixed by stabilizer S, the states $|c\rangle \in C_q$ are fixed by $g_1, \ldots, g_{N-K}$ with corresponding eigenvalues $\text{eig}(g_i) = 1$ $(i = 1,2, \ldots, N-K)$, meaning that $C_q \subset C(\lambda = 0 \ldots 0)$. Since both C_q and $C(\lambda = 0 \ldots 0)$ are 2^K-dimensional, it must be that $C_q = C(\lambda = 0 \ldots 0)$. Therefore, the encoded CB states are the eigenkets $\{|\lambda = 0\ldots0;\bar{\boldsymbol{b}}\rangle\}$.

7.3.3 Quantum-check matrix and syndrome equation

In the previous section, we established the connection between quantum and classical codes but did not explain how this connection is derived, which is the topic of this subsection. The following two theorems are important in establishing the connection between classical and quantum codes.

Theorem 1. Let E be an error and S the stabilizer group for a quantum stabilizer code C_q. If S contains an element that anticommutes with E, then for all $|c\rangle$, $|c'\rangle \in C_q$, $E|c\rangle$ is orthogonal to $|c'\rangle$:

$$\langle c'|E|c\rangle = 0. \tag{7.31}$$

This theorem is quite straightforward to prove. Since $|c\rangle, |c'\rangle \in C_q$ and the error E anticommutes with $s \in S$ ($\{E, s\} = 0$), we can write

$$E|c\rangle = Es|c\rangle = -sE|c\rangle. \tag{7.32}$$

By multiplying with $\langle c'|$ from the left side, we obtain

$$\langle c'|E|c\rangle = -\langle c'|sE|c\rangle = -\langle c'|E|c\rangle. \tag{7.33}$$

By solving (Eq. 7.33) for $\langle c'|E|c\rangle$, we obtain $\langle c'|E|c\rangle = 0$.

Theorem 2. Let E be an error and C_q be a quantum stabilizer code with generators $g_1, \ldots, g_{N-K}$. The image $E(C_q)$ under E is $C(\lambda)$, where $\lambda = \lambda_1 \ldots \lambda_{N-K}$, such that

$$\lambda_i = \begin{cases} 0, & [E, g_i] = 0 \\ 1, & \{E, g_i\} = 0 \end{cases} \quad (i = 1, \ldots, N-K) \tag{7.34}$$

This theorem can be proved by observing the action of generator g_i and an error operator E on $|c\rangle \in C_q$:

$$g_i E|c\rangle = (-1)^{\lambda_i} E g_i|c\rangle = (-1)^{\lambda_i} E|c\rangle \tag{7.35}$$

It is clear from Eq. (7.35) that for $\lambda_i = 0$, E and g_i commute since $g_i E = E g_i$. On the other hand, for $\lambda_i = 1$, from Eq. (7.35), we obtain $g_i E = -E g_i$, which means that E and g_i anticommute ($\{E, g_i\} = 0$). These two cases can be written in one equation as follows:

$$\lambda_i = \begin{cases} 0, & [E, g_i] = 0 \\ 1, & \{E, g_i\} = 0 \end{cases} \quad (i = 1, \ldots, N-K),$$

therefore proving Eq. (7.34). What remains is to prove that the image $E(C_q)$ equals $C(\boldsymbol{\lambda}) = \{|\boldsymbol{\lambda}; \overline{\boldsymbol{\delta}}\rangle : \overline{\boldsymbol{\delta}} \in F_2^K\}$. The ket $E|c\rangle$ can be represented in terms of eigenkets $|\boldsymbol{\lambda}; \overline{\boldsymbol{a}}\rangle$ as follows:

$$E|c\rangle = \sum_{\boldsymbol{\lambda}'} \sum_{\boldsymbol{a}} \alpha(\boldsymbol{\lambda}'; \overline{\boldsymbol{\delta}}) |\boldsymbol{\lambda}'; \overline{\boldsymbol{a}}\rangle, \tag{7.36}$$

where $\alpha(\boldsymbol{\lambda}'; \overline{\boldsymbol{a}})$ are projections along basis kets $|\boldsymbol{\lambda}'; \overline{\boldsymbol{a}}\rangle$. Let us now employ the property we just derived and observe the simultaneous action of g_i and E on $|c\rangle$:

$$g_i E|c\rangle = \sum_{\boldsymbol{\lambda}'} \sum_{\overline{\boldsymbol{a}}} \alpha(\boldsymbol{\lambda}'; \overline{\boldsymbol{a}}) g_i |\boldsymbol{\lambda}'; \overline{\boldsymbol{a}}\rangle = \sum_{\boldsymbol{\lambda}} \sum_{\overline{\boldsymbol{a}}} (-1)^{\lambda_i} \alpha(\boldsymbol{\lambda}; \overline{\boldsymbol{a}}) |\boldsymbol{\lambda}; \overline{\boldsymbol{a}}\rangle. \tag{7.37}$$

It is clear from Eqs. (7.37) and (7.35), in which the right side of the equation is the only function of eigenevalues $(-1)^{\lambda_i}$, that $E|c\rangle$ is equal to

$$E|c\rangle = \sum_{\overline{\boldsymbol{a}}} \alpha(\boldsymbol{\lambda}; \overline{\boldsymbol{a}}) |\boldsymbol{\lambda}; \overline{\boldsymbol{a}}\rangle, \tag{7.38}$$

and therefore, $E(C_q) = C(\boldsymbol{\lambda})$.

Because $I(C_q) = C(0 \ldots . 0)$, where I is the identity operator, it is clear that $\boldsymbol{\lambda} = \lambda_1 \ldots \lambda_{N-K}$ can be used as the error syndrome of E, denoted $S(E)$. Therefore, the *error syndrome* is defined as $S(E) = \lambda_1 \ldots \lambda_{N-K}$, where λ_i is defined by Eq. (7.34). Based on the discussion from Section 7.2, we can represent the generator g_i as a binary vector of length $2N$: $\boldsymbol{g}_i = (\boldsymbol{a}_i|\boldsymbol{b}_i)$, where the row vector $\boldsymbol{a}_i$ ($\boldsymbol{b}_i$) is obtained from the operator representation by replacing the X-operators (Z-operators) by 1's and identity operators by 0's. The binary representations of all generators can be used to create the *quantum-check matrix* $\boldsymbol{A}$ obtained by writing the binary representation of g_i as the ith row of $(\boldsymbol{A})_i$ as follows:

$$\boldsymbol{A} = \begin{pmatrix} (\boldsymbol{A})_1 \\ \vdots \\ (\boldsymbol{A})_{N-K} \end{pmatrix}, (\boldsymbol{A})_i = \boldsymbol{g}_i = (\boldsymbol{a}_i|\boldsymbol{b}_i) \tag{7.39}$$

By representing the error operator E as binary error row-vector $\boldsymbol{e} = (\boldsymbol{c}|\boldsymbol{d})$, but now with $\boldsymbol{c}$ ($\boldsymbol{d}$) obtained from error operator E by replacing the Z-operators (X-operators) with 1's and identity operators by 0's, the syndrome equation can be written as

$$S(E) = \boldsymbol{e}\boldsymbol{A}^{\mathrm{T}}, \tag{7.40}$$

which is very similar to classical error correction.

7.3.4 Necessary and sufficient conditions for quantum error correction coding

The quantum error correction circuit determines the error syndrome with the help of ancilla qubits. The measured value of syndrome S determines a set of errors $\boldsymbol{E}_S = (E_1, E_2, \ldots)$ with syndromes equal to the measured value $S(E_i) = S$, $\forall\ E_i \in \boldsymbol{E}_S$. Analogous to classical error correction, this set of errors $\boldsymbol{E}_S$ having the same syndrome S can be called the *coset*. Among different candidate errors, we chose the most probable one, say E_S, for the expected error. We then perform a recovery operation by simply applying $E^{\dagger}{}_S$. If the actual error differs from true error E_S, the resulting state $E^{\dagger}{}_S\, E|s\rangle$ will be from C_q but different from the uncorrupted state $|s\rangle$. The following two conditions should be satisfied during the design of a quantum code: (1) the encoded CB states must be chosen carefully so that the environment cannot distinguish among different CBs, and (2) the corrupted images of codewords must be orthogonal among themselves. These two conditions can be formulated as a theorem that provides the *necessary and sufficient conditions* for a quantum code C_q to correct a given set of errors $\{\boldsymbol{E}\}$.

Theorem 3 (Necessary and sufficient conditions for QECCs) [14]: The code C_q is an $\boldsymbol{E}$-error correcting code if and only if $\forall\,|\bar{i}\rangle, |\bar{j}\rangle\,(\bar{i} \neq \bar{j})$ and $\forall\ E_a, E_b \in \boldsymbol{E}$, and thus, the following two conditions are simultaneously satisfied:

$$\langle\bar{i}|E_a^{\dagger}E_b|\bar{i}\rangle = \langle\bar{j}|E_a^{\dagger}E_b|\bar{j}\rangle \quad \langle\bar{i}|E_a^{\dagger}E_b|\bar{j}\rangle = 0. \tag{7.41}$$

The first condition indicates that the action of environment is similar to that of all CBs, so the channel cannot distinguish among different CBs. The second condition indicates that the images of codewords are orthogonal to each other. The proof of this theorem is left as a homework problem.

7.3.5 A quantum stabilizer code for phase-flip channel (revisited)

The various quantum error models considered thus far assume (1) errors on different qubits are independent, (2) single qubit errors are equally likely, and (3) the single qubit-error probability is the same for all qubits. Several such models, including the phase-flip channel model, were introduced in the previous section. The phase-flip channel generates eight possible errors on the three qubits as shown in Table 7.4.

At least one of the two generators must anticommute with each of the single-qubit errors $\{E_1, E_2, E_3\}$, justifying the following selection of generators:

$$g_1 = X_1X_2 \quad g_2 = X_1X_3.$$

The error syndrome can be easily calculated from Theorem 2 by

$$S(E) = \lambda_1 \ldots \lambda_{N-K}; \quad \lambda_i = \begin{cases} 0, & [E, g_i] = 0 \\ 1, & \{E, g_i\} = 0 \end{cases},$$

Table 7.4 Syndrome lookup table for phase-flip channel.

Error E	Error probability $P_e(E)$	Error syndrome $S(E)$
$E_0 = I$	$(1-p)^3$	00
$E_1 = Z_1$	$p(1-p)^2$	11
$E_2 = Z_2$	$p(1-p)^2$	10
$E_3 = Z_3$	$p(1-p)^2$	01
$E_4 = Z_1Z_2$	$p^2(1-p)$	01
$E_5 = Z_1Z_3$	$p^2(1-p)$	10
$E_6 = Z_2Z_3$	$p^2(1-p)$	11
$E_7 = Z_1Z_2Z_3$	p^3	00

and is shown in the third column of Table 7.4. The expected error determined from the syndrome and corresponding recovery operator are provided in Table 7.5, in which out of several candidate error operators with the same syndrome, the error with the lowest weight is selected.

The stabilizer group is the set of elements of the form $s(\mathbf{p}) = g_1^{p1}\, g_2^{p2}$, where $p_1,p_2 = 0,1$. By simply varying p_i (0 or 1), we obtain the following stabilizer:

$$S = \{I, X_1X_2, X_1X_3, X_2X_3\}.$$

The single-qubit CB states can be chosen as the eigenkets of X:

$$|0\rangle \equiv |\text{eig}(X) = +1\rangle \quad |1\rangle \equiv |\text{eig}(X) = -1\rangle$$

The stabilizer S must fix the codespace C_q, including the encoded CB states:

$$|0\rangle \rightarrow |\overline{0}\rangle = |000\rangle \quad |1\rangle \rightarrow |\overline{1}\rangle = |111\rangle$$

This mapping is consistent with the no-cloning theorem claim that arbitrary kets cannot be cloned, which does not prevent the cloning of orthogonal kets.

The *failure probability* for error correction P_f is the probability that one of the nonexpected errors $\{E_4,E_5,E_6,E_7\}$ occurs, which is

$$P_f = 3p^2(1-p) + p^3.$$

Table 7.5 Syndrome lookup table and recovery operators for phase-flip channel.

Most probable error E_S	Error syndrome $S(E)$	Recovery operator R_S
$E_{00} = I$	00	$R_{00} = I$
$E_{01} = Z_3$	01	$R_{01} = Z_3$
$E_{10} = Z_2$	10	$R_{10} = Z_2$
$E_{11} = Z_1$	11	$R_{11} = Z_1$

Example: Let the error syndrome be $S = 10$, while the actual error is $E_5 = Z_1Z_3$. Based on the LUT (see Table 7.5), we choose $E_{10} = Z_2$ so that the corresponding recovery operator is $R_{10} = E_{10}^{\dagger} = Z_2$. The transmitted codeword can be represented as $|c\rangle = a|\overline{0}\rangle + b|\overline{1}\rangle$. The action of the channel leads to the state $E_5|c\rangle$, while the simultaneous action of error E_5 and recovery operator R_{10} leads to $|\psi\rangle = E_{10}^{\dagger}E_5|c\rangle = Z_1Z_2Z_3|c\rangle$. Because the action of the resulting operator is as follows,

$$Z\left\{\begin{array}{l}|+1\rangle\\|-1\rangle\end{array}\right. = \left\{\begin{array}{l}|-1\rangle\\|+1\rangle\end{array}\right. \Rightarrow E_{10}^{\dagger}E_5\left\{\begin{array}{l}|\overline{0}\rangle\\|\overline{1}\rangle\end{array}\right. = Z_1Z_2Z_3\left\{\begin{array}{l}|\overline{0}\rangle\\|\overline{1}\rangle\end{array}\right. = \left\{\begin{array}{l}|\overline{1}\rangle\\|\overline{0}\rangle\end{array}\right.$$

the final state $|\psi_f\rangle = a|\overline{1}\rangle + b|\overline{0}\rangle$ is different from the original state, and thus, the error correction has failed to correct the error operator but has been able to return the received state back to the code space.

7.3.6 Distance properties of quantum error correction codes

A QECC is defined by the encoding operation of mapping K qubits into N qubits. The QECC $[N,K]$ can be interpreted as a 2^K-dimensional subspace C_q of N-qubit Hilbert space H_2^N, together with corresponding recovery operation R. As mentioned above, this subspace (C_q) is called the code space; the kets belonging to C_q are known as codewords; and the encoded CB kets are called the basis codewords. The basis for single-qubit errors is given by $\{I,X,Y,Z\}$, as described above. Because any error, discrete or continuous, can be represented as a linear combination of the basis errors, the linear combination of correctable errors will also be a correctable error. The basis for N-qubit quantum errors is obtained by forming all possible direct products:

$$\begin{aligned}&E = j^l O_1 \otimes \cdots \otimes O_N = j^{l'} X(\boldsymbol{a})Z(\boldsymbol{b});\ \boldsymbol{a} = a_1 \cdots a_N;\ \boldsymbol{b} = b_1 \cdots b_N;\ a_i, b_i = 0,1;\ l, l' = 0,1,2,3\\&X(\boldsymbol{a}) \equiv X_1^{a_1} \otimes \cdots \otimes X_N^{a_N};\ Z(\boldsymbol{b}) \equiv Z_1^{b_1} \otimes \cdots \otimes Z_N^{b_N};\ O_i \in \{I,X,Y,Z\}\end{aligned} \tag{7.42}$$

The *weight* of an error operator $E(\boldsymbol{a},\boldsymbol{b})$ is defined to be number of qubits different from identity operator. Necessary and sufficient conditions for a QECC to correct a set of errors $E = \{E_p\}$, given by Eq. (7.41), can be compressed as follows:

$$\langle\overline{i}|E_p^{\dagger}E_q|\overline{j}\rangle = C_{pq}\delta_{ij}, \tag{7.43}$$

where the matrix elements C_{pq} satisfy the condition $C_{pq} = C^*_{qp}$, so the square matrix $\boldsymbol{C} = (C_{pq})$ is a Hermitian. A QECC for which matrix $\boldsymbol{C}$ is singular is said to be *degenerate*. If we interpret $E_p^{\dagger}E_q$ as a new error operator E, Eq. (7.43) can be rewritten as

$$\langle\overline{i}|E|\overline{j}\rangle = C_E\delta_{ij}. \tag{7.44}$$

We say that a QECC has a *distance D* if all errors of weight less than D satisfy the equation above, and there exists at least one error of weight D to violate it. In other words, the QECC distance is the weight of the smallest weight D of error E that cannot be detected by the code. Similarly to classical codes, we can relate the distance D to the error correction capability t as follows: $D \geq 2t + 1$. Namely, since $E = E_p^{\dagger}E_q$ the weight of error operator E will be $\text{wt}\left(E_p^{\dagger}E_q\right) = 2t$. If we are only interested in detecting the errors but not correcting them, the error detection capability d is related to distance D by $D \geq d+1$. Since we are interested only in detection of errors, we can set $E_p = I$ to obtain $\mathrm{wt}\left(E_p^\dagger E_q\right) = \mathrm{wt}(E_q) = d$. If we are interested in simultaneously detecting d errors and correcting t errors, the distance of the code must be $D \geq d+t+1$. The following theorem can be used to determine whether a given quantum code C_q of quantum distance D is degenerate.

Theorem 4. The quantum code C_q of distance D is a degenerate code if and only if its stabilizer S contains an element with weight less than D (excluding the identity element).

The theorem can be proved as follows. If the code C_q is a degenerate code of distance D, there will exist two correctable errors E_1, E_2 such that their action on a CB codeword is the same: $E_1|\bar{i}\rangle = E_2|\bar{i}\rangle$. By multiplying with $E_2^\dagger$ from the left we obtain: $E_2^\dagger E_1|\bar{i}\rangle = |\bar{i}\rangle$, which indicates that the error $E_2^\dagger E_1 \in S$. Since any correctable error satisfies $\langle\bar{i}|E|\bar{j}\rangle = C_E\delta_{ij}$, we obtain $\langle\bar{i}|E_2^\dagger E_1|\bar{j}\rangle = C_{12}\delta_{ij}$. Because the C_q code has a distance D it is clear that $\mathrm{wt}\left(E_2^\dagger E_1\right) < \mathrm{D}$. On the other hand, if there exists an $s \in S$ with $\mathrm{wt}(s) < D$, we can find another $s_a \in S$ so that $s_a s = s_b \in S$. By multiplying both sides with $s_a^\dagger$ we obtain $s = s_a^\dagger s_b$. From definition of stabilizer codes we know that $s|\bar{i}\rangle = |\bar{i}\rangle$, $\forall\ s \in S$. By expressing $s = s_a^\dagger s_b$, we obtain $s_a^\dagger s_b|\bar{i}\rangle = |\bar{i}\rangle$, which after multiplying with s_a from the left becomes $s_b|\bar{i}\rangle = s_a|\bar{i}\rangle$, which is equivalent to $(s_a - s_b)|\bar{i}\rangle = 0$. Since the matrix S_{ab} is singular, the code C_q is degenerate.

7.3.7 Calderbank–Shor–Steane codes (revisited)

CSS codes are constructed from two classical binary codes C and C' satisfying the following three properties:

1. C and C' are (n,k,d) and (n,k',d') codes, respectively;
2. $C' \subset C$; and
3. C and $C'^\perp$ are both t-error correcting codes.

The code construction partitions C into the cosets of C',

$C = C' \cup (c_1 + C') \cup \ldots \cup (c_N + C')$, where $c_1, c_2, \ldots, c_N \in C$, and N is the number of cosets determined by (from Lagrange's theorem) $N = 2^k/2^{k'} = 2^{k-k'}$. The *basis codewords* are obtained by identifying each one with corresponding cosset $|\bar{c}_i\rangle \Leftrightarrow c_i + C'$, so

$$|\bar{c}_i\rangle = \frac{1}{\sqrt{2^{k'}}}\sum_{c'\in C'}|c_i + c'\rangle;\quad i = 1,\cdots,N \tag{7.45}$$

If $v, w \in C$ so that $v - w = d \in C'$ (belong to the same cosset of C'), then $|\bar{v}\rangle = |\bar{w}\rangle$ because

$$|\bar{v}\rangle = \frac{1}{\sqrt{2^{k'}}}\sum_{c'\in C'}|v + c'\rangle \overset{v-w=d}{\widehat{=}} \frac{1}{\sqrt{2^{k'}}}\sum_{c'\in C'}|w + d + c'\rangle \overset{d+c'=d'}{\widehat{=}} \frac{1}{\sqrt{2^{k'}}}\sum_{d'\in C'}|w + d'\rangle = |\bar{w}\rangle \tag{7.46}$$

The code space C_q is a subspace of $H^n{}_2$ spanned by the basis codewords and therefore is $N = 2^{k-k'}$-dimensional.

Example: CSS [7,3,1] *code.* For Steane [7,3,3] code, C is the Hamming (7,4,3) and C' is the (7,3,4) maximum-length code:

$$H(C)=H(C'_{\perp})=\begin{bmatrix}0&0&0&1&1&1&1\\0&1&1&0&0&1&1\\1&0&1&0&1&0&1\end{bmatrix}$$

Based on Eqs. (7.12), (7.39), and the H-matrix above, we obtain the following generators:

$$g_1=X_4X_5X_6X_7 \quad g_2=X_2X_3X_6X_7 \quad g_3=X_1X_3X_5X_7$$

$$g_4=Z_4Z_5Z_6Z_7 \quad g_5=Z_2Z_3Z_6Z_7 \quad g_6=Z_1Z_3Z_5Z_7$$

The number of cossets is $N=2^{k-k'}=2$, and the corresponding cossets are

$(0000000+C')=\{000\,0\,000,\ 011\,0\,011,\ 101\,0\,101,\ 110\,0\,110,\ 000\,1\,111,\ 011\,1\,100,\ 101\,1\,010,\ 110\,1\,001\}$

$(1111111+C')=\{111\,1\,111,\ 100\,1\,100, 010\,1\,010,\ 001\,1\,001,\ 111\,0\,000,\ 100\,0\,011,\ 010\,0\,101,\ 001\,0\,110\}$

Therefore, based on (Eq. 7.45), we can represent the basis codewords by

$$|\overline{0}\rangle=\frac{1}{\sqrt{2^3}}[|0000000\rangle+|0110011\rangle+|1010101\rangle+|1100110\rangle+|0001111\rangle+|0111100\rangle$$
$$+|1011010\rangle+|1101001\rangle]$$

$$|\overline{1}\rangle=\frac{1}{\sqrt{2^3}}[|1111111\rangle+|1001100\rangle+|0101010\rangle+|0011001\rangle+|1110000\rangle+|1000011\rangle$$
$$+|0100101\rangle+|0010110\rangle]$$

It is straightforward to show that the generators g_1, ... ,g_6 fix the basis codewords.

By definition, a distance D of QECC is the weight of smallest error not satisfying Eq. (7.44). We can easily find that $\langle\overline{0}|X_1X_2X_3|\overline{1}\rangle=1$. Since the weight of error $X_1X_2X_3$ is 3, the distance of the code is 3, and the error correction capability is 1. It can be shown that the Steane [1,3,7] code is nondegenerate since the stabilizer S does not contain any element of weight smaller than 3.

7.3.8 Encoding and decoding circuits of quantum stabilizer codes

This section briefly describes implementing encoders and decoders for quantum stabilizer codes. An efficient encoding and decoding of quantum stabilizer codes is described in the next chapter, which is devoted to stabilizer codes. Here we provide very simple descriptions of encoders and decoders for completeness of presentation. In particular, decoders are quite easy to implement using stabilizer formalism. The error detector can be obtained by concatenation of circuits corresponding to different stabilizers. The transmission error can be identified by an intersection of corresponding syndrome measurements. Let us observe the implementation of stabilizer $S_a=X_1X_4X_5X_6$. The syndrome quantum circuit for measurement of stabilizer S_a is shown in Fig. 7.12A and B. In Fig. 7.12A the

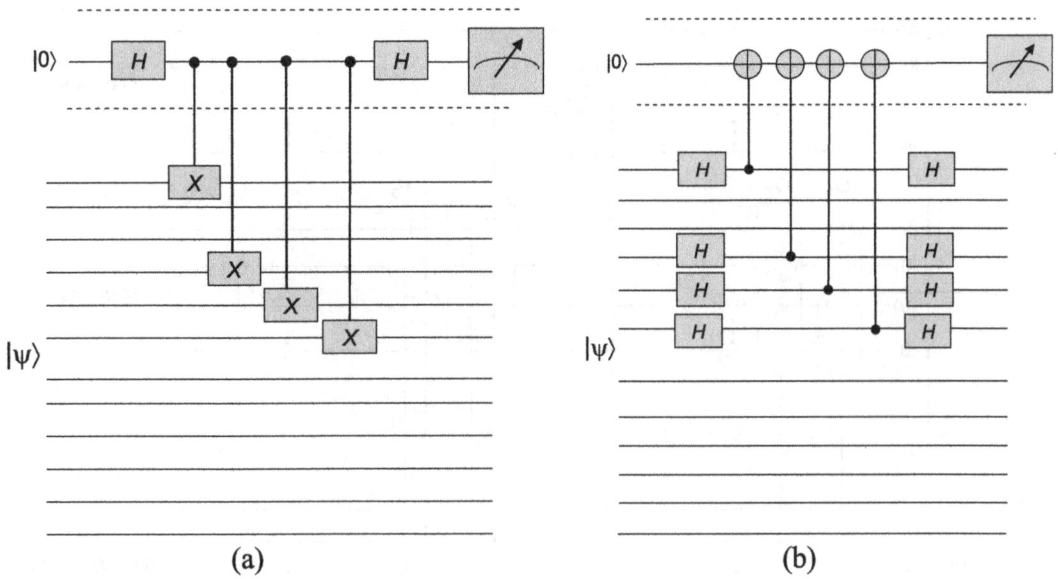

FIGURE 7.12

The syndrome quantum circuit for stabilizer $S_a = X_1X_4X_5X_6$: (A) based on H- and controlled-X gates, and (B) based on H- and CNOT-gates.

quantum syndrome implementation circuit is based on Hadamard- (H-) and controlled-X gates, while in Fig. 7.12B, the corresponding implementation is based on H- and CNOT-gates only. The initial and final Hadamard gates are applied on the ancilla. We perform the measurement on the ancilla qubit only on a $\{|0\rangle,|1\rangle\}$-basis, which gives the outcome 0 if S_a has the outcome +1 and the outcome 1 when S_a has the outcome −1. This can be explained as follows [12]:

$$(I\otimes H)S_a^c(I\otimes H)|\psi\rangle|0\rangle = (I\otimes H)S_a^c|\psi\rangle(|0\rangle+|1\rangle)/\sqrt{2} = (I\otimes H)(|\psi\rangle|0\rangle + S_a|\psi\rangle|1\rangle)/\sqrt{2}$$

$$= [|\psi\rangle(|0\rangle+|1\rangle) + S_a|\psi\rangle(|0\rangle-|1\rangle)]/2 = [(I+S_a)|\psi\rangle|0\rangle + (I-S_a)|\psi\rangle|1\rangle]/2 \tag{7.47}$$

Because measuring the ancilla projects $|\psi\rangle$ on the eigenkets of S_a, we will obtain outcome 0 when S_a has the outcome +1 and 1 when S_a has the outcome −1. In Eq. (7.47), we used superscript c to denote the control operation; for $c = 0$, the action S_a is excluded.)

We now describe how to implement the encoders for a dual-containing code defined by a full rank matrix $\boldsymbol{H}$ with $N > 2M$, based on proposal due to MacKay et al [12]. We first need to transform the $\boldsymbol{H}$-matrix by Gauss elimination into the following form: $\widetilde{\boldsymbol{H}} = [\boldsymbol{I}_M|\boldsymbol{P}_{M\times(N-M)}]$, where $\boldsymbol{I}$ is the identity matrix of size $M \times M$ and $\boldsymbol{P}$ is the binary matrix of size $M \times (N - M)$ of full rank. We further transform the $\boldsymbol{P}$-matrix also by Gaussian elimination into the following form: $\widetilde{\boldsymbol{P}} = [\boldsymbol{I}_M|\boldsymbol{Q}_{M\times(N-2M)}]$, where $\boldsymbol{Q}$ is the binary matrix of size $M \times (N - 2M)$. Clearly, for arbitrary string $\boldsymbol{f}$ of length $K = N-2M$, $[0|\boldsymbol{Qf}|\boldsymbol{f}]$ is a codeword of $C(\boldsymbol{H})$. The encoder now can be implemented in two stages, as shown in Fig. 7.13, which is modified from Ref. [12].

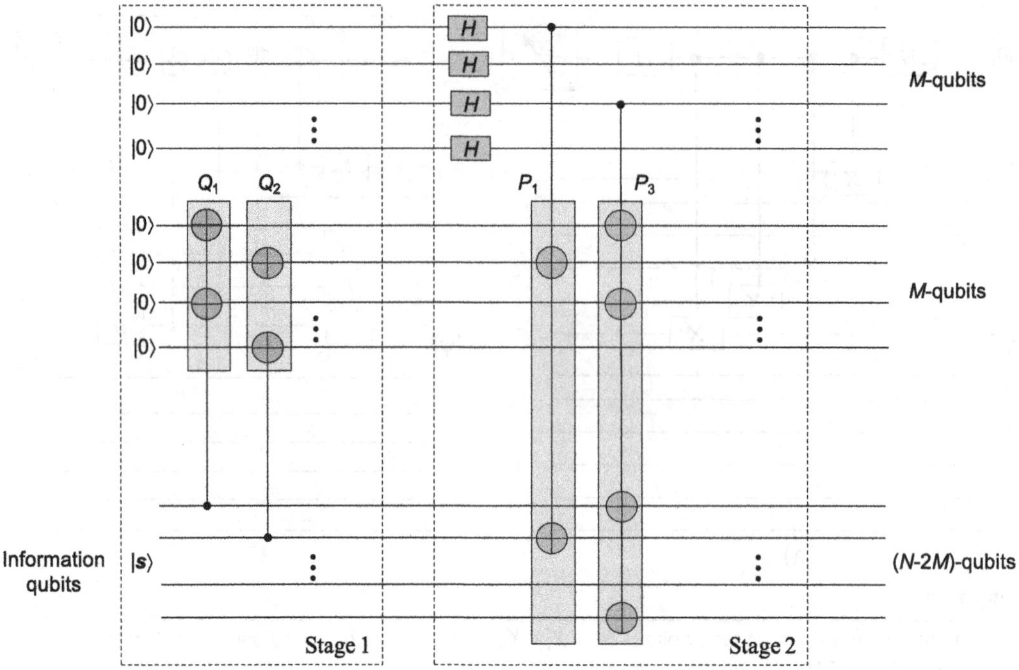

FIGURE 7.13

Encoder implementation for dual-containing codes. The block Q_k in the first stage corresponds the kth column of submatrix $\boldsymbol{Q}$. This block is controlled by the kth information qubit and executed only when the kth information qubit is $|1\rangle$. The CNOT-gates are placed according to the nonzero positions in Q_k. In the second stage, the rows of the $\boldsymbol{P}$-matrix are conditionally executed based on the content of the H-gates output. The kth Hadamard-gate controls the set of gates corresponding to the kth row of the $\boldsymbol{P}$-matrix. The CNOT-gates are placed in accordance with the nonzero-positions in P_k.

The first stage performs the following mapping:

$$|\mathbf{0}\rangle_M|\mathbf{0}\rangle_M|\boldsymbol{s}\rangle_K \rightarrow |\mathbf{0}\rangle_M|\boldsymbol{Qs}\rangle_M|\boldsymbol{s}\rangle_K, \tag{7.48}$$

where $|\boldsymbol{s}\rangle_K$ is the K-qubits' information state, $\boldsymbol{s}$ is the binary string of length K, and $|\mathbf{0}\rangle_M = \underbrace{|0\rangle\cdots|0\rangle}_{M \text{ times}}$. In other words, the information qubits serve as control qubits, while middle M ancilla qubits as target qubits. The control operations are dictated by columns in $\boldsymbol{Q}$-matrix. In second stage, the first M-qubits from stage 1 (ancillae) are processed by Hadamard-gates mapping the state $|\mathbf{0}\rangle_M$ to

$$|\mathbf{0}\rangle_M \rightarrow \underbrace{\frac{|0\rangle+|0\rangle}{\sqrt{2}}\otimes\cdots\otimes\frac{|0\rangle+|0\rangle}{\sqrt{2}}}_{M \text{ times}} = \frac{1}{2^{M/2}}\sum_{\boldsymbol{v}}|\boldsymbol{v}\rangle_M, \tag{7.49}$$

where vector $\boldsymbol{v}$ runs over all possible binary M-tuples. Let us now introduce the binary vector $\boldsymbol{y}$ to denote the concatenation of binary strings $\boldsymbol{Qs}$ and $\boldsymbol{s}$ in Eq. (7.48). From Fig. 7.13 is clear that the purpose of stage 2 is to conditionally executes the row operators of $\widetilde{\boldsymbol{H}} = [\boldsymbol{I}_M|\boldsymbol{P}_{M\times(N-M)}]$ on target qubits $|\boldsymbol{y}\rangle_{N-M}$. The qubits $|\boldsymbol{v}\rangle_M$ serve as control qubits. Note that this stage is very similar to classical encoders of systematic linear block codes (LBCs). Therefore, the operation of final stage is to perform the following mapping:

$$\frac{1}{2^{M/2}}\sum_{\boldsymbol{v}}|\boldsymbol{v}\rangle_M|\boldsymbol{y}\rangle_{N-M} \rightarrow \frac{1}{2^{M/2}}\sum_{\boldsymbol{v}}\left(\prod_{m=1}^{M}P_m^c\right)|\boldsymbol{v}\rangle_M|\boldsymbol{y}\rangle_{N-M} = \frac{1}{2^{M/2}}\sum_{\boldsymbol{x}\in C^{\perp}(H)}|\boldsymbol{x}+\boldsymbol{y}\rangle_N, \boldsymbol{y}\in C(\boldsymbol{H}) \quad (7.50)$$

The most general codeword of the form

$$|\psi\rangle = \sum_{\boldsymbol{y}\in C(\boldsymbol{H})}\alpha_y \sum_{\boldsymbol{x}\in C^{\perp}(H)}|\boldsymbol{x}+\boldsymbol{y}\rangle_N, \quad (7.51)$$

can be obtained if we start from $\sum_{s}\alpha_s|\boldsymbol{s}\rangle_K$ instead of $|\boldsymbol{s}\rangle_K$.

7.4 Important quantum coding bounds

This section describes several important quantum coding bounds, including Hamming, Gilbert–Varshamov, and Singleton. We also discuss quantum weight enumerators and the quantum MacWilliams identity.

7.4.1 Quantum Hamming bound

The quantum Hamming bound for an $[N,K]$ QECC of error correction capability t is given by

$$\sum_{j=0}^{t}\binom{N}{j}3^j 2^K \leq 2^N, \quad (7.52)$$

where t is the error correction capability, K is the information word length, and N is the codeword length. This inequality is quite straightforward to prove. The number of information words can be found as 2^K, the number of error locations is N chooses j, and the number of possible errors $\{X,Y,Z\}$ at every location is 3^j. The total number of errors for all codewords that can be corrected cannot be larger than the code space, which is 2^N-dimensional. Analogous to classical codes, the QECCs satisfying the Hamming inequality with equality sign can be called the *perfect* codes. For $t=1$, the quantum Hamming bound becomes: $(1+3N)2^K \leq 2^N$. For $[N,1]$ quantum code with $t=1$, the quantum Hamming bound is simply $2(1+3N) \leq 2^N$. The smallest possible N to satisfy the Hamming bound is $N=5$, which represents the perfect code. This cyclic code has already been introduced in Section 7.2. It is interesting to note that the quantum Hamming bound is identical to classical Hamming bound for q-ary LBCs by setting $q=4$, corresponding to cardinality of set of errors $\{I,X,Y,Z\}$. This is consistent with connection we established between QECCs and classical codes over GF(4).

The asymptotic quantum Hamming bound can be obtained by letting $N \to \infty$. For very large N, the last term in summation (Eq. 7.52) dominates and we can write $\binom{N}{t} 3^j 2^K \le 2^N$. By taking the $\log_2(\bullet)$ from both sides of inequality, we obtain $\log_2 \binom{N}{t} + t\log_2 3 + K \le N$. Using the approximation $\log_2 \binom{N}{t} \simeq N h_2(t/N)$, where $h_2(p)$ is the binary entropy function $h_2(p) = -p \log p - (1-p)\log(1-p)$, we obtain the following *asymptotic quantum Hamming bound*:

$$\frac{K}{N} \le 1 - h_2(t/N) - \frac{t}{N}\log_2 3. \tag{7.53}$$

7.4.2 Quantum Gilbert–Varshamov bound

Let us consider an $[N,K]$ QECC with distance D. Following the analogy with classical q-ary LBC that we established in the previous subsection, we expect the quantum Gilbert–Varshamov bound to be the same as the classical one for $q = 4$:

$$\sum_{j=0}^{D-1} \binom{N}{j} 3^j 2^K \ge 2^N. \tag{7.54}$$

Because all basis codewords and all errors E from Pauli group G_N with $\mathrm{wt}(E) < D$ must satisfy the equation

$$\langle \bar{i}|E|\bar{j}\rangle = C_E \delta_{ij}, \tag{7.55}$$

the number of such errors is $N_E = \sum_{j=0}^{D-1} \binom{N}{j} 3^j$. The following derivation is due to Gottesman [8] (see also [14]). Consider the state $|\psi_1\rangle$ satisfying (Eq. 7.55) for all possible errors E, which can be used as a basis codeword. Let us observe now the space orthogonal to both $|\psi_1\rangle$ and $E|\psi_1\rangle$. The dimensionality of this subspace is $2^N - N_E$. Now consider the state $|\psi_2\rangle$ from this subspace satisfying (Eq. 7.55) and determine the subspace orthogonal to both $E|\psi_1\rangle$ and $E|\psi_2\rangle$. The dimensionality of this subspace is $2^N - 2N_E$. We can iterate this procedure i ($<2^K$) times. In ith step the corresponding subspace will have dimensionality $2^N - iN_E > 0$, where $i < 2^K$. The inequality $2^N - iN_E > 0$ can be rewritten by substituting the expression N_E as follows:

$$\sum_{j=0}^{D-1} \binom{N}{j} 3^j i < 2^N; i < 2^K \tag{7.56}$$

Since i is strictly smaller than 2^K for $i = 2^K$ we obtain the inequality (Eq. 7.54). Similarly to quantum Hamming bound, the quantum Gilbert–Varshamov bound is identical to classical

Gilbert–Varshamov bound for q-ary LBCs by setting $q = 4$ (corresponding to cardinality of set of single-qubit errors $\{I,X,Y,Z\}$).

The asymptotic quantum Gilbert–Varshamov bound can be determined by following the similar procedure as for asymptotic quantum Hamming bound to obtain

$$\frac{K}{N} \geq 1 - h_2(d/N) - \frac{d}{N}\log_2 3. \tag{7.57}$$

By combining the asymptotic quantum Hamming and Gilbert–Varshamov bounds, we obtain the following upper and lower bound for the quantum code rate $R = K/N$:

$$1 - h_2(D/N) - \frac{D}{N}\log_2 3 \leq \frac{K}{N} \leq 1 - h_2(t/N) - \frac{t}{N}\log_2 3 \tag{7.58}$$

7.4.3 Quantum Singleton (Knill–Laflamme) bound [18]

In Chapter 2, we derived the classical Singleton bound for (n,k,d) LBCs as follows:

$$d \leq n - k + 1 \Leftrightarrow n - k \geq d - 1 = 2t + 1 - 1 = 2t. \tag{7.59}$$

For the quantum $[N,K]$ codes obtained by CSS construction, we have $2M = 2(n-k)$ stabilizer conditions applied to N qubit states, so the corresponding Singleton bound will be $2M = 2(n-k) \geq 4t$. Because in CSS construction $N-2M$ qubits are encoded into N qubits, the quantum Singleton bound becomes

$$N - K \geq 4t \Leftrightarrow N - K \geq 2(D-1) \Leftrightarrow \frac{K}{N} \leq 1 - \frac{2}{N}(D-1). \tag{7.60}$$

Analogous to classical codes, the quantum codes satisfying the quantum Singleton bound with equality sign can be called the *maximum distance separable* quantum codes. By combining the asymptotic Gilbert–Varshamov and Singleton bounds, we obtain the following bounds for quantum code rate:

$$1 - h_2(D/N) - \frac{D}{N}\log_2 3 \leq \frac{K}{N} \leq 1 - \frac{2}{N}(D-1). \tag{7.61}$$

For entanglement-assisted-like QECCs [17,23], the Singleton bound can be derived as follows. Consider $[N,K,D]$ QECC and assume that K Bell states are shared between the source and destination. We further encode the transmitter portion of Bell states only. The corresponding codeword can be represented by the following form:

$$\overbrace{\mathbf{d}}^{K}\ \overbrace{\mathbf{a}}^{D-1}\ \overbrace{\mathbf{b}}^{D-1}\ \overbrace{\mathbf{c}}^{N-2(D-1)}, K = 2(D-1), \tag{7.62}$$

where $\mathbf{d}$ is the receiver portion of Bell states, which stays unencoded. The von Neumann entropy of density operator ρ, introduced in Chapter 3, is defined by $S(\rho) = -\mathrm{Tr}\rho \log \rho$. We know from quantum information theory chapter that $S(\rho)$ vanishes for nonentangled pure state, and it can be as large as $\log N_D$ for maximally entangled state for two N_D-state systems $|\psi\rangle = \frac{1}{\sqrt{N_D}}\sum_i |ii\rangle$, where $|ij\rangle$ is the CB of a composite system. An important property of the von-Neumann entropy is the *subadditivity property*. Let a composite system S be a bipartite system composed of component subsystems S_1 and S_2. The subadditivity property can be stated as $S(\rho(S_1S_2)) \leq S(\rho(S_1)) + S(\rho(S_2))$, with the equality sign being satisfied if the component susbsystems are independent of each other: $\rho(S_1S_2) = \rho(S_1) \otimes \rho(S_2)$. It is clear from Eq. (7.62) that

$$S(\rho(\boldsymbol{da})) = S(\rho(\boldsymbol{bc})) \text{ and } S(\rho(\boldsymbol{db})) = S(\rho(\boldsymbol{ac})). \tag{7.63}$$

From subadditivity property it follows that

$$S(\rho(\boldsymbol{BC} = \boldsymbol{bc})) \leq S(\rho(\boldsymbol{b})) + S(\rho(\boldsymbol{c})) \text{ and } S(\rho(\boldsymbol{ac})) \leq S(\rho(\boldsymbol{a})) + S(\rho(\boldsymbol{c})). \tag{7.64}$$

Since D and A (D and B) independent of each other, we obtain

$$S(\rho(\boldsymbol{da})) = S(\rho(\boldsymbol{d})) + S(\rho(\boldsymbol{a})) \text{ and } S(\rho(\boldsymbol{db})) = S(\rho(\boldsymbol{d})) + S(\rho(\boldsymbol{b})). \tag{7.65}$$

By properly combining Eqs. (7.63)–(7.65), we obtain

$$S(\rho(\boldsymbol{d})) \leq S(\rho(\boldsymbol{b})) + S(\rho(\boldsymbol{c})) - S(\rho(\boldsymbol{a})) \text{ and } S(\rho(\boldsymbol{d})) \leq S(\rho(\boldsymbol{a})) + S(\rho(\boldsymbol{c})) - S(\rho(\boldsymbol{b})). \tag{7.66}$$

By summing two inequalities in Eq. (7.66) and dividing by 2, we obtain the Singleton bound:

$$\log_2 2^K \leq \log_2 2^{N-2(D-1)} \Leftrightarrow K \leq N - 2(D-1) \Leftrightarrow N - K \geq 2(D-1). \tag{7.67}$$

7.4.4 Quantum weight enumerators and quantum MacWilliams identity

We have shown in Chapter 2 that in classical error correction, the codeword weight distribution can be used in determination of word error probability and some other properties of the code. Similar ideas can be adopted in quantum error correction. Let A_w be the number of stabilizer S elements of weight w, and B_w be the number of elements of the same weight in centralizer of S, denoted $C(S)$. The polynomials $A(z) = \Sigma_w A_w z^w$ and $B(z) = \Sigma_w B_w z^w$ ($A_0 = B_0 = 1$) can be used as the *weight enumerators* of S and $C(S)$, respectively.

Let F be a finite field GF(q), where q is a prime power, and let F^N be a vector space of dimension N over F. A *linear code* C_F of length N over F is a subspace of F^N. Let $C_F^\perp$ denote a dual code of C_F and let the elements of F be denoted $\omega_0 = 0, \omega_1, \ldots, \omega_{q-1}$. The *composition* of a vector $\boldsymbol{v} \in F^N$ is defined to be comp($\boldsymbol{v}$) $= \boldsymbol{s} = (s_0, s_1, \ldots, s_{q-1})$, where s_i is the number of coordinates of $\boldsymbol{v}$ equal to ω_i. Clearly $\sum_{i=0}^{q-1} s_i = N$. The *Hamming weight* of vector $\boldsymbol{v}$, denoted wt($\boldsymbol{v}$), is the number of nonzero coordinates, namely wt($\boldsymbol{v}$) $= \sum_{i=1}^{q-1} s_i(\boldsymbol{v})$. Let A_i be the number of vectors in C_F having wt($\boldsymbol{v}$) $= i$. Then $\{A_i\}$ will be the weight enumerator of C_F. Similarly, let $\{B_i\}$ denote the weight numerator of dual code. The weight enumerators are related by the following MacWilliams identity [20,21]:

$$\sum_{i=0}^{n} B_i z^i = \frac{1}{|C_F|} \sum_{i=0}^{n} A_i (1 + (q-1)z)^{n-i} (1-z)^i = \frac{1}{|C_F|} (1 + (q-1)z)^n \sum_{i=0}^{n} A_i \left(\frac{1-z}{1+(q-1)z} \right)^i. \tag{7.68}$$

The expression (Eq. 7.68) can be rewritten in terms of polynomials introduced above as follows:

$$B(z) = \frac{1}{|C_F|} (1 + (q-1)z)^n A\left(\frac{1-z}{1+(q-1)z} \right). \tag{7.69}$$

We have already shown above that both quantum Hamming and Gilbert–Varshamov bounds can be obtained as special case of classical nonbinary codes for $q = 4$ (corresponding to the cardinality of set $\{I,X,Y,Z\}$). Therefore, the quantum MacWilliams identity for $[N,K]$ quantum code can be obtained from Eq. (7.69) by setting $q = 4$:

$$B(z) = \frac{1}{2^{N-K}}(1+3z)^n A\left(\frac{1-z}{1+3z}\right). \tag{7.70}$$

By matching the coefficients of z^i in Eq. (7.68) and by setting $q = 4$ we establish the relationship between B_i and A_i:

$$B_i = \frac{1}{2^{N-K}} \sum_{w=0}^{N} \left[\sum_{j=0}^{i} (-1)^j 3^{i-j} \binom{w}{j} \binom{N-w}{i-j} \right] A_w. \tag{7.71}$$

The weight distribution of classical RS (n,k,d) code with symbols from GF(q) is given by

$$A_w = \binom{n}{w}(q-1)\sum_{j=0}^{w-d}(-1)^j \binom{w-1}{j} q^{w-d-j}. \tag{7.72}$$

If the quantum code is derived from RS code, the weight distribution A_w can be found in closed form by setting $q = 4$ in Eq. (7.72).

A reader interested in rigorous derivation of Eqs. (7.70) and (7.71) and some other quantum bounds is referred to Refs. [8,14]. In the next section, we introduce the operator-sum representation, which is a powerful tool in studying various quantum channels. Namely, we can describe the action of a quantum channel by the so-called superoperator or quantum operation expressed in terms of channel operation elements. Several important quantum channel models will be then discussed using this framework. This study is useful in quantum information theory to determine the quantum channel capacity and in quantum code design. It is also relevant in the study of QECC performance over various quantum channels.

7.5 Quantum operations (superoperators) and quantum channel models

In this section, we are concerned with the operator-sum representation of a quantum operation (also known as a superoperator) and with various quantum channel models, including the depolarizing, amplitude damping, and generalized amplitude damping channels.

7.5.1 Operator-sum representation

Let the composite system C be composed of quantum register Q and environment E. This kind of system can be modeled as a closed quantum system. Because the composite system is closed, its dynamic is unitary, and final state is specified by a unitary operator U as follows $U(\rho \otimes \varepsilon_0)U^\dagger$, where ρ is a density operator of initial state of quantum register Q, and ε_0 is the initial density operator of the environment E. The reduced density operator of Q upon interaction ρ_f can be obtained by tracing out the environment:

$$\rho_f = \mathrm{Tr}_E\left[U(\rho \otimes \varepsilon_0)U^\dagger\right] \equiv \xi(\rho) \tag{7.73}$$

The transformation (mapping) of initial density operator ρ to the final density operator ρ_f, denoted $\xi : \rho \rightarrow \rho_f$, given by Eq. (7.73), is often called the *superoperator* or *quantum operation*. The final density operator can be expressed in so-called *operator-sum representation* as follows:

$$\rho_f = \sum_k E_k \rho E_k^\dagger, \tag{7.74}$$

where E_k are the operation elements for the superoperator. The operator-sum representation is suitable to represent the action of channel, evaluate performance of QECCs, and to study the quantum channel capacity, which is subject of quantum information theory chapter. To derive Eq. (7.74) we perform spectral decomposition of environment initial state:

$$\varepsilon_0 = \sum_l \lambda_l |\phi_l\rangle\langle\phi_l|; \quad |\phi_l\rangle = \text{eigenkets}(\varepsilon_0), \tag{7.75}$$

and use the definition of the trace operation given by $\text{Tr}(X) = \sum_i \langle a^{(i)}|X|a^{(i)}\rangle$. From properties of trace operation we know that it is independent on the basis, and we chose for the basis the orthonormal basis of environmental Hilbert space H_E: $\{|e_m\rangle\}$. Therefore, using the spectral decomposition and trace definition, the final density operator becomes

$$\rho_f = \sum_m \langle e_m|U\{\rho \otimes \varepsilon_0\}U^\dagger|e_m\rangle = \sum_{l,m} \lambda_l \langle e_m|U\{\rho \otimes |\phi_l\rangle\langle\phi_l|\}U^\dagger|e_m\rangle = \sum_{l,m} E_{lm}\rho E_{lm}^\dagger,$$

$$E_{lm} = \sqrt{\lambda_l}\langle e_m|U|\phi_l\rangle, \tag{7.76}$$

where E_{lm} are the operation elements of the superoperator. Using single subscript k instead of double subscript lm we obtain the operator-sum representation (Eq. 7.74). The operator-sum representation can be used in classification of quantum operations into two categories: (1) *trace-preserving* when $\text{Tr}\xi(\rho) = \text{Tr}\ \rho = 1$ and (2) *non-trace-preserving* when $\text{Tr}\xi(\rho) < 1$.

Starting from the trace-preserving condition,

$$\text{Tr}\rho = \text{Tr}\xi(\rho) = \text{Tr}\left[\sum_k E_k \rho E_k^\dagger\right] = \text{Tr}\left[\rho \sum_k E_k E_k^\dagger\right] = 1,$$

we obtain

$$\sum_k E_k E_k^\dagger = I. \tag{7.77}$$

For the non-trace-preserving quantum operation, Eq. (7.77) is not satisfied, and informally we can write $\sum_k E_k E_k^\dagger < I$.

Note that in the derivation above, we assumed that the quantum register and environment were initially nonentangled and that the environment was in a pure state; this does not reduce the generality of discussion. Observe again the situation in which the quantum register Q and its environment E are initially nonentangled, with the corresponding density operators being ρ and $\varepsilon_0 = |e_0\rangle\langle e_0|$, respectively. Upon interaction, we perform the *measurement* on observable $M = \sum_m \mu_m P_m$ of E, where $\{\mu_m\}$

are the eigenvalues of M, and P_m is the projection operator onto the subspace of states of E with eigenvalues μ_m. For the outcome μ_m, the final (normalized) state of the composite system is given by

$$\rho_{tot}^k = \frac{P_k U(\rho \otimes |e_0\rangle\langle e_0|) U^\dagger P_k}{\mathrm{Tr}(P_k U(\rho \otimes |e_0\rangle\langle e_0|) U^\dagger P_k)}. \tag{7.78}$$

The denominator can be simplified by applying the tracing per environment first:

$$\mathrm{Tr}(P_k U(\rho \otimes |e_0\rangle\langle e_0|) U^\dagger P_k) = \mathrm{Tr}_Q \mathrm{Tr}_E (P_k U(\rho \otimes |e_0\rangle\langle e_0|) U^\dagger P_k) = \mathrm{Tr}_Q\left(E_k \rho E_k^\dagger\right), E_k = \langle \mu_k | U | e_0 \rangle, \tag{7.79}$$

and by substituting this result in Eq. (7.78), we obtain

$$\rho_{tot}^k = \frac{P_k U(\rho \otimes |e_0\rangle\langle e_0|) U^\dagger P_k}{\mathrm{Tr}_Q\left(E_k \rho E_k^\dagger\right)}. \tag{7.80}$$

In practice, we are interested in the (normalized) reduced density operator for the quantum register Q, which can be obtained by tracing out the environment:

$$\rho^k = \frac{\mathrm{Tr}_E (P_k U(\rho \otimes |e_0\rangle\langle e_0|) U^\dagger P_k)}{\mathrm{Tr}_Q\left(E_k \rho E_k^\dagger\right)} = \frac{E_k \rho E_k^\dagger}{\mathrm{Tr}_Q\left(E_k \rho E_k^\dagger\right)}. \tag{7.81}$$

In Chapter 3, we learned that the probability of obtaining μ_k from the measurement of M can be determined by $\Pr(\mu_k) = \mathrm{Tr}(\rho_f P_k)$. Therefore, the *probability of measurement* being μ_k can be obtained from

$$P(k) = \Pr(\mu_k) = \mathrm{Tr}\left[P_k U(\rho \otimes |e_0\rangle\langle e_0|) U^\dagger\right] \overbrace{\underbrace{=}_{\mathrm{Tr}AB=\mathrm{Tr}BA}}^{P_k^2=P_k} \mathrm{Tr}\left[P_k U(\rho \otimes |e_0\rangle\langle e_0|) U^\dagger P_k\right] = \mathrm{Tr}_Q\left(E_k \rho E_k^\dagger\right) \tag{7.82}$$

If the measurement outcome is not observed, the composite system will be in a mixed state:

$$\rho_{tot} = \sum_k P(k) \rho_{tot}^k \tag{7.83}$$

The reduced density operator upon measurement will then be

$$\rho_Q = \sum_k P(k) \mathrm{Tr}_E\left(\rho_{tot}^k\right) = \sum_k P(k) \rho_k = \sum_k E_k \rho E_k^\dagger \tag{7.84}$$

Therefore, the representation of ρ_Q is the same as that of the operator-sum representation. In the derivation of Eq. (7.74), we did not perform any measurement; nevertheless, we obtained the same result. This conclusion can be formulated as the ***principle of implicit measurement***: Once a subsystem E of composite system C has finished interacting with the rest of the composite system $Q = C-E$, the reduced density operator for Q will not be affected by any operations that we carry out solely on E. As a consequence of this principle, we can interpret the quantum operation of the environment to the quantum register as performing the maping $\rho \to \rho_k$ with probability $P(k)$.

Without loss of generality, let us assume that quantum register Q and the environment E are initially in product state $\rho\otimes\varepsilon_0$, and ε_0 is in the mixed state $\varepsilon_0=\sum_l \lambda_l\,|\phi_l\rangle\,\langle\phi_l|$ $(\mathrm{tr}(\varepsilon_0)=\sum_l \lambda_l=1)$. Further, Q and E undergo the interaction described by U, and upon interaction, we perform the measurement, which leaves the state of E in subspace $S\subset H_E$, where H_E is the Hilbert space of the environment. The projection operator P_E associated with S and the superoperator can be represented in terms of the orthonormal basis for S $\{|g_m\rangle\}$ as follows:

$$P_E=\sum_m |g_m\rangle\langle g_m|. \tag{7.85}$$

Let $S^{\perp}$ denote the dual space with orhtonormal basis $\{|h_n\rangle\}$. Since the Hilbert space of environment H_E can be written as $S\oplus S^{\perp}=H_E$, the combined basis sets $\{|g_m\rangle\}$ and $\{|h_n\rangle\}$ can be used to create the basis for H_E. The reduced density operator for quantum register after the measurement will be

$$\xi(\rho)=\mathrm{Tr}_E\left(P_E U(\rho\otimes\varepsilon_0)U^{\dagger}P_E\right)=\sum_{l,m} E_{l,m}\rho E_{l,m}^{\dagger},\quad E_{lm}=\sqrt{\lambda_l}\langle g_m|U|\phi_l\rangle. \tag{7.86}$$

The probability that after the measurement the state of E is found in the subspace S is then as follows:

$$P(S)=\mathrm{Tr}_Q(\xi(\rho)). \tag{7.87}$$

Because $P(S)$ is the probability $0\le \mathrm{Tr}_Q(\xi(\rho))\le 1$, and the quantum operation is non-trace-preserving. If we do not observe the outcome of the measurement the corresponding setting is called the *nonselective dynamics*, otherwise, it is called the *selective dynamics*. For nonselective dynamics is clear that $P_E=I_E$ and the set $\{|g_m\rangle\}$ spans H_E (meaning that $H_E=S$). Because of this fact we expect that completeness relationship involving the operation elements E_{lm} from Eq. (7.84) is satisfied:

$$\sum_{lm} E_{lm}^{\dagger}E_{lm}=\sum_{lm}\lambda_l\langle\phi_l|U^{\dagger}|g_m\rangle\langle g_m|U|\phi_l\rangle=\sum_l \lambda_l\langle\phi_l|U^{\dagger}U|\phi_l\rangle \overset{U^{\dagger}U=I_Q\otimes I_E}{\overbrace{=}}$$

$$\sum_l \lambda_l\langle\phi_l|\phi_l\rangle I_Q=I_Q\sum_l \lambda_l \overset{\sum_l \lambda_l=\mathrm{Tr}\varepsilon_0=1}{\overbrace{=}} I_Q. \tag{7.88}$$

The quantum operation now is clearly trace-preserving and nothing has been learned about the final state of the environment. Therefore, the nonselective dynamics corresponds to the trace-preserving quantum operations.

In selective dynamics, we observe the outcome of the measurement of E, so now $P_E\neq I_E$ and $S\subset H_E$. Since $P(S)+P(S^{\perp})=1$, the trace of quantum operation is $\mathrm{Tr}(\xi(\rho))=P(S)$ $=1-P(S^{\perp})<1$ and the quantum operation is not-trace preserving. Therefore, the selective dynamics of quantum operations is a non-trace-preserving and we gain some knowledge about the final state of the environment. We have already mentioned above that the Hilbert space of environment H_E can be written as $S\oplus S^{\perp}=H_E$, and the combined basis set $\{|g_m\rangle\}$ of S and basis set $\{|h_n\rangle\}$ of $S^{\perp}$ can be used as the basis for H_E. We also defined earlier the elementary operations corresponding to S as $E_{lm}=\sqrt{\lambda_l}\langle g_m|U|\phi_l\rangle$. In similar fashion we define the element operations corresponding to $S_{\perp}$ as

$H_{ln} = \sqrt{\lambda_l}\langle h_n|U|\phi_l\rangle$. We can show that the completeness relation is now satisfied when both element operations E_{lm} and H_{ln} are involved:

$$\sum_{lm} E_{lm}^{\dagger}E_{lm} + \sum_{\text{ln}} H_{\text{ln}}^{\dagger}H_{\text{ln}} = \sum_{lm} \lambda_l\langle\phi_l|U^{\dagger}|g_m\rangle\langle g_m|U|\phi_l\rangle + \sum_{ln} \lambda_l\langle\phi_l|U^{\dagger}|h_n\rangle\langle h_n|U|\phi_l\rangle$$

$$= \sum_{l} \lambda_l\langle\phi_l|U^{\dagger}\left\{\sum_{m}|g_m\rangle\langle g_m| + \sum_{n}|h_n\rangle\langle h_n|\right\}U|\phi_l\rangle = \sum_{l} \lambda_l\langle\phi_l|U^{\dagger}U|\phi_l\rangle = I_Q \quad (7.89)$$

This derivation justifies our previous claim that for nonpreserving operations, we can write informally that $\sum_{lm} E_{lm}E_{lm}^{\dagger} < I$.

It is clear from discussion above that the operator-sum representation is base dependent. By choosing a different basis for environment and a different basis for *S* from those discussed above, we will obtain a different operator-sum representation. Therefore, the operator-sum representation is not the unique one. We also expect that we should be able to transform one set of elementary operations used in the operator-sum representation to another set of elementary operations by properly choosing the unitary transformation. In the reminder of this section we describe several important quantum channels by employing, just introduced, operator-sum representation.

7.5.2 Depolarizing channel

The depolarizing channel (see Fig. 7.14) with probability $1-p$ leaves the qubit as it is, while with probability p moves the initial state into $\rho_f = I/2$ that maximizes the von-Neumann entropy $S(\rho) = -\text{Tr}\ \rho \log \rho = 1$. The properties describing the model can be summarized as follows:

1. Qubit errors are independent.
2. Single-qubit errors (*X, Y, Z*) are equally likely.
3. All qubits have the same single-error probability $p/4$.

Operation elements E_i of channel model should be selected as follows:

$$E_0 = \sqrt{1 - 3p/4I}; E_1 = \sqrt{p/4X}; E_1 = \sqrt{p/4X}; E_1 = \sqrt{p/4Z} \quad (7.90)$$

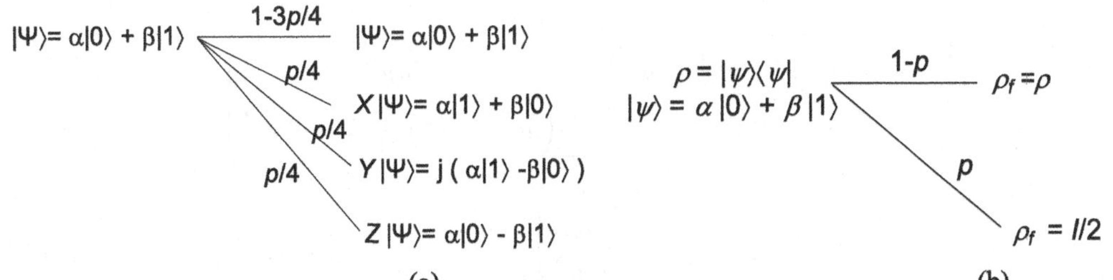

FIGURE 7.14

Depolarizing channel model: (A) Pauli operator description and (B) density operator description.

The action of depolarizing channel is to perform the following mapping: $\rho \rightarrow \xi(\rho) = \sum_i E_i \rho E_i^\dagger$, where ρ is the initial density operator. Without loss of generality we will assume that initial state was pure $|\psi\rangle = a|0\rangle + b|1\rangle$, so

$$\rho = |\psi\rangle\langle\psi| = (a|0\rangle + b|1\rangle)(\langle 0|a^* + \langle 1|b^*)$$
$$= |a|^2\begin{bmatrix}1 & 0\\ 0 & 0\end{bmatrix} + ab^*\begin{bmatrix}0 & 1\\ 0 & 0\end{bmatrix} + a^*b\begin{bmatrix}0 & 0\\ 1 & 0\end{bmatrix} + |b|^2\begin{bmatrix}0 & 0\\ 0 & 1\end{bmatrix} = \begin{bmatrix}|a|^2 & ab^*\\ a^*b & |b|^2\end{bmatrix}. \tag{7.91}$$

The resulting quantum operation can be represented using operator sum-representation as follows:

$$\xi(\rho) = \sum_i E_i \rho E_i^\dagger = \left(1 - \frac{3p}{4}\right)\rho + \frac{p}{4}(X\rho X + Y\rho Y + Z\rho Z)$$

$$= \left(1 - \frac{3p}{4}\right)\begin{bmatrix}|a|^2 & ab^*\\ a^*b & |b|^2\end{bmatrix} + \frac{p}{4}\begin{bmatrix}0 & 1\\ 1 & 0\end{bmatrix}\begin{bmatrix}|a|^2 & ab^*\\ a^*b & |b|^2\end{bmatrix}\begin{bmatrix}0 & 1\\ 1 & 0\end{bmatrix} + \frac{p}{4}\begin{bmatrix}1 & 0\\ 0 & -1\end{bmatrix}\begin{bmatrix}|a|^2 & ab^*\\ a^*b & |b|^2\end{bmatrix}$$
$$\times \begin{bmatrix}1 & 0\\ 0 & -1\end{bmatrix}$$

$$+\frac{p}{4}\begin{bmatrix}0 & -j\\ j & 0\end{bmatrix}\begin{bmatrix}|a|^2 & ab^*\\ a^*b & |b|^2\end{bmatrix}\begin{bmatrix}0 & -j\\ j & 0\end{bmatrix} = \left(1 - \frac{3p}{4}\right)\begin{bmatrix}|a|^2 & ab^*\\ a^*b & |b|^2\end{bmatrix} + \frac{p}{4}\begin{bmatrix}|b|^2 & a^*b\\ ab^* & |a|^2\end{bmatrix}$$

$$+\frac{p}{4}\begin{bmatrix}|a|^2 & -ab^*\\ -a^*b & |b|^2\end{bmatrix} + \frac{p}{4}\begin{bmatrix}|b|^2 & -a^*b\\ -ab^* & |a|^2\end{bmatrix}$$

$$= \begin{bmatrix}\left(1 - \frac{p}{2}\right)|a|^2 + \frac{p}{2}|b|^2 & (1-p)ab^*\\ (1-p)a^*b & \frac{p}{2}|a|^2 + \left(1 - \frac{p}{2}\right)|b|^2\end{bmatrix}$$

$$= \begin{bmatrix}(1-p)|a|^2 + \frac{p}{2}\left(|a|^2 + |b|^2\right) & (1-p)ab^*\\ (1-p)a^*b & \frac{p}{2}\left(|a|^2 + |b|^2\right) + (1-p)|b|^2\end{bmatrix}$$

$$\overset{|a|^2+|b|^2}{=} (1-p)\begin{bmatrix}|a|^2 & ab^*\\ a^*b & |b|^2\end{bmatrix} + \frac{p}{2}I = (1-p)\rho + \frac{p}{2}I \tag{7.92}$$

The first line in Eq. (7.92) corresponds to model shown in Fig. 7.14A and the last line corresponds to model shown in Fig. 7.14B. It is clear from Eqs. (7.91) and (7.92) that $\text{Tr}\xi(\rho) = \text{Tr}(\rho) = 1$ meaning that this superoperator is trace-preserving. Note that the depolarizing channel model in some other books/papers can slightly be different from the one described here.

7.5.3 Amplitude damping channel

In certain quantum channels the errors X, Y, and Z do not occur with the same probability. In amplitude damping channel, the operation elements are given by

$$E_0 = \begin{pmatrix} 1 & 0 \\ 0 & \sqrt{1-\varepsilon^2} \end{pmatrix}, \quad E_1 = \begin{pmatrix} 0 & \varepsilon \\ 0 & 0 \end{pmatrix} \tag{7.93}$$

The *spontaneous emission* is an example of a physical process that can be modeled using the amplitude damping channel model. If $|\psi\rangle = a\,|0\rangle + b\,|1\rangle$ is the initial qubit sate $\left(\rho = \begin{bmatrix} |a|^2 & ab^* \\ a^*b & |b|^2 \end{bmatrix}\right)$, the effect of amplitude damping channel is to perform the following mapping:

$$\rho \to \xi(\rho) = E_0 \rho E_0^\dagger + E_1 \rho E_1^\dagger = \begin{pmatrix} 1 & 0 \\ 0 & \sqrt{1-\varepsilon^2} \end{pmatrix} \begin{bmatrix} |a|^2 & ab^* \\ a^*b & |b|^2 \end{bmatrix} \begin{pmatrix} 1 & 0 \\ 0 & \sqrt{1-\varepsilon^2} \end{pmatrix}$$

$$+ \begin{pmatrix} 0 & \varepsilon \\ 0 & 0 \end{pmatrix} \begin{bmatrix} |a|^2 & ab^* \\ a^*b & |b|^2 \end{bmatrix} \begin{pmatrix} 0 & 0 \\ \varepsilon & 0 \end{pmatrix}$$

$$= \begin{pmatrix} |a|^2 & ab^*\sqrt{1-\varepsilon^2} \\ a^*b\sqrt{1-\varepsilon^2} & |b|^2(1-\varepsilon^2) \end{pmatrix} + \begin{pmatrix} |b|^2\varepsilon^2 & 0 \\ 0 & 0 \end{pmatrix} = \begin{pmatrix} |a|^2 + \varepsilon^2|b|^2 & ab^*\sqrt{1-\varepsilon^2} \\ a^*b\sqrt{1-\varepsilon^2} & |b|^2(1-\varepsilon^2) \end{pmatrix} \tag{7.94}$$

Probabilities $P(0)$ and $P(1)$ that E_0 and E_1 occur are given by

$$P(0) = \mathrm{Tr}\left(E_0 \rho E_0^\dagger\right) = \mathrm{Tr}\begin{pmatrix} |a|^2 & ab^*\sqrt{1-\varepsilon^2} \\ a^*b\sqrt{1-\varepsilon^2} & |b|^2(1-\varepsilon^2) \end{pmatrix} = 1 - \varepsilon^2|b|^2 \quad P(1) = \mathrm{Tr}\left(E_1 \rho E_1^\dagger\right)$$

$$= \mathrm{Tr}\begin{pmatrix} |b|^2\varepsilon^2 & 0 \\ 0 & 0 \end{pmatrix} = \varepsilon^2|b|^2 \tag{7.95}$$

The corresponding amplitude damping channel model is shown in Fig. 7.15.

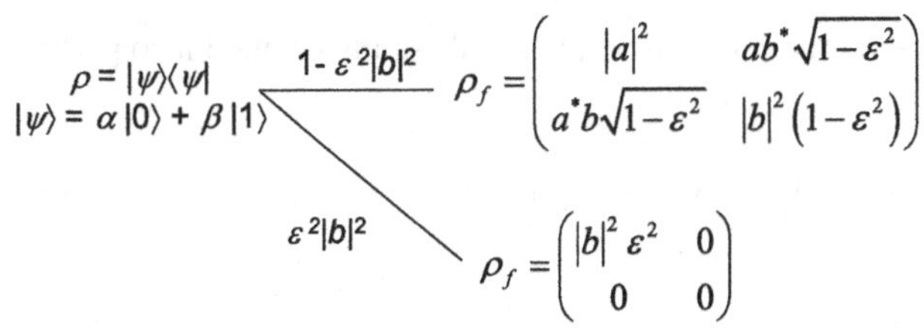

FIGURE 7.15

Amplitude damping channel model.

The density operator can for two-level system be represented as [15] $\rho = (I + \boldsymbol{R} \cdot \boldsymbol{\sigma})/2$, where $\boldsymbol{\sigma} = [X \quad Y \quad Z]$ and $\boldsymbol{R} = [R_x \quad R_y \quad R_z]$ is the Bloch vector, whose components can be determined by $R_x = \mathrm{Tr}(\rho X),\ R_y = \mathrm{Tr}(\rho Y),\ R_z = \mathrm{Tr}(\rho Z)$. Using this result we can determine the Bloch vector $\boldsymbol{R}_\rho$ for ρ by

$$R_{\rho,x} = \mathrm{Tr}(\rho X) = \mathrm{Tr}\left(\begin{bmatrix} |a|^2 & ab^* \\ a^*b & |b|^2 \end{bmatrix}\begin{bmatrix} 0 & 1 \\ 1 & 0 \end{bmatrix}\right) = ab^* + a^*b = 2\mathrm{Re}(a^*b) \quad R_{\rho,y} = 2\mathrm{Im}(a^*b)$$

$$R_{\rho,z} = |a|^2 - |b|^2 \tag{7.96}$$

and $\boldsymbol{R}_\xi$ for $\xi(\rho)$ by

$$R_{\xi,x} = \mathrm{Tr}(\rho_f X) = \mathrm{Tr}\left(\begin{pmatrix} |a|^2 + \varepsilon^2|b|^2 & ab^*\sqrt{1-\varepsilon^2} \\ a^*b\sqrt{1-\varepsilon^2} & |b|^2(1-\varepsilon^2) \end{pmatrix}\begin{bmatrix} 0 & 1 \\ 1 & 0 \end{bmatrix}\right)$$

$$= (ab^* + a^*b)\sqrt{1-\varepsilon^2} = R_{\rho,x}\sqrt{1-\varepsilon^2} \quad R_{\xi,y} = R_{\rho,y}\sqrt{1-\varepsilon^2} \quad R_{\xi,z} = R_{\rho,z}(1-\varepsilon^2) + \varepsilon^2 \tag{7.97}$$

Example 1. If the initial state was $|1\rangle$ the Bloch vectors $\boldsymbol{R}_\rho$ and $\boldsymbol{R}_\xi$ for ρ and $\xi(\rho)$ are given respectively by

$$\boldsymbol{R}_\rho = -\hat{z} \quad \boldsymbol{R}_\xi = (-1 + 2\varepsilon^2)\hat{z}.$$

Clearly, the initial amplitude of Bloch vector is reduced. The error introduced by the channel can be interpreted as $X + jY$.

Example 2. On the other hand, if the initial state was $|0\rangle$ the $\xi(\rho)$ is given by $\xi(\rho) = |0\rangle\langle 0|$, suggesting that initial state was not changed at all. Therefore, $\Pr(|1\rangle \to |0\rangle) = \varepsilon^2$.

7.5.4 Generalized amplitude damping channel

In generalized amplitude damping channel, the operation elements are given by

$$E_0 = \sqrt{p}\begin{pmatrix} 1 & 0 \\ 0 & \sqrt{1-\gamma} \end{pmatrix}, \quad E_1 = \sqrt{p}\begin{pmatrix} 0 & \sqrt{\gamma} \\ 0 & 0 \end{pmatrix}, \quad E_2 = \sqrt{1-p}\begin{pmatrix} \sqrt{1-\gamma} & 0 \\ 0 & 1 \end{pmatrix},$$
$$E_3 = \sqrt{1-p}\begin{pmatrix} 0 & 0 \\ \sqrt{\gamma} & 0 \end{pmatrix} \tag{7.98}$$

If $|\psi\rangle = a|0\rangle + b|1\rangle$ is the initial qubit sate, the effect of amplitude damping channel is to perform the following mapping:

$$\rho \to \xi(\rho) = \xi(\rho) = \frac{I}{2} + \frac{2p-1}{2}Z + \frac{1}{2}\left[R_{\rho,x}\sqrt{1-\gamma}X + R_{\rho,y}\sqrt{1-\gamma}Y + R_{\rho,z}(1-\gamma)Z\right], \tag{7.99}$$

where the initial Bloch vector is given by $R_\rho = (R_{\rho,x}, R_{\rho,y}, R_{\rho,z}) = \left(2\mathrm{Re}(a^*b), 2\mathrm{Im}(a^*b), |a|^2 - |b|^2\right)$.
The operation elements E_i $(i = 0,1,2,3)$ occur with probabilities

$$P(0) = \mathrm{Tr}\left(E_0\rho E_0^\dagger\right) = p\left[1 - \frac{\gamma}{2}(1 - R_{\rho,z})\right] \quad P(1) = \mathrm{Tr}\left(E_1\rho E_1^\dagger\right) = p\frac{\gamma}{2}(1 - R_{\rho,z})$$
$$P(3) = \mathrm{Tr}\left(E_3\rho E_3^\dagger\right) = (1-p)\left[1 - \frac{\gamma}{2}(1 + R_{\rho,z})\right] \quad P(4) = \mathrm{Tr}\left(E_4\rho E_4^\dagger\right) = (1-p)\frac{\gamma}{2}(1 + R_{\rho,z}) \tag{7.100}$$

The quantum errors considered thus far are uncorrelated. If the error probability of the correlated errors drops sufficiently rapidly with the number of errors, correlated errors with more than t errors are unlikely, and we can use quantum error correction to fix up to t errors. Errors that "knock" the qubit state out of this two-dimensional Hilbert space are known as *leakage errors*.

7.6 Summary

This chapter was devoted to quantum error correction concepts ranging from an intuitive description to a rigorous mathematical framework. After an introduction to Pauli operators, we described basic quantum codes, such as the three-qubit flip code, the three-qubit phase flip code, Shor's nine-qubit code, stabilizer codes, and CSS codes. We then formally introduced quantum error correction, including quantum error correction mapping, quantum error representation, stabilizer group definition, quantum-check matrix representation, and the quantum syndrome equation. We further provided the necessary and sufficient conditions for quantum error correction, discussed distance properties and error correction capability, and revisited CSS codes. Section 7.4 was devoted to important quantum coding bounds including the quantum Hamming, Gilbert–Varshamov, and Singleton bounds. In the same section, we discussed quantum weight enumerators and quantum MacWilliams identities. We also discussed quantum superoperators and various quantum channels, including the quantum depolarizing, amplitude damping, and generalized amplitude damping channels.

References

[1] A.R. Calderbank, P.W. Shor, Good quantum error-correcting codes exist, Phys. Rev. 54 (1996) 1098–1105.
[2] A.M. Steane, Error correcting codes in quantum theory, Phys. Rev. Lett. 77 (1996) 793.
[3] A.M. Steane, Simple quantum error-correcting codes, Phys. Rev. 54 (6) (Dec. 1996) 4741–4751.
[4] A.R. Calderbank, E.M. Rains, P.W. Shor, N.J.A. Sloane, Quantum error correction and orthogonal geometry, Phys. Rev. Lett. 78 (3) (Jan. 1997) 405–408.
[5] E. Knill, R. Laflamme, Concatenated Quantum Codes, 1996. Available at: http://arxiv.org/abs/quant-ph/9608012.
[6] R. Laflamme, C. Miquel, J.P. Paz, W.H. Zurek, Perfect quantum error correcting code, Phys. Rev. Lett. 77 (1) (July 1996) 198–201.
[7] D. Gottesman, Class of quantum error correcting codes saturating the quantum Hamming bound, Phys. Rev. 54 (Sept. 1996) 1862–1868.
[8] D. Gottesman, Stabilizer Codes and Quantum Error Correction, PhD Dissertation, California Institute of Technology, Pasadena, CA, 1997.

[9] R. Cleve, D. Gottesman, Efficient computations of encoding for quantum error correction, Phys. Rev. 56 (July 1997) 76–82.
[10] A.Y. Kitaev, Quantum error correction with imperfect gates, in: Quantum Communication, Computing, and Measurement, Plenum Press, New York, 1997, pp. 181–188.
[11] C.H. Bennett, D.P. DiVincenzo, J.A. Smolin, W.K. Wootters, Mixed-state entanglement and quantum error correction, Phys. Rev. 54 (6) (Nov. 1996) 3824–3851.
[12] D.J.C. MacKay, G. Mitchison, P.L. McFadden, Sparse-graph codes for quantum error correction, IEEE Trans. Inf. Theor. 50 (2004) 2315–2330.
[13] M.A. Neilsen, I.L. Chuang, Quantum Computation and Quantum Information, Cambridge University Press, Cambridge, 2000.
[14] F. Gaitan, Quantum Error Correction and Fault Tolerant Quantum Computing, CRC Press, 2008.
[15] I.B. Djordjevic, Quantum Information Processing, Quantum Computing, and Quantum Error Correction: An Engineering Approach, second ed., Elsevier/Academic Press, 2021.
[16] G.D. Forney Jr., M. Grassl, S. Guha, Convolutional and tail-biting quantum error-correcting codes, IEEE Trans. Inf. Theor. 53 (March 2007) 865–880.
[17] M.-H. Hsieh, Entanglement-Assisted Coding Theory, PhD Dissertation, University of Southern California, Aug. 2008.
[18] E. Knill, R. Laflamme, Theory of quantum error-correcting codes, Phys. Rev. 55 (1997) 900.
[19] J. Preskill, Ph219/CS219 Quantum Computing (Lecture Notes), Caltech, 2009. Available at: http://theory.caltech.edu/people/preskill/ph229/.
[20] F.J. MacWilliams, N.J.A. Sloane, J.-M. Goethals, The MacWilliams identities for nonlinear codes, The Bell System Techn. J 51 (4) (Apr. 1972) 803–819.
[21] F.J. MacWilliams, N.J.A. Sloane, The Theory of Error Correcting Codes, North-Holland Mathematical Library, New York, 1977.
[22] A.R. Calderbank, E.M. Rains, P.W. Shor, N.J.A. Sloane, Quantum error correction via codes over GF(4), IEEE Trans. Inf. Theor. 44 (1998) 1369–1387.
[23] I. Devetak, T.A. Brun, M.-H. Hsieh, Entanglement-assisted quantum error-correcting codes, in: V. Sidoravičius (Ed.), *New Trends in Mathematical Physics*, Selected Contributions of the XVth International Congress on Mathematical Physics, Springer, 2009, pp. 161–172.
[24] I.B. Djordjevic, Quantum LDPC codes from balanced incomplete block designs, IEEE Commun. Lett. 12 (May 2008) 389–391.
[25] I.B. Djordjevic, Photonic quantum dual-containing LDPC encoders and decoders, IEEE Photon. Technol. Lett. 21 (13) (July 1, 2009) 842–844.
[26] I.B. Djordjevic, Photonic entanglement-assisted quantum low-density parity-check encoders and decoders, Opt. Lett. 35 (9) (May 1, 2010) 1464–1466.

CHAPTER

Quantum stabilizer codes and beyond

8

Chapter outline

8.1 Stabilizer codes

Quantum stabilizer codes [1−51] are based on the stabilizer concept. Before the stabilizer code is formally introduced, the Pauli group definition from the previous chapter is revisited. A *Pauli operator on N qubits* has the form $cO_1O_2\ldots O_N$, where each $O_i \in \{I, X, Y, Z\}$, and $c = j^l$ $(l = 1,2,3,4)$. The set of Pauli operators on N-qubits forms the *multiplicative Pauli group* G_N. It was shown earlier that the basis for single-qubit errors is given by $\{I,X,Y,Z\}$, while the basis for N-qubit errors is obtained by forming all possible direct products as follows:

$$E = O_1 \otimes \cdots \otimes O_N;\quad O_i \in \{I, X, Y, Z\}. \tag{8.1}$$

Quantum Communication, Quantum Networks, and Quantum Sensing. https://doi.org/10.1016/B978-0-12-822942-2.00006-6

The N-qubit error-basis can be transformed into multiplicative Pauli group G_N if it is allowed to premultiply E by a j^l ($l = 0,1,2,3$) factor. By noticing that $Y = -jXZ$, any error in the Pauli group of N-qubit errors can be represented by

$$E = j^{\lambda}X(\boldsymbol{a})Z(\boldsymbol{b});\; \boldsymbol{a} = a_1 \cdots a_N;\; \boldsymbol{b} = b_1 \cdots b_N;\; a_i, b_i = 0, 1;\; \lambda = 0, 1, 2, 3$$
$$X(\boldsymbol{a}) \equiv X_1^{a_1} \otimes \cdots \otimes X_N^{a_N};\; Z(\boldsymbol{b}) \equiv Z_1^{b_1} \otimes \cdots \otimes Z_N^{b_N} \tag{8.2}$$

In Eq. (8.2), the subscript i ($i = 0,1,\ldots,N$) is used to denote the location of the qubit to which operator X_i (Z_i) is applied; a_i (b_i) takes the value 1 (0) if the ith X (Z) operator is to be included (excluded). Therefore, to uniquely determine the error operator (up to the phase constant $j^{\lambda} = 1$, j, -1, $-$j for $l' = 0,1,2,3$), it is sufficient to specify the binary vectors $\boldsymbol{a}$ and $\boldsymbol{b}$.

The action of Pauli group G_N on a set W is further studied. The **stabilizer** of an element $w \in W$, denoted S_w, is the set of elements from G_N that fix w: $S_w = \{g \in G_N | gw = w\}$. It is straightforward to show that S_w forms the group. Let S be the largest abelian subgroup of G_N that fixes all elements from quantum code C_Q, commonly called the stabilizer group. The **stabilizer group** $S \subseteq G_N$ is constructed from a set of $N-K$ operators $g_1,\ldots,g_{N\text{-}K}$, also known as *generators* of S. Generators have the following properties: (1) they commute among each other, (2) they are unitary and Hermitian, and (3) they have the order of 2 because $g^2{}_i = I$. Any element $s \in S$ can be written as a unique product of the powers of the generators:

$$s = g_1^{c_1} \cdots g_{N-K}^{c_{N-K}},\; c_i \in \{0, 1\};\; i = 1, \cdots, N - K \tag{8.3}$$

When c_i equals 1 (0), the ith generator is included (excluded) from the product of the generators above. Therefore, the elements from S can be labeled by simple binary strings of length $N-K$, namely $\boldsymbol{c} = c_1 \ldots c_{N-K}$. The eigenvalues of generators g_i, denoted $\{\text{eig}(g_i)\}$, can be found starting from property 3 as follows:

$$|\text{eig}(g_i)\rangle = I|\text{eig}(g_i)\rangle = g_i^2|\text{eig}(g_i)\rangle = \text{eig}^2(g_i)|\text{eig}(g_i)\rangle \tag{8.4}$$

Eq. (8.4) shows that $\text{eig}^2(g_i) = 1$, meaning that the eigenvalues are given by $\text{eig}(g_i) = \pm 1 = (-1)^{\lambda_i}; \lambda_i = 0, 1$.

The quantum stabilizer code C_Q, with parameters $[N,K]$, can now be defined as the unique subspace of Hilbert space H_2^N fixed by the elements from stabilizer S of C_Q as follows:

$$C_Q = \bigcap_{s \in S}\{|c\rangle \in H_2^N | s|c\rangle = |c\rangle\}. \tag{8.5}$$

Note that S cannot include $-I$ and jI because the corresponding code space C_Q will be trivial. Namely, if $-I$ is included in S, then it must fix arbitrary codeword $-I|c\rangle = |c\rangle$, which can be satisfied only when $|c\rangle = 0$. Since $(jI)^2 = -I$, the same conclusion can be made about jI. Because the original space for C_Q is 2^N-dimensional space, it is necessary to determine N commuting operators that specify a unique state ket $|\psi\rangle \in H_2^N$. The 2^N simultaneous eigenkets of $\{g_1, g_2, \cdots, g_{N-K}; \overline{Z}_1, \overline{Z}_2, \cdots, \overline{Z}_K\}$ can be used as a basis for H_2^N. Namely, if the subset $\{\overline{Z}_i\}$ is chosen so that each element from this subset commutes with all generators $\{g_i\}$ as well as among all elements of the subset itself, the elements from the set above can be used as the basis kets. This is the subject of the next section. In the rest of this section, some important properties of the stabilizer group and stabilizer codes are studied.

For quantum error correction codes (QECCs), it is sufficient in most cases to observe the quotient group G_N/C, $C = \{\pm I, \pm jI\}$. Namely, since it is not possible in quantum mechanics to unambiguously distinguish two states that differ only in a global phase constant, the coset $EC = \{\pm E, \pm jE\}$ can be considered a single error. Eq. (8.2) shows that any error from Pauli group G_N can be uniquely described by two binary vectors $\boldsymbol{a}$ and $\boldsymbol{b}$, each with length N. The number of such errors is therefore $2^{2N} \times 4$, where the factor 4 originates from global phase constant j^l ($l = 0,1,2,3$). Therefore, the order of G_N is 2^{2N+2}, while the order of quotient group G_N/C is $2^{2N+2}/4 = 2^{2N}$.

Two important *properties* of N-qubit errors are

(i) For $\forall$ $E_1, E_2 \in G_N$, $[E_1,E_2] = 0$ or $\{E_1,E_2\} = 0$ is valid. In other words, the elements from the Pauli group either commute or anticommute.
(ii) For $\forall$ $E \in G_N$, (1) $E^2 = \pm I$, (2) $E^\dagger = \pm E$, and (3) $E^{-1} = E^\dagger$ is valid.

Property (i) can be proved because operators X_i and Z_j commute if $i \neq j$ and anticommute when $i = j$. Using the error representation of Eq. (8.2), the error E_1E_2 can be represented as

$$E_1E_2 = j^{l_1+l_2}X(\boldsymbol{a}_1)Z(\boldsymbol{b}_1)X(\boldsymbol{a}_2)Z(\boldsymbol{b}_2) = j^{l_1+l_2}(-1)^{\boldsymbol{a}_1\boldsymbol{b}_2+\boldsymbol{b}_1\boldsymbol{a}_2}X(\boldsymbol{a}_2)Z(\boldsymbol{b}_2)X(\boldsymbol{a}_1)Z(\boldsymbol{b}_1)$$

$$= (-1)^{\boldsymbol{a}_1\boldsymbol{b}_2+\boldsymbol{b}_1\boldsymbol{a}_2}E_2E_1 = \begin{cases} E_2E_1, \boldsymbol{a}_1\boldsymbol{b}_2 + \boldsymbol{b}_1\boldsymbol{a}_2 = 0 \bmod 2 \\ -E_2E_1, \boldsymbol{a}_1\boldsymbol{b}_2 + \boldsymbol{b}_1\boldsymbol{a}_2 = 1 \bmod 2 \end{cases} \tag{8.6}$$

where the notation $\boldsymbol{ab} = \sum_i a_ib_i$ denotes the dot product. The first claim of property (ii) can be easily proved from Eq. (8.2) as follows:

$$E^2 = j^{2l'}X(\boldsymbol{a})Z(\boldsymbol{b})X(\boldsymbol{a})Z(\boldsymbol{b}) = (-1)^{l'+\boldsymbol{ab}}X(\boldsymbol{a})X(\boldsymbol{a})Z(\boldsymbol{b})Z(\boldsymbol{b}) = (-1)^{l'+\boldsymbol{ab}}I = \pm I \tag{8.7}$$

where the $Z^2 = X^2 = I$ property of Pauli operators is used. The second claim of property (ii) can be proved again, as defined in Eq. (8.2), by

$$E^\dagger = \left(j^{l'}X(\boldsymbol{a})Z(\boldsymbol{b})\right)^\dagger = (-j)^{l'}Z^\dagger(\boldsymbol{b})X^\dagger(\boldsymbol{a}) = (-j)^{l'}Z(\boldsymbol{b})X(\boldsymbol{a}) = (-j)^{l'}(-1)^{\boldsymbol{ab}}X(\boldsymbol{a})Z(\boldsymbol{b}) = \pm E \tag{8.8}$$

where the Hermitian property of Pauli matrices $X^\dagger = X$, $Z^\dagger = Z$ is used. Finally, claim (3) of property (ii) can be proved by Eq. (8.8) as follows:

$$EE^\dagger = E(\pm E) = \pm I \tag{8.9}$$

indicating that $E^{-1} = E^\dagger$.

In the previous chapter, the concept of an error syndrome was introduced as follows. Let C_Q be a quantum stabilizer code with generators $g_1, g_2, \ldots, g_{N-K}$ and let $E \in G_N$ be an error. The error syndrome for error E is defined as the bit string $S(E) = [\lambda_1\lambda_2 \ \ldots \ \lambda_{N-K}]^{\mathrm{T}}$ with component bits being determined by

$$\lambda_i = \begin{cases} 0, & [E, g_i] = 0 \\ 1, & \{E, g_i\} = 0 \end{cases} \quad (i = 1, \ldots, N-K) \tag{8.10}$$

Similarly, as in syndrome decoding for classical error correction, it is expected that errors with a nonvanishing syndrome are detectable. In Chapter 7 (Section 7.3, Theorem 1), it was proved that if an error E anticommutes with some element from stabilizer group S, the image of the computational basis (CB) codeword is orthogonal to another CB codeword, so the following can be written:

$$\langle \bar{i}|E|\bar{j}\rangle = 0. \tag{8.11}$$

It is also known from Chapter 7 that

$$\langle \bar{i}|E|\bar{j}\rangle = C_E\delta_{ij} \tag{8.12}$$

Therefore, Eqs. (8.11) and (8.12) make it clear that $C_E = 0$, which indicates that errors with nonvanishing syndromes are correctable.

Let the set of errors $E = \{E_m\}$ from G_N, for which $S(E^+{}_m E_n) \neq 0$, be observed. From Theorem 1 of Section 7.3, it is clear that

$$\langle \bar{i}|E_m^\dagger E_n|\bar{j}\rangle = 0 \tag{8.13}$$

From Eq. (8.12), it can be concluded that $C_{mn} = 0$, indicating that this set of errors can be corrected.

The third case of interest is errors E with vanishing syndrome $S(E) = 0$. Clearly, these errors commute with all generators from S because the stabilizer is Abelian, $S \subseteq C(S)$, where $C(S)$ is the **centralizer** of S, which is the set of errors $E \in G_N$ that commute with all elements from S. There are two options for error E: (1) $E \in S$, in which case there is no need to correct the error since it already fixes all the codewords; and (2) $E \in C(S) - S$, in which case the error is not detectable. This observation can formally be proved as follows. Given that $E \in C(S) - S$, $s \in S$, $|c\rangle \in C_Q$, the following is valid:

$$sE|c\rangle = Es|c\rangle = E|c\rangle. \tag{8.14}$$

Because $E \notin S$, $E|c\rangle = |c'\rangle \neq |c\rangle$ is also valid, and the error E results in a codeword different from the original. Without loss of generality, an erroneous codeword is assumed to be a basis codeword $|c'\rangle = |\bar{i}\rangle$ that can be expanded in terms of basis codewords, as can any other codewords, as follows:

$$E|\bar{i}\rangle = \sum_j p_j|\bar{j}\rangle \neq |\bar{i}\rangle. \tag{8.15}$$

By multiplying with a dual-basis codeword $\langle \bar{k}|$ from the left, the following is obtained:

$$\langle \bar{k}|E|\bar{i}\rangle = p_k \neq 0 \ \ (k \neq i) \tag{8.16}$$

For error E to be detectable, Eq. (8.12) requires p_k to be 0 when k is different from i, while Eq. (8.16) claims that $p_k \neq 0$, which indicates that error E is not detectable.

Because $C(S)$ is the subgroup of G_N, we know that its cosets can be used to partition G_N as follows:

$$EC(S) = \{Ec | c \in C(S)\} \tag{8.17}$$

From the Lagrange theorem, the following is true:

$$|G_N| = |C(S)|[G_N : C(S)] = 2^{2N+2}. \tag{8.18}$$

Based on the syndrome definition of Eq. (8.10), it is clear that with an error syndrome binary vector of length $N-K$, 2^{N-K} different errors can be identified, which means that the index of $C(S)$ in G_N is $[G_N : C(S)] = 2^{N-K}$. From Eq. (8.18), the following is obtained:

$$|C(S)| = |G_N|/[G_N : C(S)] = 2^{N+K+2}. \tag{8.19}$$

Finally, since $C = \{\pm I, \pm jI\}$, the quotient group $C(S)/C$ order is

$$|C(S)\,/\,C| = 2^{N+K}. \tag{8.20}$$

Similarly, as in classical error correction, where the different codeword errors in a set have the same syndrome, it is expected that errors $E_1, E_2 \in G_N$ belonging to the same coset $EC(S)$ will have the same syndrome $S(E) = \boldsymbol{\lambda} = [\lambda_1 \ldots \lambda_{N\text{-}K}]^T$. This observation can be expressed as the following theorem.

Theorem 1: The set of errors $\{E_i| \; E_i \in G_N\}$ have the same syndrome $\boldsymbol{\lambda} = [\lambda_1 \ldots \lambda_{N-K}]^T$ if and only if they belong to the same coset $EC(S)$ given by Eq. (8.17).

Proof. The theorem can be proved without loss of generality by observing two arbitrary errors E_1 and E_2. First, by assuming that these two errors have the same syndrome, namely $S(E_1) = S(E_2) = \boldsymbol{\lambda}$, it can be proved that they belong to the same coset. For each generator g_i, the following is valid:

$$E_1E_2g_i = (-1)^{\lambda_i}E_1g_iE_2 = (-1)^{2\lambda_i}g_iE_1E_2 = g_iE_1E_2, \tag{8.21}$$

which means that E_1E_2 commutes with g_i and therefore $E_1E_2 \in C(S)$. From claim 2 of property (ii), we know that $E_1^\dagger = \pm E_1$, and since $E_1E_2 \in C(S)$, $E_1^\dagger E_2 = c \in C(S)$. By multiplying $E_1^\dagger E_2 = c$ with E_1 from the left, $E_2 = E_1c$ is obtained, which means that $E_2 \in E_1C(S)$. Since E_2 belongs to both $E_2C(S)$ and $E_1C(S)$ (and there is no intersection between equivalence classes), it can be concluded that $E_1C(S) = E_2C(S)$, meaning that errors E_1 and E_2 belong to the same coset. Theorem 1 can be proved by completing the opposite side of the claim. Namely, by assuming that two errors E_1 and E_2 belong to the same coset, it can be proved that they have the same syndrome. Since $E_1, E_2 \in E_1C(S)$ from the coset definition of Eq. (8.17), it is clear that $E_2 = E_1c$ and therefore $E_1^\dagger E_2 = c \in C(S)$, indicating that $E_1^\dagger E_2$ commutes with all generators from S. From claim 2 of property (ii), it is clear that $E_1^\dagger = \pm E_1$, meaning that E_1E_2 must commute with all generators from S, which can be written as

$$[E_1E_2, g_i] = 0, \; \forall g_i \in S. \tag{8.22}$$

Let the syndromes corresponding to E_1 and E_2 be denoted by $S(E_1) = [\lambda_1 \cdots \lambda_{N-K}]^T$ and $S(E_2) = [\lambda'_1 \cdots \lambda'_{N-K}]^T$, respectively. Further, the commutative relations between E_1E_2 and generator g_i are established as

$$E_1E_2g_i = (-1)^{\lambda'_i}E_1g_iE_2 = (-1)^{\lambda_i+\lambda'_i}g_iE_1E_2. \tag{8.23}$$

The commutative relation of Eq. (8.22) is clearly $\lambda_i + \lambda'_i = 0 \bmod 2$, which is equivalent to $\lambda'_i = \lambda_i \bmod 2$, therefore proving that errors E_1 and E_2 have the same syndrome.

From the discussion at the beginning of the section, it is clear that nondetectable errors belong to $C(S)-S$, and the distance D of a QECC can be defined as the lowest weight among the weights of all elements from $C(S)-S$. The distance of QECC is related to the error correction capability of the codes, as discussed in the section on QECC distance properties in the previous chapter. In the same section, the concept of *degenerate* codes is introduced as a group of codes for which the matrix $\boldsymbol{C}$ with elements from Eq. (8.12) is singular. In the same chapter, a useful theorem is proved that can be used as a relatively simple check on QECC degeneracy. Namely, this theorem claims that a quantum stabilizer code with distance D is a degenerate code if and only if its stabilizer S contains at least one element (not counting the identity) weighing less than D. Therefore, two errors E_1 and E_2 are degenerate if $E_1E_2 \in S$. It can be shown that for nondegenerate quantum stabilizer codes, linearly independent correctable errors have different syndromes. Let E_1 and E_2 be two arbitrary, linearly independent correctable errors with the corresponding syndromes $S(E_1)$ and $S(E_2)$. Because these two errors are correctable, their weight in $E^{\dagger}{}_1E_2$ must be smaller than the distance of the code, and since the code is nondegenerate, clearly $E_1^{\dagger}E_2 \notin S, E_1^{\dagger}E_2 \notin C(S) - S$. Since $E_1^{\dagger}E_2$ belongs to neither S nor $C(S)-S$, it must belong to $G_N-C(S)$ and anticommute with at least one element, say g, from S. If E_1 anticommutes with g, E_2 must commute with it and vice versa, so $S(E_1) \neq S(E_2)$. From Eq. (8.11), it is clear that $\langle \bar{i}|E_1^{\dagger}E_2|\bar{j}\rangle = 0$, and therefore, because of Eq. (8.12), $C_{12} = 0$. In conclusion, only degenerate codes have $\langle \bar{i}|E_1^{\dagger}E_2|\bar{j}\rangle \neq 0$.

In some books and papers, the description of stabilizer quantum codes is based on the normalizer concept instead of the centralizer concept employed in this section. Namely, the normalizer $N_G(S)$ of stabilizer S is defined as the set of errors $E \in G_N$ that fix S under conjugation; in other words, $N_G(S) = \{E \in G_N \mid ESE^{\dagger} = S\}$. These two concepts are equivalent since it can be proved that $N_G(S) = C(S)$. To prove this claim, let us observe an element c from $C(S)$. Then for every element $s \in S$, $csc^{\dagger} = scc^{\dagger} = s$ is valid, which indicates that $C(S) \subseteq N(S)$. Let us now observe an element n from $N(S)$. Using the property (i), that the elements n and s either commute or anticommute, we have $nsn^{\dagger} = \pm snn^{\dagger} = \pm s$. Since $-I$ cannot be an element from S, clearly $nsn^{\dagger} = s$, or equivalently, $ns = sn$, meaning that s and n commute and therefore $N(S) \subseteq C(S)$. The only way to satisfy both $C(S) \subseteq N(S)$ and $N(S) \subseteq C(S)$ is by setting $N(S) = C(S)$, therefore proving that these two representations of quantum stabilizer codes are equivalent.

8.2 Encoded operators

An $[N,K]$ QECC introduces an encoding map U from a 2^K-dimensional Hilbert space, denoted H_2^K, to 2^N-dimensional Hilbert space, denoted H_2^N, as follows:

$$|\delta\rangle \in H_2^K \rightarrow |c\rangle = U|\delta\rangle \quad O \in G_K \rightarrow \overline{O} = UOU^{\dagger} \tag{8.24}$$

In particular, Pauli operators X and Z are mapped to

$$X_i \rightarrow \overline{X}_i = UX_iU^{\dagger} \quad \text{and} \quad Z_i \rightarrow \overline{Z}_i = UZ_iU^{\dagger}, \tag{8.25}$$

respectively. We showed in the previous section that arbitrary N-qubit error could be written in terms of the X- and Z-containing operators as given by Eq. (8.2). In a similar fashion, the unencoded operator O can be represented in terms of the X- and Z-containing operators as follows:

$$O = j^{\lambda}X(\boldsymbol{a})Z(\boldsymbol{b}); \ \boldsymbol{a} = a_1 \cdots a_K; \boldsymbol{b} = b_1 \cdots b_K; a_i, b_i \in \{0,1\} \tag{8.26}$$

The encoding given by Eq. (8.24) maps the operator O to the encoded operator $\overline{O}$ as follows:

$$\begin{aligned}
\overline{O} &= U\left[j^{\lambda}X(\boldsymbol{a})Z(\boldsymbol{b})\right]U^{\dagger} = j^{\lambda}UX_1^{a_1}U^{\dagger}UX_2^{a_2}\cdots U^{\dagger}UX_N^{a_N}U^{\dagger}UZ_1^{b_1}U^{\dagger}UZ_2^{b_2}\cdots U^{\dagger}UZ_N^{b_N}U^{\dagger} \\
&= j^{\lambda}\left(UX_1^{a_1}U^{\dagger}\right)\left(UX_2^{a_2}U^{\dagger}\right)\cdots U^{\dagger}\left(UX_N^{a_N}U^{\dagger}\right)\left(UZ_1^{b_1}U^{\dagger}\right)UZ_2^{b_2}\cdots U^{\dagger}\left(UZ_N^{b_N}U^{\dagger}\right) = \\
&= j^{\lambda}\underbrace{\left(\overline{X}_1\right)^{a_1}\left(\overline{X}_2\right)^{a_2}\cdots\left(\overline{X}_N\right)^{a_N}}_{\overline{X}(\boldsymbol{a})}\underbrace{\left(\overline{Z}_1\right)^{b_1}\cdots\left(\overline{Z}_N\right)^{b_N}}_{\overline{Z}(\boldsymbol{b})} = j^{\lambda}\overline{X}(\boldsymbol{a})\overline{Z}(\boldsymbol{b}).
\end{aligned} \tag{8.27}$$

Using Eq. (8.27), we can easily prove the following commutative properties:

$$\begin{aligned}[\overline{X}_i,\overline{X}_j] &= [\overline{Z}_i,\overline{Z}_j] = 0\\ [\overline{X}_i,\overline{Z}_j] &= 0(i\neq j)\\ \{\overline{X}_i,\overline{Z}_i\} &= 0\end{aligned} \tag{8.28}$$

It can also be proved that $\overline{X}_i, \overline{Z}_i$ commute with the generators g_i from S, indicating that encoded Pauli operators belong to $C(S)$.

The decoding map $U^\dagger$ returns the codewords to unencoded kets, namely $U^\dagger|c\rangle = |\delta\rangle; |c\rangle \in H_2^N, |\delta\rangle \in H_2^K$. To demonstrate this, let us start with the definition of a stabilizer,

$$s|c\rangle = |c\rangle, \forall s \in S \tag{8.29}$$

and apply $U^\dagger$ on both sides of Eq. (8.29) to obtain

$$U^\dagger s|c\rangle = U^\dagger|c\rangle. \tag{8.30}$$

By inserting $UU^\dagger = I$ between s and $|c\rangle$, we obtain

$$U^\dagger s\left(UU^\dagger\right)|c\rangle = U^\dagger|c\rangle \Leftrightarrow \left(U^\dagger sU\right)|\delta\rangle = |\delta\rangle. \tag{8.31}$$

Namely, since $U^\dagger : |c\rangle \in H_2^N \rightarrow |\delta\rangle \in H_2^K$, it is clear that stabilizer elements must be mapped to identity operators $I_K \in G_K$, meaning that $U^\dagger sU = I_k$. To better characterize the decoding mapping, we will prove that centralizer $C(S)$ is generated by encoded Pauli operators $\{\overline{X}_i, \overline{Z}_i | i = 1, \cdots, K\}$ and generators of S, $\{g_i$: $i = 1,\ldots,N-K\}$. To prove this claim, let us consider the set of W of 2^{N+K+2} operators generated by all possible powers of g_k, $\overline{X}_l$, and $\overline{Z}_m$ as follows:

$$\begin{aligned}W &= \left\{\hat{O}(\boldsymbol{a},\boldsymbol{b},\boldsymbol{c}) = j^\lambda(\overline{X}_1)^{a_1}\cdots(\overline{X}_K)^{a_K}(\overline{Z}_1)^{b_1}\cdots(\overline{Z}_K)^{b_K}(g_1)^{c_1}\cdots(g_{N-K})^{c_{N-K}}\right\}\\ &= \{j^\lambda\overline{X}(\boldsymbol{a})\overline{Z}(\boldsymbol{b})g(\boldsymbol{c})\}; \boldsymbol{a} = a_1\cdots a_K; \boldsymbol{b} = b_1\cdots b_K; \boldsymbol{c} = c_1\cdots c_{N-K}; g(\boldsymbol{c}) = (g_1)^{c_1}\cdots(g_{N-K})^{c_{N-K}}\end{aligned} \tag{8.32}$$

From commutative relations and Eq. (8.28), we conclude that the encoded operators $\overline{O}$ commute with the generators of stabilizer S, and therefore, $\overline{O} \in C(S)$. From Eq. (8.32), we see that an operator $\hat{O}(\boldsymbol{a},\boldsymbol{b},\boldsymbol{c})$ can be represented as the product of operators $\overline{O}$ and $g(\boldsymbol{c})$. Since $\overline{O} \in C(S)$ and $g(\boldsymbol{c})$ contain generators of stabilizer S, clearly $\hat{O}(\boldsymbol{a},\boldsymbol{b},\boldsymbol{c}) \in C(S)$. The cardinality of set W is $|W| = 2^{K+K+N-K+2} = 2^{N+K+2}$, and it is the same as $|C(S)|$—see (Eq. 8.19)—leading us to conclude that $C(S) = W$. Let us now observe an error E from $C(S)$; the decoding mapping performs the following action:

$$\begin{aligned}\overline{E} &\rightarrow U^\dagger\overline{E}U = U^\dagger\left[j^\lambda\overline{X}(\boldsymbol{a})\overline{Z}(\boldsymbol{b})\overline{g}(\boldsymbol{c})\right]U\\ &= j^\lambda U^\dagger(\overline{X}_1)^{a_1}UU^\dagger\cdots(\overline{X}_K)^{a_K}UU^\dagger(\overline{Z}_1)^{b_1}UU^\dagger\cdots(\overline{Z}_K)^{b_K}UU^\dagger(g_1)^{c_1}UU^\dagger\cdots UU^\dagger(g_{N-K})^{c_{N-K}}U^\dagger\\ &= j^\lambda\left(U^\dagger UX_1^{a_1}U^\dagger U\right)\cdots\left(U^\dagger UX_K^{a_K}U^\dagger U\right)\left(U^\dagger UZ_1^{b_1}U^\dagger U\right)\cdots\left(U^\dagger UZ_K^{b_K}U^\dagger U\right)\left(U^\dagger g_1^{c_1}U\right)\cdots\left(U^\dagger g_{N-k}^{c_{N-K}}U\right)\end{aligned} \tag{8.33}$$

Because stabilizer elements are mapped to identity during demapping—that is, $U^{\dagger}sU = I_k$— Eq. (8.33) becomes

$$U^{\dagger}\overline{E}U = j^{\lambda}\left(X_1^{a_1}\right)\cdots\left(X_K^{a_K}\right)\left(Z_1^{b_1}\right)\cdots\left(Z_K^{b_K}\right) = j^{\lambda}X(\boldsymbol{a})Z(\boldsymbol{b}). \tag{8.34}$$

Therefore, decoding mapping essentially maps from *C*(*S*) to G_K. The image of this mapping is G_K, and based on Eq. (8.31), we know that the kernel of mapping is *S*. In other words, we can write

$$\mathrm{Im}\left(U^{\dagger}\right) = G_K \quad \text{and} \quad \mathrm{Ker}\left(U^{\dagger}\right) = S. \tag{8.35}$$

It can also be shown that decoding mapping $U^{\dagger}$ is a homomorphism from $C(S) \rightarrow G_K$. Namely, from Eq. (8.34), decoding mapping is *onto* mapping (surjective). Now let us observe two operators from *C*(*S*), denoted $\overline{O}_1$ and $\overline{O}_2$. Decoding leads to

$$U^{\dagger} : \overline{O}_1 \rightarrow U^{\dagger}\overline{O}_1U = U^{\dagger}UO_1U^{\dagger}U = O_1, U^{\dagger} : \overline{O}_2 \rightarrow O_2. \tag{8.36}$$

The decoding mapping of the product of $\overline{O}_1$ and $\overline{O}_2$ is clearly

$$U^{\dagger} : \overline{O}_1\overline{O}_2 \rightarrow U^{\dagger}\overline{O}_1\overline{O}_2U = U^{\dagger}UO_1U^{\dagger}UO_2U^{\dagger}U = O_1O_2. \tag{8.37}$$

Since the image of the product of operators equals the product of the images of corresponding operators, the decoding mapping is a homomorphism. Let us now apply the fundamental homomorphism theorem to decode homomorphism $U^{\dagger} : C(S) \rightarrow G_K$, which claims that the quotient group $C(S)/\mathrm{Ker}\left(U^{\dagger}\right) = C(S)/S$ is isomorphic to G_K. Thus, we can write

$$C(S)\,/\,S \cong G_K. \tag{8.38}$$

The two isomorphic groups have the same order, and from the previous section, we know that the order of the unencoded group is $|G_K| = 2^{2K+2}$, so we conclude that

$$|C(S)\,/\,S| = |G_K| = 2^{2K+2}. \tag{8.39}$$

In the previous section, we showed that an error $E \in C(S){-}S$ is not detectable by the QECC scheme because these errors perform mapping from one codeword to another and can therefore be used to implement encoded operations. The errors *E* and *Es* ($s \in S$) essentially perform the same encoded operation, which means that encoded operations can be identified with cosets of *S* in *C*(*S*). This observation is consistent with isomorphism given by Eq. (8.38). Since the same vectors ***a*** and ***b*** uniquely determine both *O* and $\overline{O}$ (see Eqs. 8.26 and 8.27), and because of the isomorphism of Eq. (8.38), we can identify encoded operators $\overline{O}$ with cosets of quotient group *C*(*S*)/*S*. Therefore, we have established bijection between the encoded operations and the cosets of *C*(*S*)/*S*. In conclusion, the encoded Pauli operators $\{\overline{X}_i, \overline{Z}_j | i,j = 1, \cdots, K\}$ must belong to different cosets. Finally, we show that the quotient group can be represented by

$$C(S)/S = \cup\left\{\overline{X}_mS, \overline{Z}_nS : m,n = 1, \ldots, K\right\} \wedge \left\{j^l : l = 0,1,2,3\right\}. \tag{8.40}$$

This representation can be established by labeling the cosets with the help of Eq. (8.27) as follows:

$$\overline{O}S = j^l\overline{X}(\boldsymbol{a})\overline{Z}(\boldsymbol{b})S = j^l\left(\overline{X}_1S\right)^{a_1}\cdots\left(\overline{X}_kS\right)^{a_K}\left(\overline{Z}_1S\right)^{b_1}\cdots\left(\overline{Z}_1S\right)^{b_K}. \tag{8.41}$$

Indeed, the cosets have forms similar to those given in Eq. (8.40).

[5,1,3] Code Example (Cyclic and perfect quantum code). If the generators are given by

$$g_1 = X_1Z_2Z_3X_4,\ g_2 = X_2Z_3Z_4X_5,\ g_3 = X_1X_3Z_4Z_5,\ g_4 = Z_1X_2X_4Z_5,$$

the stabilizer elements can be generated as follows: $s = g_1^{c_1} \cdots g_4^{c_4},\ c_i \in \{0,1\}$. The encoded Pauli operators are given by $\overline{X} = X_1X_2X_3X_4X_5, \overline{Z} = Z_1Z_2Z_3Z_4Z_5$. Since $g_1\overline{X} = Y_2Y_3X_5 \in C(S) - S$, $\mathrm{wt}(g_1\overline{X}) = 3$, and the distance is $D = 3$. Since the smallest weight of stabilizer elements is 4 and thus larger than D, the code is clearly nondegenerate. Because the Hamming bound is satisfied with equality, the code is perfect. The basis codewords are given by

$$|\overline{0}\rangle = \sum_{s \in S} s|0000\rangle$$
$$|\overline{1}\rangle = \overline{X}|\overline{0}\rangle$$

8.3 Finite geometry representation

The quotient group G_N/C, where $C = \{\pm I, \pm jI\}$ as we discussed above, is the normal subgroup of G_N, and it can therefore be constructed from its cosets, with coset multiplication being defined as $(E_1C)(E_2C) = (E_1E_2)C$, where E_1 and E_2 are two error operators. We showed in the previous chapter and Section 8.2 that bijection between G_N/C and $2N$-dimensional Hilbert space F_2^{2N} could be established by Eq. (8.2). Eq. (8.6) shows that the two error operators either commute or anticommute. We can modify the commutative relation of Eq. (8.6) as follows:

$$E_1E_2 = \begin{cases} E_2E_1, \boldsymbol{a}_1\boldsymbol{b}_2 + \boldsymbol{b}_1\boldsymbol{a}_2 = 0 \bmod 2 \\ -E_2E_1, \boldsymbol{a}_1\boldsymbol{b}_2 + \boldsymbol{b}_1\boldsymbol{a}_2 = 1 \bmod 2 \end{cases} = \begin{cases} E_2E_1, \boldsymbol{a}_1\boldsymbol{b}_2 + \boldsymbol{b}_1\boldsymbol{a}_2 = 0 \\ -E_2E_1, \boldsymbol{a}_1\boldsymbol{b}_2 + \boldsymbol{b}_1\boldsymbol{a}_2 = 1 \end{cases}, \tag{8.42}$$

where the dot product is now observed per mod 2: $\boldsymbol{a}_1\boldsymbol{b}_2 = a_{11}b_{21} + \cdots a_{1N}b_{2N} \bmod 2$. The inner product of the type $\boldsymbol{a}_1\boldsymbol{b}_2 + \boldsymbol{b}_1\boldsymbol{a}_2$ is known as a *symplectic (twisted) inner product* and can be denoted $v(E_1) \odot v(E_2)$, where $v(E_i) = [\boldsymbol{a}_i|\boldsymbol{b}_i]$ $(i = 1,2)$ is the vector representation of E_i. The addition operation in a symplectic inner product is also per mod 2, meaning that it performs mapping of two vectors from F_2^{2N} into $F_2 = \{0,1\}$. Based on Eq. (8.42) and symplectic inner product notation, the commutativity relation can be written as

$$v(E_1) \odot v(E_2) = \begin{cases} 0, [E_1, E_2] = 0 \\ 1, \{E_1, E_2\} = 0 \end{cases}. \tag{8.43}$$

Important *properties* of the symplectic inner product are

(i) self-orthogonality: $\boldsymbol{v} \odot \boldsymbol{v} = 0$,
(ii) symmetry: $\boldsymbol{u} \odot \boldsymbol{v} = \boldsymbol{v} \odot \boldsymbol{u}$, and
(iii) bilinearity.

$$(\boldsymbol{u} + \boldsymbol{v}) \odot \boldsymbol{w} = \boldsymbol{u} \odot \boldsymbol{w} + \boldsymbol{v} \odot \boldsymbol{w} \quad \boldsymbol{u} \odot (\boldsymbol{v} + \boldsymbol{w}) = \boldsymbol{u} \odot \boldsymbol{v} + \boldsymbol{u} \odot \boldsymbol{w}, \quad \forall \boldsymbol{u}, \boldsymbol{v}, \boldsymbol{w} \in F_2^{2N}$$

We earlier defined the weight of an N-dimensional operator as the number of nonidentity operators. Given the fact that an N-dimensional operator E can be represented in vector (finite geometry) representation by $v(E) = [a_1 \ldots a_N|b_1b_2 \ldots b_N]$, the weight of a vector $v(E)$, denoted wt($v(E)$), sometimes called *symplectic weight* and denoted swt($v(E)$), is equal to the number of components i for which $a_i = 1$ and/or $b_i = 1$; in other words, $\mathrm{wt}([\boldsymbol{a}|\boldsymbol{b}]) = |\{i|(a_i,b_i) \neq (0,0)\}|$. The weight in finite geometry representation is equal to the weight in operator form representation, defined as the number of non-identity tensor product components in $E = cO_1 \ldots O_N$ (where $O_i \in \{I,X,Y,Z\}$ and $c = j^\lambda$, $\lambda = 0,1,2,3$); that is, $\mathrm{wt}(E) = |\{O_i \neq I\}|$. In similar fashion, the (symplectic) distance between two vectors $v(E_1)=(\boldsymbol{a}_1|\boldsymbol{b}_1)$ and $v(E_2)=(\boldsymbol{a}_2|\boldsymbol{b}_2)$ can be defined as the weight of their sum; that is, $D(v(E_1),v(E_2)) = \mathrm{wt}(v(E_1)+v(E_2))$. For example, let us observe the operators $E_1 = X_1Z_2Z_3X_4$ and $E_2 = X_2Z_3Z_4$. The corresponding finite geometry representations are $v(E_1) = (1001|0110)$ and $v(E_2) = (0100|0011)$, the weights are 4 and 3, respectively, and the distance between them is

$$D(v(E_1),v(E_2)) = \mathrm{wt}(v(E_1)+v(E_2)) = \mathrm{wt}(1101|0101) = 3.$$

The *dual* QECC can be introduced similarly to classical error correction from Chapter 2. The QECC, as described in Section 8.1, can be represented in terms of stabilizer S. Any element s from S can be represented as a vector $v(s) \in F_2^{2N}$. The stabilizer generators

$$g_i = j^l X(\boldsymbol{a}_i)Z(\boldsymbol{b}_i);\, i = 1,\ldots N-K \tag{8.44}$$

can be represented using finite geometry formalism by

$$\boldsymbol{g}_i \doteq v(g_i) = (\boldsymbol{a}_i|\boldsymbol{b}_i). \tag{8.45}$$

The *dual* (null) space $S_\perp$ can then be defined as the set of vectors orthogonal to stabilizer generator vectors; in other words, $\{v_d \in F_2^{2n} : v_d \odot v(s) = 0\}$. Because of Eq. (8.43), $S_\perp$ is the image of $C(S)$.

We now turn our attention to the syndrome representation using finite geometry representation. An error operator E from G_N can be represented using finite geometry interpretation as follows:

$$E \in G_N \rightarrow \boldsymbol{v}_E = v(E) = (\boldsymbol{a}_E|\boldsymbol{b}_E) \in F_2^{2N}. \tag{8.46}$$

The syndrome components can be determined based on Eqs. (8.10) and (8.43) by

$$S(E) = [\lambda_1\lambda_2\cdots\lambda_{N-K}]^{\mathrm{T}};\, \lambda_i = \boldsymbol{v}_E \odot \boldsymbol{g}_i,\, i = 1,\cdots,N-K. \tag{8.47}$$

Note that the ith component of syndrome λ_i can be calculated by matrix multiplication instead of the symplectic product as follows:

$$\begin{aligned}\lambda_i = \boldsymbol{v}_E \odot \boldsymbol{g}_i = \boldsymbol{b}_E\boldsymbol{a}_i + \boldsymbol{a}_E\boldsymbol{b}_i = (\boldsymbol{b}_E|\boldsymbol{a}_E)\begin{pmatrix}\boldsymbol{a}_i\\ \boldsymbol{b}_i\end{pmatrix} = \boldsymbol{v}_E\boldsymbol{J}\boldsymbol{g}_i^T = \boldsymbol{v}_E\boldsymbol{J}(\boldsymbol{A})_i^T;\, (\boldsymbol{A})_i = \boldsymbol{g}_i = (\boldsymbol{a}_i|\boldsymbol{b}_i);\\ i = 1,\cdots,N-K\end{aligned} \tag{8.48}$$

where $\boldsymbol{J}$ is the matrix used to permute the location of the X- and Z-containing operators in $\boldsymbol{v}_E$, namely

$$\boldsymbol{v}_E\boldsymbol{J} = (\boldsymbol{a}_E|\boldsymbol{b}_E)\begin{pmatrix}\boldsymbol{0}_N & \boldsymbol{I}_N\\ \boldsymbol{I}_N & \boldsymbol{0}_N\end{pmatrix} = (\boldsymbol{b}_E|\boldsymbol{a}_E),\; \boldsymbol{J} = \begin{pmatrix}\boldsymbol{0}_N & \boldsymbol{I}_N\\ \boldsymbol{I}_N & \boldsymbol{0}_N\end{pmatrix},$$

with $\boldsymbol{0}_N$ being the $N\times N$ all-zero matrix and $\boldsymbol{I}_N$ being the $N\times N$ identity matrix. Eq. (8.48) allows us to represent the symplectic inner product as matrix multiplication by defining the (quantum) check-matrix as follows:

$$\boldsymbol{A} = \begin{pmatrix}(\boldsymbol{A})_1\\ (\boldsymbol{A})_2\\ \vdots\\ (\boldsymbol{A})_{N-K}\end{pmatrix}, \tag{8.49}$$

where each row is the finite geometry representation of generator g_i. We can represent Eqs. (8.47) and (8.48) as follows:

$$S(E) = [\lambda_1 \lambda_2 \cdots \lambda_{N-K}]^{\mathrm{T}} = \boldsymbol{v}_E \boldsymbol{J} \boldsymbol{A}^T. \tag{8.50}$$

The syndrome in Eq. (8.50) is very similar to the classical error correction syndrome equation (see Chapter 2).

The encoded Pauli operators can be represented using finite geometry formalism as follows:

$$v(\overline{X}_i) = \left(\boldsymbol{a}(\overline{X}_i)\middle|\boldsymbol{b}(\overline{X}_i)\right) \quad v(\overline{Z}_i) = \left(\boldsymbol{a}(\overline{Z}_i)\middle|\boldsymbol{b}(\overline{Z}_i)\right). \tag{8.51}$$

The commutative properties of Eq. (8.28) can now be written as

$$\begin{aligned}
&v(\overline{X}_m) \odot \boldsymbol{g}_n = 0; \quad n = 1, \cdots, N-K \\
&v(\overline{X}_m) \odot v(\overline{X}_n) = 0; \quad n = 1, \cdots, K \\
&v(\overline{X}_m) \odot v(\overline{Z}_n) = 0; \quad n \neq m \\
&v(\overline{X}_m) \odot v(\overline{Z}_m) = 1
\end{aligned} \tag{8.52}$$

From Eq. (8.52), it is clear that to completely characterize the encoded Pauli operators represented with finite geometry, we must solve the system of $N-K$ equations given by Eq. (8.52). With $2N$ unknowns, we have $N-K$ degrees of freedom, which can be used to simplify encoder and decoder implementations as described in the following sections. With $N-K$ degrees of freedom, the Pauli-encoded operators $\overline{X}_i$ can be chosen 2^{N-K} ways, which equals the number of cosets in $C(S)$.

Example: [4,2,2] code. The generators and encoded Pauli operators are given by

$$\begin{aligned}
&g_1 = X_1 Z_2 Z_3 X_4 \quad g_2 = Y_1 X_2 X_3 Y_4 \\
&\overline{X}_1 = X_1 Y_3 Y_4 \quad \overline{Z}_1 = Y_1 Z_2 Y_3 \\
&\overline{X}_2 = X_1 X_3 Z_4 \quad \overline{Z}_2 = X_2 Z_3 Z_4
\end{aligned}$$

Corresponding finite geometry representations are given by

$$\begin{aligned}
&v(g_1) = (1001|0110) \quad v(g_2) = (1111|1001) \\
&v(\overline{X}_1) = (1011|0011) \quad v(\overline{X}_2) = (1010|0001) \\
&v(\overline{Z}_1) = (1010|1110) \quad v(\overline{Z}_2) = (0100|0011)
\end{aligned}$$

Finally, the quantum-check matrix is given by

$$\boldsymbol{A} = \begin{pmatrix} 1001|0110 \\ 1111|1001 \end{pmatrix}$$

Before concluding this section, we show how to represent an arbitrary $s \in S$ using finite geometry formalism. Based on generator representation Eq. (8.44), we can express an arbitrary element s from S by

$$\begin{aligned}
s &= (g_1)^{c_1} \cdots (g_{N-K})^{c_{N-K}} = \prod_{i=1}^{N-K} g_i^{c_i} = \prod_{i=1}^{N-K} \left(j^{\lambda} X(\boldsymbol{a}_i) Z(\boldsymbol{b}_i)\right)^{c_i} \\
&= j^{\lambda'} \prod_{i=1}^{N-K} \prod_k (X_k)^{c_i a_k} (Z_k)^{c_i b_k} = \mathrm{j}^{\lambda'} \mathrm{X}\left(\sum_{i=1}^{N-K} c_i \boldsymbol{a}_i\right) \mathrm{Z}\left(\sum_{i=1}^{N-K} c_i \boldsymbol{b}_i\right),
\end{aligned} \tag{8.53}$$

The corresponding finite geometry representation based on Eq. (8.45) is as follows:

$$v(s) = v_s = \sum_{i=1}^{N-K} c_i g_i. \tag{8.54}$$

Therefore, the set of vectors $\{g_i: i = 1,\ldots,N-K\}$ span a linear subspace of F_2^{2N}, which represents the image of the stabilizer S. Eq. (8.54) is very similar to classical linear block encoding from Chapter 2.

8.4 Standard form of stabilizer codes

The quantum check matrix of QECC C_q, denoted by A_q, can be written based on Eq. (8.49) as follows:

$$\mathbf{A} = \begin{pmatrix} g_1 \\ \vdots \\ g_{N-K} \end{pmatrix} = \left(\begin{array}{ccc|ccc} a_{1,1} & \cdots & a_{1,N} & b_{1,1} & \cdots & b_{1,N} \\ & \vdots & & & \vdots & \\ a_{N-K,1} & \cdots & a_{N-K,N} & b_{N-K,1} & \cdots & b_{N-K,N} \end{array}\right) = (\mathbf{A}'|\mathbf{B}'), \tag{8.55}$$

where the ith row corresponds to finite geometry representation of the generator g_i—see Eq. (8.45). Based on linear algebra, we make the following two observations:

- Observation 1: swapping the mth and nth qubits in the quantum register corresponds to the mth and nth columns swapped within both submatrices $\mathbf{A}$ and $\mathbf{B}$.
- Observation 2: adding the nth row of $\mathbf{A}_q$ to the mth row ($m \neq n$) maps $\mathbf{g}_m \rightarrow \mathbf{g}_m + \mathbf{g}_n$ and generator $g_m \rightarrow g_m g_n$, so the codewords and stabilizer are invariant to this change.

Therefore, we can perform the Gauss–Jordan elimination to put the quantum-check matrix into the following form:

$$\mathbf{A}_q = \left(\begin{array}{cc|cc} \overbrace{\mathbf{I}}^{r} & \overbrace{\mathbf{A}}^{N-r} & \overbrace{\mathbf{B}}^{r} & \overbrace{\mathbf{C}}^{N-r} \\ \mathbf{0} & \mathbf{0} & \mathbf{D} & \mathbf{E} \end{array}\right) \begin{array}{l} \}r \\ \}N-K-r \end{array};\ r = \text{rank}(\mathbf{A}) \tag{8.56}$$

In the second step, we further perform Gauss-Jordan elimination on submatrix $\mathbf{E}$ to obtain

$$\mathbf{A}_q = \left(\begin{array}{ccc|ccc} \overbrace{\mathbf{I}}^{r} & \overbrace{\mathbf{A}_1}^{N-K-r-s} & \overbrace{\mathbf{A}_2}^{K+s} & \overbrace{\mathbf{B}}^{r} & \overbrace{\mathbf{C}_1}^{N-K-r} & \overbrace{\mathbf{C}_2}^{K+s} \\ \mathbf{0} & \mathbf{0} & \mathbf{0} & \mathbf{D}_1 & \mathbf{I} & \mathbf{E}_2 \\ \mathbf{0} & \mathbf{0} & \mathbf{0} & \mathbf{D}_2 & \mathbf{0} & \mathbf{0} \end{array}\right) \begin{array}{l} \}r \\ \}N-K-r-s \\ \}s \end{array} \tag{8.57}$$

Since the last s generators do not commute with the first r operators, we must remove them, which leads to the following *standard form of the quantum-check matrix*:

$$A_q = \left(\begin{array}{ccc|ccc} \overbrace{I_r}^{r} & \overbrace{A_1}^{N-K-r} & \overbrace{A_2}^{K} & \overbrace{B}^{r} & \overbrace{C_1}^{N-K-r} & \overbrace{C_2}^{K} \\ 0 & 0 & 0 & D & I_{N-K-r} & E \end{array} \right) \begin{array}{l} \}r \\ \}N-K-r \end{array} \tag{8.58}$$

Example: The standard form for Steane's code can be obtained as follows:

$$A_q = \left(\begin{array}{c|c} 0001111 & 0000000 \\ 0110011 & 0000000 \\ 1010101 & 0000000 \\ 0000000 & 0001111 \\ 0000000 & 0110011 \\ 0000000 & 1010101 \end{array} \right) \sim \left[\begin{array}{ccccccc|ccccccc} 1&0&0&0&1&1&1&0&0&0&0&0&0&0 \\ 0&1&0&1&0&1&1&0&0&0&0&0&0&0 \\ 0&0&1&1&1&1&0&0&0&0&0&0&0&0 \\ 0&0&0&0&0&0&0&1&0&1&1&0&0&1 \\ 0&0&0&0&0&0&0&0&1&1&0&1&0&1 \\ 0&0&0&0&0&0&0&1&1&1&0&0&1&0 \end{array} \right]$$

We now turn our attention to the *standard form* representation for the *encoded Pauli operators*. Based on the previous section, encoded Pauli operators $\overline{X}_i$ can be represented using finite geometry representation as follows:

$$v(\overline{X}_i) = \left(\begin{array}{ccc|ccc} \overbrace{u_1(i)}^{r} & \overbrace{u_2(i)}^{N-K-r} & \overbrace{u_3(i)}^{K} & \overbrace{v_1(i)}^{r} & \overbrace{v_2(i)}^{N-K-r} & \overbrace{v_3(i)}^{K} \end{array} \right). \tag{8.59}$$

Since encoded Pauli operators have $N-K$ degrees of freedom, as shown in the previous section, we can set arbitrary $N-K$ components of Eq. (8.59) to zero:

$$v(\overline{X}_i) = \left(\begin{array}{ccc|ccc} \mathbf{0} & \overbrace{u_2(i)}^{N-K-r} & \overbrace{u_3(i)}^{K} & \overbrace{v_1(i)}^{r} & \mathbf{0} & \overbrace{v_3(i)}^{K} \end{array} \right). \tag{8.60}$$

Because $\overline{X}_i \in C(S)$, it must commute with all generators of S. Based on Eq. (8.50), for matrix multiplication $v(\overline{X}_i)A_q^T$, we must change the positions of the X- and Z-containing operators so that the commutativity is expressed as

$$\begin{pmatrix} \mathbf{0} \\ \hline \mathbf{0} \end{pmatrix} = (\, v_1(i) \quad \mathbf{0} \quad v_3(i) \,|\, \mathbf{0} \quad u_2(i) \quad u_3(i) \,) \left(\begin{array}{ccc|ccc} I & A_1 & A_2 & B & C_1 & C_2 \\ \mathbf{0} & \mathbf{0} & \mathbf{0} & D & I & E \end{array} \right)^{\mathrm{T}}. \tag{8.61}$$

After matrix multiplication in Eq. (8.61), we obtain the following two sets of linear equations:

$$\begin{gathered} v_1(i) + A_2 v_3(i) + C_1 u_2(i) + C_2 u_3(i) = 0 \\ u_2(i) + E\ u_3(i) = 0 \end{gathered} \tag{8.62}$$

For convenience, let us write down the encoded Pauli operators as the following matrix:

$$\overline{X}=\begin{pmatrix} v(\overline{X}_1) \\ \vdots \\ v(\overline{X}_K) \end{pmatrix}=\left(\mathbf{0}\quad u_2\quad u_3\,|\,v_1\quad \mathbf{0}\quad v_3\right);u_2=\begin{pmatrix} u_{2,1}(1) & \cdots & u_{2,N-K-r}(1) \\ \vdots & & \vdots \\ u_{2,1}(K) & \cdots & u_{2,N-K-r}(K) \end{pmatrix}$$

$$u_3=\begin{pmatrix} u_{3,1}(1) & \cdots & u_{3,K}(1) \\ \vdots & & \vdots \\ u_{3,1}(K) & \cdots & u_{3,K}(K) \end{pmatrix};v_1=\begin{pmatrix} v_{1,1}(1) & \cdots & v_{1,r}(1) \\ \vdots & & \vdots \\ v_{1,1}(K) & \cdots & v_{1,r}(K) \end{pmatrix},v_3=\begin{pmatrix} v_{3,1}(1) & \cdots & v_{3,K}(1) \\ \vdots & & \vdots \\ v_{3,1}(K) & \cdots & v_{3,K}(K) \end{pmatrix}, \tag{8.63}$$

where each row represents one of the encoded Pauli operators. Based on Eq. (8.28), we conclude that the encoded Pauli operators $\overline{X}_i$ must commute among each other, which in matrix representation can be written as

$$\mathbf{0}=\left(v_1\quad \mathbf{0}\quad v_3\,|\,\mathbf{0}\quad u_2\quad u_3\right)\left(\mathbf{0}\quad u_2\quad u_3\,|\,v_1\quad \mathbf{0}\quad v_3\right)^{\mathrm{T}}=v_3u_3^{\mathrm{T}}+u_3v_3^{\mathrm{T}}. \tag{8.64}$$

By solving the system of Eqs. (8.62) and (8.64), we obtain the following solution:

$$u_3=I,v_3=\mathbf{0},u_2=E,v_1=C_1E+C_2. \tag{8.65}$$

Based on Eq. (8.63), we derive the standard form of encoded Pauli operators $\overline{X}_i$:

$$\overline{X}=\left(\mathbf{0}\quad E^T\quad I\,\middle|\,\left(E^TC_1^T+C_2^T\right)\quad \mathbf{0}\quad \mathbf{0}\right). \tag{8.66}$$

The standard form for the encoded Pauli operators $\overline{Z}_i$ can be derived similarly. Let us first introduce the matrix representation of encoded Pauli operators $\overline{Z}_i$, similarly to Eq. (8.63):

$$\overline{Z}=\begin{pmatrix} v(\overline{Z}_1) \\ \vdots \\ v(\overline{Z}_K) \end{pmatrix}=\left(\mathbf{0}\quad u_2'\quad u_3'\,|\,v_1'\quad \mathbf{0}\quad v_3'\right) \tag{8.67}$$

The encoded Pauli operators $\overline{Z}_i$ must commute with generators g_i, which in matrix form can be represented by

$$\begin{pmatrix} \mathbf{0} \\ \hline \mathbf{0} \end{pmatrix}=\left(v_1'\quad \mathbf{0}\quad v_3'\,|\,\mathbf{0}\quad u_2'\quad u_3'\right)\left(\begin{array}{ccc|ccc} I & A_1 & A_2 & B & C_1 & C_2 \\ \mathbf{0} & \mathbf{0} & \mathbf{0} & D & I & E \end{array}\right)^{\mathrm{T}}, \tag{8.68}$$

which leads to the following set of linear equations:

$$\begin{aligned} v_1'+A_2v_3'+C_1u_2'+C_2u_3'&=\mathbf{0} \\ u_2'+E\ u_3'&=\mathbf{0}. \end{aligned} \tag{8.69}$$

Further, the encoded Pauli operators $\overline{Z}_i$ commute with $\overline{X}_i$ $(j \neq i)$ but anticommute for $j = i$—see Eq. (8.28)—which in matrix form can be represented by

$$I = \begin{pmatrix} \mathbf{0} & u'_2 & u'_3 \,|\, v'_1 & \mathbf{0} & v'_3 \end{pmatrix} \begin{pmatrix} v_1 & \mathbf{0} & \mathbf{0} \,|\, \mathbf{0} & u_2 & I \end{pmatrix}^T = v'_3, \tag{8.70}$$

and the corresponding solution is $v'_3 = I$. Finally, the $\overline{Z}_i$ commute among themselves, which in matrix form can be written by

$$\mathbf{0} = \begin{pmatrix} \mathbf{0} & u'_2 & u'_3 \,|\, v'_1 & \mathbf{0} & I \end{pmatrix} \begin{pmatrix} v'_1 & \mathbf{0} & I \,|\, \mathbf{0} & u'_2 & u'_3 \end{pmatrix}^{\mathrm{T}} = u'_3 + \left(u'_3\right)^{\mathrm{T}}, \tag{8.71}$$

whose solution is $u'_3 = \mathbf{0}$. By substituting the solutions for Eqs. (8.70) and (8.71), we obtain

$$v'_1 = A_2, u'_2 = \mathbf{0}. \tag{8.72}$$

Based on the solutions of Eqs. (8.70)–(8.72), from Eq. (8.67), we derive the standard form of the encoded Pauli operators $\overline{Z}_i$ as

$$\overline{Z} = \begin{pmatrix} \mathbf{0} & \mathbf{0} & \mathbf{0} \,|\, A_2^T & \mathbf{0} & I \end{pmatrix}. \tag{8.73}$$

Example. Standard form of [5,1,3] code. The generators of this code are

$$g_1 = X_1Z_2Z_3X_4, g_2 = X_2Z_3Z_4X_5, g_3 = X_1X_3Z_4Z_5, \text{and } g_4 = Z_1X_2X_4Z_5.$$

The standard form of quantum-check matrix can be obtained by Gauss elimination:

$$A_q = \left(\begin{array}{ccccc|ccccc} 1 & 0 & 0 & 1 & 0 & 0 & 1 & 1 & 0 & 0 \\ 0 & 1 & 0 & 0 & 1 & 0 & 0 & 1 & 1 & 0 \\ 1 & 0 & 1 & 0 & 0 & 0 & 0 & 0 & 1 & 1 \\ 0 & 1 & 0 & 1 & 0 & 1 & 0 & 0 & 0 & 1 \end{array}\right) \sim \left(\begin{array}{ccccc|ccccc} 1 & 0 & 0 & 0 & 1 & 1 & 1 & 0 & 1 & 1 \\ 0 & 1 & 0 & 0 & 1 & 0 & 0 & 1 & 1 & 0 \\ 0 & 0 & 1 & 0 & 1 & 1 & 1 & 0 & 0 & 0 \\ 0 & 0 & 0 & 1 & 1 & 1 & 0 & 1 & 1 & 1 \end{array}\right).$$

From Eq. (8.58), we conclude that

$$r = 4, A_1 = 0, C_1 = 0, E = 0, B = \begin{pmatrix} 1 & 1 & 0 & 1 \\ 0 & 0 & 1 & 1 \\ 1 & 1 & 0 & 0 \\ 1 & 0 & 1 & 1 \end{pmatrix}, A_2 = \begin{pmatrix} 1 \\ 1 \\ 1 \\ 1 \end{pmatrix}, C_2 = \begin{pmatrix} 1 \\ 0 \\ 0 \\ 1 \end{pmatrix}$$

From Eq. (8.66), the standard form for encoded Pauli X-operators is

$$\overline{X} = \begin{pmatrix} \mathbf{0} & E^T & I \,|\, \left(E^T C_1^T + C_2^T\right) & \mathbf{0} & \mathbf{0} \end{pmatrix} = \begin{pmatrix} 0 & 0 & 0 & 0 & 1 \,|\, 1 & 0 & 0 & 1 & 0 \end{pmatrix}$$

From Eq. (8.73), the standard form for encoded Pauli Z-operators is

$$\overline{Z} = \begin{pmatrix} \mathbf{0} & \mathbf{0} & \mathbf{0} \,|\, A_2^T & \mathbf{0} & I \end{pmatrix} = \begin{pmatrix} 0 & 0 & 0 & 0 & 0 \,|\, 1 & 1 & 1 & 1 & 1 \end{pmatrix}$$

Using $XZ = -jY$, we obtain the following generators (by ignoring the global phase constant) from the standard form of A_q:

$$g_1 = Y_1Z_2Z_4Y_5, g_2 = X_2Z_3Z_4X_5, g_3 = Z_1Z_2X_3X_5 \text{ and } g_4 = Z_1Z_3Y_4Y_5.$$

The encoded Pauli operators based on the above can be written in operator form as

$$\overline{X} = Z_1 Z_4 X_5, \quad \overline{Z} = Z_1 Z_2 Z_3 Z_4 Z_5.$$

8.5 Efficient encoding and decoding

This section is devoted to efficient encoder and decoder implementations as initially introduced by Gottesman and Cleve [7–10] (see also [15]).

8.5.1 Efficient encoding

Here we study an efficient implementation of an encoder for an [N,K] quantum stabilizer code. The unencoded K-qubit CB states can be obtained as simultaneous eigenkets of Pauli Z_i-operators from

$$\begin{aligned} &Z_i|\delta_1 \cdots \delta_K\rangle = (-1)^{\delta_i}|\delta_1 \cdots \delta_k\rangle; \delta_i = 0, 1; i = 1, \cdots, K \\ &|\delta_1 \cdots \delta_k\rangle = X_1^{\delta_1} \cdots X_K^{\delta_K}|0 \cdots 0\rangle_K \end{aligned} \tag{8.74}$$

The encoding operator U maps:

- the unencoded K-qubit CB ket $|\delta_1 \cdots \delta_K\rangle$ to N-qubit basis codeword $|\overline{\delta_1 \cdots \delta_K}\rangle = U|\delta_1 \cdots \delta_K\rangle$ and
- the single-qubit Pauli operator X_i to the N-qubit-encoded operator $\overline{X}_i$; that is, $X_i \to \overline{X}_i = UX_iU^\dagger$.

It follows from the above and Eq. (8.74) that

$$\begin{aligned} |\overline{\delta_1 \cdots \delta_k}\rangle &= U|\delta_1 \cdots \delta_k\rangle = U\left(X_1^{\delta_1} \cdots X_K^{\delta_K}\right)\left(U^\dagger U\right)|0 \cdots 0\rangle_K = \left[U\left(X_1^{\delta_1} \cdots X_K^{\delta_K}\right)U^\dagger\right]\underbrace{U|0 \cdots 0\rangle_K}_{|\overline{0 \cdots 0}\rangle} \\ &= \left(\overline{X}_1\right)^{\delta_1} \cdots \left(\overline{X}_K\right)^{\delta_K}|\overline{0 \cdots 0}\rangle. \end{aligned} \tag{8.75}$$

By defining the basis codeword by

$$|\overline{0 \cdots 0}\rangle = \sum_{s \in S} s|0 \cdots 0\rangle_N, \tag{8.76}$$

we can simplify the implementation of the encoder. Namely, it can be shown by mathematical induction that

$$\sum_{s \in S} s = \prod_{i=1}^{N-K}(I_N + g_i). \tag{8.77}$$

For $N-K = 1$, we have only one generator g, so the left-hand side (LHS) of Eq. (8.77) becomes

$$\sum_{s \in S} s = \sum_{i=0}^{1} g^i = I + g. \tag{8.78}$$

Let us assume that Eq. (8.77) is correct for $N-K=M$, and we can prove that the same equation is also valid for $M+1$ as follows:

$$\sum_{s\in S} s = \sum_{c_1=0;\cdots;c_M=0;c_{M+1}=0}^{1} g_1^{c_1}\cdots g_M^{c_M} g_{M+1}^{c_{M+1}} = \prod_{i=1}^{M}(I_N+g_i)\sum_{c_{M+1}=0}^{1} g_{M+1}^{c_{M+1}} = \prod_{i=1}^{M}(I_N+g_i)\cdot(I_N+g_{M+1})$$
$$= \prod_{i=1}^{M+1}(I_N+g_i). \tag{8.79}$$

Using Eq. (8.77), the basis codeword given by Eq. (8.76) can be rewritten as

$$|\overline{0\cdots0}\rangle = \prod_{i=1}^{N-K}(I_N+g_i)|0\cdots0\rangle_N. \tag{8.80}$$

Since $|\overline{\delta_1\cdots\delta_K}\rangle = (\overline{X}_1)^{\delta_1}\cdots(\overline{X}_K)^{\delta_K}|\overline{0\cdots0}\rangle$, using Eq. (8.80), the N-qubit basis codeword can be obtained as

$$|\overline{\delta_1\cdots\delta_K}\rangle = \prod_{i=1}^{N-K}(I_N+g_i)\overline{X}_1^{\delta_1}\cdots\overline{X}_K^{\delta_K}|0\cdots0\rangle_N. \tag{8.81}$$

We will use Eq. (8.81) to efficiently implement the encoder with the help of ancilla qubits. Namely, the encoding operation can be represented as the following mapping:

$$|0\cdots0\delta_1\cdots\delta_K\rangle \equiv |0\cdots0\rangle_{N-K}\otimes|\delta_1\cdots\delta_K\rangle \to |\overline{\delta_1\cdots\delta_K}\rangle. \tag{8.82}$$

Based on Eq. (8.81), we first determine the action of the encoded Pauli operators $\overline{X}_i$:

$$\overline{X}_1^{\delta_1}\cdots\overline{X}_K^{\delta_K}|0\cdots0\rangle_N = \breve{U}_1\cdots\breve{U}_K|0\cdots0\delta_1\cdots\delta_k\rangle, \tag{8.83}$$

where $\breve{U}_i$ are controlled operations, the actions of which are determined below based on the standard form of $\overline{X}_i$:

$$v(\overline{X}_i) = \left(\overbrace{\mathbf{0}}^{r}\ \ \overbrace{\mathbf{u}_2(i)}^{N-K-r}\ \ \overbrace{\mathbf{u}_3(i)}^{K}\ \middle|\ \overbrace{\mathbf{v}_1(i)}^{r}\ \ \mathbf{0}\ \ \mathbf{0}\right);\mathbf{u}_3(i)=(0\cdots1_i\cdots0) \tag{8.84}$$

By introducing the following two definitions,

$$S_x[\mathbf{u}_2(i)] \equiv X_{r+1}^{u_{2,1}(i)}\cdots X_{N-K}^{u_{2,N-K-r}(i)};\ S_z[\mathbf{v}_1(i)] \equiv Z_1^{v_{1,1}(i)}\cdots Z_r^{v_{1,r}(i)}, \tag{8.85}$$

we can express the action of encoded Pauli operator $\overline{X}_i$ based on Eqs. (8.84) and (8.85) as follows:

$$\overline{X}_i = S_x[\mathbf{u}_2(i)]S_z[\mathbf{v}_1(i)]X_{N-K+i}. \tag{8.86}$$

In a special case, for $i=K$, Eq. (8.86) becomes

$$\overline{X}_K = S_x[\mathbf{u}_2(K)]S_z[\mathbf{v}_1(K)]X_N, \tag{8.87}$$

so the action of $\overline{X}_K^{\delta_K}$ on $|0\cdots0\rangle_N$ is as follows:

$$\overline{X}_K^{\delta_k}|0\cdots0\rangle_N \overbrace{\underbrace{=}_{S_z[v_1(K)]|0\cdots1\rangle_N=|0\cdots1\rangle_N}}^{X_N|0\cdots0\rangle_N=|0\cdots1\rangle_N} \begin{cases} |0\cdots0\rangle_N, & \delta_K=0 \\ S_x[u_2(K)]|0\cdots1\rangle_N, & \delta_K=1 \end{cases} = \{S_x[u_2(K)]\}^{\delta_K}|0\cdots\delta_K\rangle_N. \tag{8.88}$$

The overall action of $\overline{X}_1^{\delta_1}\cdots\overline{X}_K^{\delta_K}$ on $|0\cdots0\rangle_N$ can be obtained iteratively as

$$\overline{X}_1^{\delta_1}\cdots\overline{X}_K^{\delta_K}|0\cdots0\rangle_N = \prod_{i=1}^{K}\{S_x[u_2(i)]\}^{\delta_i}|0\cdots0\delta_1\cdots\delta_K\rangle_N. \tag{8.89}$$

Clearly, the action of operators in Eq. (8.83) is

$$\breve{U}_i = \{S_x[u_2(i)]\}^{\delta_i}. \tag{8.90}$$

Therefore, the action of $\breve{U}_i$ is a controlled-$S_x[u_2(i)]$ operation controlled by information qubit δ_i associated with the $(N-K+i)$th qubit. The qubits affected by $\breve{U}_i$ operation are the target qubits at positions $r+1$ to $N-K$. The number of required two-qubit gates is $\leq K(N-K-r)$.

To determine N-qubit basis codewords based on Eq. (8.81), we must apply the operator $G = \prod_{i=1}^{N-K}(I_N+g_i)$ to $\overline{X}_1^{\delta_1}\cdots\overline{X}_K^{\delta_K}|0\cdots0\rangle_N$. The action of G can be determined more simply by the following factorization:

$$G = \prod_{i=1}^{N-K}(I_N+g_i) = G_1G_2;\, G_1 = \prod_{i=1}^{r}(I_N+g_i),\, G_2 = \prod_{i=r+1}^{N-K}(I_N+g_i). \tag{8.91}$$

Since the qubits at positions qubits $r+1$ to $N-K$ are affected by $\breve{U}_i$, G_1 and $\breve{U}_i$ operations commute (the order in which we apply them does not matter), and we can write

$$|\overline{\delta_1\cdots\delta_K}\rangle = G_1G_2\overline{X}_1^{\delta_1}\cdots\overline{X}_K^{\delta_K}|0\cdots0\rangle_N = G_1\overline{X}_1^{\delta_1}\cdots\overline{X}_K^{\delta_K}G_2|0\cdots0\rangle_N. \tag{8.92}$$

From the standard form of the quantum-check matrix of Eq. (8.58), it is clear that G_2 is composed of Z-operators only, and since $Z_i|0\cdots0\rangle_N = |0\cdots0\rangle_N$, state $|0\cdots0\rangle_N$ is invariant for operation G_2, and we can write

$$|\overline{\delta_1\cdots\delta_K}\rangle = G_1\overline{X}_1^{\delta_1}\cdots X_K^{\delta_K}|0\cdots0\rangle_N = G_1\breve{U}_T|0\cdots0\delta_1\cdots\delta_K\rangle,\quad \breve{U}_T = \breve{U}_1\cdots\breve{U}_K \tag{8.93}$$

To complete the encoder implementation, we must determine the action of G_1, which contains the factors like $(I+g_i)$. From the standard form of Eq. (8.58), we know that the finite geometry representation of generator g_i is given by

$$g_i = (0\cdots01_i0\cdots0A_1(i)A_2(i)|B(i)C_1(i)C_2(i)). \tag{8.94}$$

From Eq. (8.94), it is clear that the ith position operator X is always present, while the presence of operator Z depends on the content of $B_i(i)$, so we can express g_i as follows:

$$\underline{g_i = T_i X_i Z_i^{B_i(i)}}; i=1, \cdots, r \tag{8.95}$$

where T_i is the operator derived from g_i by removing all operators associated with the ith qubit.

Example. Generator g_1 for [5,1,3] code is given by $g_1 = Y_1 Z_2 Z_4 Y_5$. By removing out all operators associated with the first qubit, we obtain $T_1 = Z_2 Z_4 Y_5$.

The action of $(I + g_i)$ on $|\psi\rangle = \overline{X}_1^{\delta_1} \cdots \overline{X}_K^{\delta_K} |0 \cdots 0\rangle_N$ based on Eq. (8.95) is given by

$$(I + g_i)|\psi\rangle = \breve{U}_T|0\cdots 0\delta_1 \cdots \delta_K\rangle + T_i X_i Z_i^{B_i(i)} \breve{U}_T |0\cdots 0\delta_1 \cdots \delta_K\rangle. \tag{8.96}$$

Eq. (8.95) shows that the action of g_i affects the qubits on positions 1 to r, while $\breve{U}_T$ affect qubits from r+1 to N–K—see Eqs. 8.85 and 8.90—indicating that these operators commute $Z_i^{B_i(i)} \breve{U}_T = \breve{U}_T Z_i^{B_i(i)}$. Because the state $|0\rangle$ is invariant on the action of Z, the state $|0\cdots 0\delta_1 \cdots \delta_K\rangle$ is also invariant on the action of $Z_i^{B_i(i)}$; that is, $Z_i^{B_i(i)} |0\cdots 0\delta_1 \cdots \delta_K\rangle = |0\cdots 0\delta_1 \cdots \delta_K\rangle$. On the other hand, the action of the X-operator on state $|0\rangle$ is to perform qubit flipping to $|1\rangle$, so we can write $X_i|0\cdots 0\delta_1 \cdots \delta_K\rangle = |0\cdots 01_i 0\cdots 0\delta_1 \cdots \delta_K\rangle$ Therefore, the overall action of $(I + g_i)$ on $|\psi\rangle$ is

$$(I + g_i)|\psi\rangle = \breve{U}_T|0\cdots 0\delta_1 \cdots \delta_K\rangle + T_i \breve{U}_T |0\cdots 1_i \cdots 0\delta_1 \cdots \delta_K\rangle. \tag{8.97}$$

Eq. (8.97) can be expressed in terms of Hadamard gate action on the ith qubit:

$$H_i|\psi\rangle = \breve{U}_T H_i |0\cdots 0_i \cdots 0\delta_1 \cdots \delta_K\rangle \overbrace{=}^{H_i|\delta_i\rangle = \frac{1}{\sqrt{2}}\left[|0\rangle + (-1)^{\delta_i}|1\rangle\right],\ \delta_i = 0,1} \breve{U}_T\{|0\cdots 0_i \cdots 0\delta_1 \cdots \delta_K\rangle + |0\cdots 1_i \cdots 0\delta_1 \cdots \delta_K\rangle\}. \tag{8.98}$$

Further, let us apply the operator-controlled T_i, that is, $T_i^{\alpha_i}$ $(\alpha_i = 0, 1)$, on Eq. (8.98) from the left to obtain

$$T_i^{\alpha_i} H_i|\psi\rangle = \breve{U}_T|0\cdots 0_i \cdots 0\delta_1 \cdots \delta_K\rangle + T_i \breve{U}_T |0\cdots 1_i \cdots 0\delta_1 \cdots \delta_k\rangle, \tag{8.99}$$

which is the same as the action of $I + g_i$ given by Eq. (8.97). Therefore, we conclude that

$$(I + g_i)|\psi\rangle = T_i^{\alpha_i} H_i |\psi\rangle, \tag{8.100}$$

meaning that the action of $(I + g_i)$ on $|\psi>$ is to first apply Hadamard gate H_i on $|\psi>$ followed by the controlled T_i gate (T_i-gate controls the qubit at position i). Since T_i acts on no more than N–1 qubits and controlled T_i gates on no more than N–1 two-qubit gates, plus application of the one-qubit Hadamard gate H_i, the number of gates needed for the operation of Eq. (8.100) is (N–1) + 1 gates.

Based on the definition of G_1 operation (see Eq. 8.91), we must iterate the action $(I + g_i)$ per i = 1,…, r to obtain the basis codewords:

$$|\overline{\delta_1 \cdots \delta_k}\rangle = \underbrace{\prod_{i=1}^{N-K} (I + g_i)}_{\prod_{i=1}^{r} T_i^{\alpha_i} H_i} \underbrace{\overline{X}_1^{\delta_1} \cdots \overline{X}_k^{\delta_k}}_{\prod_{m=1}^{K} \breve{U}_m} |0\cdots 0\rangle_N = \left(\prod_{i=1}^{r} T_i^{\alpha_i} H_i\right)\left(\prod_{m=1}^{K} \breve{U}_m\right)|0\cdots 0\delta_1 \cdots \delta_K\rangle. \tag{8.101}$$

From Eq. (8.101), it is clear that the number of required Hadamard gates is r, the number of two-qubit gates implementing controlled T_i gates is $r(N-1)$, and the number of controlled $S_x[\mathbf{u}_2(i)]$ gates (operating on qubits $r+1$ to $N-K$) is $K(N-K-r)$, so the number of total gates is not larger than

$$N_{\text{gates}} = r[(N-1)+1] + K(N-K-r) = (K+r)(N-K). \tag{8.102}$$

Since the number of gates is a linear function of codeword length N, this implementation can be called "efficient." Based on Eqs. (8.90) and (8.85), we can rewrite Eq. (8.101) in a form suitable for implementation:

$$|\delta_1 \cdots \delta_k\rangle = \left(\prod_{i=1}^{r} T_i^{\alpha_i} H_i\right)\left(\prod_{m=1}^{K}\{S_x[\mathbf{u}_2(m)]\}^{\delta_m}\right)|0\cdots 0\delta_1\cdots\delta_K\rangle, \quad S_x[\mathbf{u}_2(m)] = X_{r+1}^{u_{2,1}(m)}\cdots X_{N-K}^{u_{2,N-K-r}(m)} \tag{8.103}$$

Example. Let us now study the implementation of the encoding circuit for [5,1,3] quantum stabilizer code; the same code was observed in the previous section. The generators derived from standard from representation were found to be

$$g_1 = Y_1Z_2Z_4Y_5, g_2 = X_2Z_3Z_4X_5, g_3 = Z_1Z_2X_3X_5 \text{ and } g_4 = Z_1Z_3Y_4Y_5.$$

The encoded Pauli-X operator was found as

$$\overline{X} = Z_1Z_4X_5,$$

and the corresponding finite geometry representation is

$$v(\overline{X}) = (0\ \ 0\ \ 0\ \ 0\ \ 1|1\ \ 0\ \ 0\ \ 1\ \ 0).$$

Since $r=4$, $N=5$, and $K=1$, from Eq. (8.84) it is clear that $N-K-r=0$, so $\mathbf{u}_2=0$, meaning that $S_x[u_2]=I$. From Eq. (8.103), we see that we must determine the T_i operators, which can be determined from the generators g_i by taking out the ith qubit term:

$$T_1 = Z_2Z_4Y_5, T_2 = Z_3Z_4X_5, T_3 = Z_1Z_2X_5 \text{ and } T_4 = Z_1Z_3Y_5.$$

Based on Eq. (8.103), the basis codewords can be obtained as

$$|\delta\rangle = \left[\prod_{i=1}^{4}(T_i)^{\alpha_i}H_i\right]|0000\delta\rangle,$$

wherein α_i auxiliary qubits after the H_i gate control the remaining qubits (described by T_i), and the corresponding encoder implementation is shown in Fig. 8.1.

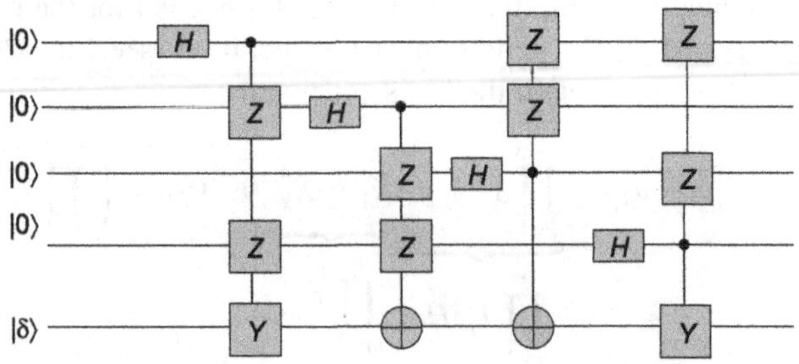

FIGURE 8.1

Efficient implementation of the encoding circuit for [5,1,3] quantum stabilizer code. Since the action of Z in $T_i^{\alpha i}$ ($i = 1,2$) is trivial, it can be omitted.

Example. The generators for the [8,3,3] quantum stabilizer code are given by

$$g_1 = X_1X_2X_3X_4X_5X_6X_7X_8, g_2 = Z_1Z_2Z_3Z_4Z_5Z_6Z_7Z_8, g_3 = X_2X_4Y_5Z_6Y_7Z_8, g_4 = X_2Z_3Y_4X_6Z_7Y_8 \text{ and }$$
$$g_5 = Y_2X_3Z_4X_5Z_6Y_8.$$

The corresponding quantum-check matrix based on the generators above is given by

$$A = \left(\begin{array}{cccccccc|cccccccc} 1 & 1 & 1 & 1 & 1 & 1 & 1 & 1 & 0 & 0 & 0 & 0 & 0 & 0 & 0 & 0 \\ 0 & 0 & 0 & 0 & 0 & 0 & 0 & 0 & 1 & 1 & 1 & 1 & 1 & 1 & 1 & 1 \\ 0 & 1 & 0 & 1 & 1 & 0 & 1 & 0 & 0 & 0 & 0 & 0 & 1 & 1 & 1 & 1 \\ 0 & 1 & 0 & 1 & 0 & 1 & 0 & 1 & 0 & 0 & 1 & 1 & 0 & 0 & 1 & 1 \\ 0 & 1 & 1 & 0 & 1 & 0 & 0 & 1 & 0 & 1 & 0 & 1 & 0 & 1 & 0 & 1 \end{array}\right)$$

By Gauss elimination, we can put the quantum-check matrix above in the standard form of Eq. (8.58) as follows:

$$A = \left(\begin{array}{cccc|c|ccc|cccc|c|ccc} 1 & 0 & 0 & 0 & 1 & 1 & 1 & 0 & 0 & 1 & 0 & 0 & 1 & 1 & 0 & 1 \\ 0 & 1 & 0 & 0 & 1 & 1 & 0 & 1 & 0 & 0 & 1 & 0 & 1 & 0 & 1 & 1 \\ 0 & 0 & 1 & 0 & 1 & 0 & 1 & 1 & 0 & 1 & 0 & 1 & 1 & 0 & 1 & 0 \\ 0 & 0 & 0 & 1 & 0 & 1 & 1 & 1 & 0 & 0 & 1 & 1 & 1 & 1 & 0 & 0 \\ 0 & 0 & 0 & 0 & 0 & 0 & 0 & 0 & 1 & 1 & 1 & 1 & 1 & 1 & 1 & 1 \end{array}\right)$$

From code parameters and Eq. (8.58), it is clear that $r = 4, K = 3, N - K - r = 1$ and

$$A_1 = \begin{pmatrix} 1 \\ 1 \\ 1 \\ 0 \end{pmatrix}, A_2 = \begin{pmatrix} 1 & 1 & 0 \\ 1 & 0 & 1 \\ 0 & 1 & 1 \\ 1 & 1 & 1 \end{pmatrix}$$

$$B = \begin{pmatrix} 0 & 1 & 0 & 0 \\ 0 & 0 & 1 & 0 \\ 0 & 1 & 0 & 1 \\ 0 & 0 & 1 & 1 \end{pmatrix}, C_1 = \begin{pmatrix} 1 \\ 1 \\ 1 \\ 1 \end{pmatrix}, C_2 = \begin{pmatrix} 1 & 0 & 1 \\ 0 & 1 & 1 \\ 0 & 1 & 0 \\ 1 & 0 & 0 \end{pmatrix}, D = (1 \quad 1 \quad 1 \quad 1), E = (1 \quad 1 \quad 1)$$

Based on the standard form of Eq. (8.66) for encoded Pauli-X operators, we obtain

$$\overline{X} = \left(\mathbf{0} \quad E^T \quad I \,\middle|\, (E^T C_1^T + C_2^T) \quad \mathbf{0} \quad \mathbf{0}\right) = \left(\begin{array}{cccc|c|ccc|cccc|c|ccc} 0 & 0 & 0 & 0 & 1 & 1 & 0 & 0 & 0 & 1 & 1 & 0 & 0 & 0 & 0 & 0 \\ 0 & 0 & 0 & 0 & 1 & 0 & 1 & 0 & 1 & 0 & 0 & 1 & 0 & 0 & 0 & 0 \\ 0 & 0 & 0 & 0 & 1 & 0 & 0 & 1 & 0 & 0 & 1 & 1 & 0 & 0 & 0 & 0 \end{array}\right)$$

Based on the standard form of Eq. (8.73) for encoded Pauli-Z operators, we obtain

$$\overline{Z}=\begin{pmatrix}\mathbf{0} & \mathbf{0} & \mathbf{0} \,|\, A_2^T & \mathbf{0} & I\end{pmatrix}=\left(\begin{array}{cccccccc|cccccccc}0&0&0&0&0&0&0&0&1&1&0&1&0&1&0&0\\0&0&0&0&0&0&0&0&1&0&1&1&0&0&1&0\\0&0&0&0&0&0&0&0&0&1&1&1&0&0&0&1\end{array}\right)$$

From the encoded Pauli-X matrix above and Eqs. (8.84) and (8.103), it is clear that

$$u_2=\begin{pmatrix}1\\1\\1\end{pmatrix},\quad S_x[u_2(1)]=X_5,\ \ S_x[u_2(2)]=X_5,\ \ S_x[u_2(3)]=X_5$$

The generators derived from the standard quantum-check matrix are obtained as

$$g_1=X_1Z_2Y_5Y_6X_7Z_8,\ g_2=X_2Z_3Y_5X_6Z_7Y_8, g_3=Z_2X_3Z_4Y_5Y_7X_8,$$
$$g_4=Z_3Y_4Z_5Y_6X_7X_8,\ \text{and}\ g_5=Z_1Z_2Z_3Z_4Z_5Z_6Z_7Z_8.$$

The corresponding T_i operators are obtained from generators g_i by omitting the ith term:

$$T_1=Z_2Y_5Y_6X_7Z_8, T_2=Z_3Y_5X_6Z_7Y_8, T_3=Z_2Z_4Y_5Y_7X_8, T_4=Z_3Z_5Y_6X_7X_8.$$

Based on Eq. (8.103), we obtain the efficient encoder implementation shown in Fig. 8.2.

To further simplify encoder implementation, we can modify the standard form of Eq. (8.58) by further Gauss elimination to obtain

$$A_q=\left(\begin{array}{ccc|ccc}\overbrace{I_r}^{r} & \overbrace{A_1}^{N-K-r} & \overbrace{A_2}^{K} & \overbrace{B}^{r} & \overbrace{0}^{N-K-r} & \overbrace{C}^{K}\\ 0 & 0 & 0 & D & I_{N-K-r} & E\end{array}\right)\begin{array}{l}\}r\\ \}N-K-r\end{array} \tag{8.104}$$

The corresponding encoded Pauli operators can be represented by

$$\overline{X}=\begin{pmatrix}\mathbf{0} & E^T & I_K \,|\, C^T & \mathbf{0} & \mathbf{0}\end{pmatrix},\ \overline{Z}=\begin{pmatrix}\mathbf{0} & \mathbf{0} & \mathbf{0} \,|\, A_2^T & \mathbf{0} & I_K\end{pmatrix}. \tag{8.105}$$

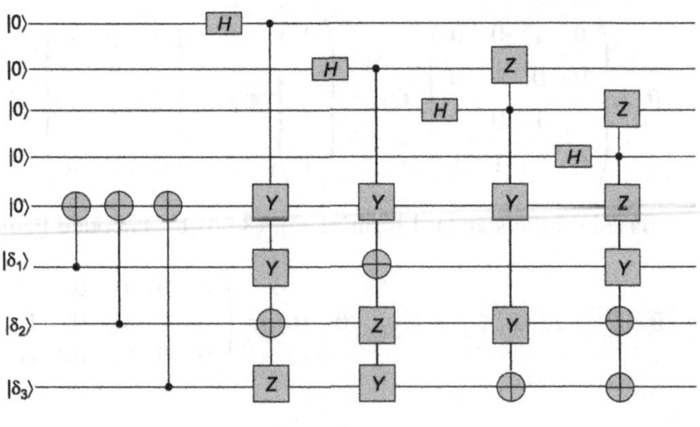

FIGURE 8.2

Efficient implementation of the encoding circuit for [8,3,3] quantum stabilizer code.

From Eq. (8.104), it is clear that generators can be represented using finite geometry representation by

$$g_i = (0 \quad \cdots \quad 0 \quad 1_i \quad 0 \quad \cdots \quad a_{r+1} \quad \cdots \quad a_N | \; b_1 \quad \cdots \quad b_r \quad 0 \quad \cdots \quad 0 \quad b_{N-K+1} \quad \cdots \quad b_N). \tag{8.106}$$

Based on Eq. (8.105), the ith encoded Pauli-X operator can be represented by

$$v(\overline{X}_i) = (0 \quad \cdots \quad 0 \quad a'_{r+1} \quad \cdots \quad a'_{N-K} \quad 0 \quad \cdots \quad 0 \quad 1_{N-K+i} \quad 0 \quad \cdots \quad 0 | \; b'_1 \quad \cdots \quad b'_r \quad 0 \quad \cdots \quad 0). \tag{8.107}$$

Based on Eqs. (8.103), (8.106), and (8.107), we show the two-stage encoder configuration in Fig. 8.3. The first stage is related to controlled-X operations in which the ith information qubit [or equivalently $(N-K+i)$th codeword qubit] is used to control ancilla qubits at positions $r+1$ to $N-K$ based on Eq. (8.107). In the second stage, the ith ancilla qubit, upon application of the Hadamard gate, is used to control qubits at codeword locations $i+1$ to N based on Eq. (8.106).

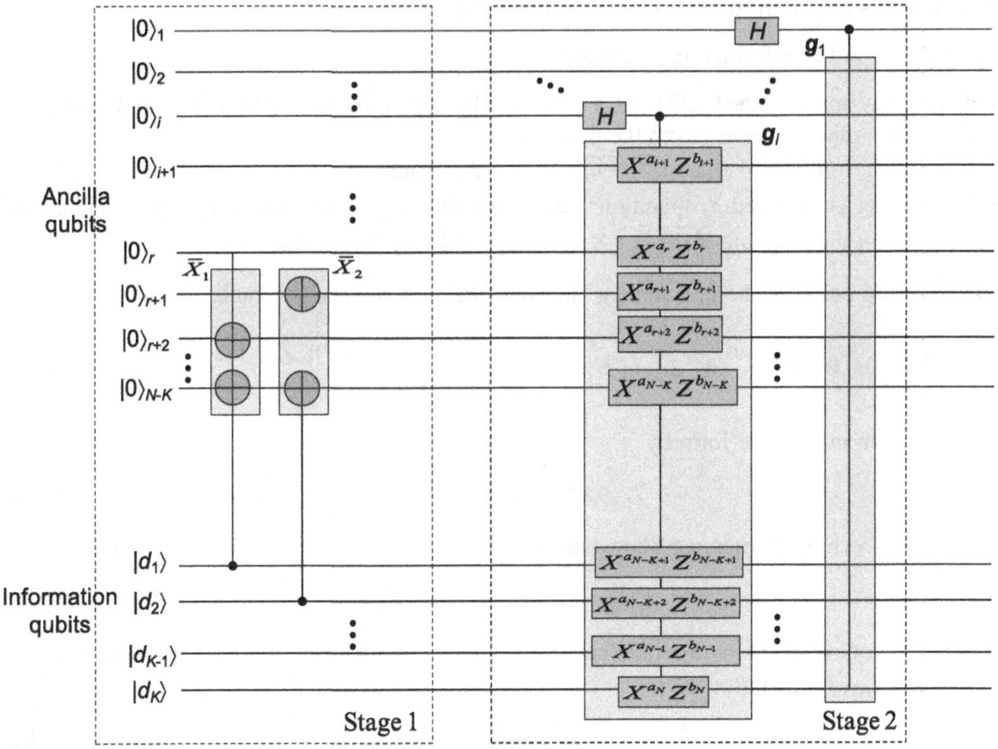

FIGURE 8.3

Efficient encoder implementation based on the standard form given by Eq. (8.104).

8.5.2 Efficient decoding

The straightforward approach to decoding is to implement the decoding circuit as an encoding circuit with the gates in reverse order. However, this approach is not necessarily optimal in terms of gate utilization. Here we describe an efficient implementation from Gottesman [8] (see also [15]). Gottesman's approach is to introduce K-ancilla qubits being in state $|0\ldots0\rangle_K$ and perform decoding as the following mapping:

$$|\psi_{\text{in}}\rangle = |\overline{\delta_1\cdots\delta_K}\rangle \otimes |0\cdots0\rangle_K \rightarrow |\psi_{\text{out}}\rangle = U_{\text{decode}}|\psi_{\text{in}}\rangle = U_{\text{decode}}|\overline{\delta_1\cdots\delta_K}\rangle \otimes |0\cdots0\rangle_K = |\overline{0\cdots0}\rangle \otimes |\delta_1\cdots\delta_K\rangle. \tag{8.108}$$

With respect to Eq. (8.108), the following two remarks are important:

- *Remark 1.* U_{decode} is a linear operator; it is sufficient to concentrate on decoding the basis codewords.
- *Remark 2.* The decoding procedure, in addition to decoded state $|\delta_1\ldots\delta_K\rangle$, returns N qubits in state $|\overline{0\cdots0}\rangle$ that are needed in the encoding process. This occurs from applying the operator G on ket $|0\ldots0\rangle_N$. Therefore, by implementing the encoder and decoder in the transceiver by applying Gottesman's approach, we can save Nr gates in the encoding process.

Decoding can be performed in two stages:

(1) by properly applying the CNOT gates to the ancilla qubits, we put the ancillae into decoded state $|\delta_1\ldots\delta_K\rangle$; that is, we perform the mapping $|\psi_{\text{in}}\rangle = |\overline{\delta_1\cdots\delta_K}\rangle \otimes |0\cdots0\rangle_K \rightarrow |\overline{\delta_1\cdots\delta_K}\rangle \otimes |\delta_1\cdots\delta_K\rangle$, and

(2) by applying a controlled-X_i operation to each encoded qubit i conditioned on the ith ancilla qubit, we perform the mapping $|\overline{\delta_1\cdots\delta_K}\rangle \otimes |\delta_1\cdots\delta_K\rangle \rightarrow |\overline{0\cdots0}\rangle \otimes |\delta_1\cdots\delta_K\rangle$

The starting point for the first stage is the standard form of encoded Pauli operators $\overline{Z}_i$:

$$\overline{\mathbf{Z}} = (\mathbf{0} \quad \mathbf{0} \quad \mathbf{0} \,|\, A_2^T \quad \mathbf{0} \quad I), v(\overline{Z}_i) = (v_l), \quad v_l = \begin{cases} A_{2,l}(i); \; l = 1, \cdots, r \\ \delta_{l,N-K+i}; l = r+1, \cdots, N \end{cases}, \tag{8.109}$$

which are given in operator form by

$$\overline{Z}_i = Z_1^{A_{2,1}(i)} \cdots Z_r^{A_{2,r}(i)} Z_{N-K+i}. \tag{8.110}$$

From the previous chapter, we know that the basis codeword $|\overline{\delta_1\cdots\delta_K}\rangle$ is the eigenket of $\overline{Z}_i$:

$$\overline{Z}_i|\overline{\delta_1\cdots\delta_K}\rangle = (-1)^{\delta_i}|\overline{\delta_1\cdots\delta_K}\rangle. \tag{8.111}$$

The basis codeword $|\overline{\boldsymbol{\delta}}\rangle = |\overline{\delta_1\cdots\delta_K}\rangle$ can be expanded in terms N-qubit CB kets $\{|\boldsymbol{d}\rangle = |d_1\ldots d_N\rangle \,|\, d_i = 0,1; \; i = 1,\ldots,N\}$ as follows:

$$|\overline{\boldsymbol{\delta}}\rangle = |\overline{\delta_1\cdots\delta_K}\rangle = \sum_{\boldsymbol{d} \in F_2^N} C_{\overline{\delta}}(\boldsymbol{d})|d_1\cdots d_N\rangle. \tag{8.112}$$

Let us now apply Eq. (8.111) on Eq. (8.112) to obtain

$$(-1)^{\delta_i}|\overline{\delta}\rangle = \overline{Z}_i|\overline{\delta}\rangle = \sum_{d\in F_2^n} C_{\overline{\delta}}(d)\overline{Z}_i|d_1\cdots d_N\rangle = \sum_{d\in F_2^N} C_{\overline{\delta}}(d)(-1)^{v(\overline{Z}_i)\cdot d}|d_1\cdots d_N\rangle$$

$$= (-1)^{v(\overline{Z}_i)\cdot d}\sum_{d\in F_2^N} C_{\overline{\delta}}(d)|d_1\cdots d_N\rangle = (-1)^{v(\overline{Z}_i)\cdot d}|\overline{\delta}\rangle. \tag{8.113}$$

From Eq. (8.113), it is obvious that

$$\delta_i = v(\overline{Z}_i)\cdot d. \tag{8.114}$$

We indicated above that in the first stage of the decoding procedure, we must properly apply CNOT gates on the ancillae, and it is therefore important to determine the action of the X_{a_i} gate on $|\overline{\delta_1\cdots\delta_K}\rangle\otimes|0\cdots0\rangle_K$. By applying the expansion Eq. (8.113), we obtain

$$X_{a_1}^{\delta_1}|\overline{\delta_1\cdots\delta_K}\rangle\otimes|0\cdots0\rangle_K = \sum_{d\in F_2^N} C_{\overline{\delta}}(d)|d_1\cdots d_N\rangle\otimes\left[X_{a_1}^{\delta_1}|0\cdots0\rangle_K\right]$$

$$\overset{\delta_i=v(\overline{Z}_1)\cdot d}{\overbrace{=}} \sum_{d\in F_2^N} C_{\overline{\delta}}(d)|d_1\cdots d_N\rangle\otimes X_{a_1}^{v(\overline{Z}_1)\cdot d}|0\cdots0\rangle_K. \tag{8.115}$$

From Eq. (8.115), it is clear that we must determine the action of $X_{a_1}^{v(\overline{Z}_1)\cdot d}$ to $|0\rangle$:

$$X_{a_1}^{v(\overline{Z}_1)\cdot d}|0\rangle = X_{a_1}^{v_1(\overline{Z}_1)d_1}\cdots X_{a_1}^{v_N(\overline{Z}_1)d_N}|0\rangle = |v(\overline{Z}_1)\cdot d\rangle = |\delta_1\rangle. \tag{8.116}$$

Therefore, the action of $X_{a_1}^{\delta_1}$ gate on $|\overline{\delta_1\cdots\delta_K}\rangle\otimes|0\cdots0\rangle_K$ is

$$X_{a_1}^{\delta_1}|\overline{\delta_1\cdots\delta_k}\rangle\otimes|0\cdots0\rangle_K = |\overline{\delta_1\cdots\delta_K}\rangle\otimes|\delta_1\cdots0\rangle_K. \tag{8.117}$$

By applying a similar procedure on the remaining ancilla qubits, we obtain

$$\prod_{i=1}^{K} X_{a_i}^{\delta_i}|\overline{\delta_1\cdots\delta_K}\rangle\otimes|0\cdots0\rangle_K = |\overline{\delta_1\cdots\delta_K}\rangle\otimes|\delta_1\cdots\delta_K\rangle. \tag{8.118}$$

In the second stage of decoding, as indicated above, we must perform the mapping $|\overline{\delta_1\cdots\delta_K}\rangle\otimes|\delta_1\cdots\delta_K\rangle\to|\overline{0\cdots0}\rangle\otimes|\delta_1\cdots\delta_K\rangle$ This mapping can be implemented by applying a controlled-X_i operation to the ith encoded qubit, which serves as the target qubit. The ith ancilla qubit serves as the control qubit. This action can be described as follows:

$$(\overline{X}_1)^{\delta_1}|\overline{\delta_1\delta_2\cdots\delta_K}\rangle\otimes|\delta_1\delta_2\cdots\delta_K\rangle = |\overline{(\delta_1\oplus\delta_1)\delta_2\cdots\delta_K}\rangle\otimes|\delta_1\delta_2\cdots\delta_K\rangle = |\overline{0\delta_2\cdots\delta_K}\rangle\otimes|\delta_1\cdots\delta_K\rangle. \tag{8.119}$$

Further, we perform a similar procedure on the remaining encoded qubits:

$$\prod_{i=1}^{K}(\overline{X}_i)^{\delta_i}|\overline{\delta_1\cdots\delta_K}\rangle\otimes|\delta_1\cdots\delta_K\rangle = |\overline{0\cdots0}\rangle\otimes|\delta_1\cdots\delta_K\rangle. \tag{8.120}$$

The overall decoding process, by applying Eqs. (8.118) and (8.120), can be represented by

$$\prod_{i=1}^{K}\left(\overline{X}_i\right)^{\delta_i}\prod_{l=1}^{K}X_{a_l}^{\delta_l}\left|\overline{\delta_1\cdots\delta_K}\right\rangle\otimes|0\cdots0\rangle_K=\left|\overline{0\cdots0}\right\rangle\otimes|\delta_1\cdots\delta_K\rangle. \tag{8.121}$$

The total number of gates for decoding based on Eq. (8.121) can be found as

$$K(r+1)+K(N-K+1)=K(N-K+r+2)\leq(K+r)(N-K) \tag{8.122}$$

because $\overline{Z}_i$ acts on r+1 qubits (see Eq. (8.109)), and there are K such actions based on Eq. (8.121). Controlled-X_i operations require no more than N−K+1 two-qubit gates, and there are K such actions (see Eq. (8.121)). On the other hand, the number of gates required in reverse encoding is $(K+r)(N-K)$— see Eq. (8.102). Therefore, the decoder implemented based on Eq. (8.121) is more efficient than decoding based on efficient reverse encoding.

Example. Decoding circuit for [4,2,2] quantum stabilizer code can be obtained based on Eq. (8.121). The starting points are generators of the code

$$g_1=X_1Z_2Z_3X_4 \text{ and } g_2=Y_1X_2X_3Y_4.$$

The corresponding quantum-check matrix and its standard from (obtained Gaussian elimination) are given by

$$\boldsymbol{A}=\left(\begin{array}{cccc|cccc}1&0&0&1&0&1&1&0\\1&1&1&1&1&0&0&1\end{array}\right)\sim\left(\begin{array}{cccc|cccc}1&0&0&1&0&1&1&0\\0&1&1&0&1&1&1&1\end{array}\right)$$

The encoded Pauli operators in finite geometry form are given by

$$\overline{X}=\left(\mathbf{0}\quad \boldsymbol{E}^T\quad I\,\middle|\,(\boldsymbol{E}^T\boldsymbol{C}_1^T+\boldsymbol{C}_2^T)\quad \mathbf{0}\quad \mathbf{0}\right)=\left(\begin{array}{cccc|cccc}0&0&1&0&1&1&0&0\\0&0&0&1&0&1&0&0\end{array}\right)$$

$$\overline{Z}=\left(\mathbf{0}\quad \mathbf{0}\quad \mathbf{0}\,\middle|\,\boldsymbol{A}_2^T\quad \mathbf{0}\quad I\right)=\left(\begin{array}{cccc|cccc}0&0&0&0&0&1&1&0\\0&0&0&0&1&0&0&1\end{array}\right)$$

The corresponding operator representation of encoded Pauli operators are

$$\overline{X}_1=Z_1Z_2X_3,\overline{X}_2=Z_2X_4.$$
$$\overline{Z}_1=Z_2Z_3,\overline{Z}_2=Z_1Z_4.$$

The decoding circuit obtained using Eq. (8.121) is shown in Fig. 8.4. The decoding circuit has two stages. In the first stage, we apply CNOT gates on ancillae, with the control states being the encoded

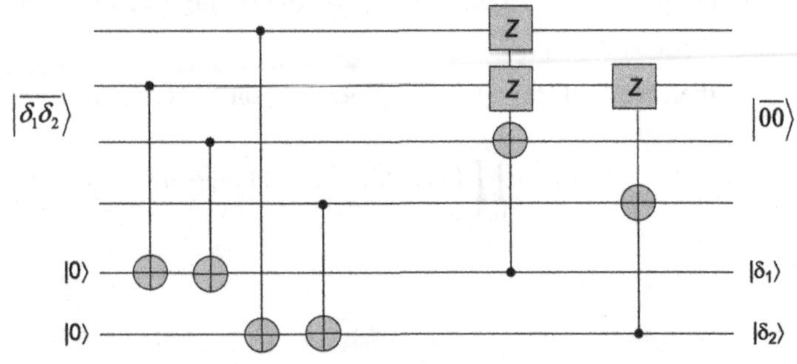

FIGURE 8.4

Efficient implementation of decoding circuit for [4,2,2] quantum stabilizer code.

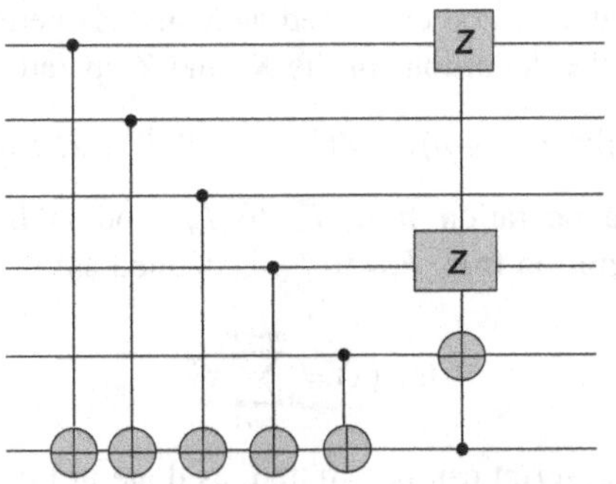

FIGURE 8.5

Efficient implementation of decoding circuit for [5,1,3] quantum stabilizer code.

states. The encoded Pauli state $\overline{Z}_1 = Z_2Z_3$ indicates that the first ancilla qubit is controlled by encoded kets at positions 2 and 3. The encoded Pauli state $\overline{Z}_2 = Z_1Z_4$ indicates that the second ancilla qubit is controlled by encoded kets at positions 1 and 4. The second stage is implemented based on encoded Pauli operators $\overline{X}_i$. The ancilla qubits now serve as control qubits and encoded qubits as target qubits. The encoded Pauli state $\overline{X}_1 = Z_1Z_2X_3$ indicates that the first ancilla qubit controls the encoded kets at positions 1, 2, and 3, and the corresponding controlled gates are Z at positions 1 and 2, and X at position 3. The encoded Pauli state $\overline{X}_2 = Z_2X_4$ indicates that the second ancilla qubit controls the encoded kets at positions 2 and 4, while the corresponding controlled gates are Z at position 2, and X at position 4.

Example. The decoding circuit for [5,1,3] quantum stabilizer code can be obtained similarly to the previous example, and the starting points are generators of the code:

$$g_1 = Y_1Z_2Z_4Y_5, g_2 = X_2Z_3Z_4Z_5, g_3 = Z_1Z_2X_3X_5 \text{ and } g_4 = Z_1Z_3Y_4Y_5.$$

Following a procedure similar to the example above, the standard form encoded Pauli operators are given by

$$\overline{Z} = Z_1Z_2Z_3Z_4Z_5 \text{ and } \overline{X} = Z_1Z_4X_5.$$

The corresponding decoding circuit is shown in Fig. 8.5.

8.6 Nonbinary stabilizer codes

The previous sections considered quantum stabilizer codes represented by a finite geometry over F_2^{2N}, $F_2 = \{0,1\}$. Because these codes are defined over F_2^{2N}, they are called "binary" stabilizer codes. This section concerns stabilizer codes defined over F_q^{2N}, where $q = p^l$ is a prime power (p is a prime, and $l \geq 1$ is an integer) [23–28]. This class of codes can be called the "nonbinary" stabilizer codes. Although many definitions and properties from the previous sections are applicable here, certain modifications are required. First, we operate on q-ary quantum digits; analogous to qubits, they can be called "qudits." Second, we must extend the definitions of quantum gates to qudits. In the previous chapter, we saw that arbitrary qubit error could be represented in terms of Pauli operators $\{I,X,Y,Z\}$. We have also seen that the Y operator can be expressed as X and Z operators. A similar strategy can be applied here. We must extend the definitions of the X- and Z-operators to qudits as follows:

$$X(a)|x\rangle = |x+a\rangle, \quad Z(b) = \omega^{\mathrm{tr}(bx)}|x\rangle; x, a, b \in F_q \tag{8.123}$$

where tr(•) denotes the trace operation from F_q to F_p, and ω is a pth root of unity, namely $\omega = \exp(j2\pi/p)$. The trace operation from F_{q^m} to F_q is defined as

$$\mathrm{tr}_{q^m}(x) = \sum_{i=0}^{m-1} x^{q^i}. \tag{8.124}$$

If F_q is a prime field, the subscript can be omitted, as done in Eq. (8.123).

Before proceeding further with nonbinary quantum stabilizer codes, we believe it is convenient to introduce several additional *nonbinary quantum gates* that are useful in describing arbitrary quantum operations on qudits. The nonbinary quantum gates are shown in Fig. 8.6, which also describes the action of the gates. The F-gate corresponds to the discrete Fourier transform gate. Its action on ket $|0\rangle$ is the superposition of all basis kets having the same probability amplitude:

$$F|0\rangle = \frac{1}{\sqrt{q}} \sum_{u \in F_q} |u\rangle. \tag{8.125}$$

An operator A can be represented by $A = \sum_{i,j} A_{ij}|i\rangle\langle j|$. Based on Fig. 8.6, we determine the action of operator $FX(b)F^{\dagger}$ as follows:

$$FX(a)F^{\dagger} = \frac{1}{\sqrt{q}} \sum_{i,j \in F_q} \omega^{\mathrm{tr}(ij)} |i\rangle\langle j| \sum_{x \in F_q} \omega |x+a\rangle\langle x| \frac{1}{\sqrt{q}} \sum_{l,k \in F_q} \omega^{-\mathrm{tr}(lk)} |l\rangle\langle k| = \frac{1}{q} \sum_{i,j \in F_q} \omega^{\mathrm{tr}(ia)} \underbrace{\sum_{x \in F_q} \omega^{\mathrm{tr}(ix - xl)}}_{1, i=l \quad 0, i \neq l} |i\rangle\langle l|$$

$$= \sum_{i \in F_q} \omega^{\mathrm{tr}(ia)} |i\rangle\langle i| = Z(a) \tag{8.126}$$

Using the basic nonbinary gates shown in Fig. 8.6 and Eq. (8.123), we can perform more complicated operations, as illustrated in Fig. 8.7. These circuits are used later in the section to efficiently implement quantum nonbinary encoders and decoders. Note that the circuit at the bottom of Fig. 8.7 is obtained by concatenating the circuits at the top.

It can be shown that the set of errors $\varepsilon = \{X(a)X(b)|a, b \in F_q\}$ satisfies the following properties: (1) it contains the identity operator, and (2) $\mathrm{tr}\left(E_1^{\dagger}E_2\right) = 0 \quad \forall E_1, E_2 \in \varepsilon$, and (3) $\forall E_1, E_2 \in \varepsilon$: $E_1E_2 = cE_3, E_3 \in \varepsilon, c \in F_q$. This set forms an error basis for the set of $q \times q$ matrices and is sometimes

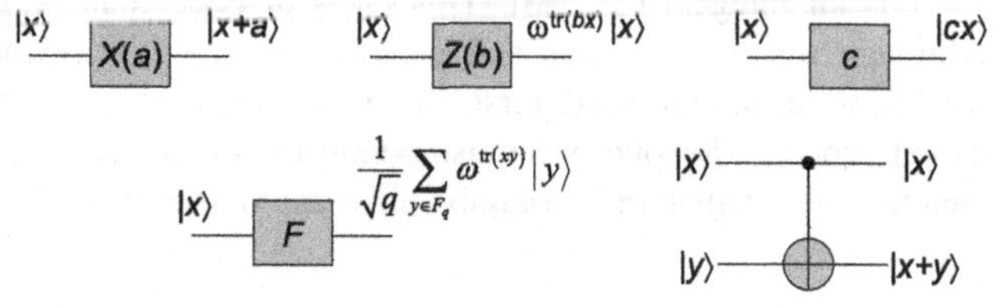

FIGURE 8.6

Basic nonbinary quantum gates.

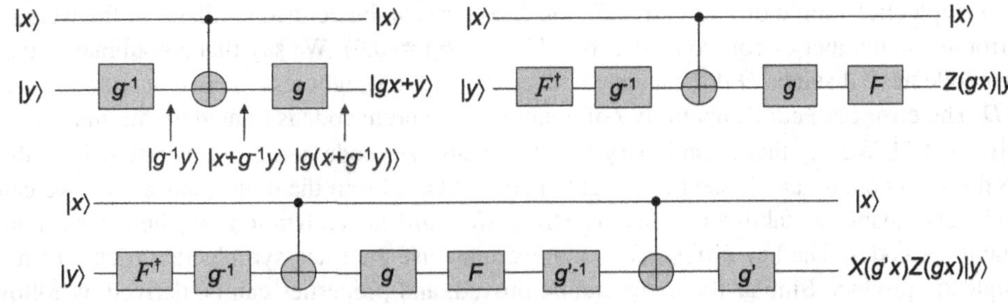

FIGURE 8.7

Nonbinary quantum circuits derived from basic nonbinary gates.

called a "nice error basis." It can also be shown that the following two properties hold: (1) $X(a)Z(b) = \omega^{-tr(ab)}Z(b)X(a)$ and (2) $X(a+a')Z(b+b') = \omega^{-tr(a'b)}X(a)Z(b)X(a')Z(b')$.

Example. The 4-ary nice error basis over $F_4 = \{0, 1, \alpha, \overline{\alpha}\}$ can be obtained as the tensor product of the Pauli basis:

$$X(0) = II, X(1) = IX, X(\alpha) = XI, X(\overline{\alpha}) = XX,$$
$$Z(0) = II, Z(1) = ZI, Z(\alpha) = ZZ, Z(\overline{\alpha}) = IZ.$$

To determine the nice basis error on N-qudits, we introduce the notation $X(\boldsymbol{a}) = X(a_1)\otimes\cdots\otimes X(a_N)$ and $Z(\boldsymbol{b}) = Z(b_1)\otimes\cdots\otimes Z(b_N)$, where $\boldsymbol{a}=(a_1,\cdots,a_N)$, $\boldsymbol{b}=(b_1,\cdots,b_N)$ and $a_i,\ b_i \in F_q$. The set $\varepsilon_N = \left\{X(\boldsymbol{a})X(\boldsymbol{b})\middle|\boldsymbol{a},\boldsymbol{b}\in F_q^N\right\}$ is the nice error basis defined over F_q^{2N}. Similarly to the Pauli multiplicative group, we can define the *error group* by

$$G_N = \left\{\omega^c X(\boldsymbol{a})Z(\boldsymbol{b})\middle|\boldsymbol{a},\boldsymbol{b}\in F_q^N, c\in F_q\right\}. \tag{8.127}$$

Let S be the largest abelian subgroup of G_N that fixes all elements from quantum code C_Q, called the stabilizer group. The $[N,K]$ nonbinary stabilizer code C_Q is defined as the K-dimensional subspace of the N-qudit Hilbert space H_q^N as follows:

$$C_Q = \bigcap_{s\in S}\left\{|c\rangle \in H_q^N\middle| s|c\rangle = |c\rangle\right\}. \tag{8.128}$$

The definition of nonbinary stabilizer codes is a straightforward generalization of the quantum stabilizer code over H_2^N. Therefore, the similar properties, definitions, and theorems introduced in the previous sections apply here. For example, two errors $E_1 = \omega^{c_1}X(\boldsymbol{a}_1)Z(\boldsymbol{b}_1), E_2 = \omega^{c_2}X(\boldsymbol{a}_2)Z(\boldsymbol{b}_2)\in G_N$ commute if and only if their trace symplectic product vanishes; that is,

$$\mathrm{tr}(\boldsymbol{a}_1\boldsymbol{b}_2 - \boldsymbol{a}_2\boldsymbol{b}_1) = 0. \tag{8.129}$$

From error property 2, namely $X(a)Z(b)X(a')Z(b') = \omega^{\mathrm{tr}(a'b)}X(a+a')Z(b+b')$, it is straightforward to verify that $E_1E_2 = \omega^{\mathrm{tr}(\boldsymbol{a}_2\boldsymbol{b}_1)}X(\boldsymbol{a}_1+\boldsymbol{a}_2)Z(\boldsymbol{b}_1+\boldsymbol{b}_2)$ and $E_2E_1 = \omega^{\mathrm{tr}(\boldsymbol{a}_1\boldsymbol{b}_2)}X(\boldsymbol{a}_1+\boldsymbol{a}_2)$ $Z(\boldsymbol{b}_1+\boldsymbol{b}_2)$. Clearly, $E_1E_2 = E_2E_1$ only when $\omega^{\mathrm{tr}(\boldsymbol{a}_2\boldsymbol{b}_1)} = \omega^{\mathrm{tr}(\boldsymbol{a}_1\boldsymbol{b}_2)}$, which is equivalent to Eq. (8.129).

The (symplectic) weight of qudit error $E=\omega^c X(\boldsymbol{a})Z(\boldsymbol{b})$ can be defined similarly to the weight of a qubit error as the number of components i for which $(a_i,b_i)\neq(0,0)$. We say that a nonbinary quantum stabilizer code has a distance D if it can detect all errors with a weight less than D, but none that are of weight D. The error correction capability t of nonbinary quantum code is related to minimum distance by $t=\lfloor(D-1)/2\rfloor$. We say that a nonbinary quantum stabilizer code is nondegenerate if its stabilizer group S does not contain an element of weight smaller than t. From the definitions above, we can see that nonbinary quantum stabilizer codes are straightforward generalizations of their corresponding qubit stabilizer codes. The key difference is that we must use the trace-symplectic product instead of the symplectic product. Similar theorems can be proved, and properties can be derived by following similar procedures considering the differences outlined above. On the other hand, quantum hardware implementation is more challenging. For example, instead of using Pauli X- and Z-gates, we must use the $X(a)$- and $Z(b)$-gates shown in Fig. 8.6. We can also define the standard form, but instead of using F_2 during Gauss elimination, we must perform Gauss elimination in F_q. For example, we can determine the syndrome based on syndrome measurements, as illustrated in the previous sections. Let the generator $\boldsymbol{g}_i$ be given as follows:

$$\boldsymbol{g}_i=[\boldsymbol{a}_i|\boldsymbol{b}_i]=[0 \ \cdots \ 0 \ a_i \ \cdots \ a_N \ | \ 0 \ \cdots \ 0 \ b_i \ \cdots \ b_N]\in F_q^{2N}, a_i, b_i\in F_q. \tag{8.130}$$

The quantum circuit shown in Fig. 8.8 will provide a nonzero measurement if a detectible error does not commute with a multiple of $\boldsymbol{g}_i$. By comparing the corresponding qubit syndrome decoding circuits from the previous chapter, we conclude that the F-gate for nonbinary quantum codes has a role similar to that of the Hadamard gate for codes over F_2. Note that the syndrome circuit is a generalization of the bottom circuit shown in Fig. 8.7. From Fig. 8.8, the input to the stabilizer circuit is clearly $E|\psi\rangle$, where E is the qudit error and $|\psi\rangle$ is the codeword; thus, the stabilizer circuit performs the following mapping:

$$|0\rangle E|\psi\rangle\rightarrow\sum_{u\in F_q}F^\dagger|u\rangle X(u\boldsymbol{a}_i)Z(u\boldsymbol{b}_i)E|\psi\rangle. \tag{8.131}$$

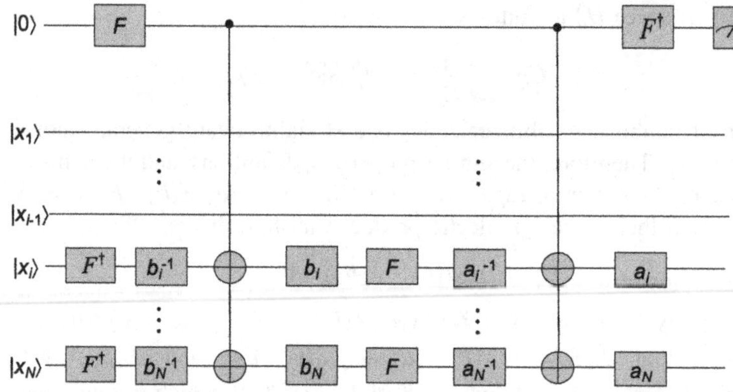

FIGURE 8.8

Nonbinary syndrome decoding circuit corresponding to generator $\boldsymbol{g}_i$.

By ignoring the complex phase constant, the qudit error E can be represented as $E = X(\boldsymbol{a})Z(\boldsymbol{b})$. Based on Eq. (8.129), it is clear that $X(\boldsymbol{a}_i)Z(\boldsymbol{b}_i)E = \omega^{\mathrm{tr}(\boldsymbol{a}\boldsymbol{b}_i - \boldsymbol{a}_i\boldsymbol{b})}EX(\boldsymbol{a}_i)Z(\boldsymbol{b}_i)$. Based on Fig. 8.6, we can show that $X(u\boldsymbol{a}_i)Z(u\boldsymbol{b}_i)E = \omega^{\mathrm{tr}(u(\boldsymbol{a}\boldsymbol{b}_i - \boldsymbol{a}_i\boldsymbol{b}))}EX(\boldsymbol{a}_i)Z(\boldsymbol{b}_i)$, and by substituting in Eq. (8.131), we obtain

$$|0\rangle E|\psi\rangle \to \sum_{u\in F_q} F^\dagger|u\rangle X(u\boldsymbol{a}_i)Z(u\boldsymbol{b}_i)E|\psi\rangle = \sum_{u\in F_q} F^\dagger|u\rangle \omega^{\mathrm{tr}(u(\boldsymbol{a}\boldsymbol{b}_i - \boldsymbol{a}_i\boldsymbol{b}))}EX(\boldsymbol{a}_i)Z(\boldsymbol{b}_i)|\psi\rangle. \tag{8.132}$$

Since $X(\boldsymbol{a}_i)Z(\boldsymbol{b}_i)|\psi\rangle = |\psi\rangle$ and $X(\boldsymbol{a}_i)Z(\boldsymbol{b}_i)\in S$, Eq. (8.132) becomes

$$|0\rangle E|\psi\rangle \to \left[\sum_{u\in F_q} F^\dagger|u\rangle \omega^{\mathrm{tr}(u(\boldsymbol{a}\boldsymbol{b}_i - \boldsymbol{a}_i\boldsymbol{b}))}\right]E|\psi\rangle = \left[\sum_{u\in F_q}\sum_{v\in F_q} \omega^{-\mathrm{tr}(uv)}\omega^{\mathrm{tr}(u(\boldsymbol{a}\boldsymbol{b}_i - \boldsymbol{a}_i\boldsymbol{b}))}|v\rangle\right]E|\psi\rangle$$

$$= \sum_{v\in F_q}|v\rangle \underbrace{\sum_{u\in F_q}\omega^{\mathrm{tr}(u(\boldsymbol{a}\boldsymbol{b}_i - \boldsymbol{a}_i\boldsymbol{b}) - uv)}}_{1,\ \text{for } v=\boldsymbol{a}\boldsymbol{b}_i - \boldsymbol{a}_i\boldsymbol{b}} E|\psi\rangle = |\boldsymbol{a}\boldsymbol{b}_i - \boldsymbol{a}_i\boldsymbol{b}\rangle E|\psi\rangle. \tag{8.133}$$

Eq. (8.133) indicates that the measurement gives the ith syndrome component $\lambda_i = \mathrm{tr}(\boldsymbol{a}\boldsymbol{b}_i - \boldsymbol{a}_i\boldsymbol{b})$. By $N-K$ measurements on corresponding generators $\boldsymbol{g}_i$ ($i = 1,\ldots,N-K$), we obtain the syndrome $S(E)$:

$$S(E) = [\lambda_1\lambda_2\cdots\lambda_{N-K}]^{\mathrm{T}};\ \lambda_i = \mathrm{tr}(\boldsymbol{a}\boldsymbol{b}_i - \boldsymbol{a}_i\boldsymbol{b}), i = 1,\cdots,N-K. \tag{8.134}$$

Syndrome Eq. (8.134) is similar to Eq. (8.47), and similar conclusions can be made. The correctable qudit error maps the code space to the q^K-dimensional subspace of q^N-dimensional Hilbert space. Since there are $N-K$ generators or equivalent syndrome positions, there are q^{N-K} different cosets. All qudit errors belonging to the same coset have the same syndrome. By selecting the most probable qudit error for the coset representative, typically the lowest weight error, we can uniquely identify the qudit error and consequently perform the error correction action.

We can define the projector P onto the code space C_Q as follows:

$$P = \frac{1}{|S|}\sum_{s\in S} s. \tag{8.135}$$

Clearly, $Ps = s$ for every $s \in S$. Further,

$$P^2 = \frac{1}{|S|}\sum_{s\in S} Ps = \frac{1}{|S|}\sum_{s\in S} s = P. \tag{8.136}$$

Since $s^\dagger \in S$, it is also valid that $P^\dagger = P$, indicating that P is an orthogonal projector. The dimensionality of C_Q is $\dim C_Q = q^N/|S|$.

Quantum bounds similar to those described in the previous chapter can be derived for nonbinary stabilizer codes as well. For example, the Hamming bound for the nonbinary quantum stabilizer code $[N,K,D]$ (where K is the number of information qudits, and N is the codeword length) over F_q is given by

$$\sum_{i=0}^{\lfloor (D-1)/22 \rfloor} \binom{N}{i} (q^2-1)^i q^K \le q^N. \tag{8.137}$$

This inequality is obtained from the inequality (7.52) in Chapter 7 by substituting the term 3^i with $(q^2-1)^i$, and it can be proved using a similar methodology. The number of information qudits is now q^K, and the number of possible codewords is q^N.

8.7 Subsystem codes

Subsystem codes [27–32] represent a generalization of the nonbinary stabilizer codes of the previous section. They can also be considered a generalization of decoherence-free subspaces [33,34] and noiseless subsystems [35]. The key idea behind subsystem codes is to decompose the quantum code C_Q as the tensor product of two subsystems A and B as $C_Q = A \otimes B$. The information qudits belong to subsystem A and noninformation qudits, also known as gauge qudits, belong to subsystem B. We are concerned only with the errors introduced in subsystem A.

In the previous chapter, we discussed that both noise $\mathfrak{E}$ and recovery $\mathfrak{R}$ processes could be described as the quantum operations $\mathfrak{E}, \mathfrak{R}$: $L(H) \rightarrow L(H)$, where $L(H)$ is the space of linear operators on Hilbert space H. These mappings can be represented in terms of operator-sum representation, $\mathfrak{E}(\rho) = \sum_i E_i \rho E_i^\dagger, E_i \in L(H), \mathfrak{E} = \{E_i\}$. Given that quantum code C_Q is a subspace of H, we say that the set of errors $\mathfrak{E}$ is correctable if a recovery operation $\mathfrak{R}$ exists such that $\mathfrak{R}\mathfrak{E}(\rho) = \rho$ for any state ρ from $L(C_Q)$. In terms of subsystem codes, we say that a recovery operation exists such that for any ρ^A from $L(A)$ and ρ^B from $L(B)$, it is also valid that $\mathfrak{R}\mathfrak{E}(\rho^A \otimes \rho^B) = \rho^A \otimes \rho'^B$, where the state ρ'^B is not relevant. The necessary and sufficient condition for the set of errors $\mathfrak{E} = \{E_m\}$ to be correctable is that $PE_m^\dagger E_n P = I^A \otimes g_{mn}^B, \forall m, n$, where P is the projector on code subspace. Such a set of errors is called a correctable set of errors. Clearly, the linear combination of errors from $\mathfrak{E}$ is also correctable, so it makes sense to observe the correctable set of errors as linear space with a properly chosen operator basis. If we are concerned with quantum registers/systems composed of N-qubits, the corresponding error operators are Pauli operators. The Pauli group on N-qubits, G_N, was introduced in Section 8.1. The stabilizer formalism can also be used in describing the stabilizer subsystem codes. Stabilizer subsystem codes are determined by a subgroup of G_N that contains the element jI, called the gauge group $\mathcal{G}$, and by a stabilizer group S that is properly chosen such that $S' = j^\lambda S$ is the center of $\mathcal{G}$, denoted $Z(\mathcal{G})$ (that is, the set of elements from $\mathcal{G}$ that commute with all elements from $\mathcal{G}$). We are concerned with the decomposition of H as $H = C \oplus C^\perp = (H_A \otimes H_B) \oplus C^\perp$, where $C^\perp$ is the dual of $C = H_A \otimes H_B$. The gauge operators are chosen so that they act trivially on subsystem A but generate the full algebra of subsystem B. The information is encoded on subsystem A, while subsystem B is used to absorb the effects of gauge operations. The Pauli operators for K logical qubits are obtained from the isomorphism $N(\mathcal{G})/S \simeq G_K$, where $N(\mathcal{G})$ is the normalizer of $\mathcal{G}$. On the other hand, subsystem B consists of R gauge qubits recovered from isomorphism $\mathcal{G}/S \simeq G_N$, where $N = K + R + s$, $s \geq 0$. If $\widetilde{X}_1$ and $\widetilde{Z}_1$ represent the images of X_i and Z_i under automorphism U of G_N, the stabilizer can be described by $S = \langle \widetilde{Z}_1, \cdots, \widetilde{Z}_s \rangle; R + s \leq N; R, s \geq 0$. The gauge group can be specified by $\mathcal{G} = \langle jI, \widetilde{Z}_1, \cdots, \widetilde{Z}_{s+R}, \widetilde{X}_{s+1}, \cdots, \widetilde{X}_R \rangle$. The images must satisfy the commutative relations Eq. (8.28). The logical (encoded) Pauli operators are then $\overline{X}_1 = \widetilde{X}_{s+R+1}, \overline{Z}_1 = \widetilde{Z}_{s+R+1}, \cdots, \overline{X}_K = \widetilde{X}_N, \overline{Z}_K = \widetilde{Z}_N$. The detectable errors are elements of $G_N - N(S)$ and the undetectable errors in $N(S) - \mathcal{G}$. Undetectable errors are related to the logical Pauli operators because $N(S)/\mathcal{G} \simeq N(\mathcal{G})/S'$. Namely, if $n \in N(S)$, $g \in \mathcal{G}$ exists such that $ng \in N(\mathcal{G})$, and if $g' \in \mathcal{G}$ such that $ng' \in N(\mathcal{G})$, $gg' \in \mathcal{G} \cap N(\mathcal{G}) = S'$. The distance D of this code is defined as the minimum weight among undetectable errors. The subsystem code encodes K qubits into N-qubit codewords and has R gauge qubits and therefore can be denoted the $[N,K,R,D]$ code.

We turn our attention now to subsystem codes defined as a subspace of q^N-dimensional Hilbert space H_q^N. For a and b from F_q, we define the unitary operators (qudit errors) $X(a)$ and $Z(b)$ as follows:

$$X(a)|x\rangle = |x+a\rangle, \quad Z(b) = \omega^{\mathrm{tr}(bx)}|x\rangle; \quad x, a, b \in F_q; \omega = e^{j2\pi/p} \tag{8.138}$$

As expected, since nonbinary stabilizer subsystems codes are generalizations of nonbinary stabilizer codes, similar qudit errors occur. The trace operation is defined by Eq. (8.124). The qudit *error group* is defined by

$$G_N = \left\{\omega^c X(\boldsymbol{a})Z(\boldsymbol{b})\middle|\boldsymbol{a},\boldsymbol{b}\in F_q^N, c\in F_p\right\}; \quad \boldsymbol{a} = (a_1\cdots a_N), \boldsymbol{b} = (b_1\cdots b_N); a_i, b_i \in F_q. \tag{8.139}$$

Let C_Q be a quantum code such that $H = C_Q \oplus C_Q^{\perp}$. The $[N,K,R,D]$ subsystem code over F_q is defined as the decomposition of code-space C_Q into a tensor product of two subsystems A and B such that $C_Q = A \otimes B$, where the dimensionality of A equals $\dim A = q^K$, $\dim B = q^R$, and all errors of weight less than $D_{\min}$ on subsystem A can be detected. What is interesting about this class of codes is that when constructed from classical codes, the corresponding classical codes do not need to be dual-containing (self-orthogonal). Note that the subsystem codes can also be defined over F_{q^2}.

For $[N,K,R,D_{\min}]$ stabilizer subsystem codes over F_2, it can be shown that the centralizer of S is given by

$$C_{G_N}(S) = \langle \mathcal{G}, \overline{X}_1, \overline{Z}_1, \cdots, \overline{X}_K, \overline{Z}_K \rangle. \tag{8.140}$$

Since $C_Q = A \otimes B$, $\dim A = 2^K$ and $\dim B = 2^R$. From Eq. (8.135), we know that stabilizer S can be used as a projector to C_Q. The dimensionality of quantum code defined by stabilizer S is 2^{K+R}. The stabilizer S, therefore, defines an $[N, K+R, D]$ stabilizer code. Based on Eq. (8.140), we conclude that the image operators $\widetilde{Z}_i, \widetilde{X}_i; i = s+1, \cdots, R$ behave as encoded operators on gauge qubits, while $\overline{Z}_i, \overline{X}_i$ act on information qubits. In total, we have a set of $2(K+R)$ encoded operators of $[N, K+R, D]$ stabilizer codes given by

$$\{\overline{X}_1, \overline{Z}_1, \cdots, \overline{X}_K, \overline{Z}_K, \widetilde{X}_{s+1}, \widetilde{Z}_{s+1}, \cdots, \widetilde{X}_{s+R}, \widetilde{Z}_{s+R}\}. \tag{8.141}$$

Therefore, with Eq. (8.141), we have just established the connection between stabilizer and subsystem codes. Since subsystem codes are more flexible for design, they can be used to design new stabilizer code classes. More importantly, Eq. (8.141) provides an opportunity to implement encoders and decoders for subsystem codes using approaches similar to those developed for stabilizer codes, such as standard form formalism. For example, subsystem code [4,1,1,2] is described by stabilizer group $S = \langle \widetilde{Z}_1, \widetilde{Z}_2 \rangle$ and gauge group $\mathcal{G} = \langle S, \widetilde{X}_3, \widetilde{Z}_3, jI \rangle$, where $\widetilde{Z}_1 = X_1X_2X_3X_4, \widetilde{Z}_2 = Z_1Z_2Z_3Z_4$, $\widetilde{Z}_3 = Z_1Z_2$, and $\widetilde{X}_3 = X_2X_4$. The encoded operators are given by $\overline{X}_1 = X_2X_3$ and $\overline{Z}_1 = Z_2Z_4$. The corresponding [4,2] stabilizer code based on Eq. (8.141) is described by generators $g_1 = \widetilde{X}_3, g_2 = \widetilde{Z}_3, g_3 = \overline{X}_1$, and $g_4 = \overline{Z}_1$. In Fig. 8.9, we show two possible encoder versions of the same [4,1,1,2] subsystem code. Note that the state of the gauge qubits is irrelevant as far as information qubits are concerned. They can be chosen randomly. In Fig. 8.9B, we show that by properly setting the gauge qubits, we can simplify encoder implementation.

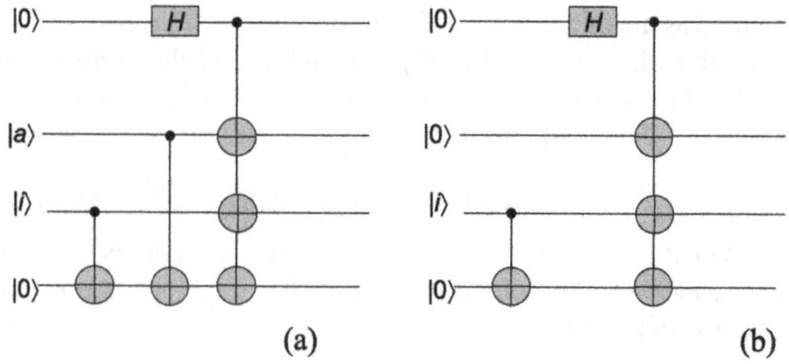

FIGURE 8.9

Encoder implementation of [4,1,1,2] subsystem code: (A) gauge qubit $|a\rangle$ has arbitrary value, and (B) gauge qubit is set to $|0\rangle$.

Based on Eq. (8.141) and its surrounding text description, we conclude that with the help of stabilizer S and gauge G groups of subsystem code, we can create a stabilizer S_A, whose encoder can encode for subsystem code provided that gauge qubits are initialized to $|0\rangle$-qubit. Clearly, $S_A \supseteq S$, indicating that the space stabilized by S_A is the subspace of $C_Q = A \otimes B$, which is stabilized by S. Since $|S_A|/|S| = 2^R$, the dimension of the subspace stabilized by S_A is $2^{K+R-R} = 2^K$. If the subsystem stabilizer and gauge groups are given by $S = \langle \widetilde{Z}_1, \cdots, \widetilde{Z}_s \rangle; s = N - K - R$, and $\mathcal{G} = \langle jI, S, \widetilde{Z}_{s+1}, \cdots, \widetilde{Z}_{s+R}, \widetilde{X}_{s+1}, \cdots, \widetilde{X}_{s+R} \rangle$, respectively, then the stabilizer group of S_A is given by $S_A = \langle S, \widetilde{Z}_{s+1}, \cdots, \widetilde{Z}_{s+R} \rangle$. The quantum-check matrix for S_A can be put in the standard form of Eq. (8.104) as follows:

$$A(S_A) = \left(\begin{array}{ccc|ccc} \overbrace{I_r}^{r} & \overbrace{A_1}^{N-K-r} & \overbrace{A_2}^{K} & \overbrace{B}^{r} & \overbrace{0}^{N-K-r} & \overbrace{C}^{K} \\ 0 & 0 & 0 & D & I_{N-K-r} & E \end{array} \right) \begin{array}{l} \}r \\ \}N-K-r \end{array}, r = \text{rank}(A). \quad (8.142)$$

The encoded Pauli operators of S_A are given by

$$\overline{X} = (0 \;\; E^T \;\; I_K | C^T \;\; 0 \;\; 0), \; \overline{Z} = (0 \;\; 0 \;\; 0 | A_2^T \;\; 0 \;\; I_K). \quad (8.143)$$

The encoder circuit for the [*N*,*K*,*R*,*D*] subsystem code can be implemented as shown in Fig. 8.3, wherein all ancilla qubits are initialized to the $|0\rangle$-qubit.

To illustrate, let us study the encoder implementation of [9,1,4,3] Bacon−Shor code [36,37] (see also [27]). The stabilizer S and gauge G groups of this code are given by

$$S = \langle \widetilde{Z}_1 = X_1X_2X_3X_7X_8X_9, \widetilde{Z}_2 = X_4X_5X_6X_7X_8X_9, \widetilde{Z}_3 = Z_1Z_3Z_4Z_6Z_7Z_9, \widetilde{Z}_4 = Z_2Z_3Z_5Z_6Z_8Z_9 \rangle$$
$$\mathcal{G} = \langle S, \mathcal{G}_X, \mathcal{G}_Z \rangle, \mathcal{G}_X = \langle \widetilde{X}_5 = X_2X_5, \widetilde{X}_6 = X_3X_6, \widetilde{X}_7 = X_5X_8, \widetilde{X}_8 = X_6X_9 \rangle,$$
$$\mathcal{G}_Z = \langle \widetilde{Z}_5 = Z_1Z_3, \widetilde{Z}_6 = Z_4Z_6, \widetilde{Z}_7 = Z_2Z_3, \widetilde{Z}_8 = Z_5Z_6 \rangle.$$

The stabilizer S_A can be formed by adding the images from $\mathcal{G}_Z$ to S to get $S_A = \langle S, \mathcal{G}_Z \rangle$, and the corresponding encoder is shown in Fig. 8.10A. The stabilizer S_A can also be obtained by $\mathcal{G}_X$ to S to get $S_B = \langle S, \mathcal{G}_X \rangle$, with the corresponding encoder shown in Fig. 8.10B. Clearly, the encoder from Fig. 8.10B has a larger number of Hadamard gates but a smaller number of CNOT-gates, which are much more difficult for implementation. Therefore, the gauge qubits provide flexibility in encoder implementation.

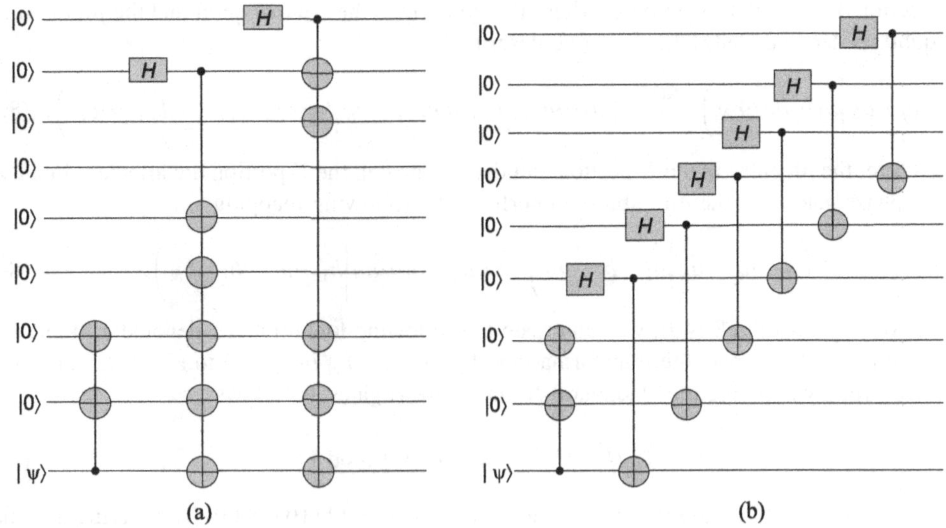

FIGURE 8.10

Encoder for [9,1,4,3] subsystem code (or equivalently [9,1,3] stabilizer code) for S_A given by (A) $S_A = \langle S, \mathcal{G}_Z \rangle$ and (B) $S_B = \langle S, \mathcal{G}_X \rangle$.

We further describe another encoder implementation method due to Grassl et al. [38] (see also [27]), also known as the *conjugation method*. The key idea of this method is to start with quantum-check matrix $\boldsymbol{A}$, given by Eq. (8.142), and transform it into the following form:

$$\boldsymbol{A}' = [\boldsymbol{0} \quad \boldsymbol{0} \mid \boldsymbol{I}_{N-K} \quad \boldsymbol{0}]. \tag{8.144}$$

During this transformation, we must memorize the actions we performed, which are equivalent to the gates being employed on information and ancilla qubits to perform this transformation. Note that the quantum-check matrix of Eq. (8.144) corresponds to the canonical quantum codes described in the previous chapter, which perform the following simple mapping:

$$|\psi\rangle \rightarrow \underbrace{|0\rangle \otimes \cdots \otimes |0\rangle}_{N-K \text{ times}} |\psi\rangle = |\boldsymbol{0}\rangle_{N-K} |\psi\rangle. \tag{8.145}$$

In this encoding mapping, $N-K$ ancilla qubits are appended to K information qubits. The corresponding encoded Pauli operators are given by

$$\overline{\boldsymbol{X}} = (\boldsymbol{0} \quad \boldsymbol{I}_K \mid \boldsymbol{0} \quad \boldsymbol{0}), \ \overline{\boldsymbol{Z}} = (\boldsymbol{0} \quad \boldsymbol{0} \mid \boldsymbol{0} \quad \boldsymbol{I}_K). \tag{8.146}$$

Let us observe one particular generator of the quantum-check matrix, say $\boldsymbol{g}=(a_1,\ldots,a_N| \, b_1,\ldots,b_N)$. The action of the Hadamard gate on the ith qubit is to swap the positions of a_i and b_i as follows, where we have underlined the affected qubits for convenience:

$$\left(a_1 \cdots \underline{a_i} \cdots a_N \middle| b_1 \cdots \underline{b_i} \cdots b_N \right) \xrightarrow{H} \left(a_1 \cdots \underline{b_i} \cdots a_N \middle| b_1 \cdots \underline{a_i} \cdots b_N \right). \tag{8.147}$$

The action of the CNOT gate on **g**, where the ith qubit is the control qubit and the jth qubit is the target qubit, is denoted $\text{CNOT}^{i,j}$ and can be described as

$$\left(a_1 \cdots \underline{a_j} \cdots a_N \middle| b_1 \cdots \underline{b_i} \cdots b_N\right) \xrightarrow{\text{CNOT}^{i,j}} \left(a_1 \cdots a_{j-1} \underline{a_j + a_i} a_{j+1} \cdots a_N \middle| b_1 \cdots b_{i-1} \underline{b_i + b_j} b_{i+1} \cdots b_N\right). \quad (8.148)$$

Therefore, the jth entry in the X-portion and the ith entry in the Z-portion are affected. Finally, the action of phase gate P on the ith qubit is to perform the following mapping:

$$\left(a_1 \cdots \underline{a_i} \cdots a_N \middle| b_1 \cdots \underline{b_i} \cdots b_N\right) \xrightarrow{P} \left(a_1 \cdots \underline{a_i} \cdots a_N \middle| b_1 \cdots \underline{a_i + b_i} \cdots b_N\right). \quad (8.149)$$

Based on Eqs. (8.147–8.149), we can create the following lookup table, denoted as Table 8.1.

Based on the table above, the transformation of **g**=$(a_1,\ldots,a_N| b_1,\ldots,b_N)$ to **g'**=(a'$_1$,…,a'$_N$|0,…,0) can be achieved by applying the Hadamard H and phase P gates as follows:

$$\bigotimes_{i=1}^{N} H^{\bar{a}_i b_i} P^{a_i b_i}, \quad \bar{a}_i = a_i + 1 \bmod 2. \quad (8.150)$$

For example, **g**=(10010|01110) can be transformed to **g'**=(11110|00000) by applying a sequence of gates $H_2H_3P_4$ based on Table 8.1 or Eq. (8.150).

Based on Eq. (8.148), the error operator $\boldsymbol{e}=(a_1,\ldots,a_i = 1,\ldots, a_N|\ 0,\ldots,0)$ can be converted to $\boldsymbol{e}'=(0,\ldots,a'_i = 1,0 \ldots 0|0 \ldots 0)$ by applying the following sequence of gates:

$$\prod_{m=1, m\neq n}^{N} (CNOT^{m,n})^{a_n}. \quad (8.151)$$

For example, the error operator $\boldsymbol{e}$=(11110|00000) can be transformed to e'=(0100|0000) by the application sequence of the CNOT gates: $\text{CNOT}^{2,1}\text{CNOT}^{2,3}\text{CNOT}^{2,4}$.

Therefore, the first stage is to convert the Z-portion of the stabilizer matrix to an all-zero submatrix by applying the H- and P-gates using Eq. (8.150). In the second stage, we convert every row of the stabilizer matrix of the first stage (**a'**|**0**) to $(0 \ldots a'_i = 10 \ldots 0|0 \ldots 0)$ using Eq. (8.151). After the second stage, the stabilizer matrix has the form $(\boldsymbol{I}_{N-K}0|\boldsymbol{00})$. In the third stage, we transform the obtained stabilizer matrix to $(\boldsymbol{00}|\boldsymbol{I}_{N-K}\boldsymbol{0})$ by applying the H-gates first to the $N-K$ qubits. The final form of

Table 8.1 Lookup table of actions of gates *I*, *H*, and *P* on the *i*th qubit. By the action of its corresponding gate, the *i*th qubit (a_i,b_i) is transformed into (c,d).

(a_i,b_i)	Gate	(c,d)
(0,0)	I	(0,0)
(1,0)	I	(1,0)
(0,1)	H	(1,0)
(1,1)	P	(1,0)

the stabilizer matrix is that of canonical code, which appends $N-K$ ancilla qubits in the $|0\rangle$state to K information qubits; that is, it performs the mapping of Eq. (8.145). During this three-stage transformation, encoded Pauli operators are transformed into the form given by Eq. (8.146). Let us denote the gate sequence applied during three-stage transformation as $\{U\}$. By applying this gate sequence in reverse order, we obtain the encoding circuit of the corresponding stabilizer code.

Let us now apply the conjugation method to subsystem codes. The key difference for stabilizer codes is that we must observe the whole gauge group instead. One option is to create the stabilizer group S_A based on gauge group $\mathcal{G} = <S,\mathcal{G}_Z,\mathcal{G}_X>$ by either $S_A = <S,\mathcal{G}_Z>$ or $S_B = <S,\mathcal{G}_X>$ and then apply the conjugation method. The second option is to transform the gauge group, represented using finite geometry formalism, as follows:

$$\mathcal{G} = \begin{bmatrix} \underline{S} \\ \underline{\mathcal{G}_Z} \\ \mathcal{G}_X \end{bmatrix} \xrightarrow{\{U\}} \mathcal{G}\prime = \begin{bmatrix} \mathbf{0} & \mathbf{0} & \mathbf{0} & I_s & \mathbf{0} & \mathbf{0} \\ \mathbf{0} & \mathbf{0} & \mathbf{0} & \mathbf{0} & I_R & \mathbf{0} \\ \mathbf{0} & I_R & \mathbf{0} & \mathbf{0} & \mathbf{0} & \mathbf{0} \end{bmatrix}. \tag{8.152}$$

With transformation by Eq. (8.152), we performed the following mapping:

$$|\phi\rangle|\psi\rangle \rightarrow \underbrace{|0\rangle \otimes \cdots \otimes |0\rangle}_{s=N-K-R \text{ times}} |\phi\rangle|\psi\rangle = |\mathbf{0}\rangle_s|\phi\rangle|\psi\rangle, \tag{8.153}$$

where the quantum state $|\psi\rangle$ corresponds to information qubits, while the quantum state $|\phi\rangle$ represents the gauge qubits. Therefore, with transformation by Eq. (8.152), we obtain the *canonical subsystem code*—that is, the subsystem code in which $s = N-K-R$ ancilla qubits in the $|0\rangle$state are appended to information and gauge qubits. By applying the sequence of gates $\{U\}$ in Eq. (8.152) in the opposite order, we obtain the encoder of subsystem code.

As an illustrative example, let us again observe the subsystem [4,1,1,2] code, whose gauge group can be represented using the finite geometry representation as follows:

$$\mathcal{G} = \left[\begin{array}{cccc|cccc} 1 & 1 & 1 & 1 & 0 & 0 & 0 & 0 \\ 0 & 0 & 0 & 0 & 1 & 1 & 1 & 1 \\ \hline 0 & 0 & 0 & 0 & 0 & 0 & 1 & 1 \\ 0 & 1 & 0 & 1 & 0 & 0 & 0 & 0 \end{array}\right].$$

By applying a sequence of transformations, we can represent the [4,1,1,2] subsystem code in canonical form:

$$\xrightarrow{U_1 = CNOT^{1,2}CNOT^{1,3}CNOT^{1,4}} \left[\left[\begin{array}{cccc|cccc} 1 & 0 & 0 & 0 & 0 & 0 & 0 & 0 \\ 0 & 0 & 0 & 0 & 0 & 1 & 1 & 1 \\ \hline 0 & 0 & 0 & 0 & 0 & 0 & 1 & 1 \\ 0 & 1 & 0 & 1 & 0 & 0 & 0 & 0 \end{array}\right]\right] \xrightarrow{U_2 = H_2H_3H_4} \left[\left[\begin{array}{cccc|cccc} 1 & 0 & 0 & 0 & 0 & 0 & 0 & 0 \\ 0 & 1 & 1 & 1 & 0 & 0 & 0 & 0 \\ \hline 0 & 0 & 1 & 1 & 0 & 0 & 0 & 0 \\ 0 & 0 & 0 & 0 & 0 & 1 & 0 & 1 \end{array}\right]\right]$$

$$\xrightarrow{U_3 = CNOT^{2,3}CNOT^{2,4}} \left[\left[\begin{array}{cccc|cccc} 1 & 0 & 0 & 0 & 0 & 0 & 0 & 0 \\ 0 & 1 & 0 & 0 & 0 & 0 & 0 & 0 \\ \hline 0 & 0 & 1 & 1 & 0 & 0 & 0 & 0 \\ 0 & 0 & 0 & 0 & 0 & 0 & 0 & 1 \end{array}\right]\right] \xrightarrow{U_4 = CNOT^{4,3}} \left[\left[\begin{array}{cccc|cccc} 1 & 0 & 0 & 0 & 0 & 0 & 0 & 0 \\ 0 & 1 & 0 & 0 & 0 & 0 & 0 & 0 \\ \hline 0 & 0 & 0 & 1 & 0 & 0 & 0 & 0 \\ 0 & 0 & 0 & 0 & 0 & 0 & 0 & 1 \end{array}\right]\right]$$

$$\xrightarrow{U_5 = H_1H_2} \left[\left[\begin{array}{cccc|cccc} 0 & 0 & 0 & 0 & 1 & 0 & 0 & 0 \\ 0 & 0 & 0 & 0 & 0 & 1 & 0 & 0 \\ \hline 0 & 0 & 0 & 1 & 0 & 0 & 0 & 0 \\ 0 & 0 & 0 & 0 & 0 & 0 & 0 & 1 \end{array}\right]\right].$$

Now starting from canonical codeword $|0\rangle|0\rangle|\psi\rangle|\phi\rangle$ and applying the sequence of transformations $\{U_i\}$ in the opposite order, we obtain the encoder shown in Fig. 8.11A. An alternative encoder can be obtained by swapping control and target qubits of U_3 and U_4 to obtain the encoder shown in Fig. 8.11B.

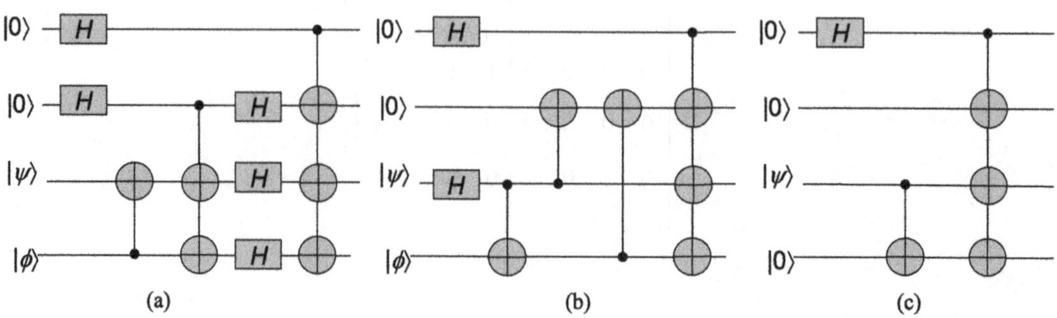

FIGURE 8.11

Encoder for [4,1,1,2] subsystem code by conjugation method: (A) first version, (B) second version, and (C) optimized version.

By setting the gauge qubit to $|0\rangle$ and performing several simple modifications, we obtain the optimized encoder circuit shown in Fig. 8.11C.

Before concluding this section, we provide the following theorem, which allows us to establish the connection between classical codes and quantum subsystem codes. The proof of the theorem is left as a homework problem,

Theorem 2. Let C be a classical linear subcode of F_2^{2N} and D denote its subcode $D=C\cap C^{\perp}$. If $x=|C|$ and $y=|D|$, a subsystem code $C_Q=A\otimes B$ exists in which $\dim A=q^N/(xy)^{1/2}$ and $\dim B=(x/y)^{1/2}$. The minimum distance of subsystem A is given by

$$D_{\min}=\begin{cases}\mathrm{wt}((C+C^{\perp})-C)=\mathrm{wt}(D^{\perp}-C), D^{\perp}\neq C\\ \mathrm{wt}(D^{\perp}), D^{\perp}=C\end{cases},\quad C^{\perp}=\left\{u\in F_q^{2N}\middle| u\odot v=0\ \forall v\in C\right\} \tag{8.154}$$

The minimum distance of $[N,K,R]$ subsystem code $D_{\min}$ given by Eq. (8.154) is more challenging to determine than that of the stabilizer code because of the set difference. The definition can be relaxed, and we can define the minimum distance as $D_m=\mathrm{wt}(C)$. Further, we say that the subsystem code is nondegenerate (pure) if $D_m\geq D_{\min}$. The Hamming and Singleton bounds can now be determined by simply observing the corresponding bounds of the nonbinary stabilizer codes. Since $\dim A=q^K$, $\dim B=q^R$, and the Singleton bound of $[N,K,R,D_{\min}]$ subsystem code is given by

$$K+R\leq N-2D_{\min}+2. \tag{8.155}$$

The corresponding Hamming quantum bound of subsystem code is given based on Eq. (8.137) by

$$\sum_{i=0}^{\lfloor (D_{\min}-1)/2\rfloor}\binom{N}{i}(q^2-1)^i q^{K+R}\leq q^N. \tag{8.156}$$

8.8 Topological codes

Topological quantum error-correction codes [41–51] are typically defined on a two-dimensional lattice, with the quantum parity-check being geometrically local. The locality of the parity-check is of crucial importance since the syndrome measurements are easy to implement when the qubits involved in syndrome verification are proximate to each other. In addition, the possibility to implement the universal quantum gates topologically yields to the increased interest in topological quantum codes. On the other hand, someone may argue that quantum low-density parity-check (LDPC) codes can be designed so that quantum parity checks are only local. In addition, through the concept of subsystem codes, a portion of qubits can be converted to qubits not carrying any encoded information but instead serving as gauge qubits. The gauge qubits can be used to "absorb" the effect of errors. Moreover, the subsystem codes allow syndrome measurements with smaller number qubit interaction. The combination of the locality of topological codes and a small number of interactions leads us to another generation of QECCs, known as the topological subsystem codes, revealed by Bombin [44].

The topological codes on square lattice such as Kitaev's toric code [41,42], quantum lattice code with a boundary [43], surface codes, and planar codes [44] are easier to design and implement

[45–47]. These basic codes on qubits can be generalized to higher alphabets [48,49], resulting in quantum topological codes on qudits. In this section, we are concerned only with quantum topological codes on qubits; the corresponding topological codes on qudits can be obtained in a manner similar to that provided in Section 8.7 (see also [48,49]).

Kitaev's toric code [41,42] is defined on a square lattice with periodic boundary conditions, meaning that the lattice has a torus topology as shown in Fig. 8.12. The qubits are associated with the edges of the lattice. For an $m \times m$ square lattice on the torus, there are $N = 2m^2$ qubits. For each vertex v and plaquette (or face) p—see Fig. 8.13 (right)—we associate the stabilizer operators as follows:

$$A_v = \bigotimes_{i \in n(v)} X_i \quad \text{and} \quad B_p = \bigotimes_{i \in n(p)} Z_i, \tag{8.157}$$

where X_i and Z_i denote the corresponding Pauli X- and Z-operators on position i. With $n(v)$, we denote the vertex v neighborhood, that is, the set of edges incident to vertex v; while with $n(p)$, we denote the plaquette p neighborhood, that is, the set of edges encircling the face p. (We use the symbol $\otimes$ to denote the tensor product, as we did earlier.) Based on Pauli operator properties, it is clear that operators A_v and B_p mutually commute. Namely, the commutation of operators A_v among themselves is trivial to prove. Since the A_v and B_p operators have either zero or two common edges—see Fig. 8.12 (right)—they commute (an even number of anticommuting Pauli operators results in commuting stabilizer operators). The operators A_v and B_p are Hermitian and have the eigenvalues 1 and -1. Let H_2^N denote the Hilbert space, where $N = 2m^2$. The toric code space C_Q can be defined as follows:

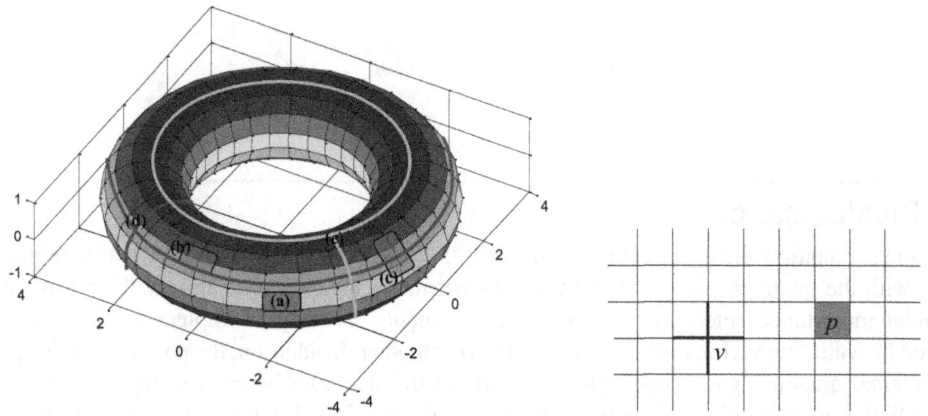

FIGURE 8.12

The square lattice on the torus. The qubits are associated with the edges. The figure on the right shows an enlarged portion of the square lattice on the torus. The Z-containing operators represent strings on the lattice, while the X-containing operators represent strings on the dual lattice. The trivial elementary loop (cycle) denoted (a) corresponds to a certain B_p operator, while the trivial elementary cycle denoted (b) corresponds to a certain A_v operator on the lattice dual. The trivial loops can be obtained as products of individual elementary loops—see, for example, the cycle denoted (c). The nontrivial loops on the dual of lattice, such as (d), or on the lattice, such as (e), correspond to the generators from the set of encoded (logical) Pauli operators.

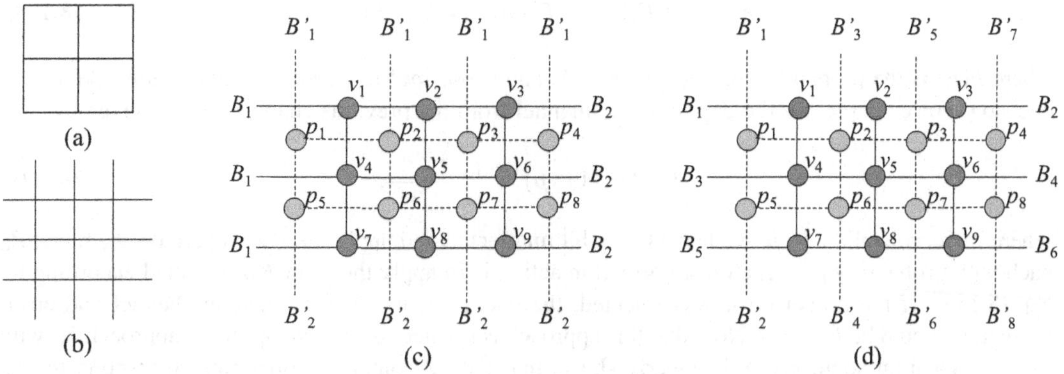

FIGURE 8.13

Quantum codes derived from square lattice on the plane: (A) *z*-type boundary, (B) *x*-type boundary, and (C) 2 × 3 lattice-based code and (D) a generalization of topological code (C). B_1 and B_2 denote the original lattice boundaries, while B'_1 and B'_2 denote the dual lattice boundaries.

Modified from Ref. A.Y. Kitaev, Fault-tolerant quantum computation by anyons, Ann. Phys. 303 (1) (2003) 2–30

$$C_Q = \left\{|c\rangle \in H_2^N \middle| A_v|c\rangle = |c\rangle, B_p|c\rangle = |c\rangle;\ \ \forall v,p\right\}. \tag{8.158}$$

Eq. (8.158) (see the content inside the brackets) represents the eigenvalue equation, and since A_v and B_p mutually commute, they have common eigenkets. The stabilizer group is defined by $S = \langle A_v, B_p \rangle$.

Since $\prod_v A_v = I, \prod_p B_p = I$ there are $M = 2m^2 - 2$ independent stabilizer operators. The number of information qubits can be determined by $K = N - M = 2$, and the code space dimensionality is $\dim C_Q = 2^{N-K} = 4$. This toric code clearly has a low quantum code rate. Another interesting property to note is that the stabilizers A_v contain only *X*-operators, while the stabilizers B_p contain only *Z*-operators. This toric code represents a particular instance of the Calderbank–Shor–Steane codes described in Chapter 7. The encoder and decoder can, therefore, be equivalently implemented as described in Section 8.5. We saw above that commutation of A_v and B_p arises from the fact that these two operators have either zero or two common edges, which is equivalent to the classical code with the property that any two rows overlap in an even number of positions, representing a dual-containing code.

Let us now observe the error E acting on codeword $|c\rangle$, resulting in state $E|c\rangle$, which is not necessarily the eigenket with eigenvalue $+1$ of vertex and plaquette operators. If E and A_p commute, ket $E|c\rangle$ will be the $+1$ eigenket; otherwise, it will be the -1 eigenket. A similar conclusion applies to B_p. By performing the measurements on each of the vertex A_v and plaquette B_p operators, we obtain the syndrome pair $s = (s_v, s_p)$, where $s_i = \pm 1$, $i \in \{v, p\}$. The intersection of syndrome pairs gives us the set of errors having the same syndrome, namely the coset (the set of errors that differ in an element from stabilizer S). Of all possible coset errors, we choose the most probable, which is typically the one having the lowest weight. We can also consider stabilizer elements individually instead of in pairs, and the corresponding error syndrome can be obtained as explained in the previous chapter or Sections 8.1 and 8.3. Another approach is to select the most likely error compatible with the syndrome, namely

$$E_{\mathrm{ML}} = \underset{E \in \mathfrak{S}}{\operatorname{argmax}} P(E), \mathfrak{S} = \{E | EA_v = s_v A_v E, EB_p = s_p B_p E\}, \tag{8.159}$$

where *P*(*E*) is the probability of a given error *E*, and subscript ML denotes the maximum-likelihood decision. For example, for the depolarizing channel from the previous chapter, *P*(*E*) is given by

$$P(E) = (1-p)^{N-\mathrm{wt}(E)} \frac{p^{\mathrm{wt}(E)}}{3}, \tag{8.160}$$

where with probability $1-p$, we leave the qubit unaffected and apply the Pauli operators *X*, *Y*, and *Z*, each with probability *p*/3. The error correction action is to apply the error E_{ML} selected according to Eq. (8.159). If the proper error was selected, the overall action of the quantum channel and error correction would be $E^2 = \pm I$. Note that this approach is not necessarily the optimum approach, as will become apparent soon. Fig. 8.12 clearly shows that only *Z*-containing operators correspond to the strings on the lattice, while only *X*-containing operators correspond to the strings on the dual lattice. The plaquette operators are represented as elementary cycles (loops) on the lattice, while the vertex operators are represented by elementary loops on the dual lattice. Trivial loops can be obtained as the products of elementary loops. Therefore, the plaquette operators generate the group of homologically trivial loops on the torus, while the vertex operators generate homologically trivial loops on the dual lattice. Two error operators E_1 and E_2 have the same effect on a codeword $|c\rangle$ if E_1E_2 contains only homologically trivial loops, and thus, we say these two errors are homologically equivalent. This concept is very similar to the error coset concept introduced in the previous chapter. Four independent operators exist that perform the mapping of stabilizer *S* to itself. These operators commute with all operators from *S* (both A_v's and B_p's) but do not belong to *S*, and they are known as encoded (logical) Pauli operators. The encoded Pauli operators correspond to the homologically nontrivial loops of the *Z*- and *X*-types, as shown in Fig. 8.12. Let *L* denote the set of logical Pauli operators. Because the operators $l \in L$ and ls for every $s \in S$ are equivalent, only homology classes (cosets) are important. The corresponding maximum-likelihood error operator selection can be performed as follows:

$$l_{\mathrm{ML}} = \underset{l \in L}{\operatorname{argmax}} \sum_{s \in S} P(E = lsR(s)), \tag{8.161}$$

where *R*(*s*) is the representative error from the coset corresponding to syndrome *s*. The error correction action is performed by applying the operator $l_{\mathrm{ML}}R(s)$. Note that the algorithm above can be computationally extensive. An interesting algorithm of the complexity of $\mathcal{O}(m^2 \log m)$ can be found in Refs. [46,47]. Since we have already established the connection between quantum LDPC codes and torus codes, we can use the LDPC decoding algorithms described in Chapter 9 instead. We now determine the encoded (logical) Pauli operators following the description given by Bravyi and Kitaev [44]. Let us consider an operator of the form

$$\overline{Y}(c, c') = \prod_{i \in c} Z_i \prod_{j \subset c'} X_j, \tag{8.162}$$

where *c* is an *I*-cycle (loop) on the lattice and *c*' is an *I*-cycle on the dual lattice. It is straightforward to show that $\overline{Y}(c, c')$ commutes with all stabilizers and thus performs the mapping of code subspace C_Q to itself. Since this mapping depends on homology classes (cosets) of *c* and *c*', $\overline{Y}([c], [c'])$ This denotes the encoded Pauli operators. It can be shown that $\overline{Y}([c], [c'])$ forms a linear basis for $L(C_Q)$. The logical operators $\overline{X}_1(0,[c'_1]),\overline{X}_2(0,[c'_2]),\overline{Z}_1([c_1],0)$, and $\overline{Z}_2([c_2],0)$, where c_1,c_2 are the cycles on the original lattice and c'_1,c'_2 are the cycles on the dual lattice, can be used as generators of $L(C_Q)$.

We now turn our attention to the quantum topological codes on a *lattice with boundary* introduced by Bravyi and Kitaev [44]. Instead of dealing with lattices on the torus, here we consider the finite square lattice on the plane. Two types of boundaries can be introduced: the z-type boundary shown in Fig. 8.13A and the x-type boundary shown in Fig. 8.13B. In Fig. 8.13C, we provide an illustrative example [44] for a 2×3 lattice.

The simplest form of quantum boundary code can be obtained by simply alternating x-type and z-type boundaries, as illustrated in Fig. 8.13C. Since an $n\times m$ lattice has $(n+1)(m+1)$ horizontal edges and nm vertical edges, by associating the edges with qubits, the corresponding codeword length becomes $N=2\,nm+n+m+1$. The stabilizers can be formed in a fashion similar to toric codes. The vertex and plaquette operators can be defined in a fashion similar to Eq. (8.157), and the code subspace can be introduced by Eq. (8.158). The free ends of edges do not contribute to stabilizers. Note that the inner vertex and plaquette stabilizers are of weight 4, while the outer stabilizers are of weight 3. This quantum code is an example of an irregular quantum stabilizer code. The number of plaquette and vertex stabilizers is determined by $n(m+1)$ and $(n+1)m$, respectively.

For example, as shown in Fig. 8.13C, the vertex stabilizers are given by

$$\begin{aligned}
&A_{v_1}=X_{B_1v_1}X_{v_1v_2}X_{v_1v_4}, A_{v_2}=X_{v_1v_2}X_{v_2v_3}X_{v_2v_5}, A_{v_3}=X_{v_2v_3}X_{v_3B_2}X_{v_3v_6},\\
&A_{v_4}=X_{B_1v_4}X_{v_4v_5}X_{v_1v_4}X_{v_4v_7}, A_{v_5}=X_{v_4v_5}X_{v_5v_6}X_{v_2v_5}X_{v_5v_8}, A_{v_6}=X_{v_5v_6}X_{v_6B_2}X_{v_3v_6}X_{v_6v_9},\\
&A_{v_7}=X_{B_1v_7}X_{v_7v_8}X_{v_4v_7}, A_{v_8}=X_{v_7v_8}X_{v_8v_9}X_{v_5v_8}, A_{v_9}=X_{v_8v_9}X_{v_9B_2}X_{v_6v_9}.
\end{aligned}$$

On the other hand, the plaquette stabilizers are given by

$$\begin{aligned}
&B_{p_1}=Z_{B'_1p_1}Z_{p_1p_5}Z_{p_1p_2}, B_{p_2}=Z_{B'_1p_2}Z_{p_2p_6}Z_{p_1p_2}Z_{p_2p_3}, B_{p_3}=Z_{B'_1p_3}Z_{p_3p_7}Z_{p_2p_3}Z_{p_3p_4}, B_{p_4}=Z_{B'_1p_4}Z_{p_4p_8}Z_{p_3p_4},\\
&B_{p_5}=Z_{p_1p_5}Z_{p_5B'_2}Z_{p_5p_6}, B_{p_6}=Z_{p_2p_6}Z_{p_6B'_2}Z_{p_5p_6}Z_{p_6p_7}, B_{p_7}=Z_{p_3p_7}Z_{p_7B'_2}Z_{p_6p_7}Z_{p_7p_8}, B_{p_8}=Z_{p_4p_8}Z_{p_8B'_2}Z_{p_7p_8}.
\end{aligned}$$

From this example, it is clear that the number of independent stabilizers is given by $N-K=2\,nm+n+m$, so the number of information qubits of this code is only $K=1$. The encoded (logical) Pauli operators can also be obtained by Eq. (8.162), but now with *I*-cycle c (c') defined as a string that ends on the boundary of the lattice (dual lattice). An error E is undetectable if it commutes with stabilizers—that is, it can be represented as a linear combination of operators given by Eq. (8.162). Since the distance is defined as the minimum weight of error that cannot be detected, *I*-cycle c (c') has the length $n+1$ $(m+1)$, and the minimum distance is determined by $D=\min(n+1,m+1)$. The parameters of this quantum code are $[2\,nm+n+m+1,1,\min(n+1,m+1)]$. Clearly, by letting one dimension go to infinity, the minimum distance becomes infinitely large as the quantum code rate tends to zero.

This class of codes can be generalized in several ways: (1) by setting $B_1=B_2=B$, we obtain the square lattice on the *cylinder*, (2) by setting $B_1=B_2=B'_1=B'_2=B$, we obtain the square lattice on the *sphere*, and (3) by assuming that free edges end at different boundaries, we obtain a stronger code, since the number of cycles of length 4 in the corresponding bipartite graph representation (see Chapter 9 for definition) is smaller. The 2×3-based lattice code belonging to class 3 is shown in Fig. 8.13D. Note that the lattice does not need to be square/rectangular, and we can further generalize from this fact. Several examples of such codes are provided in Refs. [43,44]. By employing the good properties of topological and subsystem codes, we can design subsystem topological codes [45].

8.9 Surface codes

Surface codes are closely related to the quantum topological codes on the boundary introduced by Bravyi and Kitaev [43,44]. This section is devoted to this class of quantum codes, given their popularity and possible application in quantum information processing and quantum communications [52–58]. Various versions of surface codes exist; here, we describe two that are quite popular. The surface code is defined on a two-dimensional lattice, such as the one shown in Fig. 8.14, with the qubits clearly indicated in Fig. 8.14A. Stabilizers of plaquette type can be defined as shown in Fig. 8.14B. Each plaquette stabilizer denoted by X (Z) is composed of Pauli X (Z)-operators on qubits located in the intersection of the edges of the corresponding plaquette. As an illustration, the plaquette stabilizer denoted X and related to qubits 1 and 2 is X_1X_2. Similarly, the plaquette stabilizer denoted Z and related to qubits 5, 6, 8, and 9 is $Z_5Z_6Z_8Z_9$. To simplify the notation, we can use the representation in Fig. 8.14C, where the shaded plaquettes correspond to all X-containing operators' stabilizers, while the white plaquettes correspond to all Z-containing operators' stabilizers. Weight-2 stabilizers are allocated around the perimeter, while weight-4 stabilizers are located within the interior. If we denote the set of all blue (white) plaquettes with $\{B_p\}$ ($\{W_p\}$), we can define the stabilizer on plaquette p as follows:

$$S_p = \bigotimes_{i\in p} \sigma_i, \quad \sigma_i = \begin{cases} X_i, \text{ for } p\in B_p \\ Z_i, \text{ for } p\in W_p \end{cases}. \tag{8.163}$$

All stabilizers commute and have the eigenvalues +1 and −1. The surface code can now be defined in a fashion similar to Eq. (8.158), namely

$$C_{SC} = \{|c\rangle \in H_2^N \,|\, S_p|c\rangle = |c\rangle; \ \forall p\}. \tag{8.164}$$

Given that the surface code space is two-dimensional, it encodes a single logical qubit.

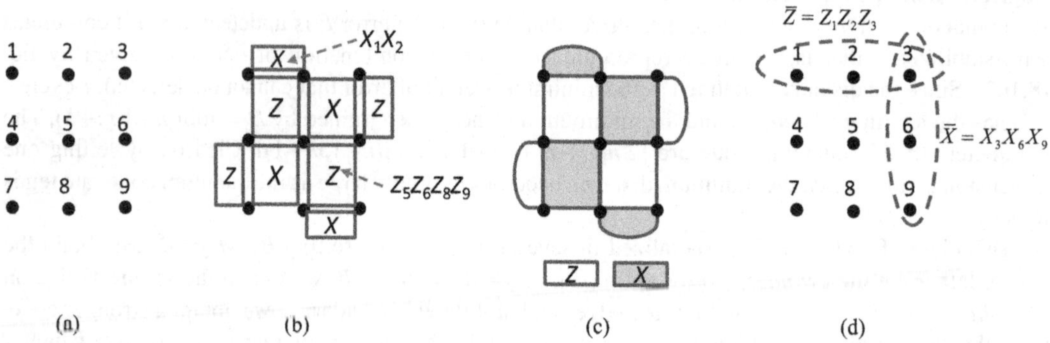

FIGURE 8.14

Illustration of a surface code: (A) the qubits are located in the lattice positions, (B) all-X and all-Z plaquette operators, (C) popular representation of surface codes in which the X- and Z-plaquettes are clearly indicated, and (D) logical operators.

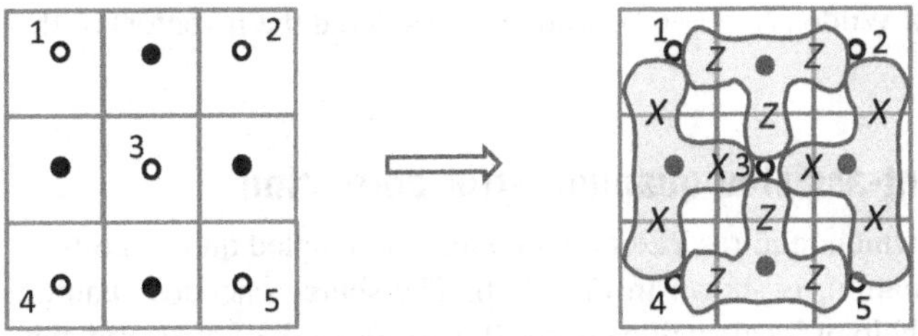

FIGURE 8.15

Another interpretation of the surface code.

The logical operators preserve the code space, and as such, manipulate the logical qubit state. For surface code, the logical operators are run over both sides of the lattice as shown in Fig. 8.14D and can be represented as $\overline{Z} = Z_1Z_2Z_3, \overline{X} = X_3X_6X_9$. Clearly, the logical operator spans two boundaries of the same type. The codeword length is the product of the side lengths expressed as the number of qubits, and for the surface code from Fig. 8.14, we have $N = L_x \times L_z = 3\times3 = 9$. In addition, the number of information qubits is $K = 1$. The minimum distance of this code is determined as the minimum side length, that is, $d = \min(L_x, L_z) = 3$, indicating that this code can correct a single qubit error. Let E denote the error operator applied on codeword $|c\rangle$—that is, we can write $E|c\rangle$. Let us assume that this error operator represents the single physical qubit Pauli flip. $E|c\rangle$ will anticommute with all codewords $|c'\rangle$, that is, $\langle c'|E|c\rangle = -1$. We can perform the syndrome measurements by placing ancilla qubits in the middle of each plaquette. The intersection of syndromes can help identify the most probable error operator. Given that the surface codes are topological, the decoding algorithm described in the previous section also applies here.

Let us provide another description of a surface code. Let us now place the qubits on the dual lattice, as illustrated in Fig. 8.15. There are two types of qubits: quiescent states denoted by open (white) circles and measurement qubits denoted by filled (black) circles. We manipulate the quiescent states to perform the desired quantum operation. The measurement qubits serve the role of ancilla qubits and facilitate syndrome measurements. This example has five data qubits (quiescent states) and four measurement qubits. The measurement qubits are associated with either all-X or all-Z stabilizers. From Fig. 8.15 (right), we can read out the following stabilizers for this surface code: $Z_1Z_2Z_3$, $X_1X_3X_4$, $X_2X_3X_5$, and $Z_3Z_4Z_5$. Since the codeword length is $N = 5$, and the number of stabilizer generators is $N-K = 4$, we conclude that the number of information qubits is $K = N-(N-K) = 1$ and that the surface code dimensionality is $2^K = 2$.

8.10 Entanglement-assisted quantum codes

In this section, entanglement-assisted (EA) QECCs [20,59–63] are described. These QECCs use preexisting entanglements between the transmitter and receiver to improve transmission reliability. EA quantum codes are a generalization of superdense coding [39,59]. They can also be considered particular instances of the subsystem codes [29–31] described in Section 8.7. The key advantage of EA quantum codes over CSS codes is that they do not require the classical codes from which they are derived to be dual-containing. Namely, arbitrary classical code can be used to design EA quantum codes, provided that a preentanglement exists between the source and destination. This chapter uses the notation of Brun, Devetak, Hsieh, and Wilde [20,59–63], who are considered the inventors of this class of quantum codes.

8.10.1 Principles of entanglement-assisted quantum error correction

The block-scheme of EA quantum code, which requires a certain number of entangled qubits (ebits) to be shared between the source and destination, is shown in Fig. 8.16. The source encodes quantum information in K-qubit state $|\psi\rangle$ with the help of local ancilla qubits $|0\rangle$ and source-half of shared ebits (e ebits) into N-qubits and then sends the encoded qubits over a noisy quantum channel (such as a free-space or fiber-optic channel). The receiver performs decoding on all qubits ($N+e$ qubits) to diagnose the channel error and performs a unitary recovery operation to reverse the action of the channel. Note that the channel has no effect on the receiver's half of the shared ebits. By omitting the ebits, the conventional quantum coding scheme is obtained.

As discussed in previous sections, both noise $\mathfrak{N}$ and recovery $\mathfrak{R}$ processes can be described as quantum operations $\mathfrak{N},\mathfrak{R}$: $L(H) \rightarrow L(H)$, where $L(H)$ is the space of linear operators on Hilbert space H. These mappings can be represented in terms of operator-sum representation:

$$\mathfrak{N}(\rho) = \sum\nolimits_i E_i \rho E_i^\dagger, E_i \in L(H), \mathfrak{N} = \{E_i\}. \tag{8.165}$$

Given that the quantum code C_Q is a subspace of H, we say that a set of errors $\mathfrak{E}$ is correctable if there is a recovery operation $\mathfrak{R}$ such that $\mathfrak{R}\mathfrak{N}(\rho) = \rho$ for any state ρ from $L(C_Q)$. Further, each error operator E_i can be expanded in terms of Pauli operators:

$$E_i = \sum_{[\boldsymbol{a}|\boldsymbol{b}]\in F_2^{2N}} \alpha_{i,[\boldsymbol{a}|\boldsymbol{b}]} E_{[\boldsymbol{a}|\boldsymbol{b}]},\ E_{[\boldsymbol{a}|\boldsymbol{b}]} = j^\lambda X(\boldsymbol{a})Z(\boldsymbol{b});\ \boldsymbol{a} = a_1 \cdots a_N; \boldsymbol{b} = b_1 \cdots b_N; a_i, b_i = 0,1; \lambda = 0,1,2,3$$
$$X(\boldsymbol{a}) \equiv X_1^{a_1} \otimes \cdots \otimes X_N^{a_N}; Z(\boldsymbol{b}) \equiv Z_1^{b_1} \otimes \cdots \otimes Z_N^{b_N} \tag{8.166}$$

An $[N,K,e;D]$ EA-QECC consists of (1) an encoding map,

$$\hat{U}_{\text{enc}}: L^{\otimes K} \otimes L^{\otimes e} \rightarrow L^{\otimes N}, \tag{8.167}$$

and (2) decoding trace-preserving mapping,

$$\mathfrak{D}: L^{\otimes N} \otimes L^{\otimes e} \rightarrow L^{\otimes K}, \tag{8.168}$$

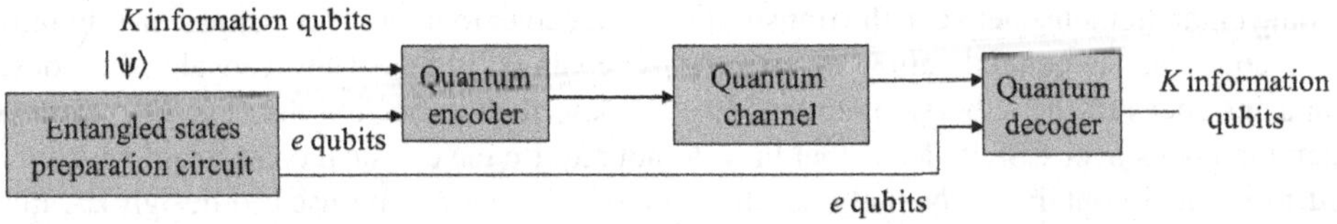

FIGURE 8.16

A generic entanglement-assisted quantum error correction scheme.

such that the composition of encoding mapping, channel transformation, and decoding mapping is the identity mapping:

$$\mathcal{D} \circ \mathcal{N} \circ \breve{U}_{\text{enc}} \circ \breve{U}_{\text{app}} = I^{\otimes K}; I: L \rightarrow L; \quad U_{\text{app}}|\psi\rangle = |\psi\rangle|\Phi^{\otimes e}\rangle \tag{8.169}$$

where U_{app} is an operation that appends the transmitter half of the maximum-entangled state. $|\Phi^{\otimes e}\rangle$

8.10.2 Entanglement-assisted canonical quantum codes

The simplest EA quantum code, the EA canonical code, performs the following trivial encoding operation:

$$\hat{U}_c: |\psi\rangle \rightarrow |\mathbf{0}\rangle_l \otimes |\Phi^{\otimes e}\rangle \otimes |\psi\rangle_K; l = N - K - e \tag{8.170}$$

The canonical EA code, shown in Fig. 8.17, is, therefore, an extension of canonical code presented in Chapter 7, in which e ancilla qubits are replaced by e maximally entangled kets shared between the transmitter and receiver. Since the number of information qubits is K and codeword length is N, the number of remaining ancilla qubits in the $|0\rangle$-state is $l = N-K-e$.

The EA canonical code can correct the following set of errors:

$$\mathcal{N}_{\text{canonical}} = \{X(\boldsymbol{a})Z(\boldsymbol{b}) \otimes Z(\boldsymbol{b}_{\text{eb}})X(\boldsymbol{a}_{\text{eb}}) \otimes X(\alpha(\boldsymbol{a}, \boldsymbol{b}_{\text{eb}}, \boldsymbol{a}_{\text{eb}}))Z(\beta(\boldsymbol{a}, \boldsymbol{b}_{\text{eb}}, \boldsymbol{a}_{\text{eb}})) : \boldsymbol{a}, \boldsymbol{b} \in F_2^l; \boldsymbol{b}_{\text{eb}}, \boldsymbol{a}_{\text{eb}} \in F_2^e\}$$
$$\alpha, \beta: F_2^l \times F_2^e \times F_2^e \rightarrow F_2^K \tag{8.171}$$

where the Pauli operators $X(\boldsymbol{c})$ ($\boldsymbol{c} \in \{\boldsymbol{a}, \boldsymbol{a}_{\text{eb}}, \alpha(\boldsymbol{a}, \boldsymbol{b}_{\text{eb}}, \boldsymbol{a}_{\text{eb}})\}$) and $Z(\boldsymbol{d})$ ($\boldsymbol{d} \in \{\boldsymbol{b}, \boldsymbol{b}_{\text{eb}}, \beta(\boldsymbol{a}, \boldsymbol{b}_{\text{eb}}, \boldsymbol{a}_{\text{eb}})\}$) are introduced in Eq. (8.166).

To prove the claim given by Eq. (8.171), let us observe one particular error E_c from the set of errors in Eq. (8.171) and study its action on the transmitted codeword $|\mathbf{0}\rangle_l \otimes |\Phi^{\otimes e}\rangle \otimes |\psi\rangle_K$:

$$X(\boldsymbol{a})Z(\boldsymbol{b})|\mathbf{0}\rangle_l \otimes \left(Z(\boldsymbol{b}_{\text{eb}})X(\boldsymbol{a}_{\text{eb}}) \otimes I^{R_x}\right)|\Phi^{\otimes e}\rangle \otimes X(\alpha(\boldsymbol{a}, \boldsymbol{b}_{\text{eb}}, \boldsymbol{a}_{\text{eb}}))Z(\beta(\boldsymbol{a}, \boldsymbol{b}_{\text{eb}}, \boldsymbol{a}_{\text{eb}}))|\psi\rangle. \tag{8.172}$$

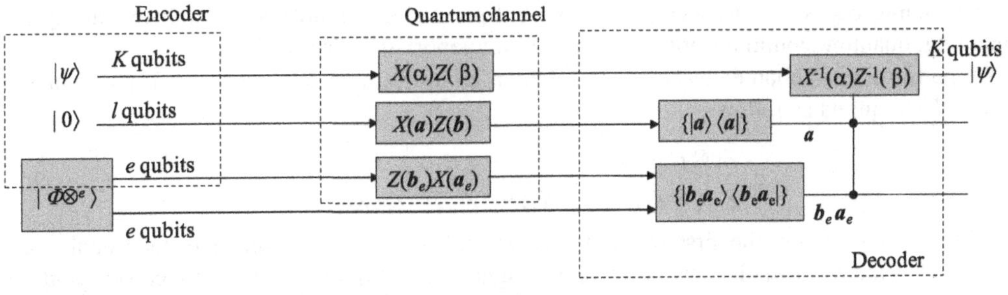

FIGURE 8.17

An entanglement-assisted canonical code.

The action of operator $X(\boldsymbol{a})Z(\boldsymbol{b})$ on state $|\boldsymbol{0}\rangle_l$ is given by $X(\boldsymbol{a})Z(\boldsymbol{b})|\boldsymbol{0}\rangle_l \overset{Z(\boldsymbol{b})|\boldsymbol{0}\rangle_l=|\boldsymbol{0}\rangle_l}{\widehat{=}} X(\boldsymbol{a})|\boldsymbol{0}\rangle_l = |\boldsymbol{a}\rangle$. Meanwhile, the action of $Z(\boldsymbol{b}_{\text{eb}})X(\boldsymbol{a}_{\text{eb}})$ on the maximum entangled state is $\left(Z(\boldsymbol{b}_{\text{eb}})X(\boldsymbol{a}_{\text{eb}}) \otimes I^{R_x}\right)\left|\Phi^{\otimes e}\right\rangle = |\boldsymbol{b}_{\text{eb}}, \boldsymbol{a}_{\text{eb}}\rangle$, with I^{Rx} denoting the identity operators applied on the receiver half of the maximum-entangled state. Finally, the action of the channel on the information portion of the codeword is given by $X(\alpha(\boldsymbol{a}, \boldsymbol{b}_{\text{eb}}, \boldsymbol{a}_{\text{eb}}))Z(\beta(\boldsymbol{a}, \boldsymbol{b}_{\text{eb}}, \boldsymbol{a}_{\text{eb}}))|\psi\rangle = |\psi'\rangle$. The vector $(\boldsymbol{a}\ \boldsymbol{b}\ \boldsymbol{b}_{\text{eb}}\ \boldsymbol{a}_{\text{eb}})$ uniquely specifies the error operator E_c and is called the syndrome vector. Since state $|\boldsymbol{0}\rangle_l$ is invariant on the action of $Z(\boldsymbol{b})$, vector $\boldsymbol{b}$ can be omitted from the syndrome vector. To obtain the $\boldsymbol{a}$-portion of the syndrome, we must perform simultaneous measurements on $Z(\boldsymbol{d}_i)$, where $\boldsymbol{d}_i=(0\cdots01_i0\cdots0)$. To obtain the $(\boldsymbol{b}_{\text{eb}}\ \boldsymbol{a}_{\text{eb}})$-portion of the syndrome vector, we must perform measurements on $Z(\boldsymbol{d}_1)Z(\boldsymbol{d}_1)$, ..., $Z(\boldsymbol{d}_e)Z(\boldsymbol{d}_e)$ for $\boldsymbol{b}_{\text{eb}}$ and $X(\boldsymbol{d}_1)X(\boldsymbol{d}_1)$, ..., $X(\boldsymbol{d}_e)X(\boldsymbol{d}_e)$ for $\boldsymbol{a}_{\text{eb}}$. Once the syndrome vector is determined, we apply $X^{-1}(\alpha(\boldsymbol{a}, \boldsymbol{b}_e, \boldsymbol{a}_e))Z^{-1}(\beta(\boldsymbol{a}, \boldsymbol{b}_e, \boldsymbol{a}_e))$ to undo the action of the channel. From syndrome measurements, it is clear that we have applied exactly the same measurements as in the superdense coding, indicating that EA codes can indeed be interpreted as generalizations of superdense coding (see Ref. [39] and Section 8.8 of [61]). To avoid the need for measurement, we can perform the following controlled unitary decoding:

$$U_{\text{canocnical, dec}} = \sum_{\boldsymbol{a}, \boldsymbol{b}_{\text{eb}}, \boldsymbol{a}_{\text{eb}}} |\boldsymbol{a}\rangle\langle \boldsymbol{a}| \otimes |\boldsymbol{b}_{\text{eb}}, \boldsymbol{a}_{\text{eb}}\rangle\langle \boldsymbol{b}_{\text{eb}}, \boldsymbol{a}_{\text{eb}}| \otimes X^{-1}(\alpha(\boldsymbol{a}, \boldsymbol{b}_{\text{eb}}, \boldsymbol{a}_{\text{eb}}))Z^{-1}(\beta(\boldsymbol{a}, \boldsymbol{b}_{\text{eb}}, \boldsymbol{a}_{\text{eb}})). \quad (8.173)$$

The EA error correction can also be described using the stabilizer formalism interpretation. Let $S_{\text{canonical}}$ be the nonabelian group generated by

$$\begin{aligned} S_{\text{canonical}} &= \left\langle S_{\text{canonical},I}, S_{\text{canonical},E} \right\rangle;\ S_{\text{canonical},I} = \langle Z_1, \cdots, Z_l \rangle, \\ S_{\text{canonical},E} &= \langle Z_{l+1}, \cdots, Z_{l+e}, X_{l+1}, \cdots, X_{l+e} \rangle, \end{aligned} \quad (8.174)$$

where the isotropic subgroup is denoted $S_{\text{canonical},I}$, and the symplectic subgroup of anticommuting operators is denoted $S_{\text{canonical},E}$. The representation of Eq. (8.174) has certain similarities with the subsystem codes from Section 8.7. Namely, by interpreting $S_{\text{canonical},I}$ as the stabilizer group and $S_{\text{canonical},E}$ as the gauge group, it turns out that EA codes represent a generalization of subsystem codes. Moreover, subsystem codes do not require entanglement between the transmitter and receiver. On the other hand, EA codes with few ebits have reasonable encoder and decoder complexity compared with dual-containing codes, and as such, they represent promising candidates for fault-tolerant quantum computing, quantum communication, and quantum teleportation applications.

We perform an Abelian extension of nonabelian group $S_{\text{canonical}}$, denoted $S_{\text{canonical,ext}}$, which now acts on $N + e$ qubits as follows:

$$\begin{aligned} &Z_1 \otimes I, \cdots, Z_l \otimes I \\ &Z_{l+1} \otimes Z_1, X_{l+1} \otimes X_1, \cdots, Z_{l+e} \otimes Z_e, X_{l+e} \otimes X_e \end{aligned} \quad (8.175)$$

Of the $N + e$ qubits, the first N qubits correspond to the transmitter, and additional e qubits correspond to the receiver. The operators on the right side of Eq. (8.175) act on receiver qubits. This extended group is abelian and can be used as a stabilizer to fix the code subspace $C_{\text{canonical,EA}}$. Such an obtained stabilizer group can be called the EA stabilizer group. The number of information qubits of this stabilizer can be determined by $K=N-l-e$, so the code can be denoted $[N,K,e,D]$, where D is the distance of the code.

Before concluding this subsection, it is useful to identify the correctable set of errors. Analogous with stabilizer codes, we expect the correctable set of EA QECC errors defined by $S_{\text{canonical}}=<S_{\text{canonical},I}, S_{\text{canonical},E}>$ to be given by

$$\mathfrak{N}_{\text{canonical}}=\left\{E_m\middle|\forall E_1,E_2\Rightarrow E_2^{\dagger}E_1\in S_{\text{canonical},I}\cup\left(G_N-C(S_{\text{canonical}})\right)\right\}, \tag{8.176}$$

where $C(S_{\text{canonical}})$ is the centralizer of $S_{\text{canonical}}$ (that is, the set of errors that commute with all elements from $S_{\text{canonical}}$). We showed above (see Eq. 8.172 and the corresponding text) that every error can be specified by the syndrome ($\boldsymbol{a}\ \boldsymbol{b}_{\text{eb}}\ \boldsymbol{a}_{\text{eb}}$). If two errors E_1 and E_2 have the same syndrome ($\boldsymbol{a}\ \boldsymbol{b}_{\text{eb}}\ \boldsymbol{a}_{\text{eb}}$), $E_2^{\dagger}E_1$ will have the all-zeros syndrome, and such an error clearly belongs to $S_{\text{canonical},I}$. An error such as $E_2^{\dagger}E_1$ is trivial, as it fixes all codewords. If two errors have different syndromes, then the syndrome for $E_2^{\dagger}E_1$, which has the form ($\boldsymbol{a}\ \boldsymbol{b}_{\text{eb}}\ \boldsymbol{a}_{\text{eb}}$), indicates that such an error can be corrected provided that it does not belong to $C(S_{\text{canonical}})$. Namely, when an error belongs to $C(S_{\text{canonical}})$, its action results in another codeword different from the original codeword, so such an error is undetectable.

8.10.3 General entanglement-assisted quantum codes

From group theory, if V is an arbitrary subgroup of G_N of order 2^M, a set of generators $\{\overline{X}_{p+1},\cdots,\overline{X}_{p+q};\overline{Z}_1,\cdots,\overline{Z}_{p+q}\}$ exists that satisfies commutation properties similar to those of Eq. (8.28), namely

$$\left[\overline{X}_m,\overline{X}_n\right]=\left[\overline{Z}_m,\overline{Z}_n\right]=0\ \forall m,n;\ \left[\overline{X}_m,\overline{Z}_n\right]=0\ \forall m\neq n;\ \left\{\overline{X}_m,\overline{Z}_n\right\}=0\ \forall m=n \tag{8.177}$$

We also know that the unitary equivalent observables, A and UAU^{-1}, have identical spectra. Thus, we can establish a bijection between the two groups V and S that preserves their commutation relations so that $\forall\ v\in V$. A corresponding $s\in S$ exists such that $v=UsU^{-1}$ (up to a general phase constant). Using these results, we can establish a bijection between the canonical EA quantum code described by $S_{\text{canonical}}=\langle S_{\text{canonical},I},S_{\text{canonical},E}\rangle$ and a general EA code characterized by $S=\langle S_I,S_E\rangle$, with $S=US_{\text{canonical}}U^{-1}$ (or equivalently $S_{\text{canonical}}=U^{-1}SU$), where U is a corresponding unitary operator.

Based on the discussion above, we can generalize the results from the previous subsection as follows. Given a general group $S=<S_I, S_E>$ ($|S_I|=2^{N-K-e}$ and $|S_E|=2^{2e}$), there is an $[N,K,e;D]$ EA-QECC code, illustrated in Fig. 8.18 and denoted C_{EA}, defined by an encoding–decoding pair $(\mathfrak{E},\mathfrak{D})$ with the following properties:

1. The EA code C_{EA} can correct any error from the following set of errors $\mathfrak{N}$:

$$\mathfrak{N}=\left\{E_m\middle|\forall E_1,E_2\Rightarrow E_2^{\dagger}E_1\in S_I\cup\left(G_N-C(S)\right)\right\}. \tag{8.178}$$

2. The code-space C_{EA} is a simultaneous eigenspace of the Abelian extension of S, denoted S_{ext}.
3. To decode, the error syndrome is obtained by simultaneously measuring the observables from S_{ext}.

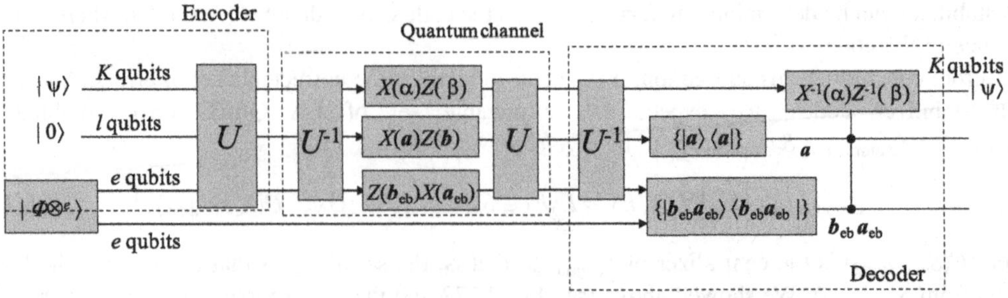

FIGURE 8.18

An entanglement-assisted code obtained as a generalization of the canonical entanglement-assisted code.

Because the commutation relations of S are the same as those of the EA canonical code $C_{\text{canonical,EA}}$, we can find a unitary operator U such that $S_{\text{canonical}} = USU^{-1}$. Based on Fig. 8.18, encoding mapping can be represented as a composition of the unitary operator U and encoding mapping of the canonical code $\mathcal{E}_{\text{canonical}}$, namely $\mathcal{E} = U \circ \mathcal{E}_{\text{canonical}}$. Similarly, decoding mapping can be represented by $\mathcal{D} = \mathcal{D}_{\text{canonical}} \circ \hat{U}^{-1}$, $\hat{U} = \text{extension}(U)$. Further, the composition of encoding, error, and decoding mapping would result in the identity operator

$$\mathcal{D} \circ \mathcal{N} \circ \mathcal{E} = \mathcal{D}_{\text{canonical}} \circ \hat{U}^{-1} \circ \hat{U} \mathcal{N}_{\text{canonical}} \hat{U}^{-1} \circ \hat{U} \mathcal{E} = I^{\otimes K} \tag{8.179}$$

The set of errors given by Eq. (8.178) is the correctable set of errors because $C_{\text{canonical,EA}}$ is a simultaneous eigenspace of $S_{\text{canonical,ext}}$, and $S_{\text{ext}} = US_{\text{canonical, ext}}U^{-1}$, and by definition, $C_{\text{EA}} = U(C_{\text{canonical,EA}})$, so we conclude that C_{ext} is a simultaneous eigenspace of S_{ext}. We learned in the previous subsection that the decoding operation of $D_{\text{canonical}}$ involves (1) measuring the set of generators of $S_{\text{canonical,ext}}$ yielding the error syndrome according to the error E_c and (2) performing a recovery operation to undo the action of channel error E_c. The decoding action of C_{EA} is equivalent to measuring $S_{\text{ext}} = US_{\text{canonical, ext}}U^{-1}$, followed by the recovery operation UE_cU^{-1} and U^{-1} to undo the encoding.

The distance D of the EA code can be introduced similarly to that of a stabilizer code. The distance D of the EA code is defined as the minimum weight among undetectable errors. In other words, we may say that the EA quantum code has a distance D if it can detect all errors with weights less than D but none with weights of D or more. The error correction capability t of the EA quantum code is related to minimum distance D by $t = \lfloor (D-1)/2 \rfloor$. The weight of a qubit error $[\mathbf{u}|\mathbf{v}]$ ($u,v \in F_2^N$) can be defined in a fashion similar to that for stabilizer codes, that is, as the number of components i for which $(u_i,b_i) \neq (0,0)$ or equivalently $\text{wt}([\mathbf{u}|\mathbf{v}]) = \text{wt}(\mathbf{u} + \mathbf{v})$, where "+" denotes bitwise addition mod 2.

The EA quantum bounds are similar to the quantum coding bounds presented in Chapter 8. For example, the EA quantum Singleton bound is given by

$$N + e - K \geq 2(D-1). \tag{8.180}$$

8.10.4 Entanglement-assisted quantum error correction codes derived from classical quaternary and binary codes

We now turn our attention to establishing the connection between EA codes and quaternary classical codes, which are defined over $F_4 = \mathrm{GF}(4) = \{0,1,\omega,\omega^2 = \varpi\}$. As a reminder, the addition and multiplication tables for F_4 are given below.

+	0	ϖ	1	ω
0	0	ϖ	1	ω
ϖ	ϖ	0	ω	1
1	1	ω	0	ϖ
ω	ω	1	ϖ	0

×	0	ϖ	1	ω
0	0	0	0	0
ϖ	0	ω	ϖ	1
1	0	ϖ	1	ω
ω	0	1	ω	ϖ

The addition table for binary 2-tuples, $F_2^2 = \{00,01,10,11\}$ is given by

+	00	01	11	10
00	00	01	11	10
01	01	00	10	11
11	11	10	00	01
10	10	11	01	00

Further, from Chapter 7, we know that the multiplication table of Pauli matrices is given by

×	*I*	X	Y	Z
I	*I*	X	Y	Z
X	X	*I*	jZ	-jY
Y	Y	-jZ	*I*	jX
Z	Z	jY	-jX	*I*

If we ignore the phase factor in Pauli matrices multiplication, we can establish the following correspondences between G_N, F_2^2, and F_4:

G	F_2^2	F_4
I	00	0
X	10	ω
Y	11	1
Z	01	ϖ

It is easily shown that the mapping f: $F_4 \rightarrow F_2^2$ is an isomorphism. Therefore, if a classical code (n,k,d) C_4 over GF(4) exists, an $[N,K,e,D] = [n,2k-n+e,e;D]$ EA-QECC exists for some nonnegative integer e that can be described by the following quantum-check matrix:

$$A_{\mathrm{EA}} = f\left(\widetilde{\boldsymbol{H}}_4\right);\; \widetilde{\boldsymbol{H}}_4 = \begin{pmatrix} \omega \boldsymbol{H}_4 \\ \varpi \boldsymbol{H}_4 \end{pmatrix}, \tag{8.181}$$

where $\boldsymbol{H}_4$ denotes the parity-check matrix of the classical code (n,k,d) over F_4. The symplectic product of binary vectors from Eq. (8.181) is equal to the trace of the product of their GF(4) representations as follows:

$$\boldsymbol{h}_m \odot \boldsymbol{h}_n = \operatorname{tr}\left(f^{-1}(\boldsymbol{h}_m)\overline{f^{-1}(\boldsymbol{h}_n)}\right), \quad \operatorname{tr}(w) = w + \overline{w}, \tag{8.182}$$

where $\boldsymbol{h}_m$ $(\boldsymbol{h}_n)$ is the mth (nth) row of $\boldsymbol{A}_{\mathrm{EA}}$, and the overbar denotes the conjugation. The symplectic product of $\boldsymbol{A}_{\mathrm{EA}}$ is given by

$$\operatorname{tr}\left\{\begin{pmatrix}\omega \boldsymbol{H}_4 \\ \overline{\omega}\boldsymbol{H}_4\end{pmatrix}\begin{pmatrix}\omega \boldsymbol{H}_4 \\ \overline{\omega}\boldsymbol{H}_4\end{pmatrix}^{\dagger}\right\} = \operatorname{tr}\left\{\begin{pmatrix}\omega \boldsymbol{H}_4 \\ \overline{\omega}\boldsymbol{H}_4\end{pmatrix}\begin{pmatrix}\overline{\omega}\boldsymbol{H}_4^{\dagger} & \omega \boldsymbol{H}_4^{\dagger}\end{pmatrix}\right\} = \operatorname{tr}\left\{\begin{pmatrix}\boldsymbol{H}_4\boldsymbol{H}_4^{\dagger} & \overline{\omega}\boldsymbol{H}_4\boldsymbol{H}_4^{\dagger} \\ \omega \boldsymbol{H}_4\boldsymbol{H}_4^{\dagger} & \boldsymbol{H}_4\boldsymbol{H}_4^{\dagger}\end{pmatrix}\right\}$$

$$= \operatorname{tr}\left\{\begin{pmatrix}1 & \overline{\omega} \\ \omega & 1\end{pmatrix} \otimes \boldsymbol{H}_4\boldsymbol{H}_4^{\dagger}\right\} = \begin{pmatrix}1 & \overline{\omega} \\ \omega & 1\end{pmatrix} \otimes \boldsymbol{H}_4\boldsymbol{H}_4^{\dagger} + \begin{pmatrix}1 & \omega \\ \overline{\omega} & 1\end{pmatrix} \otimes \overline{\boldsymbol{H}}_4\boldsymbol{H}_4^{\mathrm{T}} \tag{8.183}$$

The rank of $\widetilde{\boldsymbol{H}}_4\widetilde{\boldsymbol{H}}_4^{\dagger}$ does not change when we perform the similarity transformation $\boldsymbol{B}\boldsymbol{A}\widetilde{\boldsymbol{H}}_4\boldsymbol{H}_4^{\dagger}\boldsymbol{A}^{\dagger}\boldsymbol{B}^{\dagger}$, where $\boldsymbol{A} = \begin{bmatrix}1 & \overline{\omega} \\ \omega & 1\end{bmatrix} \otimes \boldsymbol{I}, \boldsymbol{B} = \begin{bmatrix}1 & 0 \\ 1 & 1\end{bmatrix} \otimes \boldsymbol{I}$, to obtain

$$\boldsymbol{B}\boldsymbol{A}\widetilde{\boldsymbol{H}}_4\boldsymbol{H}_4^{\dagger}\boldsymbol{A}^{\dagger}\boldsymbol{B}^{\dagger} = \begin{bmatrix}\overline{\boldsymbol{H}}_4\boldsymbol{H}_4^{\mathrm{T}} & \boldsymbol{0} \\ \boldsymbol{0} & \boldsymbol{H}_4\boldsymbol{H}_4^{\dagger}\end{bmatrix} = \overline{\boldsymbol{H}}_4\boldsymbol{H}_4^{\mathrm{T}} \otimes \boldsymbol{H}_4\boldsymbol{H}_4^{\dagger}. \tag{8.184}$$

The rank of $\boldsymbol{B}\boldsymbol{A}\widetilde{\boldsymbol{H}}_4\boldsymbol{H}_4^{\dagger}\boldsymbol{A}^{\dagger}\boldsymbol{B}^{\dagger}$ is then obtained by

$$\operatorname{rank}\left(\boldsymbol{B}\boldsymbol{A}\widetilde{\boldsymbol{H}}_4\boldsymbol{H}_4^{\dagger}\boldsymbol{A}^{\dagger}\boldsymbol{B}^{\dagger}\right) = \operatorname{rank}\left(\overline{\boldsymbol{H}}_4\boldsymbol{H}_4^{\mathrm{T}} \otimes \boldsymbol{H}_4\boldsymbol{H}_4^{\dagger}\right) = \operatorname{rank}\left(\overline{\boldsymbol{H}}_4\boldsymbol{H}_4^{\mathrm{T}}\right) + \operatorname{rank}\left(\boldsymbol{H}_4\boldsymbol{H}_4^{\dagger}\right)$$

$$= 2\operatorname{rank}\left(\boldsymbol{H}_4\boldsymbol{H}_4^{\dagger}\right) = 2e, e = \operatorname{rank}\left(\boldsymbol{H}_4\boldsymbol{H}_4^{\dagger}\right). \tag{8.185}$$

Eq. (8.185) indicates that the number of required ebits is $e = \operatorname{rank}\left(\boldsymbol{H}_4\boldsymbol{H}_4^{\dagger}\right)$, so the parameters of this EA code based on Eq. (8.181) are $[N,K,e] = [n, n + e - 2(n-k), e] = [n, 2k - n + e, e]$.

Example. The EA code derived from the (4,2,3) 4-ary classical code described by the parity-check matrix

$$\boldsymbol{H}_4 = \begin{pmatrix}1 & 1 & 0 & 1 \\ 1 & \omega & 1 & 0\end{pmatrix}$$

has parameters [4,1,1,3], and based on Eq. (8.186), the corresponding quantum-check matrix is given by

$$A_{EA} = f\left(\widetilde{\boldsymbol{H}}_4\right) = f\left(\begin{bmatrix}\omega \boldsymbol{H}_4 \\ \overline{\omega}\boldsymbol{H}_4\end{bmatrix}\right) = f\left(\begin{bmatrix}\omega & \omega & 0 & \omega \\ \omega & \overline{\omega} & \omega & 0 \\ \overline{\omega} & \overline{\omega} & 0 & \overline{\omega} \\ \overline{\omega} & 1 & \overline{\omega} & 0\end{bmatrix}\right) = \begin{pmatrix}X & X & I & X \\ X & Z & X & I \\ Z & Z & I & Z \\ Z & Y & Z & I\end{pmatrix}$$

In the remainder of this subsection, we describe how to relate EA quantum codes with classical binary codes. Let $\boldsymbol{H}$ be a binary parity-check matrix of dimension $(n\text{-}k)\times n$. We show that the corresponding EA-QECC has the parameters $[n,2k-n+e;e]$, where $e=\text{rank}(\boldsymbol{H}\boldsymbol{H}^{\text{T}})$. The quantum-check matrix of CSS-like EA quantum code can be represented as follows:

$$\boldsymbol{A}_{\text{EA}}=\left(\begin{array}{c|c}\boldsymbol{H} & \boldsymbol{0}\\ \boldsymbol{0} & \boldsymbol{H}\end{array}\right). \tag{8.186}$$

The rank of the symplectic product of $\boldsymbol{A}_{\text{AE}}$ based on Eq. (8.187) is given by

$$\begin{aligned}\text{rank}\{\boldsymbol{A}_{\text{EA}}\odot\boldsymbol{A}_{EA}^{\text{T}}\}&=\text{rank}\left\{\begin{pmatrix}\boldsymbol{H}\\ \boldsymbol{0}\end{pmatrix}\begin{pmatrix}\boldsymbol{0} & \boldsymbol{H}^{\text{T}}\end{pmatrix}+\begin{pmatrix}\boldsymbol{0}\\ \boldsymbol{H}\end{pmatrix}\begin{pmatrix}\boldsymbol{H}^{\text{T}} & \boldsymbol{0}\end{pmatrix}\right\}\\ &=\text{rank}\left\{\begin{pmatrix}0 & \boldsymbol{H}\boldsymbol{H}^{\text{T}}\\ 0 & 0\end{pmatrix}+\begin{pmatrix}0 & 0\\ \boldsymbol{H}\boldsymbol{H}^{\text{T}} & 0\end{pmatrix}\right\}=\text{rank}\{\boldsymbol{H}\boldsymbol{H}^{\text{T}}+\boldsymbol{H}\boldsymbol{H}^{\text{T}}\}=2\text{rank}\{\boldsymbol{H}\boldsymbol{H}^{\text{T}}\}\\ &=2e,\ e=\text{rank}\{\boldsymbol{H}\boldsymbol{H}^{\text{T}}\}.\end{aligned} \tag{8.187}$$

Eq. (8.187) illustrates that the number of required qubits is $e=\text{rank}(\boldsymbol{H}\,\boldsymbol{H}^{\text{T}})$, so the parameters of this EA code are $[N,K,e]=[n,n+e-2(n-k),e]=[n,2k-n+e,e]$. Therefore, dual-containing codes can be considered EA codes for which $e=0$.

The EA code given by Eq. (8.186) can be generalized as follows:

$$\boldsymbol{A}_{\text{EA}}=(\boldsymbol{A}_x|\boldsymbol{A}_z), \tag{8.188}$$

where $(n\text{-}k)\times n$ submatrix $\boldsymbol{A}_{\text{x}}$ $(\boldsymbol{A}_{\text{z}})$ contains only X-operators (Z-operators). The rank of the symplectic product of $\boldsymbol{A}_{\text{AE}}$ can be obtained as

$$\text{rank}\{\boldsymbol{A}_{\text{EA}}\odot\boldsymbol{A}_{EA}^{\text{T}}\}=\text{rank}\{\boldsymbol{A}_x\boldsymbol{A}_z^{\text{T}}+\boldsymbol{A}_z\boldsymbol{A}_x^{\text{T}}\}=2e,\quad e=\text{rank}\{\boldsymbol{A}_x\boldsymbol{A}_z^{\text{T}}+\boldsymbol{A}_z\boldsymbol{A}_x^{\text{T}}\}/2. \tag{8.189}$$

The parameters of this EA code are $[N,K,e]=[n,n+e\text{-}(n\text{-}k),e]=[n,k+e,e]$.

Another generalization of EA code defined by Eq. (8.186) can be introduced by

$$\boldsymbol{A}_{\text{EA}}=\left(\begin{array}{c|c}\boldsymbol{H}_x & \boldsymbol{0}\\ \boldsymbol{0} & \boldsymbol{H}_z\end{array}\right), \tag{8.190}$$

where $(n\text{-}k_{\text{x}})\times n$ submatrix $\boldsymbol{H}_{\text{x}}$ contains only X-operators, while $(n\text{-}k_{\text{z}})\times n$ submatrix $\boldsymbol{H}_{\text{z}}$ contains only Z-operators. The rank of the symplectic product of $\boldsymbol{A}_{\text{AE}}$ is given by

$$\begin{aligned}\text{rank}\{\boldsymbol{A}_{\text{EA}}\odot\boldsymbol{A}_{EA}^{\text{T}}\}&=\text{rank}\left\{\begin{pmatrix}\boldsymbol{H}_x\\ \boldsymbol{0}\end{pmatrix}\begin{pmatrix}\boldsymbol{0} & \boldsymbol{H}_z^{\text{T}}\end{pmatrix}+\begin{pmatrix}\boldsymbol{0}\\ \boldsymbol{H}_z\end{pmatrix}\begin{pmatrix}\boldsymbol{H}_x^{\text{T}} & \boldsymbol{0}\end{pmatrix}\right\}=\text{rank}\left\{\begin{pmatrix}0 & \boldsymbol{H}_x\boldsymbol{H}_z^{\text{T}}\\ 0 & 0\end{pmatrix}\right.\\ &\left.+\begin{pmatrix}0 & 0\\ \boldsymbol{H}_z\boldsymbol{H}_x^{\text{T}} & 0\end{pmatrix}\right\}=\text{rank}\{\boldsymbol{H}_x\boldsymbol{H}_z^{\text{T}}+\boldsymbol{H}_z\boldsymbol{H}_x^{\text{T}}\}=2e,\quad e=\text{rank}\{\boldsymbol{H}_x\boldsymbol{H}_z^{\text{T}}\}.\end{aligned} \tag{8.191}$$

The EA code parameters are $[N,K,e;D] = [n,n + e-(n-k_x)-(n-k_y),\ e;\ \min(d_x,d_z)] = [n,k_x + k_y-n + e,\ e;\ \min(d_x,d_z)]$.

Encoders and decoders for EA codes can be implemented in a fashion similar to that for subsystem codes, as described in Section 8.7: (1) using standard format formalism [8,10] and (2) using the conjugation method of Grassl et al. [38] (see also [27,64]).

8.11 Summary

This chapter has been devoted to stabilizer codes and their relatives. Stabilizer codes were introduced in Section 8.1. Their basic properties were discussed in Section 8.2, and encoded operations were introduced in the same section (Section 8.2). In Section 8.3, the finite geometry interpretation of stabilizer codes was introduced. This representation was used in Section 8.4 to introduce the standard form of the stabilizer code. The standard form is a basic representation for efficient encoder and decoder implementations, as discussed in Section 8.5. Section 8.6 was devoted to nonbinary stabilizer codes generalized from the previous sections. Subsystem codes were introduced in Section 8.7; also discussed is the efficient encoding and decoding of those codes. An important class of quantum codes, namely topological codes, was discussed in Section 8.8. Section 8.9 was devoted to surface codes. Finally, EA codes were discussed in Section 8.10.

References

[1] A.R. Calderbank, P.W. Shor, Good quantum error-correcting codes exist, Phys. Rev. A 54 (1996) 1098–1105.

[2] A.M. Steane, Error correcting codes in quantum theory, Phys. Rev. Lett. 77 (1996) 793.

[3] A.M. Steane, Simple quantum error-correcting codes, Phys. Rev. A 54 (6) (1996) 4741–4751.

[4] A.R. Calderbank, E.M. Rains, P.W. Shor, N.J.A. Sloane, Quantum error correction and orthogonal geometry, Phys. Rev. Lett. 78 (3) (1997) 405–408.

[5] E. Knill, R. Laflamme, Concatenated Quantum Codes, 1996. Available at: http://arxiv.org/abs/quant-ph/9608012.

[6] R. Laflamme, C. Miquel, J.P. Paz, W.H. Zurek, Perfect quantum error correcting code, Phys. Rev. Lett. 77 (1) (1996) 198–201.

[7] D. Gottesman, Class of quantum error correcting codes saturating the quantum Hamming bound, Phys. Rev. A 54 (1996) 1862–1868.

[8] D. Gottesman, Stabilizer Codes and Quantum Error Correction (Ph.D. Dissertation), California Institute of Technology, Pasadena, CA, 1997.

[9] I.B. Djordjevic, Quantum Information Processing, Quantum Computing, and Quantum Error Correction: An Engineering Approach, second ed., Elsevier/Academic Press, 2021.

[10] R. Cleve, D. Gottesman, Efficient computations of encoding for quantum error correction, Phys. Rev. A 56 (1997) 76–82.

[11] A.Y. Kitaev, Quantum error correction with imperfect gates, in: Quantum Communication, Computing, and Measurement, Plenum Press, New York, 1997, pp. 181–188.

[12] C.H. Bennett, D.P. DiVincenzo, J.A. Smolin, W.K. Wootters, Mixed-state entanglement and quantum error correction, Phys. Rev. A 54 (6) (1996) 3824–3851.

[13] D.J.C. MacKay, G. Mitchison, P.L. McFadden, Sparse-graph codes for quantum error correction, IEEE Trans. Inf. Theor. 50 (2004) 2315–2330.

[14] M.A. Neilsen, I.L. Chuang, Quantum Computation and Quantum Information, Cambridge University Press, Cambridge, 2000.

[15] F. Gaitan, Quantum Error Correction and Fault Tolerant Quantum Computing, CRC Press, 2008.

[16] G.D. Forney Jr., M. Grassl, S. Guha, Convolutional and tail-biting quantum error-correcting codes, IEEE Trans. Inf. Theor. 53 (2007) 865–880.

[17] A.R. Calderbank, E.M. Rains, P.W. Shor, N.J.A. Sloane, Quantum error correction via codes over GF(4), IEEE Trans. Inf. Theor. 44 (1998) 1369–1387.

[18] I.B. Djordjevic, Quantum LDPC codes from balanced incomplete block designs, IEEE Commun. Lett. 12 (2008) 389–391.

[19] I.B. Djordjevic, Photonic quantum dual-containing LDPC encoders and decoders, IEEE Photon. Technol. Lett. 21 (13) (2009) 842–844.

[20] I.B. Djordjevic, Photonic entanglement-assisted quantum low-density parity-check encoders and decoders, Opt. Lett. 35 (9) (2010) 1464–1466.

[21] S.A. Aly, A. Klappenecker, P.K. Sarvepalli, On quantum and classical BCH codes, IEEE Trans. Inf. Theor. 53 (3) (2007) 1183–1188.

[22] S.A. Aly, A. Klappenecker, P.K. Sarvepalli, Primitive quantum BCH codes over finite fields, in: Proc. Int. Symp. Inform. Theory 2006 (ISIT 2006), 2006, pp. 1114–1118.

[23] A. Ashikhmin, E. Knill, Nonbinary quantum stabilizer codes, IEEE Trans. Inf. Theor. 47 (7) (2001) 3065–3072.

[24] A. Ketkar, A. Klappenecker, S. Kumar, P.K. Sarvepalli, Nonbinary stabilizer codes over finite fields, IEEE Trans. Inf. Theor. 52 (11) (2006) 4892–4914.

[25] J.-L. Kim, J. Walker, Nonbinary quantum error-correcting codes from algebraic curves, Discrete Math. 308 (14) (2008) 3115–3124.

[26] P.K. Sarvepalli, A. Klappenecker, Nonbinary quantum Reed-Muller codes, in: Proc. Int. Symp. Inform. Theory 2005 (ISIT 2005), 2005, pp. 1023–1027.

[27] P.K. Sarvepalli, Quantum Stabilizer Codes and Beyond (Ph.D. Dissertation), Texas A&M University, August 2008.

[28] S.A.A. Aly Ahmed, Quantum Error Control Codes (Ph.D. Dissertation), Texas A&M University, May 2008.

[29] S.A. Aly, A. Klappenecker, Subsystem code constructions, in: Proc. ISIT 2008, Canada, Toronto, July 6–11, 2008, pp. 369–373.

[30] S. Bravyi, Subsystem codes with spatially local generators, Phys. Rev. A 83 (2011) 012320.

[31] A. Klappenecker, P.K. Sarvepalli, On subsystem codes beating the quantum Hamming or Singleton Bound, Proc. Math. Phys. Eng. Sci. 463 (2087) (2007) 2887–2905.

[32] A. Klappenecker, P.K. Sarvepalli, Clifford Code Construction of Operator Quantum Error Correcting Codes, April 2006 quant-ph/0604161.

[33] D.A. Lidar, I.L. Chuang, K.B. Whaley, Decoherence-free subspaces for quantum computation, Phys. Rev. Lett. 81 (12) (1998) 2594–2597.

[34] P.G. Kwiat, A.J. Berglund, J.B. Altepeter, A.G. White, Experimental verification of decoherence-free subspaces, Science 290 (2000) 498–501.

[35] L. Viola, E.M. Fortunato, M.A. Pravia, E. Knill, R. Laflamme, D.G. Cory, Experimental realization of noiseless subsystems for quantum information processing, Science 293 (2001) 2059–2063.

[36] D. Bacon, Operator quantum error correcting subsystems for self-correcting quantum memories, Phys. Rev. A 73 (1) (2006) 012340.

[37] D. Bacon, A. Casaccino, Quantum error correcting subsystem codes from two classical linear codes, in: Proc. *Forty-Fourth Annual Allerton Conference on Communication, Control, and Computing*, 2006, pp. 520–527. Monticello, Illinois.

[38] M. Grassl, M. Rötteler, T. Beth, Efficient quantum circuits for non-qubit quantum error-correcting codes, Int. J. Found. Comput. Sci. 14 (5) (2003) 757–775.
[39] A. Harrow, Superdense coding of quantum states, Phys. Rev. Lett. 92 (18) (2004) 187901.
[40] E. Knill, R. Laflamme, Theory of quantum error-correcting codes, Phys. Rev. A 55 (1997) 900.
[41] J. Preskill, Ph219/CS219 Quantum Computing (Lecture Notes), Caltech, 2009. Available at: http://theory.caltech.edu/people/preskill/ph229/.
[42] A. Kitaev, Topological quantum codes and anyons, in: Quantum Computation: A Grand Mathematical Challenge for the Twenty-First Century and the Millennium (Washington, DC, 2000), Volume 58 of Proc. Sympos. Appl. Math., Amer. Math. Soc., Providence, RI, 2002, pp. 267–272.
[43] A.Y. Kitaev, Fault-tolerant quantum computation by anyons, Ann. Phys. 303 (1) (2003) 2–30. Available at: http://arxiv.org/abs/quant-ph/9707021.
[44] S.B. Bravyi, A.Y. Kitaev, Quantum Codes on a Lattice With Boundary, 1998. Available at, http://arxiv.org/abs/quant-ph/9811052.
[45] H. Bombin, Topological subsystem codes, Phys. Rev. A 81 (2010) 032301.
[46] G. Duclos-Cianci, D. Poulin, Fast decoders for topological subsystem codes, Phys. Rev. Lett. 104 (2010) 050504.
[47] G. Duclos-Cianci, D. Poulin, A renormalization group decoding algorithm for topological quantum codes, in: Proc. 2010 IEEE Information Theory Workshop-ITW 2010, August 30, 2010–September3, 2010 (Dublin, Ireland).
[48] M. Suchara, S. Bravyi, B. Terhal, Constructions and noise threshold of topological subsystem codes, J. Phys. Math. Theor. 44 (2011) 26. Paper no. 155301.
[49] P. Sarvepalli, Topological color codes over higher alphabet, in: Proc. 2010 IEEE Information Theory Workshop-ITW 2010, 2010 (Dublin, Ireland).
[50] F. Hernando, M.E. O'Sullivan, E. Popovici, S. Srivastava, Subfield-subcodes of generalized toric codes, in: Proc. *ISIT 2010*, June 13–18, 2010, pp. 1125–1129. Austin, Texas, U.S.A.
[51] H. Bombin, Topological order with a twist: ising anyons from an abelian model, Phys. Rev. Lett. 105 (2010) 030403.
[52] A.G. Fowler, M. Mariantoni, J.M. Martinis, A.N. Cleland, Surface codes: towards practical large-scale quantum computation, Phys. Rev. A 86 (3) (2012) 032324.
[53] D. Litinski, A game of surface codes: large-scale quantum computing with lattice surgery, Quantum 3 (2019) 128.
[54] A.J. Landahl, C. Ryan-Anderson, Quantum Computing by Color-Code Lattice Surgery, 2014. Available at, https://arxiv.org/abs/1407.5103.
[55] C. Horsman, A.G. Fowler, S. Devitt, R.V. Meter, Surface code quantum computing by lattice surgery, New J. Phys. 14 (2012) 123011.
[56] A.G. Fowler, C. Gidney, Low Overhead Quantum Computation Using Lattice Surgery, 2018. Available at: https://arxiv.org/abs/1808.06709.
[57] R. Raussendorf, D.E. Browne, H.J. Briegel, Measurement-based quantum computation with cluster states, Phys. Rev. A 68 (2003) 022312.
[58] S. Bravyi, A. Kitaev, Universal quantum computation with ideal Clifford gates and noisy ancillas, Phys. Rev. A 71 (2) (2005) 022316.
[59] T. Brun, I. Devetak, M.H. Hsieh, Correcting quantum errors with entanglement, Science 314 (2006) 436–439.
[60] M.-H. Hsieh, I. Devetak, T. Brun, General entanglement-assisted quantum error correcting codes, Phys. Rev. A 76 (2007) 062313.
[61] M.-H. Hsieh, Entanglement-Assisted Coding Theory (Ph.D. Dissertation), University of Southern California, August 2008.

[62] I. Devetak, T.A. Brun, M.-H. Hsieh, "Entanglement-assisted quantum error-correcting codes, in: V. Sidoravičius (Ed.), *New Trends in Mathematical Physics*, Selected Contributions of the XVth International Congress on Mathematical Physics, Springer, 2009, pp. 161–172.
[63] M.-H. Hsieh, W.-T. Yen, L.-Y. Hsu, High performance entanglement-assisted quantum LDPC codes need little entanglement, IEEE Trans. Inf. Theory 57 (3) (2011) 1761–1769.
[64] M.M. Wilde, Quantum Coding with Entanglement (Ph.D. Dissertation), University of Southern California, August 2008.

Further reading

[1] M.M. Wilde, T.A. Brun, Optimal entanglement formulas for entanglement-assisted quantum coding, Phys. Rev. A 77 (2008). Paper 064302.
[2] M.M. Wilde, H. Krovi, T.A. Brun, Entanglement-assisted quantum error correction with linear optics, Phys. Rev. A 76 (2007) 052308.
[3] M.M. Wilde, T.A. Brun, Protecting quantum information with entanglement and noisy optical modes, Quant. Inf. Process. 8 (2009) 401–413.
[4] M.M. Wilde, T.A. Brun, Entanglement-assisted quantum convolutional coding, Phys. Rev. A 81 (2010). Paper 042333.
[5] M.M. Wilde, D. Fattal, Nonlocal quantum information in bipartite quantum error correction, Quant. Inf. Process. 9 (2010) 591–610.
[6] Y. Fujiwara, D. Clark, P. Vandendriessche, M. De Boeck, V.D. Tonchev, Entanglement-assisted quantum low-density parity-check codes, Phys. Rev. A 82 (4) (2010). Paper 042338.
[7] C.H. Bennett, S.J. Wiesner, Communication via one- and two-particle operators on Einstein-Podolsky-Rosen states, Phys. Rev. Lett. 69 (1992) 2881–2884.
[8] M.M. Wilde, Quantum-shift-register circuits, Phys. Rev. A 79 (6) (2009). Paper 062325.
[9] S. Taghavi, T.A. Brun, D.A. Lidar, Optimized entanglement-assisted quantum error correction, Phys. Rev. A 82 (4) (2010). Paper 042321.
[10] D. Bacon, A. Casaccino, Quantum error correcting subsystem codes from two classical linear codes, in: Proc. *Forty-Fourth Annual Allerton Conference*, Allerton House, UIUC, Illinois, USA, September 27–29, 2006, pp. 520–527.
[11] D. Kribs, R. Laflamme, D. Poulin, Unified and generalized approach to quantum error correction, Phys. Rev. Lett. 94 (18) (2005) paper 180501.
[12] M.A. Nielsen, D. Poulin, Algebraic and information-theoretic conditions for operator quantum error correction, Phys. Rev. A 75 (2007). Paper 064304.
[13] T. Camara, H. Ollivier, J.-P. Tillich, A class of quantum LDPC codes: construction and performances under iterative decoding, in: Proc. ISIT 2007, June 24–29, 2007, pp. 811–815.
[14] S. Lin, S. Zhao, A class of quantum irregular LDPC codes constructed from difference family, in: Proc. ICSP, October 24–28, 2010, pp. 1585–1588.
[15] W. Bosma, J.J. Cannon, C. Playoust, The magma algebra system I: the user language, J. Symbolic Comput. 4 (1997) 235–266.

CHAPTER

Quantum low-density parity-check codes

9

Chapter outline

9.1 Classical low-density parity-check codes

Codes on graphs, such as turbo codes [1] and low-density parity-check (LDPC) codes [2−14], have revolutionized communications and are becoming standard in many applications. LDPC codes, invented by Gallager in the 1960s, are linear block codes for which the parity-check matrix has a low density of 1's [4]. LDPC codes have recently generated great interest within the coding community [2−21], and this has resulted in a great deal of understanding of the various aspects of LDPC codes and their decoding process. An iterative LDPC decoder based on the sum-product algorithm (SPA) has been shown to achieve performance as close to the Shannon limit as 0.0045 dB [20]. The inherently low complexity of this decoder opens up avenues for its use in various high-speed applications. Because of their low decoding complexity, LDPC codes are intensively studied for quantum applications as well.

If the parity-check matrix has a low density of 1's, and the number of 1's per row and per column are both constant, the code is said to be a *regular LDPC* code. To facilitate implementation at high speed, we prefer regular rather than irregular LDPC codes. The graphical representation of LDPC

Quantum Communication, Quantum Networks, and Quantum Sensing. https://doi.org/10.1016/B978-0-12-822942-2.00004-2

codes, known as bipartite (Tanner) graph representation, is helpful in the efficient description of LDPC decoding algorithms. A *bipartite (Tanner) graph* is a graph whose nodes may be separated into two classes (*variable* and *check* nodes), and where *undirected edges* may only connect two nodes not residing in the same class. The Tanner graph is drawn according to the following rule: check (function) node c is connected to variable (bit) node v whenever element h_{cv} in a parity-check matrix $\boldsymbol{H}$ is a 1. In an $m \times n$ parity-check matrix, there are $m = n - k$ check nodes and n variable nodes.

Example 1. As an illustrative example, consider the $\boldsymbol{H}$-matrix of the following code:

$$\boldsymbol{H} = \begin{bmatrix} 1 & 0 & 1 & 0 & 1 & 0 \\ 1 & 0 & 0 & 1 & 0 & 1 \\ 0 & 1 & 1 & 0 & 0 & 1 \\ 0 & 1 & 0 & 1 & 1 & 0 \end{bmatrix}$$

For any valid codeword $x = [x_0, x_1, \ldots, x_{n-1}]$, the checks used to decode the codeword are written as

- Equation (c_0): $x_0 + x_2 + x_4 = 0 \pmod 2$
- Equation (c_1): $x_0 + x_3 + x_5 = 0 \pmod 2$
- Equation (c_2): $x_1 + x_2 + x_5 = 0 \pmod 2$
- Equation (c_3): $x_1 + x_3 + x_4 = 0 \pmod 2$

The bipartite graph (Tanner graph) representation of this code is given in Fig. 9.1A. The circles represent the bit (variable) nodes, while squares represent the check (function) nodes. For example, the variable nodes x_0, x_2, and x_4 are involved in Eq. (c_0) and therefore connected to the check node c_0.

A closed path in a bipartite graph comprising l edges that closes back on itself is called a *cycle of length l*. The shortest cycle in the bipartite graph is called the *girth*. The girth influences the minimum distance of LDPC codes, correlates the extrinsic LLRs, and therefore affects decoding performance. Large-girth LDPC codes are preferable because the large girth increases the minimum distance and decorrelates the extrinsic info in the decoding process. To improve the iterative decoding performance, we must avoid cycles of length 4 and preferably of 6 as well. To check for the existence of short cycles, one has to search over the $\boldsymbol{H}$-matrix for the patterns shown in Fig. 9.1B and C.

In the rest of this section, we describe a method for designing large-girth quasi-cyclic (QC) LDPC codes (Subsection 9.1.1); and an efficient and simple variant of SPA suitable for use in optical communications, namely the min-sum-with-correction term algorithm (Subsection 9.1.2). We also evaluate the bit error rate (BER) performance of QC-LDPC codes against concatenated codes and turbo-product codes (TPCs) (Subsection 9.1.3). Nonbinary LDPC codes are discussed in Subsection 9.1.4. The LDPC code designs are reviewed in Subsection 9.1.5.

9.1.1 Large-girth quasi-cyclic binary low-density parity-check codes

Based on Tanner's bound for the minimum distance of an LDPC code [22]

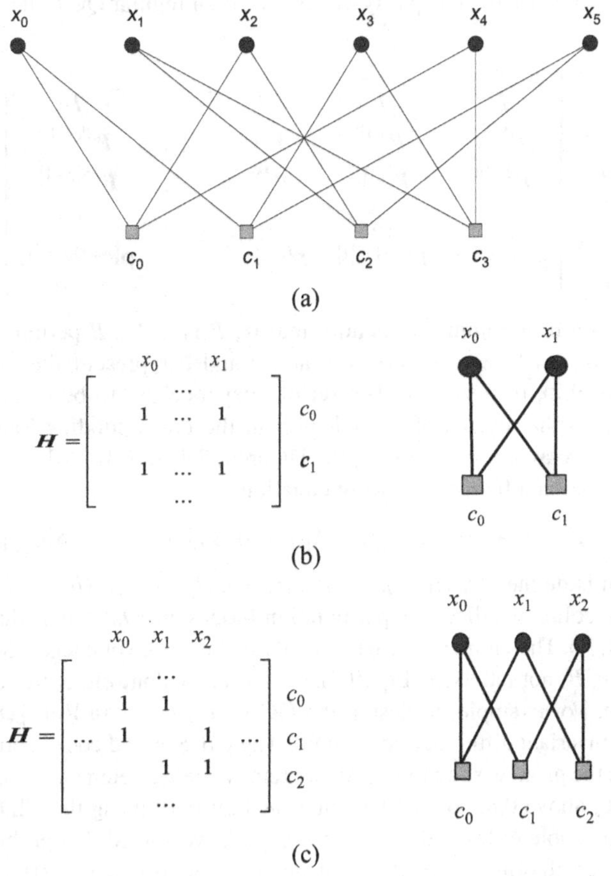

FIGURE 9.1

(A) Bipartite graph of (6, 2) code described by ***H*** matrix above. Cycles in a Tanner graph: (B) cycle of length 4 and (C) cycle of length 6.

$$d \geq \begin{cases} 1 + \dfrac{w_c}{w_c - 2}\left((w_c - 1)^{\lfloor (g-2)/4 \rfloor} - 1\right), & g/2 = 2m + 1 \\ 1 + \dfrac{w_c}{w_c - 2}\left((w_c - 1)^{\lfloor (g-2)/4 \rfloor} - 1\right) + (w_c - 1)^{\lfloor (g-2)/4 \rfloor}, & g/2 = 2m \end{cases} \tag{9.1}$$

(where g and w_c denote the girth of the code graph and the column weight, respectively, and d stands for the minimum distance of the code), it follows that large girth leads to an exponential increase in the minimum distance, provided that the column weight is at least 3. ($\lfloor\,\rfloor$ denotes the largest integer less than or equal to the enclosed quantity.) For example, the minimum distance of girth-10 codes with column weight $r = 3$ is at least 10. The parity-check matrix of regular QC LDPC codes [23,24] can be represented by

$$H = \begin{bmatrix} I & I & I & \cdots & I \\ P^{S[0]} & P^{S[1]} & P^{S[2]} & \cdots & P^{S[c-1]} \\ P^{2S[0]} & P^{2S[1]} & P^{2S[2]} & \cdots & P^{2S[c-1]} \\ \cdots & \cdots & \cdots & \cdots & \cdots \\ P^{(r-1)S[0]} & P^{(r-1)S[1]} & P^{(r-1)S[2]} & \cdots & P^{(r-1)S[c-1]} \end{bmatrix}, \tag{9.2}$$

where I is $B \times B$ (B is a prime number) identity matrix, P is a $B \times B$ permutation matrix given by $P = (p_{ij})_{B\times B}$, $p_{i,i+1} = p_{B,1} = 1$ (zero otherwise), and r and c represent the number of block-rows and block-columns in (9.2), respectively. The set of integers S are to be carefully chosen from the set {0, 1, ..., $B - 1$} so that cycles of short length in the corresponding Tanner (bipartite) graph representation of (9.2) are avoided. According to Theorem 2.1 in Ref. [24], we must avoid cycles of length $2k$ ($k = 3$ or 4) defined by the following equation:

$$S[i_1]j_1 + S[i_2]j_2 + \cdots + S[i_k]j_k = S[i_1]j_2 + S[i_2]j_3 + \cdots + S[i_k]j_1 \bmod B, \tag{9.3}$$

where the closed path is defined by (i_1, j_1), (i_1, j_2), (i_2, j_2), (i_2, j_3), ..., (i_k, j_k), (i_k, j_1) with the pair of indices denoting row-column indices of permutation-blocks in (9.2) such that $l_m \neq l_{m+1}$, $l_k \neq l_1$ ($m = 1, 2, \ldots, k$; $l \in \{i, j\}$). Therefore, we have to identify the sequence of integers $S[i] \in \{0, 1, \ldots, B-1\}$ ($i = 0, 1, \ldots, r-1$; $r < B$) not satisfying Eq. (9.3), which can be done either by computer search or in a combinatorial fashion. For example, to design the QC LDPC codes in Ref. [18], we introduced the concept of the cyclic-invariant difference set (CIDS). The CIDS-based codes come naturally as girth-6 codes, and to increase the girth, we had to selectively remove certain elements from a CIDS. The design of LDPC codes of rate above 0.8, column weight 3, and girth 10 using the CIDS approach is a challenging and still open problem. Instead, in our paper [23], we solved this problem by developing an efficient computer search algorithm. We add an integer at a time from the set {0, 1, ..., $B - 1$} (not used before) to the initial set S and check if Eq. (9.3) is satisfied. If Eq. (9.3) is satisfied, we remove that integer from set S and continue our search with another integer from set {0, 1, ..., $B - 1$} until we exploit all the elements from {0, 1, ..., $B - 1$}. The code rate of these QC codes, R, is lower-bounded by

$$R \geq \frac{|S|B - rB}{|S|B} = 1 - r/|S|, \tag{9.4}$$

and the codeword length is $|S|B$, where $|S|$ denotes the cardinality of set S. For a given code rate R_0, the number of elements from S to be used is $\lfloor r/(1 - R_0) \rfloor$. With this algorithm, LDPC codes with arbitrary rates can be designed.

Example 2. By setting $B = 2311$, the set of integers to be used in (9.2) is obtained as $S = \{1, 2, 7, 14, 30, 51, 78, 104, 129, 212, 223, 318, 427, 600, 808\}$. The corresponding LDPC code has rate $R_0 = 1{-}3/15 = 0.8$, column weight 3, girth 10, and length $|S|B = 15 \cdot 2311 = 34{,}665$. In the example above, the initial set of integers was $S = \{1, 2, 7\}$, and the set of rows to be used in (9.2) is {1, 3, 6}. Using a different initial set will result in a different set from that obtained above.

Example 3. By setting $B = 269$, the set S is obtained as $S = \{0, 2, 3, 5, 9, 11, 12, 14, 27, 29, 30, 32, 36, 38, 39, 41, 81, 83, 84, 86, 90, 92, 93, 95, 108, 110, 111, 113, 117, 119, 120, 122\}$. If 30 integers are used, the corresponding LDPC code has rate $R_0 = 1-3/30 = 0.9$, column weight 3, girth 8, and length $30 \cdot 269 = 8070$.

9.1.2 Decoding of binary low-density parity-check codes

This subsection describes the min-sum with correction term decoding algorithm [25,26]. It is a simplified version of the original algorithm proposed by Gallager [4]. Gallager proposed a near-optimal iterative decoding algorithm for LDPC codes that computes the distributions of the variables to calculate the a posteriori *probability* (APP) of a bit v_i of a codeword $v = [v_0, v_1, \ldots, v_{n-1}]$ to be equal to 1, given a received vector $y = [y_0, y_1, \ldots, y_{n-1}]$. This iterative decoding scheme engages the passing of the extrinsic info back and forth among the c-nodes and the v-nodes over the edges to update the distribution estimation. Each iteration in this scheme is composed of two half-iterations. In Fig. 9.2, we illustrate both the first and the second halves of an iteration of the algorithm. As an example, in Fig. 9.2A, we show the message sent from v-node v_i to the c-node c_j. The v_i-node collects the information from the channel (y_i sample) in addition to extrinsic info from other c-nodes connected to the v_i-node, processes them, and sends the extrinsic info (not already available info) to c_j. This extrinsic info contains the information about the probability $\Pr(c_i = b|y_0)$, where $b \in \{0, 1\}$. This is performed in all c-nodes connected to v_i-node. On the other hand, Fig. 9.2B shows the extrinsic info sent from c-node c_i to the v-node v_j that contains information about $\Pr(c_i$ equation is satisfied $| \mathbf{y})$. This is done repeatedly to all the c-nodes connected to the v_i-node.

After this intuitive description, we describe the min-sum-with-correction-term algorithm in more detail [25] because of its simplicity and suitability for high-speed implementation. Generally, we can either compute *the* APP $\Pr(v_i|\mathbf{y})$ or the APP ratio $l(v_i) = \Pr(v_i = 0|\mathbf{y})/\Pr(v_i = 1|\mathbf{y})$, which is also referred to as the likelihood ratio. In the log-domain version of the sum-product algorithm, we replace these likelihood ratios with log-likelihood ratios (LLRs) because the probability domain includes many multiplications that lead to numerical instabilities, whereas the computation using LLR computation involves addition only. Moreover, the log-domain representation is more suitable for finite precision representation. Thus, we compute LLRs by $L(v_i) = \log[\Pr(v_i = 0|\mathbf{y})/\Pr(v_i = 1|\mathbf{y})]$. For the final decision, if $L(v_i) > 0$, we decide in favor of 0, and if $L(v_i) < 0$, we decide in favor of 1. To further explain the algorithm, we introduce the following notations from MacKay [27] (see also [25]):

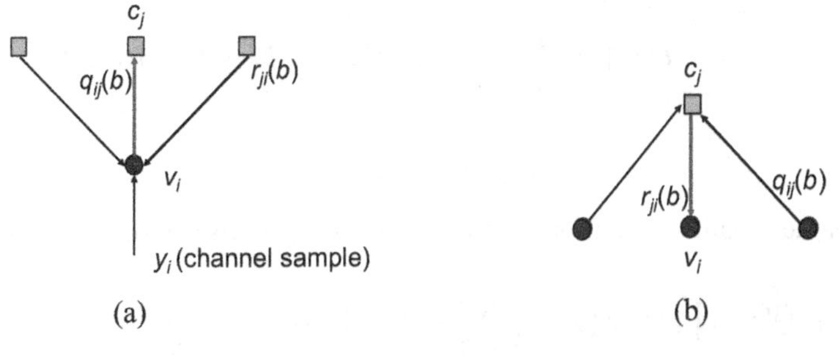

FIGURE 9.2

Illustration of the half-iterations of the sum-product algorithm: (A) first half-iteration: extrinsic info sent from v-nodes to c-nodes, and (B) second half-iteration: extrinsic info sent from c-nodes to v-nodes.

$V_j = \{v\text{-nodes connected to } c\text{-node } c_j\}$
$V_j\backslash i = \{v\text{-nodes connected to } c\text{-node } c_j\}\backslash\{v\text{-node } v_i\}$
$C_i = \{c\text{-nodes connected to } v\text{-node } v_i\}$
$C_i\backslash j = \{c\text{-nodes connected to } v\text{-node } v_i\}\backslash\{c\text{-node } c_j\}$
$M_v(\sim i) = \{\text{messages from all } v\text{-nodes except node } v_i\}$
$M_c(\sim j) = \{\text{messages from all } c\text{-nodes except node } c_j\}$
$P_i = \Pr(v_i = 1|y_i)$
$S_i =$ event that the check equations involving c_i are satisfied
$q_{ij}(b) = \Pr(v_i = b|S_i, y_i, M_c(\sim j))$
$r_{ji}(b) = \Pr(\text{check equation } c_j \text{ is satisfied}|v_i = b, M_v(\sim i))$

In the log-domain version of the sum-product algorithm, all the calculations are performed in the log-domain as follows:

$$L(v_i) = \log\left[\frac{Pr(v_i = 0|y_i)}{Pr(v_i = 1|y_i)}\right], L(r_{ji}) = \log\left[\frac{r_{ji}(0)}{r_{ji}(1)}\right], L\left(q_{ji}\right) = \log\left[\frac{q_{ji}(0)}{q_{ji}(1)}\right]. \tag{9.5}$$

The algorithm starts with the initialization step where we set $L(v_i)$ as follows:

$$\begin{aligned} &L(v_i) = (-1)^{y_i}\log\left(\frac{1-\varepsilon}{\varepsilon}\right), \quad \text{for BSC} \\ &L(v_i) = 2y_i/\sigma^2, \quad \text{for binary input AWGN} \\ &L(v_i) = \log\left(\frac{\sigma_1}{\sigma_0}\right) - \frac{(y_i-\mu_0)^2}{2\sigma_0^2} + \frac{(y_i-\mu_1)^2}{2\sigma_1^2}, \quad \text{for BA}-\text{AWGN} \\ &L(v_i) = \log\left(\frac{Pr(v_i = 0|y_i)}{Pr(v_i = 1|y_i)}\right), \text{for abritrary channel} \end{aligned} \tag{9.6}$$

where ε is the probability of error in the binary symmetric channel (BSC), σ^2 is the variance of the Gaussian distribution of the additive white Gaussian noise (AWGN), and μ_j and σ_j^2 ($j = 0, 1$) represent the mean and variance of the Gaussian process corresponding to the bits $j = 0, 1$ of a binary asymmetric (BA)-AWGN channel. After initialization of $L(q_{ij})$, we calculate $L(r_{ji})$ as follows:

$$\begin{aligned} L(r_{ji}) &= L\left(\sum_{i'\in V_j\backslash i} b_{i'}\right) = L(\ldots \oplus b_k \oplus b_l \oplus b_m \oplus b_n \ldots) \\ &= \ldots L_k \boxplus L_l \boxplus L_m \boxplus L_n \boxplus \ldots \end{aligned} \tag{9.7}$$

where $\oplus$ denotes the modulo-2 addition, and $\boxplus$ denotes a pairwise computation defined by

$$L_1 \boxplus L_2 = \prod_{k=1}^{2} \text{sign}(L_k)\cdot\phi\left(\sum_{k=1}^{2}\phi(|L_k|)\right), \quad \phi(x) = -\log\tanh(x/2) \tag{9.8}$$

Upon calculation of $L(r_{ji})$, we update as

$$L\left(q_{ij}\right)=L(v_i)+\sum_{j'\in C_i\backslash j}L\left(r_{j'i}\right),L(Q_i)=L(v_i)+\sum_{j\in C_i}L\left(r_{ji}\right) \tag{9.9}$$

Finally, the decision step is as follows:

$$\widehat{v}_i=\begin{cases}1, & L(Q_i)<0\\ 0, & \text{otherwise}\end{cases} \tag{9.10}$$

If the syndrome equation $\widehat{\boldsymbol{v}}\boldsymbol{H}^T=\boldsymbol{0}$ is satisfied or the maximum number of iterations is reached, we stop; otherwise, we recalculate $L(r_{ji})$ and update $L(q_{ij})$ and $L(Q_i)$ and check again. It is important to set the number of iterations high enough to ensure that most codewords are decoded correctly and low enough not to affect processing time. It is important to mention that the decoder for good LDPC codes requires a smaller number of iterations to guarantee successful decoding.

The **Gallager log-domain SPA** can be formulated as follows:

1. Initialization: For $j=0, 1, \ldots, n-1$; initialize the messages to be sent from v-node i to v-node j to channel LLRs, namely $L\left(q_{ij}\right)=L(v_i)$.
2. The c-node update rule: For $j=0, 1, \ldots, n-k-1$; compute $L\left(r_{ji}\right)=\left(\prod_{i'\in R_j\backslash i}\alpha_{i'j}\right)\phi\left[\sum_{i'\in R_j\backslash i}\phi\left(\beta_{i'j}\right)\right]$, where $\alpha_{ij}=\text{sign}\left[L\left(q_{ij}\right)\right]$, $\beta_{ij}=\left|L\left(q_{ij}\right)\right|$, and $\phi(x)=-\log\tanh(x/2)=\log[(e^x+1)/(e^x-1)]$
3. The v-node update rule: For $i=0, 1, \ldots, n-1$; set $L\left(q_{ij}\right)=L(v_i)+\sum_{j'\in C_i\backslash j}L\left(r_{j'i}\right)$ for all c-nodes for which $h_{ji}=1$.
4. Bit decisions: Update $L(Q_i)$ $(i=0, \ldots, n-1)$ by $L(Q_i)=L(v_i)+\sum_{j\in C_i}L\left(r_{ji}\right)$ and set $\widehat{v}_i=1$ when $L(Q_i)<0$ (otherwise, $\widehat{v}_i=0$). If $\widehat{\boldsymbol{v}}\boldsymbol{H}^T=0$ or a predetermined number of iterations has been reached, then stop; otherwise, go to step 1.

Because the c-node update rule involves log and tanh functions, it is computationally intensive, and many approximations exist. Very popular of these is the **min-sum-plus-correction-term approximation** [6]. Namely, it can be shown that the "box-plus" operator $\boxplus$ can be calculated by

$$L_1\boxplus L_2=\prod_{k=1}^{2}\text{sign}(L_k)\cdot\min(|L_1|,|L_2|)+c(x,y), \tag{9.11}$$

where $c(x,y)$ denotes the correction factor defined by

$$c(x,y)=\log[1+\exp(-|x+y|)]-\log[1+\exp(-|x-y|)], \tag{9.12}$$

commonly implemented as a look-up table.

9.1.3 Bit error rate performance of binary low-density parity-check codes

The results of simulations for an AWGN channel model are given in Fig. 9.3, where we compare large-girth LDPC codes (Fig. 9.3A) with RS codes, concatenated RS codes, TPCs, and other classes of

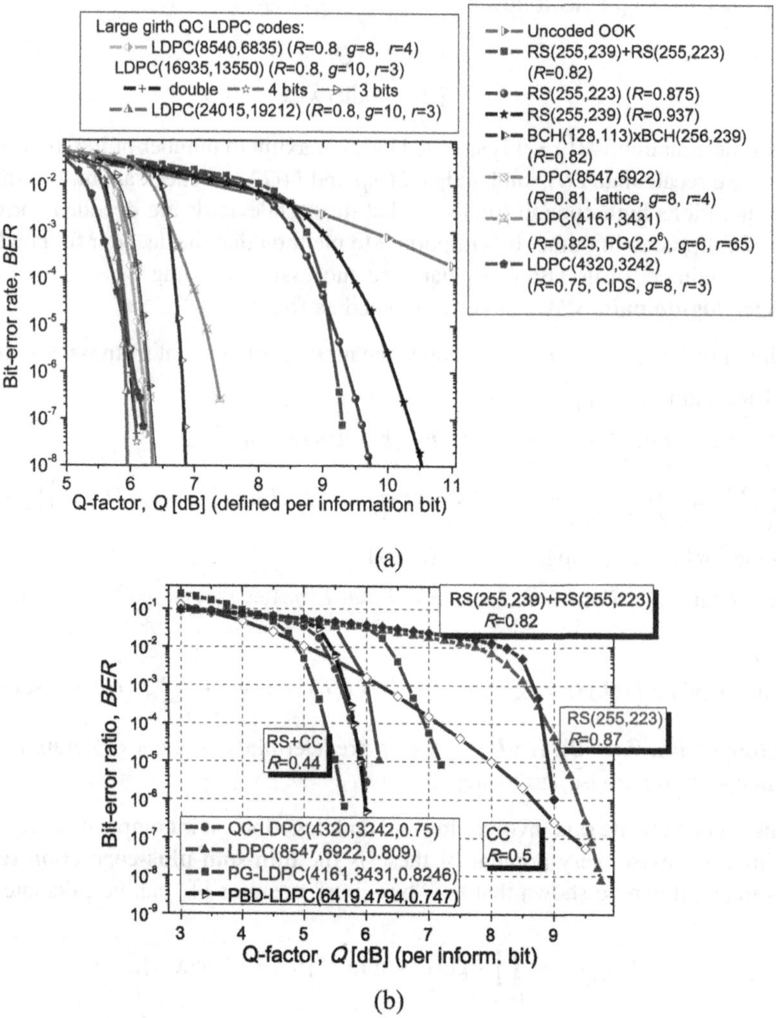

FIGURE 9.3

(A) Large girth quasi-cyclic low-density parity-check (QC-LDPC) codes against RS codes, concatenated RS codes, TPCs, and girth-6 LDPC codes on an addditive white Gaussian noise (AWGN) channel model, and (B) LDPC codes versus convolutional, concatenated RS, and concatenation of convolutional and RS codes on an AWGN channel. The number of iterations in the sum-product-with-correction-term algorithm was set to 25. The Q-factor is defined as $Q = (\mu_1 - \mu_0)/(\sigma_1 + \sigma_0)$, where μ_i is the mean value corresponding to bit i ($i = 0, 1$) and σ_i is the corresponding standard deviation.

LDPC codes. In all simulation results in this section, we maintained double precision. For the LDPC(16935, 13550) code, we also provided 3- and 4-bit fixed-point simulation results (see Fig. 9.3A). Our results indicate that the 4-bit representation has performance comparable to double-precision representation, whereas the 3-bit representation performs 0.27 dB worse than double-precision representation at a BER of $2 \cdot 10^{-8}$. The girth-10 LDPC(24015, 19212) code of rate 0.8 outperforms the concatenation RS(255, 239) + RS(255, 223) (of rate 0.82) by 3.35 dB and RS(255, 239) by 4.75 dB, both at a BER of 10^{-7}. The same LDPC code outperforms projective geometry (PG) $(2,2^6)$-based LDPC(4161, 3431) (of rate 0.825) of girth 6 by 1.49 dB at BER of 10^{-7} and outperforms CIDS-based LDPC(4320, 3242) of rate 0.75 and girth-8 LDPC codes by 0.25 dB. At a BER of 10^{-10}, it outperforms lattice-based LDPC(8547, 6922) of rate 0.81 and girth-8 LDPC code by 0.44 dB, and BCH(128, 113) × BCH(256, 239) TPC of rate 0.82 by 0.95 dB. The net coding gain at a BER of 10^{-12} is 10.95 dB. In Fig. 9.3B, various LDPC codes are compared with RS (255, 223) code, concatenated RS code of rate 0.82, and convolutional code (CC) (of constraint length 5). It can be seen that LDPC codes, both regular and irregular, offer much better performance than hard-decision codes. It should be noted that pairwise balanced design (PBD) [58] based irregular LDPC code of rate 0.75 is only 0.4 dB away from the concatenation of convolutional-RS codes (denoted in Fig. 9.3B as RS + CC) with a significantly lower code rate of $R = 0.44$ at BER of 10^{-6}. As expected, irregular LDPC codes (black colored curves) outperform regular LDPC codes.

A primary problem in decoder implementation for large-girth *binary* LDPC codes is excessive codeword length. To solve this problem, we consider *nonbinary* LDPC codes over $GF(2^m)$. By designing codes over higher-order fields, we aim to achieve coding gains comparable to binary LDPC codes but for shorter codeword lengths. Also note that 4-ary LDPC codes (over GF(4)) can be straightforwardly related to entanglement-assisted quantum LDPC codes. Moreover, properly designed classical nonbinary LDPC codes can be used as nonbinary stabilizer codes.

9.1.4 Nonbinary low-density parity-check codes

The parity-check matrix H of a nonbinary QC-LDPC code can be organized as an array of submatrices of equal size as in (9.13), where $H_{i,j},\ 0 \leq i < \gamma,\ 0 \leq j < \rho$, is a $B \times B$ submatrix in which each row is a cyclic shift of the row preceding it. This modular structure can be exploited to facilitate hardware implementation of the decoders of QC-LDPC codes [7,8]. Furthermore, the QC nature of their generator matrices enables encoding of QC-LDPC codes to be performed in linear time using simple shift-register-based architectures [8,11].

$$\boldsymbol{H} = \begin{bmatrix} \boldsymbol{H}_{0,0} & \boldsymbol{H}_{0,1} & \cdots & \boldsymbol{H}_{0,\rho-1} \\ \boldsymbol{H}_{1,0} & \boldsymbol{H}_{1,1} & \cdots & \boldsymbol{H}_{1,\rho-1} \\ \vdots & \vdots & \ddots & \vdots \\ \boldsymbol{H}_{\gamma-1,0} & \boldsymbol{H}_{\gamma-1,1} & \cdots & \boldsymbol{H}_{\gamma-1,\rho-1} \end{bmatrix}. \tag{9.13}$$

If we select entries of $\boldsymbol{H}$ from the binary field GF(2) as in the previous section, the resulting QC-LDPC code is a binary LDPC code. On the other hand, if the selection is made from the Galois field of q elements denoted by GF(q), we obtain a q-ary QC-LDPC code. To decode binary LDPC codes, as described above, the iterative message-passing algorithm is referred to as the SPA. For nonbinary

LDPC codes, a variant of the SPA known as the q-ary SPA (QSPA) is used [9]. When the field order is a power of 2, i.e., $q = 2^m$, where m is an integer and $m \geq 2$, a fast Fourier transform (FFT)-based implementation of QSPA, referred to as FFT-QSPA, significantly reduces the computational complexity of QSPA. FFT-QSPA is further analyzed and improved in Ref. [10]. A mixed-domain FFT-QSPA implementation, in short MD-FFT-QSPA, aims to reduce hardware implementation complexity by transforming multiplications in the probability domain into additions in the log domain whenever possible. It also avoids the instability issues commonly faced in probability-domain implementations.

Following the code design discussed above, we generated (3,15)-regular, girth-8 LDPC codes over the fields GF(2^p), where $0 \leq p \leq 7$. All the codes had a code rate (R) of at least 0.8 and hence an overhead OH $= (1/R - 1)$ of 25% or less. We compared the BER performance of these codes against each other and against some other well-known codes, such as RS(255, 239), RS(255, 223) and concatenated RS(255, 239) + RS(255, 223) codes; and BCH(128, 113) × BCH(256, 239) TPC. We used the binary AWGN (BI-AWGN) channel model in our simulations and set the maximum number of iterations to 50. In Fig. 9.4A, we present the BER performance of the set of nonbinary LDPC codes discussed above. Using the figure, we conclude that when we fix the girth of a nonbinary regular, rate-0.8 LDPC code at 8, increasing the field order above 8 exacerbates BER performance. In addition to having better BER performance than codes over higher-order fields, codes over GF(4) have smaller decoding complexities when decoded using the MD-FFT-QSPA algorithm since the complexity of this

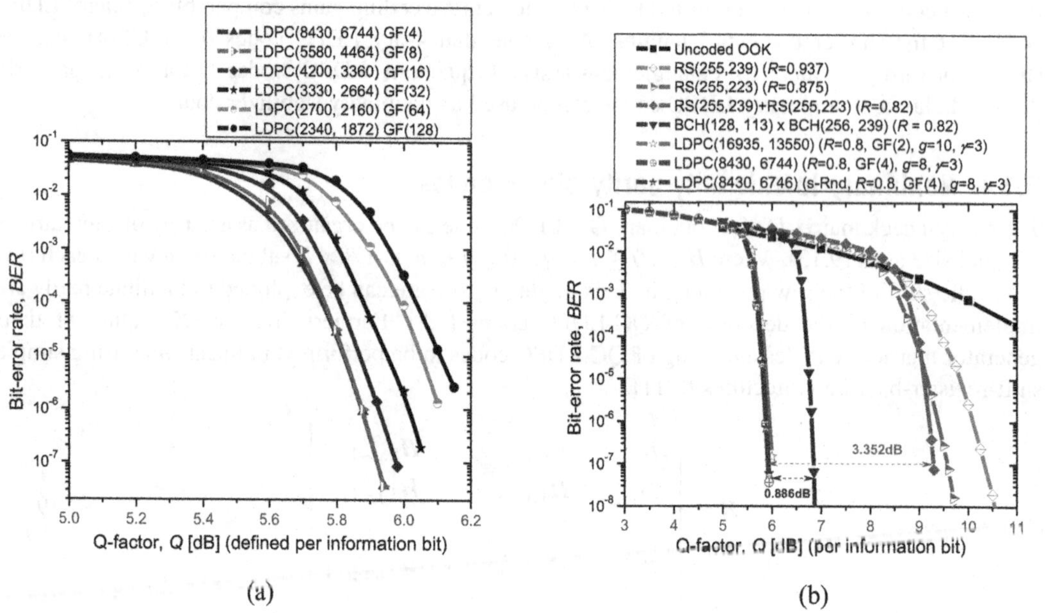

FIGURE 9.4

(A) Comparison of nonbinary, (3,15)-regular, girth-8 low-density parity-check (LDPC) codes over BI-AWGN channel. (B) Comparison of 4-ary (3,15)-regular, girth-8 LDPC codes; a binary, girth-10 LDPC code, three RS codes, and a TPC code.

algorithm is proportional to the field order. Thus, we focus our attention on nonbinary, regular, rate-0.8, girth-8 LDPC codes over GF(4) in the rest of the section.

In Fig. 9.4B, we compare the BER performance of the LDPC(8430, 6744) code over GF(4) discussed in Fig. 9.4A, with that of the RS(255, 239) code, RS(255, 223) code, RS(255, 239) + RS(255, 223) concatenation code, and BCH(128, 113) × BCH(256, 239) TPC. We observe that the LDPC code over GF(4) outperforms all of these codes by a significant margin. In particular, it provides an additional coding gain of 3.363 and 4.401 dB at BER of 10^{-7} when compared with the concatenation code RS(255, 239) + RS(255, 223) and the RS(255, 239) code, respectively. Its coding gain improvement over BCH(128, 113) × BCH(256, 239) TPC is 0.886 dB at BER of 4×10^{-8}. We also presented in Fig. 9.4B a competitive, binary, (3,15)-regular, LDPC(16935, 13550) code proposed in Ref. [23]. We can see that the 4-ary, (3,15)-regular, girth-8 LDPC(8430, 6744) code beats the bit-length-matched binary LDPC code with a margin of 0.089 dB at BER of 10^{-7}. More importantly, the complexity of the MD-FFT-QSPA used for decoding the nonbinary LDPC code is lower than the min-sum-with-correction-term algorithm [6] used for decoding the corresponding binary LDPC code. When the MD-FFT-QSPA is used for decoding a (γ,ρ)-regular q-ary LDPC($N/\log q$, $K/\log q$) code, which is bit-length-matched to a (γ,ρ)-regular binary LDPC(N,K) code, the complexity is given by $(M/\log q)$ $2\rho q(\log q + 1 - 1/(2\rho))$ additions, where $M = N - K$ is the number of check nodes in the binary code. On the other hand, to decode the bit-length-matched binary counterpart using min-sum-with-correction-term algorithm [6], one needs $15M(\rho - 2)$ additions. Thus, a (3,15)-regular 4-ary LDPC code requires 91.28% of the computational resources required for decoding a (3,15)-regular binary LDPC code of the same rate and bit length.

9.1.5 Low-density parity-check code design

The most obvious way to design LDPC codes is to construct a low-density parity-check matrix with prescribed properties. Some important designs among others include: (i) Gallager codes (semirandom construction) [4], (ii) MacKay codes (semirandom construction) [26], (iii) finite-geometry-based LDPC codes [17,30,31], (iv) combinatorial-design-based LDPC codes [19], (v) QC (array, block-circulant) LDPC codes [2,3,18,23,24] (and references therein), (vi) irregular LDPC codes [31], and (vii) Tanner codes [22]. Quasi-cyclic LDPC codes of large girth were discussed in Section 9.1.1. The rest of this section briefly describes several LDPC codes classes, namely Gallager, Tanner, and MacKay codes. Other code classes, commonly referred to as structured (because they have regular structures in their parity-check matrices that can be exploited to facilitate hardware implementation of encoders and decoders), are described in corresponding sections related to quantum LDPC codes. Note that the basic design principles for both classical and quantum LDPC codes are very similar, and the same designs already used for classical LDPC code design can be used to design quantum LDPC codes. Therefore, various code designs from Chapter 17 of Ref. [21], with certain modifications can be used for quantum LDPC codes design as well.

9.1.5.1 Gallager codes

The $\boldsymbol{H}$-matrix for Gallager code has the following general form:

$$\boldsymbol{H} = [\boldsymbol{H}_1 \quad \boldsymbol{H}_2 \quad \cdots \quad \boldsymbol{H}_{w_c}]^{\mathrm{T}} \tag{9.14}$$

where $\boldsymbol{H}_1$ is $p \times p \cdot w_r$ matrix of row weight w_r, and $\boldsymbol{H}_i$ are column-permuted versions of $\boldsymbol{H}_1$-submatrix. The row-weight of H is w_r, and column-weight is w_c. The permutations are carefully chosen to avoid the cycles of length 4. The H-matrix is obtained by computer search.

9.1.5.2 Tanner codes and generalized low-density parity-check codes

In Tanner codes [22], each bit-node is associated with a code bit, and each check-node is associated with a subcode whose length is equal to the degree of a node, as illustrated in Fig. 9.5. Note that the so-called generalized LDPC codes [15,32–35] were inspired by Tanner codes.

Example 4. Let us consider the following example:

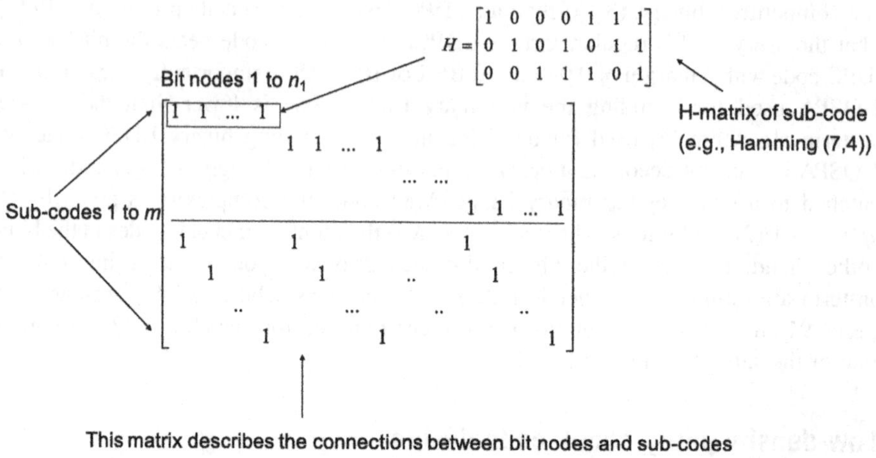

The Tanner code design in this example is performed in two stages. We first design the global code by starting from an identity matrix $I_{m/2}$ and replacing every nonzero element with n_1 ones and every zero element by n_1 zeros. The lower submatrix is obtained by concatenating the identity matrices In_1. In the second stage, we substitute all-one row-vectors (of length n_1) by the parity-check matrix of local linear block code (n_1, k_1), such as Hamming, BCH, or RM code. The resulting parity-check matrix is used as the parity-check matrix of an LDPC code.

The GLDPC codes can be constructed in similar fashion. To construct a GLDPC code, one can replace each single parity-check equation of a global LDPC code by the parity-check matrix of a simple linear block code, known as the constituent (local) code, and this construction is proposed by

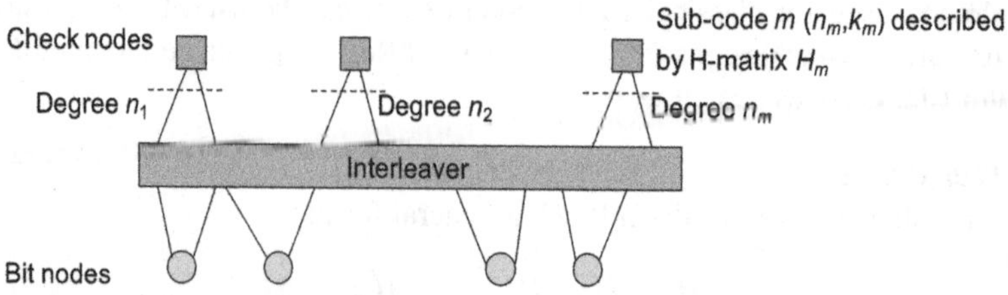

FIGURE 9.5

An illustration of Tanner codes.

Lentmaier and Zigangirov [33], and we will refer to this construction as the LZ-GLDPC code construction. One illustrative example is shown in Fig. 9.6. In another construction proposed by Boutros et al. in Ref. [32], referred here as B-GLDPC code construction, the parity-check matrix, **H**, is a sparse matrix partitioned into W submatrices $\boldsymbol{H}_1, \ldots, \boldsymbol{H}_W$. The $\boldsymbol{H}_1$ submatrix is a block-diagonal matrix generated from an identity matrix by replacing the ones by a parity-check matrix $\boldsymbol{H}_0$ of a local code of codeword-length n and dimension k:

$$\boldsymbol{H}=\left[\boldsymbol{H}_1^T \quad \boldsymbol{H}_2^T \quad \ldots \quad \boldsymbol{H}_W^T\right]^T \quad \boldsymbol{H}_1=\begin{bmatrix} \boldsymbol{H}_0 & & 0 \\ & \boldsymbol{H}_0 & \\ & & \cdots \\ 0 & & \boldsymbol{H}_0 \end{bmatrix} \tag{9.15}$$

Each submatrix $\boldsymbol{H}_j$ in (9.15) is derived from $\boldsymbol{H}_1$ by random column permutations. The code rate of a GLDPC code is lower bounded by

$$R=K/N \geq 1-W(1-k/n), \tag{9.16}$$

where K and N denote the dimension and the codeword-length of a GLDPC code, W is the column weight of a global LDPC code, and k/n is the code rate of a local code (k and n denote the dimension and the codeword-length of a local code). The GLDPC codes can be classified as follows [15]: (i) GLDPC codes with algebraic local codes of short length, such as Hamming codes, BCH codes, RS codes, or Reed-Muller codes, (ii) GLDPC codes for which the local codes are high-rate regular or irregular LDPC codes with large minimum distance, and (iii) fractal GLDPC codes in which the local code is, in fact, another GLDPC code. For more details on LZ-GLDPC and B-GLDPC-like codes and their generalization-fractal GLDPC codes (a local code is another GLDPC code), the interested reader is referred to Refs. [15,35] (and references therein).

9.1.5.3 MacKay codes

Following MacKay (1999) [26] below are listed several ways to generate the sparse matrices in order of increasing algorithm complexity (not necessarily improved performance):

1. $\boldsymbol{H}$-matrix is generated by starting from an all-zero matrix and randomly inverting w_c (not necessarily distinct bits) in each column. The resulting LDPC code is irregular code.
2. $\boldsymbol{H}$-matrix is generated by randomly creating weight-w_c columns.

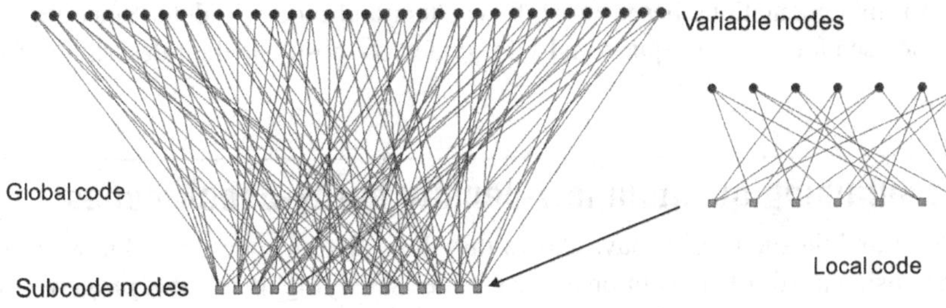

FIGURE 9.6

Illustration of construction of Lentmaier–Zigangirov generalized low-density parity-check (LZ-GLDPC) codes.

3. $\boldsymbol{H}$-matrix is generated with weight-w_c columns and uniform row weight (as nearly as possible).
4. $\boldsymbol{H}$-matrix is generated with weight-w_c columns, weight-w_r rows, and no two columns having overlap larger than one.
5. $\boldsymbol{H}$-matrix is generated as in (9.4), and short cycles are avoided.
6. $\boldsymbol{H}$-matrix is generated as in (9.5) and can be represented as $\boldsymbol{H} = [\boldsymbol{H}_1|\boldsymbol{H}_2]$, where $\boldsymbol{H}_2$ is invertible or at least has a full rank.

The construction via (9.5) may lead to an $\boldsymbol{H}$-matrix that is not of full rank. Nevertheless, it can be put in the following form by column swapping and Gauss-Jordan elimination:

$$\boldsymbol{H} = \begin{bmatrix} \boldsymbol{P}^T & \boldsymbol{I} \\ 0 & 0 \end{bmatrix}, \tag{9.17}$$

and by eliminating the all-zero submatrix, we obtain the parity-check matrix of systematic LDPC code:

$$\widetilde{\boldsymbol{H}} = \begin{bmatrix} \boldsymbol{P}^T & \boldsymbol{I} \end{bmatrix}. \tag{9.18}$$

Among various construction algorithms listed above, the construction (9.5) will be described in more detail.

Outline of Construction Algorithm (5)

1. Chose code parameters n, k, w_c, w_r, and g (minimum cycle length-girth). The resulting $\boldsymbol{H}$-matrix will be an $m \times n$ $(m = n - k)$ matrix with w_c ones per column, and w_r ones per row.
2. Set the column counter to $i_c = 0$.
3. Generate a weight-w_c column vector and place it in i_c-th column of the $\boldsymbol{H}$-matrix.
4. If the weight of each row $\leq w_r$, the overlap between any two columns ≤ 1, and all cycle-lengths are $\geq g$, then increment the counter $i_c = i_c + 1$.
5. If $i_c = n$, stop; else go to step 3.

This algorithm could take hours to run with no guarantee of regularity of the $\boldsymbol{H}$-matrix. Moreover, it may not finish at all, and we need to restart the search with another set of parameters. Richard and Urbanke (in 2001) proposed a linear complexity in length technique based on $\boldsymbol{H}$-matrix [28].

An alternative approach to simplify encoding is to design the codes via algebraic, geometric, or combinatorics methods [24], [17,29,30]. The $\boldsymbol{H}$-matrix of those designs can be put in cyclic or QC form, leading to implementations based on shift registers and mod-2 adders. Since these classes of codes can be also used for design of quantum LDPC codes, we postpone their description for incoming sections.

9.2 Dual-containing quantum low-density parity-check codes

Calderbank, Shor and Steane [36,37] have shown that the quantum codes, now known as CSS codes, can be designed using a pair of conventional linear codes satisfying the *twisted property*—that is, one of the codes includes the dual of another code. This class of quantum codes has already been studied in

Chapter 8. Among CSS codes, particularly simple are those based on dual-containing codes [38], whose (quantum) check matrix can be represented by [38]:

$$A = \left[\begin{array}{c|c} H & 0 \\ \hline 0 & H \end{array}\right], \tag{9.19}$$

where $HH^{\mathrm{T}} = 0$, which is equivalent to $C^{\perp}(H) \subset C(H)$, $C(H)$ is the code having H as the parity-check matrix, and $C^{\perp}(H)$ is its corresponding dual code. It has been shown in Ref. [38] that the requirement $HH^{\mathrm{T}} = \mathbf{0}$ is satisfied when rows of H have an even number of 1s and any two of them overlap by an even number of 1s. The LDPC codes satisfying these two requirements in Ref. [38] were designed by exhaustive computer search. in Ref. [39], they were designed as codes over GF(4) by identifying the Pauli operators I, X, Y, Z with elements from GF(4), while in Ref. [40], they were designed in QC fashion. In what follows, we will show how to design the dual-containing LDPC codes using the combinatorial objects known as balanced incomplete block designs (BIBDs) [46–50]. Note that the theory behind BIBDs is well known (see Ref. [41]), and BIBDs of unity index have already been used to design LDPC codes of girth-6 [42]. Note, however, that dual-containing LDPCs are girth-4 LDPC codes and can be designed based on BIBDs with an even index.

A balanced incomplete block design, denoted BIBD(v,b,r,k,λ), is a collection of subsets (also known as *blocks*) of a set $V$of size v, with a size of each subset being k, so that: (i) each pair of elements (also known as *points*) occurs in *exactly* λ of the blocks, and (ii) every point occurs in exactly r blocks. The BIBD parameters satisfy the following two conditions [41], (a) $vr = bk$, and (b) $\lambda(v-1) = r(k-1)$. Because the BIBD parameters are related (conditions a and b), it is sufficient to identify just three of them: v, k, and λ. It can be easily verified [42] that a point-block incident matrix represents a parity-check matrix H of an LDPC code of the code rate R lower bounded by $R \geq [b\text{-rank}(H)]/b$, where b is the codeword length, and with rank( ) we denoted the rank of the parity-check matrix. The parameter k corresponds to the column weight, r to the row weight, and v to the number of parity-checks. The corresponding quantum code rate is lower bounded by $R_Q \geq [b - 2\cdot\text{rank}(H)]/b$. By selecting the index of BIBD $\lambda = 1$, the parity-check matrix has a girth of at least 6. For a classical LDPC code to be applicable in quantum error-correction the following two conditions are to be satisfied [38]: (1) the LDPC code must contain its dual or equivalently any two rows of the parity-check matrix must have even overlap and the row weight must be even ($HH^{\mathrm{T}} = 0$), and (2) the code must have rate greater than 1/2. The BIBDs with even index λ satisfy condition (1). The parameter λ corresponds to the number of ones in which two rows overlap.

Example 5. The parity-check matrix from BIBD(7 ,7, 4, 4, 2) = {{1, 2, 4, 6}, {2, 6, 3, 7}, {3, 5, 6, 1}, {4, 3, 2, 5}, {5, 1, 7, 2}, {6, 7, 5, 4}, {7, 4, 1, 3}}, given as H_1-matrix below, has the rank 3, the even overlap between any two rows, and row weight is even as well (or equivalently $H_1 H_1^{\mathrm{T}} = 0$). Fig. 9.7 shows the equivalence between this BIBD, parity-check matrix H, and the Tanner graph. Namely, the blocks correspond to the bit-nodes and provide the position of nonzero elements in corresponding columns. For example, block $B_1 = \{1, 2, 4, 6\}$ corresponds to the bit-node v_1 and it is involved in parity-checks c_1, c_2, c_4 and c_6. Therefore, we establish edges between variable-node v_1 and set check-nodes c_1, c_2, c_4 and c_6. The B_1 block gives also the positions of nonzero elements in the first column of the parity-check matrix.

Note that so-called λ-configurations (in definition of BIBD, the word exactly from condition (i) is replaced by *at most*) are not applicable here. However, the H-matrix from BIBD of unity index can be converted to H-matrix satisfying condition (1) by adding a column with all-ones, which is

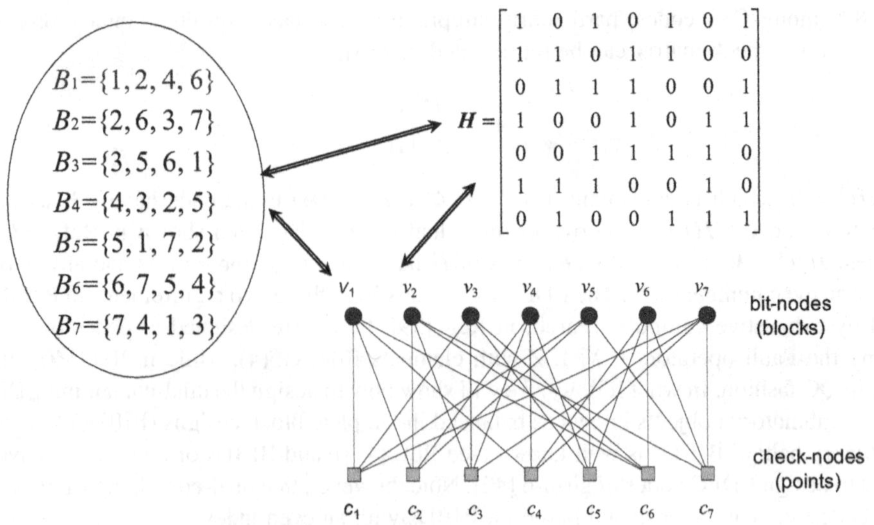

FIGURE 9.7

Equivalence between balanced incomplete block design (BIBD), parity-check matrix ***H***, and Tanner graph.

equivalent to adding an additional block to unity index BIBD having all elements from V. For example, the parity-check matrix from BIBD(7, 7, 3, 3, 1), after the addition of a column with all ones, is given as $\boldsymbol{H}_2$-matrix below, and satisfies the condition $\boldsymbol{H}_2\boldsymbol{H}_2^{\mathrm{T}} = 0$:

$$\boldsymbol{H}_1 = \begin{bmatrix} 1 & 0 & 1 & 0 & 1 & 0 & 1 \\ 1 & 1 & 0 & 1 & 1 & 0 & 0 \\ 0 & 1 & 1 & 1 & 0 & 0 & 1 \\ 1 & 0 & 0 & 1 & 0 & 1 & 1 \\ 0 & 0 & 1 & 1 & 1 & 1 & 0 \\ 1 & 1 & 1 & 0 & 0 & 1 & 0 \\ 0 & 1 & 0 & 0 & 1 & 1 & 1 \end{bmatrix} \quad \boldsymbol{H}_2 = \begin{bmatrix} 1 & 1 & 1 & 0 & 0 & 0 & 0 & 1 \\ 1 & 0 & 0 & 1 & 1 & 0 & 0 & 1 \\ 1 & 0 & 0 & 0 & 0 & 1 & 1 & 1 \\ 0 & 1 & 0 & 1 & 0 & 1 & 0 & 1 \\ 0 & 1 & 0 & 0 & 1 & 0 & 1 & 1 \\ 0 & 0 & 1 & 1 & 0 & 0 & 1 & 1 \\ 0 & 0 & 1 & 0 & 1 & 1 & 0 & 1 \end{bmatrix}$$

The following method, due to Bose [43], is a powerful method to design many *different BIBDs with desired index* λ. Let S be a set of elements. Associate to each element u from S n symbols $u_1, u_2, \ldots, u_n$. Let sets $S_1, \ldots, S_t$ satisfy the following three conditions: (i) every set S_i $(i = 1, \ldots, t)$ contains k symbols (the symbols from the same set are different from one another); (ii) among kt symbols in t sets exactly r symbols belong to each of n classes $(nr = kt)$; and (iii) the differences from t sets are symmetrically repeated so that each repeats λ times. If s is an element from S, from each S_i set we are able to form another set $S_{i,s}$ by adding s to S_i keeping the class number (subscript) unchanged; then sets $S_{i,s}$ $(i = 1, \ldots, t; s \in S)$ represent a (mn, nt, r, k, λ) BIBD. By observing the elements from BIBD blocks as position of ones in corresponding columns of a parity-check matrix, the code rate of an LDPC code such obtained is lower bounded by $R \geq 1 - m/t$ ($R_Q \geq 1 - 2\,m/t$), and the codeword length is determined by $b = nt$.

In the rest of this section, we introduce several constructions employing the method of Bose.

Construction 1: If $6t + 1$ is a prime or prime power and θ is a primitive root of GF($6t + 1$), then the following t initial sets $S_i = (0, \theta^i, \theta^{2t+i}, \theta^{4t+i})$ ($i = 0, 1, \ldots, t - 1$) form a BIBD($6t + 1$, $t(6t + 1)$, $4t$, 4, 2). The BIBD is formed by adding the elements from GF($6t + 1$) to the initial blocks S_i. Because the index of BIBD is even ($\lambda = 2$), and row weight $r = 4t$ is even, the corresponding LDPC code is a dual-containing code ($\boldsymbol{HH}^{\mathrm{T}} = \mathbf{0}$). The quantum code rate for this construction, and constructions 2 and 3 as well, is lower bounded by $R_Q \geq (1 - 2/t)$. The BIBD(7, 7 ,4, 4, 2) given above is obtained using this construction method. For $t = 30$ a dual-containing LDPC(5430, 5249) code is obtained. The corresponding quantum LDPC code has the rate 0.934, which is significantly higher than any of the codes introduced in Ref. [38] ($R_Q = 1/4$).

Construction 2: If $10t + 1$ is a prime or prime power and θ is a primitive root of GF($10t + 1$), then the following t initial sets $S_i = (\theta^i, \theta^{2t+i}, \theta^{4t+i}, \theta^{6t+i})$ form a BIBD($10t + 1$, $t(10t + 1)$, $4t$, 4, 2). For example, for $t = 24$, the dual-containing LDPC(5784, 5543) code is obtained, and the corresponding CSS code has the rate 0.917.

Construction 3: If $5t + 1$ is a prime or prime power and θ is a primitive root of GF($5t + 1$), then the following t initial sets ($\theta^i, \theta^{2t+i}, \theta^{4t+i}, \theta^{6t+i}, \theta^{8t+i}$) form a BIBD($5t$+1, $t(5t$+1), $5t$, 5, 4). Note that parameter t is to be even for LDPC code to satisfy condition (1). For example, for $t = 30$, the dual-containing LDPC(4530, 4379) code is obtained, and corresponding CSS LDPC code has the rate 0.934.

Construction 4: If $2t + 1$ is a prime or prime power and θ is a primitive root of GF($2t + 1$), then the following $5t + 2$ initial sets,

$$
\begin{aligned}
&\left(\theta_1^i, \theta_1^{t+i}, \theta_3^{i+a}, \theta_3^{t+i+a}, 0_2\right) \quad (i = 0, 1, \ldots, t-1)\\
&\left(\theta_2^i, \theta_2^{t+i}, \theta_4^{i+a}, \theta_4^{t+i+a}, 0_3\right) \quad (i = 0, 1, \ldots, t-1)\\
&\left(\theta_3^i, \theta_3^{t+i}, \theta_5^{i+a}, \theta_5^{t+i+a}, 0_4\right) \quad (i = 0, 1, \ldots, t-1)\\
&\left(\theta_4^i, \theta_4^{t+i}, \theta_1^{i+a}, \theta_1^{t+i+a}, 0_5\right) \quad (i = 0, 1, \ldots, t-1)\\
&\left(\theta_5^i, \theta_5^{t+i}, \theta_2^{i+a}, \theta_2^{t+i+a}, 0_1\right) \quad (i = 0, 1, \ldots, t-1)\\
&(0_1, 0_2, 0_3, 0_4, 0_5), (0_1, 0_2, 0_3, 0_4, 0_5)
\end{aligned}
$$

form a BIBD($10t$+5, ($5t$+2), ($2t$+1), $5t$+2, 5, 2). Similarly, as in the previous construction, the parameter t will be even. The quantum code rate is lower bounded by $R_Q \geq [1 - 2 \cdot 5/(5t + 2)]$, and the codeword length is determined by ($5t + 2$) ($2t + 1$). For $t = 30$, the dual-containing LDPC(5490, 5307) is obtained, and corresponding quantum LDPC codes has the rate 0.934. The following two constructions are obtained by converting unity index BIBD into $\lambda = 2$ BIBD.

Construction 5: If $12t + 1$ is a prime or prime power and θ is a primitive root of GF($12t + 1$), then the following t initial sets $(0, \theta^i, \theta^{4t+i}, \theta^{8t+i})$ ($i = 0, 2, \ldots, 2t - 2$) form a BIBD($12t + 1$, $t(12t + 1)$, $4t$, 4, 1). To convert this unity index BIBD into $\lambda = 2$ BIBD we have to add an additional block (1, ..., $12t + 1$).

Construction 6: If $20t+1$ is a prime or prime power and θ is a primitive root of GF($20t+1$), then the following t initial sets $(\theta^i, \theta^{4t+i}, \theta^{8t+i}, \theta^{12t+i}, \theta^{16t+i})$ $(i=0, 2, \ldots, 2t-2)$ form a BIBD($20t+1$, $t(20t+1)$, $5t$, 5, 1). Similarly as we did in the previous construction, to convert this unity index BIBD into $\lambda=2$ BIBD, we must add an additional block (1, ..., $20t+1$).

Construction 7: If $2(2\lambda+1)t+1$ is a prime power, and θ is a primitive root of GF[$2(2\lambda+1)t+1$], then the following t initial sets $S_i=(\theta^i, \theta^{2t+i}, \theta^{4t+i}, \ldots, \theta^{4\lambda t+i})$ $(i=0, 1, \ldots, t-1)$ form BIBD($2(2\lambda+1)t+1$,$t[2(2\lambda+1)t+1]$, $(2\lambda+1)t$, $2\lambda+1$, λ). The BIBD is formed by adding the elements from GF[$2(2\lambda+1)t+1$] to the initial blocks S_i. For any even index λ and even parameter t (the row weight is even), the corresponding LDPC code is a dual-containing code ($\boldsymbol{HH}^{\mathrm{T}}=\boldsymbol{0}$). The quantum code rate for this construction is lower bounded by $R_Q \geq (1-2/t)$, and the minimum distance is lower bounded by $d_{\min} \geq 2\lambda+2$. For any odd index λ design, we must add an additional block (1, 2, ..., $2(2\lambda+1)t+1$) so that the row weight of $\boldsymbol{H}$ becomes even.

Construction 8: If $2(2\lambda-1)t+1$ is a prime power, and θ is a primitive root of GF[$2(2\lambda-1)t+1$], then the following t initial sets $S_i=(0, \theta^i, \theta^{2t+i}, \ldots, \theta^{(4\lambda-1)t+i})$ form BIBD($2(2\lambda-1)t+1$, [$2(2\lambda-1)$ $t+1$]t, $2\lambda t$, 2λ, λ). For any even index λ, the corresponding LDPC code is dual-containing code. The quantum LPDC code rate is lower bounded by $R_Q \geq (1-2/t)$, and the minimum distance is lower bounded by $d_{\min} \geq 2\lambda+1$.

Construction 9: If $(\lambda-1)t$ is a prime power, and θ is a primitive root of GF[$(\lambda-1)t+1$], then the following t initial sets $(0, \theta^i, \theta^{t+i}, \ldots, \theta^{(\lambda-2)t+i})$ form BIBD[$(\lambda-1)t+1$, $((\lambda-1)t+1)t$, λt, λ, λ]. Again for an even index λ, a quantum LDPC code of rate $R_Q \geq (1-2/t)$ is obtained whose minimum distance is lower bounded by $d_{\min} \geq \lambda+1$. Similarly, as in the two previous constructions, for any odd index λ design, we must add an additional block (1, 2, ..., $(\lambda-1)t$).

Construction 10: If $2k-1$ is a prime power, and θ is a primitive root of GF($2k-1$), then the following initial sets

$$\left(0, \theta^i, \theta^{i+2}, \ldots, 0^{i+2k-4}\right), \quad \left(\infty, \theta^{i+1}, \theta^{i+3}, \ldots, 0^{i+2k-3}\right); (i=0,1)$$

form BIBD($2k$, $4(2k-1)$, $2(2k-1)$, k, $2(k-1)$). For an even k, the quantum code rate is lower bounded by $R_Q \geq [1-1/(2k-1)]$, the codeword length is determined by $4(2k-1)$, and the minimum distance is lower bounded by $d_{\min} \geq k+1$.

We performed the simulations for error-correcting performance of BIBD-based dual-containing codes described above as a function of noise level by Monte Carlo simulations. We simulated the classical BSC to be compatible with current literature [38–40]. The results of the simulations are shown in Fig. 9.8 for 30 iterations in the sum-product-with-correction-term algorithm described in the previous section. We simulated dual-containing LDPC codes with high rates and moderate lengths so that corresponding quantum LDPC code has a rate around 0.9. BER curves correspond to the $C/C^{\perp}$ case and are obtained by counting the errors only on those codewords from C not belonging to $C^{\perp}$. The codes from BIBD with index $\lambda=2$ outperform the codes with index $\lambda=4$. The codes derived from unity-index BIBDs by adding all 1 columns outperform the codes derived from BIBDs of even index. In the simulations presented in Fig. 9.8, codes with parity-check matrices with column weight $k=4$ or 5 are observed. The code from projective geometry (PG) is an exception. It is based on BIBD(s^2+s+1, $s+1$, 1), where s is a prime power; in our example, parameter s was set to 64. The LDPC(4162, 3432) code based on this BIBD outperforms other codes; however, the code rate is lower

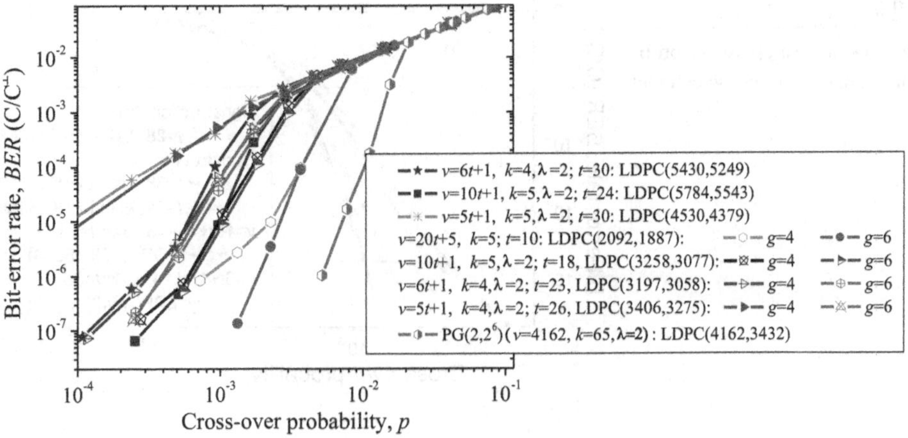

FIGURE 9.8

Bit error rates against crossover probability on binary symmetric channel.

($R = 0.825$), and the column weight is large (65). For more details on PG codes and secondary structures developed from them, the interested reader is referred to the next section.

To improve the BER performance of proposed high-rate sparse dual-containing codes we employed an efficient algorithm due to Sankaranarayanan and Vasic [44], for removing the cycles of length 4 in corresponding bipartite graph. As shown in Fig. 9.8, this algorithm can significantly improve the BER performance, especially for weak (index four BIBD-based) codes. For example, with LDPC(3406, 3275) code the BER of 10^{-5} can be achieved at cross-over probability $1.14 \cdot 10^{-4}$ if sum-product-with-correction-term algorithm is employed ($g = 4$ curve), while the same BER can be achieved at cross-over probability $6.765 \cdot 10^{-4}$ when the algorithm proposed in Ref. [44] is employed ($g = 6$ curve). Note that this algorithm modifies the parity-check matrix by adding the auxiliary variables and checks, so that the 4 cycles are removed in the modified parity-check matrix. The algorithm attempts to minimize the required number of auxiliary variable/check nodes while removing the 4-cycles. The modified parity-check matrix is used only in decoding phase, while the encoder remains the same.

In Fig. 9.9, we further compare BER performance of three quantum LDPC codes of quantum rate above 0.9, which are designed by employing Constructions 7 and 8: (i) quantum LDPC(7868, 7306, 0.9285, $\geq$6) code from Construction 7 by setting $t = 28$ and $\lambda = 2$, (ii) quantum LDPC(8702, 8246, 0.946, $\geq$5) code from Construction 8 by setting $t = 38$ and $\lambda = 2$, and (iii) quantum LDPC(8088, 7416, 0.917, $\geq$9) from Construction 8 by setting $t = 24$ and $\lambda = 4$.

For comparison purposes, two curves for quantum LDPC codes from Fig. 9.8 are also plotted. The BIBD codes from Constructions 7 and 8 outperform the codes from previous constructions for BERs around 10^{-7}. The code from BIBD with index $\lambda = 4$ and Construction 8 outperforms the codes with index $\lambda = 2$ from both constructions because it has a larger minimum distance. The code with index $\lambda = 2$ from Construction 7 outperforms the corresponding code from Construction 8.

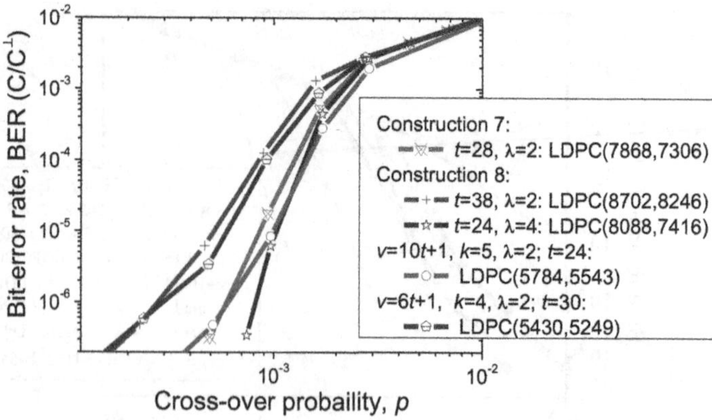

FIGURE 9.9

Bit error rates against crossover probability on a binary symmetric channel.

9.3 Entanglement-assisted quantum low-density parity-check codes

The EA-QECCs [51–57] make use of pre-existing entanglement between transmitter and receiver to improve the reliability of transmission. Dual-containing LDPC codes themselves have a number of advantages compared with other classes of quantum codes, because of sparseness of their quantum parity-check matrices, which leads to small number of interactions required to perform decoding. However, the dual-containing LDPC codes are in fact girth-4 codes, and the existence of short cycles deteriorates the error performance by correlating the extrinsic information in decoding process. By using the EA-QECC concept, arbitrary classical code can be used as quantum code. Namely, if classical codes are not dual containing, they correspond to a set of stabilizer generators that do not commute. By providing the entanglement between source and destination, the corresponding generators can be embedded into larger set of commuting generators, which gives a well-defined code space.

By using the LDPC codes of girth at least 6 we can dramatically improve the performance and reliability of current QKD schemes. The number of ebits required for EA-QECC is determined by $e = \text{rank}(\boldsymbol{HH}^{\mathrm{T}})$, as we described in previous chapter. Note that for dual-containing codes $\boldsymbol{HH}^{\mathrm{T}} = 0$ meaning that the number of required ebits is zero. Therefore, the minimum number of ebits in an EA-QECC is 1, and here we are concerned with the design of LDPC codes of girth $g \geq 6$ with $e = 1$ or reasonable small e. For example, an EA quantum LDPC code of quantum check matrix $\boldsymbol{A}$, given below has $\text{rank}(\boldsymbol{HH}^{\mathrm{T}}) = 1$ and girth 6:

$$\boldsymbol{A} = \begin{pmatrix} \boldsymbol{H} & 0 \\ 0 & \boldsymbol{H} \end{pmatrix} \quad \boldsymbol{H} = \begin{bmatrix} 1 & 0 & 1 & 1 & 1 \\ 1 & 1 & 0 & 0 & 1 \\ 0 & 1 & 1 & 1 & 0 \end{bmatrix}. \tag{9.20}$$

Because two Pauli operators on n-qubits commute if and only if there is an even number of places in which they differ (neither of which is the identity I operator), we can extend the generators in A by adding $e = 1$ column so that they can be embedded into a larger Abelian group, the procedure is known as Abelianization in abstract algebra. For example, by adding $(0, 1, 1)^{\mathrm{T}}$ column to $\boldsymbol{H}$-matrix above newly obtained code is dual-containing quantum code, and corresponding quantum-check matrix is given by:

$$A' = \begin{pmatrix} \boldsymbol{H}' & \boldsymbol{0} \\ \boldsymbol{0} & \boldsymbol{H}' \end{pmatrix} \quad \boldsymbol{H}' = \left(\begin{array}{ccccc|c} 1 & 0 & 1 & 1 & 1 & 0 \\ 1 & 1 & 0 & 0 & 1 & 1 \\ 0 & 1 & 1 & 1 & 0 & 1 \end{array}\right). \tag{9.21}$$

The last column in $\boldsymbol{H}'$ corresponds to the ebit, the qubit that is shared between the source and destination, which is illustrated in Fig. 9.10. The source encodes quantum information in state $|\psi\rangle$ with the help of local ancilla qubits $|0\rangle$ and source-half of shared ebits, and then sends the encoded qubits over a noisy quantum channel (say free-space optical channel or optical fiber). The receiver performs decoding on all qubits to diagnose the channel error and performs a recovery unitary operation to reverse the action of the channel. Note that channel does not affect at all the receiver's half of shared ebits. The encoding of quantum LDPC codes is very similar to the different classes of EA block-codes as described in previous chapter. Since the encoding of EA LDPC is not different from that of any other EA block-code, in the rest of this section we deal with the design of EA LDPC codes instead. The following theorem can be used for design of EA codes from BIBDs that require only one ebit to be shared between source and destination.

Theorem 1: Let $\boldsymbol{H}$ be a $b \times v$ parity-check matrix of an LDPC code derived from a BIBD(v, k, 1) of odd r. The rank of $\boldsymbol{HH}^{\mathrm{T}}$ is equal to 1, while the corresponding EA LDPC code of CSS type has the parameters [v, $v - 2b + 1$] and requires one ebit to be shared between source and destination.

Proof: Because any two rows or columns in $\boldsymbol{H}$ of size $b \times v$, derived from BIBD(v, k, 1) overlap in *exactly* one position, by providing that row weight r is odd, the matrix $\boldsymbol{HH}^{\mathrm{T}}$ is all-one matrix. The rank of all-one matrix $\boldsymbol{HH}^{\mathrm{T}}$ is 1. Therefore, LDPC codes from BIBDs of unity index ($\lambda = 1$) and odd r have $e = \mathrm{rank}(\boldsymbol{HH}^{\mathrm{T}}) = 1$, while the number of required ebits to be shared between source and destination is $e = 1$. If the EA code is put in CSS form, then the parameters of EA codes will be [v, $v - 2b + 1$].

Because the girth of codes derived by employing Theorem 1 is 6, they can significantly outperform quantum dual-containing LDPC codes. Note that certain blocks in so-called λ-configurations [41] have the overleap of zero and, therefore, cannot be used for design of EA codes with $e = 1$. Shrikhande [58] has shown that the generalized Hadamard matrices, affine resolvable BIBDs, group-divisible designs and orthogonal arrays of strength two [Shrikhande] [68] are all related; and because for these combinatorial objects $\lambda = 0$ or 1, they are not suitable for design of EA codes of $e = 1$. Note that if the quantum check matrix is put into the following format $\boldsymbol{A} = [\boldsymbol{H}|\boldsymbol{H}]$ then the corresponding EA code has

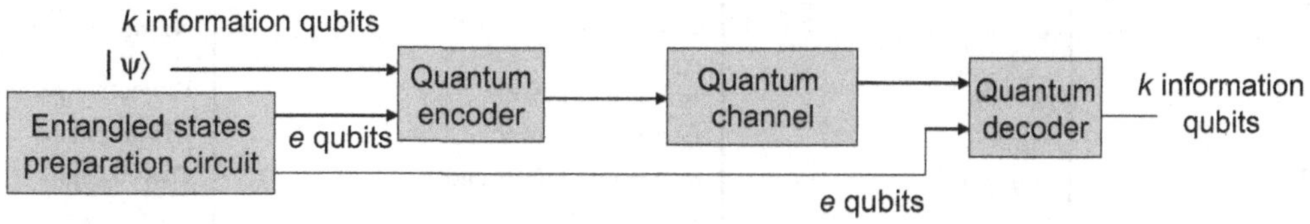

FIGURE 9.10

The operating principle of entanglement-assisted quantum codes.

parameters $[v, v-b+1]$. This design can be generalized by employing two BIBDs having the same parameter v, as explained in Theorem below.

Theorem 2: Let $\boldsymbol{H}_1$ be a $b_1 \times v$ parity-check matrix of an LDPC code derived from a BIBD(v, k_1, 1) of odd r_1 and $\boldsymbol{H}_2$ be a $b_2 \times v$ parity-check matrix of an LDPC code derived from a BIBD(v, k_2, 1) of odd r_2. The number of ebits required in corresponding EA code with quantum check matrix $\boldsymbol{A} = [\boldsymbol{H}_1|\boldsymbol{H}_2]$ is $e = \text{rank}(\boldsymbol{H}_1\boldsymbol{H}_2^{\text{T}} + \boldsymbol{H}_2\boldsymbol{H}_1^{\text{T}})/2 = 1$, while the corresponding EA LDPC code has the parameters $[v, v-b+1]$.

The proof of this theorem is a straightforward generalization of the proof of Theorem 1. Below we describe three designs belonging to the class of codes from Theorem 1.

Design 1 (Steiner triple system): If $6t+1$ is a prime power and θ is a primitive root of GF($6t+1$), then the following t initial blocks $(\theta^i, \theta^{2t+i}, \theta^{4t+i})$ ($i = 0, 1, \ldots, t-1$) form BIBD($6t+1$, 3, 1) with $r = 3t$. The BIBD is formed by adding the elements from GF($6t+1$) to the initial blocks. The corresponding LDPC code has rank($\boldsymbol{HH}^{\text{T}}$) = 1 and girth of 6.

Design 2 (projective planes): A finite projective plane of order q, say PG(2, q), is an (q^2+q+1, $q+1$, 1) BIBD, $q \geq 2$ and q is a power of prime [17,17,21,30,53]. The finite projective plane-based geometries are summarized in Table 9.1. The point set of the design consists of all the *points* on PG(2, q), and the block set of the design consist of all *lines* on PG(2, q). The points and lines of a finite projective plane satisfy the following set of axioms: (1) every line consists of the same number of points, (2) any two points on the plane are connected by a unique line, (3) any two lines on the plane intersect at a unique point, and (4) a fixed number of lines pass through any point on the plane. Since any two lines on a projective plane intersect at a unique point, there are no parallel lines on the plane ($\lambda = 1$). The incidence matrix (parity-check matrix of corresponding LDPC codes) of such a design, for $q = 2^s$, is cyclic and hence, any row of the matrix can be obtained by cyclic-shifting (right or left) another row of the matrix. Since the row weight $q+1 = 2^s+1$ is odd, the corresponding LDPC code has rank($\boldsymbol{HH}^{\text{T}}$) = 1. The minimum distance of codes from projective planes, affine planes, oval designs, and unitals [45] is at least 2^s+2, 2^s+1, $2^{s-1}+1$ and $2^{s/2}+2$, respectively. Corresponding ranks of $\boldsymbol{H}$-matrices are given respectively by Ref. [45] 3^s+1, 3^s, $3^s - 2^s$, and $2^{3s/2}$. Note, however, that affine plane-based codes have rank($\boldsymbol{HH}^{\text{T}}$) > 1, because they contain parallel lines. The code rate of classical projective plane codes is given by $R = (q^2+q+1-\text{rank}(\boldsymbol{H}))/(q^2+q+1) = (q^2+q-3^s)/(q^2+q+1)$, while the quantum code rate of corresponding quantum code is given by

$$R_Q = (q^2+q+1-2\text{rank}(\boldsymbol{H})+\text{rank}(\boldsymbol{HH}^{\text{T}}))/(q^2+q+1) = (q^2+q-2\cdot 3^s)/(q^2+q+1). \tag{9.22}$$

Table 9.1 Finite projective plane 2-(v, k, 1)-based geometries.

k	v	Parameter	Name
$q+1$	q^2+q+1	q- a prime power	Projective planes
q	q^2	q- a prime power	Affine planes
$q/2$	$q(q-1)/2$	q- a prime power and even	Oval designs
$\sqrt{q}+1$	$q\sqrt{q}+1$	q- a prime power and square	Unitals

Design 3 (m-dimensional projective geometries): The finite projective geometries [41,58] PG(m, p^s) are constructed using ($m+1$)-tuples $x=(x_0, x_1, \ldots, x_m)$ of elements x_i from GF(p^s) (p is a prime, s is a positive integer), not all simultaneously equal to zero, called points. Two ($m+1$)-tuples $\boldsymbol{x}$ and $y=(y_0, y_1, \ldots, y_m)$ represent the same point if $y=\lambda x$, where λ is a nonzero element from GF(p^s). Therefore, each point can be represented on p^s-1 ways (an equivalence class). The number of points in PG(m, p^s) is given by

$$v=\left[p^{(m+1)s}-1\right]/(p^s-1). \tag{9.23}$$

The points (equivalence classes) can be represented by $\left[\alpha^i\right]=\left\{\alpha^i, \beta\alpha^i, \ldots, \beta^{p^s-2}\alpha^i\right\}(0\le i\le v)$, where $\beta=\alpha^v$. Let $[\alpha^i]$ and $[\alpha^j]$ be two distinct points in PG(m, p^s), then the line passing through them consists of points of the form $[\lambda_1\alpha^i+\lambda_2\alpha^j]$, where λ_1, $\lambda_2\in$ GF(p^s). Because $[\lambda_1\alpha^i+\lambda_2\alpha^j]$ and $[\beta^k\lambda_1\alpha^i+\beta^k\lambda_2\alpha^j]$ represent the same point, each line in PG(m, p^s) consists of $k=(p^{ms}-1)/(p^s-1)$ points. The number of lines intersecting at a given point is given by k, and the number of lines in m-dimensional PG is given by

$$b=\left[p^{s(m+1)}-1\right](p^{sm}-1)/\left[(p^{2s}-1)(p^s-1)\right]. \tag{9.24}$$

The parity-check matrix is obtained as the incidence matrix $\boldsymbol{H}=(h_{ij})_{b\times v}$ with rows corresponding to the lines and columns to the points, and columns being arranged in the following order: $[\alpha^0], \ldots, [\alpha^v]$. This class of codes are sometimes called type I projective geometry codes [21]. The $h_{ij}=1$ if the jth point belongs to the ith line, and zero otherwise. Each row has weight p^s+1, and each column has the weight k. Any two rows or columns have exactly one "1" in common, and provided that p^s is even, the row weight (p^s+1) will be odd, and the corresponding LDPC code will have rank($\boldsymbol{HH}^\mathrm{T}$) = 1. The quantum code rate of corresponding codes will be given by

$$R_Q=\left(v-2\mathrm{rank}(\boldsymbol{H})+\mathrm{rank}\left(\boldsymbol{HH}^\mathrm{T}\right)\right)/v. \tag{9.25}$$

By defining a parity-check matrix as an incidence matrix with lines corresponding to columns and points to rows, the corresponding LDPC code will have rank($\boldsymbol{HH}^\mathrm{T}$) = 1 by providing that a row weight k is odd. These types of PG codes are sometimes called type II PG codes [21].

Design 4 (Finite geometry codes on m-flats) [45]: In projective geometric terminology, an ($m+1$)-dimensional element of the set is referred to as an *m-flat*. For examples, *points* are 0-flats, *lines* are 1-flats, *planes* are 2-flats, and *m-dimensional* spaces are ($m-1$)-flats. An l-dimension flat is said to be contained in an m-dimension flat if it satisfies a set-theoretic containment. Since any m-flat is a finite vector subspace, it is justified to talk about a basis set composed of basis elements independent in GF(p) (p is a prime). In the following discussion, the basis element of a 0-flat is referred to as a point, and generally, this should cause no confusion because this basis element is representative of the 0-flat. Using the above definitions and linear algebraic concepts, it is straightforward to count the number of m-flats, defined by a basis with $m+1$ points, in PG(n, q). The number of such flats, say $N_{PG}(m, n, q)$, is the number of ways of choosing $m+1$ independent points in PG(n, q) divided by the number of ways of choosing $m+1$ independent points in any m-flat:

$$N_{PG}(m,n,q)=\frac{(q^{n+1}-1)(q^{n+1}-q)(q^{n+1}-q^2)\cdots(q^{n+1}-q^m)}{(q^{m+1}-1)(q^{m+1}-q)(q^{m+1}-q^2)\cdots(q^{m+1}-q^m)}=\prod_{i=0}^{m}\frac{(q^{n+1-i}-1)}{(q^{m+1-i}-1)}. \tag{9.26}$$

Hence, the number of 0-flats in PG(n, q) is $N_{PG}(0,n,q) = \frac{q^{n+1}-1}{q-1}$, and the number of 1-flats is

$$N_{PG}(1,n,q) = \frac{(q^{n+1}-1)(q^n-1)}{(q^2-1)(q-1)}. \tag{9.27}$$

When $n = 2$, the number of points is equal to the number of lines, and this agrees with the dimensions of point-line incidence matrices of projective plane codes introduced above.

An algorithm to construct an m-flat in PG(n, q) begins by recognizing that the elements of GF(q^{n+1}) can be used to represent points of PG(n, q) in the following manner [45,59]. If α is the primitive element of GF(q^{n+1}), then α^v, where $v = (q^{n+1}-1)/(q-1)$, is a primitive element of GF(q). Each one of the first v powers of α can be taken as the basis of one of the 0-flats in PG(n, q). In other words, if α^i is the basis of a 0-flat in PG(n, q), then every α^k, such that $i \equiv k \bmod v$, is contained in the subspace. In a similar fashion, a set of $m+1$ powers of α that are independent in GF(q) forms a basis of an m-flat. If $\alpha^{s_1}, \alpha^{s_2}, \ldots, \alpha^{s_{m+1}}$ is a set of $m+1$ basis elements of an m-flat, then any point in the flat can be written as

$$\zeta_i = \sum_{j=1}^{m+1} \varepsilon_{ij}\alpha^{s_j}, \tag{9.28}$$

where the ε_{ij} are chosen such that no two vectors $< \varepsilon_{i1}, \varepsilon_{i2}, \ldots, \varepsilon_{i(m+1)} >$ are linear multiples over GF(q). Now every ζ_i can be equivalently written as a power of the primitive element α.

To construct an LDPC code, we generate all m-flats and l-flats in PG(n, q) using the method described above. An incidence matrix of m-flats and l-flats in PG(n, q) is a binary matrix with $N_{PG}(m, n, q)$ rows and $N_{PG}(l, n, q)$ columns. The (i, j)-th element of the incidence matrix $\boldsymbol{A}_{PG}(m,l,n,q)$ is a 1 only if the jth l-flat is contained in the ith m-flat of the geometry. An LDPC code is constructed by considering the incidence matrix or its transpose as the parity-check matrix of the code. It is widely accepted that the performance of an LDPC code under an iterative decoder is deteriorated by the presence of four cycles in the Tanner graph of the code. To avoid four cycles, we impose an additional constraint that l should be 1 less than m. If l is 1 less than m, then no two distinct m-flats have more than one l-dimensional subspace in common. This guarantees that the girth (length of the shortest cycle) of the graph is 6.

In Fig. 9.11, we show the code length as a function of quantum code rate for EA LDPC codes derived from designs 2 and 3 obtained for different values of parameter $s = 1-9$. The results of the simulations are shown in Fig. 9.12 for 30 iterations in the sum-product algorithm. The EA quantum LDPC codes from Design 3 are compared with the dual-containing quantum LDPC codes discussed in the previous chapter. We can see that EA LDPC codes significantly outperform dual-containing LDPC codes.

Before concluding this section, we provide a very general design method [29] based on the theory of mutually orthogonal Latin squares (MOLS) constructed using the MacNeish–Mann theorem [58]. This design allows us to determine a corresponding code for an arbitrary composite number q.

Design 5 (iterative decodable codes based on MacNeish–Mann theorem) [29]: Let the prime decomposition of an integer q be

$$q = p_1^{s_1} p_2^{s_2} \cdots p_m^{s_m} \tag{9.29}$$

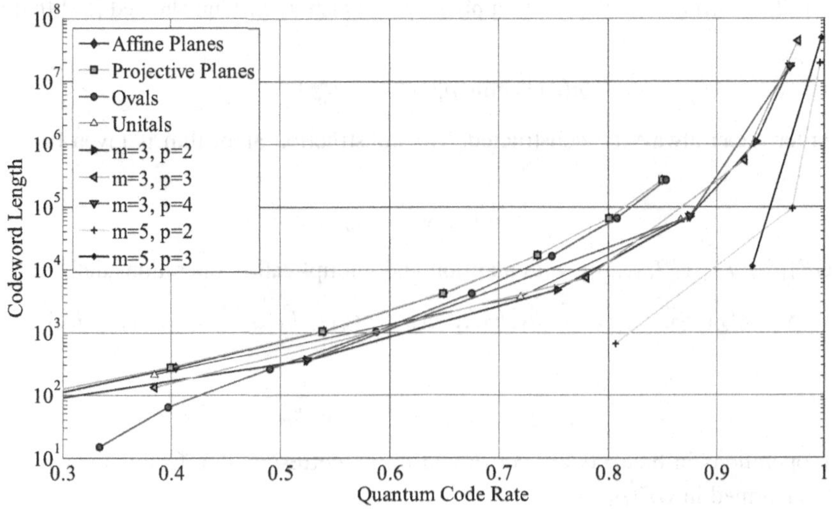

FIGURE 9.11

Entaglement asssited (EA) quantum low-density parity-check codes from *m*-dimensional projective geometries.

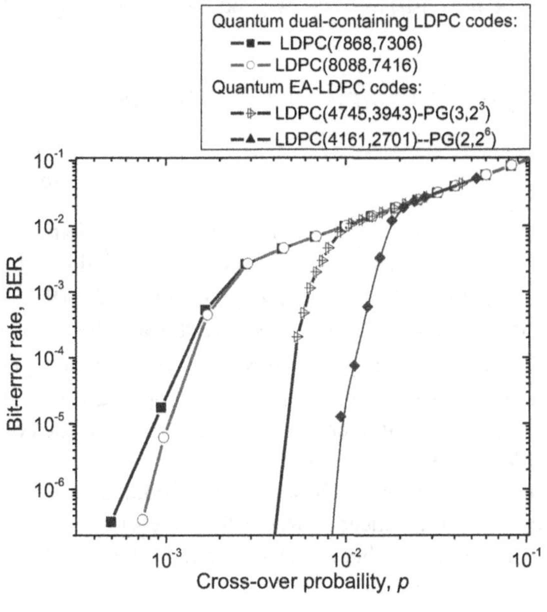

FIGURE 9.12

Bit error rates against crossover probability on a binary symmetric channel.

with p_i ($i = 1, 2, \ldots, m$) being the prime numbers. MacNeish and Mann showed [58] that a complete set of

$$n(q) = \min\left(p_1^{s_1}, p_2^{s_2}, \ldots, p_m^{s_m}\right) - 1 \tag{9.30}$$

MOLS of order q can always be constructed. The construction algorithm is given below. Consider m-tuples

$$\gamma = (g_1, g_2, \ldots, g_m), \tag{9.31}$$

where $g_i \in GF\left(p_i^{s_i}\right)$, $i = 1, 2, \ldots, m$, with addition and multiplication operations defined by

$$\gamma_1 + \gamma_2 = (g_1, g_2, \ldots, g_m) + (h_1, h_2, \ldots, h_m) = (g_1 + h_1, g_2 + h_2, \ldots, g_m + h_m) \tag{9.32}$$

and

$$\gamma_1 \gamma_2 = (g_1 h_1, g_2 h_2, \ldots, g_m h_m), \tag{9.33}$$

wherein the operations in brackets are performed in the corresponding Galois fields (e.g., $g_1 h_1$ and $g_1 + h_1$ are performed in $GF\left(p_1^{s_1}\right)$).

Denote the elements from $GF\left(p_i^{s_i}\right)$, $i = 1, 2, .., m$ as $0, g_i^{(1)} = 1, g_i^{(2)}, \ldots, g_i^{\left(p_i^{s_i} - 1\right)}$. The following elements possess multiplicative inverses:

$$\gamma_i = \left(g_1^{(i)}, g_2^{(i)}, \ldots, g_m^{(i)}\right), \quad i = 1, 2, \ldots, n(q), \tag{9.34}$$

since they are composed of nonzero elements of corresponding Galois fields, and $\gamma_i - \gamma_j$ for $i \neq j$ as well. Denote the first element as $\gamma_0 = (0, 0, \ldots, 0)$, the next $n(q)$ as in (9.34), and the rest $q - n(q) - 1$ elements as

$$\gamma_j = \left(g_1^{(i_1)}, g_2^{(i_2)}, \ldots, g_m^{(i_m)}\right), j = n(q) + 1, \ldots, q - 1; i_1 = 0, 1, \ldots, p_1^{s_1} - 1; i_2 = 0, 1, \ldots, p_2^{s_2} - 1; \ldots;$$
$$i_m = 0, 1, \ldots, p_m^{s_m} - 1, \tag{9.35}$$

with the combinations of components leading to (9.34) being excluded. The $n(q)$ arrays of elements, MOLS, L_i, $i = 1, 2, \ldots, n(q)$ are further formed with (j,k)-th cell being filled by $\gamma_j + \gamma_i \gamma_k$; $i = 1, 2, \ldots, n(q); 0 \leq j, k \leq q - 1$. (The multiplication $\gamma_i \gamma_k$ is performed using (9.33), and the addition $\gamma_j + \gamma_i \gamma_k$ using (9.32).) To establish the connection between MOLS and BIBD, we introduce the following one-to-one mapping l: L->V, with L being the MOLS and V being the integer set of q^2 elements (*points*). The simplest example of such mapping is the linear mapping $l(x, y) = q(x-1) + y$; $1 \leq x, y \leq q$ (therefore, $1 \leq l(x,y) \leq q^2$). (The numbers $l(x,y)$ are referred to as the cell *labels*.) Each L_i ($i = 1, \ldots, n(q)$) defines a set of q *blocks* (*lines*) $B_i = \{l(x, y) | L_i(x, y) = s, 1 \leq s \leq q\}$. Each line s ($1 \leq s \leq q$) in B_i contains the labels of element s from L_i. These blocks are equivalent to the sets of lines of slopes $0 \leq s \leq q-1$ in the lattice design we introduced in Ref. [19]. Labels of every line of a design specify the positions of 1's in a row of the parity-check matrix. Namely, the *incidence matrix* of a design is a $qn(q) \times q^2$ matrix $\boldsymbol{H} = (h_{ij})$ defined by

$$h_{ij} = \begin{cases} 1 & \text{if the } i-\text{th block contains the } j-\text{th point} \\ 0 & \text{otherwise} \end{cases} \tag{9.36}$$

and represents the *parity-check matrix*. The code length is q^2, the code rate is determined by $R = (q^2 - \text{rank}(\boldsymbol{H}))/q^2$, and it may be estimated as $R \cong (q^2 - n(q)q)/q^2 = 1 - n(q)/q$. The quantum code rate of corresponding quantum code is given by

$$R_Q = (q^2 - 2\text{rank}(\boldsymbol{H}) + \text{rank}(\boldsymbol{H}\boldsymbol{H}^{\mathrm{T}}))/q^2 \simeq (q^2 - 2n(q)q + \text{rank}(\boldsymbol{H}\boldsymbol{H}^{\mathrm{T}}))/q^2. \tag{9.37}$$

9.4 Iterative decoding of quantum low-density parity-check codes

In Chapters 7–8, we defined the quantum error correction code (QECC), which encodes K qubits into N qubits, denoted as $[N,K]$, by an encoding mapping U from the K-qubit Hilbert space H_2^K onto a 2^K-dimensional subspace C_q of the N-qubit Hilbert space H_2^N. The $[N,K]$ quantum stabilizer code C_Q has been defined as the unique subspace of Hilbert space H_2^N that is fixed by the elements from stabilizer S of C_Q as follows:

$$C_Q = \bigcap_{s \in S} \{|c\rangle \in H_2^N \,|\, s|c\rangle = |c\rangle\}. \tag{9.38}$$

Any error in Pauli group G_N of N-qubit errors can be represented by

$$\begin{aligned} &E = j^{\lambda} X(\boldsymbol{a}) Z(\boldsymbol{b}); \;\; \boldsymbol{a} = a_1 \cdots a_N; \;\; \boldsymbol{b} = b_1 \cdots b_N; \;\; a_i, b_i = 0, 1; \;\; \lambda = 0, 1, 2, 3 \\ &X(\boldsymbol{a}) \equiv X_1^{a_1} \otimes \cdots \otimes X_N^{a_N}; \;\; Z(\boldsymbol{b}) \equiv Z_1^{b_1} \otimes \cdots \otimes Z_N^{b_N} \end{aligned} \tag{9.39}$$

An important *property* of N-qubit errors is: $\forall \, E_1, E_2 \in G_N$ the following is valid: $[E_1, E_2] = 0$ or $\{E_1, E_2\} = 0$. In other words, the elements from Pauli group either commute or anticommute:

$$E_1 E_2 = j^{l_1 + l_2} X(\boldsymbol{a}_1) Z(\boldsymbol{b}_1) X(\boldsymbol{a}_2) Z(\boldsymbol{b}_2) = j^{l_1 + l_2} (-1)^{\boldsymbol{a}_1 \boldsymbol{b}_2 + \boldsymbol{b}_1 \boldsymbol{a}_2} X(\boldsymbol{a}_2) Z(\boldsymbol{b}_2) X(\boldsymbol{a}_1) Z(\boldsymbol{b}_1)$$

$$= (-1)^{\boldsymbol{a}_1 \boldsymbol{b}_2 + \boldsymbol{b}_1 \boldsymbol{a}_2} E_2 E_1 = \begin{cases} E_2 E_1, \boldsymbol{a}_1 \boldsymbol{b}_2 + \boldsymbol{b}_1 \boldsymbol{a}_2 = 0 \text{ mod} 2 \\ -E_2 E_1, \boldsymbol{a}_1 \boldsymbol{b}_2 + \boldsymbol{b}_1 \boldsymbol{a}_2 = 1 \text{ mod} 2 \end{cases} \tag{9.40}$$

In Chapters 7–8, we introduced the concept of syndrome of an error as follows. Let C_Q be a quantum stabilizer code with generators $g_1, g_2, \ldots, g_{N-K}$ and let $E \in G_N$ be an error. The error syndrome for error E is defined the bit string $S(E) = [\lambda_1, \lambda_2, \ldots, \lambda_{N-K}]^{\mathrm{T}}$ with component bits being determined by

$$\lambda_i = \begin{cases} 0, [E, g_i] = 0 \\ 1, \{E, g_i\} = 0 \end{cases} \quad (i = 1, \ldots, N-K) \tag{9.41}$$

We showed in Chapter 8 that the set of errors $\{E_i | E_i \in G_N\}$ have the same syndrome $\boldsymbol{\lambda} = [\lambda_1 \ldots \lambda_{N-K}]^{\mathrm{T}}$ if an only if they belong to the same coset $EC(S) = \{Ec | c \in C(S)\}$, where $C(S)$ is the *centralizer* of S (that is the set of errors $E \in G_N$ that commute with all elements from S). The Pauli encoded operators have been introduced by the following mappings:

$$X_i \rightarrow \overline{X}_i = UX_{N-K+i}U^{\dagger} \quad \text{and} \quad Z_i \rightarrow \overline{Z}_i = UZ_{N-K+i}U^{\dagger}; i = 1, \cdots, K \tag{9.42}$$

where U is a Clifford operator acting on N qubits (an element of Clifford group-$N(G_N)$, where N is normalizer of G_N). By using this interpretation, the encoding operation is represented by

$$U(|\mathbf{0}\rangle_{N-K} \otimes |\psi\rangle), \quad |\psi\rangle \in H_2^K \tag{9.43}$$

An equivalent definition of syndrome elements is given by Ref. [60]:

$$\lambda_i = \begin{cases} 1, & Eg_i = g_iE \\ -1, & Eg_i = -g_iE \end{cases} \quad (i = 1, \ldots, N-K) \tag{9.44}$$

This definition can simplify verification of commutativity of the two errors $E = E_1 \otimes \ldots \otimes E_N$ and $F = F_1 \otimes \ldots \otimes F_N$, where $E_i, F_i \in \{I, X, Y, Z\}$, as follows:

$$EF = \prod_{i=1}^{N} E_i F_i. \tag{9.45}$$

The Pauli channels can be described using the following error model, which has already been discussed in Section 7.5:

$$\xi(\rho) = \sum_{E \in G_N} p(E) E \rho E^{\dagger}, \quad p(E) \geq 0, \sum_{E \in G_N} p(E) = 1. \tag{9.46}$$

A memoryless channel model can be defined in similar fashion to the discrete memoryless channel models for classical codes (see Chapter 2) as follows [60]:

$$p(E = E_1 \cdots E_N) = p(E_1) \cdots p(E_N), E \in G_N. \tag{9.47}$$

The depolarizing channel is a very popular channel model for which $p(I) = 1 - p$, $p(X) = p(Y) = p(Z) = p/3$, for certain depolarizing strength $0 \leq p \leq 1$.

The probability of E error given syndrome λ can be calculated by

$$p(E|\lambda) \sim p(E) \prod_{i=1}^{N-K} \delta_{\lambda_i, Eg_i}. \tag{9.48}$$

The most likely error E_{opt} given the syndrome can be determined by

$$E_{\text{opt}}(\lambda) = \underset{E \in G_N}{\arg\max}\, p(E|\lambda). \tag{9.49}$$

Unfortunately, this approach is an NP-hard problem. We can instead use the qubit-wise most likely error evaluation as follows [60]:

$$E_{q,\text{opt}}(\lambda) = \underset{E_q \in G_1}{\arg\max}\, p_q(E_q|\lambda), \quad p_q(E_q|\lambda) = \sum_{E_1, \cdots, E_{q-1}, E_{q+1}, \cdots, E_N \in G_1} p(E_1 \cdots E_N|\lambda). \tag{9.50}$$

Note that this marginal optimum does not necessarily coincide with the global optimum.

It is clear that from Eq. (9.50), that this marginal optimization involves summation of many terms exponential in N. The SPA similar to that discussed in Section 9.1 can be used here. For this purpose we can also use *bipartite graph* representation of stabilizers. Namely we can identify stabilizers with function nodes and qubits with variable nodes. As an illustrative example let us consider (5, 1) quantum cyclic code generated by the following four stabilizers: $g_1 = X_1Z_2Z_3X_4$, $g_2 = X_2Z_3Z_4X_5$, $g_3 = X_1X_3Z_4X_5$, $g_4 = Z_1X_2X_4Z_5$. The corresponding quantum-check matrix is given by

$$A = \begin{bmatrix} x \quad z \\ 10010|01100 \\ 01001|00110 \\ 10100|00011 \\ 01010|10001 \end{bmatrix}$$

The corresponding bipartite graph is shown in Fig. 9.13, where we used dashed lines to denote the X-operator action on particular qubits and solid lines do denote the action of Z-operators. This is an example of *regular* quantum stabilizer code, with qubit node degree 3 and stabilizer node degree 4. For example, the qubit q_1 is involved in stabilizers g_1, g_3 and g_4 and therefore, there exist the edges between this qubit and corresponding stabilizers. The weight of edge is determined by action of corresponding Pauli operator on observed qubit.

Let us denote the messages to be sent from qubit q to generator (check) node c as $m_{q\to c}$, which are concerned probability distributions over $E_q \in G_1 = \{I, X, Y, Z\}$ and represent an array of four positive numbers, one for each value of E_q. Let us denote the neighborhood of q by $n(q)$ and neighborhood of c by $n(c)$. The initialization is performed as follows $m_{q\to c}(E_q) = p_q(E_q)$, where $p_q(E_q)$ is given by (9.50). Each generator ("check") node c sends the message to its neighbor qubit-node q, by properly combining the extrinsic information from other neighbors as follows:

$$m_{c\to q}(E_q) \sim \sum_{E_{q'},\ q'\in n(c)\backslash q} \left(\delta_{\lambda_c, E_c g_c} \prod_{q'\in n(c)\backslash q} m_{q'\to c}(E_{q'}) \right), \tag{9.51}$$

where $n(c)\backslash q$ denotes all neighbors of generator (check)-node c except q. The proportionality factor can be determined from normalization condition: $\sum_{E_i} m_{c\to q} = 1$. Clearly, the message $m_{c\to q}$ is related to the syndrome component λ_c associated to generator g_c and extrinsic information collected from all

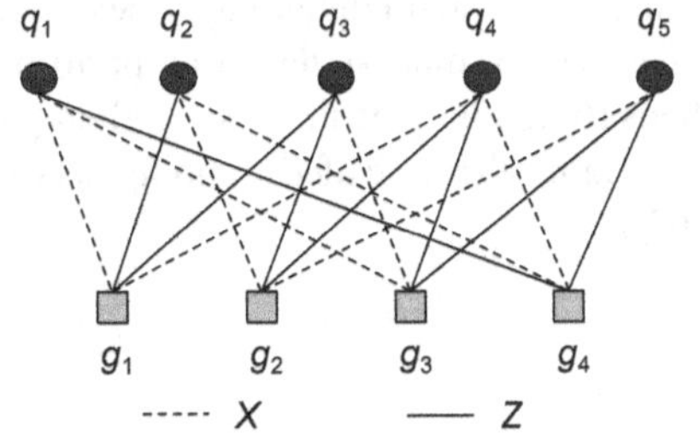

FIGURE 9.13

The bipartite graph of (5, 1) quantum cyclic code with generators $g_1 = X_1Z_2Z_3X_4$, $g_2 = X_2Z_3Z_4X_5$, $g_3 = X_1X_3Z_4X_5$, and $g_4 = Z_1X_2X_4Z_5$.

neighbors of c except q. This step can be called, in analogy with classical LDPC codes, as *generator (check)-node update rule*.

The message to be sent from qubit q to generator (check) c is obtained by properly combining extrinsic information (message) from all generator-neighbors of q, say c', except generator-node c:

$$m_{q\rightarrow c}(E_q) \sim p_q(E_q) \prod_{c' \in n(q)\backslash c} m_{c'\rightarrow q}(E_q), \tag{9.52}$$

where $n(q)\backslash c$ denotes all neighbors of qubit-node q except c. Clearly, $m_{q\rightarrow c}$ is a function of prior probability $p(E_q)$ and extrinsic messages collected from all neighbors of q except c. The proportionality constant in (9.52) can also be obtain from normalization condition. Again, in analogy with classical LDPC codes, this step can be called *qubit-node update rule*.

The steps (9.51) and (9.52) are iterated until a valid codeword is obtained or predetermined number of iteration is reached. Now we have to calculate the beliefs $b_q(E_q)$, that is, the marginal conditional probability estimation $p_q(E_q|\lambda)$, as follows:

$$b_q(E_q) \sim p_q(E_q) \prod_{c' \in n(q)} m_{c'\rightarrow q}(E_q). \tag{9.53}$$

Clearly, the beliefs are determined by taking the extrinsic messages of all neighbors of qubit q into account and the prior probability $p_q(E_q)$. The recovery operator can be obtained as tensor product of qubit-wise maximum-belief Pauli operators:

$$E_{\mathrm{SPA}} = \otimes_{q=1}^{N} E_{\mathrm{SPA},q}, \quad E_{\mathrm{SPA},q} = \arg\max_{E_q \in G_1} \{b_q(E_q)\}. \tag{9.54}$$

If the bipartite graph is a tree than the beliefs will converge to true conditional marginal probabilities $b_q(E_q) = p_q(E_q|\lambda)$ in number of steps equal to the tree's depth. However, in the presence of cycles, the beliefs will not converge to the correct conditional marginal probabilities in general. We will perform iterations until trivial error syndrome is obtained, that is until $E_{\mathrm{SPA}} g_i = \lambda_i \ (\forall\, i = 1, \ldots, N-K)$, or until the predetermined number of iterations has been reached. This algorithm is applicable to both dual-containing and EA quantum codes.

The quantum SPA algorithm above is essentially very similar to the classical probability-domain SPA. In Section 9.1, we described the log-domain version of this algorithm, which is more suitable for hardware implementation as the multiplication of probabilities is replaced by addition of reliabilities.

The cycles of length 4, which exist in dual containing codes, deteriorate the decoding BER performance. There exist various methods to improve decoding performance, such as freezing, random perturbation, and collision methods [60]. Someone may use method [44], which was already described in Section 9.2. In addition, by avoiding the so-called trapping sets the BER performance of LDPC decoders can be improved [61,62].

9.5 Spatially coupled quantum low-density parity-check codes

The starting point can be the quasi cyclic (QC)-LPDC code with the template parity-check matrix given by

$$H_{QC} = \begin{bmatrix} I & I & I & \dots & I \\ B^{S[0]} & B^{S[1]} & B^{S[2]} & \dots & B^{S[c-1]} \\ B^{2S[0]} & B^{2S[1]} & B^{2S[2]} & \dots & B^{2S[c-1]} \\ \dots & \dots & \dots & \dots & \dots \\ B^{(r-1)S[0]} & B^{(r-1)S[1]} & B^{(r-1)S[2]} & \dots & B^{(r-1)S[c-1]} \end{bmatrix}, \quad B = \begin{bmatrix} 0 & 1 & 0 & \dots & 0 \\ 0 & 0 & 1 & \dots & 0 \\ \dots & \dots & \dots & \dots & \dots \\ 0 & 0 & 0 & \dots & 1 \\ 1 & 0 & 0 & \dots & 0 \end{bmatrix}_{b \times b} \tag{9.55}$$

where I and B are identity and permutation matrices of size $b \times b$, and integers $S[i] \in \{0, 1, \dots, b-1\}$ ($i = 0, 1, \dots, r-1$; $r < b$) are properly chosen to satisfy the girth (the largest cycle in corresponding bipartite graph representation of H_{QC}) constrains, as described in Ref. [63]. To memorize the parity-check matrix of this QC-LDPC code, for each block-row we need to story just the position of the first nonzero entry in the first row of corresponding permutation submatrix, all other rows are block-cyclic permutations of the first row. By using this QC-LDPC code as a template design, we create a spatially coupled (SC)-LDPC code as illustrated in Fig. 9.14 (left). The codeword length of this SC-LDPC code will be $b \times (l \times c - m \times (l-1))$, where l is the number of coupled template QC-LDPC codes and m is the coupling length expressed in terms of number of blocks. Because there are $r \times c \times l$ nonempty submatrices in the parity-check matrix of the SC-LDPC code, we can introduce the layer index (l.i.) and thus reduce the memory requirements as illustrated in Fig. 9.14 (right). In such a way we do not need to memorize the all-zeros submatrices. For full-rank parity-check matrix of the template QC-LDPC code, the code rate of SC-LDPC code is given by

$$R = 1 - \frac{rl}{lc - m(l-1)}. \tag{9.56}$$

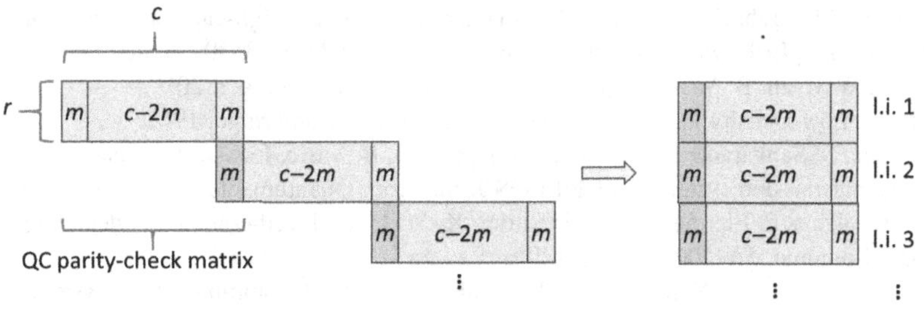

FIGURE 9.14

Parity-check matrix of spatially coupled low-density parity-check code (LDPC) design derived from the template quasi-cyclic-LDPC code.

Therefore, for fixed l, by increasing the coupling length m we can reduce the code rate and thus improve the error-correction capability of the code. For the FPGA implementation of decoders for SC-LDPC codes, the interested reader is referred to our recent paper [64]. In a more general SC design, where the coupled code layers are not the same, which is known as *time-variant* SC-LDPC code, the memory size will be c times larger than that of the template QC-LDPC code.

By ensuring that the following condition is satisfied for the SC-LDPC code:

$$\boldsymbol{H}_{SC}\boldsymbol{H}_{SC}^{T}=0, \tag{9.57}$$

and the corresponding quantum LDPC code will belong to the class of CSS codes. The quantum-check matrix of the quantum SC-LDPC code will be

$$\boldsymbol{A}_{SC}=\left[\begin{array}{c|c}\boldsymbol{H}_{SC} & 0\\ 0 & \boldsymbol{H}_{SC}\end{array}\right], \tag{9.58}$$

and can be decoded using the SPA similar to that provided in the previous section.

The SC-LDPC code design can straightforwardly be generalized to the convolutional LDPC code, such as the one introduced in Ref. [65]. Similarly to the classical counterparts [66], the quantum SC-LDPC codes outperform corresponding quantum block LDPC codes as shown in Ref. [67].

9.6 Summary

This chapter has been devoted to quantum LDPC codes. In Section 9.1, we described classical LDPC codes, their design, and decoding algorithms. In Section 9.2, we described dual-containing quantum LDPC codes. Various design algorithms were described. Section 9.3 was devoted to entanglement-assisted quantum LDPC codes. Various classes of finite geometry codes were described. In Section 9.4, we described the probabilistic sum-product algorithm based on the quantum-check matrix. The quantum SC LDPC codes were discussed briefly in Section 9.5.

References

[1] W.E. Ryan, Concatenated convolutional codes and iterative decoding, in: J.G. Proakis (Ed.), Wiley Encyclopedia of Telecommunications, John Wiley & Sons, 2003.

[2] I.B. Djordjevic, M. Arabaci, L. Minkov, Next generation FEC for high-capacity communication in optical transport networks, IEEE/OSA J. Lightw. Technol. 27 (2009) 3518–3530 (Invited Paper.).

[3] I. Djordjevic, W. Ryan, B. Vasic, Coding for Optical Channels, Springer, 2010.

[4] R.G. Gallager, Low Density Parity Check Codes, MIT Press, Cambridge, 1963.

[5] I.B. Djordjevic, S. Sankaranarayanan, S.K. Chilappagari, B. Vasic, Low-density parity-check codes for 40 Gb/s optical transmission systems, IEEE/LEOS J. Sel. Top. Quantum Electron. 12 (4) (2006) 555–562.

[6] J. Chen, A. Dholakia, E. Eleftheriou, M. Fossorier, X.-Y. Hu, Reduced-complexity decoding of LDPC codes, IEEE Trans. Commun. 53 (2005) 1288–1299.

[7] I.B. Djordjevic, B. Vasic, Nonbinary LDPC codes for optical communication systems, IEEE Photon. Technol. Lett. 17 (10) (October 2005) 2224–2226.

[8] M. Arabaci, I.B. Djordjevic, R. Saunders, R.M. Marcoccia, Non-binary quasi-cyclic LDPC based coded modulation for beyond 100-Gb/s transmission, IEEE Photon. Technol. Lett. 22 (6) (March 2010) 434–436.

[9] M.C. Davey, Error-correction Using Low-Density Parity-Check Codes, Ph.D. dissertation, Univ. Cambridge, Cambridge, U.K., 1999.

[10] D. Declercq, M. Fossorier, Decoding algorithms for nonbinary LDPC codes over GF(q), IEEE Trans. Commun. 55 (4) (April 2007) 633–643.

[11] M. Arabaci, I.B. Djordjevic, R. Saunders, R.M. Marcoccia, Polarization-multiplexed rate-adaptive non-binary-LDPC-coded multilevel modulation with coherent detection for optical transport networks, Opt Express 18 (3) (2010) 1820–1832.

[12] M. Arabaci, I.B. Djordjevic, An alternative FPGA implementation of decoders for quasi-cyclic LDPC codes, in: Proc. TELFOR 2008, November 2008, pp. 351–354.

[13] M. Arabaci, I.B. Djordjevic, R. Saunders, R. Marcoccia, Non-binary LDPC-coded modulation for high-speed optical metro networks with back propagation, in: Proc. SPIE Photonics West 2010, OPTO: Optical Communications: Systems and Subsystems, Optical Metro Networks and Short-Haul Systems II, 23–28 January, 2010. San Francisco, California, USA, Paper no. 7621-17.

[14] M. Arabaci, Nonbinary-LDPC-Coded Modulation Schemes for High-Speed Optical Communication Networks, PhD Dissertation, University of Arizona, November 2010.

[15] I.B. Djordjevic, O. Milenkovic, B. Vasic, Generalized low-density parity-check codes for optical communication systems, IEEE/OSA J. Lightw. Technol. 23 (May 2005) 1939–1946.

[16] B. Vasic, I.B. Djordjevic, R. Kostuk, Low-density parity check codes and iterative decoding for long haul optical communication systems, IEEE/OSA J. Lightw. Technol. 21 (February 2003) 438–446.

[17] I.B. Djordjevic, et al., Projective plane iteratively decodable block codes for WDM high-speed long-haul transmission systems, IEEE/OSA J. Lightw. Technol. 22 (3) (March 2004) 695–702.

[18] O. Milenkovic, I.B. Djordjevic, B. Vasic, Block-circulant low-density parity-check codes for optical communication systems, IEEE/LEOS J. Sel. Top. Quantum Electron. 10 (2004) 294–299.

[19] B. Vasic, I.B. Djordjevic, Low-density parity check codes for long haul optical communications systems, IEEE Photon. Technol. Lett. 14 (8) (August 2002) 1208–1210.

[20] S. Chung, et al., On the design of low-density parity-check codes within 0.0045 dB of the Shannon Limit, IEEE Commun. Lett. 5 (February 2001) 58–60.

[21] S. Lin, D.J. Costello, Error Control Coding: Fundamentals and Applications, second ed., Prentice-Hall, Inc., USA, 2004.

[22] R.M. Tanner, A recursive approach to low complexity codes, IEEE Tans. Information Theory IT-27 (September 1981) 533–547.

[23] I.B. Djordjevic, L. Xu, T. Wang, M. Cvijetic, Large girth low-density parity-check codes for long-haul high-speed optical communications, in: Proc. OFC/NFOEC, IEEE/OSA, San Diego, CA, 2008. Paper no. JWA53.

[24] M.P.C. Fossorier, Quasi-cyclic low-density parity-check codes from circulant permutation matrices, IEEE Trans. Inf. Theor. 50 (2004) 1788–1793.

[25] W.E. Ryan, An introduction to LDPC codes, in: B. Vasic (Ed.), CRC Handbook for Coding and Signal Processing for Recording Systems, CRC Press, 2004.

[26] H. Xiao-Yu, E. Eleftheriou, D.-M. Arnold, A. Dholakia, Efficient implementations of the sum-product algorithm for decoding of LDPC codes," in Proc, IEEE Globecom. 2 (November 2001), 1036-1036E.

[27] D.J.C. MacKay, Good error correcting codes based on very sparse matrices, IEEE Trans. Inf. Theor. 45 (1999) 399–431.

[28] T. Richardson, A. Shokrollahi, R. Urbanke, Design of capacity approaching irregular low-density parity-check codes, IEEE Trans. Inf. Theor. 47 (2) (February 2001) 619–637.

[29] I.B. Djordjevic, B. Vasic, MacNeish-Mann theorem based iteratively decodable codes for optical communication systems, IEEE Commun. Lett. 8 (8) (August 2004) 538–540.

[30] Y. Kou, S. Lin, M.P.C. Fossorier, Low-density parity-check codes based on finite geometries: a rediscovery and new results, IEEE Trans. Inf. Theor. 47 (7) (November 2001) 2711–2736.

[31] I.B. Djordjevic, S. Sankaranarayanan, B. Vasic, Irregular low-density parity-check codes for long haul optical communications, IEEE Photon. Technol. Lett. 16 (1) (January 2004) 338–340.

[32] J. Boutros, O. Pothier, G. Zemor, Generalized low density (Tanner) codes, in: Proc. 1999 IEEE Int. Conf. Communication (ICC 1999), vol. 1, 1999, pp. 441–445.

[33] M. Lentmaier, K.S. Zigangirov, On generalized low-density parity check codes based on Hamming component codes, IEEE Commun. Lett. 3 (8) (August 1999) 248–250.

[34] T. Zhang, K.K. Parhi, "High-performance, low-complexity decoding of generalized low-density parity-check codes, Proc. IEEE Globecom. Conf. 1 (2001) 181–185.

[35] I.B. Djordjevic, L. Xu, T. Wang, M. Cvijetic, GLDPC codes with Reed-Muller component codes suitable for optical communications, IEEE Commun. Lett. 12 (Sept. 2008) 684–686.

[36] A.R. Calderbank, P.W. Shor, Good quantum error-correcting codes exist, Phys. Rev. A 54 (1996) 1098–1105.

[37] A.M. Steane, Error correcting codes in quantum theory, Phys. Rev. Lett. 77 (1996) 793.

[38] D.J.C. MacKay, G. Mitchison, P.L. McFadden, Sparse-graph codes for quantum error correction, IEEE Trans. Inf. Theor. 50 (2004) 2315–2330.

[39] T. Camara, H. Ollivier, J.-P. Tillich, Constructions and Performance of Classes of Quantum LDPC Codes, 2005 quant-ph/0502086.

[40] M. Hagiwara, H. Imai, Quantum Quasi-Cyclic LDPC Codes, 2007.

[41] I. Anderson, Combinatorial Designs and Tournaments, Oxford University Press, Oxford, 1997.

[42] B. Vasic, O. Milenkovic, Combinatorial constructions of low-density parity-check codes for iterative decoding, IEEE Trans. Inf. Theor. 50 (June 2004) 1156–1176.

[43] R.C. Bose, On the construction of balanced incomplete block designs, Ann. Eugen. 9 (1939) 353–399.

[44] S. Sankaranarayanan, B. Vasic, Iterative decoding of linear block codes: a parity-check orthogonalization approach, IEEE Trans. Inf. Theor. 51 (September 2005) 3347–3353.

[45] S. Sankaranarayanan, I.B. Djordjevic, B. Vasic, Iteratively decodable codes on *m*-flats for WDM high-speed long-haul transmission, J. Lightwave Technol. 23 (November 2005) 3696–3701.

[46] I.B. Djordjevic, Quantum LDPC codes from balanced incomplete block designs, IEEE Commun. Lett. 12 (May 2008) 389–391.

[47] I.B. Djordjevic, Photonic quantum dual-containing LDPC encoders and decoders, IEEE Photon. Technol. Lett. 21 (13) (July 1, 2009) 842–844.

[48] I.B. Djordjevic, On the photonic implementation of universal quantum gates, Bell states preparation circuit, quantum relay and quantum LDPC encoders and decoders, IEEE Photon. J. 2 (1) (February 2010) 81–91.

[49] I.B. Djordjevic, Photonic implementation of quantum relay and encoders/decoders for sparse-graph quantum codes based on optical hybrid, IEEE Photon. Technol. Lett. 22 (19) (October 1, 2010) 1449–1451.

[50] I.B. Djordjevic, Cavity quantum electrodynamics based quantum low-density parity-check encoders and decoders, Paper no. 7948-38, in: SPIE Photonics West 2011, Advances in Photonics of Quantum Computing, Memory, and Communication IV, The Moscone Center, San Francisco, California, USA, 22–27 January 2011.

[51] T. Brun, I. Devetak, M.-H. Hsieh, Correcting quantum errors with entanglement, Science 314 (2006) 436–439.

[52] M.-H. Hsieh, I. Devetak, T. Brun, General entanglement-assisted quantum error correcting codes, Phys. Rev. A 76 (6) (December 19, 2007) 062313-1–062313-7.

[53] I.B. Djordjevic, Photonic entanglement-assisted quantum low-density parity-check encoders and decoders, Opt. Lett. 35 (9) (May 1, 2010) 1464–1466.

[54] M.-H. Hsieh, Entanglement-Assisted Coding Theory, PhD Dissertation, University of Southern California, August 2008.

[55] I. Devetak, T.A. Brun, M.-H. Hsieh, "Entanglement-assisted quantum error-correcting codes, in: V. Sidoravičius (Ed.), New Trends in Mathematical Physics, Selected Contributions of the XVth International Congress on Mathematical Physics, Springer, 2009, pp. 161−172.
[56] M.-H. Hsieh, W.-T. Yen, L.-Y. Hsu, High performance entanglement-assisted quantum LDPC codes need little entanglement, IEEE Trans Inform Theory 57 (3) (March 2011) 1761−1769.
[57] M.M. Wilde, Quantum Coding with Entanglement, PhD Dissertation, University of Southern California, August 2008.
[58] D. Raghavarao, Constructions and Combinatorial Problems in Design of Experiments, Dover Publications, Inc., New York, 1988.
[59] I.F. Blake, R.C. Mullin, The Mathematical Theory of Coding, Academic Press, New York, 1975.
[60] D. Poulin, Y. Chung, On the iterative decoding of sparce quantum codes, Quant. Inf. Comput. 8 (10) (2008) 987−1000.
[61] Z. Zhang, L. Dolecek, B. Nikolic, V. Anantharam, M. Wainwright, Design of LDPC decoders for improved low error rate performance: quantization and algorithm choices, IEEE Trans. Commun. 57 (11) (November 2009) 3258−3268.
[62] M. Ivkovic, S.K. Chilappagari, B. Vasic, Eliminating trapping sets in low-density parity-check codes by using Tanner graph covers, IEEE Tans. Inform. Theory 54 (8) (August 2008) 3763−3768.
[63] I.B. Djordjevic, L. Xu, T. Wang, M. Cvijetic, Large girth low-density parity-check codes for long-haul high-speed optical communications, in: Optical Fiber Communication Conference/National Fiber Optic Engineers Conference, OSA Technical Digest (CD), Optical Society of America, 2008 paper JWA53.
[64] X. Sun, I.B. Djordjevic, FPGA implementation of rate-adaptive spatially-coupled LDPC codes suitable for optical communications, Opt Express 27 (3) (2019) 3422−3428.
[65] J. Felstrom, K.S. Zigangirov, Time-varying periodic convolutional codes with low-density parity-check matrix, IEEE Trans. Inf. Theor. 45 (6) (1999) 2181−2191.
[66] D.J. Costello, et al., A comparison between LDPC block and convolutional codes, in: IEEE Information Theory and Applications Workshop, 2006.
[67] I. Andriyanova, D. Mauricey, J.-P. Tillich, Spatially coupled quantum LDPC codes, in: Proc. 2012 Information Theory Workshop, 2012, pp. 327−331.
[68] S.S. Shrikhande, Generalized Hadamard matrices and orthogonal arrays of strength two, Can. J. Math. 16 (1964) 736−740.

CHAPTER

Quantum networking

10

Chapter outline

Quantum Communication, Quantum Networks, and Quantum Sensing. https://doi.org/10.1016/B978-0-12-822942-2.00001-7

10.1 Quantum communications networks and the quantum Internet

Quantum information processing (QIP) opens up new avenues for several applications, including high-performance computing, high-precision sensing, and secure communications [1–3]. Among the various QIP attributes, entanglement represents a unique feature, enabling quantum computers to solve problems that are numerically intractable for classical computers [4], enabling quantum-enhanced sensors with measurement sensitivities exceeding the classical limit [5], and providing certifiable security for data transmissions whose security is guaranteed by the laws of quantum mechanics rather than the unproven assumptions used in cryptography based on computational security [2,3].

Preshared entanglement can be used to (1) improve classical channel capacity [6–12], (2) enable secure communications, (3) improve sensor sensitivity, (4) enable distributed quantum computing [13], (5) enable entanglement-assisted (EA) distributed sensing [5], and (6) enable provably secure quantum computer access [14], to mention a few applications. To distribute entangled signal-idler photon pairs, either a fiber-based quantum network or satellite-to-ground links can be used [11,12]. Unfortunately, we cannot ignore dispersion effects, channel attenuation, and phase noise in a fiber-based quantum network. Moreover, entangled states must be stored in quantum memories before being used. Given that they are imperfect, it is necessary to use quantum error correction (QEC) to deal with decoherence effects. On the other hand, the satellite-to-ground link will experience diffraction loss, atmospheric turbulence effects, scattering effects, and background radiation. Clearly, the entanglement distribution cannot be considered ideal in either case. Moreover, it has been shown in Ref. [11] that under imperfect preshared entanglement distribution, EA communication can be inferior to classical communication, depending on the parameters of the distribution channel. Therefore, we must use EA communication in those scenarios that truly benefit from employing it.

This section describes the basics of quantum networking and associated problems. The quantum network represents a network of nodes interconnected by quantum links. The quantum links could be heterogeneous, including fiber-optics links, free-space optical (FSO) links, and satellite links. The quantum communication network (QCN) is a system of quantum nodes in which an uninterrupted quantum channel can be established between any two nodes. When quantum nodes are equipped with quantum computers capable of exchanging qubits (or qudits) over the QCN by teleportation, the corresponding QCN is commonly referred to as the *quantum Internet*. Fig. 10.1 illustrates an envisioned heterogeneous QCN architecture suitable for long-distance entanglement distribution among multiple QIP nodes. Various quantum interconnects (QuICs) and quantum interfaces are needed to implement such a quantum Internet infrastructure. To interconnect and entangle distributed QIP devices, the stationary quantum information must be retrieved on demand from quantum memories and converted into flying (mobile) quantum information by heterogeneous QuIC modules. The photonic quantum information then transmits to the destination through various genres of QuIC modules via heterogeneous fibers, ground-to-satellite, and FSO links. Another QuIC module then converts the received flying quantum information back to stationary quantum information for local processing. Therefore, based on this illustrative example, we conclude that the *quantum interface* can be defined as a quantum module or subsystem allowing to connect the stationary and flying qubits to establish a quantum channel between any two remote nodes.

While QuICs have been individually validated over specific types of quantum links, a full-stack integration of QuICs into a unified, heterogeneous QCN remains elusive at this point. Two

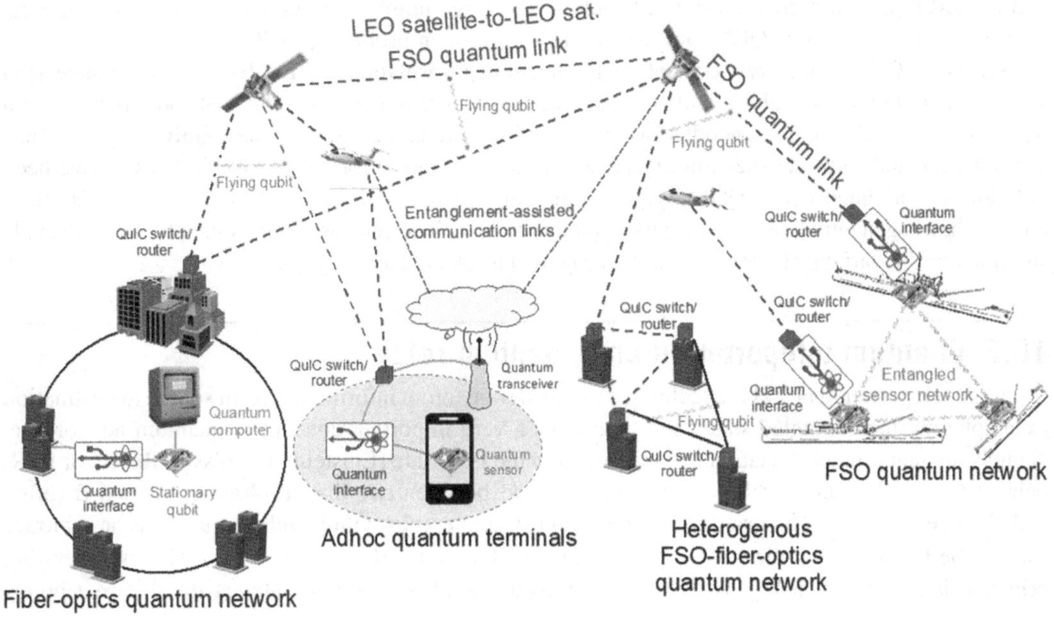

FIGURE 10.1

Envisioned wide-area quantum communication network empowered by heterogeneous quantum interconnects, including quantum switches and routers, transceivers, quantum interfaces, and quantum memories. *FSO*, free-space optical; *LEO*, low Earth orbit.

outstanding challenges must be solved before such a global QCN (or quantum Internet) can be realized. First, the distribution of entanglement over long distances has been an outstanding challenge due to *photon losses*; i.e., quantum signals cannot be amplified without introducing additional noise that degrades or destroys the transmitted entanglement. Hence, quantum communication (QuCom) calls for fundamentally distinct loss-mitigation mechanisms to establish long-range entanglement. In this regard, quantum repeaters have been pursued over the last decade to overcome the exponentially low entanglement distribution rate versus transmission distance in optical fibers. Despite encouraging advances of using silicon vacancy defects in diamond quantum memory to surpass the repeaterless key-generation bound [15], several technological hurdles, including the scalability of quantum devices, the indistinguishability of emitted photons, and practical QEC, have to be overcome before the development of fully functional quantum repeaters for long-distance QuCom. As an alternative solution, satellites have been leveraged as relays for QuCom over thousands of kilometers by virtue of the more favorable quadratic scaling of photon loss versus the distance in FSO links. Remarkably, quantum key distribution [16], quantum teleportation [17], and entanglement distribution [18] have been demonstrated in a ground-to-satellite QuCom testbed across several thousand kilometers, showing tremendous potential for building up satellite-mediated wide-area QuCom networks via FSO links. At present, while QuCom is individually validated over a specific type of quantum point-to-point link, a QCN that enables full integration of diverse QIP devices into a unified QuCom network must still be developed. In particular, the distribution of a large number of quantum states in multiaccess environments over various QCN topologies remains an open problem [19,20].

Secondly, QIP devices rest on stationary quantum information carried by, e.g., solid-state spin states, superconducting qubits, and ultracold atoms, whereas entanglement distribution hinges on flying quantum information encoded on photons. To interconnect QIP devices, QuIC modules must convert back and forth between stationary and flying quantum information. While QuICs have been individually validated over specific types of quantum links, a full-stack integration of QuICs into a unified, heterogeneous QuCom network remains elusive. The purpose of this chapter is to provide solutions to the various challenges that heterogeneous QCNs currently face.

10.2 Quantum teleportation and quantum relay

Quantum teleportation [1,22] is a technique to transfer quantum information from source to destination by employing the entangled states and represents a very important concept in quantum networking. Namely, in quantum teleportation, entanglement in the Bell state (Einstein–Podolsky–Rosen, or EPR, pair) is used to transport arbitrary quantum state $|\psi\rangle$ between two distant observers A and B (often called Alice and Bob). The quantum teleportation system employs three qubits; qubit 1 is an arbitrary state to be teleported, while qubits 2 and 3 are in Bell state $|B_{00}\rangle=(|00\rangle+|11\rangle)/\sqrt{2}$. The operating principle is illustrated in Fig. 10.2, where the evolution of three-qubit states at the different points

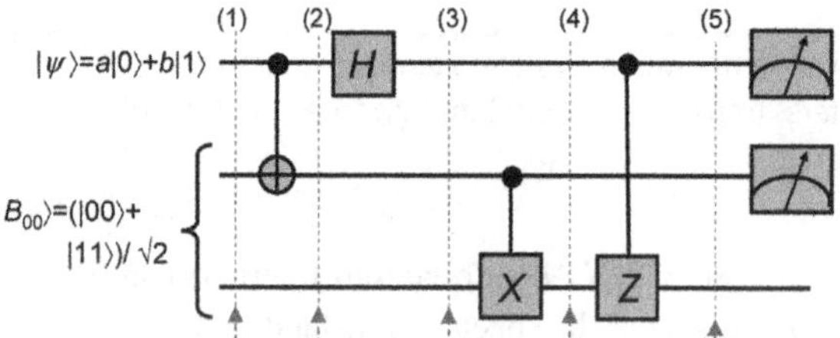

$$|\psi_1\rangle=|\psi\rangle|B_{00}\rangle=(a|0\rangle+b|1\rangle)(|00\rangle+|11\rangle)/\sqrt{2}$$
$$=(a|000\rangle+a|011\rangle+b|100\rangle+b|111\rangle)/\sqrt{2}$$
$$|\psi_2\rangle=CNOT_{12}|\psi_1\rangle=(a|000\rangle+a|011\rangle+b|110\rangle+b|101\rangle)/\sqrt{2}$$
$$|\psi_3\rangle=H_1|\psi_2\rangle$$
$$=\left[a(|0\rangle+|1\rangle)|00\rangle+a(|0\rangle+|1\rangle)|11\rangle+b(|0\rangle-|1\rangle)|10\rangle+b(|0\rangle-|1\rangle)|01\rangle\right]/2$$
$$|\psi_4\rangle=CNOT_{23}|\psi_3\rangle$$
$$=\left[a(|0\rangle+|1\rangle)|00\rangle+a(|0\rangle+|1\rangle)|10\rangle+b(|0\rangle-|1\rangle)|11\rangle+b(|0\rangle-|1\rangle)|01\rangle\right]/2$$
$$=\left[a|000\rangle+a|100\rangle+a|010\rangle+a|110\rangle+b|011\rangle-b|111\rangle+b|001\rangle-b|101\rangle\right]/2$$
$$|\psi_5\rangle=C_{13}(Z)|\psi_4\rangle$$
$$=\left[a|000\rangle+a|100\rangle+a|010\rangle+a|110\rangle+b|011\rangle+b|111\rangle+b|001\rangle+b|101\rangle\right]/2$$
$$=\left[|00\rangle\underbrace{(a|0\rangle+b|1\rangle)}_{|\psi\rangle}+|01\rangle\underbrace{(a|0\rangle+b|1\rangle)}_{|\psi\rangle}+|10\rangle\underbrace{(a|0\rangle+b|1\rangle)}_{|\psi\rangle}+|11\rangle\underbrace{(a|0\rangle+b|1\rangle)}_{|\psi\rangle}\right]/2$$

FIGURE 10.2

Explaining the quantum teleportation principle.

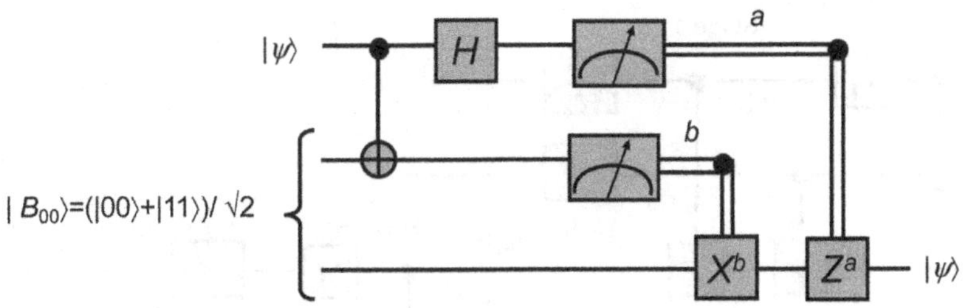

FIGURE 10.3

Quantum teleportation scheme performing the measurement in the middle of the circuit.

(1)–(5) is provided as well. On the destination side, we perform the measurement on the first two qubits, and the possible results are $|00\rangle$, $|01\rangle$, $|10\rangle$, and $|11\rangle$. Whatever the content results, the third qubit is always $|\psi\rangle$, which is the same as the desired state to be teleported.

Another alternative quantum teleportation scheme is shown in Fig. 10.3. In this version, we perform the measurement in the middle of the circuit; based on the measurement results, we conditionally execute *X*- and *Z*-operators. We can interpret this circuit as the application of the principle of deferred measurement.

To extend the total teleportation distance, *quantum repeaters* [22], entanglement purification [23], and *quantum relays* [24–26] can be used. The key obstacle quantum repeater and relay for implementation is the no-cloning theorem. However, we can use the quantum teleportation principle described above to implement the quantum relay. One such quantum relay, based on the quantum teleportation circuit from Fig. 10.2, is shown in Fig. 10.4.

Applying the principle of implicit measurement, we can represent the circuit shown in Fig. 10.4 as provided in Fig. 10.5. Clearly, Stage 1 is the same as Fig. 10.2, while Stage 2 is the same as Fig. 10.3. Both stages serve as quantum teleportation circuits. So the content of qubit 3 at the output of Stage 1 is

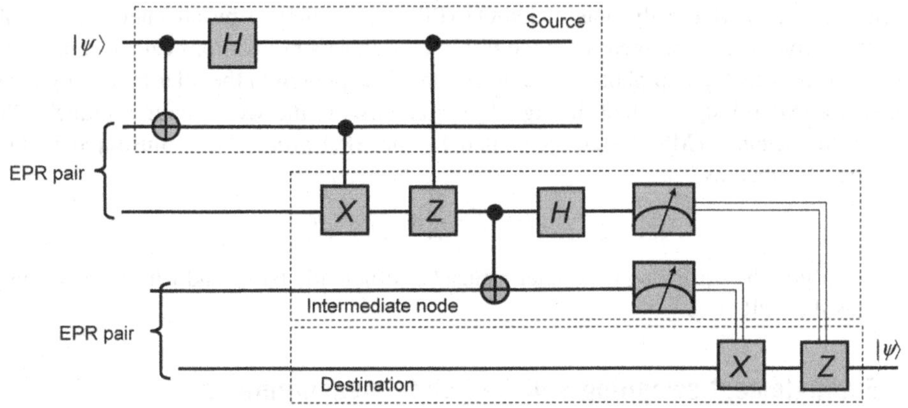

FIGURE 10.4

Quantum relay.

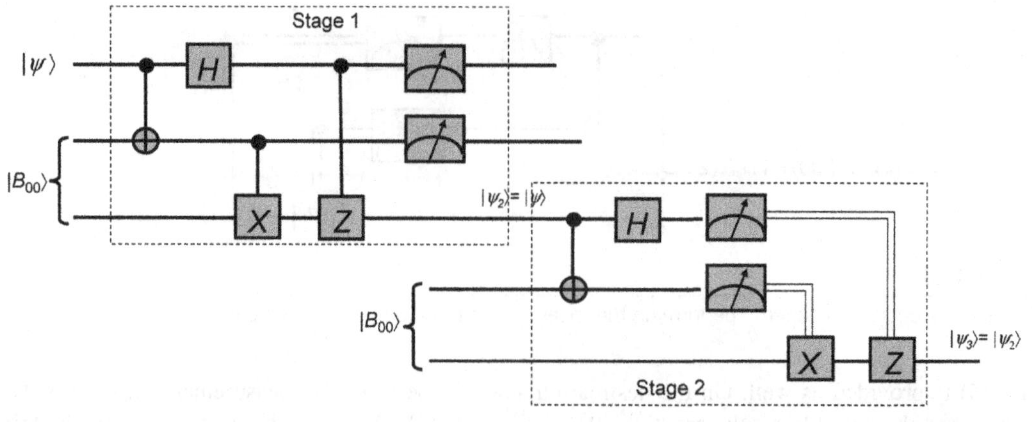

FIGURE 10.5

Quantum relay and principle of implicit measurement.

$|\psi_2\rangle = |\psi\rangle$. On the other hand, as Stage 2 is a teleportation circuit, the content of qubit 6 at the output of Stage 2 will be the same as the qubit 3 input of Stage 2—that is, we can write $|\psi_3\rangle = |\psi_2\rangle = |\psi\rangle$. Therefore, the overall action of the quantum circuit is to teleport the quantum state $|\psi\rangle$ from qubit 1 to qubit 6.

10.3 Entanglement distribution

Entanglement distribution is a key technique enabling quantum communication and quantum networking by employing the quantum teleportation concept. The generation of entangled states has already been discussed in Chapter 5 (see Section 5.9), here we repeat some basic ideas for completeness of presentation. In Fig. 10.6(left), we illustrate the well-known heralded generation of single-photon states through parametric downconversion (PDC). The crystal exhibiting second-order nonlinearity, such as periodically poled LiNbO3 (PPLN), is pumped by an intense beam of light. Thanks to the second-order nonlinearity of the PPLN crystal, a pair of photons called signal s and idler i are generated from each pump photon. The detection of the photon in the idler branch by the single-photon detector (SPD) indicates that the signal photon exists in the second output branch. The two-mode squeezed-vacuum (TMSV) state generated by the SPD process and illustrated in Fig. 10.6 (right) can be described as

$$|\psi\rangle = e^{za^\dagger b^\dagger - z^* ab}|0,0\rangle, z = re^{j\theta}, \tag{10.1}$$

where $z = r\exp(j\theta)$. The squeezing parameter r must be sufficiently large, and when $r \to \infty$, the TMSV state becomes the EPR state.

10.3.1 Entanglement swapping and Bell state measurements

The basic technique to entangle two systems that have previously not been interacting is *entanglement swapping (ES)*, which is described in Fig. 10.7. Two TMSVs are first generated by the SPD process to obtain

$$|\psi_{12}\rangle = e^{za_1^\dagger a_2^\dagger - z^* a_1 a_2}|0,0\rangle, \quad |\psi_{34}\rangle = e^{za_3^\dagger a_4^\dagger - z^* a_3 a_4}|0,0\rangle. \tag{10.2}$$

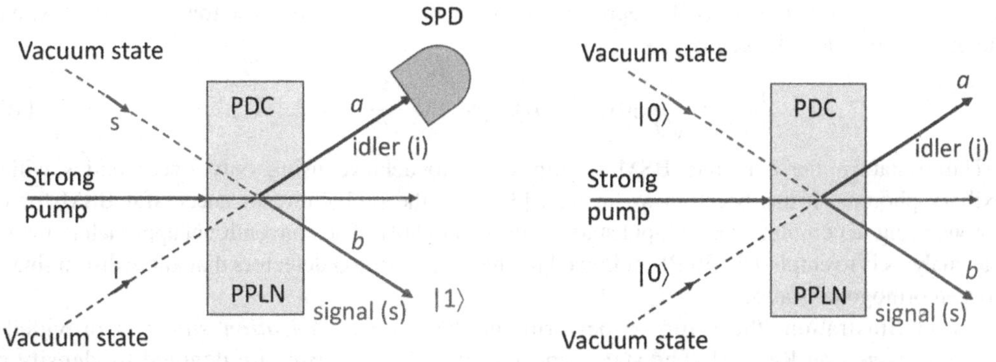

FIGURE 10.6

The heralded generation of the single photon states and two-mode squeezed vacuum state $|\psi\rangle$. *SPD*: single-photon detector, *PDC*: parametric downconversion, *PPLN*: periodically poled lithium niobate crystal.

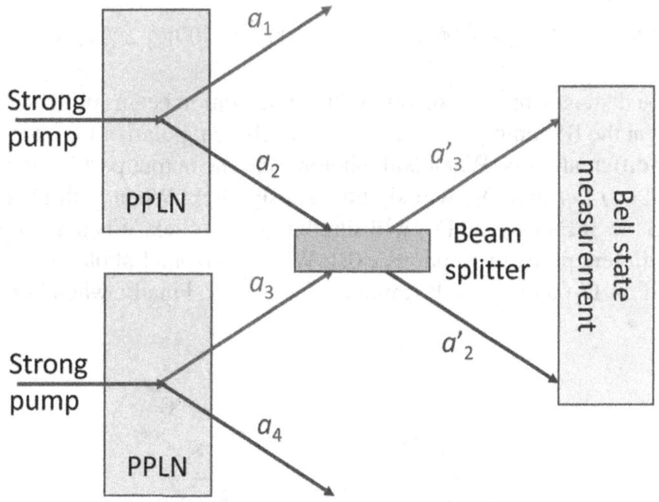

FIGURE 10.7

Entanglement swapping.

We then interact photons represented by annihilation operators a_2 and a_3 on a beam splitter (BS), followed by *Bell state measurement* (BSM). By doing so we effectively perform the projection of photons a_1 and a_4 onto a different entangled state related to the four Bell states so that photons a_1 and a_4 become entangled even though they were not initially entangled.

BSMs play a key role in photonic quantum computation and communication, including quantum teleportation [21], superdense coding [27], quantum swapping [28], quantum repeaters [29], and

QKD [30–33]. A *complete BSM* represents a projection of any two-photon state to maximally entangled Bell states, defined by

$$|\psi^{\pm}\rangle = \frac{1}{\sqrt{2}}(|01\rangle \pm |10\rangle), \quad |\phi^{\pm}\rangle = \frac{1}{\sqrt{2}}(|00\rangle \pm |11\rangle). \tag{10.3}$$

Unfortunately, the complete BSM is impossible to achieve using only linear optics without auxiliary photons. It has been shown in Ref. [34] that the probability of successful BSM for two photons being in complete mixed input states is limited to 50%. The conventional approach to the Bell state analysis is to employ 50:50 BS, followed by the single-photon detectors that allow discrimination between orthogonal states.

As an illustration, the setup to perform the *BSM on polarization states* is provided in Fig. 10.8, based on Ref. [2]. The states emitted by Alice and Bob are denoted by density operators ρ_A and ρ_B, respectively. When photons at BS inputs 1 and 2 are of the same polarization, that is $|HH\rangle = |00\rangle$ or $|VV\rangle = |11\rangle$, according to Hong-Ou-Mendel effect [35,36], described in next section, they will appear simultaneously at either output port 3 or port 4, and the Bell states $|\phi^{\pm}\rangle$ are transformed to

$$|\phi^{\pm}\rangle = \frac{1}{\sqrt{2}}(|00\rangle_{12} \pm |11\rangle_{12}) \overset{BS}{\rightarrow} |\phi^{\pm}_{\text{out}}\rangle = \frac{j}{2\sqrt{2}}(|00\rangle_{33} + |00\rangle_{44} \pm |11\rangle_{33} \pm |11\rangle_{44}), \tag{10.4}$$

and clearly cannot be distinguished by corresponding polarization beam splitters (PBSs). On the other hand, when photons at the BS input ports 1 and 2 are of different polarizations, they can exit either the same output port or different ports. When both photons exit the output port 3, after the PBS the single photon detectors (SPDs) D_{10} and D_{11} will simultaneously click. When both photons exit the output port 4, after the PBS the SPDs D_{20} and D_{21} will simultaneously click. When input photons of different polarizations exit different ports, two options exist. When horizontal photon exits port 3 and vertical photon port 4, the SPDs D_{11} and D_{20} will simultaneously click. Finally, when horizontal photon exits

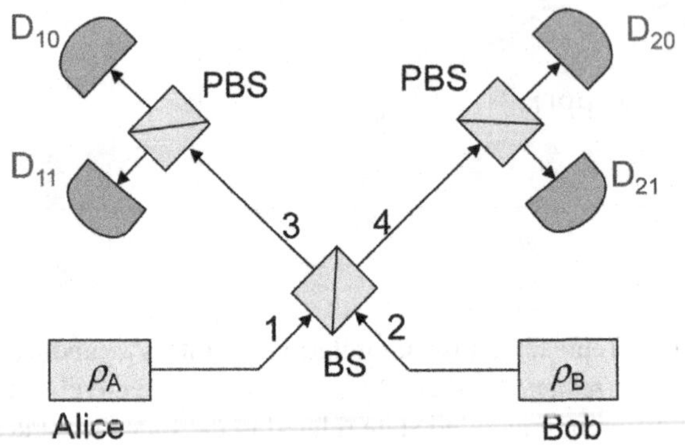

FIGURE 10.8

Setup to perform Bell state measurements on polarization states.

port 4 and vertical photon port 3, the SPDs D_{10} and D_{21} will simultaneously click. Given that the Bell states $|\psi^{\pm}\rangle$ are transformed (by the BS) to

$$\begin{aligned} |\psi^{+}\rangle &= \frac{1}{\sqrt{2}}(|01\rangle_{12} + |10\rangle_{12}) \stackrel{BS}{\rightarrow} |\psi_{out}^{+}\rangle = j(|01\rangle_{33} + |10\rangle_{44}), \\ |\psi^{-}\rangle &= \frac{1}{\sqrt{2}}(|01\rangle_{12} - |10\rangle_{12}) \stackrel{BS}{\rightarrow} |\psi_{out}^{-}\rangle = |01\rangle_{34} - |10\rangle_{34}, \end{aligned} \tag{10.5}$$

which can be distinguished by PBSs and SPDs.

Based on the discussion above, we conclude that the projection on $|\psi^{+}\rangle$ occurs when both SPDs after any of two PBSs simultaneously click. In other words, when either D_{10} and D_{11} or D_{20} and D_{21} simultaneously click, the projection on $|\psi^{+}\rangle$ is identified. Similarly, when either D_{10} and D_{21} or D_{20} and D_{21} simultaneously click, the projection on $|\psi^{-}\rangle$-state is identified. In conclusion, when ideal SPDs are used, BSM efficiency is ½. If the SPDs have nonideal detection efficiency η, the BSM efficiency will be $0.5\eta^2$.

10.3.2 Hong–Ou–Mendel effect

Two photons belonging to different systems can be entangled by the BS through the Hong–Ou–Mendel (HOM) effect [35,36], which is one of the most well-known two-photon interference phenomena. It has found applications in linear optical quantum computing [36] and quantum communications [37], among others. In Fig. 10.9A, we illustrate BS action on a single photon. Assuming that amplitude reflection and transmission coefficients are denoted by A_r and A_t, respectively, the BS operator can be represented by

$$\begin{aligned} U_{BS} &= A_t|d_2\rangle\langle l| + A_r|d_1\rangle\langle l| + A_r|d_2\rangle\langle r| + A_t|d_1\rangle\langle r|, \\ &|A_r|^2 + |A_t|^2 = 1, A_tA_r^* + A_rA_t^* = 0, \end{aligned} \tag{10.6}$$

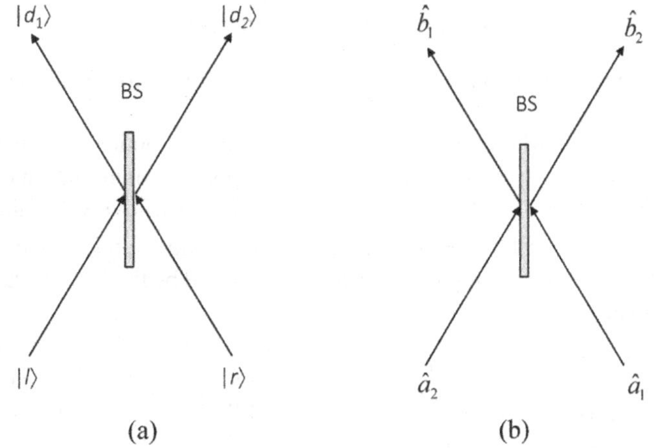

FIGURE 10.9

(A) Beam splitter (BS) operation and (B) illustration of Hon–Ou–Mendel effect for two photons.

where we use the $|l\rangle$ state ($|r\rangle$ state) to denote the photon coming from the left (right) side of the BS and $|d_i\rangle$ ($i = 1,2$) to denote the corresponding BS outputs. The reflection and transmission amplitude coefficients can be parameterized by $A_r = \cos\theta$, $A_t = \text{j}\ \sin\theta$, and the BS operator action can also be represented as

$$U_{BS} = \cos\,\theta(|d_1\rangle\langle l| + |d_2\rangle\langle r|) + j\,\sin\,\theta(|d_2\rangle\langle l| + |d_1\rangle\langle r|). \tag{10.7}$$

In similar fashion, the PBS action on either the horizontal photon $|\text{H}\rangle$ or the vertical photon $|\text{V}\rangle$ coming from the left or right side of the PBS can be described by

$$U_{PBS} = A_r\left|H_l^{(\text{out})}\right\rangle\left\langle H_l^{(\text{in})}\right| + A_r\left|H_r^{(\text{out})}\right\rangle\left\langle H_r^{(\text{in})}\right| + A_t\left|V_l^{(\text{out})}\right\rangle\left\langle V_r^{(\text{in})}\right| + A_t\left|V_r^{(\text{out})}\right\rangle\left\langle V_l^{(\text{in})}\right|. \tag{10.8}$$

In other words, the horizontal photon is reflected by the PBS, while the vertical photon goes through the PBS. Finally, the action of the half-wave plate at 45° can be described as

$$U_{\lambda/2,45^o} = \left|V^{(\text{out})}\right\rangle\left\langle H^{(\text{in})}\right| + \left|H^{(\text{out})}\right\rangle\left\langle V^{(\text{in})}\right|. \tag{10.9}$$

In other words, the horizontal photon gets converted to the vertical photon and vice versa.

Let us now observe two photons at BS inputs, as shown in Fig. 10.9A. Using the formalism given by Eq. (10.6), we can represent the joint output state of the BS as follows:

$$\begin{aligned}|\psi\rangle_{\text{out}} &= (U_{BS}|l\rangle)\otimes(U_{BS}|r\rangle) = (A_r|d_1\rangle + A_t|d_2\rangle)(A_r|d_2\rangle + A_t|d_1\rangle)\\ &= A_r^2|d_1\rangle|d_2\rangle + A_rA_t|d_1\rangle|d_1\rangle + A_tA_r|d_2\rangle|d_2\rangle + A_t^2|d_2\rangle|d_1\rangle.\end{aligned} \tag{10.10}$$

Now by replacing $A_r = \cos\theta$, $A_t = \text{j}\sin\theta$, we obtain

$$|\psi\rangle_{\text{out}} = \cos^2\,\theta|d_1\rangle|d_2\rangle + j\,\sin\,\theta\,\cos\,\theta|d_1\rangle|d_1\rangle + j\,\sin\,\theta\,\cos\,\theta|d_2\rangle|d_2\rangle - \sin^2\,\theta|d_2\rangle|d_1\rangle. \tag{10.11}$$

For a 50:50 BS, $\cos\theta = \sin\theta = 1/\sqrt{2}$, so the first and last term cancel each other, which is commonly referred to as the *two-photon destructive interference*, and the output state becomes

$$|\psi\rangle_{\text{out}} = \frac{j}{2}|d_1\rangle|d_1\rangle + \frac{j}{2}|d_2\rangle|d_2\rangle, \tag{10.12}$$

indicating that both photons arrive together at the same output port. This effect is commonly referred to as the HOM effect.

Let us now explain the HOM effect by employing the quantum optics formalism we introduced in Chapter 5 (related to Gaussian transformations). The corresponding annihilation operators of photons at the input of the BS (see Fig. 10.9B) are denoted $\hat{a}_1$ and $\hat{a}_2$, respectively. Let the photons at the BS output be represented by the annihilation operators denoted $\hat{b}_1$ and $\hat{b}_2$, respectively. In the Heisenberg picture, the annihilation operators of input modes are transformed by the BS according to the following Bogoliubov transformation:

$$\begin{bmatrix}\hat{b}_1\\ \hat{b}_2\end{bmatrix} = \begin{bmatrix}\hat{U}a_1\hat{U}^\dagger\\ \hat{U}a_2\hat{U}^\dagger\end{bmatrix} = \begin{bmatrix}\sqrt{T} & \sqrt{1-T}\\ -\sqrt{1-T} & \sqrt{T}\end{bmatrix}\begin{bmatrix}\hat{a}_1\\ \hat{a}_2\end{bmatrix} = \begin{bmatrix}t & r\\ -r & t\end{bmatrix}\begin{bmatrix}\hat{a}_1\\ \hat{a}_2\end{bmatrix}. \tag{10.13}$$

So the output state of the BS splitter will be

$$|\Psi_{\text{out}}\rangle = \hat{U}|1\rangle_1|1\rangle_2 = \hat{U}\hat{a}_1^\dagger|0\rangle_1\hat{a}_2^\dagger|0\rangle_2 = \underbrace{\hat{U}\hat{a}_1^\dagger\hat{U}^\dagger}_{\hat{b}_1^\dagger}\underbrace{\hat{U}\hat{a}_2^\dagger\hat{U}^\dagger}_{\hat{b}_2^\dagger}\underbrace{\hat{U}|0\rangle_1|0\rangle_2}_{|0\rangle_1|0\rangle_2}$$

$$\begin{aligned}
&= \left(t\hat{a}_1^\dagger + r\hat{a}_2^\dagger\right)\left(-r\hat{a}_1^\dagger + t\hat{a}_2^\dagger\right)|0\rangle_1|0\rangle_2 \\
&= \left[-tr\left(\hat{a}_1^\dagger\right)^2 + t^2\hat{a}_1^\dagger\hat{a}_2^\dagger - r^2\hat{a}_2^\dagger\hat{a}_1^\dagger + rt\left(\hat{a}_2^\dagger\right)^2\right]|0\rangle_1|0\rangle_2 \\
&= -tr\sqrt{2}|2\rangle_1|0\rangle_2 + (t^2 - r^2)|1\rangle_1|1\rangle_2 + rt\sqrt{2}|0\rangle_1|2\rangle_2,
\end{aligned} \tag{10.14}$$

where we use a different type of BS (consistent with Chapter 5) and use $|n\rangle$ ($n = 0,1,2$) to denote the number state. For a 50:50 BS, we have $t = r = 1/\sqrt{2}$, so the two photons at different outputs cancel, and the output state becomes

$$|\Psi_{\text{out}}\rangle = -\frac{1}{\sqrt{2}}|2\rangle_1|0\rangle_2 + \frac{1}{\sqrt{2}}|0\rangle_1|2\rangle_2, \tag{10.15}$$

which represents the two-photon NOON state. In other words, two photons arrive together at the same output port with the same probability.

10.3.3 Continuous variable quantum teleportation

The entanglement swapping scenario and teleportation, in which an intermediate node distributes the entanglement, are illustrated in Fig. 10.10. In this scenario, both Bob and Alice possess CV states.

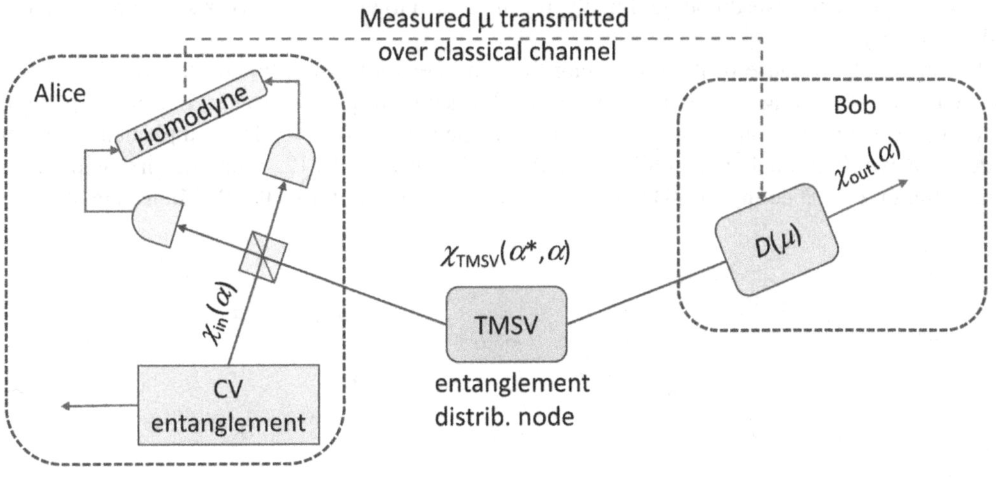

FIGURE 10.10

Entanglement swapping and teleportation with an intermediate node distributing entanglement.

Alice mixes her CV mode with the TSMV mode on the BS. To characterize this input, we can use the characteristic function, defined as $\chi_{in}(\alpha) = \langle D(\alpha) \rangle$, where $D(\alpha)$ is the displacement operator. The TMSV state characteristic function can be determined by $\chi_{TMSV}(\alpha^*, \alpha) = \exp\left[-|\alpha|^2 \exp(-2r)\right]$, where r is the squeezing parameter. Alice performs homodyne detection on the BS outputs to obtain μ and transmits it over a classical channel to Bob, who uses it to perform the displacement operator $D(\mu)$ on his qubit from the TMSV pair. Bob's characteristic function is given by $\chi_{out}(\alpha) = \chi_{in}(\alpha)\chi_{TMSV}(\alpha^*, \alpha)$. Clearly, only when $r \to \infty$ Bob will be able perfectly to recover Alice's transmitted state, since then $\chi_{TMSV}(\alpha^*, \alpha) \to 1$.

10.3.4 Quantum network coding

Quantum network coding allows simultaneous quantum teleportation between the quantum repeaters that are far from each other and can be used to reduce the cost of large-scale ES deployment [38–41].

The concept of network coding was introduced in 1978 in the context of satellite communication in which two ground users communicated with the help of a satellite to XOR the input streams and broadcast the combined stream. Clearly, this concept is similar to the popular butterfly network shown in Fig. 10.11 and reintroduced in the networking framework in 2000 [42]. In the butterfly network, two source nodes S_1 and S_2 would like to distribute information a and b to two destination nodes D_1 and D_2 with the help of two intermediate nodes I_1 and I_2. At the same time, the source node S_1 (S_2) distributes the available information to destination node D_1 (D_2). Both source nodes forward available information to the first intermediate node, which XORs it and forwards it to the second intermediate node. The second intermediate node forwards the received information from I_1 to both destination nodes. The destination node D_1 has information originating from S_1 already available, and to get the information from S_2, it XORs the information obtained from S_1 and I_2 to get $a \oplus (a \oplus b) = b$. In a similar fashion, the destination node D2 recovers the information sent by S2 by $b \oplus (a \oplus b) = a$. The linear network coding is suitable for broadcasting and multicasting applications. It can also be used to improve resilience to eavesdropping. Finally, it can be used to improve network scalability, efficiency, and throughput.

Unfortunately, because of the no-cloning theorem, network coding is not directly extendable to quantum network coding, as has been shown by Hayashi et al. [38]. However, in Ref. [39], Hayashi showed that when two source nodes share preentanglement, it is possible to implement a quantum butterfly network through teleportation, as illustrated in Fig. 10.12, with the first experimental demonstration described in Ref. [41]. The source node S_1 (S_2) performs the BSM on state $|\psi_1\rangle$ ($|\psi_2\rangle$)

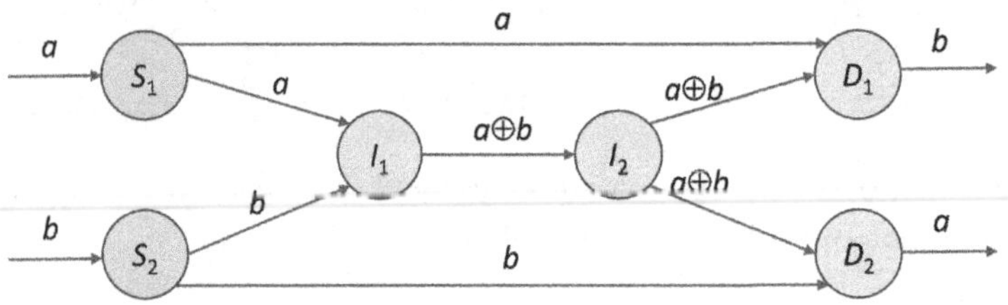

FIGURE 10.11

Illustrating classical network coding on a butterfly network.

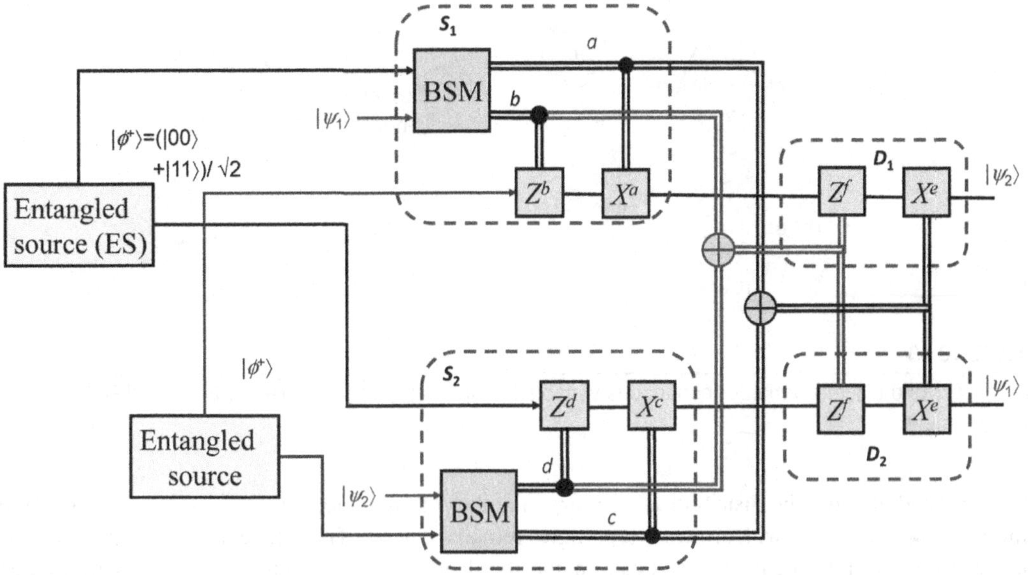

FIGURE 10.12

Illustrating quantum network coding on butterfly network. *BSM*: Bell state measurement.

and one qubit of the $|\phi^+\rangle$ entangled state. Based on the results of BSM *ab* (*cd*), the source node S_1 (S_2) applies the unitary operation X^aZ^b on the second qubit in $|\phi^+\rangle$. The intermediate node XORs *a* and *c* (*b* and *d*) to obtain $e = a \oplus c$ and $f = b \oplus d$. The destination nodes D_1 and D_2 apply the unitary gate X^eZ^f to recover the respective qubits $|\psi_2\rangle$ and $|\psi_1\rangle$. This concept is straightforwardly extendable to higher-complexity quantum networks.

10.4 Engineering entangled states and hybrid continuous-variable–discrete-variable quantum networks

This section describes how to engineer entangled states employing photon addition and subtraction modules, described in Section 10.4.2, for hybrid CV–DV quantum networks.

10.4.1 Hybrid continuous-variable–discrete-variable quantum networks

To take advantage of quantum resources for QIP, distributed quantum computing, quantum networking, and distributed quantum sensing, it is necessary to interface quantum systems based on different information encodings (either discrete or continuous). To illustrate, most current approaches for quantum computers are based on discrete-variable (DV) encoding of information, while continuous-variable (CV) bosonic quantum systems are known to be more suitable for QCNs and future quantum Internet. Furthermore, hybrid DV–CV QCNs can enable the deterministic teleportation of DV states [43]. Even though entanglement distribution over point-to-point links has been

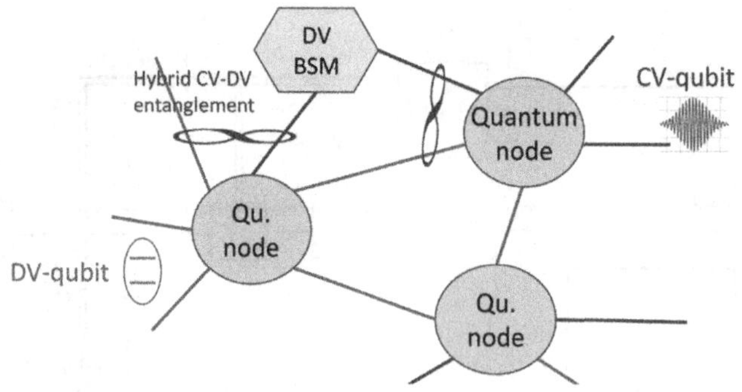

FIGURE 10.13

Vision of hybrid discrete-variable–continuous-variable transparent quantum network. *BSM*: Bell state measurement.

demonstrated already, the distribution of a large number of entangled states in multiaccess and multihop quantum networks with arbitrary topologies remains elusive. The efficient and robust interfacing between CV and DV quantum nodes can enable the distribution of quantum states over transparent hybrid DV–CV multihop, multiuser networks with arbitrary network topology. Thus, developing the means to connect DV and CV optical quantum systems will enable a major step forward in optical QuICs and networks. The vision of the hybrid CV–DV quantum network is illustrated in Fig. 10.13. Nodes composed of transmitters and receivers capable of generating DV, CV, and hybrid DV–CV entangled states are used as the backbone of the quantum network. Bell-state measurements are then used to perform teleportation and entanglement swapping operations for the CV-to-DV and DV-to-CV information transfer and interconnection between different types of nodes. Hybrid DV–CV systems represent a new paradigm in the encoding of information for various quantum applications [43–48]. An interface between DV–CV optical systems is needed to optimally interconnect various types of DV quantum systems over a long distance. The generation of hybrid DV–CV entangled states, their teleportation, and entanglement swapping through entangling measurements are described in the upcoming subsections.

10.4.2 Photon addition and photon subtraction modules

The photon addition module is based on PDC, as introduced in Fig. 10.6. By replacing the vacuum state in the input signal port with a coherent state $|\alpha\rangle$, we are able to add a single photon to the coherent state to get $a^\dagger|\alpha\rangle$, where $a^\dagger$ is the creation operator, and this process is illustrated in Fig. 10.14.

A BS can be used to perform a single-photon subtraction, as illustrated in Fig. 10.15. The operation of a BS can be described by the following unitary transformation $\hat{U} = \exp\left[j\theta\left(\hat{a}^\dagger\hat{b} + \hat{a}\hat{b}^\dagger\right)\right]$, and for a small θ, the action on input state $|\alpha\rangle|0\rangle$ will be

$$\begin{aligned}|\psi\rangle_{\text{out}} &= \exp\left[j\theta\left(\hat{a}^\dagger\hat{b} + \hat{a}\hat{b}^\dagger\right)\right]|\alpha\rangle|0\rangle \cong |\alpha\rangle|0\rangle + j\theta\left(\hat{a}^\dagger\hat{b} + \hat{a}\hat{b}^\dagger\right)|\alpha\rangle|0\rangle \\ &= |\alpha\rangle|0\rangle + j\theta(\hat{a}|\alpha\rangle)|1\rangle.\end{aligned} \tag{10.16}$$

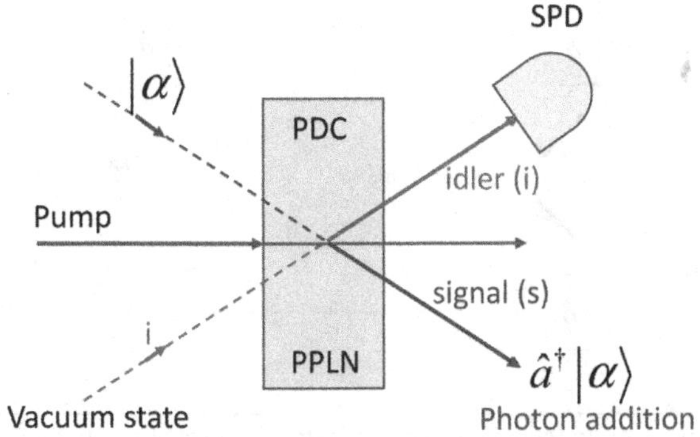

FIGURE 10.14

Illustration of a single-photon addition. *SPD*: single-photon detector, *PDC*: parametric downconversion, *PPLN*: periodically poled lithium niobate crystal.

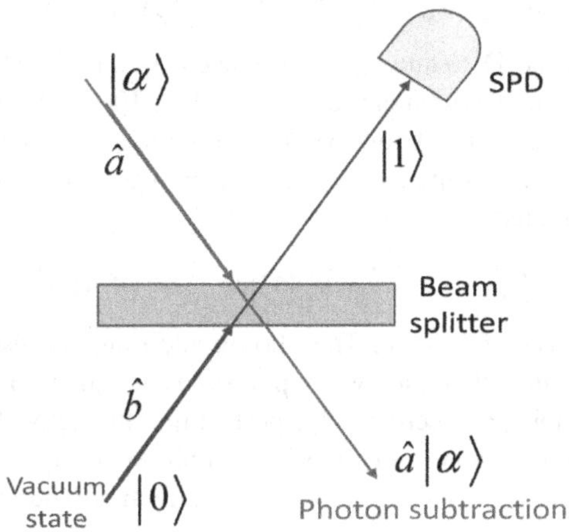

FIGURE 10.15

Illustration of a single-photon subtraction by a beam splitter.

Detection of a photon by the SPD heralds the single-photon subtraction at other output ports.

The photon addition and subtraction are two basic entanglement engineering operations to be used in the creation of multipartite entangled states, hybrid CV–DV entangled states, teleportation, and entanglement swapping.

10.4.3 Generation of hybrid discrete-variable–continuous-variable entangled states

The photon addition modules can be used to entangle two independent DV quantum states, as illustrated in Fig. 10.16. When the photon gets detected on SPD1 (SPD2), we do not know whether it

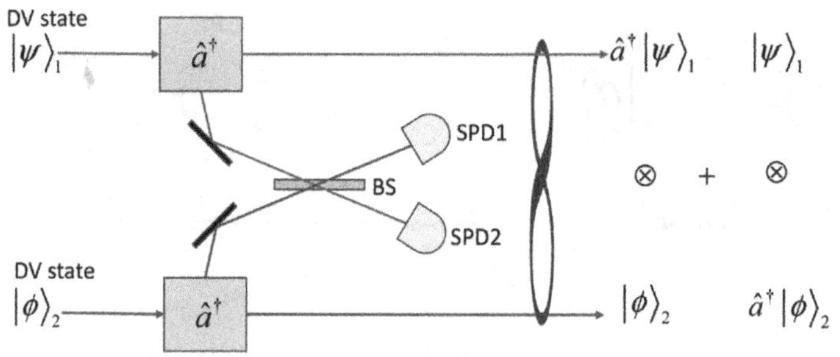

FIGURE 10.16

Entangling two independent discrete-variable quantum states by delocalized photon addition.

originated from the upper or lower photon addition module. This inability to distinguish between the two possibilities indicates that the following quantum state is entangled:

$$|\Psi\rangle_{out} = 2^{-1/2}\left[\left(\hat{a}^\dagger|\psi\rangle\right)|\phi\rangle + |\psi\rangle\left(\hat{a}^\dagger|\phi\rangle\right)\right]. \tag{10.17}$$

In similar fashion, when the DV state $|\phi\rangle_2$ is replaced by the CV state $|\alpha\rangle_2$, we will be able to generate the hybrid DV–CV state, which is illustrated in Fig. 10.17). Similarly, when the photon gets detected on SPD1 (SPD2), we do not know whether it originated from the upper or lower photon addition module. This inability to distinguish between the two possibilities indicates that the following hybrid DV–CV state is entangled:

$$|\Psi\rangle_{out} = 2^{-1/2}\left[\left(\hat{a}^\dagger|\psi\rangle\right)|\alpha\rangle + |\psi\rangle\left(\hat{a}^\dagger|\alpha\rangle\right)\right], \tag{10.18}$$

which represents the hybrid DV–CV state. The photon addition can also be applied to multipartite entangled states. Let us assume that the two input states to quantum circuity in Fig. 10.16 are multipartite with M and N qubits, respectively. By performing the delocalized photon addition on the Mth qubit from the top multipartite state and on the first qubit on the bottom multipartite state; once the photons have been simultaneously detected on SPDs, we effectively herald the multipartite state with $(M + N)$ qubits (Fig. 10.17).

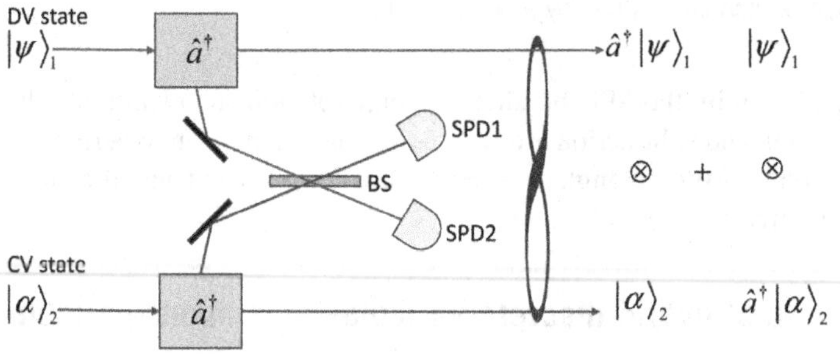

FIGURE 10.17

Generation of entangled hybrid discrete-variable–continuous-variable state by delocalized photon addition.

10.4.4 Hybrid continuous-variable–discrete-variable state teleportation and entanglement swapping through entangling measurements

To perform entanglement swapping between DV only and hybrid DV–CV nodes, we can perform a two-photon subtraction approach, as shown in Fig. 10.18. When the SPD1 (SPD2) clicks, we do not know whether the signal originates from photon subtracted from a DV state $|\phi\rangle_2$ or a DV state $|\psi\rangle_3$, and this uncertainty entangles DV state $|\psi\rangle_1$ and CV state $|\alpha\rangle_4$, effectively performing entanglement swapping.

The entanglement swapping scenario and teleportation in which an intermediate node distributes the entanglement is illustrated in Fig. 10.19. In this scenario, Bob possesses a CV state, while Alice a hybrid DV–CV state. Alice mixes her CV mode (from the hybrid state) with the TSMV mode on a BS. Alice performs the homodyne detection on BS outputs to obtain μ and transmits it over the classical channel to Bob, who uses it to perform the displacement operator $D(\mu)$ on his qubit from the TMSV pair.

Very often, Schrödinger's cat states are used to represent CV qubits [44,48], as follows: $|\psi_{CV}\rangle = N_{\pm}(|\alpha\rangle \pm |-\alpha\rangle)$, where $N_{\pm}$ is the normalization factor. The cat states are typically obtained as the approximation of single-mode squeezed vacuum states for properly chosen squeezing parameter r, such as $r = 0.18$ [44]. Unfortunately, such generated cat states have little tolerance for loss. To overcome this problem, the cat states will be used only for reference. Other CV states that are more tolerant to loss effects are currently under investigation. For instance, the *entangled-photon holes* introduced by Franson [49] exhibit good tolerance to loss and amplification [50] and can be generated as shown in Fig. 10.20. Two-photon subtraction can be represented by $|\psi\rangle = \hat{a}\hat{b}S_2(z)|0,0\rangle$, where

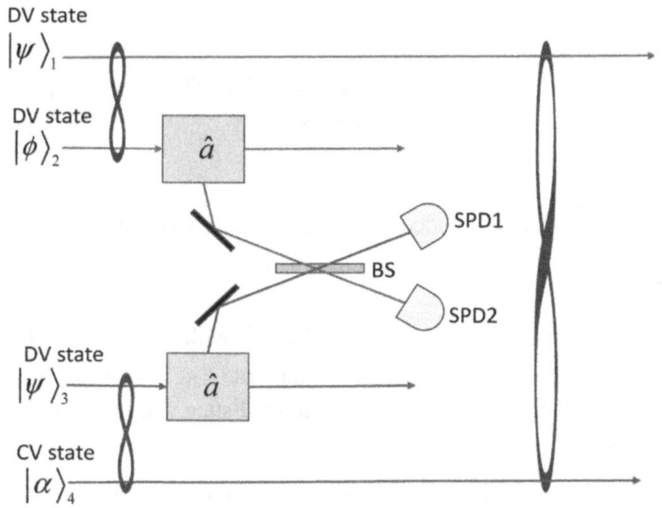

FIGURE 10.18

Entanglement swapping between discrete-variable only node and hybrid discrete-variable–continuous-variable node by two-photon subtraction.

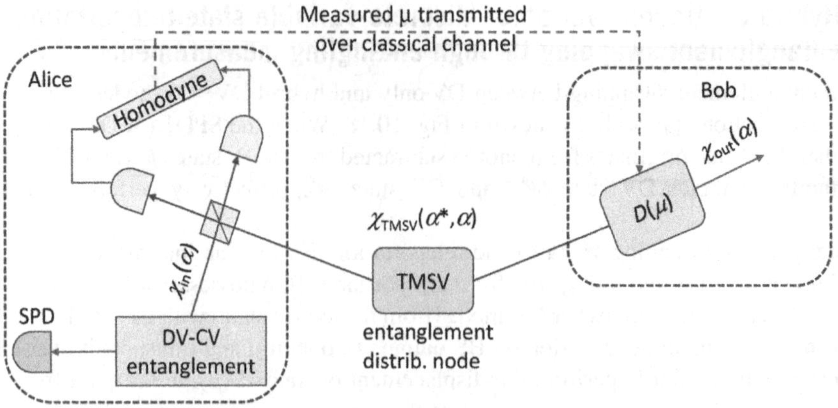

FIGURE 10.19

Entanglement swapping and teleportation in which an intermediate node distributes the entanglement.

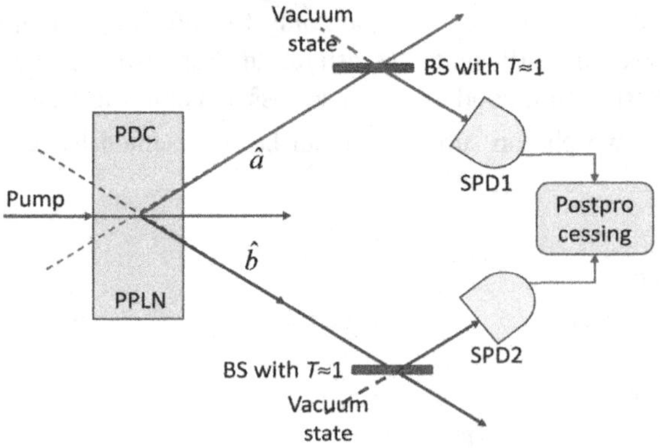

FIGURE 10.20

Generation of entangled-photon holes-embedded two-mode squeezed-vacuum states.

$S_2(z)$ is a two-mode squeezing operator $S_2(z) = e^{z\hat{a}^\dagger \hat{b}^\dagger - z^* \hat{a}\hat{b}}$, with $z = r\exp(j\theta)$. If we introduce a new basis $\hat{a}_+ = 2^{-1/2}(\hat{a} + \hat{b}), \hat{a}_- = 2^{-1/2}(\hat{a} - \hat{b})$, we can write $S_2(z)$ as the product of two single-mode squeezing operators $S_\pm(\pm z)$. Now a two-photon subtracted state can be represented in the form that reminds us of the cat state,

$$|\psi\rangle = (\hat{a}_+^2 S_+|0\rangle)(S_-|0\rangle) - (S_+|0\rangle)(\hat{a}_-^2 S_-|0\rangle) \tag{10.19}$$

but it does not require the squeezing parameter to be low.

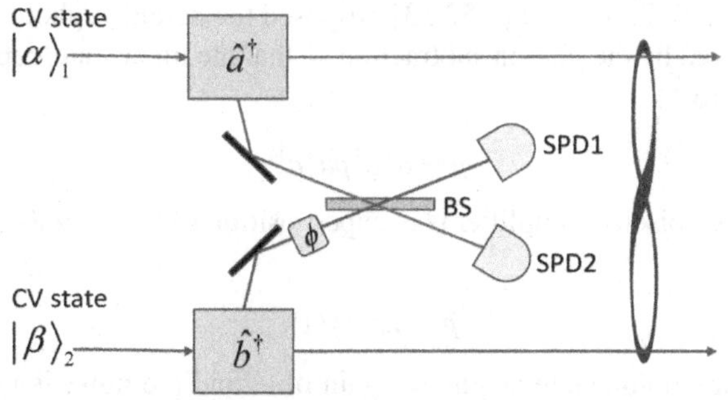

FIGURE 10.21

Generation of macroscopic entangled continuous-variable–continuous-variable states.

10.4.5 Generation of entangled macroscopic light states

A recent paper has shown that macroscopic light states as CV states can be entangled [51]. This is a very interesting alternative, since macroscopic light states are very tolerant of channel attenuation. The corresponding scheme to generate entangled CV–CV macroscopic states is shown in Fig. 10.21. The output state can be represented by $|\psi\rangle = \mathfrak{N}^{-1/2}(\hat{a}^\dagger|\alpha\rangle_1|\beta\rangle_2 + e^{j\phi}|\alpha\rangle_1\hat{b}^\dagger|\beta\rangle_2)$, where $\mathfrak{N}$ is the normalization factor. The phase shift needs to be properly chosen to ensure that the output macroscopic states are entangled. Now, using the following property of the displacement operator $\hat{a}^\dagger D(\alpha) = D(\alpha)(\hat{a}^\dagger + \alpha^*)$, we can rewrite the output states as follows:

$$|\psi\rangle = \mathfrak{N}^{-1/2}D_1(\alpha)D_2(\beta)(|1\rangle|0\rangle + e^{j\phi}|0\rangle|1\rangle) + \mathfrak{N}^{-1/2}\alpha^*(1+e^{j\phi})|\alpha\rangle|\beta\rangle. \tag{10.20}$$

Clearly, the first state is an entangled state, and the second state is a separable state. Now by setting $\phi = \pi$, the second term becomes zero, and the scheme shown in Fig. 10.21 can indeed entangle the macroscopic CV states that are tolerant to losses, thus allowing the distances between nodes to be significantly increased.

10.4.6 Noiseless amplification

To extend the transmission distance between quantum nodes, noiseless amplification can potentially be used [52–62]. The amplification process can typically be represented by

$$a \to b = \sqrt{G}a + n;\ \langle n\rangle = 0, \langle n^\dagger n\rangle \neq 0 \tag{10.21}$$

where b is the amplifier output mode, G is the amplifier gain, while n is the noise mode. Clearly, the signal-to-noise ratio becomes deteriorated since

$$\langle b^\dagger b\rangle = Ga^\dagger a + \sqrt{G}(a^\dagger n + n^\dagger a) + n^\dagger n, \tag{10.22}$$

indicating the amplified mode is affected by signal noise and noise terms.

To solve this problem, Bellini's group [52,53] proposed the heralded photon amplifier in which the photon addition is followed by the photon subtraction so that the input state, represented by the density operator ρ, get mapped to

$$\rho \rightarrow a\left(a^{\dagger} \rho a\right) a^{\dagger}. \tag{10.23}$$

When the input of the noiseless amplifier is in superposition state $\rho = \alpha\,|0\rangle\langle 0| + \beta\,|1\rangle\langle 1|$, the input state is mapped to

$$\rho \rightarrow 2\beta|1\rangle\langle 1|, \tag{10.24}$$

indicating that the single-photon state acquires a gain of 2, and the noise is not added.

10.5 Cluster state-based quantum networking

We move our attention to different approaches enabling the next generation of QCNs. We start with cluster state-based quantum networking, the concept we introduced in Refs. [3,60].

10.5.1 Cluster states and cluster state processing

The cluster states [54,55] belong to the class of graph states [56,57], which also include Bell states, GHZ states, W-states, and various other entangled states. When cluster C is defined as a connected subset on a d-dimensional lattice, it obeys the set of eigenvalue equations:

$$S_a|\phi\rangle_C = |\phi\rangle_C, \quad S_a = X_a \bigotimes_{b \in N(a)} Z_b, \tag{10.25}$$

where S_a are *stabilizer operators* with $N(a)$ denoting the neighborhood of $a \in C$. The cluster state defined by the equation above is also known as the *graph state*, denoted by $|G\rangle$. Namely, we can define the cluster state using the graph $G=(V,E)$ representation with V being the set of vertices, corresponding to qubits in quantum computing or nodes in networking applications; and E being the set of edges, corresponding to the controlled-phase gates. The graph state $|G\rangle$ can be expanded using the computational basis as follows:

$$|G\rangle = \frac{1}{2^N} \sum_{\boldsymbol{x} \in \{0,1\}^N} (-1)^{\boldsymbol{x}^{\mathrm{T}} \boldsymbol{A} \boldsymbol{x}} |\boldsymbol{x}\rangle, \quad \boldsymbol{x} = x_1 \cdots x_N \in \{0,1\}^N, \tag{10.26}$$

where $\boldsymbol{A}$ is the adjacency matrix. For vertices $a,b \in V$ representing the end points of an edge, the corresponding element in adjacency matrix is given by

$$A_{a,b} = \begin{cases} 1, \{a,b\} \in E \\ 0, \ \text{otherwise} \end{cases} \tag{10.27}$$

For weighted graphs, we also need to associate the weight with edge $\{a,b\}$, that is, $A_{a,b} = w_{a,b}$. The neighborhood of vertex a, denoted as $N(a)$, is defined as the set of vertices adjacent to it—that is, $N(a) = \{b \in V | \{a,b\} \in E\}$. The number of neighbors related to a is called the *degree* of the vertex a. An ordered list of vertices $a_1 = a, a_2, \ldots, a_n = b$, wherein a_i and a_{i+1} are adjacent (for all i) is called the $\{a,b\}$-path. The graph that has an $\{a,b\}$-path for every $a,b \in V$ is called the *connected graph*.

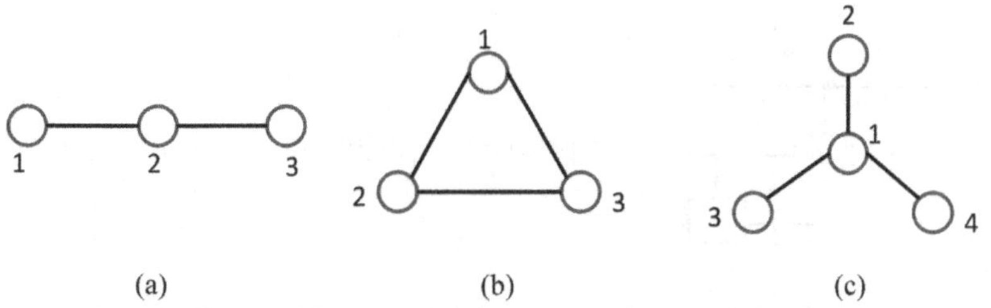

FIGURE 10.22

Illustration of graph states: (A) three-vertex path, (B) triangle of three vertices, and (C) the star graph.

As an illustration, Fig. 10.22 provides three examples of graph states: (a) three-vertex path, (b) triangle of three vertices, and (c) the star graph corresponding to the $|\mathrm{GHZ}\rangle$-state on $N=4$ qubits. The neighbors of vertex 1 in the star graph are $\{2,3,4\}$. The corresponding adjacency matrices are as follows:

$$\boldsymbol{A}_a=\begin{bmatrix}0&1&0\\1&0&1\\0&1&0\end{bmatrix},\boldsymbol{A}_b=\begin{bmatrix}0&1&1\\1&0&1\\1&1&0\end{bmatrix},\boldsymbol{A}_c=\begin{bmatrix}0&1&1&1\\1&0&0&0\\1&0&0&0\\1&0&0&0\end{bmatrix}.$$

Based on Eq. (10.25), the corresponding stabilizers for the three-vertex path are *XZI*, *ZXZ*, and *IZX*, while for the triangle-graph the stabilizers are *XZZ*, *ZXZ*, and *ZZX*. The two-qubit cluster state has stabilizers *XZ* and *ZX*, the corresponding state can be represented by

$$|\phi\rangle_2=2^{-1/2}(|0+\rangle+|1-\rangle), \tag{10.28}$$

and is related to the Bell state (up to corresponding local transformations). On the other hand, the cluster state corresponding to the three-vertex path is given by

$$|\phi\rangle_3=2^{-1/2}(|{+}0+\rangle+|{-}1-\rangle), \tag{10.29}$$

and is related to the GHZ-state (up to corresponding local transformations). The vertices in the graph represent $|+\rangle$ states, while the edges represent the controlled-phase gates.

The corresponding quantum gates-based model for the three-vertex path graph is provided in Fig. 10.23. The action of the controlled-Z gate on initial state $|+\rangle|+\rangle$ can be described by

$$C_Z^{(1,2)}|+\rangle|+\rangle=2^{-1/2}(|0\rangle|+\rangle+|1\rangle|-\rangle). \tag{10.30}$$

The output of this circuit is the cluster state $|\phi\rangle_3$, since

$$\begin{aligned}|\Psi_{\text{out}}\rangle&=2^{-1/2}C_Z^{(2,3)}(|0\rangle_1|+\rangle_2+|1\rangle_1|-\rangle_2)|+\rangle_3=2^{-1/2}C_Z^{(2,3)}(|+\rangle_1|0\rangle_2+|-\rangle_1|1\rangle_2)|+\rangle_3\\&=2^{-1/2}(|+\rangle_1|0\rangle_2|+\rangle_3+|-\rangle_1|1\rangle_2|-\rangle_3)=|\phi_3\rangle.\end{aligned} \tag{10.31}$$

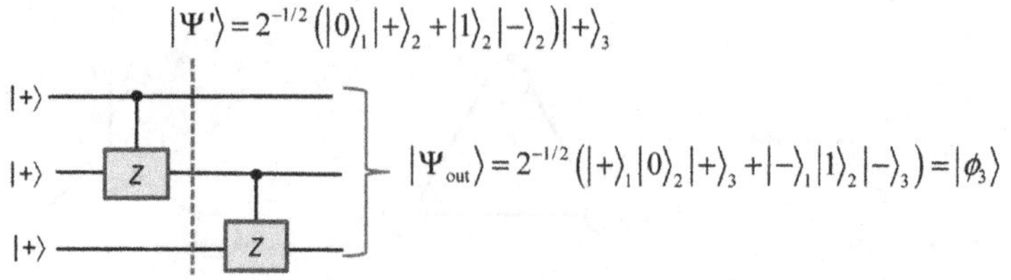

FIGURE 10.23

The quantum circuit corresponding to the three-qubit linear cluster state (three-vertex path graph).

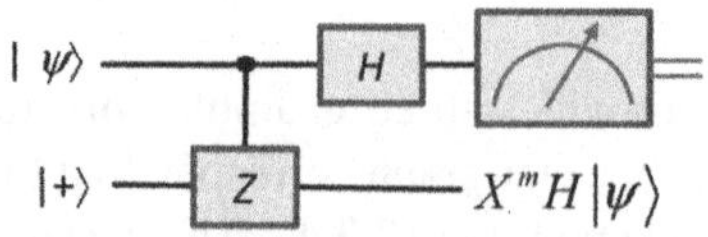

FIGURE 10.24

The quantum circuit to perform one-bit teleportation.

One of the driving forces for one-way quantum computing, measurement-based quantum computing, and distributed quantum computing is the one-bit teleportation [3,64,65], with corresponding quantum circuit provided in Fig. 10.24, in which the first qubit is arbitrary state $|\psi\rangle$, while the second qubit is the $|+\rangle$-state. When we perform the measurement on the first qubit in diagonal basis $\{|+\rangle,|-\rangle\}$, the state of the second qubit will be $X^m H|\psi\rangle$, where m is the result (outcome) of the measurement (on the first qubit). Therefore, by performing the $C_Z(|\psi\rangle \otimes |+\rangle)$ and measuring the first qubit in X-basis, we effectively teleported the quantum state from physical qubit 1 to qubit 2. Let us now apply this concept to the graph shown in Fig. 10.25. By applying the C_Z-gate between qubits 0 and 1, followed by the X-measurements on qubits 0, 1, and 2, the resulting quantum state of the third qubit will be

$$X^{m_2} \underbrace{HX^{m_1}H}_{Z^{m_1}} X^{m_0} H|\psi\rangle_3 = X^{m_2} Z^{m_1} X^{m_0} H|\psi\rangle_3, \tag{10.32}$$

where m_i are outcomes of measurements on qubits 0, 1, and 2; and we employed the following identity $HXH = Z$.

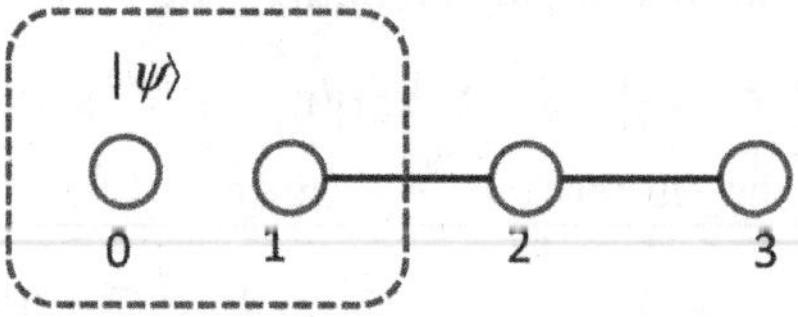

FIGURE 10.25

The illustration of teleportation of quantum state though linear chain of qubits. We apply the C_Z-gate between qubits 0 and 1, followed by the X-measurements on qubits 0, 1, and 2.

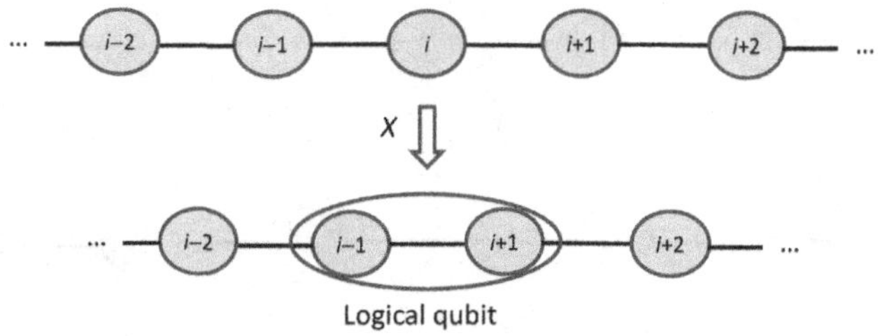

FIGURE 10.26

Illustrating the role of *X*-measurement in linear chain cluster.

Let us now observe the linear chain cluster state and perform the measurement on a qubit i located between $i-1$ and $i+1$ qubits [3]. The X-measurement on the i-the qubit will have the following two effects—(1) removing the i-th qubit from the cluster and (2) joining the neighboring qubits $i-1$ and $i+1$ into a single joint logical qubit—as illustrated in Fig. 10.26. The single logic qubit will represent 0 as $|++\rangle$ and 1 as $|--\rangle$, which is similar to repetition coding.

We now study the role of Z-measurements. Suppose we want to eliminate the qubit a in the cluster provided in Fig. 10.27. Let us apply the Z-measurement on qubit a and denote the outcome of the measurement as m_a. Given that the measurement Z commutes with C_Z-gates, we can first apply the C_Z-gate between a and 2 qubits and then perform the Z-measurement on qubit a before applying C_Z-gates between 1 and 2 as well as 2 and 3. So the action of C_Z-gate between qubits a and 2 will be $C_Z^{(1,2)}|+\rangle_a|+\rangle_2 = |0\rangle_a|+\rangle_2 + |1\rangle_a|-\rangle_2$. Now by performing the Z-measurement on qubit a, the qubit a will be removed and the state of qubit 2 will be $|(-1)^{ma}\rangle$. The remaining qubits in the cluster will be in the state $|+\rangle_1|(-1)^{ma}\rangle_2|+\rangle_3$. Let us now observe the linear chain cluster state as in Fig. 10.28 and let us perform the Z-measurement on qubit i, located between $i-1$ and $i+1$ qubits. The Z-measurement on

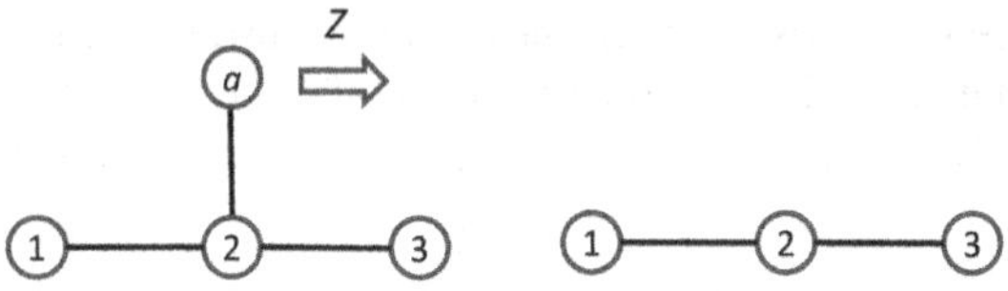

FIGURE 10.27

The role of Z-measurements is to remove the unwanted qubits in the cluster, such as the qubit *a*.

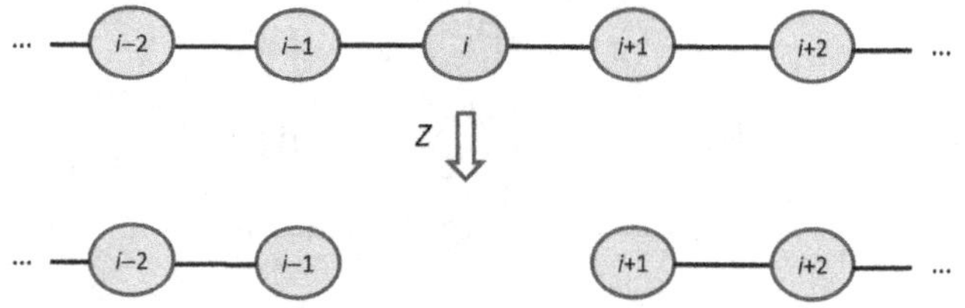

FIGURE 10.28

Illustrating the role of *Z*-measurement in linear chain cluster.

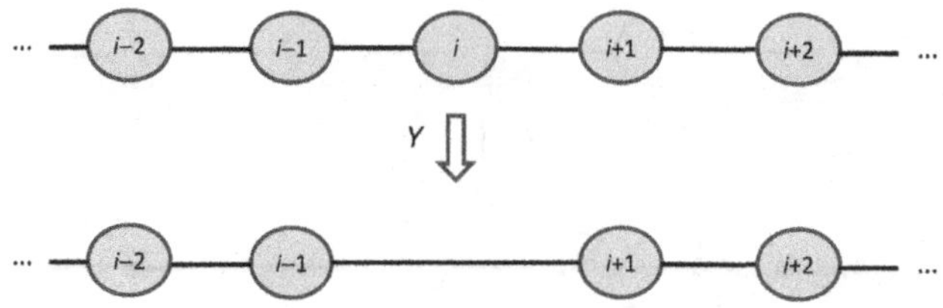

FIGURE 10.29

Illustrating the role of *Y*-measurement in linear chain cluster.

the *i*-the qubit will have the following two effects: (1) breaking the connections between the *i*-th qubit and its neighbors and (2) removing the *i*-th qubit from the cluster. The Z-measurements should also be applied in the final stage of the one-way quantum computing on remaining qubits to obtain the classical output results.

Finally, let us consider the role of *Y*-measurement on the linear chain cluster state shown in Fig. 10.29. We are interested in the effect of the *Y*-measurement on qubit *i*, located between $i-1$ and $i+1$ qubits. The *Y*-measurement on the *i*-the qubit will have the following two effects: (1) removal of the *i*-th qubit from the cluster and (2) linking of the neighboring qubits $i-1$ and $i+1$, as illustrated in Fig. 10.29.

10.5.2 Cluster state-based quantum networks

To create a 2-D cluster state, among others, we can employ the approach proposed by Gilbert et al. [60], which utilizes linear states generated by SPDC, local unitaries, and type I fusion to create the desired 2-D cluster state. The type I fusion is illustrated in Fig. 10.30, based on ref. [60]. The vertical photon is reflected by the polarization beam splitter (PBS), while the horizontal photon is transmitted through the PBS. Given the probabilistic nature of the PBS, with the photons present at both left and right input ports, there are four possible outcomes, each occurring with a probability of 0.25. Two outcomes correspond to the desired fusion operators, and the success probability of the fusion is 0.5. When a single photon is detected by the detector, the successful fusion is declared. The procedure to

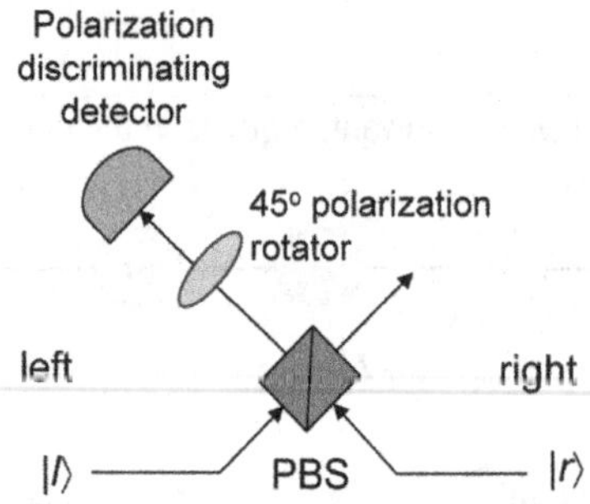

FIGURE 10.30

Type I fusion.

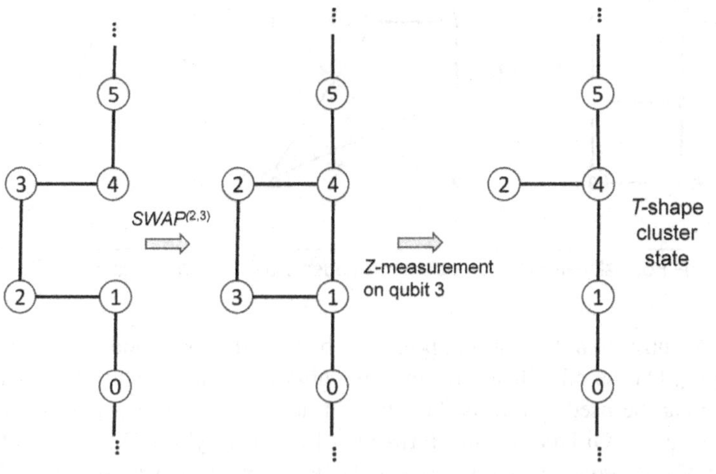

FIGURE 10.31

T-shape cluster state creation using Gilbert's approach.

create a T-shape cluster state is described in Fig. 10.31. To create the box-cluster state, we start with a four-qubit linear cluster state, relabel qubits 2 and 3, and apply the Hadamard gates to qubits 2 and 3, which effectively establishes the bond between qubits 1 and 4. Essentially, relabeling the qubits is equivalent to the SWAP gate action. To create the box-on-chain cluster state, we start with a longer linear chain of qubits and apply the same approach as in a box-state creation. Two T-shape cluster states can be fused together to get the *H*-shape cluster state, *etc.* One of the key issues of the fusion approach is its probabilistic nature, and it will succeed with a certain probability. To solve this problem, we can employ the photon addition concept introduced in Fig. 10.16, wherein the quantum states to be entangled are now cluster states. The key advantage over type I fusion is that creating a desired cluster state with the photon addition concept becomes a deterministic process. Another key advantage of this concept is that the qubits in the cluster do not have to be DV only. Namely, the photon addition concept to create the desired cluster state can be applied to all possible types of qubits, including DV, CV, and hybrid DV−CV.

Once a 2-D cluster state of nodes is created, wherein the qubits are located at different nodes in QCN, we must use properly selected Y and Z measurements to create EPR pairs between two arbitrary nodes in the quantum network [54]. As an illustration, Fig. 10.32 shows a 2-D cluster state with nine nodes. To establish EPR pairs between nodes 3 and 7 and nodes 1 and 9, we first disconnect the 4−7 and 2−3 links, then perform Y measurements on nodes 8, 5, 6, followed by Z-measurement on node 2 and Y-measurement on node 4 to form two desired Bell states. Given that the 2-D cluster state is universal, it is possible to use the same network architecture for both QCN and distributed quantum computing. Moreover, when several 2-D cluster states are run in parallel on the same network nodes, we will be able to reconfigure the QCN as needed with the help of the software-defined network (SDN) concept.

To further extend the transmission distance, QEC can also be applied on hybrid DV−CV states. To take advantage of both DV and CV "worlds," we can first encode the CV and DV states before entangling them into a hybrid encoded CV−DV codeword. This approach is compatible with DV- and CV-only

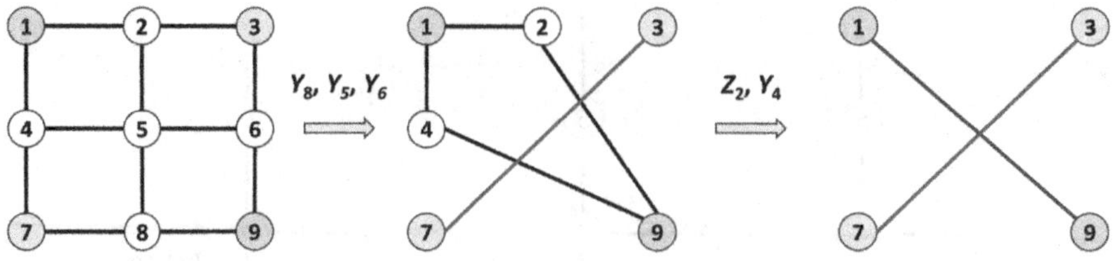

FIGURE 10.32

Establishing Einstein–Podolsky–Rosen pairs between nodes 1 and 9 and nodes 3 and 7.

nodes. For the CV subsystem, the starting point can be the GKP-two-mode squeezed code and GKP-stabilizer code [61]. On the other hand, for the DV subsystem, the quantum low-density parity-check (QLDPC) coding can be used, given its flexibility in design and the fact that many stabilizer codes can be interpreted using QLDPC coding formalism [3]. This hybrid QEC concept is illustrated in Fig. 10.33. The CV QEC (on the top) is based on GKP-two-mode squeezed-state code, in which two-mode squeezing (TMS) operation, denoted by $U_{CV}(G)$, is used to entangle an arbitrary CV state and a GKP grid state. On the other hand, in DV encoder K information qubits are encoded into an N-qubit codeword with the help of ancilla qubits; see Ref. [3] for additional details. The DV and CV codewords get entangled by applying a delocalized photon addition module, as described in text related to Fig. 10.14. The quantum channel is modeled as the additive white Gaussian noise (AWGN), which introduces the complex Gaussian displaced perturbation to the annihilation operator, effectively performing the mapping $\hat{a} \rightarrow \hat{a} + (\varepsilon_I + j\varepsilon_Q)/2$, with ε_I and ε_Q being zero-mean Gaussian random variables of variance σ^2. Once both CV modes get transmitted over the AWGN channels, another conjugate TMS operation takes place, and the overall action can be represented in terms of effective displacements:

$$\begin{aligned} &U_{CV}^{\dagger}(G)\left[U_p(\varepsilon_p)U_{GKP}(\varepsilon_{GKP})\right]U_{CV}^{\dagger}(G) = \\ &U_p\left(\varepsilon_p\sqrt{G} + \varepsilon_{GKP}\sqrt{G-1}\right)U_{GKP}\left(\varepsilon_{GKP}\sqrt{G} + \varepsilon_p\sqrt{G-1}\right). \end{aligned} \tag{10.33}$$

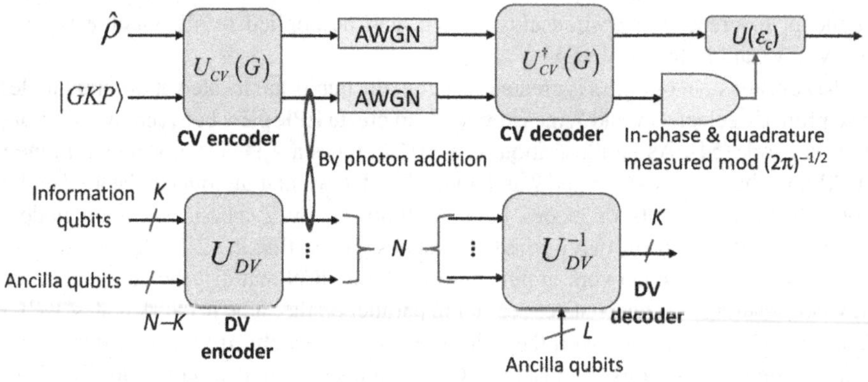

FIGURE 10.33

Illustrating hybrid continuous-variable–discrete-variable quantum error correction.

Clearly, for sufficient overall gain $G >> 1$, the effective displacements of arbitrary and GKP states are highly correlated. Given that, on a GKP state, we can simultaneously measure both quadratures (I and Q) per mod $(2\pi)^{-1/2}$, and with high precision, after estimating the GKP displacement error, we can use it to undo the action of the channel on arbitrary state ρ. The quantum errors introduced on the DV subsystem, even when continuous, can be represented in terms of Pauli operators. Because the DV and CV subsystems are entangled, the DV subsystem might experience additional errors caused by AWGN displacements on the CV subsystem, and since the QEC methods are significantly different, it makes sense to complete the CV error correction before the DV error correction takes place. The subsystem coding approach [3] is also applicable in hybrid QEC, in which the information is encoded, for instance, on the DV subsystem, such as in distributed quantum computing, and the other (CV) subsystem is used to absorb the ambient errors. Finally, for distributed quantum computing, it also makes sense to apply concatenated coding, in which DV information is protected first, followed by DV to CV conversion to improve tolerance to the channel loss and CV error correction.

10.6 Surface code-based and quantum low-density parity-check code-based quantum networking

Here we describe a quantum-error-correction-based implementation of a QCN with surface codes (SCs) [63]. The surface code [63–66] is defined on a 2-D lattice with information (data) qubits denoted by solid circles, with one particular example provided in Fig. 10.34.

There are two types of stabilizers (quantum parity-check equations): (1) stabilizers containing all Z Pauli operators, denoted by white plaquettes in Fig. 10.34, and (2) stabilizers containing all X Pauli operators, denoted by shaded plaquettes. To illustrate, the plaquette stabilizer related to qubits 1 and 2 will be X_1X_2. On the other hand, the plaquette stabilizer related to qubits 1, 2, 4, and 5 will be $Z_1Z_2Z_4Z_5$. The minimum distance of the surface code will be the minimum side length, that is, $d = \min(2,3) = 2$. The codeword length is determined as the product of side lengths, that is, $n = 2 \times 3 = 6$. There are five stabilizers, $n-k = 5$, and therefore, the number of information (data) qubits will simply be $n-(n-k) = 1$.

The logical qubits of SC will be located at different nodes, as illustrated in Fig. 10.35. Each SC will be on a $d \times d$ grid, with neighboring nodes connected via d wavelengths in the telecommunication C and L-bands, transmitted using single-mode fibers. By performing the product $Z \otimes Z$ measurement,

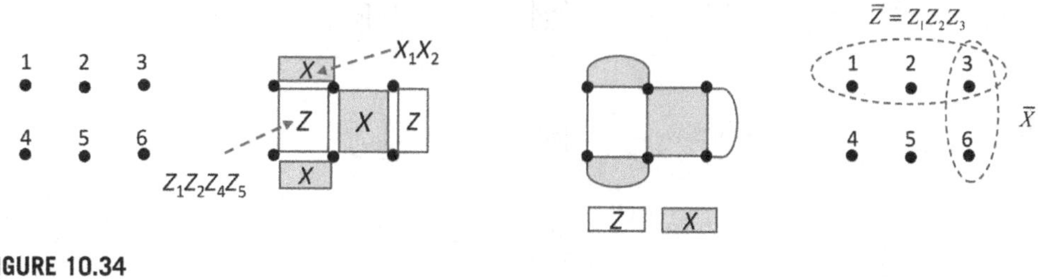

FIGURE 10.34

Illustrating the surface code concept. The minimum distance of this code is $d = \min(2,3) = 2$, and the codeword length is $n = 2 \times 3 = 6$. The number of stabilizers (plaquettes) is $n-k = 5$. Finally, the number of data (information) qubits is simply $n-(n-k) = k = 1$.

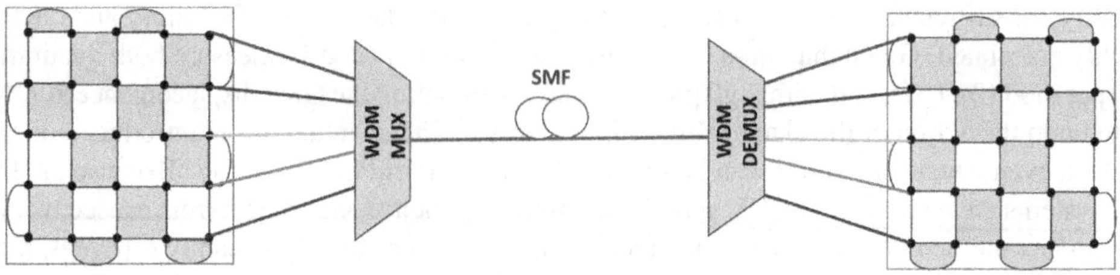

FIGURE 10.35

Simplified description of connecting two logical qubits from two neighboring nodes by *d* wavelengths.

logical qubits will create the EPR pairs. The measurement results will pass to the SDN controller, which will record the exact number of EPR pairs created. For the desired Q-LAN, corresponding product measurements must be simultaneous.

An example of the QCN with five nodes with a reconfigurable arrangement is illustrated in Fig. 10.36. Each node has an SC patch for the corresponding logical qubits. By performing the simultaneous $Z \otimes Z$ product measurements between logical qubits q_1 and q_2, q_2 and q_3, q_3 and q_4, q_4 and q_5, q_5 and q_1, we can entangle nodes 1 to 5 to create a ring QCN, as illustrated in Fig. 10.36A, which is most suitable for investigations of multihop transmission. On the other hand, for the 5-node mesh network shown in Fig. 10.36B, by performing the simultaneous $Z \otimes Z$ measurements and simultaneous $X \otimes X$ measurements, we can entangle five qubits into the mesh configuration. Finally, by disconnecting the outer links in Fig. 10.36B, we obtain the star configuration in Fig. 10.36C. The exact maximum entangled state between nodes in the QCN is determined by providing the measurement

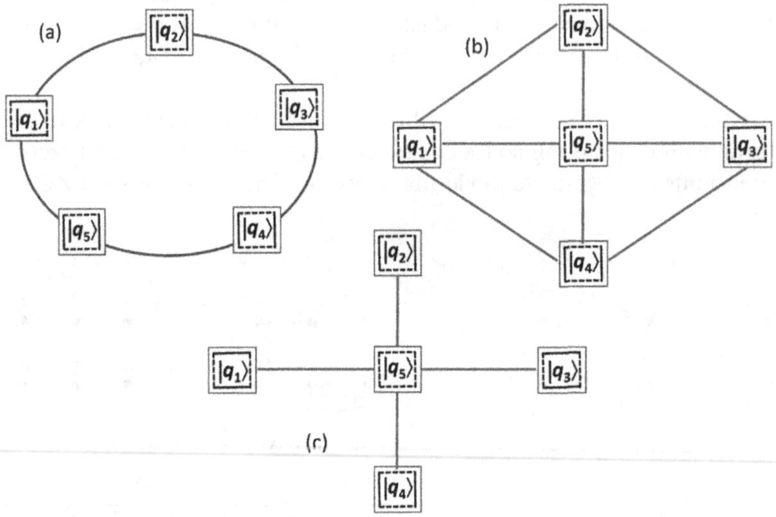

FIGURE 10.36

Illustrative 5-node: ring (A), mesh (B), and star (C) QCNs.

results to the SDN control plane. This approach will entangle the logical qubits in an arbitrary network topology. By equipping every node with multiple qubit patches, the QCN will be able to simultaneously perform quantum networking, distributed computing, and other distributed quantum information science applications.

To extend the transmission distance between neighboring nodes in the SC-code-based QCN, QEC-based repeaters are typically used. So far, QEC-based repeaters have been based on 2-D QEC, such as Calderbank−Shor−Steane (CSS) code-based repeaters [64] and surface-code-based repeaters [65]. Unfortunately, CSS codes are essentially girth-4 QLDPC codes, with poor error-correction performance [3]. On the other hand, the surface codes of [65] introduce large latency and are not compatible with the SC-based QCN. In Ref. [63], we proposed a different approach, whereby we interpret an intermediate node as an SC patch and apply the patch deformation approach suggested by Litinski [67], thus extending the logical qubit to two spatially separated patches. This is illustrated in Fig. 10.37, based on a recent publication [63]. In this example, three wavelengths are needed to connect remote patches. Once the logical qubit is extended to the intermediate node, we further perform product $X \otimes Z$ measurements to entangle the logical qubits q_1 and q_2. This approach is applicable to several intermediate nodes, thus offering the potential to significantly extend the distance between any two desired nodes in the QCN.

For completeness of presentation, we illustrate the lattice surgery in Fig. 10.38, applied on the product $Z \otimes Z$ between adjacent patches. In *lattice surgery* [67,68], we modify the configuration as shown in Fig. 10.38 (right) by introducing the patches with dark-red edges, after which we measure the stabilizers for d cycles to get the outcome of measurements and split again. This represents the

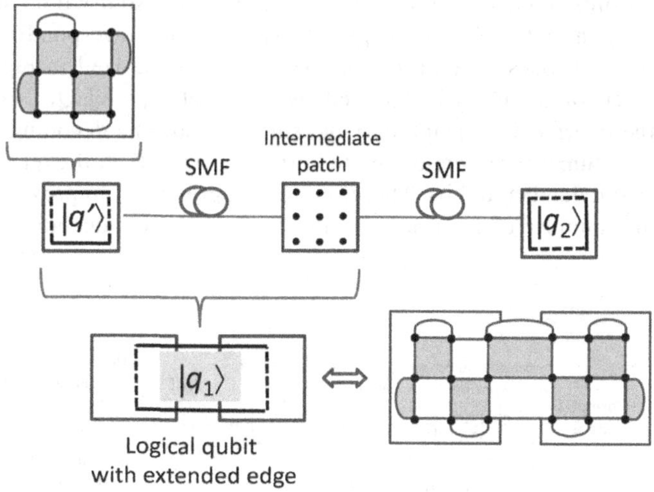

FIGURE 10.37

Extending the distance between two nodes in QCN by creating a logical qubit spanning two spatially separated SC patches.

After I.B. Djordjevic, Surface-codes-based quantum communication networks. Special Issue on Physical-Layer Security, Quantum Key Distribution, and Post-Quantum Cryptography, Entropy 22 (9) (2020) 1059.

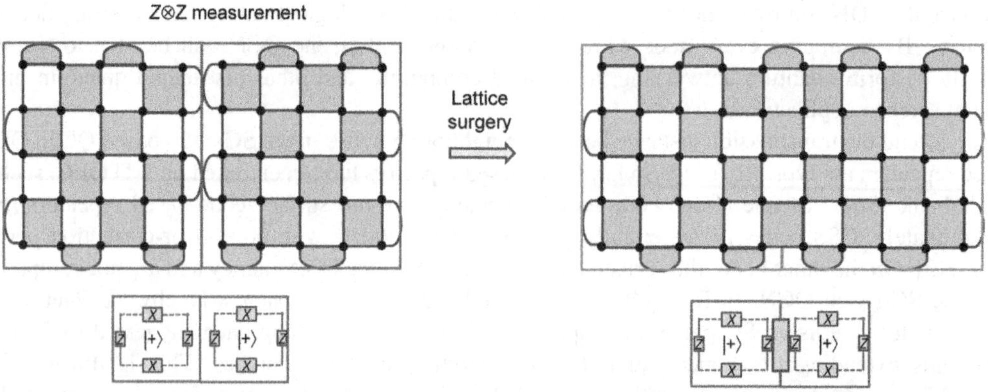

FIGURE 10.38

Illustrating the lattice surgery procedure for the measurement on the product $Z \otimes Z$.

After I.B. Djordjevic, Surface-codes-based quantum communication networks. Special Issue on Physical-Layer Security, Quantum Key Distribution, and Post-Quantum Cryptography, Entropy 22 (9) (2020) 1059.

way to introduce entanglement between two adjacent patches. Namely, we start with the state $|++\rangle = 0.5(|00\rangle + |11\rangle + |01\rangle + |10\rangle)$ and perform the measurement on $Z \otimes Z$ operator. If the result of the measurement is +1 the qubits end up in state $2^{-1/2}(|00\rangle + |11\rangle)$; otherwise, in state $2^{-1/2}$ $(|01\rangle + |10\rangle)$. In either case, the qubits are maximally entangled.

To extend the transmission distance between nodes, we can also use the QEC-based quantum networking employing the QLDPC coding [3,69], wherein the corresponding QLDPC code is composed of multiple subcodes. The proposed concept is summarized in Fig. 10.39. An information state composed of K qubits is encoded by systematic $[N,K]$ QLDPC code to get the codeword $|\boldsymbol{q}_t\rangle$. At the intermediate quantum node, simple syndrome decoding is used to identify the most probable quantum error operator and to correct it. After that, the reencoding takes place by inserting additional redundant qubits. Therefore, we progressively improve the error correction strength based on the number of intermediate nodes. The QLDPC decoding takes place at the destination node.

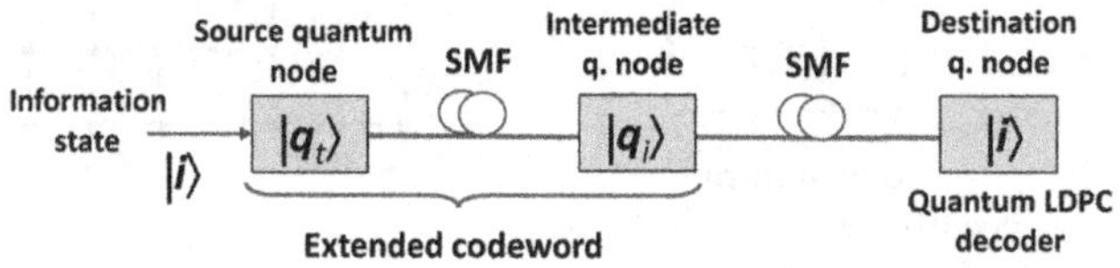

FIGURE 10.39

Extending the distance between two nodes in a QCN by reconfigurable QLDPC code, wherein for each intermediate node, extra redundant qubits are added. In each intermediate node, simple syndrome decoding is used to identify the most probable error operator and to correct it before the reencoding takes place. The QLDPC decoding takes place at the destination node.

10.7 Entanglement-assisted communication and networking

As discussed earlier, entanglement represents a unique resource for QIP. Quantum communication (QuCom) is the cornerstone to fully exploit the power of entanglement. QuCom enables the distribution of entanglement at a distance to interconnect quantum devices to endow quantum advantages in sensing, communication, and information gathering. To illustrate, the preshared entanglement can be used as in Refs. [11,12] to (1) enable secure communications, (2) beat the classical channel capacity [6,10], (3) beat the standard quantum limit for sensing applications, (4) enable distributed quantum computing, and (5) enable EA distributed sensing [5]. To distribute the entangled signal-idler photon pairs, either the fiber-based quantum network or satellite-to-ground links can be used, which is illustrated in Fig. 10.40. The optimum encoding for EA communication has been known for decades [6], while the design of an optimum receiver is still an open problem. It has been shown in Ref. [6] that with the two-mode Gaussian states, EA capacity can be achieved. In Ref. [10], the authors proposed to employ multiple feedforward sum-frequency generation sections in the receiver based on [70]. This scheme was shown to be suitable for use in quantum binary discrimination problems to detect the target in a highly noisy environment [70] but does not represent an EA channel capacity-achieving scheme. To come close to the EA channel capacity, the authors in Ref. [10] applied repetition coding over 10^6 bosonic modes, thus occupying the whole C and L bands as well as the portion of the S-band. To solve this problem, the author in Ref. [12] proposed the optimum encoding based on Gaussian modulation (GM) of the signal photon in a TMSV state. In this chapter, we also review recent quantum detector proposals employing the optical parametric amplifier (OPA) as the basic building block in corresponding joint receivers. One of the receivers is based on optical phase conjugation (OPC). By moving the OPC block to the transmitter side where the signal photon is brighter rather than on the receiver side in Ref. [10], when the signal photon has low brightness and is affected by noise, we can employ the conventional homodyne receivers used in classical optical communications. This approach leads to improvements in bandwidth (spectral) efficiencies compared with the case when the OPC is performed on the receiver side.

10.7.1 Entanglement-assisted communication networks

In EA communication networking, we are interested in constructing an EA communication network in which any two users can communicate at rate higher than classical channel capacity.

The key issue is to distribute the entangled states across the network and store them in the quantum memories, which are not widely available. To distribute the entangled signal-idler photon pairs, either the fiber-based quantum network or satellite-to-ground links can be used, which is illustrated in Fig. 10.40. Entangled states get distributed by the fiber-optics-based network as shown in Fig. 10.40A, which are further stored in corresponding quantum memories. Alice uses her signal photon to impose the classical data to be transmitted over a noisy and lossy quantum channel. Bob detects the transmitted signal with the help of the idler photon. The satellites can also be used in entanglement distribution, which is illustrated in Fig. 10.40B. Given that satellite-to-ground links are more tolerable to the atmospheric turbulence effects compared with ground-to-satellite links, the usage of satellite-to-ground links to distribute entanglement represents a favorable scenario.

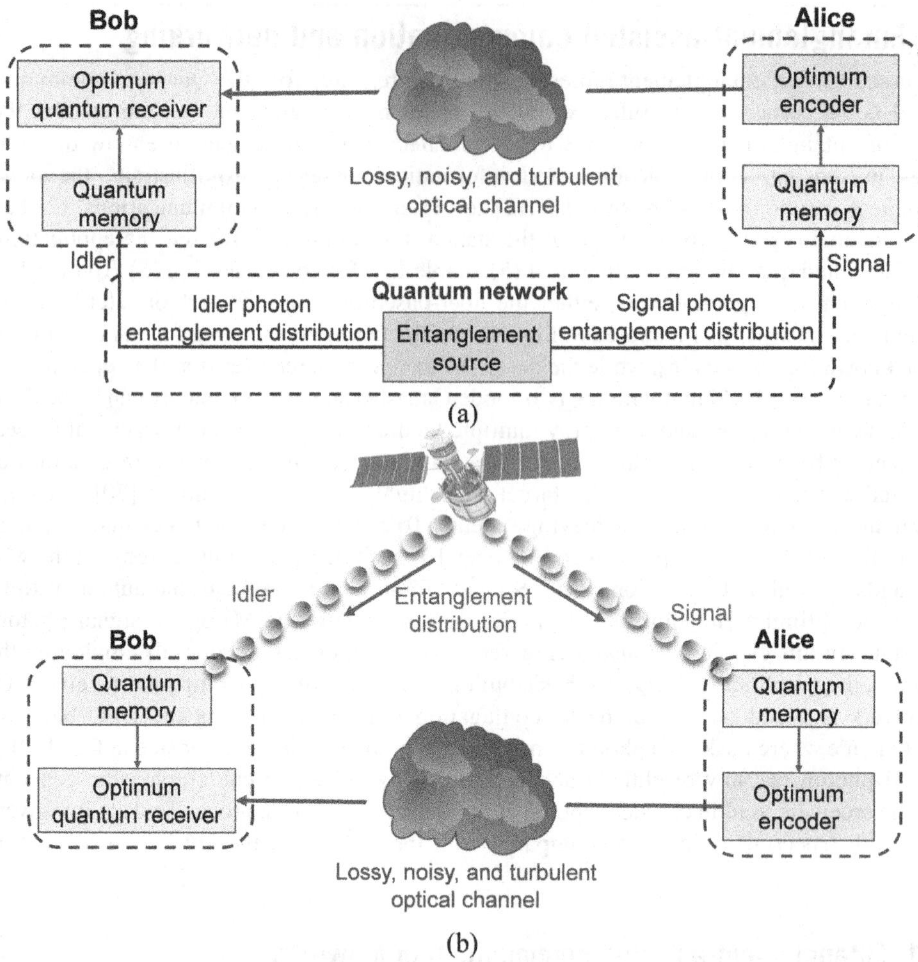

FIGURE 10.40

Entanglement-assisted classical optical communication concept: (A) quantum fiber-optics network-based entanglement distribution and (B) satellite-based entanglement distribution.

Each signal-idler pair is a TMSV state generated by spontaneous parametric downconversion (SPDC), and can be represented in Fock basis as

$$|\psi\rangle_{s,i} = (N_s + 1)^{-1/2} \sum\nolimits_{n=0}^{\infty} [N_s/(N_s + 1)]^{n/2} |n\rangle_s |n\rangle_i, \tag{10.34}$$

with the mean photon number being $N_s = \left\langle \hat{a}_s^\dagger \hat{a}_s \right\rangle = \left\langle \hat{a}_i^\dagger \hat{a}_i \right\rangle$, with corresponding signal and idler creation operators denoted by $\hat{a}_s^\dagger$ and $\hat{a}_i^\dagger$, respectively. The phase-sensitive cross-correlation $C_{si} = \langle \hat{a}_s \hat{a}_i \rangle = \sqrt{N_s(N_s + 1)}$ is used to specify the signal-idler entanglement. Clearly, when $N_s << 1$, we have that $C_{si} \approx \sqrt{N_s}$ which is larger than the classical limit N_s. The TMSV state is a pure maximally entangled zero-mean Gaussian state with Wigner covariance matrix being:

$$\Sigma_{TMSV} = \begin{bmatrix} (2N_s+1)\mathbf{1} & 2\sqrt{N_s(N_s+1)}\mathbf{Z} \\ 2\sqrt{N_s(N_s+1)}\mathbf{Z} & (2N_s+1)\mathbf{1} \end{bmatrix}, \tag{10.35}$$

where $\mathbf{1}$ is the identity matrix, and $\mathbf{Z} = \text{diag}(1,-1)$ is the Pauli Z-matrix. The SPDC-based entangled source represents a broadband source with $D = T_m W$ i.i.d. signal-idler mode pairs, where W denotes the phase-matching bandwidth and T_m is the measurement interval.

The corresponding model for EA classical communication is provided in Fig. 10.41, where the preshared entanglement is distributed with the help of two channels: signal channel, denoted by Φ_s, and the idler channel, denoted by Φ_i. We assume that QEC is applied to protect the quantum states stored in quantum memories against decoherence effects. The Alice-to-Bob channel is modeled by the single-mode thermal lossy Bosonic channel model

$$\hat{a}_s = \sqrt{T}\hat{a}'_s + \sqrt{1-T}\hat{a}_b, \tag{10.36}$$

where T is the transmissivity of the main channel, while $\hat{a}_b$ is a thermal (background) mode with the mean photon number being $N_b/(1-T)$. This channel can also be interpreted as a zero-mean AWGN channel with a power-spectral density of N_b and attenuation coefficient being T.

To improve the bandwidth efficiency of the repetition phase encoding scheme over a large number of bosonic modes [10], author in Refs. [11,12] proposed to apply the Gaussian modulation on single-photon of the TMSV state, which is illustrated in Fig. 10.41. Alice performs the Gaussian modulation on the signal photon with the help of an electrooptical I/Q modulator by generating the I and Q coordinates in the electrical domain and using them as RF inputs to the I/Q modulator, wherein Gaussian samples are properly scaled to account for the I/Q modulator insertion loss such that the average number of transmitted signal photons per mode is equal to $N_s = \langle (\hat{a}'_s)^\dagger \hat{a}'_s \rangle$.

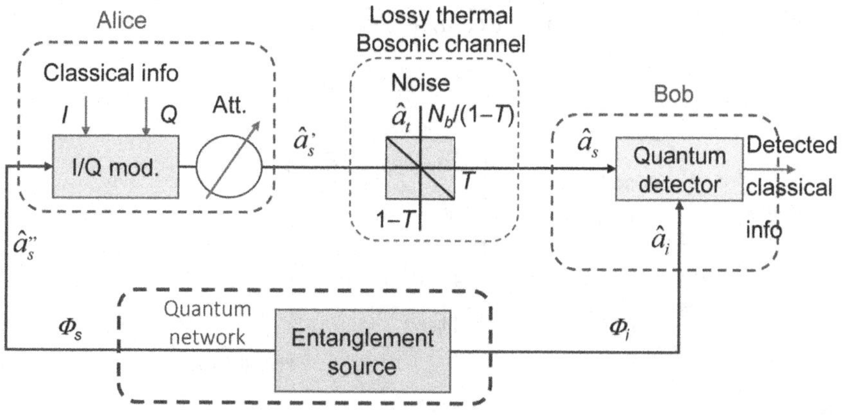

FIGURE 10.41

Entanglement-assisted communication system model. I/Q mod: electrooptical I/Q modulator, Att.: attenuator.

Given that the action of the BSs describing the quantum channel can be represented by the matrix $BS(\tau) = \begin{bmatrix} \sqrt{\tau}\mathbf{1} & \sqrt{1-\tau}\mathbf{1} \\ -\sqrt{1-\tau}\mathbf{1} & \sqrt{\tau}\mathbf{1} \end{bmatrix}$, to determine the covariance matrix after the BS in the entanglement distribution channel (see Fig. 10.41), we need to apply the following symplectic operation [1–3] by

$$BS_c(T) = \mathbf{1} \oplus BS(T) = \begin{bmatrix} \mathbf{1} & \mathbf{0} & \mathbf{0} \\ \mathbf{0} & \sqrt{T}\mathbf{1} & \sqrt{1-T}\mathbf{1} \\ \mathbf{0} & -\sqrt{1-T}\mathbf{1} & \sqrt{T}\mathbf{1} \end{bmatrix} \tag{10.37}$$

on the input covariance matrix $\mathbf{\Sigma}_{TMSV}$ to obtain

$$\mathbf{\Sigma}' = BS_c(T) \begin{bmatrix} \mathbf{\Sigma}_{TMSV} & \mathbf{0} \\ \mathbf{0} & \sigma'^2\mathbf{1} \end{bmatrix} [BS_c(T)]^T, \tag{10.38}$$

where the variance of the thermal state is $N_b/(1-T)$ thermal photons. By keeping Alice and Bob submatrices and ignoring the rest, we obtain the following covariance matrix:

$$\mathbf{\Sigma}_{AB} = \begin{bmatrix} (2N_s+1)\mathbf{1} & 2\sqrt{TN_s(N_s+1)}\mathbf{Z} \\ 2\sqrt{TN_s(N_s+1)}\mathbf{Z} & (2N_s''+1)\mathbf{1} \end{bmatrix}, \tag{10.39}$$

where $N_s'' = N_sT + N_b$.

10.7.2 Nonlinear receivers for entanglement-assisted communication systems

The EA receiver often employs the OPA as the basic building block. A single *OPA-based receiver* is provided in Fig. 10.42, where the gain is selected to be $G = 1 + \varepsilon$, $\varepsilon << 1$. The idler output of the OPA is given by [71]

$$\hat{A}_i = \sqrt{G}\,\hat{a}_i + \sqrt{G-1}\hat{a}_s^\dagger. \tag{10.40}$$

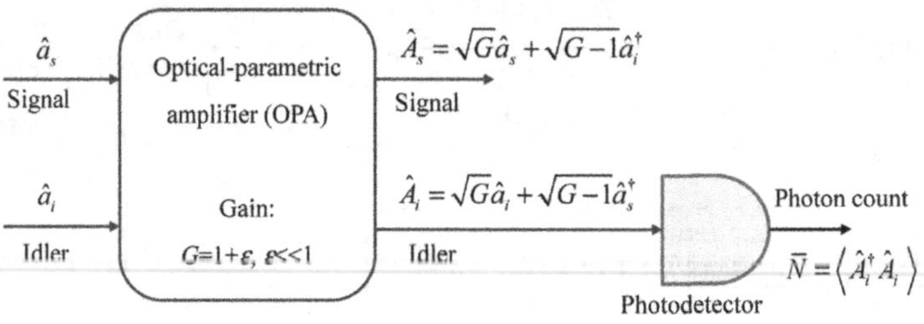

FIGURE 10.42

The optical phase conjugation-based EA receiver. The photons are detected at the port where the idler is amplified.

Assuming that M-ary PSK is imposed by the I/Q modulator, the signal mode at the output modulator $\hat{a}'_s$ is related to the input mode of the modulator $\hat{a}''_s$ by $\hat{a}'_s = e^{j\varphi}\hat{a}''_s$ where φ is the phase shift introduced by the modulator. The photodetector output operator is given by

$$\begin{aligned}\hat{i} &= R\hat{A}_i^\dagger \hat{A}_i = R\left(\sqrt{G}\ \hat{a}_i^\dagger + \sqrt{G-1}\ \hat{a}_s\right)\left(\sqrt{G}\ \hat{a}_i + \sqrt{G-1}\ \hat{a}_s^\dagger\right) \\ &= R\left[G\hat{a}_i^\dagger \hat{a}_i + \sqrt{G(G-1)}\hat{a}_i^\dagger e^{-j\varphi}\left(\hat{a}''_s\right)^\dagger + \sqrt{G(G-1)}\, e^{j\varphi}\hat{a}''_s \hat{a}''_i + (G-1)\underbrace{\hat{a}''_s\left(\hat{a}''_s\right)^\dagger}_{\left(\hat{a}''_s\right)^\dagger \hat{a}''_s + 1}\right],\end{aligned} \tag{10.41}$$

where R is the photodiode responsivity. Without loss of generality and given that we are concerned with fundamental limits, in the rest of this section, we will assume that $R = 1$ A/W. The expected value of the photocurrent operator is related to the photon count average by

$$\begin{aligned}\langle \hat{i} \rangle &= \left\langle \hat{A}_i^\dagger \hat{A}_i \right\rangle = GN_i + e^{-j\varphi}\sqrt{G(G-1)}C_{si} + e^{j\varphi}\sqrt{G(G-1)}C_{si} + (G-1)\left(N''_s + 1\right) \\ &= GN_s + 2\sqrt{G(G-1)}C_{si}\cos\varphi + (G-1)\left(N''_s + 1\right),\end{aligned} \tag{10.42}$$

where we used that $N_i = N_s$ for the TMSV state. The second line of this expression matches Eq. (10.17) provided in Ref. [10]. Clearly, the photocurrent average is proportional to cosφ, and we can detect the transmitted phase, but they are also two shot noise terms.

Given that the OPA receiver is not suitable for balanced detection, we can instead use the OPC receiver shown in Fig. 10.43. Here we use the OPA to nonlinearly interreact the signal mode $\hat{a}_s$ with the vacuum mode $\hat{a}_v$ to get the following output at the idler port $\hat{a}_c = \sqrt{G}\ \hat{a}_v + \sqrt{G-1}\ \hat{a}_s^\dagger$, which for gain $G = 2$ is simply $\hat{a}_c = \sqrt{2}\ \hat{a}_v + \hat{a}_s^\dagger$. Now by performing the mixing of the a_c-mode with the

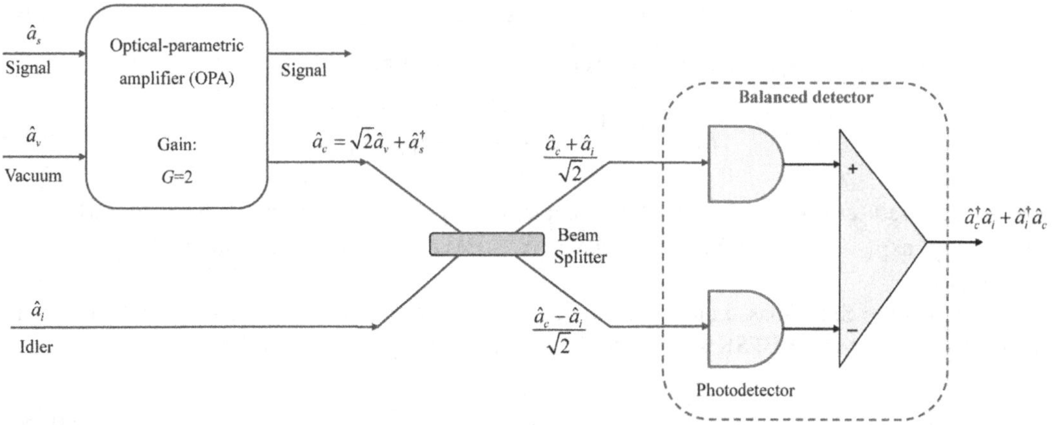

FIGURE 10.43

The operating principle of the optical phase conjugation receiver.

idler mode on a balanced beam splitter (BBS) followed by the balanced detection (BD), we obtain the following BD photocurrent operator:

$$\begin{aligned}\hat{i}_{BD} &= \frac{1}{2}\left(\hat{a}_c^\dagger + \hat{a}_i^\dagger\right)(\hat{a}_c + \hat{a}_i) - \frac{1}{2}\left(\hat{a}_c^\dagger - \hat{a}_i^\dagger\right)(\hat{a}_c - \hat{a}_i) \\ &= \hat{a}_c^\dagger \hat{a}_i + \hat{a}_i^\dagger \hat{a}_c,\end{aligned} \tag{10.43}$$

and after substituting the expression for the a_c-mode, we obtain

$$\hat{i}_{BD} = \sqrt{2}\hat{a}_v^\dagger \hat{a}_i + \hat{a}_s \hat{a}_i + \sqrt{2}\hat{a}_i^\dagger \hat{a}_v + \hat{a}_i^\dagger \hat{a}_s^\dagger. \tag{10.44}$$

For M-ary PSK, we have that $\hat{a}_s' = e^{j\varphi}\hat{a}_s''$, and given that the vacuum mode and the idler mode are uncorrelated, we obtain the following expression for the expectation of the BD photocurrent operator:

$$\langle \hat{i}_{BD} \rangle = e^{j\phi} \underbrace{\langle \hat{a}_s'' \hat{a}_i \rangle}_{C_{si}} + e^{-j\phi} \underbrace{\left\langle \hat{a}_i^\dagger \hat{a}_s^\dagger \right\rangle}_{C_{si}} = 2C_{si} \cos\varphi, \tag{10.45}$$

thus canceling the noise terms in the single-detector case in Eq. (10.42). By ignoring the terms originating from the vacuum state, the variance of the BD photocurrent operator will be

$$\begin{aligned}\mathrm{Var}\left(\hat{i}_{BD}\right) &= \langle \hat{i}_{BD}^2 \rangle - \langle \hat{i}_{BD} \rangle^2 \\ &= N_i N_s'' + (N_i + 1)\left(N_s'' + 1\right) + 2C_{si}^2 \cos(2\varphi) - 4C_{si}^2 \cos^2\varphi.\end{aligned} \tag{10.46}$$

For binary PSK (BPSK), the expression for variance simplifies to

$$\mathrm{Var}^{(\mathrm{BPSK})}\left(\hat{i}_{BD}\right) = N_i N_s'' + (N_i + 1)\left(N_s'' + 1\right) - 2C_{si}^2. \tag{10.47}$$

Further, in a low-brightness, highly noisy regime where $N_s << 1$ and $N_b >> 1$, we obtain the following expression for variance:

$$\mathrm{Var}^{(\mathrm{BPSK})}\left(\hat{i}_{BD}\right) \approx (2N_s + 1)N_b. \tag{10.48}$$

The corresponding expression for bit error probability will be

$$P_b^{(\mathrm{BPSK})} = \frac{1}{2}\mathrm{erfc}\left(\sqrt{\frac{4TN_s(N_s + 1)}{(2N_s + 1)N_b}}\right), \tag{10.49}$$

where we used erfc(•) to denote the complementary error function, defined by $\mathrm{erfc}(u) = (2/\sqrt{\pi})\int_u^\infty \exp(-x^2)dx$. For $N_s << 1$ and $N_b >> 1$, this expression agrees with the Eqs. (C1), (C2) of Ref. [10].

Given that the SPDC-based entangled source is a broadband source, by employing D modes all modulated with the same BPSK signal, the bit error probability can be reduced:

$$P_b^{(\mathrm{BPSK})}(D) = \frac{1}{2}\mathrm{erfc}\left(\sqrt{\frac{4DTN_s(N_s + 1)}{(2N_s + 1)N_b}}\right). \tag{10.50}$$

The corresponding binary symmetric channel capacity per mode will be

$$C_{\mathrm{OPC}}^{(\mathrm{BPSK})}=\frac{1}{D}\left[1-h\left(P_b^{(\mathrm{BPSK})}(D)\right)\right], \tag{10.51}$$

where $h(x)$ is the binary entropy function, defined by $h(x) = -x\log_2 x - (1-x)\log_2(1-x)$.

10.7.3 Entanglement-assisted communication with optical phase conjugation on the transmitter side

If we do not ignore the variance terms originating from the vacuum state in the OPC receiver (Fig. 10.43), we obtain the following expression for the BD photocurrent operator variance:

$$\mathrm{Var}\left(\hat{i}_{BD}\right)=2(N_i+1)+4N_i+N_iN_s''+(N_i+1)\left(N_s''+1\right)+2C_{si}^2\cos(2\varphi)-4C_{si}^2\cos^2\varphi. \tag{10.52}$$

By moving the OPA to the transmitter side by ensuring that $N_s >> N_\nu$, where the signal photon brightness is much higher, the output at the idler output port of the OPA will be just phase-conjugated signal mode. Moreover, the SPDC generator and OPA can be integrated on the same chip. Alternatively, similarly to fiber-optics communications [72], we can use periodically poled $LiNbO_3$ (PPLN) waveguide to realize the phase-conjugation by second harmonic generation (SHG) and difference frequency generation (DFG). Though the SHG pump frequency ω_p is doubled and parallel through DFG, the second harmonic photon $2\omega_p$ interacts with input signal mode ω_s to get the phase-conjugated signal at $\omega_{s,\mathrm{PC}} = 2\omega_p-\omega_s$. By moving the OPC to the transmitter side, we can use a balanced detection scheme for classical communication for EA detection, which is illustrated in Fig. 10.44, where we mix the phase-conjugated signal and idler modes directly on BBS. The BD photocurrent operator is now simply

$$\hat{i}_{BD}=\hat{a}_s\hat{a}_i+\hat{a}_i^\dagger\hat{a}_s^\dagger, \tag{10.53}$$

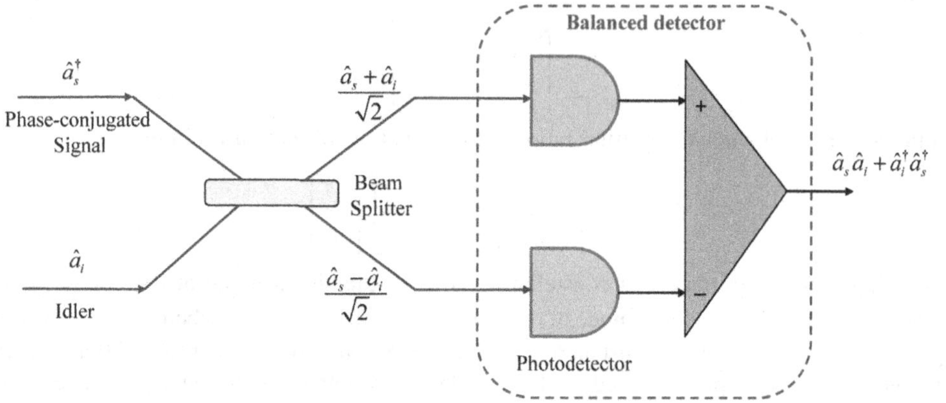

FIGURE 10.44

The balanced beam splitter-based homodyne-balanced detection receiver is used as an entanglement-assisted receiver.

and the expectation of the BD photocurrent operator, for M-ary PSK with $\widehat{a}_s' = e^{j\varphi}\widehat{a}_s''$ and $T = 1$, is given by

$$\langle \widehat{i}_{BD} \rangle = e^{j\phi} \underbrace{\langle \widehat{a}_s'' \widehat{a}_i \rangle}_{C_{si}} + e^{-j\phi} \underbrace{\langle \widehat{a}_i^\dagger \widehat{a}_s^\dagger \rangle}_{C_{si}} = 2C_{si} \cos \varphi, \tag{10.54}$$

which is identical to that of the OPC receiver (see Eq. 10.45). Clearly, in this classical coherent detection receiver, we use the phase-conjugated signal photon at the incoming optical signal port and the idler photon at the local oscillator (LO) laser signal port. On the other hand, the variance of the BD photocurrent operator variance is given by

$$\mathrm{Var}(\widehat{i}_{BD}) = \langle \widehat{i}_{BD}^2 \rangle - \langle \widehat{i}_{BD} \rangle^2 = N_i N_s'' + (N_i + 1)(N_s'' + 1) + 2C_{si}^2 \cos(2\varphi) - 4C_{si}^2 \cos^2 \varphi, \tag{10.55}$$

and compared with Eq. (10.52), it does not have the terms originating from the vacuum state.

Given that this receiver is identical to the coherent detection receiver, in which instead of the LO laser signal, the idler mode is used, different coherent detection receivers described in Ref. [73] (see also Ref [12]) are directly applicable in EA communications. Now we describe an optical hybrid (OH) based joint balanced detection scheme, shown in Fig. 10.45, because it is suitable for implementation in integrated optics and as such, is compatible with quantum nanophotonics. In Fig. 10.45A, we provide the configuration of 2 × 2 optical hybrid-based BD, while in Fig. 10.45B, we provide the corresponding integrated optics implementation of 2 × 2 OH. The OH has two-phase trimmers to introduce the phase shifts ϕ_1 and ϕ_2, respectively. The optical hybrid is composed of two input and two output Y-junctions, and the scattering matrix can be described by

$$\mathbf{S} = \begin{bmatrix} e^{j\phi_1}\sqrt{1-\kappa} & \sqrt{1-\kappa} \\ \sqrt{\kappa} & e^{j\phi_2}\sqrt{\kappa} \end{bmatrix} \tag{10.56}$$

where κ is the power splitting ratio of Y-junctions, and this matrix transforms input signal and idler modes to

$$\begin{bmatrix} \widehat{A}_s \\ \widehat{A}_i \end{bmatrix} = \mathbf{S} \begin{bmatrix} \widehat{a}_s^\dagger \\ \widehat{a}_i \end{bmatrix}. \tag{10.57}$$

In particular, for the power splitting ratio $\kappa = 1/2$ this transformation is simply

$$\begin{bmatrix} \widehat{A}_i \\ \widehat{A}_i \end{bmatrix} = \frac{1}{\sqrt{2}} \begin{bmatrix} e^{j\phi_1} & 1 \\ 1 & e^{j\phi_2} \end{bmatrix} \begin{bmatrix} \widehat{a}_s^\dagger \\ \widehat{a}_i \end{bmatrix}. \tag{10.58}$$

Assuming that an arbitrary 2-D constellation is used, the signal mode at the output of the I/Q modulator $\widehat{a}_s'$ is related to the input mode of the modulator $\widehat{a}_s''$ by $\widehat{a}_s' = s\widehat{a}_s''$, where $s = s_I + js_Q$ is the 2-D signal constellation point. For instance, for M-ary PSK, we have $s = \exp(j\varphi)$. After that, optical phase conjugation takes place. Based on Fig. 10.45A, we obtain the following balanced detector, photocurrent operator:

$$\widehat{i}_{BD} = \frac{1}{2} s \left(e^{-j\phi_1} - e^{j\phi_2}\right) a_s'' a_i + \frac{1}{2} s^\dagger \left(e^{j\phi_1} - e^{-j\phi_2}\right) a_i^\dagger \left(a_s''\right)^\dagger. \tag{10.59}$$

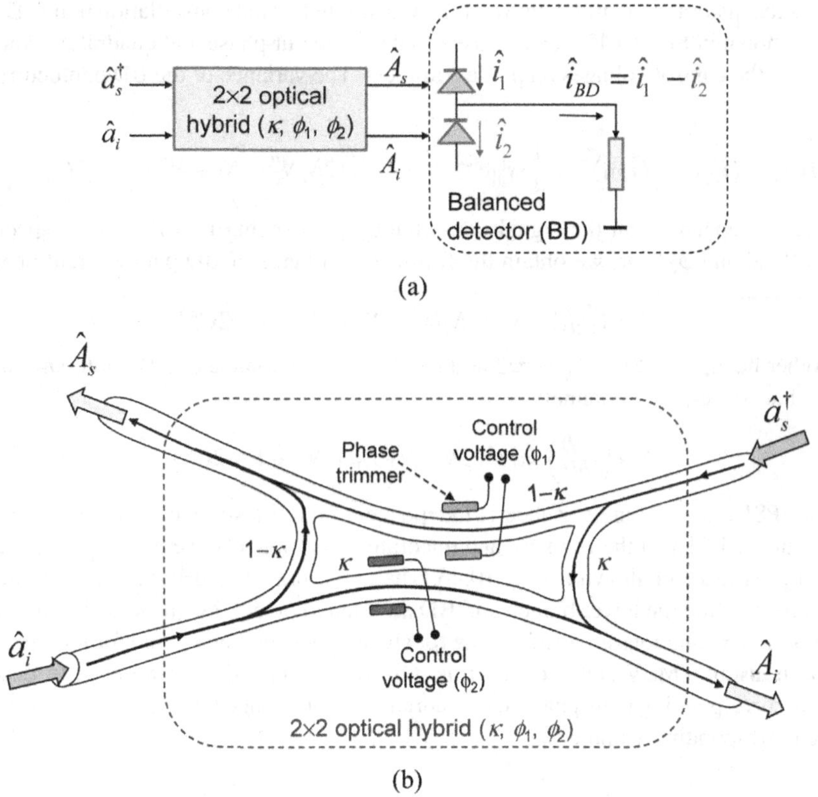

FIGURE 10.45

The 2 × 2 optical hybrid-based EA balanced detection receiver: (A) receiver configuration and (B) 2 × 2 optical hybrid implementation in integrated optics.

The expectation of photocurrent operator is given by

$$\langle \hat{i}_{BD} \rangle = \frac{1}{2} s C_{si} \left(e^{-j\phi_1} - e^{j\phi_2}\right) + \frac{1}{2} s^{\dagger} \left(e^{j\phi_1} - e^{-j\phi_2}\right) C_{si}. \tag{10.60}$$

In a special case, by setting $\phi_1 = 0$ rad and $\phi_2 = \pi$, we obtain the following mean of BD photocurrent operator:

$$\left\langle \hat{i}_{BD}^{(I)} \right\rangle = 2C_{si} \underbrace{\mathrm{Re}\{s\}}_{s_I} = 2C_{si} s_I, \tag{10.61}$$

thus detecting the in-phase component s_I of transmitted signal constellation point. Clearly, for M-ary PSK Re{s} = cosφ, thus obtaining Eq. (10.54). On the other hand, by setting $\phi_1 = \pi/2$ and $\phi_2 = \pi/2$, we obtain the following mean of BD photocurrent operator:

$$\left\langle \hat{i}_{BD}^{(Q)} \right\rangle = 2C_{si} \underbrace{\mathrm{Im}\{s\}}_{s_Q} = 2C_{si} s_Q, \tag{10.62}$$

thus detecting the quadrature component s_Q of the transmitted signal constellation point. Therefore, the joint receiver shown in Fig. 10.45A can be used to determine in-phase and quadrature components by properly setting the control voltages on phase trimmers. The variance of the BD photocurrent operator is given by

$$\mathrm{Var}(\hat{i}_{BD}) = \langle \hat{i}_{BD}^2 \rangle - \langle \hat{i}_{BD} \rangle^2 = \frac{1}{4} ss^\dagger \left| e^{j\phi_1} - e^{-j\phi_2} \right|^2 \left(2N_i N_s'' + N_i + N_s'' + 1 - 2C_{si}^2 \right), \tag{10.63}$$

which is clearly dependent on the signal constellation point being transmitted. In special case, by setting $\phi_1 = 0$ rad and $\phi_2 = \pi$, we obtain the following variance of BD photocurrent operator:

$$\mathrm{Var}\left(\hat{i}_{BD}^{(I)}\right) = ss^\dagger \left(2N_i N_s'' + N_i + N_s'' + 1 - 2C_{si}^2 \right). \tag{10.64}$$

On the other hand, by setting $\phi_1 = \pi/2$ and $\phi_2 = \pi/2$, the variance of BD photocurrent operator is the same as that for in-phase channel:

$$\mathrm{Var}\left(\hat{i}_{BD}^{(Q)}\right) = ss^\dagger \left(2N_i N_s'' + N_i + N_s'' + 1 - 2C_{si}^2 \right). \tag{10.65}$$

For M-ary PSK, $ss^\dagger = 1$ and variance is independent on a transmitted signal constellation point.

To simultaneously detect the in-phase and quadrature components, we need to use the 2×4 optical hybrid-based joint receiver shown in Fig. 10.46. The upper branch and BD are used to detect the in-phase component, while the lower branch and BD are used to detect the quadrature component. Given that two additional Y-junctions are needed, one for signal mode and the second for idler mode, we need to adjust the transmit TMSV state so that Eqs. (10.61) and (10.62) can be used to describe mean photocurrents corresponding to in-phase and quadrature components, while Eqs. (10.64) and (10.65) to describe the corresponding variances.

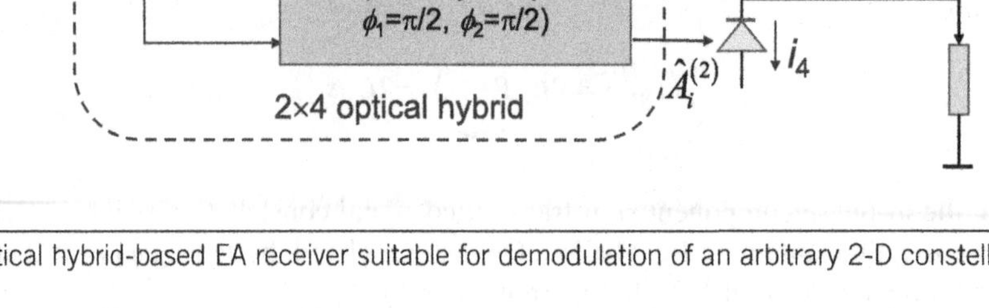

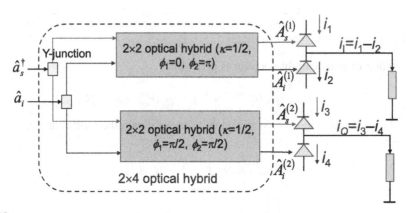

FIGURE 10.46

The 2 × 4 optical hybrid-based EA receiver suitable for demodulation of an arbitrary 2-D constellation.

For the Gaussian modulation (GM) we described above and the OH-based EA demodulator shown in Fig. 10.46, the corresponding channel capacity will be

$$C = \sum_s p_s \log_2\left(1 + \frac{4TN_s(N_s+1)|s|^2}{2ss^\dagger\left(2N_iN_s'' + N_i + N_s'' + 1 - 2C_{si}^2\right)}\right)$$
$$= \sum_s p_s \log_2\left(1 + \frac{2TN_s(N_s+1)|s|^2}{ss^\dagger\left(2N_iN_s'' + N_i + N_s'' + 1 - 2TN_s(N_s+1)\right)}\right), \tag{10.66}$$

where p_s is the prior probability of symbol s and $N_s'' = N_sT + N_b$.

The *Holevo capacity*, representing the quantum limit of classical capacity, is determined by [6].

$$C_{\mathrm{Hol}} = g(TN_s + N_b) - g(N_b), \tag{10.67}$$

where $g(x) = (x+1)\log_2(x+1) - x\log_2 x$.

On the other hand, the *EA capacity* is given by [11,12].

$$C_{\mathrm{EA}} = g(N_s) + g(N_s'') - \left[g\left(\frac{\nu_+ - 1}{2}\right) + g\left(\frac{\nu_- - 1}{2}\right)\right], \tag{10.68}$$

where symplectic eigenvalues are defined by

$$\nu_\mp = \sqrt{(N_s + N_s'' + 1)^2 - 4TN_s(N_s+1)} \mp (N_s'' - N_s). \tag{10.69}$$

Given that according to the uncertainty principle, both in-phase and quadrature components of a Gaussian state cannot be simultaneously measured with the complete precision, for homodyne detection, the information is encoded on a single quadrature so that the average number of received photons is $4TN_s$, while the average number of noise photons is $2N_b+1$, and the corresponding classical capacity for homodyne detection is $C_{\mathrm{hom}} = 0.5\log_2[1 + 4TN_s/(2N_b+1)]$. On the other hand, in heterodyne detection, both quadratures are used so that the average number of received signal photons will be $0.5 \times 0.5 \times 4TN_s = TN_s$ (one-half comes from splitting to two quadratures and the second half from heterodyne splitting), while the average number of noise photons per quadrature is $(2N_b + 1)/2 + 1/2 = N_b + 1$. The corresponding heterodyne channel capacity will be $C_{\mathrm{het}} = \log_2[1 + TN_s/(N_b+1)]$. To achieve the channel capacity in the classical case, we need to use the Gaussian modulation by generating samples from two uncorrelated zero-mean Gaussian sources and, with the help of an arbitrary waveform generator, impose them on the optical carrier using an I/Q modulator. To achieve the Holevo capacity, we need to use the Gaussian state. For instance, the coherent state with GM can achieve the Holevo capacity.

We first evaluate the capacity improvement of the EA communication scheme, with OPC on the transmitter side and GM, over Holevo capacity. For channel transmissivity $T = 0.1$, the average number of background photons $N_b = 20$, and the average signal photon number $N_s = 10^{-3}$, from Fig. 10.47 we conclude that this EA scheme (with GM and OPC on the transmitter side) outperforms the Holevo capacity by > 1.95 times for the number of signal-idler Bosonic modes up to $D = 94$. On the other hand, the BPSK based EA scheme, with OPC on the transmitter side, employing receiver shown in Fig. 10.46 outperforms the Holevo capacity >1.2 times for the number of signal-idler modes

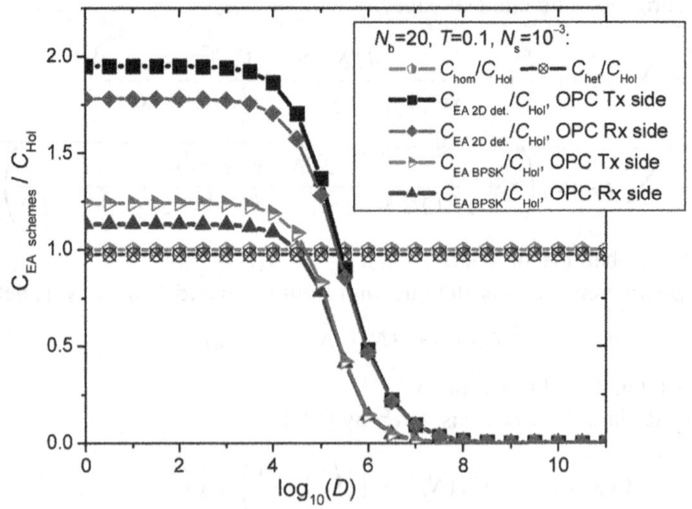

FIGURE 10.47

The capacity improvements of EA scheme with a 2-D detector and OPC on transmitter side over Holevo capacity versus the number of signal-idler modes for $N_s = 10^{-3}$, $N_b = 20$, and $T = 0.1$.

up to 6075. The EA scheme with OPC on the transmitter side for up to 3000 modes outperforms the corresponding scheme with OPC on the receiver side by > 1.1 times.

By close inspection of Fig. 10.47, contrary to Ref. [10], we conclude that, when the OPC is placed on the transmitter side, a large number of signal-idler modes is not required; moreover, the single signal-idler pair is sufficient. This motivates to study how the capacity of the proposed scheme with GM compares with the EA capacity, and the corresponding results are summarized in Fig. 10.48 by setting the simulation parameters as follows: channel transmissivity $T = 0.1$ and the average number of background photons $N_b = 20$. Evidently, the capacity of the EA scheme, with GM and OPC on the transmitter side, is significantly better than EA BPSK, homodyne, and heterodyne capacities.

To summarize, the EA scheme employing the 2-D EA demodulator and OPC on the transmitter side significantly outperforms EA BPSK, homodyne, heterodyne, and Holevo capacities. At the same time, the EA communication scheme in which OPC is placed on the transmitter side outperforms the corresponding scheme when OPC is used on the receiver side.

10.8 Summary

This chapter has been devoted to discussing how to build a future quantum network. The basic concepts of QCNs and the quantum Internet are introduced in Section 10.1, followed by the introduction of quantum teleportation and quantum relay concepts in Section 10.2. Entanglement distribution, a key technique enabling quantum networking by employing the quantum

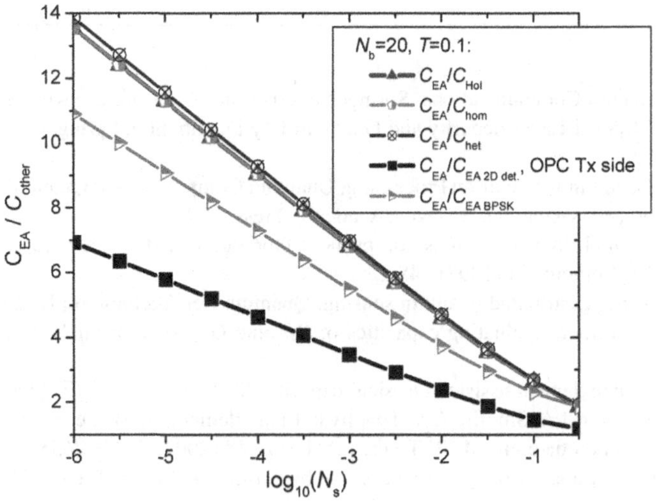

FIGURE 10.48

The EA capacity improvements over Holevo, homodyne, heterodyne, and proposed EA scheme (with OPC on transmitter side) against the average number of signal photons for $N_b = 20$ and $T = 0.1$.

teleportation concept, has been described in detail in Section 10.3. A brief description of how to generate entangled states is followed by descriptions of entanglement swapping, Bell state measurements, and the Hong–Ou–Mendel effect in Subsections 10.3.1 and 10.3.2. CV quantum teleportation has been described in Subsection 10.3.3, followed by the basics of quantum network coding in Subsection 10.3.4. Section 10.4 has been devoted to description how to engineer the entangled states by photon addition and photon subtraction. The hybrid CV–DV networking concept has been introduced in Subsection 10.4.1. After the description of photon subtraction and addition concepts in Subsection 10.4.2, we have described how to apply them to build a hybrid CV–DV quantum network. The following topics on hybrid quantum networks have been described: generation of hybrid DV–CV entangled states (in Subsection 10.4.3), hybrid CV–DV states teleportation (Subsection 10.4.4), and entanglement swapping through entangling measurements (Subsection 10.4.4). Next, the generation of entangled macroscopic light states has been discussed in Subsection 10.4.5, given that macroscopic light states are much more tolerant to channel loss than DV states. The section ends with the noiseless amplification in Subsection 10.4.6, which represents a possible approach to extend the transmission distance between the quantum nodes. The focus of the chapter has been then moved to the cluster state-based quantum networking in Section 10.5. The cluster states concept has been introduced first, followed by the cluster state processing, both in Subsection 10.5.1. The cluster state-based quantum networking concept has been described in Subsection 10.5.2 as both quantum Internet and distributed quantum computing enabling technology. In Section 10.6, surface code-based and QLDPC code-based quantum networking concepts have been described. Section 10.7 has been devoted to EA communication and networking.

References

[1] G. Cariolaro, Quantum Communications, Springer International Publishing, Switzerland, 2015.

[2] I.B. Djordjevic, Physical-Layer Security and Quantum Key Distribution, Springer International Publishing, Switzerland, 2019.

[3] I.B. Djordjevic, Quantum Information Processing, Quantum Computing, and Quantum Error Correction: An Engineering Approach, second ed., Elsevier/Academic Press, 2021.

[4] P.W. Shor, Polynomial-time algorithms for prime factorization and discrete logarithms on a quantum computer, SIAM J. Comput. 26 (1997) 1484.

[5] Z. Zhang, Q. Zhuang, Distributed quantum sensing, Quantum Sci. Technol. 6 (4) (2021) 043001.

[6] A.S. Holevo, R.F. Werner, Evaluating capacities of Bosonic Gaussian channels, Phys. Rev. 63 (3) (2001) 032312.

[7] A.S. Holevo, On entanglement assisted classical capacity, J. Math. Phys. 43 (9) (2002) 4326–4333.

[8] C.H. Bennett, P.W. Shor, J.A. Smolin, A.V. Thapliyal, Entanglement-assisted capacity of a quantum channel and the reverse Shannon theorem, IEEE Trans. Inf. Theor. 48 (2002) 2637–2655.

[9] M. Sohma, O. Hirota, Capacity of a channel assisted by two-mode squeezed states, Phys. Rev. 68 (2) (2003) 022303.

[10] H. Shi, Z. Zhang, Q. Zhuang, Practical route to entanglement-assisted communication over noisy Bosonic channels, Phys. Rev. Appl. 13 (3) (2020) 034029.

[11] I.B. Djordjevic, On entanglement assisted classical optical communications, IEEE Access 9 (March 17, 2021) 42604–42609.

[12] I.B. Djordjevic, Quantum receivers for entanglement assisted classical optical communications, article ID 7500114, IEEE Photonics J. 13 (3) (June 2021) (Early posting: 27 April 2021).

[13] R.V. Meter, S.J. Devitt, The path to scalable distributed quantum computing, IEEE Computer 49 (9) (2016).

[14] A.M. Childs, Secure assisted quantum computation, Quant. Inf. Comput. 5 (6) (2005) 456–466.

[15] M.K. Bhaskar, et al., Experimental demonstration of memory-enhanced quantum communication, Nature 580 (2020) 60.

[16] S.-K. Liao, et al., Satellite-to-ground quantum key distribution, Nature 549 (2017) 43.

[17] J.-G. Ren, et al., Ground-to-satellite quantum teleportation, Nature 549 (2017) 70.

[18] J. Yin, et al., Satellite-based entanglement distribution over 1200 kilometers, Science 356 (1140) (2017).

[19] Entanglement Management and Control in Transparent Optical Quantum Networks, Funding Opportunity Announcement (FOA) number: DE-FOA-0002476.

[20] Report of the DOE Quantum Internet Blueprint Workshop, "From Long-Distance Entanglement to Building a Nationwide Quantum Internet", 2020.

[21] C.H. Bennett, G. Brassard, C. Crépeau, R. Jozsa, A. Peres, W.K. Wootters, Teleporting an unknown quantum state via dual classical and Einstein-Podolsky-Rosen channels, Phys. Rev. Lett. 70 (13) (1993) 1895–1899.

[22] H.-J. Briegel, W. Dür, J.I. Cirac, P. Zoller, Quantum repeaters: the role of imperfect local operations in quantum communication, Phys. Rev. Lett. 81 (26) (1998) 5932–5935.

[23] J.-W. Pan, C. Simon, Č. Brunker, A. Zeilinger, Entanglement purification for quantum communication, Nature 410 (April 26, 2001) 1067–1070.

[24] I.B. Djordjevic, Photonic implementation of quantum relay and encoders/decoders for sparse-graph quantum codes based on optical hybrid, IEEE Photon. Technol. Lett. 22 (19) (October 1, 2010) 1449–1451.

[25] I.B. Djordjevic, On the photonic implementation of universal quantum gates, Bell states preparation circuit and quantum LDPC encoders and decoders based on directional couplers and HNLF, Opt Express 18 (8) (April 1, 2010) 8115–8122.

[26] I.B. Djordjevic, On the photonic implementation of universal quantum gates, Bell states preparation circuit, quantum relay and quantum LDPC encoders and decoders, IEEE Photon. J. 2 (1) (Feb. 2010) 81–91.
[27] K. Mattle, H. Weinfurter, P.G. Kwiat, A. Zeilinger, Dense coding in experimental quantum communication, Phys. Rev. Lett. 76 (25) (1996) 4656.
[28] M. Źukowski, A. Zeilinger, M.A. Horne, A.K. Ekert, "Event-ready-detectors" Bell experiment via entanglement swapping, Phys. Rev. Lett. 71 (26) (1993) 4287.
[29] N. Sangouard, C. Simon, H. De Riedmatten, N. Gisin, Quantum repeaters based on atomic ensembles and linear optics, Rev. Mod. Phys. 83 (1) (2011) 33.
[30] H.-K. Lo, M. Curty, B. Qi, Measurement-device-independent quantum key distribution, Phys. Rev. Lett. 108 (13) (2012) 130503.
[31] C.C.W. Lim, C. Portmann, M. Tomamichel, R. Renner, N. Gisin, Device-independent quantum key distribution with local Bell test, Phys. Rev. X 3 (3) (2013) 031006.
[32] H.-L. Yin, et al., Measurement-device-independent quantum key distribution over a 404 km optical fiber, Phys. Rev. Lett. 117 (2016) 190501.
[33] R. Valivarthi, I. Lucio-Martinez, A. Rubenok, P. Chan, F. Marsili, V.B. Verma, M.D. Shaw, J.A. Stern, J.A. Slater, D. Oblak, S.W. Nam, W. Tittel, Efficient Bell state analyzer for time-bin qubits with fast-recovery WSi superconducting single photon detectors, Opt Express 22 (2014) 24497–24506.
[34] N. Lütkenhaus, J. Calsamiglia, K.-A. Suominen, Bell measurements for teleportation, Phys. Rev. 59 (1999) 3295–3300.
[35] Z.-Y.J. Ou, Quantum Optics for Experimentalists, World Scientific, New Jersey-London-Singapore, 2017.
[36] E. Knill, R. Laflamme, G.J. Milburn, A scheme for efficient quantum computation with linear optics, Nature 409 (2001) 46–52.
[37] A. Zeilinger, Experiment and the foundations of quantum physics, in: B. Bederson (Ed.), More Things in Heaven and Earth, Springer, New York, NY, 1999, pp. 482–498.
[38] M. Hayashi, K. Iwama, H. Nishimura, R. Raymond, S. Yamashita, Quantum network coding, in: T. Wolfgang, W. Pascal (Eds.), STACS, Springer, Berlin, Heidelberg, 2007, pp. 610–621.
[39] M. Hayashi, Prior entanglement between senders enables perfect quantum network coding with modification, Phys. Rev. 76 (2007) 040301.
[40] H. Kobayashi, F.L. Gall, H. Nishimura, M. Rötteler, Constructing quantum network coding schemes from classical nonlinear protocols, in: Proc 2011 IEEE International Symposium on Information Theory Proceedings, 2011, pp. 109–113.
[41] H. Lu, et al., Experimental quantum network coding, Npj Quantum Information 5 (October 23, 2019) 89.
[42] R. Ahlswede, N. Cai, S.-Y.R. Li, R.W. Yeung, Network information flow, IEEE Trans. Inf. Theor. 46 (4) (2000) 1204–1216.
[43] S. Takeda, T. Mizuta, M. Fuwa, P. van Loock, A. Furusawa, Deterministic quantum teleportation of photonic quantum bits by a hybrid technique, Nature 500 (2013) 315.
[44] A.E. Ulanov, D. Sychev, A.A. Pushkina, I.A. Fedorov, A.I. Lvovsky, Quantum teleportation between discrete and continuous encodings of an optical qubit, Phys. Rev. Lett. 118 (2017) 160501.
[45] S.-W. Lee, H. Jeong, Near-deterministic quantum teleportation and resource-efficient quantum computation using linear optics and hybrid qubits, Phys. Rev. 87 (2013) 022326.
[46] S. Takeda, et al., Entanglement swapping between discrete and continuous variables, Phys. Rev. Lett. 114 (2015) 100501.
[47] K. Huang, et al., Engineering optical hybrid entanglement between discrete- and continuous-variable states, New J. Phys. 21 (2019) 083033.
[48] G. Guccione, et al., Connecting heterogeneous quantum networks by hybrid entanglement swapping, Sci. Adv. 6 (22) (May 29, 2020) eaba4508.
[49] J.D. Franson, Entangled photon holes, Phys. Rev. Lett. 96 (9) (March 6, 2006) 090402.

[50] J.D. Franson, Quantum communication using entangled photon holes, in: Frontiers in Optics 2012/Laser Science XXVIII, OSA Technical Digest (Online), Optical Society of America, 2012 paper FTh2C.1.

[51] N. Biagi, L.S. Constanzo, M. Bellini, A. Zavatta, Entangling macroscopic light states by delocalized photon addition, Phys. Rev. Lett. 124 (3) (January 24, 2020) 033604.

[52] A. Zavatta, J. Fiurášek, M. Bellini, A high-fidelity noiseless amplifier for quantum light states, Nat. Photonics 5 (2011) 52–56.

[53] C.N. Gagatsos, J. Fiurášek, A. Zavatta, M. Bellini, N.J. Cerf, Heralded noiseless amplification and attenuation of non-Gaussian states of light, Phys. Rev. 89 (6) (June 9, 2014) 062311.

[54] I.B. Djordjevic, On global quantum communication networking, Entropy 22 (8) (July 29, 2020) 831.

[55] D. McMahon, Quantum Computing Explained, John Wiley & Sons, Inc., Hoboken, NJ, 2008.

[56] O. Gühne, G. Tóth, P. Hyllus, H.J. Briegel, Bell inequalities for graph states, Phys. Rev. Lett. 95 (12) (2005) 120405.

[57] M. Hein, W. Dür, J. Eisert, R. Raussendorf, M. van den Nest, H.J. Briegel, Entanglement in graph states and its applications, in: G. Casati, et al. (Eds.), International School of Physics Enrico Fermi (Varenna, Italy), Course CLXII, Quantum Computers, Algorithms and Chaos, 2006, pp. 115–218.

[58] M.A. Neilsen, Cluster-state quantum computation, Rep. Math. Phys. 57 (1) (2006) 147–161.

[59] A. Mantri, T. Demarie, J. Fitzsimons, Universality of quantum computation with cluster states and (X, Y)-plane measurements, Sci. Rep. 7 (2017) 42861.

[60] G. Gilbert, et al., Efficient construction of photonic quantum-computational clusters, Phys. Rev. 73 (June 2006) 064303.

[61] K. Noh, S.M. Girvin, L. Jiang, Encoding an oscillator into many oscillators, Phys. Rev. Lett. 125 (2020) 080503.

[62] J. Park, J. Joo, A. Zavatta, M. Bellini, H. Jeong, Efficient noiseless linear amplification for light fields with larger amplitudes, Opt Express 24 (2016) 1331–1346.

[63] I.B. Djordjevic, Surface-codes-based quantum communication networks, Special Issue on Physical-Layer Security, Quantum Key Distribution, and Post-Quantum Cryptography, Entropy 22 (9) (Sept. 2020) 1059.

[64] L. Jiang, J.M. Taylor, K. Nemoto, W.J. Munro, R. Van Meter, M.D. Lukin, Quantum repeater with encoding, Phys. Rev. 79 (#3) (March 20, 2009). Art. 032325.

[65] A.G. Fowler, D.S. Wang, C.D. Hill, T.D. Ladd, R. Van Meter, L.C.L. Hollenberg, Surface code quantum communication, Phys. Rev. Lett. 104 (#18) (May 6, 2010). Art. 180503.

[66] I.B. Djordjevic, QKD-enhanced cybersecurity protocols, IEEE Photonics J. 13 (2) (April 2021) 7600208.

[67] D. Litinski, Game of surface codes: large-scale quantum computing with lattice surgery, Quantum 3 (Feb. 2019) 128.

[68] C. Horsman, A.G. Fowler, S. Devitt, R.V. Meter, Surface code quantum computing by lattice surgery, New J. Phys. 14 (2012) 123011.

[69] I.B. Djordjevic, Quantum LDPC codes from balanced incomplete block designs, IEEE Commun. Lett. 12 (May 2008) 389–391.

[70] Q. Zhuang, et al., Optimum mixed-state discrimination for noisy entanglement-enhanced sensing, Phys. Rev. Lett. 118 (2017) 040801.

[71] Z.-Y. Jeff Ou, Quantum Optics for Experimentalists, World Scientific Publishing Co. Pte. Ltd., 2017.

[72] S.L. Jansen, et al., Long-haul DWDM transmission systems employing optical phase conjugation, IEEE J. Sel. Top. Quant. Electron. 12 (2006) 505–520.

[73] I.B. Djordjevic, Advanced Optical and Wireless Communications Systems, Springer International Publishing AG, Cham, Switzerland, 2018.

Further reading

[1] A. Harrow, Superdense coding of quantum states, Phys. Rev. Lett. 92 (18) (May 7, 2004) 187901, 187901-187904.

[2] D. Gottesman, I.L. Chuang, Demonstrating the viability of universal quantum computation using teleportation and single-qubit operations, Nature 402 (November 25, 1999) 390–393.

[3] C.K. Hong, Z.Y. Ou, L. Mandel, Measurement of subpicosecond time intervals between two photons by interference, Phys. Rev. Lett. 59 (1987) 2044.

[4] M. Celebiler, G. Stette, On increasing the down-link capacity of a regenerative satellite repeater in point-to-point communications, Proc. IEEE 66 (1) (Jan. 1978) 98–100.

[5] A. Zavatta, S. Viciani, M. Bellini, Quantum-to-Classical transition with single-photon-added coherent states of light, Science 306 (2004) 660.

[6] A. Zavatta, V. Parigi, M.S. Kim, M. Bellini, Subtracting photons from arbitrary light fields: experimental test of coherent state invariance by single-photon annihilation, New J. Phys. 10 (2008) 123006.

CHAPTER

Quantum sensing and quantum radars

11

Chapter outline

11.1 Quantum phase estimation

We expect that the photon number operator $\widehat{N}$ and phase operator $\widehat{\phi}$ should represent the canonically conjugate operators, so their commutator yields to the uncertainty relation, which imposes physical limits on phase measurements, as follows:

$$\Delta\phi \ \Delta N \geq 1, \tag{11.1}$$

where $\Delta\phi$ and ΔN represent the uncertainties in the phase and photon number, respectively. Let us observe the coherent state represented by

$$|\alpha\rangle = e^{-|\alpha|^2/2} \sum_n \frac{\alpha^n}{\sqrt{n!}} |n\rangle, \tag{11.2}$$

Quantum Communication, Quantum Networks, and Quantum Sensing. https://doi.org/10.1016/B978-0-12-822942-2.00007-8

where $\alpha = \alpha_I + j\alpha_Q$ is a complex number, and $|n\rangle$ are the Fock (photon number) states, which are the eigenkets of the photon number operator; that is, $\widehat{N}|n\rangle = n|n\rangle$. The expected value of the number operator is

$$\overline{N} = \langle\alpha|\widehat{N}|\alpha\rangle = \underbrace{\langle\alpha|a^\dagger a|\alpha\rangle}_{\langle\alpha|\alpha^* \ \alpha|\alpha\rangle} = |\alpha|^2. \tag{11.3}$$

On the other hand, the expected value of the square of the number operator is

$$\langle\widehat{N}^2\rangle = \langle\alpha|\widehat{N}^2|\alpha\rangle = \langle\alpha|\widehat{N}\widehat{N}|\alpha\rangle = \langle\alpha|\underbrace{a^\dagger}_{\langle\alpha|\alpha^*} a \underbrace{a^\dagger a|\alpha\rangle}_{\alpha|\alpha\rangle} = |\alpha|^2\langle\alpha|\underbrace{aa^\dagger}_{a^\dagger a+\widehat{1}}|\alpha\rangle = |\alpha|^4 + |\alpha|^2$$

$$= \overline{N}^2 + \overline{N}, \tag{11.4}$$

where we use the fundamental commutation relation of the annihilation and creation operators $[a, a^\dagger] = \widehat{1}$, with $\widehat{1}$ being the identity operator.

The corresponding variance is

$$\sigma^2(\widehat{N}) = \mathrm{Var}(\widehat{N}) = \langle\widehat{N}^2\rangle - \langle\widehat{N}\rangle^2 = \overline{N}^2 + \overline{N} - \overline{N}^2 = \overline{N}, \tag{11.5}$$

and the fractional uncertainty in the photon number will be

$$\frac{\sigma(N)}{\overline{N}} = \frac{\sqrt{\overline{N}}}{\overline{N}} = \frac{1}{\sqrt{\overline{N}}} \xrightarrow{N\to\infty} 0. \tag{11.6}$$

The probability of detecting n photons is given by

$$P_n = |\langle n|\alpha\rangle|^2 = e^{-|\alpha|^2}\frac{|\alpha|^{2n}}{n!} = e^{-\overline{N}}\frac{\overline{N}^n}{n!}, \tag{11.7}$$

which represents a *Poisson distribution* with a mean value of $|\alpha|^2$.

From the Heisenberg inequality of Eq. (11.1), we obtain the following phase measurement limitation:

$$\Delta\phi \geq \frac{1}{\sqrt{\overline{N}}}, \tag{11.8}$$

which is commonly referred to as the *standard quantum limit* (SQL).

Unfortunately, since the 1960s, it has been well known that to obtain the proper commutation relations, $\widehat{\exp(j\phi)}$ must be unitary, while $\widehat{\phi}$ must be Hermitian [1–5]. To solve this problem, Dirac proposed redefining the creation and annihilation operators as follows [1]:

$$\widehat{a}^\dagger = \sqrt{\widehat{N}}e^{-j\hat{\phi}}, \quad \widehat{a} = e^{j\hat{\phi}}\sqrt{\widehat{N}}, \tag{11.9}$$

where $\widehat{\phi}$ is the Hermitian operator of the phase. From the commutation relation $[a, a^\dagger] = \widehat{1}$, we obtain

$$e^{j\hat{\phi}}\widehat{N}e^{-j\hat{\phi}} = \widehat{N} + \widehat{1}. \tag{11.10}$$

Unfortunately, the exponential operator $\widehat{\exp(j\phi)}$ is not unitary because

$$\underbrace{e^{j\hat{\phi}}}_{\widehat{a}N^{-1/2}}\underbrace{e^{-j\hat{\phi}}}_{N^{-1/2}\widehat{a}^{\dagger}} = \widehat{a}N^{-1}\widehat{a}^{\dagger} \neq \widehat{1}. \tag{11.11}$$

Barnett and Pegg constructed the unitary exponential operators in Ref. [2]:

$$e^{j\hat{\phi}} = \sum_{n=-\infty}^{\infty} |n\rangle\langle n+1|, \quad e^{-j\hat{\phi}} = \sum_{n=-\infty}^{\infty} |n+1\rangle\langle n|. \tag{11.12}$$

The exponential operator is now unitary:

$$e^{j\hat{\phi}}\left(e^{j\hat{\phi}}\right)^{\dagger} = \left(e^{j\hat{\phi}}\right)^{\dagger}e^{j\hat{\phi}} = \widehat{1}. \tag{11.13}$$

Unfortunately, negative number states do not physically exist. The most common approach to this problem is to employ the non-Hermitian Susskind–Glogower operator [3–5]:

$$\begin{aligned} \widehat{E} &= e^{j\hat{\phi}} = \left(\widehat{N}+\widehat{1}\right)^{-1/2}\widehat{a} = \left(\widehat{a}\widehat{a}^{\dagger}\right)^{-1/2}\widehat{a} = \sum_{n=0}^{\infty}|n\rangle\langle n+1|, \\ \Rightarrow \widehat{E}^{\dagger} &= e^{-j\hat{\phi}} = \widehat{a}^{\dagger}\left(\widehat{N}+\widehat{1}\right)^{-1/2} = \widehat{a}^{\dagger}\left(\widehat{a}\widehat{a}^{\dagger}\right)^{-1/2} = \sum_{n=0}^{\infty}|n+1\rangle\langle n|. \end{aligned} \tag{11.14}$$

Clearly, the following is valid:

$$\widehat{E}\widehat{E}^{\dagger} = \sum_{n=0}^{\infty}\sum_{m=0}^{\infty}|n\rangle\langle n+1|m+1\rangle\langle m| = \sum_{n=0}^{\infty}|n\rangle\langle n| = \widehat{1}. \tag{11.15}$$

Unfortunately,

$$\widehat{E}^{\dagger}\widehat{E} = \sum_{n=0}^{\infty}\sum_{m=0}^{\infty}|n+1\rangle\langle n|m\rangle\langle m+1| = \sum_{n=0}^{\infty}|n+1\rangle\langle n+1| = \widehat{1} - |0\rangle\langle 0| \neq \widehat{1}, \tag{11.16}$$

and strictly speaking, the exponential operator is not unitary. However, for the average number of photons $n \geq 1$, it is approximately unitary. On the other hand, the following two operators derived from the exponential operator are unitary:

$$\widehat{C} = \frac{\widehat{E}+\widehat{E}^{\dagger}}{2}, \quad \widehat{S} = \frac{\widehat{E}-\widehat{E}^{\dagger}}{2j}, \tag{11.17}$$

and can be used as observables. These operators are analogs of cosϕ and sinϕ and satisfy the following commutation relation:

$$\left[\widehat{C},\widehat{S}\right] = \frac{j}{2}|0\rangle\langle 0|, \tag{11.18}$$

as well as the following trigonometric-like identity:

$$\widehat{C}^2 + \widehat{S}^2 = \widehat{1} - \frac{1}{2}|0\rangle\langle 0|. \tag{11.19}$$

Because the following two commutation relations are satisfied:

$$\left[\widehat{C},\widehat{N}\right]=j\widehat{S},\ \left[\widehat{S},\widehat{N}\right]=-j\widehat{C}, \tag{11.20}$$

the corresponding uncertainty relations are

$$\Delta\widehat{N}\ \Delta\widehat{C}\geq\frac{1}{2}\left|\langle\widehat{S}\rangle\right|,\ \ \Delta\widehat{N}\ \Delta\widehat{S}\geq\frac{1}{2}\left|\langle\widehat{C}\rangle\right|. \tag{11.21}$$

11.1.1 Quantum interferometry

Let us observe the Mach–Zehnder interferometer (MZI) with two input ports (I_1 and I_2) and two output ports (O_1 and O_2), as shown in Fig. 11.1. It comprises two balanced beam splitters (BBSs) and two mirrors. A photon entering the input port I_1 has two possible pathways to reach the second BBS: the upper path and the lower path. When the photon travels along the upper path, it acquires the phase shift ϕ. The state when the photon leaves the first BBS along the upper path can be denoted $|u\rangle=|1\rangle$, with the state when the photon chooses the lower path denoted $|l\rangle=|0\rangle$. When observing the output port outcomes, we cannot specify the path along which the photon has traveled, so the state of a photon arriving at the second BBS can be described as superposition state $|\psi\rangle=2^{-1/2}(|0\rangle+|1\rangle)$ Given that the photon number n in this case cannot exceed 1, the operator $\widehat{C}$ is truncated to

$$\widehat{C}=2^{-1}(|0\rangle\langle 1|+|1\rangle\langle 0|), \tag{11.22}$$

By selecting the computational basis as CB = $\{|0\rangle=[1\ 0]^T, |1\rangle=[0\ 1]^T\}$, we conclude that the operator $\widehat{C}$ is related to the Pauli-X operator by

$$\widehat{C}=2^{-1}(|0\rangle\langle 1|+|1\rangle\langle 0|)=2^{-1}\underbrace{\begin{bmatrix}0&1\\1&0\end{bmatrix}}_{X}=2^{-1}X. \tag{11.23}$$

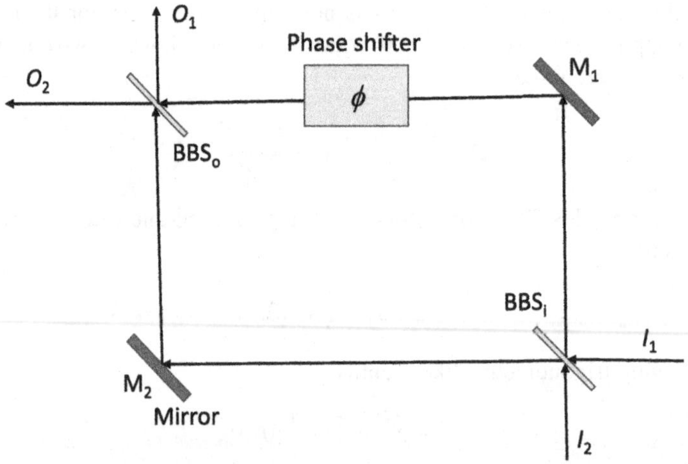

FIGURE 11.1

The Mach–Zehnder interferometer with phase shift in the upper branch. *BBS*: balanced beam splitter.

Since $X^2 = \widehat{1}$, we have $\widehat{C}^2 = 2^{-2}\widehat{1}$. By taking the mean values of $\widehat{C}$ and $\widehat{C}^2$ with respect to state $|\psi\rangle$, we obtain

$$\langle \widehat{C} \rangle = \langle \psi|\widehat{C}|\psi\rangle = 2^{-2}\big(\langle 1|e^{-j\phi} + \langle 0|)\underbrace{X(|0\rangle + e^{j\phi}|1\rangle)}_{|1\rangle + e^{j\phi}|0\rangle}\big) = 2^{-1}\underbrace{\frac{e^{j\phi}+e^{-j\phi}}{2}}_{\cos\phi} = 2^{-1}\cos\phi,$$
$$\langle \widehat{C}^2 \rangle = \langle \psi|\underbrace{\widehat{C}^2}_{2^{-2}\widehat{I}}|\psi\rangle = 1/4. \tag{11.24}$$

The uncertainty of measurement of observable $\widehat{C}$ is now

$$\Delta\widehat{C} = \sqrt{\langle \widehat{C}^2 \rangle - \langle \widehat{C} \rangle^2} = 2^{-1}\sqrt{1-\cos^2\phi} = 2^{-1}|\sin\phi|. \tag{11.25}$$

Phase responsivity is given by

$$\left|\frac{d\langle \widehat{C} \rangle}{d\phi}\right| = \frac{1}{2}|\sin\phi|. \tag{11.26}$$

When we repeat the measurement N times, we expect from statistical theory that the variance will be reduced N times and the standard deviation $N^{1/2}$ times. The overall input state can be represented by $|\psi_R\rangle = |\psi\rangle^{\otimes N} = |\psi\rangle \otimes \cdots \otimes |\psi\rangle$. The corresponding observable is given by $\widehat{C}_R = \oplus_{n=1}^{N}\widehat{C}^{(n)}$. For simplicity, let us ignore the factor ½ in $\widehat{C}$. The expected value of observable $\widehat{C}_R$ is given by

$$\langle \widehat{C}_R \rangle = \langle \psi_R|\widehat{C}_R|\psi_R\rangle = N\cos\phi. \tag{11.27}$$

Further, because $\widehat{C}^2 = \widehat{1}$, the variance of $\widehat{C}_R$ given N realizations is

$$\left(\Delta\widehat{C}_R\right)^2 = N\left(\langle \widehat{C}^2 \rangle - \langle \widehat{C} \rangle^2\right) = N(1-\cos^2\phi) = N\sin^2\phi. \tag{11.28}$$

Therefore, the uncertainty of the phase estimate can be evaluated as in Ref. [6]:

$$\Delta\phi_{SQL} = \frac{\Delta\widehat{C}_R}{\left|\frac{d\langle \widehat{C}_R \rangle}{d\phi}\right|} = \frac{N^{1/2}|\sin\phi|}{N|\sin\phi|} = \frac{1}{\sqrt{N}}. \tag{11.29}$$

11.1.2 Supersensitive regime and Heisenberg limit

When N independent photons are used to determine the estimate of phase, the uncertainty is limited by $N^{-1/2}$. When N photons used in the phase estimate are entangled, we can achieve sensitivity better than the SQL. That regime is known as the *supersensitive regime*, and we say that the corresponding quantum sensor is *supersensitive* [7].

For this purpose, we can use the NOON states, defined by

$$|\psi_N\rangle = 2^{-1/2}(|N0\rangle + |0N\rangle) = 2^{-1/2}(|N\rangle|0\rangle + |0\rangle|N\rangle). \tag{11.30}$$

which are highly entangled states often used in various quantum sensing applications. Let us introduce the following notation $|n\rangle_u\ |m\rangle_l$ to denote the quantum state of n photons in the top branch and m photons in the lower branch of the MZI. We are interested in measuring the phase difference arising when the photons are passed along the top path. Because of entanglement, the phase shifts from the N photons act collectively to introduce the phase shift $N\phi$, and the quantum state reaching the BBS_o can be represented by

$$|\psi_N\rangle = 2^{-1/2}\left(e^{jN\phi}|N0\rangle + |0N\rangle\right). \tag{11.31}$$

To measure the phase shift ϕ, the detector needs to measure the following observable:

$$\hat{O}_N = |0N\rangle\langle N0| + |N0\rangle\langle 0N|. \tag{11.32}$$

The expectation of this operator can be determined by

$$\begin{aligned}\langle\hat{O}_N\rangle &= \langle\psi_N|\hat{O}_N|\psi_N\rangle = \langle\psi_N|(|0N\rangle\langle N0| + |N0\rangle\langle 0N|)\left(|0N\rangle + e^{jN\phi}|N0\rangle\right) \\ &= \frac{1}{2}\left(\langle 0N| + \langle N0|e^{-jN\phi}\right)\left(|N0\rangle + e^{jN\phi}|0N\rangle\right) = \frac{1}{2}\left(e^{-jN\phi} + e^{jN\phi}\right) = \cos(N\phi).\end{aligned} \tag{11.33}$$

Given that

$$\begin{aligned}\hat{O}_N^2 &= (|0N\rangle\langle N0| + |N0\rangle\langle 0N|)(|0N\rangle\langle N0| + |N0\rangle\langle 0N|) \\ &= |0N\rangle\langle 0N| + |N0\rangle\langle N0| = \hat{1},\end{aligned} \tag{11.34}$$

where $\hat{1}$ is the identity operator, we determine that the expectation of $\hat{O}_N^2$ is simply $\langle\hat{O}_N^2\rangle = \langle\psi_N|\underbrace{\hat{O}_N^2}_{\hat{1}}|\psi_N\rangle = \langle\psi_N|\psi_N\rangle = 1$, so the observable variance becomes

$$\left(\Delta\hat{O}_N\right)^2 = \langle\hat{O}_N^2\rangle - \langle\hat{O}_N\rangle^2 = 1 - \cos^2(N\phi) = \sin^2(N\phi). \tag{11.35}$$

The uncertainty of the phase estimate for the NOON input state can now be evaluated by

$$\Delta\phi_{HL} = \frac{\Delta\hat{O}_N}{\left|\frac{d\langle\hat{O}_N\rangle}{d\phi}\right|} = \frac{|\sin(N\phi)|}{N|\sin(N\phi)|} = \frac{1}{N}, \tag{11.36}$$

which is known as the *Heisenberg limit*. Therefore, using the entanglement states as input states to the quantum sensor, we can outperform the SQL. Compared with unentangled states, we can improve the phase sensitivity by $N^{-1/2}$ times with maximally entangled states.

The two-photon NOON state can easily be obtained by employing the Hong–Ou–Mendel effect [8,9] (see Chapter 10). Higher order NOON states can be generated by spontaneous parametric downconversion (SPDC) [9–13] followed by the postelection process to achieve the desired entangled states [14–17].

11.2 Quantum Fisher information and quantum Cramér–Rao bound

Let us briefly review unbiased estimation and the classical Cramér–Rao bound [6,18,19].

11.2.1 Cramér–Rao bound

The Fisher information corresponds to the amount of information that a random variable X carries about the unknown parameter θ. Let $f(x|\theta)$ denote the conditional probability density function of x given θ. The *unbiased estimator* $\tilde{\theta}(x)$ is one for which the expected value of estimation error is zero regardless of value of θ; that is, we can write

$$\left\langle \tilde{\theta}(x) - \theta \middle| \theta \right\rangle = \int \left[\tilde{\theta}(x) - \theta\right] f(x|\theta)dx = 0. \tag{11.37}$$

The partial derivative of the estimation error is also zero; that is,

$$\frac{\partial}{\partial\theta} \int \left[\tilde{\theta}(x) - \theta\right] f(x|\theta)dx = 0. \tag{11.38}$$

Applying the derivative product rule, we obtain

$$\begin{aligned} &-\underbrace{\int f(x|\theta)dx}_{1} + \int \left[\tilde{\theta}(x) - \theta\right] \underbrace{\frac{\partial}{\partial\theta} f(x|\theta)}_{f(x|\theta)\frac{\partial}{\partial\theta}\log f(x|\theta)} dx = 0 \\ &\Rightarrow \int \left[\tilde{\theta}(x) - \theta\right] f(x|\theta) \frac{\partial}{\partial\theta} \log f(x|\theta)dx = 1 \end{aligned} \tag{11.39}$$

Since $f(x|\theta) = \sqrt{f(x|\theta)}\sqrt{f(x|\theta)}$ after substituting into the previous equation, and by applying the Bunyakovsky–Cauchy–Schwarz inequality, we obtain

$$\begin{aligned} 1 &= \left\{ \int \left[\tilde{\theta}(x) - \theta\right] \sqrt{f(x|\theta)}\sqrt{f(x|\theta)}\, \frac{\partial}{\partial\theta} \log f(x|\theta)dx \right\}^2 \\ &\le \underbrace{\int \left[\tilde{\theta}(x) - \theta\right]^2 f(x|\theta)dx}_{\mathrm{Var}\left(\tilde{\theta}\right)} \cdot \underbrace{\int \left[\frac{\partial}{\partial\theta} \log f(x|\theta)\right]^2 f(x|\theta)dx}_{I_F(\theta)} \\ &\le \mathrm{Var}\left(\tilde{\theta}\right) I_F(\theta), \end{aligned} \tag{11.40}$$

and the second term in Eq. (11.40) is known as the *Fisher information*. Given that the partial derivative of $\log f(x|\theta)$ is known as the *score* representing the steepness of the log-likelihood function, we can define the Fisher information as the average value of the score. Further, given that $\langle \partial \log f(x|\theta)/\partial\theta \rangle = 0$ we can also define the Fisher information as the variance of the score.

The previous inequality can be rewritten as

$$\mathrm{Var}\left(\tilde{\theta}\right) \ge \frac{1}{I_F(\theta)}, I_F(\theta) = \int \left[\frac{\partial}{\partial\theta} \log f(x|\theta)\right]^2 f(x|\theta)dx, \tag{11.41}$$

which is commonly referred to as the *Cramér–Rao bound.* In distributed classical sensing, we use M classical sensors to measure the same parameter θ, and by averaging out the outcomes of the measurements, we can reduce the variance of unbiased estimator M times while reducing the standard deviation $\sigma(\widetilde{\theta})$ by $\sqrt{M}$ times; that is,

$$\mathrm{Var}_M(\widetilde{\theta}) = \sigma_M^2(\widetilde{\theta}) \geq \frac{1}{M\, I_F(\theta)}, \tag{11.42}$$

which represents the standard limit for distributed sensing.

11.2.2 Quantum Cramér–Rao bound

Quantum Fisher information (QFI) is a quantum equivalent of the classical Fisher information. The QFI of state ρ with respect to observable $\widehat{O}$ can be defined by

$$Q_F(\rho, \widehat{O}) = 2\sum_{m,n} \frac{(\lambda_m - \lambda_n)^2}{\lambda_m + \lambda_n} \left|\langle \lambda_m|\widehat{O}|\lambda_n\rangle\right|^2, \tag{11.43}$$

where λ_k are eigenvalues of ρ and $|\lambda_k\rangle$ are corresponding eigenkets; that is, $\rho|\lambda_k\rangle = \lambda_k|\lambda_k\rangle$.

At this point, it is convenient to formulate quantum sensing as the unitary transformation of the system described by density operator ρ as illustrated in Fig. 11.2. The unknown parameter θ is embedded in a unitary transformation $\widehat{U}(\theta) = \exp(-j\widehat{H}\theta)$, with the Hermitian operator $\widehat{H}$ serving as the generator of the transformation [20]. In quantum sensing, we prepare the initial quantum state represented by the density operator ρ_0 and pass it through the quantum sensing device, represented by the unitary transformation $\widehat{U}(\theta)$, to obtain the output state ρ_θ by the following transformation:

$$\rho_\theta = \widehat{U}(\theta)\rho_0\widehat{U}^\dagger(\theta) = e^{-j\widehat{H}\theta}\rho_0 e^{j\widehat{H}\theta}. \tag{11.44}$$

In this case, QFI restricts the precision of the statistical estimation of parameter θ by

$$\mathrm{Var}_M(\widetilde{\theta}) = \sigma_M^2(\widetilde{\theta}) \geq \frac{1}{M\, Q_F(\rho, \widehat{O})}, \tag{11.45}$$

which is known as the *quantum Cramér–Rao bound*, where M denotes the number of independent measurements.

The QFI can also be defined as the expectation of the *symmetric logarithmic derivative* (SLD) square L_ρ^2; in other words,

$$Q_F(\rho, \widehat{O}) = \left\langle L_\rho^2(\widehat{O}) \right\rangle = \mathrm{Tr}\left(\rho L_\rho^2(\widehat{O})\right), \tag{11.46}$$

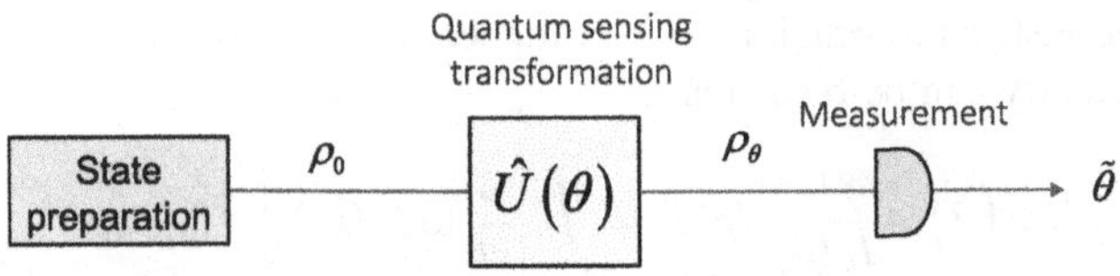

FIGURE 11.2

Quantum sensing model for parameter θ estimation.

wherein the SLD can be implicitly defined by Refs. [21,22]:

$$j\left[\rho, \widehat{O}\right] = \frac{1}{2}\left\{\rho, L_\rho\left(\widehat{O}\right)\right\}, \tag{11.47}$$

with [,] denoting the commutator and {,} the anticommutator. It can also be expressed in terms of eigenvalues λ_m and eigenkets $|\lambda_m\rangle$ of ρ by Ref. [23]:

$$L_\rho\left(\widehat{O}\right) = 2j\sum_{m,n}\frac{\lambda_m - \lambda_n}{\lambda_m + \lambda_n}\langle\lambda_m|\widehat{O}|\lambda_n\rangle|\lambda_m\rangle\langle\lambda_n|, \quad \rho = \sum_m \lambda_m|\lambda_m\rangle\langle\lambda_m|. \tag{11.48}$$

The SLD is linear in $\widehat{O}$, and when the operator is Hermitian, the SLD is also Hermitian. If state ρ is the pure state, that is, $\rho = |\psi\rangle\langle\psi|$, the SLD is simply

$$L_\rho\left(\widehat{O}\right) = 2\frac{\partial\rho}{\partial\theta} = 2j(-H\rho + \rho H), \tag{11.49}$$

so the QFI becomes

$$\begin{aligned} Q_F\left(\rho, \widehat{O}\right) &= \left\langle L_\rho^2\left(\widehat{O}\right)\right\rangle = \langle\psi|L_\rho^2\left(\widehat{O}\right)|\psi\rangle = -4\Bigg[\langle\psi|H|\psi\rangle^2\overbrace{\langle\psi\psi\rangle}^{1} - \langle\psi|H|\psi\rangle\langle\psi|H|\psi\rangle \\ &\qquad - \overbrace{\langle\psi\psi\rangle}^{1}\langle\psi|H^2|\psi\rangle\overbrace{\langle\psi\psi\rangle}^{1} + \langle\psi|H|\psi\rangle\langle\psi|H|\psi\rangle\Bigg] \\ &= 4\left[\langle\psi|H^2|\psi\rangle - \langle\psi|H|\psi\rangle^2\right] = 4\mathrm{Var}(H)_\theta. \end{aligned} \tag{11.50}$$

Therefore, the QFI for pure states is four times higher than the variance of Hamiltonian. For mixed states, the QFI is convex, provided that corresponding probabilities are parameter θ independent.

Finally, the QFI can be defined in terms of *Uhlmann fidelity* [24] $F(\rho_1, \rho_2) = \left[\mathrm{Tr}\left(\sqrt{\sqrt{\rho_1}\rho_2\sqrt{\rho_1}}\right)\right]^2$ as follows:

$$Q_F(\rho_\theta) = \lim_{\xi\to 0} 8\left\{\left[1 - \sqrt{F(\rho_\theta\rho_{\theta+\xi})}\right] / \xi\right\}. \tag{11.51}$$

11.3 Distributed quantum sensing

The conceptual schematic of the distributed quantum sensing (DQS) concept [20,25,26] is provided in Fig. 11.3. Based on the initial quantum state ρ_0, the source quantum module generates a multipartite entangled probe state shared by M sensors to scan an object of interest. The measurement results obtained by M sensors are (classically) postprocessed to determine a certain global parameter of the scanned object. On the other hand, in distributed classical sensing, the quantum state of M sensors can be separated; that is, it can be written as a tensor product $\rho_1 \otimes \cdots \otimes \rho_M$.

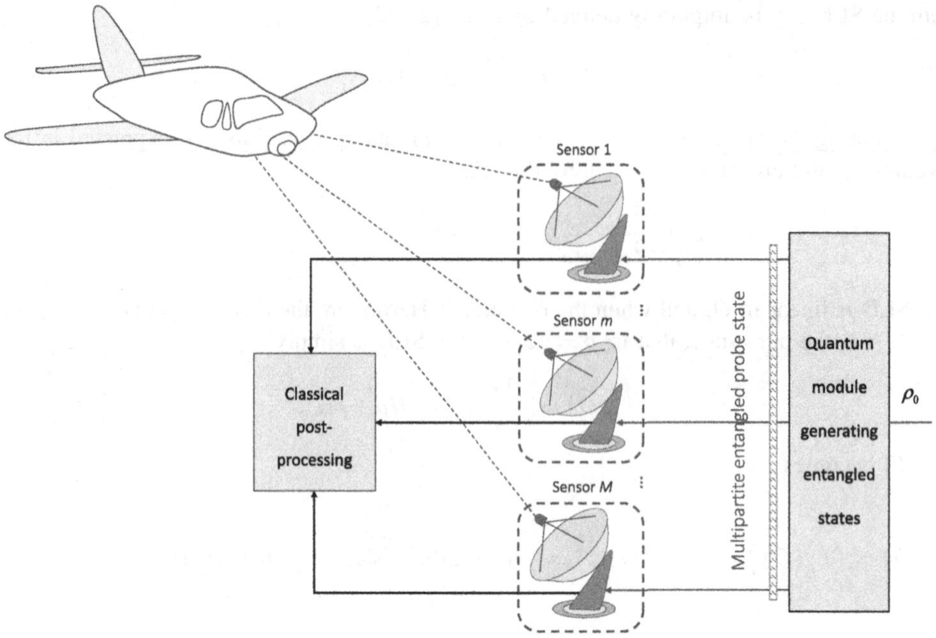

FIGURE 11.3

Distributed quantum sensing concept for entangled inputs.

The distributing sensing scenario can be described by a unitary operator being the product of M unitary operators, that is, $U(\tilde{\boldsymbol{\theta}}) = \otimes_{m=1}^{M} U(\tilde{\theta}_m)$, wherein each unitary measures one parameter. The following transformation describes the output quantum state:

$$\rho_M(\boldsymbol{\theta}) = U(\boldsymbol{\theta})\rho_M U^\dagger(\boldsymbol{\theta}), \boldsymbol{\theta} = [\theta_1, \cdots, \theta_M], \tag{11.52}$$

where ρ_M is the input entangled probe state. In DQS, we are interested in estimating the global parameter rather than all unknown parameters. This can be achieved by calculating the weighted average as follows:

$$\overline{\theta} = \sum_{m=1}^{M} w_m\theta_m = \boldsymbol{w} \cdot \boldsymbol{\theta}^{\mathrm{T}}, \boldsymbol{w} = [w_1, \cdots, w_M], w_m \geq 0, \sum_{m=1}^{M} w_m = 1. \tag{11.53}$$

In special cases when $w_m = 1/M$ and $\theta_m = \theta$, we can beat the SQL and achieve the Heisenberg limit using the entangled probe state.

Let us now observe the joint quantum sensing operation on M-probes pure state $|\psi_M\rangle$:

$$\rho_M(\theta) = U^{\otimes M}(\theta)|\psi_M\rangle\langle\psi_M|\left(U^\dagger\right)^{\otimes M}(\theta), U^{\otimes M}(\theta) = e^{-j\theta\sum_m \hat{H}_m}, \tag{11.54}$$

where $\widehat{H}_m$ generators are applied to different probes. Based on Eq. (11.50), the QFI becomes

$$Q_F(\rho(\theta)) = 4\mathrm{Var}\left(\sum\nolimits_m \widehat{H}_m\right)_{|\psi_M\rangle} = 4\left[\langle\psi_M|\left(\sum\nolimits_m \widehat{H}_m\right)^2|\psi_M\rangle - \langle\psi_M|\sum\nolimits_m \widehat{H}_m|\psi_M\rangle^2\right]. \quad (11.55)$$

When the pure state $|\psi_M\rangle$ can be represented as a tensor product of (not necessarily identical) states $|\psi_m\rangle$ $(m = 1, \ldots, M)$, that is, $|\psi_M\rangle = |\psi_1\rangle \otimes \cdots \otimes |\psi_M\rangle$, the first term in the previous equation can be represented as

$$\langle\psi_M|\left(\sum\nolimits_m \widehat{H}_m\right)^2|\psi_M\rangle = \sum\nolimits_m \langle\psi_m|\widehat{H}_m^2|\psi_m\rangle + \sum\nolimits_{m\neq m'}\sum\nolimits_{m'} \langle\psi_m|\widehat{H}_m|\psi_m\rangle\langle\psi_{m'}|\widehat{H}_{m'}|\psi_{m'}\rangle, \quad (11.56)$$

and after substituting it into the previous equation, we derive the following additivity property of the QFI for separable states:

$$Q_F\left(U^{\otimes M}(\theta)\otimes_{m=1}^{M}|\psi_m\rangle\right) = \sum\nolimits_m Q_F(U_m(\theta)|\psi_m\rangle). \quad (11.57)$$

When generators applied to different probes are identical, that is, $\widehat{H}_m = \widehat{H}$ and $|\psi_m\rangle = |\psi\rangle$, we obtain

$$Q_F\left(U^{\otimes M}(\theta)\otimes_{m=1}^{M}|\psi\rangle\right) = M\cdot Q_F(U(\theta)|\psi\rangle). \quad (11.58)$$

thus proving the quantum Cramér–Rao bound of Eq. (11.45). Eq. (11.57) is not valid for multipartite entangled states.

Let us observe the case when the probe states live in the D-dimensional space, such as atoms, *nitrogen-vacancy* centers, or *continuous-variable* systems with finite photon-number cutoff. So we are essentially concerned with the *discrete-variable* systems [27–29], with eigenkets of the generators $\widehat{H}$ being $|\lambda_n\rangle$ satisfying $\widehat{H}|\lambda_n\rangle = \lambda_n|\lambda_n\rangle$. Let the $\lambda_{\max}$ and $\lambda_{\min}$ denote the maximum and minimum eigenvalues, respectively. Each term in summation of Eq. (11.57) is limited by $Q_F(U_m(\theta)|\psi_m\rangle) \leq (\lambda_{\max} - \lambda_{\min})^2$, and the upper bound (UB) for the separable input state will be

$$Q_F\left(U^{\otimes M}(\theta)\otimes_{m=1}^{M}|\psi_m\rangle\right) \leq Q_F^{\max,C} = M(\lambda_{\max} - \lambda_{\min})^2. \quad (11.59)$$

The global generator cannot be decomposed as the sum of independent generators for the entangled probe state. The corresponding maximum and minimum eigenvalues are $M\ \lambda_{\max}$ and $M\ \lambda_{\min}$, respectively. The QFI is now upper bounded by

$$Q_F\left(U^{\otimes M}(\theta)|\psi_M\rangle\right) \leq Q_F^{\max,\mathrm{Ent.}} = (M\lambda_{\max} - M\lambda_{\min})^2 = M^2(\lambda_{\max} - \lambda_{\min})^2. \quad (11.60)$$

From the quantum Cramér–Rao bound, the precision scales are the same as $\sigma_M(\widehat{\theta}) \sim 1/M$, the Heisenberg limit. The maximum of QFI is achieved by the superposition state $|\psi_M\rangle = 2^{-1/2}\left(|\lambda_{\max}\rangle^{\otimes M} + |\lambda_{\min}\rangle^{\otimes M}\right)$.

11.3.1 Distributed quantum sensing of in-phase displacements

In this subsection, we are concerned with the in-phase displacement, and thus the generator is described by $\widehat{H} = \widehat{I}$, where $\widehat{I}$ is the in-phase operator. When expressed in a shot-noise unit (SNU), the in-phase and quadrature operators are related to the creation $a^\dagger$ and annihilation a operators by $\widehat{I} = a^\dagger + a, \widehat{Q} = j(a^\dagger - a)$. So the displacement operator $D(\alpha)$ performs the following mapping:

$$D(\boldsymbol{\alpha})\boldsymbol{x} = \boldsymbol{x} + \boldsymbol{\alpha},\ \boldsymbol{x} = \begin{bmatrix} \widehat{I} \\ \widehat{Q} \end{bmatrix}, \boldsymbol{\alpha} = \begin{bmatrix} \mathrm{Re}\{\alpha\} \\ \mathrm{Im}\{\alpha\} \end{bmatrix} = \begin{bmatrix} \alpha_I \\ \alpha_Q \end{bmatrix}. \tag{11.61}$$

In this quantum sensing protocol, the entanglement is generated with the help of linear optics and single-mode squeezed-vacuum (SMSV) states, as illustrated in Fig. 11.4, where we use identical displacements and weights.

From in-phase and quadrature operators definitions, we conclude that $\widehat{I}^2 + \widehat{Q}^2 = 4a^\dagger a + 2\widehat{1}$, so $\left[\mathrm{Var}(\widehat{I}) + \mathrm{Var}(\widehat{Q})\right]/4 - 1/2 = N_s = \langle a^\dagger a \rangle$, with N_s being the energy constraint assuming that mean values are equal to zero. From uncertainty principle and energy constraint, we conclude that $\mathrm{Var}(\widehat{I})$ is maximized when the input state is squeezed-vacuum state, with optimum QFI being

$$Q_F(U(\theta)|\psi\rangle) = 4(\sqrt{N_s} + \sqrt{N_s + 1})^2. \tag{11.62}$$

Based on Eqs. (11.57) and (11.58), we conclude that the QFI for M separable probes is upper bounded by

$$Q_F(U^{\otimes M}(\theta)|\psi_M\rangle) \le Q_F^{\max,C} = 4M(\sqrt{N_s/M} + \sqrt{N_s/M + 1})^2, \tag{11.63}$$

wherein the maximum QFI is achieved when input is a tensor product of i.i.d. SMSV states.

For an entangled input state the QFI is given by

$$Q_F(U^{\otimes M}(\theta)|\psi_M\rangle) = 4M\ \mathrm{Var}(\widehat{I})_{|\psi_M\rangle}, \widehat{I} = \sum_m M^{-1/2}\widehat{I}_m, \tag{11.64}$$

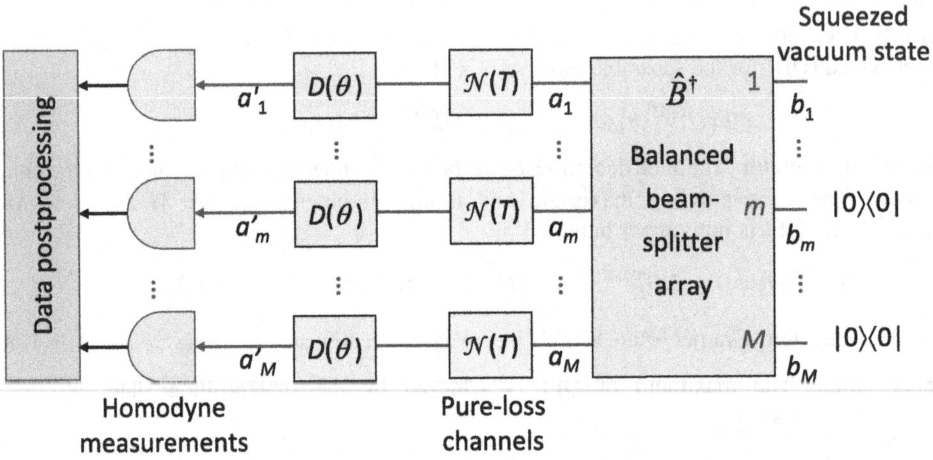

FIGURE 11.4

Distributed quantum sensing scheme to measure displacements on in-phase components with identical weights and displacements.

and the corresponding global quadrature operators satisfy the commutation relation and can be obtained by a passive linear transformation applied on input single-mode quadrature operators. Therefore, we can interpret the global quadratures as being generated by passing the single-mode inputs through a BBS array whose action is described by the operator $B^{\dagger}$. Given that energy is preserved during the beam-splitter array transformation, an approach that concentrates all energy on the global mode while maintaining the other modes as vacuum states represents the optimum solution [20]. Thus, the multimode problem is reduced to a single-mode problem, and the QFI becomes upper bounded by

$$Q_F\left(U^{\otimes M}(\theta)|\psi_M\rangle\right) \leq Q_F^{\max,\text{Ent.}} = 4M\left(\sqrt{N_s} + \sqrt{N_s+1}\right)^2 \underset{N_s \gg 1}{\rightarrow} 16MN_s. \tag{11.65}$$

Therefore, the following *unitary reduction* is valid [20]:

$$\widehat{B}^{\dagger} D^{\otimes M}(\theta)\widehat{B} = D(\sqrt{M}\,\theta)D^{\otimes(M-1)}(0). \tag{11.66}$$

From the quantum Cramér–Rao bound we obtain the following sensitivity bound:

$$\text{Var}_M\left(\widetilde{\theta}\right) = \sigma_M^2\left(\widetilde{\theta}\right) \geq \frac{1}{M\left(\sqrt{N_s} + \sqrt{N_s+1}\right)^2}, \tag{11.67}$$

wherein the equality sign is satisfied, according to Eq. (11.66), when input state on port 1 of the BBS is an SMSV state.

To account for imperfections, we use the pure-loss Bosonic channels $N\,(T_m)$ of transmissivity T_m, shown in Fig. 11.4, which is described by the Bogoliubov transform:

$$\widehat{a} \rightarrow \sqrt{T_m}\widehat{a} + \sqrt{1-T_m}\widehat{e} = \sqrt{T_m}\left(\widehat{a} + \sqrt{1/T_m - 1}\,\widehat{e}\right), \tag{11.68}$$

where $\widehat{e}$ is the annihilation operator of the environment mode. Since there are M sensors, there will be M pure-loss channels for transmissivities $(T_1, \ldots, T_M) = \boldsymbol{T}$. The entangled probed state is prepared by applying the SMSV state at input port 1 of the BBS array. Each probe mode $\widehat{a}_m$ passes through the pure-loss Bosonic channel and the displacement gate $D(\theta)$, and the in-phase (real) quadrature $\text{Re}\{\widehat{a}'_m\} = \widehat{I}'_m$ get measured by the homodyne detector to get $\text{Re}\{\theta_m\}$. The estimation of parameter θ is performed with classical data postprocessing by $\widetilde{\theta} = M^{-1}\sum_m \theta_m$. The joint state at the input of homodyne detectors of the sensor nodes can be represented by

$$\begin{aligned}
\boldsymbol{\rho}_{M,N_s,\boldsymbol{T}} &= \otimes_{m=1}^{M} D_m(\theta_m) \underbrace{\otimes_{m=1}^{M} N_m(T_m)\boldsymbol{\rho}_{M,N_s}}_{\boldsymbol{\rho}^{(\text{eqv})}_{M,N_s,\boldsymbol{T}}} \otimes_{m=1}^{M} D_m^{\dagger}(\theta_m) \\
&= \otimes_{m=1}^{M} D_m(\theta_m)\boldsymbol{\rho}^{(\text{eqv})}_{M,N_s,\boldsymbol{T}} \otimes_{m=1}^{M} D_m^{\dagger}(\theta_m),
\end{aligned} \tag{11.69}$$

where $\boldsymbol{\rho}^{(\text{eqv})}_{M,N_s,\boldsymbol{T}}$ is the density operator of the equivalent entangled state, representing the action of pure-loss channels on the entangled input state. In the absence of imperfections, the variance of the estimate is given by Eq. (11.67). By modeling the imperfections using pure-loss Bosonic channels, we must account for loss, and assuming that all channels have the same transmissivity $T_m = T$, we must multiply Eq. (11.65) by T. Given that the variance of the vacuum state is 1 (in SNUs) based on the pure loss channel model of Eq. (11.66), we must add another term for the variance of the estimator, which is $(1-T)\cdot 1/M$, to obtain the following entanglement sensitivity:

$$\sigma^2_{M,\text{Ent.}}\left(\widetilde{\theta}\right) = \frac{T}{M\left(\sqrt{N_s}+\sqrt{N_s+1}\right)^2} + \frac{1-T}{M}. \tag{11.70}$$

Using more rigorous analysis by employing the Uhlmann fidelity-based definition of QFI—see Eq. (11.51)—for the optimum Gaussian state (in this case, the SMSV state), the same entanglement sensitivity as given by the previous equation is obtained, as shown in Ref. [20].

For a separable input state, our starting point is Eq. (11.63), and by applying the same approach as for the entangled input state, we obtain the following sensitivity:

$$\sigma^2_{M}\left(\widetilde{\theta}\right) \geq \sigma^2_{M,\text{Sep.}}\left(\widetilde{\theta}\right) = \frac{T}{M\left(\sqrt{N_s/M}+\sqrt{N_s/M+1}\right)} + \frac{1-T}{M}, \tag{11.71}$$

wherein the equality sign corresponds to the product of SMSV states, which represents the optimum state in this scenario.

By close inspection of Fig. 11.4, we conclude that we can use the balanced beam-combiner array described by the operator $\widehat{B}$ and a single homodyne detector instead of using M homodyne detectors. Further, since the BBS array operator $\widehat{B}^{\dagger}$ commutes with pure-loss Bosonic channels, we come up with an equivalent circuit shown in Fig. 11.5.

Applying the unitary reduction property given by Eq. (11.66), the DQS-equivalent model can be further simplified as shown in Fig. 11.6. This model is not relevant from an implementation point of view, but only to simplify the sensitivity analysis. On the other hand, the model from Fig. 11.5 can be used as an alternative implementation with only one homodyne detector.

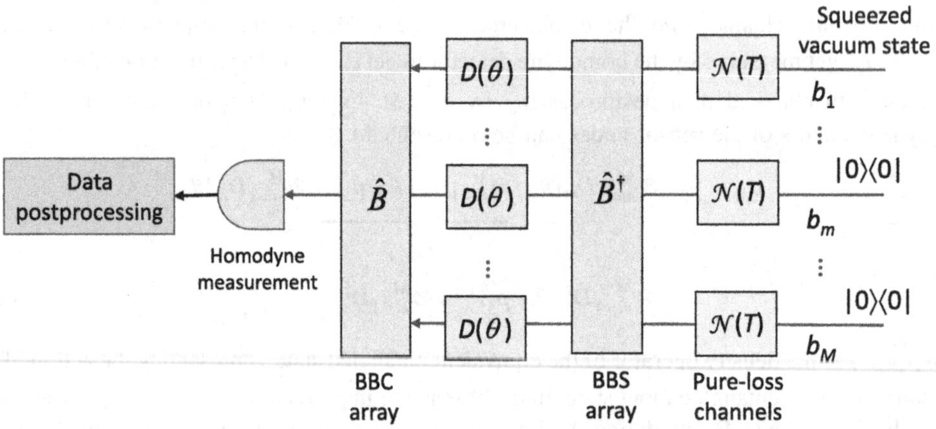

FIGURE 11.5

Equivalent model for distributed quantum sensing to measure displacements on in-phase components with identical weights and displacements.

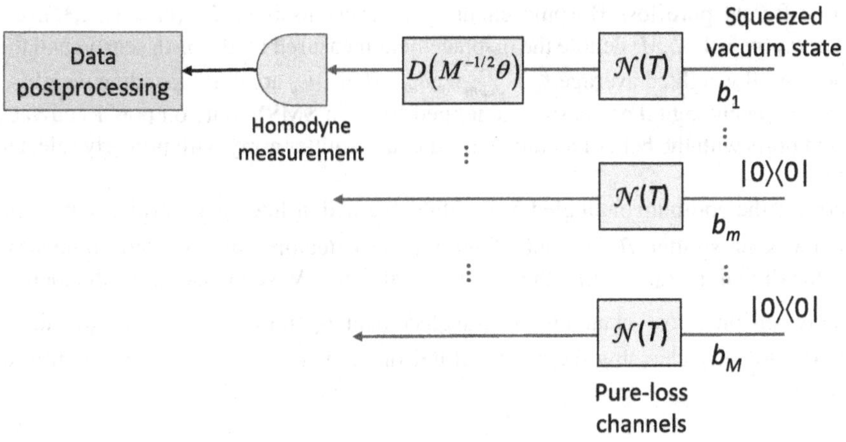

FIGURE 11.6

Simplified equivalent model for distributed quantum sensing to measure displacements on in-phase components with identical weights and displacements.

For the pure input state, we can represent the QFI through the variance of the generator; therefore, we can generalize Eq. (11.65), corresponding to the entangled input state, to obtain the UB of the QFI as follows:

$$Q_F\left(U^{\otimes M}(\theta)|\psi_M\rangle\right) \le Q_F^{(\text{Ent., UB})} = T\cdot 4M(\sqrt{N_s} + \sqrt{N_s+1})^2 + (1-T)\cdot 4M, \tag{11.72}$$

wherein the first term corresponds to the variance of the signal after the pure-loss channels, while the second term corresponds to the noise power introduced by the channels. The lower bound (LB) of the variance of the estimator will be just reciprocal of the QFI UB; that is, we can write

$$\sigma^2_{M,\text{Ent., LB}}\left(\tilde{\theta}\right) = \frac{1}{Q_F^{(\text{Ent., UB})}} = \frac{1}{T\cdot 4M(\sqrt{N_s} + \sqrt{N_s+1})^2 + (1-T)\cdot 4M}. \tag{11.73}$$

For separable input state, in a similar fashion, we can generalize Eq. (11.63) to obtain the following UB for the QFI:

$$Q_F^{(\text{Sep.,UB})} = T\cdot 4M(\sqrt{N_s/M} + \sqrt{N_s/M+1})^2 + (1-T)\cdot 4M. \tag{11.74}$$

The LB of the variance of the estimator for a separable input state is just reciprocal of the QFI UB; that is, we can write

$$\sigma^2_{M,\text{Sep., LB}}\left(\tilde{\theta}\right) = \frac{1}{Q_F^{(\text{Sep., UB})}} = \frac{1}{T\cdot 4M(\sqrt{N_s/M} + \sqrt{N_s/M+1})^2 + (1-T)\cdot 4M}. \tag{11.75}$$

11.3.2 General distributed quantum sensing of in-phase displacements

In this subsection, we are concerned with estimating the arbitrary weighted average of homodyne measurements, wherein the displacements are different, and imperfections in quantum sensors are modeled by different pure loss Bosonic channels of transmissivity T_m ($m = 1,\ldots,M$), as shown in Fig. 11.7. Let θ_m ($m = 1,\ldots,M$) denote the displacement measured by the m-th sensor, and the goal is to estimate the overall weighted average $\overline{\theta} = \sum_m w_m \theta_m$, where w_m are corresponding weights. The input modes $\widehat{a}_m$ are in an entangled probe state generated from an SMSV state on port 1 and vacuum states on other input ports with the help of an unbalanced beam-splitter array with properly selected coupling coefficients.

To determine the optimum entangled probe state, we will utilize the equivalence relations. We can first insert the beam-splitter $\widehat{B}$ in front of homodyne detectors, and we can compensate for this insertion in the digital domain so that the QFI is not affected. We can also insert the identity operator $\widehat{1} = \widehat{B}\widehat{B}^{\dagger}$ between pure loss channels and displacement operators to get an equivalent model in Fig. 11.8A. By properly choosing the beam-splitter operator $\widehat{B}$, we can perform the following transformation [20]:

$$\widehat{B}^{\dagger}\left[\otimes_{m=1}^{M} D(\theta_m)\right]\widehat{B} = D\left(\frac{\overline{\theta}}{\overline{w}}\right)\left[\otimes_{m=2}^{M} D(\theta'_m)\right], \overline{w} = \left(\sum\nolimits_m w_m^2\right)^{1/2}, \tag{11.76}$$

with other parameters θ'_m ($m = 2,\ldots,M$) independent of the parameter $\overline{\theta}$ of interest. Using this property, we created the equivalent circuit provided in Fig. 11.8B, and clearly, we need to measure the first output mode $\widehat{a}''_1$ to obtain an estimate of $\overline{\theta}$.

Based on transform Eq. (11.76) and the action of pure loss channels of Eq. (11.68), we can express the first output mode $\widehat{a}''_1$ as [20,25]

$$a''_1 = \sum_{m=1}^{M} \frac{w_m}{\overline{W}} T_m^{1/2} \widehat{a}_m + \text{vaccum terms} = \overline{T}^{1/2} \widehat{b}_1 + \text{vaccum terms}, \overline{W} = \left(\sum\nolimits_m T_m w_m^2\right)^{1/2}, \tag{11.77}$$

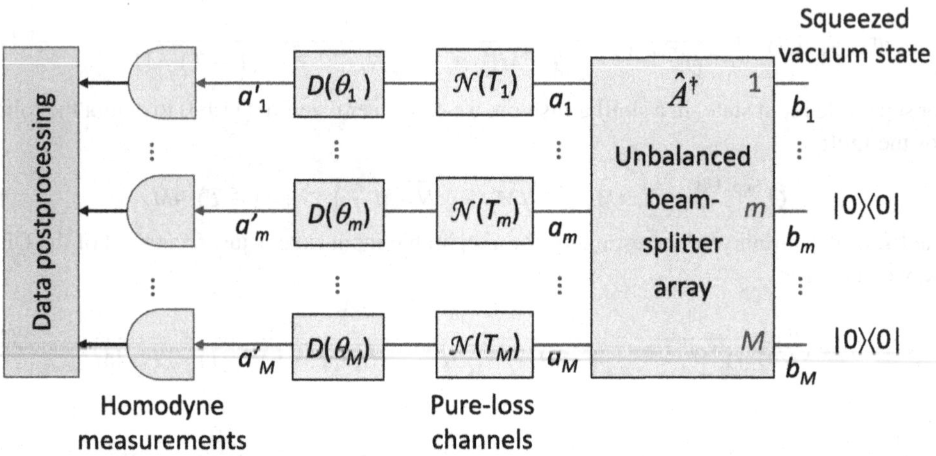

FIGURE 11.7

General distributed quantum sensing scheme to measure displacements on in-phase components.

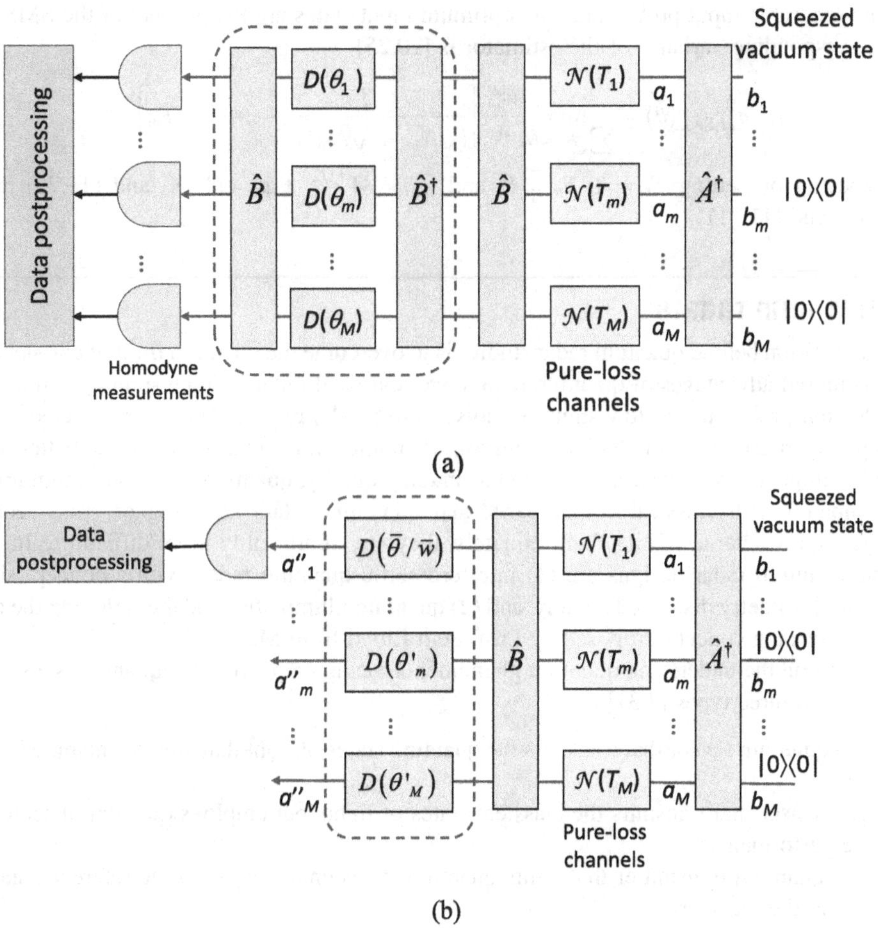

FIGURE 11.8

Equivalent model for generally distributed quantum sensing scheme to measure displacements on in-phase components.

with $\overline{T} = \sum_m T_m w_m^2/\overline{W}^2$. From Eq. (11.77), we conclude that the single-mode input $\widehat{b}_1$ is the superposition of modes $\widehat{a}_m$, that is, $\widehat{b}_1 = \sum_m T_m^{1/2}\widehat{a}_m w_m/\overline{W}$. The first output mode $\widehat{a}_1''$ is obtained by passing the input mode $\widehat{b}_1$ through a pure loss Bosonic channel of effective transmissivity $\overline{T}$. Given the energy constraint $N_s = \sum_m N_m$, the optimum Gaussian state is obtained by applying the beam-splitter array $\widehat{A}^\dagger$ on the SMSV state $\widehat{b}_1$ (on input port 1) and vacuum states on other input ports.

The estimator $\overline{\theta} = \sum_m w_m \mathrm{Re}\{\widehat{a}_m'\}$ is unbiased, and similarly to Eq. (11.70), the minimum variance of the estimator is given by

$$\sigma_{M,\mathrm{Ent.}}^2(\overline{\theta}) = \frac{\overline{w}^2}{4}\left[\frac{\overline{T}}{(\sqrt{N_s}+\sqrt{N_s+1})^2} + 1 - \overline{T}\right]. \tag{11.78}$$

For the separable input probe state, the optimum input states are the product of the SMSV states, and the corresponding variance of the estimator is [20,25]

$$\sigma^2_{M,\text{Sep.}}(\theta) = \min_{\sum_m N_m = N_S} \frac{w_m^2}{4}\left[\frac{T_m}{(\sqrt{N_m} + \sqrt{N_m+1})^2} + 1 - T_m\right]. \tag{11.79}$$

As expected by setting $T_m = T$, $\theta_m = \theta$, and $w_m = M^{-1/2}$, Eqs. (11.78) and (11.79) reduce to Eqs. (11.70) and (11.71).

11.4 Quantum radars

The key motivation behind quantum radar studies is to overcome the quantum limit of classical sensors [7]. The potential advantages of quantum radars over classical radars can be summarized as follows: better detection probability in low signal-to-noise ratio (SNR) regime, better target detection probability, improved detection through clouds and fog when microwave photons are used, better resilience to jamming, improved synthetic-aperture radar imaging quality, quantum radars are more difficult to detect compared with classical counterparts, and quantum radars have higher cross sections, to mention just a few characteristics. Unfortunately, they are significantly more difficult to implement. Two basic quantum radar designs are (1) interferometric quantum radar, with a concept similar to quantum interferometry discussed already, and (2) quantum illumination radar employing the quantum illumination sensing concept proposed by Prof. Seth Lloyd from MIT [30].

Depending on the underlying quantum phenomenon being employed, the quantum sensors can be categorized into three types [7,31]:

- *Type-1*, a quantum sensor that transmits the quantum states of light that are not entangled with the receiver.
- *Type-2*, a sensor that transmits the classical states of lights but employs quantum detectors to improve performance.
- *Type-3*, a quantum transmitter that emits quantum states entangled with the reference states available at the receiver.

The *single-photon radar*, shown in Fig. 11.9, belongs to the type 1 quantum sensors and operates similarly to classical radars but employs a single-photon pulse. This type of quantum radar has a larger radar cross section (RCS) than classical radars [7].

The quantum laser detection and ranging (*LADAR*) is a type-2 quantum sensor that employs the light detection and ranging (LIDAR) technology [7,32–34]. Since LADAR operates in visible and near-visible regimes, the laser beam cannot propagate through clouds and fogs. On the other hand, since the operating wavelengths are much lower than classical radars, LADAR has a better spectral resolution.

Entangled photon-based quantum radars belong to type-3 quantum sensors, and the corresponding operational principle is illustrated in Fig. 11.10. The entangled source generates an entangled pair of photons, the signal and idler photons. The idler photon is kept in the quantum memory of the receiver, while the signal photon is transmitted over a noisy, lossy, and atmospheric turbulent channel toward the target. The radar detects the reflected photon, and quantum correlation is exploited to improve sensitivity and target detection probability.

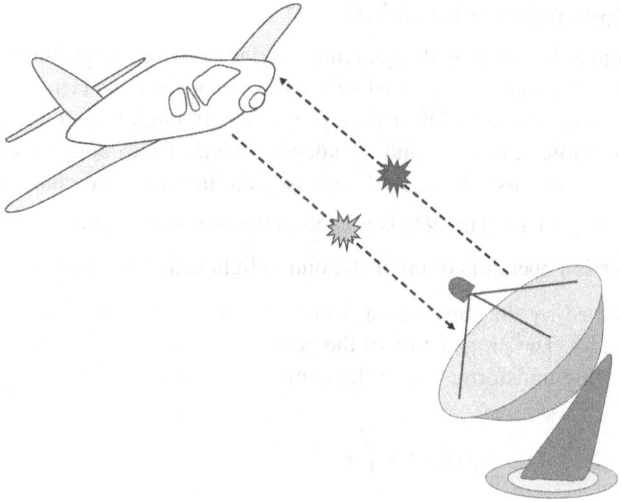

FIGURE 11.9

Single-photon (monostatic) quantum radar concept.

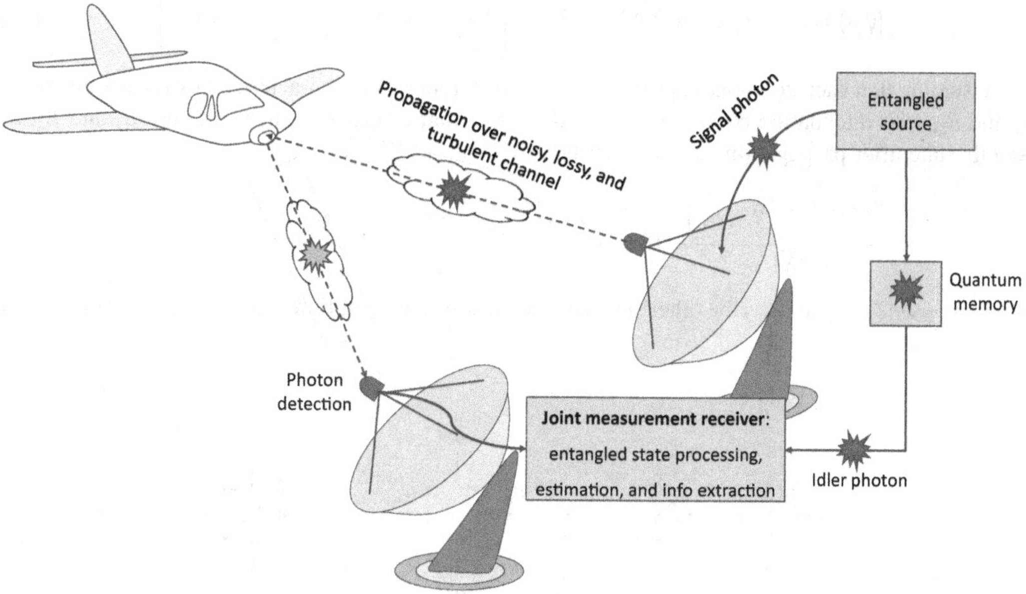

FIGURE 11.10

Entangled photon-assisted (bistatic) quantum radar concept.

11.4.1 Interferometric quantum radars

Interferometric quantum radars employ the quantum interferometry concepts introduced in Section 11.1. We showed in Section 11.1.2 that entangled NOON states can achieve the Heisenberg limit. However, in interferometric quantum radars, the NOON state propagates through the atmospheric channel and exhibits diffraction, absorption, scattering, and turbulence effects. To account for diffraction, absorption, and scattering effects, we can use the cascade of lossy thermal Bosonic channels (or beam splitters) [7,35,36], as shown in Fig. 11.11. The details related to the i-th beam splitter are provided. The $\widehat{a}_{\text{in}}^{(i)}$ and $\widehat{a}_{\text{out}}^{(i)}$ denote the annihilation operators of input and output light beam, respectively. The mode $\widehat{a}_{\text{s}}^{(i)}$ is used to describe light absorbed by the medium or scattered. We use $\widehat{b}^{(i)}$ to denote the background light introduced by the medium. The propagation of the photon by a cascade (chain) of beam splitters can be described by the following transformation of the annihilation operator [7]:

$$\widehat{a}(\omega) \rightarrow e^{-\alpha(\omega)z/2+jk(\omega)z}\widehat{a}(\omega) + j\sqrt{\alpha(\omega)}\int_0^z e^{[-\alpha(\omega)/2+jk(\omega)](z-z')}\widehat{b}(\omega)dz', k(\omega) = \frac{\omega n(\omega)}{c}, \tag{11.80}$$

where $n(\omega)$ is a frequency-dependent refraction index of the medium, and $\alpha(\omega)$ is the attenuation coefficient considering both absorption and scattering effects.

Let us represent the NOON state in terms of creation operators:

$$|\psi_N\rangle = 2^{-1/2}(|N0\rangle + |0N\rangle) = 2^{-1/2}\left[\frac{\left(a_1^\dagger\right)^N}{\sqrt{N!}}|0\rangle_1|0\rangle_2 + \frac{\left(a_2^\dagger\right)^N}{\sqrt{N!}}|0\rangle_1|0\rangle_2\right]. \tag{11.81}$$

Assuming that each component propagates over different media characterized by refraction indices n_1 and n_2, with attenuation coefficients α_1 and α_2, and that propagation distances are L_1 and L_2, the NOON state after propagation can be described by

$$|\psi_N'\rangle = \frac{e^{[-\alpha_1(\omega)/2+jk_1(\omega)]NL_1}\left(a_1^\dagger\right)^N}{\sqrt{2N!}}|0\rangle_1|0\rangle_2 + \frac{e^{[-\alpha_2(\omega)/2+jk_2(\omega)]NL_2}\left(a_2^\dagger\right)^N}{\sqrt{2N!}}|0\rangle_1|0\rangle_2 + |\psi_{\text{other}}\rangle, \tag{11.82}$$

where we use $|\psi_{\text{other}}\rangle$ to describe other states due to background radiation that gets scattered outside the NOON-basis.

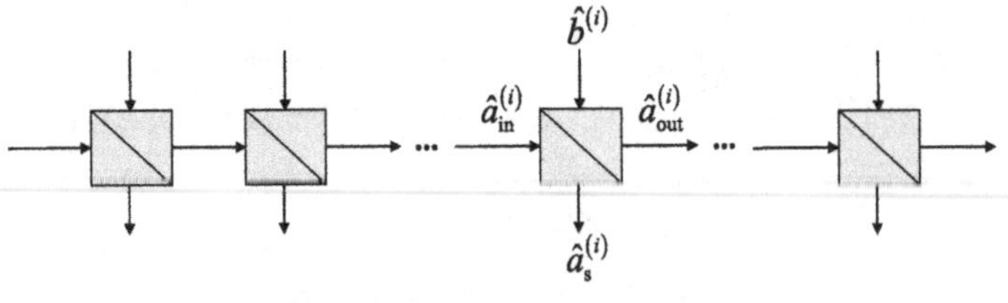

FIGURE 11.11

Modeling light attenuation effects by a cascade of lossy thermal Bosonic channels.

Let us now apply the quantum interferometry concept by inserting the multiplication factor $\exp(jN\phi)$ in front of the first term in the previous equation:

$$|\psi_N^{(\mathrm{Rx})}\rangle=\frac{1}{\sqrt{2}}e^{jN\phi}e^{[-\alpha_1(\omega)/2+jk_1(\omega)]NL_1}|N\rangle|0\rangle+\frac{1}{\sqrt{2}}e^{[-\alpha_2(\omega)/2+jk_2(\omega)]NL_2}|0\rangle|N\rangle+|\psi_{\mathrm{other}}\rangle, \tag{11.83}$$

Based on Eq. (11.33), we obtain the following expectation of the observable $\widehat{O}_N$:

$$\begin{aligned}\langle\widehat{O}_N\rangle&=\langle\psi_N^{(\mathrm{Rx})}|\widehat{O}_N|\psi_N^{(\mathrm{Rx})}\rangle=\langle\psi_N^{(\mathrm{Rx})}|(|0N\rangle\langle N0|+|N0\rangle\langle 0N|)\psi_N^{(\mathrm{Rx})}\\&=(T_1T_2)^{N/2}\cos[N(\phi-\phi_d)],\ \ \phi_d=k_2L_2-k_1L_1;T_i=e^{-\alpha_iL_i},i=1,2\end{aligned} \tag{11.84}$$

where ϕ_d denotes the dispersion phase shift, and T_i is the effective transmissivity of channel $i=1,2$. The phase responsivity is simply

$$\frac{d\langle\widehat{O}_N\rangle}{d\phi}=-(T_1T_2)^{N/2}N\sin[N(\phi-\phi_d)]. \tag{11.85}$$

Since $\widehat{O}_N^2=\widehat{1}$ [see Eq. 11.34], we determine that the expectation of $\widehat{O}_N^2$ is given by

$$\langle\widehat{O}_N^2\rangle=\langle\psi_N^{(\mathrm{Rx})}|\underbrace{\widehat{O}_N^2}_{\widehat{1}}|\psi_N^{(\mathrm{Rx})}\rangle=\langle\psi_N^{(\mathrm{Rx})}|\psi_N^{(\mathrm{Rx})}\rangle=\frac{1}{2}(T_1^N+T_2^N). \tag{11.86}$$

The variance of the observable $\widehat{O}_N$ now becomes

$$\begin{aligned}(\Delta\widehat{O}_N)^2&=\langle\widehat{O}_N^2\rangle-\langle\widehat{O}_N\rangle^2=\frac{1}{2}(T_1^N+T_2^N)-(T_1T_2)^N\cos^2[N(\phi-\phi_d)]\\&=\frac{1}{2}(T_1^N+T_2^N)-(T_1T_2)^N\{1-\sin^2[N(\phi-\phi_d)]\}.\end{aligned} \tag{11.87}$$

The uncertainty of the phase estimate for the NOON state can now be evaluated by

$$\begin{aligned}\Delta\phi&=\frac{\Delta\widehat{O}_N}{\left|\frac{d\langle\widehat{O}_N\rangle}{d\phi}\right|}=\frac{\sqrt{2^{-1}(T_1^N+T_2^N)-(T_1T_2)^N\{1-\sin^2[N(\phi-\phi_d)]\}}}{(T_1T_2)^{N/2}N|\sin[N(\phi-\phi_d)]|}\\&=\frac{\sqrt{2^{-1}(T_1^{-N}+T_2^{-N})-1+\sin^2[N(\phi-\phi_d)]}}{N|\sin[N(\phi-\phi_d)]|}.\end{aligned} \tag{11.88}$$

This expression is consistent with Ref. [37]. The Heisenberg limit can be achieved only in the absence of attenuation:

$$\lim_{T_i\to 1}\Delta\phi=\frac{1}{N}. \tag{11.89}$$

The uncertainty in the range of the target is simply

$$\Delta R\approx\frac{\lambda}{2\pi}\Delta\phi, \tag{11.90}$$

where λ is the photon wavelength.

The atmospheric turbulence is caused by variations in the refractive index of the transmission medium due to spatial variations in temperature and pressure (related to wind and solar heating) [18,38]. When the optical beam propagates through a turbulence channel, it exhibits wavefront distortions, manifest as random amplitude and phase fluctuations. In interferometric quantum radars, the transmissivities T_i are now random variables related to the gamma-gamma distribution of irradiance. To guarantee sensitivity of quantum radars, we can use adaptive optics (AO) as proposed in Ref. [37]. A linear AO system consists of three subsystems: (1) a wavefront sensor to detect the atmospheric distortions, (2) a wavefront corrector to compensate for turbulence effects, and (3) a control processor to monitor the wavefront sensor information and update the wavefront corrector [18,39]. AO-based correction can maintain supersensitivity over ranges up 5000 km [37]. The key idea behind Smith's proposal [37] is to introduce the phase shift ϕ_{AO} in the upper branch (interrogating the target), see Fig. 11.1, which is controlled by the operator. In the bistatic quantum radar scenario shown in Fig. 11.10, let us assume that the NOON states source is located on the transmitter side of the radar system, so distortions due to turbulence and attenuation effects of the lower branch (from the transmitter to receiver side of the radar system) can be neglected. The phase estimation error after correction by AO will be

$$\Delta\phi^{(\mathrm{corr})} \approx \frac{\sqrt{\frac{1}{2}\left(\frac{1}{\left(T_1^{(\mathrm{corr})}\right)^N}+1\right) - 1 + \sin^2\left[N\left(\phi+\phi_{\mathrm{AO}}-\phi_d^{(\mathrm{corr})}\right)\right]}}{N\left|\sin\left[N\left(\phi+\phi_{\mathrm{AO}}-\phi_d^{(\mathrm{corr})}\right)\right]\right|}, \tag{11.91}$$

where the corrected dispersion phase shift $\phi_d^{(\mathrm{corr})}$ and corrected transmissivity of the medium $T_1^{(\mathrm{corr})}$ are given respectively by

$$\phi_d^{(\mathrm{corr})} \approx \frac{\omega}{c} n_1^{(\mathrm{corr})} L_1, \quad T_1^{(\mathrm{corr})} = e^{-\alpha^{(\mathrm{corr})} L_1}, \tag{11.92}$$

with $n_1^{(\mathrm{corr})}$ and $\alpha_1^{(\mathrm{corr})}$ being the corrected refraction index and corrected attenuation coefficient of the medium, respectively. The corrected dispersion phase shift and corrected transmissivity expressions require the knowledge of the range of the target L_1. Smith suggested using classical radar; however, the location of quantum radar is easier to detect. An alternative would be to use the range of target determined in the absence of AO and run an iterative procedure. When ϕ_{AO} is properly chosen such that the argument of sin-function in the denominator is an odd multiple of $\pi/2$,

$$N\left(\phi+\phi_{\mathrm{AO}}-\phi_d^{(\mathrm{corr})}\right) = (2m+1)\frac{\pi}{2};\, m = 0, \pm1, \pm2, \cdots \tag{11.93}$$

the corrected estimation error becomes simply

$$\Delta\phi^{(\mathrm{corr})} \approx \frac{\sqrt{\frac{1}{2}\left(\frac{1}{\left(T_1^{(\mathrm{corr})}\right)^N}+1\right)}}{N}, \tag{11.94}$$

and in the limit $T_1^{(\mathrm{corr})} \to 1$, we reach the Heisenberg limit.

Alternatively, we can use AO to correct for aberrations in the wavefront of a single photon introduced by turbulence to improve $T_1^{(\text{corr})}$ in a fashion similar to that of Ref. [40]. This approach does not require knowledge of the target range.

11.4.2 Quantum illumination-based quantum radars

Quantum illumination is an approach proposed in Ref. [30] to improve the sensitivity of the quantum sensing systems operating in lossy and noisy regimes. Since it is not restricted to any specific wavelength, it can be applied on both LADAR and quantum radars operating in X-band (8–12 GHz). This approach does not require any interferometer; simple photon counters are sufficient. It applies to both single-photon-based and entangled photon-based quantum radars. The interaction between single photons and the target can be described by a beam splitter of reflectivity R; see the i-th beam splitter in Fig. 11.11 (clearly $R = 1-T$). The noise source is represented by the annihilation operator $\hat{b}$ injecting in average b background photons. The dimensionality of quantum states is D, and we assume that $bD << 1$.

To facilitate explanations, we adopt an approach similar to that used in Ref. [7]. For instance we use TR to denote that the target is within the range and NTR to denote that target is not present within the range. Let us first consider *nonentangled* photon-based quantum radar technology. For the *NTR* case, the emitted signal photon is lost, the radar receiver receives only noise photons, and the density operator of this state can be represented by

$$\rho_{th} = \otimes_{d=1}^{D} |d\rangle\langle d|, \tag{11.95}$$

where $|d\rangle$ represents the d-the baseket, for instance, a single photon in mode d. Since $bD << 1$, we can use the following approximation:

$$\rho_{th} \approx (1 - bD)|0\rangle\langle 0| + b\sum_{d=1}^{D} |d\rangle\langle d|, \tag{11.96}$$

with $|0\rangle$ being the vacuum state.

For the TR case, there is background radiation in addition to the signal photon represented by density operator ρ_s, and we can use the following approximation for the mixed density operator:

$$\rho_{TR} \approx R\rho_s + (1 - R)\rho_{th} = R\rho_s + (1 - R)(1 - bD)|0\rangle\langle 0| + (1 - R)b\sum_{d=1}^{D} |d\rangle\langle d|. \tag{11.97}$$

To estimate the detection probability for two hypotheses (*TR*, *NTR*), we should repeat the experiment multiple times by interrogating the target with single-photon pulses. Let us use "+" to denote the detection of the target event and "−" to denote the event that a target is not detected. So the probability of detecting the target in the *NTR* case is $P(+|NTR) = b$, while the probability of not detecting the target for the same case is $P(-|NTR) = 1-b$. The normalization condition is satisfied: $P(+|NTR)+P(-|NTR) = 1$. On the other hand, for the TR case, the probability of detecting the target will have two components:

$$P(+|TR) = R + (1 - R)b, \tag{11.98}$$

indicating that the detection could be either due to the single photon reflected from the target or background radiation. The probability of not detecting the target for the TR case is

$$P(-|TR) = 1 - P(+|TR) = 1 - R - (1-R)b = (1-R)(1-b). \tag{11.99}$$

The number of times we need to repeat the experiment to successfully detect the target depends on the *SNR*, which can be estimated by

$$SNR \approx \frac{P(+|TR)}{P(+|NTR)} = \frac{R+(1-R)b}{b} = 1 - R + \frac{R}{b} \approx \mathcal{O}\left(\frac{R}{b}\right). \tag{11.100}$$

The number of repeated experiments needed in a high SNR regime ($SNR > 1$) is $N_{\text{high}} \approx \mathcal{O}(1/R)$. On the other hand, the repeated experiments needed in a low *SNR* regime ($SNR < 1$) are [7] $N_{\text{low}} \approx \mathcal{O}(8b/R^2)$. Thus, in a low-SNR regime, $R < b$, so $N_{\text{low}} > N_{\text{high}}$ as expected.

In *entangled* photon-based quantum radar technology, we can use the entangled source generating the generalized Bell states:

$$|B_{00}\rangle = D^{-1/2}\sum\nolimits_d |d\rangle_s |d\rangle_i, \tag{11.101}$$

where subscript *s* refers to the *signal photon* and subscript *i* to the *idler (auxiliary) photon*, while the corresponding density operator is $\rho = |B_{00}\rangle\langle B_{00}|$. In the *NTR* case, the signal photon is lost, while at the same time, the idler photon is in a completely mixed state, so the density operator can be described by

$$\begin{aligned} \rho_{NTR}^{(\text{Ent.})} &\approx \rho_{th} \otimes \frac{\hat{1}_i}{D} + \mathcal{O}(b^2) = \left[(1-bD)|0\rangle_s\langle 0| + b\sum_{d=1}^{D} |d\rangle_s\langle d|\right] \otimes \frac{\hat{1}_i}{D} + \mathcal{O}(b^2) \\ &\approx \left[(1-bD)|0\rangle_s\langle 0| + b\hat{1}_s\right] \otimes \frac{\hat{1}_i}{D} + \mathcal{O}(b^2) \\ &\approx \left[\left(1 - \frac{b}{D}D\right)|0\rangle_s\langle 0| + \frac{b}{D}\hat{1}_s\right] \otimes \hat{1}_i + \mathcal{O}(b^2). \end{aligned} \tag{11.102}$$

In the TR case, on the other hand, the generalized Bell state can be detected with probability *R*, while the $\rho_{NTR}^{(\text{Ent.})}$ can be detected with probability $1-R$, so

$$\rho_{TR}^{(\text{Ent.})} = R\rho + (1-R)\rho_{NTR}^{(\text{Ent.})} + \mathcal{O}(Rb, Rb^2). \tag{11.103}$$

By close inspection of Eqs. (11.96) and (11.102), we conclude that the only difference from the nonentangled case is that we need to replace *b* with *b*/*D*. Therefore, the corresponding detection probabilities for the entangled case are simply

$$\begin{aligned} P^{(\text{Ent.})}(-|TR) &= (1-R)\left(1-\frac{b}{D}\right), P^{(\text{Ent.})}(+|TR) = R + (1-R)\frac{b}{D}, \\ P^{(\text{Ent.})}(-|NTR) &= 1 - \frac{b}{D}, P^{(\text{Ent.})}(+|NTR) = \frac{b}{D}. \end{aligned} \tag{11.104}$$

Similarly, the *SNR* for the entangled case can be obtained from Eq. (11.100) by replacing *b* with *b*/*D*:

$$SNR^{(\text{Ent.})} \approx \frac{P^{(\text{Ent.})}(+|TR)}{P^{(\text{Ent.})}(+|NTR)} = \frac{R + (1-R)\frac{b}{D}}{\frac{b}{D}} = 1 - R + \frac{RD}{b} \approx \mathcal{O}\left(\frac{RD}{b}\right). \tag{11.105}$$

The number of required signal photons to detect the presence of the target in low and high *SNR* regimes is then

$$N_{\text{high}}^{(\text{Ent.})} \approx \mathcal{O}(1/R), N_{\text{low}}^{(\text{Ent.})} \approx \mathcal{O}(8b/R^2 D). \tag{11.106}$$

The *SNR* improvement when entangled photons are used over the nonentangled case is

$$\frac{SNR^{(\text{Ent.})}}{SNR} = \frac{1 - R + \frac{RD}{b}}{1 - R + \frac{R}{b}} \approx D. \tag{11.107}$$

There is no improvement in terms of the number of required signal photons in a high-SNR regime, while the improvement in a low-SNR regime is

$$\frac{N_{\text{low}}^{(\text{Ent.})}}{N_{\text{low}}} \approx \frac{\mathcal{O}(8b/R^2 D)}{\mathcal{O}(8b/R^2)} \approx \frac{1}{D}. \tag{11.108}$$

The number of entangled qubits (also known as ebits) of the maximally entangled state is $e = \log_2 D$. The signal-to-jam ratio (*SJR*) can be approximately estimated for nonentangled quantum illumination by $SJR \approx \mathcal{O}(R/J)$, where J is the average number of photons injected by the jammer. On the other hand, for entanglement-based quantum illumination, the *SJR* can be approximated by Ref. [7]:

$$SJR^{(\text{Ent.})} = \mathcal{O}\left(\frac{R}{J} 2^e\right), \tag{11.109}$$

and the improvement in *SJR* for entanglement-based quantum illumination is 2^e-times. Therefore, the entanglement-based quantum illumination is exponentially less sensitive to jamming.

From an implementation point of view, the quantum illumination with Gaussian states is of high interest [41–45]. Namely, when the spontaneous parametric downconverter entangled-photon source is employed, while the optimum quantum measurement is performed on the receiver side; in lossy, noisy, and low-brightness regime, it provides a factor-of-four error-probability exponent improvement (6 dB) over the corresponding coherent-state transmission [41,42]. On the other hand, when SPDC generates entangled signal-photon pairs because the dimensionality is $D = 2$, then from Eq. (11.107), we expect a 3 dB improvement in the SNR. This case is considered in greater detail in the upcoming section.

11.4.3 Entanglement-assisted quantum radars

This section studies the entanglement-assisted target detection by employing the Gaussian states generated by the continuous-wave SPDC. The SPDC-based entangled source is the broadband source with $D = T_m W$ i.i.d. signal-idler mode pairs, with W being the phase-matching bandwidth and T_m being the measurement interval. Each signal-idler mode pair, with corresponding signal and idler annihilation operators being $\hat{a}_s$ and $\hat{a}_i$, is, in fact, a two-mode squeezed-vacuum (*TMSV*) state whose representation in Fock basis is given by

$$|\psi\rangle_{s,i} = \frac{1}{\sqrt{N_s + 1}} \sum_{n=0}^{\infty} \left(\frac{N_s}{N_s + 1}\right)^{n/2} |n\rangle_s |n\rangle_i, \tag{11.110}$$

where $N_s = \left\langle \hat{a}_s^\dagger \hat{a}_s \right\rangle = \left\langle \hat{a}_i^\dagger \hat{a}_i \right\rangle$ denotes the mean photon number per mode. The signal-idler entanglement is specified by the phase-sensitive cross-correlation (PSCC) $\langle \hat{a}_s \hat{a}_i \rangle = \sqrt{N_s(N_s+1)}$, which represents the quantum limit. The *TMSV* is a pure maximally entangled zero-mean Gaussian state with Wigner covariance matrix being

$$\boldsymbol{\Sigma}_{TMSV} = \begin{bmatrix} (2N_s+1)\mathbf{1} & 2\sqrt{N_s(N_s+1)}\mathbf{Z} \\ 2\sqrt{N_s(N_s+1)}\mathbf{Z} & (2N_s+1)\mathbf{1} \end{bmatrix}, \tag{11.111}$$

where $\mathbf{1}$ is the identity matrix, and $\mathbf{Z} = \text{diag}(1,-1)$ is the Pauli Z-matrix. Clearly, in a low-brightness regime $N_s << 1$, the PSCC is $\langle \hat{a}_s \hat{a}_i \rangle \approx \sqrt{N_s}$, which is much larger than the classical limit N_s. As described earlier, by going back to Fig. 11.10, the entangled source is used on the transmitter side to generate the signal photon (probe) and idler photon (local reference). The signal photon is transmitted over a noisy, lossy, and atmospheric turbulent channel toward the target. The reflected photon (the radar return) is detected by the radar receiver, and a quantum correlation between radar return and retained reference (idler photon) is exploited to improve receiver sensitivity. The interaction between the probe (signal) photon and the target can be described by a beam splitter of transmissivity *T*, see the *i*-th beam splitter in Fig. 11.11. Therefore, we can model the radar transmitter–target–radar receiver (main) channel as a lossy thermal Bosonic channel $\hat{a}_{\text{Rx}}(\varphi) = \sqrt{T}e^{j\varphi}\hat{a}_s + \sqrt{1-T}\hat{a}_{th}$, where $\hat{a}_{th}$ is a thermal state with the mean photon number being $(1-T)\left\langle \hat{a}_{th}^\dagger \hat{a}_{th} \right\rangle = N_b \gg 1$. With φ we denoted signal-mode phase shift. The idler-mode channel can be described by the lossy Bosonic channel $\hat{a}_{\text{Rx,idler}} = \sqrt{T_i}\hat{a}_i + \sqrt{1-T_i}\hat{a}_v$, where T_i is the transmissivity of the idler channel and $\hat{a}_v$ is the annihilation operator of the vacuum mode. The radar returned probe and retained reference (stored idler) can be described by the following covariance matrix:

$$\boldsymbol{\Sigma}_t = \begin{bmatrix} (2N_s+1)\mathbf{1} & 2\sqrt{TN_s(N_s+1)}\mathbf{Z}\delta_{1h} \\ 2\sqrt{TN_s(N_s+1)}\mathbf{Z}\delta_{1h} & (2N_s'+1)\mathbf{1} \end{bmatrix}, N_s' = TN_s + N_b, \tag{11.112}$$

where *t* is the target indicator. When there is no target, $h=0$, the return signal contains no probe noise but contains background noise, and the covariance matrix is diagonal. When the target is present, $h=1$, and antidiagonal terms, representing the correlation between the signal and idler, are nonzero. Since we are interested in a low-brightness regime ($N_s << 1$), we will have that $N_b >> TN_s$, and the covariance matrix get simplified to

$$\boldsymbol{\Sigma}_t \cong \begin{bmatrix} (2N_s+1)\mathbf{1} & 2\sqrt{TN_s(N_s+1)}\mathbf{Z}\delta_{1h} \\ 2\sqrt{TN_s(N_s+1)}\mathbf{Z}\delta_{1h} & (2N_b+1)\mathbf{1} \end{bmatrix}. \tag{11.113}$$

The joint measurement receiver may use the optical parametric amplifier (OPA), shown in Fig. 11.12 and based on Ref. [46], with a low gain of $G-1=\varepsilon << 1$ to obtain

$$\hat{a}(\varphi) = \sqrt{G}\ \hat{a}_{\text{Rx,idler}} + \sqrt{G-1}\ \hat{a}_{\text{Rx}}^\dagger(\varphi) \tag{11.114}$$

for each single-idler pair of a given mode.

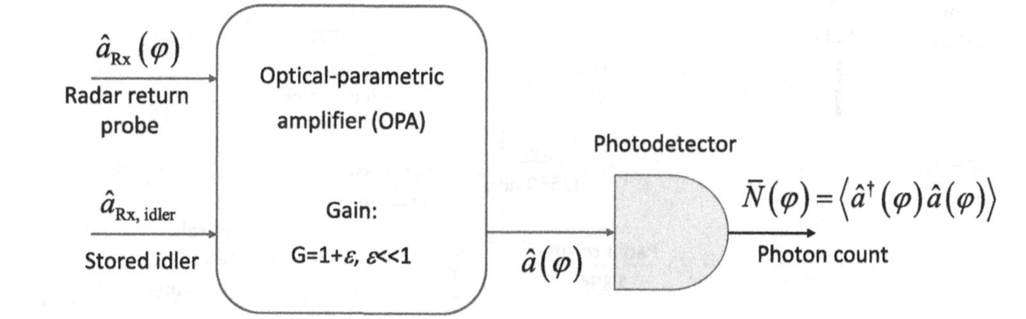

FIGURE 11.12

Operating principle of optical-parametric amplifier-based receiver. Photons are detected at port where idler is amplified.

The direct detection of the OPA has the following mean photon number:

$$\overline{N}(\varphi)=\langle\hat{a}^{\dagger}(\varphi)\hat{a}(\varphi)\rangle=\left\langle\left[\sqrt{G}\ \hat{a}^{\dagger}_{\mathrm{Rx,idler}}+\sqrt{G-1}\ \hat{a}_{\mathrm{Rx}}(\varphi)\right]\left[\sqrt{G}\ \hat{a}_{\mathrm{Rx,idler}}+\sqrt{G-1}\ \hat{a}^{\dagger}_{\mathrm{Rx}}(\varphi)\right]\right\rangle \tag{11.115}$$

The signal-to-noise ratio will be [44]

$$SNR_{EA}=\frac{4\langle(N(0)-N(\pi))\rangle^2}{\{\sigma[N(0)]+\sigma[N(\pi)]\}^2}=\frac{16TT_iT_{\mathrm{imp}}\eta DN_s}{N_b+N_{el}}, \tag{11.116}$$

where η is the detector quantum efficiency, $T_{\mathrm{imp}}\le 1$ is used to denote the system imperfections, and N_{el} is to account for post-detection electronics noise. The corresponding classical *SNR* based on homodyne detection is given by Ref. [44]:

$$SNR=\frac{8TDN_s}{N_b}. \tag{11.117}$$

Therefore, improvement in the *SNR* for an idealized OPA ($\eta=1$, $T_i=T_{\mathrm{imp}}=1$, $N_{\mathrm{el}}=0$) is

$$\frac{SNR_{EA}}{SNR}=\frac{\dfrac{16TDN_s}{N_b}}{\dfrac{8TDN_s}{N_b}}=2, \tag{11.118}$$

indicating that up to 3 dB improvement is possible when the joint measurement is performed with the help of OPA.

The simplified experimental setup used in Ref. [44] to demonstrate improved entanglement-assisted target detection over that of classical target detection is provided in Fig. 11.13. (Full details of the experimental setup can be found in Ref. [44].) The same pump laser was used for both SPDC source and OPA. The periodically poled LiNbO3 (PPLN) crystal was used to generate the photon pairs at frequencies ω_s (signal) and ω_i (idler) satisfying the energy conservation principle $\hbar(\omega_s+\omega_i)=\hbar\omega_p$, and momentum conservation principle: $\boldsymbol{k}_s+\boldsymbol{k}_i=\boldsymbol{k}_p$, which is illustrated in Fig. 11.14.

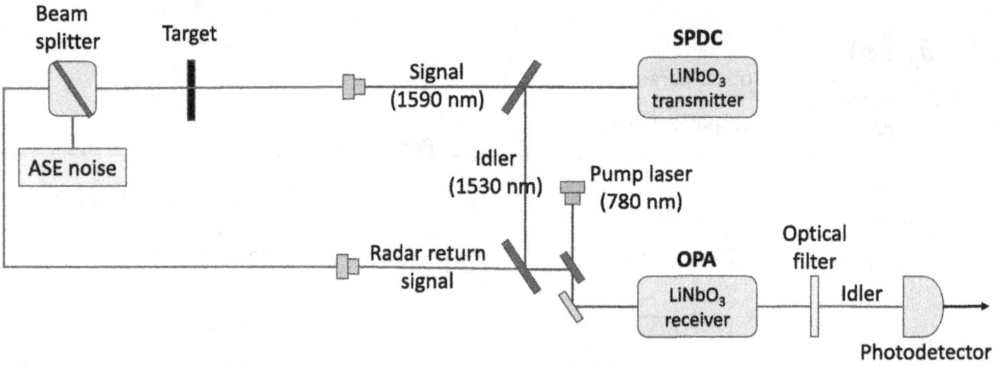

FIGURE 11.13

Setup suitable to demonstrate sensitivity improvement by quantum illumination.

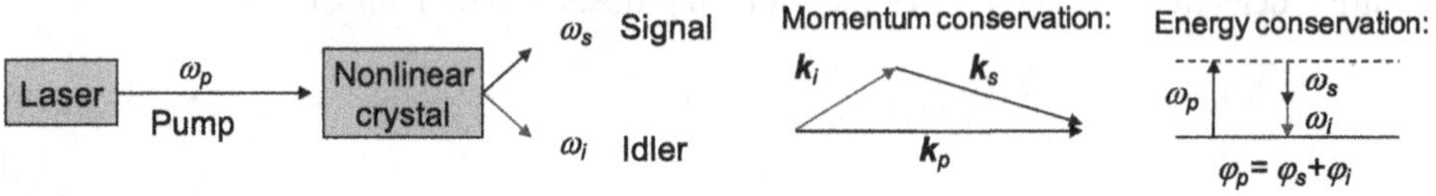

FIGURE 11.14

Illustration of the spontaneous parametric downconversion principle.

The SPDC generates the entangled probe and reference. The signal photon (probe) was sent toward the target. To emulate background noise, the amplified spontaneous emission (ASE) source, such as an EDFA, was used whose output is combined with the probe, interrogating the target, by a beam splitter. The radar return signal gets combined with the retained reference and processed by the second nonlinear PPLN crystal (the OPA), serving as the quantum receiver. The probe wavelength is then filtered out, and we measure the reference arm. SNR measurements were performed for both quantum illumination and classical homodyne detection. After all experimental imperfections, the quantum illumination beats the classical limit by $\sim$1 dB. The experiment was performed in a very noisy environment, the environment noise power was 75 dBm, and the probe was attenuated by 14 dB. Even in this noisy-lossy environment, the authors demonstrated the quantum enhancement achieved by the entanglement. Given that this experiment was conducted in a low-brightness, highly noisy regime ($N_b >> TN_s$), N_{el} can be neglected, but system imperfections, quantum receiver suboptimality, and nonideal detector quantum efficiency cannot be neglected for causing reduced improvements in the SNR. For instance, a simple photodetector might not be sufficient to accurately measure the PSCC.

To compensate for this, the authors in Ref. [45] proposed using a *sum-frequency generation* (SFG) receiver, whose conceptual diagram is provided in Fig. 11.15. The interaction Hamiltonian in SFG $\widehat{H}_I$ describes the interaction of three modes: ancillary mode $\widehat{b}$, signal mode $\widehat{a}_{s,d}$, and reference (idler) mode $\widehat{a}_{i,d}$:

$$\widehat{H}_I = \hbar g \sum_{d=1}^{D} \left(\widehat{b}^{\dagger} \widehat{a}_{s,d} \widehat{a}_{i,d} + \widehat{b} \widehat{a}_{s,d}^{\dagger} \widehat{a}_{i,d}^{\dagger} \right), \quad (11.119)$$

where g describes the interaction strength and $\hbar$ is the reduced Planck constant.

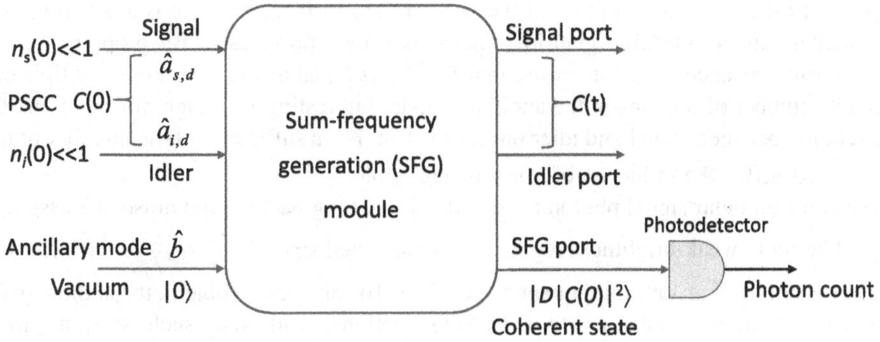

FIGURE 11.15

Conceptual scheme of sum-frequency generation receiver.

The information about PSCC is contained in $\hat{a}_{s,d}\hat{a}_{i,d}$ term. If the target is absent, the interaction Hamiltonian is zero. When the target is present, this interaction Hamiltonian will enable the coherent interaction between photons in signal and idler modes and photons in ancillary mode. By detecting the photons in ancillary mode, we can determine how much correlation was present between radar return signal photons and reference (idler) photons. The SFG has three input ports: signal, idler, and vacuum. The PSCC of signal and idler is denoted by $C(0)$. In the SFG module, the coherent conversion of correlation occurs with photons in the ancillary mode, acquiring the knowledge of signal-idler cross-correlation. The SFG output port is a coherent state $|D\,C(0)\rangle$ containing the PSCC $C(0)$ information. The assumption for proper operation is that both signal and idler are weak with the average number of signal photons $n_s(0) << 1$ and mean idler photon number $n_i(0) << 1$, which is not a correct assumption in radar quantum illumination because the radar return is strong. Let us now more formally describe the operating principle of the SFG receiver, which is based on Ref. [45]. Initially, at time instance $t = 0$, signal-idler mode pairs $\{\hat{a}_{s,d}, \hat{a}_{i,d}\}$ with frequencies $\{f_{s,d}, f_{i,d}\}$ are i.i.d. zero-mean Gaussian states, while the SFG mode $\hat{b}$ at frequency $f_b = f_{s,d} + f_{i,d}$ is in the vacuum mode. In the SFG module, the nonlinear, coherent interaction occurs, and during this interaction, the ancillary mode gets the information about signal-idler PSCC. The evolution of different parameters during SFG interaction can be described as follows [45]:

$$\begin{aligned} C(t) &= C(0)\cos(\sqrt{D}\,gt) \\ b(t) &= -j\sqrt{D}\;C(0)\sin(\sqrt{D}\,gt) \\ n_b(t) &= \left[D|C(0)|^2 + n_i(0)n_s(0)\right]\sin^2(\sqrt{D}\,gt) \\ n_s(t) &= n_s(0), n_i(t) = n_i(0), \end{aligned} \tag{11.120}$$

where we dropped the subscript d for simplicity. The assumption is that the state evolution stays in a weak-brightness, weak cross-correlation regime [45]:

$$n_s(t) = \left\langle \hat{a}^{\dagger}_{s,d}\hat{a}_{s,d}\right\rangle_t \ll 1,\;\; n_i(t) = \left\langle \hat{a}^{\dagger}_{i,d}\hat{a}_{i,d}\right\rangle_t \ll 1, |C(t)|^2 = |\langle \hat{a}_{s,d}\hat{a}_{i,d}\rangle_t|^2 \ll 1, \tag{11.121}$$

where subscript t denotes the averaging of the state at time instance t. There is a coherent oscillation between ancillary mode and all signal-idler pairs as time t progresses. By properly choosing the measurement time instance t, we can ensure that $D^{-1/2}gt$ is equal to $\pi/2$ (or an odd multiple of $\pi/2$) to get the largest number of photons in the ancillary mode. Interestingly enough, for this t, the $C(t) = 0$, so the correlation between signal and idler photons is lost. For a sufficiently large number of modes D, $D|C(0)|^2 \gg n_i(0)n_s(0)$, the influence of noise is negligible.

Given that return radar signal photon is embedded in strong background noise, the assumptions of return signal being in weak-brightness regimes is not satisfied since $\left\langle \hat{a}^{\dagger}_{s,d}\hat{a}_{s,d}\right\rangle_{t=0} = N_b \gg 1$, and the SFG receiver is not itself an optimum quantum receiver. To solve this problem, the authors in Ref. [45] proposed using multiple feed-forward (FF)-SFG sections, with one such section provided in Fig. 11.16.

The input cross-correlation between signal and idler modes in the k-the stage is denoted by $C_{si}^{(k)}$. The signal in the k-th stage is divided by a highly transmissive beam-splitter of reflectivity $\sqrt{R} \ll 1$ into a bright beam and the weak beam that goes through the FF-SFG module. The bright beam bypasses the nonlinear modules. The weak beam is processed by the two-mode squeezer of squeezing parameter r_k, denoted as $S(r_k)$, followed by SFG processing to generate the SFG mode $\hat{b}_k$, and desqueezing module $S(-r_k)$. Squeezing and desqueezing modules are used to adaptively tune the cross-correlation. The final squeezer $S(\varepsilon_k)$ is used to wipe out any remaining squeezing dependence in the cross-correlation. The evolution is terminated when the remaining cross-correlation becomes $\varepsilon << 1$. The SFG mode $\hat{b}_k$ is not entangled with any SFG outputs, while SFG outputs are in a two-mode Gaussian state. Therefore, we can use two-mode squeezing operator $S(r)$ to model the action of

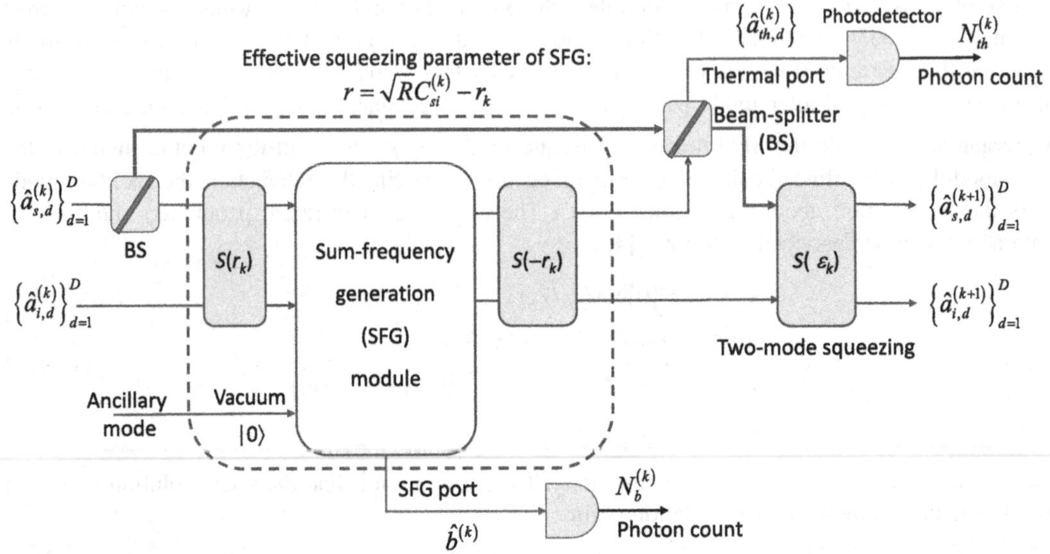

FIGURE 11.16

Conceptual diagram of feed-forward sum-frequency generation receiver.

SFG on each signal-idler pair, wherein $r = \sqrt{R}C_{si}^{(k)} - r_k$. The parameter r_k is adapted depending on the target detection $\widetilde{h}$ in previous stages, wherein $\widetilde{h} = 1$ means that the target was detected and $\widetilde{h} = 0$ means that there was no target in range. Since we adapt the FF-SFG receiver based on previous decisions, this receiver has certain similarities with the Dolinar receiver [47,48]. As discussed in the text related to the previous figure, the SFG mode contains information about signal-idler cross-correlation. In the absence of a target, the SFG mode is just the vacuum mode. Any meaningful photon count at SFG port $N_b^{(k)}$, while the previous decision was $\widetilde{h} = 0$ will indicate that there was a target detection error. Given that the number of photons lost by the signal modes entering the SFG matches the number of photons that DFG mode gains, the purpose of desqueezer $S(-r_k)$ is to ensure that the number of photons at the thermal port, obtained by combining the bright beam and weak signal beam exiting SFG, matches the number of photons at SFG port. The structure with two beam splitters is, in fact, the interferometric structure. The D-mode thermal state, composed of $\widehat{a}_{th,d}^{(k)}$ modes, will have the same mean photon number as the DFG mode $D|r|^2$. When the target is present ($h = 1$), the statistics of $N_b^{(k)} + N_{th}^{(k)}$ will match the photon number statistics of the coherent state $\left|2D|r|^2\right\rangle$. When the target is absent ($h = 0$), the statistics of $N_b^{(k)} + N_{th}^{(k)}$ are the same as for the vacuum state. Therefore, we can apply the binary coherent-state discrimination approach [6,12,45–50] to the entanglement-assisted quantum radar detection problem.

Ref. [46] shows that the concatenated FF-SFG receiver can achieve the quantum Chernoff bound for quantum illumination. It is an optimum receiver in this regime: signal is weak, while noise is strong.

11.4.4 Quantum radar equation

Let us first briefly some basic concepts from *classical radar theory.* The basic operation of the monostatic classical radar, see Fig. 11.9, is to send an electromagnetic wave toward the target, which is reflected by the target with radar return measured by the receiver. The range to the target can be determined by $R = c\Delta t/2$, where Δt is the transit time (time interval that elapses from the transmission of the pulse until its return upon reflection). Let G_{Tx} denote the gain of transmit antenna and P_{Tx} be the power of the transmitter. The power density from a uniformly radiating antenna emitting the spherical wave at distance R is given by $P_{TX}/(4\pi R^2)$. Therefore, the power density at the target is $P_{Tx}G_{Tx}/(4\pi R^2)$. The power of the reflected signal from the target can be written as

$$P_t = \frac{P_{Tx}G_{Tx}}{4\pi R^2}\sigma, \tag{11.122}$$

where σ is the RCS representing the measure of energy that the target reflects back toward the radar. The power density of the reflected signal at the receiver side, therefore, is

$$W_{Rx} = \frac{P_{Tx}G_{Tx}}{4\pi R^2}\frac{\sigma}{4\pi R^2}. \tag{11.123}$$

The power of the reflected signal at the receiver side becomes

$$P_{Rx} = W_{Rx}A_{Rx} = \frac{P_{Tx}G_{Tx}}{4\pi R^2}\frac{\sigma A_{Rx}}{4\pi R^2} = \frac{P_{Tx}G_{Tx}\sigma A_{Rx}}{(4\pi)^2 R^4}, \tag{11.124}$$

where A_{Rx} is the effective area of the receive antenna. This equation is commonly referred to as the *radar range equation*. If the medium (air) is not transparent due to the presence of clouds or fog, we need to introduce the atmospheric attenuation F, defined as $F = \exp(-\alpha R/2)$, with α being the attenuation coefficient, as follows:

$$P_{Rx} = \frac{P_{Tx} G_{Tx} \sigma A_{Rx}}{(4\pi)^2 R^4} F^4. \tag{11.125}$$

The RCS can be defined more formally by Refs. [7,51]:

$$\sigma = \lim_{R \to \infty} 4\pi R^2 \frac{|\boldsymbol{E}_{Rx}|^2}{|\boldsymbol{E}_t|^2}, \tag{11.126}$$

where $\boldsymbol{E}_t$ represents the incident electric field at the target, while $\boldsymbol{E}_{Rx}$ denotes the scattered electric field measured at the receiver side.

The quantum RCS needs to be redefined because the energy in the classical regime is proportional to the magnitude square of the amplitude, while the photon energy is proportional to the frequency. So the quantum RCS should represent an objective measure of the quantum radar visibility of the target. Given that in optical communication we typically use the intensity instead of power, we can define the *quantum RCS* as follows [7]:

$$\sigma = \lim_{R \to \infty} 4\pi R^2 \frac{\langle I_{Rx} \rangle}{\langle I_t \rangle}, \tag{11.127}$$

where I_{Rx} is the intensity of the reflected beam (from the target) measured at the receiver side, while I_t is the incident intensity of the beam at the target. Given that the power in a sphere of radius r, centered around the point source, is related to the intensity I of the uniform source by $P = I \cdot 4\pi r^2$, we conclude that density of the power W in classical radar has the similar role as the intensity for the quantum radar. Therefore, the *quantum radar equation* is very similar to the classical radar equation:

$$P_{Rx}^{(Q)} = \frac{P_{Tx}^{(Q)} \sigma A_{Rx}}{(4\pi)^2 R^4}, \tag{11.128}$$

where A_{Rx} is the receive aperture. For more rigorous treatment of the quantum RCS and radar equation, the interested reader is referred to Ref. [7]. The same author has found that for single-photon radars, quantum RCS has a sidelobe structure dictated by quantum mechanical effects. Moreover, he found that $\sigma_Q \geq \sigma$ for a rectangular target near the specular direction provided that the radar pulse contains more than one photon per pulse.

11.5 Summary

This chapter was devoted to quantum sensing and quantum radars. In Section 11.1, the quantum phase estimation problem was addressed, and the physical limits on phase measurements were described. Quantum interferometry was described in Section 11.1.1 and applied to the sparable quantum probe to derive the SQL. In Section 11.1.2, highly entangled NOON states were used as the quantum probe to achieve the Heisenberg limit. QFI was introduced in Section 11.2 and used to derive the quantum

Cramér–Rao bound, representing the lower bound for the precision of quantum sensing. In the same section, the classical Cramér–Rao bound was derived. Section 11.3 was devoted to distributed quantum sensing, in which multiple quantum sensors are used to measure a global parameter. Optimum entangled probes were shown to overcome the SQL and achieve the Heisenberg limit. DQS was employed to determine quadrature displacements. The performance limits for DQS were also derived. Section 11.4 focused on quantum radars. After a brief description of common quantum radar architectures, the following promising quantum radar technologies were described: interferometric quantum radars (Section 11.4.1), quantum illumination-based quantum radars (Section 11.4.2), and entanglement-assisted target detection (Section 11.4.3). Finally, the quantum radar equation was discussed in Section 11.4.4.

References

[1] P.A. M Dirac, The quantum theory of the emission and absorption of radiation, Proc. Roy. Soc. Lond. A 114 (767) (1927) 243–265.
[2] S.M. Barnet, D.T. Pegg, Phase in quantum optics, J. Phys. Math. Gen. 19 (18) (1986) 3849.
[3] L. Susskind, J. Glogower, Quantum mechanical phase and time operator, Physics 1 (1) (1964) 49–61.
[4] C.C. Gerry, P.L. Knight, Introductory Quantum Optics, Cambridge University Press, Cambridge, UK, 2008.
[5] D.S. Simon, Quantum sensors: improved optical measurement via specialized quantum states, J. Sens. 2016 (2016). Article ID 6051286.
[6] C.W. Helstrom, Quantum Detection and Estimation Theory, Academic Press, New York, 1976.
[7] M. Lanzagorta, Quantum Radar, Morgan and Claypool Publishers, 2012.
[8] C.K. Hong, Z.Y. Ou, L. Mandel, Measurement of subpicosecond time intervals between two photons by interference, Phys. Rev. Lett. 59 (1987) 2044.
[9] Z.-Y.J. Ou, Quantum Optics for Experimentalists, World Scientific, Hackensack-London, 2017.
[10] I.B. Djordjevic, Quantum Information Processing, Quantum Computing, and Quantum Error Correction: An Engineering Approach, second ed., Elsevier/Academic Press, 2021.
[11] I.B. Djordjevic, Physical-Layer Security and Quantum Key Distribution, Springer Nature Switzerland AG, Cham, Switzerland, 2019.
[12] G. Cariolaro, Quantum Communications, Springer International Publishing Switzerland, Cham: Switzerland, 2015.
[13] C. Weedbrook, S. Pirandola, R. García-Patrón, N.J. Cerf, T.C. Ralph, J.H. Shapiro, S. Lloyd, Gaussian quantum information, Rev. Mod. Phys. 84 (2) (Apr.-Jun. 2012) 621.
[14] M.W. Mitchell, J.S. Lundeen, A.M. Steinberg, Super-resolving phase measurements with a multiphoton entangled state, Nature 429 (6988) (2004) 161–164.
[15] P. Walther, J.-W. Pan, M. Aspelmeyer, R. Ursin, S. Gasparoni, A. Zeilinger, De Broglie wavelength of a non-local four-photon state, Nature 429 (6988) (2004) 158–161.
[16] I. Afek, O. Ambar, Y. Silberberg, High-NOON states by mixing quantum and classical light, Science 328 (5980) (2010) 879–881.
[17] Y. Israel, I. Afek, S. Rosen, O. Ambar, Y. Silberberg, Experimental tomography of NOON states with large photon numbers, Phys. Rev. 85 (2) (2012) 022115.
[18] I.B. Djordjevic, Advanced Optical and Wireless Communications Systems, Springer International Publishing, Switzerland, 2017.
[19] R.N. McDonough, A.D. Whalen, Detection of Signals in Noise, Academic Press, San Diego, 1995.
[20] Z. Zhang, Q. Zhuang, Distributed quantum sensing, Quantum Sci. Technol. 6 (4) (2021) 043001.

[21] S.L. Braunstein, C.M. Caves, Statistical distance and the geometry of quantum states, Phys. Rev. Lett. 72 (22) (1994) 3439–3443.
[22] S.L. Braunstein, C.M. Caves, G.J. Milburn, Generalized uncertainty relations: theory, examples, and Lorentz invariance, Ann. Phys. 247 (1) (1996) 135–173.
[23] M.G.A. Paris, Quantum estimation for quantum technology, Int. J. Quant. Inf. 07 (Suppl. 01) (2009) 125–137.
[24] A. Uhlmann, The "transition probability" in the state space of a *-algebra, Rep. Math. Phys. 9 (2) (1976) 273–279.
[25] Q. Zhuang, Z. Zhang, J.H. Shapiro, Distributed quantum sensing using continuous-variable multipartite entanglement, Phys. Rev. A 97 (3) (March 21 , 2018) 032329.
[26] Q. Zhuang, J. Preskill, L. Jiang, Distributed quantum sensing enhanced by continuous-variable error correction, New J. Phys. 22 (2020) 022001.
[27] V. Giovannetti, S. Lloyd, L. Maccone, Quantum-enhanced measurements: beating the standard quantum limit, Science 306 (5700) (2004) 1330–1336.
[28] M. Zwierz, C.A. Pérez-Delgado, P. Kok, General optimality of the Heisenberg limit for quantum metrology, Phys. Rev. Lett. 105 (18) (2010) 180402.
[29] M. Zwierz, C.A. Pérez-Delgado, P. Kok, Erratum: general optimality of the Heisenberg limit for quantum metrology, Phys. Rev. Lett. 107 (5) (2011) 059904, 105, 180402 (2010).
[30] S. Lloyd, Enhanced sensitivity of photodetection via quantum illumination, Science 321 (5895) (2008) 1463–1465.
[31] Harris Corporation, *Quantum Sensors Program*, Final Technical Report, AFRL-RI-RS-TR-2009-208, Aug. 2009.
[32] P. Kumar, V. Grigoryan, M. Vasilyev, Noise-Free amplification: towards quantum laser radar, in: Proc. 14th Coherent Laser Radar Conference, July 9-13, 2007. Snowmass, CO.
[33] Z. Dutton, J.H. Shapiro, S. Guha, LADAR resolution improvement using receivers enhanced with squeezed-vacuum injection and phase-sensitive amplification, J. Opt. Soc. Am. B 27 (2010) A63.
[34] J.H. Shapiro, Quantum pulse compression laser radar, in: Proc. SPIE 6603, Noise and Fluctuations in Photonics, Quantum Optics, and Communications, 2007, p. 660306.
[35] J.R. Jeffers, N. Imoto, R. Loudon, Quantum optics of traveling-wave attenuators and amplifiers, Phys. Rev. A 47 (4) (1993) 3346.
[36] R. Loudon, The Quantum Theory of Light, third ed., Oxford University Press, 2000.
[37] J.F. Smith, Quantum entangled radar theory and a correction method for the effects of the atmosphere on entanglement, in: Proc. SPIE Defense, Security, and Sensing, SPIE Quantum Information and Computation VII Conference Vol. 7342, 2009, p. 73420A. Orlando, Florida.
[38] L.C. Andrews, R.L. Philips, Laser Beam Propagation through Random Media, SPIE Press, Bellingham, Washington, 2005.
[39] R.K. Tyson, Principles of Adaptive Optics, CRC Press, Boca Raton, FL, 2015.
[40] V. Nafria, C. Cui, I.B. Djordjevic, Z. Zhang, Adaptive-optics enhanced distribution of entangled photons over turbulent free-space optical channels, in: Proc. CLEO 2021, 09–14 May 2021 paper FTu1N.2.
[41] S. Hui-Tan, B.I. Erkmen, V. Giovannetti, S. Guha, S. Lloyd, L. Maccone, S. Pirandola, J.H. Shapiro, Quantum illumination with Gaussian states, Phys. Rev. Lett. 101 (25) (2008) 253601.
[42] J.F. Shapiro, S. Lloyd, Quantum illumination versus coherent-state target detection, New J. Phys. 11 (June 2009) 063045.
[43] S. Guha, B.I. Erkmen, Gaussian-state quantum-illumination receivers for target detection, Phys. Rev. A 80 (5) (2009) 052310.
[44] Z. Zhang, S. Mouradian, F.N.C. Wong, J.H. Shapiro, Entanglement-Enhanced sensing in a lossy and noisy environment, Phys. Rev. Lett. 114 (11) (2015) 110506.

[45] Q. Zhuang, Z. Zhang, J.H. Shapiro, Optimum mixed-state discrimination for noisy entanglement-enhanced sensing, Phys. Rev. Lett. 118 (4) (2017) 040801.

[46] H. Shi, Z. Zhang, Q. Zhuang, Practical route to entanglement-assisted communication over noisy Bosonic channels, Phys. Rev. Appl. 13 (3) (2020) 034029.

[47] S.J. Dolinar, An optimum receiver for the binary coherent state quantum channel, MIT Res. Lab. Electron. Quart. Progr. Rep. 111 (1973) 115−120.

[48] A. Assalini, N. Dolla Pozza, G. Pierobon, Revisiting the Dolinar receiver through multiple-copy state discrimination theory, Phys. Rev. A 84 (2011) 022342.

[49] A. Acín, E. Bagan, M. Baig, L. Masanes, R. Muñoz-Tapia, Multiple-copy two-state discrimination with individual measurements, Phys. Rev. A 71 (3) (2005) 032338.

[50] V. Vilnrotter, C.-W. Lau, Quantum detection theory for the free-space channel, IPN Prog. Rep. 42−146 (Aug. 15, 2001) 1−34.

[51] E.F. Knott, J.F. Shaeffer, M.T. Tuley, Radar Cross Section, second ed., SciTech Publishing, 2004.

CHAPTER

Quantum machine learning

12

Chapter outline

Quantum Communication, Quantum Networks, and Quantum Sensing. https://doi.org/10.1016/B978-0-12-822942-2.00010-8

12.1 Machine learning fundamentals

Machine learning (ML) is a multidisciplinary science (employing artificial intelligence, computer science, and statistics) that studies various algorithms to extract meaningful information from available data and provide automated solutions to complex computational problems [1–3]. The strength of ML algorithms (MLAs) lies in their ability to learn from available data, and as such, ML is a data-driven rather than model-driven approach. The various learning models can be grouped into three generic categories:

- *Supervised learning*, in which the MLA employs labeled data samples. In other words, we provide a dataset to the machine together with the correct answer (label) for each data point and expect the MLA to determine key characteristics so that the next time an unknown data point is provided, the algorithm will correctly predict the corresponding outcome. Classification and regression problems belong to this category. In a *classification* problem, the machine should classify the data sequence into several distinct categories (classes). In a *regression* problem, the machine should correctly predict the values of a continuous variable. Famous algorithms in this category include K-nearest neighbors (K-NN), *support vector machines* (SVMs), artificial neural networks (ANNs), decision trees, random forest, linear regression, and logistic regression.
- *Unsupervised learning*, in which a sequence of data is given to the machine, and it is expected to determine the structure of the data. For instance, in clustering, the algorithm must find similar types of data points and group them into different clusters. The clustering algorithm can implement a spam filter to identify which e-mail represents junk e-mail. Well-known algorithms in this category include principal component analysis (PCA), K-means clustering, and hierarchical clustering.
- *Reinforcement learning* (RL), in which the machine must sense the environment and choose the appropriate action based on its current state and environment. RL is applicable to time-varying conditions in which the external situation is under constant change. It is also applicable to solving problems for which the state space is enormous. RL is popular in data processing, industrial automation, and training system development.

12.1.1 Machine learning basics

The key idea behind *supervised learning* is to use label training data to determine a predictive model that allows us to make accurate predictions about future data, as illustrated in Fig. 12.1. As an illustration, the filtering spam problem can be solved by supervised learning. In this example, we send

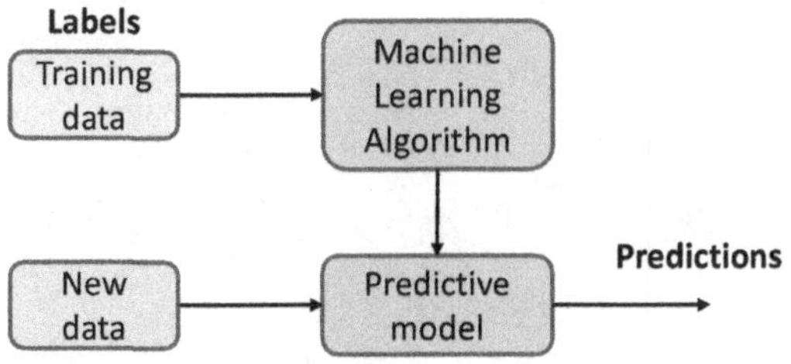

FIGURE 12.1

Supervised learning.

a series of e-mails to the MLA that are properly labeled spam or not-spam. The MLA builds a predictor model to properly classify unknown e-mails as spam or not-spam.

A supervised classification problem with discrete labels is commonly referred to as a *classification task*. In a *regression* task, the predictor can generate the continuous outcome signal, whereas in a classification task, we are concerned with determining the category to which an input belongs. A *binary classifier* determines the decision boundary to partition the corresponding vector space into two classes. In a *multiclass classifier* problem, we are concerned with partitioning the input vector space into multiple vector sets, with each set representing a different class. The binary classification example provided in Fig. 12.2 uses seven training examples labeled with a negative sign and seven labeled with a positive label. This binary classification problem is two-dimensional. We can use supervised MLA to determine the decision boundary (the classification rule), represented by the line with the positive slope. Using two independent binary classifiers, we can create a multiclass classifier with four labels. In certain situations, the decision boundary might be *nonlinear*, in which case SVMs classified as supervised learning algorithms can be used. The key idea behind an SVM is to construct a hyperplane in a larger, high-dimensional (HD) hyperspace such that the margin separation between two classes in a binary classification problem is maximized, as illustrated in Fig. 12.3. In this HD hyperspace, each data is represented as a point, and the purpose of the SVM is to construct a hyperplane maximizing the separation between the points from two classes, wherein

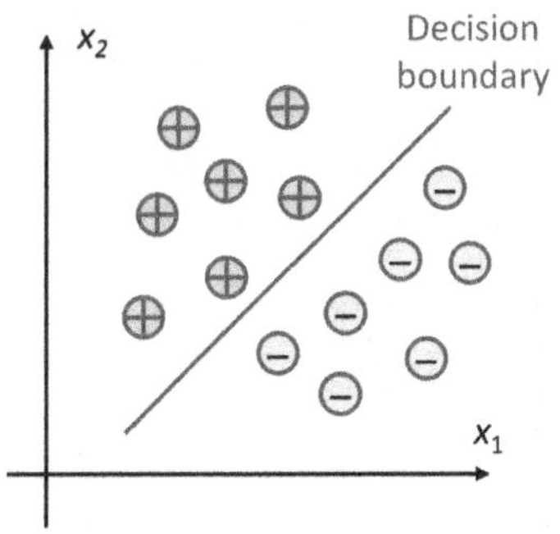

FIGURE 12.2

Binary classification problem.

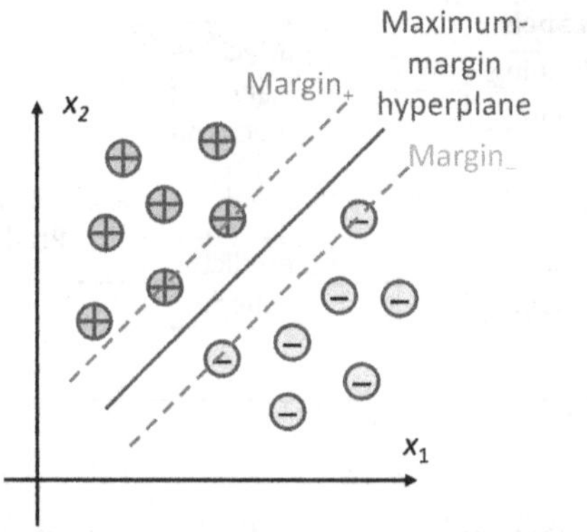

FIGURE 12.3

Optimum margin classifier to maximize separation between two classes; separation is defined as distance between two margins.

the separation is defined in terms of the distance between two margins. Therefore, we avoid the nonlinear boundary problem by mapping it into HD hyperspace, where there is more space and the hyperplane is easier to determine (the hyperplane is one dimension smaller than the hyperspace). The SVM applies to *linear separable problems* as well, for which a hyperplane exists that separates the points belonging to two different classes; in this case, transformation to an HD space is not required.

In a *regression* problem, we are concerned with determining an approximate function to accurately predict an outcome given a finite number of training examples. In contrast to the multitask classification problem, where the outcome is discrete, the regression outcome is continuous. In ML, predictor variables are commonly referred to as *features*, while response variables are called *target* variables. In this chapter, we adopt the same terminology. In a *linear regression* problem, given a feature variable x and target variable y, we are interested in drawing a straight line that minimizes the average squared distance between the data points and fitted line, as illustrated in Fig. 12.4.

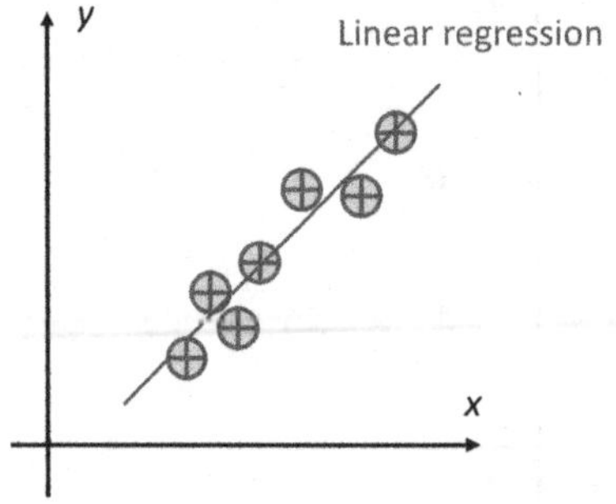

FIGURE 12.4

The linear regression problem.

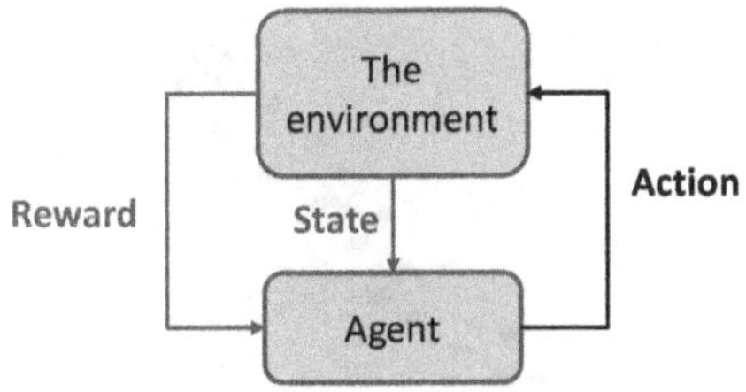

FIGURE 12.5

The reinforcement learning problem.

In RL, we are interested in developing an *agent* (system) that improves its performance and accuracy through interactions with the environment, as illustrated in Fig. 12.5. Compared with supervised learning, the feedback is not a label or value but a measure of how well the action was measured by the reward function. By interacting with the environment, the agent learns various strategies to maximize the reward. The environmental state is related to a positive or negative reward, with the reward is the overall goal. For instance, in chess, the reward could be winning the game, and the state could be the outcome of each move. The agent in this example is the chess engine, which decides a series of moves based on the state of the board (the environment).

In *unsupervised learning*, we are dealing with an unknown structure or unlabeled data sequence and exploring the data structure to extract meaningful information. In *clustering*, also known as unsupervised classification, we perform exploratory data analysis to group data points into meaningful clusters without prior knowledge of cluster memberships. A cluster represents a group of data points that share a certain degree of similarity but are dissimilar in form to the members of other classes. A popular learning algorithm from this class is the *K-means* algorithm, in which we partition the data points into K classes based on pairwise distances; an example is provided in Fig. 12.6.

We often encounter data with large dimensionality—in other words, each observation has many dimensions. In the preprocessing stage, we often try to reduce data noise to improve the predictive performance of the corresponding algorithm while simultaneously compressing the data into a smaller dimensional subspace. The corresponding unsupervised algorithm is known as a *dimensionality*

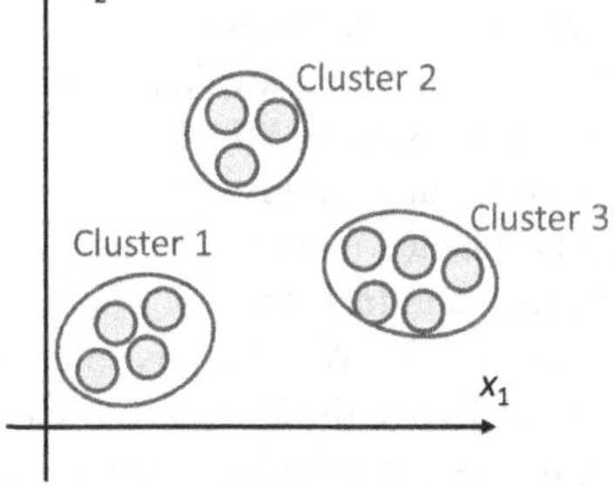

FIGURE 12.6

The clustering problem.

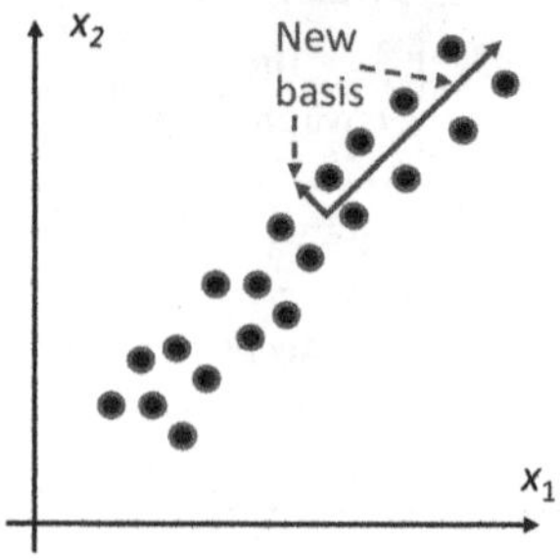

FIGURE 12.7

Principal component analysis concept.

reduction algorithm. Dimensionality reduction can also provide better visualization by projecting to a lower dimensionality space. A popular algorithm from this class is *PCA*, which can reveal geometric structures that are not obvious. When the MLA has no idea about the data structure, it must determine the structure on its own, and the intuitive way to do this is to determine the eigenvectors of the available data to obtain geometric insight into the prevailing directions in the predictive space. PCA builds on this idea by transforming the original ("natural") basis into one of lower dimensionality that is better aligned with the available data, as illustrated in Fig. 12.7. The directions in the new basis are related to the actual data structure, while the length of the basis vector is dictated by the variance in that direction. The basis vectors with small variances can be discarded, and thus we can reduce the dimensionality. In this particular example, we reduce two-dimensional (2-D) data to a single dimension. In PCA, we organize data in terms of matrix $\boldsymbol{X}=(x_{ij})_{m\times n}$, where n is the number of experiments and m denotes the number of measurement values for each experiment. Therefore, each row represents measurements of the same type. By removing the mean value for each row, we can determine the *covariance matrix* by $\boldsymbol{C}=(1/n)\boldsymbol{X}\boldsymbol{X}^{\mathrm{T}}$. To determine the principal component vectors, we need to diagonalize the covariance matrix. PCA employs the following steps during the diagonalization procedure [1]: (1) choose the direction of the maximum variance; (2) of the available directions orthogonal to the previous basis vectors, select the one with the largest variance; and (3) repeat step 2 until we run out of dimensions. The resulting basis vectors are the *principal components*. The next subsection provides additional details about PCA.

The ML predictive modeling workflow comprises [2] preprocessing, learning, evaluation, and prediction stages, as summarized in Fig. 12.8. The raw data rarely come in a form suitable for use in ML, and preprocessing represents a crucial step in ML. We can interpret the raw data as a series of objects from which we want to extract relevant features. The *feature* can be any measurable property of the phenomenon under study. Properly selected features are discriminating and help the MLA identify patterns and distinguish among different data instances. For optimum performance of the underlying MLAs, the features must be on the same scale. We often use a geometric approach to group features into a *feature vector*. The feature vectors live in the HD *feature space*. Some features might be highly correlated, and we will need to employ dimensionality reduction techniques. To evaluate MLA performance, we randomly split the available dataset into a training dataset and a test dataset. The *training dataset* is used to train and optimize the ML model, while the *test dataset* is used to evaluate and

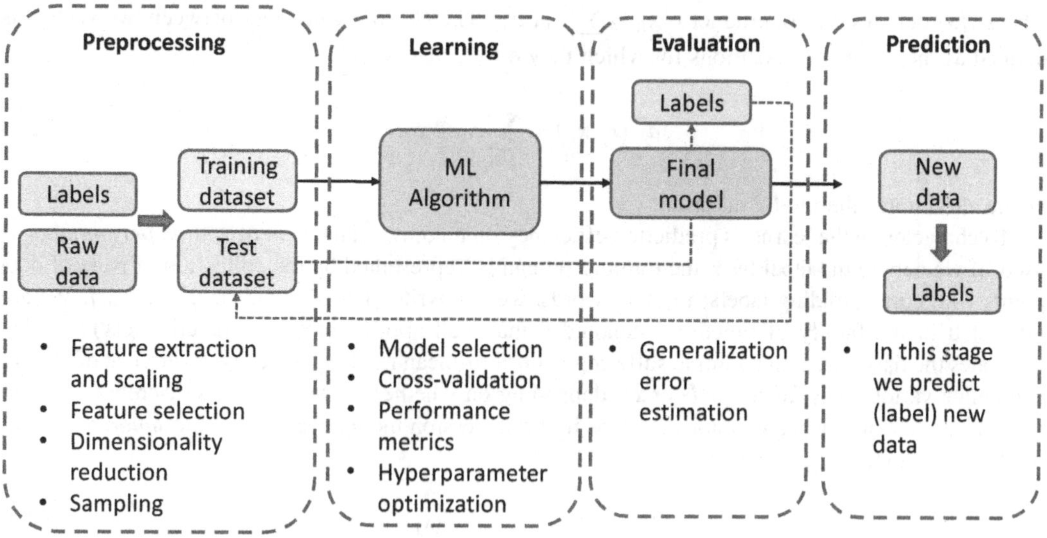

FIGURE 12.8

Road map for machine learning.

finalize the ML model. To estimate how well the predictor model works on unseen data, we evaluate the so-called generalization (test) error, and if we are happy with the performance of the predictor, we use it to predict new/future data.

A feature vector with d features can be represented as a d-dimensional column vector:

$$\boldsymbol{x}_j = \sum_{i=1}^{d} x_{ij}\boldsymbol{e}_i = \begin{bmatrix} x_{1j} \\ x_{2j} \\ \vdots \\ x_{dj} \end{bmatrix}, \tag{12.1}$$

where the i-th basis vector $\boldsymbol{e}_i$ corresponds to the i-th feature. The index j corresponds to the j-th data instance. In designing MLAs, distance functions, norms, and angles between the feature vectors are commonly used. As an illustration, the *Euclidean distance* between the j-th and k-th data instances is defined by

$$d(\boldsymbol{x}_j, \boldsymbol{x}_k) = \sqrt{\sum_{i=1}^{d} (x_{ij} - x_{ik})^2}. \tag{12.2}$$

The cosine angle between two vectors is known as the cosine *similarity*, defined as

$$\cos(\boldsymbol{x}_j, \boldsymbol{x}_k) = \frac{\boldsymbol{x}_j^{\mathrm{T}} \boldsymbol{x}_k}{\|\boldsymbol{x}_j\| \|\boldsymbol{x}_k\|}, \tag{12.3}$$

where $\boldsymbol{x}_j^{\mathrm{T}}\boldsymbol{x}_k$ denotes the dot product $\boldsymbol{x}_j^{\mathrm{T}}\boldsymbol{x}_k = \sum_{i=1}^{d} x_{ij}x_{ik}$. The *Hamming distance* between two vectors is defined as the number of positions for which they differ; that is,

$$d_H(\boldsymbol{x}_j, \boldsymbol{x}_k) = \sum_{i=1}^{d} x_{ij} \oplus x_{ik}, \tag{12.4}$$

where $\oplus$ denotes the mod-2 addition.

To characterize the learner's prediction efficiency on unknown data, *generalization performance* is used. If we denote the label by y, the training set can be represented by the collection of pairs of data points with corresponding labels; in other words, we can write $\{(\boldsymbol{x}_1,y_1), (\boldsymbol{x}_2,y_2), \ldots, (\boldsymbol{x}_n,y_n)\}$. We are interested in the family of functions, denoted f, that well approximates the function $g(\boldsymbol{x}) = y$ that generates the data based on a sample suffering from zero-mean random noise and a variance of σ^2. We can define various loss functions $L(y,f(\boldsymbol{x}))$ depending on whether the value of y is discrete or continuous. If the response y is continuous, such as in the regression model, we can use the *squared error* as the loss function:

$$L\left(y_j, f(\boldsymbol{x}_j)\right) = \left[y_j - f(\boldsymbol{x}_j)\right]^2. \tag{12.5}$$

The *absolute error* can be used instead of the squared error and is defined as

$$L\left(y_j, f(\boldsymbol{x}_j)\right) = \left|y_j - f(\boldsymbol{x}_j)\right|. \tag{12.6}$$

In binary classification problems, we can use the *0−1 loss function*, defined as

$$L\left(y_j, f(\boldsymbol{x}_j)\right) = \mathcal{I}\left(y_j \neq f(\boldsymbol{x}_j)\right), \quad \mathcal{I}(e) = \begin{cases} 1, & \text{if } e \text{ is true} \\ 0, & \text{if } e \text{ is false} \end{cases} \tag{12.7}$$

where $\mathcal{I}$ denotes the indicator function. Clearly, y_j is the true label, while $f(\boldsymbol{x}_j)$ is the predicted label. Unfortunately, as shown by Feldman et al. [4,5], classification problem optimization based on the 0−1 loss function is NP-hard, and often we use a convex approximation instead, such as the *hinge loss*, defined as

$$L\left(y_j, f(\boldsymbol{x}_j)\right) = \max\left(0, 1 - f(\boldsymbol{x}_j)y\right), \tag{12.8}$$

whose range is the set of real numbers R. Once the loss function is defined, we can introduce *training error*, also known as *empirical risk*:

$$E = \frac{1}{n}\sum_{j=1}^{n} L\left(y_j, f(\boldsymbol{x}_j)\right). \tag{12.9}$$

To measure how accurately an MLA predicts the outcomes of a predictor for previously unseen data, we use *generalization error* (GE). GE can be defined as

$$E_{GE} = \int_{x \in \mathcal{X}, y \in \mathcal{Y}} L(y, f(\boldsymbol{x}))p(y, \boldsymbol{x})d\boldsymbol{x}dy, \tag{12.10}$$

where $\boldsymbol{x}$ is the feature vector from the feature space $\mathcal{X}$, and $p(y,\boldsymbol{x})$ is the joint distribution of $\boldsymbol{x}$ and y. Typically, the training error is lower than the GE. Unfortunately, this distribution is unknown, and we cannot compute the GE. However, we can estimate it by calculating the *test error* (TE) with the testing dataset we left aside, as follows:

$$E_{TE} = \frac{1}{n_{TE}} \sum_{j=1}^{n_{TE}} L\left(y_j, f\left(\boldsymbol{x}_j\right)\right), \tag{12.11}$$

where averaging is performed over the testing set with n_{TE} instances. The TE gives an estimate of the GE, and when the testing set represents an unbiased sample from the joint distribution $p(y,\boldsymbol{x})$, it approaches the GE in the limit, that is, $\lim_{n_{TE} \to \infty} E_{TE} = E_{GE}$. It may be a good idea to bound the GE probabilistically:

$$\Pr(E_{GE} - E_{TE} \leq \varepsilon) \geq 1 - \delta, \tag{12.12}$$

where the goal is to characterize the probability $1-\delta$ that the GE is smaller than some error-bound ε, which typically depends on δ and the number of instances in the test dataset.

12.1.2 Principal component analysis

In PCA, we organize data in terms of matrix $\boldsymbol{X}' = (x'_{ij})_{m \times n}$:

$$X' = \begin{bmatrix} x'_{11} & x'_{12} & & \cdots & x'_{1n} \\ x'_{21} & x'_{22} & x'_{23} & \cdots & x'_{2n} \\ \vdots & & & & \\ & & & \ddots & \\ x'_{m1} & x'_{m2} & x'_{m3} & \cdots & x'_{mn} \end{bmatrix}, \tag{12.13}$$

where n is the number of experiments and m denotes the number of measurement values for each experiment. Therefore, each row represents measurements of the same type, and we can determine the *mean value* of each row by

$$\mu_i = \frac{1}{n} \sum_{j=1}^{n} x_{ij}; \quad i = 1, 2, \cdots, m \tag{12.14}$$

Now we create a new matrix $\boldsymbol{X}=(x_{ij})_{m\times n}$ by removing the mean value for each row, that is, $x_{ij} = x'_{ij} - \mu_i$. The mean values of the rows in $\boldsymbol{X}$ now equal 0. We further calculate the *covariance matrix* by $\boldsymbol{C}=(1/n)\boldsymbol{X}\boldsymbol{X}^{\mathrm{T}}$. To determine the principal component vectors, we must diagonalize the covariance matrix. Let us now look at the example in Fig. 12.9.

The coordinates x and y could be interpreted as the height and weight of individuals in a studied population of people. We can clearly identify two dominant (principal) directions, one represented by the green vector (light gray in print version) of length σ_s^2 (the signal variance) and the second by the red vector (dark gray in print version) of length σ_n^2 (the noise variance), which is orthogonal to the first.

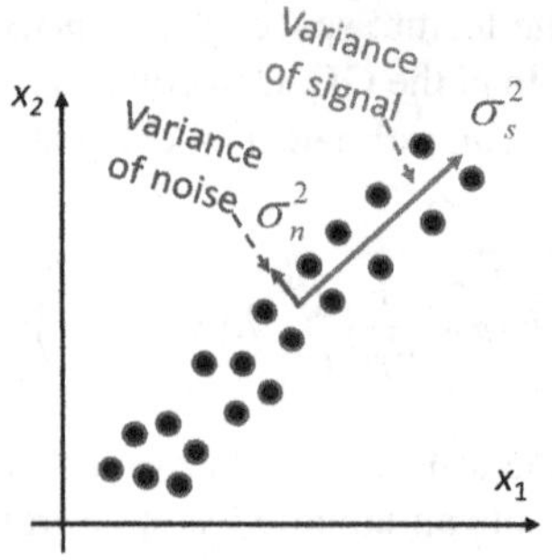

FIGURE 12.9

Principal component analysis. The direction of the largest variance does not coincide with the original ("native") basis.

Evidently $\sigma_s^2 \gg \sigma_n^2$, or in other words, the signal-to-noise ratio, defined as $SNR = \sigma_s^2/\sigma_n^2$, is much larger than 1. We can use these two vectors, well aligned with the data, as the new basis. The direction denoted with the green vector (light gray in print version) is dominant and carries the most significant information, and we can discard the red vector's (dark gray in print version) direction, thus reducing the complexity of the problem. We have essentially reduced the dimensionality of the system while minimizing the information loss. Interpreted differently, we translated and rotated the original basis to obtain a new basis better aligned with the actual data. The length of the vector in a given direction is determined by estimating the spread in the same direction. This change of basis can be represented by the unitary matrix U satisfying condition $U^{\dagger}U$, with I being the identity matrix. (For real-valued problems, the Hermitian transposition † can be replaced by the transposition T.) This change of the base corresponds to the diagonalization of the correlation matrix, which is the essence of PCA. Intuitively, we can summarize the PCA procedure as follows [1,6]: (1) choose the direction of the maximum variance; (2) of the available directions orthogonal to the previous basis vectors, select the one with the largest variance; and (3) repeat step 2 until we run out of dimensions. The basis vectors obtained from PCA are the *principal components*. PCA relies on the following three assumptions:

- *Linearity.* The basis change is a linear operation, but some problems are inherently nonlinear.
- *Directions with large variances are the most interesting.* Directions with large variances are considered to belong to the signal vector, while directions with low variances belong to the noise vector. Unfortunately, this is not true for all problems.
- *Orthogonality.* To make problems easy to solve, we assume the principal components are orthogonal, which is not true for all datasets. As an illustrative example, to explore a city using the PCA approach, we can make only 90 degrees turns. This situation is suitable for Phoenix and Tucson but not for cities with angled (curved) streets, such as Washington, DC.

Fig. 12.10 provides two illustrative examples for which PCA is not useful. In Fig. 12.10 (left), the dataset is nonorthogonal, and we do not expect PCA to be useful given that one of the three assumptions above is not satisfied. On the other hand, in Fig. 12.10 (right), known as the Ferris wheel dataset, there is no dominant direction. This problem can be solved by moving to the polar coordinate system by the transformation $x = r\cos\theta$, $y = r\sin\theta$; unfortunately, this is a nonlinear transformation.

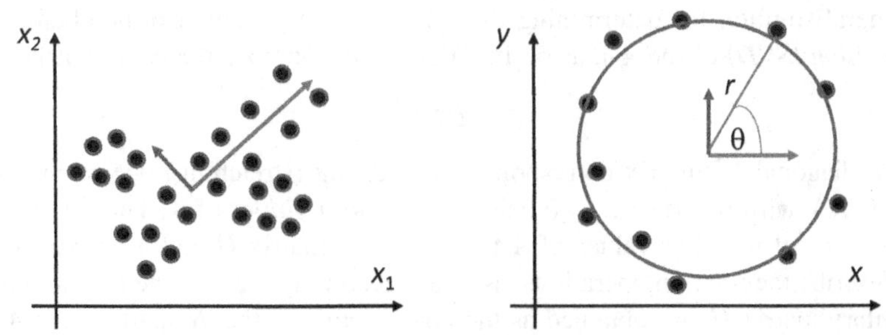

FIGURE 12.10

Datasets for which principal component analysis is not beneficial.

Problems in lower-dimensional space, particularly 2-D, are easy to solve, and we can apply the best fitting approach. Unfortunately, many problems in ML have hundreds or even thousands of components. In HD space, PCA can reveal the dataset structure and reduce the dimensionality of the problem, given that low variance dimensions can be discarded. By studying the eigenvector structure of the correlation matrix, we can make PCA efficient and practical. Given that the correlation matrix $\boldsymbol{C}$ is a square matrix of size $m \times m$, we can perform *spectral (eigenvalue) decomposition* [7–10]:

$$\boldsymbol{C} = \boldsymbol{U}\boldsymbol{D}\boldsymbol{U}^{\dagger}, \tag{12.15}$$

where $\boldsymbol{D}$ is the diagonal matrix of eigenvalues of the correlation matrix $\boldsymbol{C}$, that is, $\boldsymbol{D} = \text{diag}(\lambda_1,\ldots, \lambda_m)$, while $\boldsymbol{U}$ is the unitary matrix with columns corresponding to the orthonormalized eigenvectors of $\boldsymbol{C}$. The eigenvalues of $\boldsymbol{C}$, denoted λ_i, are solved via the *characteristic polynomial*:

$$\det(\boldsymbol{C} - \lambda\boldsymbol{I}) = 0. \tag{12.16}$$

The eigenvectors are now used as the principal components. Eigenvectors with small eigenvalues are discarded from the study, thus effectively performing *dimensionality reduction*. Unfortunately, ML problems can occur when the number of measurements m is too large, particularly in image signal-processing problems, where m can represent the number of pixels (which can be on the order of thousands). It can happen that $m >> n$, in which case the correlation matrix will be huge, and the complexity of the spectral decomposition procedure will be too high. To avoid this problem, instead of defining the correlation matrix as $\boldsymbol{C}=(1/n)\boldsymbol{X}\boldsymbol{X}^{\mathrm{T}}$, we can redefine it as $\boldsymbol{R}=(1/n)\ \boldsymbol{X}^{\mathrm{T}}\boldsymbol{X}$, which has the size $n \times n$ (and $n << m$). The eigenvalue λ, eigenvector $\boldsymbol{u}$, and correlation matrix $\boldsymbol{R}$ satisfy the characteristic equation $\boldsymbol{R}\boldsymbol{u} = \lambda\boldsymbol{u}$. If we multiply this equation by matrix $\boldsymbol{X}$ from the left, we obtain

$$\lambda\boldsymbol{X}\boldsymbol{u} = \boldsymbol{X}\underbrace{\boldsymbol{R}}_{(1/n)\boldsymbol{X}^{\mathrm{T}}\boldsymbol{X}}\boldsymbol{u} = \underbrace{\frac{1}{n}\boldsymbol{X}\boldsymbol{X}^{\mathrm{T}}}_{\boldsymbol{C}}\boldsymbol{X}\boldsymbol{u} = \boldsymbol{C}\boldsymbol{X}\boldsymbol{u}, \tag{12.17}$$

which indicates that λ is also the eigenvalue of $\boldsymbol{C}$, with the corresponding eigenvector being $\boldsymbol{X}\boldsymbol{u}$. Therefore, to avoid the high complexity of spectral decomposition of $\boldsymbol{C}$, we perform spectral decomposition of $\boldsymbol{R}$ instead and transform eigenvectors $\boldsymbol{u}$ by multiplying them with matrix $\boldsymbol{X}$ from the left, and these new eigenvectors represent the principal directions.

A more general solution for determining the principal components can be obtained by *singular value decomposition (SVD)* of the **A**-matrix. In SVD, we decompose the $m \times n$ matrix **A** as follows:

$$\boldsymbol{A} = \boldsymbol{U}\boldsymbol{\Sigma}\boldsymbol{V}^{\dagger}, \tag{12.18}$$

where the $m \times n$ diagonal $\sum$-matrix corresponds to a scaling (stretching) operation and is given by $\sum = \mathrm{diag}(\sigma_1, \sigma_2, \ldots)$, with $\sigma_1 \geq \sigma_2 \geq \ldots \geq 0$ being the *singular values* of **A**. The singular values are in fact the square roots of the eigenvalues of $\boldsymbol{AA}^{\dagger}$. The $m \times m$ matrix **U** and $n \times n$ matrix **V** are unitary matrices that describe the rotation operations, as illustrated in Fig. 12.11. The left singular vectors, the columns of unitary matrix **U**, are obtained as the eigenvectors of the Wishart matrix $\boldsymbol{AA}^{\dagger}$. The other rotation matrix **V** has the eigenvectors of $\boldsymbol{A}^{\dagger}\boldsymbol{A}$ for columns and thus represents the right singular vectors.

PCA by SVD can be summarized as follows:

- Organize the dataset into a real-valued matrix **X** of size $m \times n$ in which the mean per measurement type (row) has been subtracted.
- Scale this matrix by $\sqrt{n}$ as $\boldsymbol{X}/\sqrt{n} = \boldsymbol{Y}$. Clearly, $\boldsymbol{YY}^{\mathrm{T}}$ correspond to the covariance matrix, since $\boldsymbol{YY}^{\mathrm{T}} = (1/n)\boldsymbol{XX}^{\mathrm{T}} = \boldsymbol{C}$.
- Apply the SVD procedure to $\boldsymbol{Y} = \boldsymbol{X}/\sqrt{n}$ to obtain $\boldsymbol{Y} = \boldsymbol{U}\boldsymbol{\Sigma}\boldsymbol{V}^{\mathrm{T}}$.
- The columns in **U**, denoted $\boldsymbol{u}_i$ $(i = 1,2,\ldots,m)$, are eigenvectors of the covariance matrix **C** and thus represent the principal components.
- Determine the *scoring matrix* **S** as follows:

$$\boldsymbol{S} = \boldsymbol{U}^{\mathrm{T}}\boldsymbol{X}. \tag{12.19}$$

The j-th column of the scoring matrix is clearly

$$S_j = \left(\boldsymbol{U}^{\mathrm{T}}\boldsymbol{X}\right)_j = \begin{bmatrix} \boldsymbol{u}_1^{\mathrm{T}} \cdot \boldsymbol{X}_j \\ \boldsymbol{u}_2^{\mathrm{T}} \cdot \boldsymbol{X}_j \\ \vdots \\ \boldsymbol{u}_m^{\mathrm{T}} \cdot \boldsymbol{X}_j \end{bmatrix}, \tag{12.20}$$

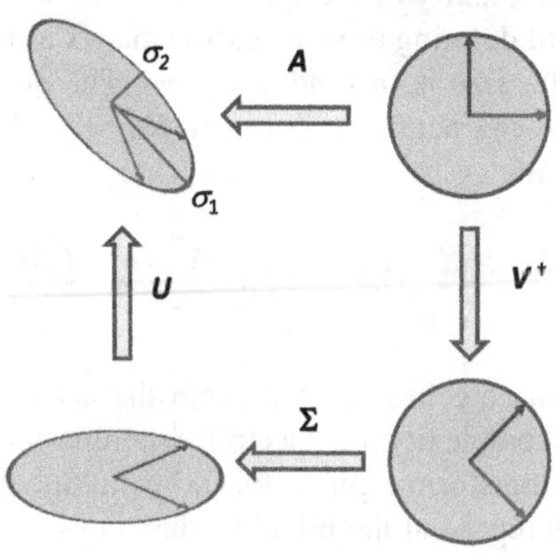

FIGURE 12.11

Singular value decomposition procedure.

with components representing projections along the principal components and X_j denoting the j-th column of matrix X. The rows in S corresponding to small eigenvalues are discarded, and system dimensionality is reduced to $D < m$.

These steps complete the *training phase*. In the *scoring phase*, for each new data vector $y' = [y'_1\ y'_2\ \ldots\ y'_m]^T$, we subtract the mean value $\mu = (1/n)\Sigma_j y'_j$ to obtain $y = [y_1\ y_2\ \ldots\ y_m]^T = [y'_1 - \mu\ \ y'_2 - \mu\ \ldots\ y'_m - \mu]^T$. The scoring phase has the following steps:

- Calculate the weight (scoring) vector for the y vector by

$$S(y) = \begin{bmatrix} u_1^T \cdot y \\ u_2^T \cdot y \\ \vdots \\ u_D^T \cdot y \end{bmatrix}. \tag{12.21}$$

- Calculate the Euclidean distances of this weight vector from all columns of the scoring matrix; that is,

$$d_j = d(S(y), S_j); \ \ j = 1, 2, \cdots, n. \tag{12.22}$$

- The score is defined as

$$\text{score}(y) = \min_j \ d_j \geq 0. \tag{12.23}$$

When we score a vector from the training dataset, we obtain the minimum possible score.

PCA has applications in image compression, data visualization, facial recognition, malware detection, image spam detection, and other areas.

12.1.3 Support vector machines

MLAs based on SVMs, which were proposed by Vapnik Lerner, and Chervonenkis in 1963–64 [11,12], belong to the class of supervised learning algorithms. The key idea behind an SVM is to construct a so-called *maximum-margin hyperplane* such that the margin separation between two classes in a binary classification problem is maximized, as illustrated in Fig. 12.12 for the 2-D case. Let us assume we are given a dataset of n points as follows:

$$\{(x_1, y_1), (x_2, y_2), \cdots, (x_n, y_n)\}; \ \ y_j \in \{-1, +1\}, \ \ x_j = \begin{bmatrix} x_{j1} & x_{j2} & \cdots & x_{jd} \end{bmatrix}^T, \tag{12.24}$$

where y_j are labels corresponding to d-dimensional real-valued data points x_j. Our goal is to derive the hyperplane that can separate the data points with the label $+1$ from the data points with the label -1. The equation of the hyperplane in d-dimensional hyperspace $\Re^d$ is given by

$$w^T \cdot x - b = 0, \ \ w = \begin{bmatrix} w_1 & w_2 & \cdots & w_d \end{bmatrix}^T, \tag{12.25}$$

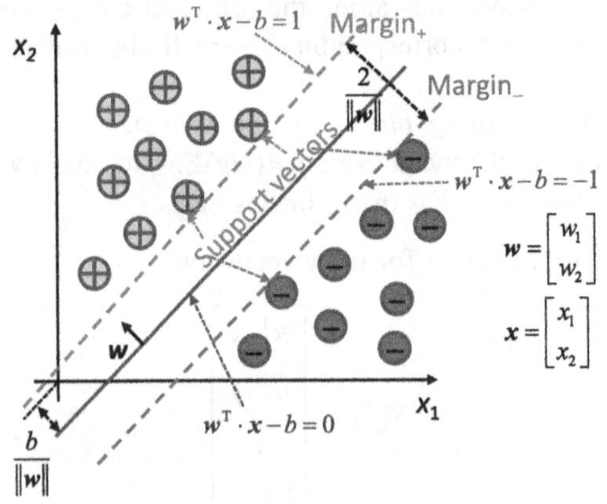

FIGURE 12.12

Defining the optimum margin classifier problem of maximizing the separation between two classes. The samples on the margins are known as support vectors.

where w is the weight vector (not necessarily normalized) orthogonal to the hyperplane, and b is the bias describing the offset from the origin. As shown in Fig. 12.12, the offset from the origin is given by $b/\|w\|$, where $\|w\|$ is the length of the normal vector w. When the training dataset is linearly separable, we can select two parallel hyperplanes that completely separate two classes of data points so that the distance is maximized, as illustrated in Fig. 12.12. The region between these two hyperplanes is known as the *margin*, and the hyperplane intersecting the margin region exactly in the middle is commonly referred to as the maximum-margin hyperplane. Parallel hyperplanes are chosen in such a way that no point stands between them, and we maximize the separation between them. These hyperplanes are described by the following equations in the hyperspace:

$$\begin{aligned} w^T \cdot x - b &= +1 \ (\text{margin}_+) \\ w^T \cdot x - b &= -1 \ (\text{margin}_-) \end{aligned} \tag{12.26}$$

All the data points lying on or above the boundary hyperplane denoted margin $_+$ belong to the class with the label $y = +1$. On the other hand, all the data points lying on or below the boundary hyperplane denoted margin $_-$ belong to the class with the label $y = -1$. The data points lying on either boundary hyperplane are known as *support vectors*. Therefore, the linear SVM is the hyperplane that separates data points with positive labels form data points with negative labels with maximum margin. The separation between boundary hyperplanes is $2/\|w\|$, as illustrated in Fig. 12.12, and by minimizing the length of the normal vector $\|w\|$, we simultaneously maximize the separation between boundary hyperplanes $2/\|w\|$. (If we set $b = 0$ and align the maximum-margin hyperplane with the x_1-axis, we conclude that the distance between the boundary hyperplanes is 2 when the normal vector is of unit length.) For the j-th training data point x_j, the output of the linear SVM is determined by

$$y_j = \text{sign}\left(w^T \cdot x_j - b\right), \quad \text{sign}(z) = \begin{cases} +1, & z > 0 \\ -1, & z < 0 \end{cases} \tag{12.27}$$

To prevent data points from falling within the margin region, we impose the following two constraints:

$$\begin{aligned} w^{T} \cdot x_j - b &\geq +1, \quad \text{for } y_j = +1 \\ w^{T} \cdot x_j - b &\leq -1, \quad \text{for } y_j = -1 \end{aligned} \tag{12.28}$$

The two constraints can be combined as follows:

$$y_j \left(w^{T} \cdot x_j - b \right) \geq 1, \quad \text{for } j = 1, 2, \cdots, n \tag{12.29}$$

The *optimization problem* can now be formulated as

$$\left(\hat{w}, \hat{b} \right) = \underset{w,b}{\arg\min} \|w\|^2, \quad \text{subject to:} \quad y_j \left(w^{T} \cdot x_j - b \right) \geq 1, \quad \forall\ j = 1, 2, \cdots, n \tag{12.30}$$

This optimization problem has linear complexity in time. When unknown data point x arrives, it is classified by the *linear SVM classifier* as follows:

$$x \rightarrow y = \text{sign}\left(\hat{w}^{T} \cdot x - \hat{b} \right). \tag{12.31}$$

We can also formulate the optimization problem by employing the *Lagrangian method*, also known as the *dual* problem, as follows:

$$\left(\hat{w}, \hat{b} \right) = \underset{w,b}{\arg\min}\ \underset{\lambda_j}{\max} \overbrace{\left\{ \frac{\|w\|^2}{2} - \sum_{j=1}^{n} \lambda_j \left[y_j \left(w^{T} \cdot x_j - b \right) - 1 \right] \right\}}^{L(w,b,\boldsymbol{\lambda})} = \underset{w,b}{\arg\min}\ \underset{\lambda_j}{\max} L(w, b, \boldsymbol{\lambda}), \tag{12.32}$$

$$\text{subject to:} \quad y_j \left(w^{T} \cdot x_j - b \right) \geq 1, \quad \forall\ j = 1, 2, \cdots, n$$

where λ_j $(j = 1,\ldots,n)$ are Lagrangian multipliers and are set to zero for all nonsupport vectors. By taking the derivatives of $L(w,b,\boldsymbol{\lambda})$ with respect to w_i and setting them to zero, we obtain the following set of equations:

$$\frac{\partial L(w, b, \boldsymbol{\lambda})}{\partial w_i} = w_i - \sum_{j=1}^{n} \lambda_j y_j x_{j,i} = 0. \tag{12.33}$$

By solving for w_i, we obtain $w_i = \sum_{j=1}^{n} \lambda_j y_j x_{j,i}$ $(i = 1, \cdots, d)$, which can be written in compact vector form as

$$w = \sum_{j=1}^{n} \lambda_j y_j x_j. \tag{12.34}$$

On the other hand, by taking the derivatives of $L(w,b,\boldsymbol{\lambda})$ with respect to b and setting them to zero, we obtain the following constraint:

$$\frac{\partial L(w, b, \boldsymbol{\lambda})}{\partial b} = \sum_{j=1}^{n} \lambda_j y_j = 0. \tag{12.35}$$

Now by taking the derivative of L with respect to λ_j and setting it to zero, we obtain

$$\frac{\partial L(\boldsymbol{w}, b, \boldsymbol{\lambda})}{\partial \lambda_j} = 1 - y_j\left(\boldsymbol{w}^{\mathrm{T}} \cdot \boldsymbol{x}_j - b\right) = 0, \tag{12.36}$$

from which we obtain

$$\boldsymbol{w}^{\mathrm{T}} \cdot \boldsymbol{x}_j - b = 1/y_j, \tag{12.37}$$

By substituting Eqs. (12.34) and (12.37) into Eq. (12.32), the optimization problems simplify to

$$\begin{aligned} \widehat{\boldsymbol{\lambda}} &= \max_{\lambda_1, \cdots, \lambda_n} \left(\sum_{j=1}^{n} \lambda_j - \frac{1}{2} \sum_{j=1}^{n} \sum_{k=1}^{n} \lambda_j \lambda_k y_j y_k \boldsymbol{x}_j^{\mathrm{T}} \cdot \boldsymbol{x}_k \right) \\ &\text{subject to: } \sum_{j=1}^{n} \lambda_j y_j = 0, \ \lambda_j \geq 0 \end{aligned} \tag{12.38}$$

Typically, only a few λ_j's are nonzero. The optimum $\boldsymbol{w}$ is now determined after substituting $\widehat{\boldsymbol{\lambda}}$ into Eq. (12.34) to obtain

$$\widehat{\boldsymbol{w}} = \sum_{j=1}^{n} \widehat{\lambda}_j y_j \boldsymbol{x}_j. \tag{12.39}$$

Offset b is finally determined after substituting Eq. (12.39) into Eq. 12.37) to obtain

$$\widehat{b} = \widehat{\boldsymbol{w}}^{\mathrm{T}} \cdot \boldsymbol{x}_j - \underbrace{1/y_j}_{y_j} = \widehat{\boldsymbol{w}}^{\mathrm{T}} \cdot \boldsymbol{x}_j - y_j. \tag{12.40}$$

To obtain a better estimate of offset b, we should average out over all support vectors by

$$\widehat{b} = \frac{1}{\left|\{j|\lambda_j \neq 0\}\right|} \sum_{j:\ \lambda_j \neq 0} \left(\widehat{\boldsymbol{w}}^{\mathrm{T}} \cdot \boldsymbol{x}_j - y_j\right), \tag{12.41}$$

where the averaging is performed over all j's for which $\lambda_j \neq 0$, while $|\{\ \dots\ \}|$ denotes the cardinality of the corresponding set.

When unknown data point $\boldsymbol{x}$ arrives, it is classified by the *dual SVM classifier* as follows:

$$\boldsymbol{x} \rightarrow y = \mathrm{sign}\left(\widehat{\boldsymbol{w}}^{\mathrm{T}} \cdot \boldsymbol{x} - \widehat{b}\right). \tag{12.42}$$

12.1.3.1 Soft margin

In certain classification problems, it might happen that several instances of a given class can be located with the points from the other class. To deal with such problems, we can introduce nonnegative *slack variables*, which specify how much a particular point deviates from the margin region. After introducing the slack variables $\zeta_j \geq 0$, we can redefine the constraints as

$$y_j\left(\boldsymbol{w}^{\mathrm{T}} \cdot \boldsymbol{x}_j - b\right) \geq 1 - \zeta_j, \ \text{ for } j = 1, 2, \cdots, n \tag{12.43}$$

which essentially enlarges the margin, commonly referred to as the *soft margin.* The primal optimization problem can now be formulated as

$$(\widehat{w},\widehat{b}) = \arg\min_{w,b}\left(\frac{1}{2}\|w\|^2 + C\sum_j \zeta_j\right), \quad \text{subject to: } y_j(w^T \cdot x_j - b) \geq 1 - \zeta_j, \ \forall\ j = 1,2,\cdots,n \tag{12.44}$$

Using the Lagrangian approach, we can formulate the optimization problem by

$$(\widehat{w},\widehat{b}) = \arg\min_{w,b}\max_{\lambda_j}\left\{\frac{\|w\|^2}{2} + C\sum_{j=1}^{n}\zeta_j - \sum_{j=1}^{n}\lambda_j\left[y_j(w^T\cdot x_j - b) - 1 + \zeta_j\right] - \sum_{j=1}^{n}\eta_j\zeta_j\right\}, \tag{12.45}$$

$$\text{subject to: } y_j(w^T\cdot x_j - b) \geq 1 - \zeta_j, \ \forall\ j = 1,2,\cdots,n$$

By repeating the procedure above with the equation obtained from the derivative with respect to multiplier η_j, that is, $\partial L/\partial\zeta_j = C - \lambda_j - \eta_j = 0$, we conclude that $\lambda_i \leq C$, and the corresponding dual optimization problem becomes

$$\widehat{\lambda} = \max_{\lambda_1,\cdots,\lambda_n}\left(\sum_{j=1}^{n}\lambda_j - \frac{1}{2}\sum_{j=1}^{n}\sum_{k=1}^{n}\lambda_j\lambda_k y_j y_k x_j^T \cdot x_k^T\right), \quad \text{subject to: } \sum_{j=1}^{n}\lambda_j y_j = 0, \ 0 \leq \lambda_j \leq C. \tag{12.46}$$

12.1.3.2 The kernel method

When the decision boundaries are nonlinear, we can apply the *kernel trick (method)* to the maximum-margin hyperplane. The key idea is to use a similar algorithm in which the dot product is replaced by a *nonlinear kernel function.* This approach allows fitting of the maximum-margin hyperplane in transformed feature space. Very often, the transformation is nonlinear and the corresponding transformed space is HD. In a nonlinear SVM, we are interested in determining the nonlinear classification rule that corresponds to a linear classification in the transformed space, while the corresponding nonlinear mapping can be represented by $x_j \rightarrow \varphi(x_j)$. The corresponding kernel function k maps the dot product of x_j and x_k as follows: $k(x_j, x_k) = \varphi(x_j)^T \cdot \varphi(x_k)$. Based on the above, the classification vector can be represented by

$$w = \sum_{j=1}^{n}\lambda_j y_j \varphi(x_j), \tag{12.47}$$

wherein the λ_j multipliers are obtained by the following optimization problem:

$$\widehat{\lambda} = \max_{\lambda_1,\cdots,\lambda_n}\left(\sum_{j=1}^{n}\lambda_j - \frac{1}{2}\sum_{j=1}^{n}\sum_{k=1}^{n}\lambda_j\lambda_k y_j y_k \underbrace{\varphi(x_j^T)\cdot\varphi(x_k^T)}_{k(x_j,x_k)}\right) = \max_{\lambda_1,\cdots,\lambda_n}\left(\sum_{j=1}^{n}\lambda_j - \frac{1}{2}\sum_{j=1}^{n}\sum_{k=1}^{n}\lambda_j\lambda_k y_j y_k k(x_j,x_k)\right),$$

$$\text{subject to: } \sum_{j=1}^{n}\lambda_j y_j = 0, \ 0 \leq \lambda_j \leq C. \tag{12.48}$$

The corresponding algorithm is known as the *coordinate descent* and is suitable for dealing with large and sparse datasets.

The offset b for the nonlinear classification can be determined using an equation similar to

$$\widehat{b} = \widehat{w}^{\mathrm{T}} \cdot \varphi(x_k) - y_k = \sum_{j=1}^{n} \lambda_j y_j \underbrace{\varphi(x_j)^T \cdot \varphi(x_j)}_{k(x_j, x_k)} - y_k = \sum_{j=1}^{n} \lambda_j y_j k(x_j, x_k) - y_k. \tag{12.49}$$

When it arrives, the unknown data point x is classified by the *dual nonlinear SVM classifier* as follows:

$$x \to y = \mathrm{sign}\left(\widehat{w}^{\mathrm{T}} \cdot \varphi(x) - \widehat{b}\right) = \mathrm{sign}\left(\sum_{j=1}^{n} \lambda_j y_j k(x_j, x) - \widehat{b}\right). \tag{12.50}$$

Popular *nonlinear kernels* include

- *Polynomial*: $k(x_j, x) = \left(x_j^{\mathrm{T}} \cdot x + c\right)^p$, $c \in \{0, 1\}$, where $c = 0$ corresponds to the homogenous case, and $c = 1$ applies to the inhomogeneous case.
- *Gaussian radial basis function*: $k(x_j, x) = \exp\left(-\gamma \|x_j, x\|^2\right)$, $\gamma > 0$. Sometimes the parameter γ is selected by $1/(2\sigma^2)$.
- *Hyperbolic tangent*: $k(x_j, x) = \tanh(c_1 x_j \cdot x + c_2)$, for some (but not every) $c_1 > 0$ and $c_2 < 0$.

12.1.4 Clustering

Clustering refers to grouping objects in a particularly meaningful way. We typically group points in a dataset that are in close proximity within the feature space. Numerous clustering algorithms exist and are classified as extrinsic or intrinsic, hierarchical or partitional, agglomerative or divisive, or hard or soft. Extrinsic clustering depends on category labels and requires data preprocessing. Intrinsic clustering, on the other hand, does not require training labels on objects. Hierarchical clustering has a parent–child structure represented with dendrograms. In partition clustering, we partition data points into disjoint clusters with no hierarchical relationship among them. In agglomerative clustering, we apply the bottom-up approach, interpreting each object as a cluster and then merging the existing clusters. In divisive clustering, we apply the top-down approach in which all objects belong to the same initial cluster, which we split into different clusters according to certain criteria. In hard clustering, clusters do not overlap; in other words, a particular object from the dataset belongs to only one cluster. In soft clustering, on the other hand, the clusters may overlap, and we associate each data point with the probabilities of being within different clusters. In this section, we describe K-means and expectation maximization (EM), belonging to the class of unsupervised learning algorithms, and K-NN algorithm, belonging to the class of supervised learning algorithms.

12.1.4.1 K-means clustering

In the *K-means clustering* algorithm, which is a hard-clustering algorithm, we partition the dataset points into K clusters based on their pairwise distances. We typically use the Euclidean distance, defined by Eq. (12.2); that is, for two data points $\boldsymbol{x}_i=(x_{i1} \ldots x_{id})$ and $\boldsymbol{x}_j=(x_{j1} \ldots x_{jd})$, the Euclidean distance is defined by $d(\boldsymbol{x}_i,\boldsymbol{x}_j) = \left[\sum_{k=1}^{d}(\boldsymbol{x}_{ik}-\boldsymbol{x}_{jk})^2\right]^{1/2}$. Other distance definitions can be used, including the Manhattan (taxicab) distance, $d_M(\boldsymbol{x}_i,\boldsymbol{x}_j) = \sum_{k=1}^{d}|\boldsymbol{x}_{ik}-\boldsymbol{x}_{jk}|$. The starting point is the dataset of points $\{\boldsymbol{x}_1,\boldsymbol{x}_2,\ldots,\boldsymbol{x}_n\}$ that we want to partition into K clusters. We then portion them into K initial clusters either at random or by applying a certain heuristic. For each cluster k, we specify the *centroid* $\boldsymbol{c}_k$ as follows:

$$\boldsymbol{c}_k = \frac{1}{N_{c_k}}\sum_{j=1}^{N_{c_k}}\boldsymbol{x}_{c_k}^{(j)}, \tag{12.51}$$

where N_{c_k} denotes the number of points in the k-th cluster, and $\boldsymbol{x}_{c_k}^{(j)}$ denotes the points in that cluster. We then perform a new partition by assigning each data point to the closest centroid. The centroids are then recalculated for each new cluster. This procedure is repeated until convergence is achieved, which occurs when the data points no longer switch clusters. The *K-means clustering algorithm* can be summarized as follows (inspired by Ref. [1]):

- *Given*: the number of clusters K and data point set $\{\boldsymbol{x}_1,\boldsymbol{x}_2,\ldots,\boldsymbol{x}_n\}$.
- *Initialization*: Partition the dataset into K initial clusters selected at random or heuristically.
- *Centroid determination step*: For each cluster, determine the centroid as a center of mass using Eq. (12.51).
- *Point reassigning step*: Assign data point $\boldsymbol{x}_i$ ($i = 1,\ldots,n$) to cluster $\boldsymbol{c}_j$ so that $d(\boldsymbol{x}_i,\boldsymbol{c}_j) \leq d(\boldsymbol{x}_i,\boldsymbol{c}_k)$ for all $k \in \{1,\ldots,K\}$.
- Repeat the previous two steps until the stopping criterion is met (the data points no longer switch clusters).

This algorithm belongs to the class of "hill-climbing" algorithms, which converge to a local minimum. To avoid this problem, we could restart the algorithm multiple times with different initial centroids and select one that minimizes overall intracluster *distortion* (or compactness) D, which is defined as

$$D = \sum_{i=1}^{n} d(\boldsymbol{x}_i,\boldsymbol{c}(\boldsymbol{x}_i)), \tag{12.52}$$

where $\boldsymbol{c}(\boldsymbol{x}_i)$ denotes the centroid to which data point $\boldsymbol{x}_i$ belongs.

A key question is how to select the optimum K. The choice of optimum K is a tricky one. If K is too large, it can overfit the dataset. We have to try with various Ks, and for each K, rerun the algorithm for different initial centroids while selecting the best result. (This is not a very scientific approach.) Ideally, we could select the K that minimizes the distortion function (12.52); unfortunately, this problem is NP-complete. Alternatively, we can use variants such as K-medoids (or K-medians) and fuzzy K-means. In K-medoids, the centroids are placed at the actual median point. In fuzzy K-means, the point can belong to multiple clusters.

12.1.4.2 Cluster quality

Another relevant question is how we can measure the *cluster quality*? We can use internal (topological) or external validation for this purpose. In internal validation, we specify cluster quality based on the clusters themselves, such as spacing within or between clusters. In external validation, we determine the quality of clusters based on the knowledge of labels. Popular topological (internal) measurements include [1] cluster correlation, similarity matrix, sum of squares error, cohesion & separation, and silhouette coefficient.

Given the dataset and the clusters, we define and determine two matrices:

- *Distance matrix* $\boldsymbol{D}=(d_{ij})_{n\times n}$, where d_{ij} is the distance between data points $\boldsymbol{x}_i$ and $\boldsymbol{x}_j$, and
- *Adjacency matrix* $\boldsymbol{A}=(a_{ij})_{n\times n}$, where $a_{ij}=1$ when the data points $\boldsymbol{x}_i$ and $\boldsymbol{x}_j$ belong to the same cluster, and $a_{ij}=0$ otherwise.

We calculate the *correlation* between these two matrices as follows:

$$\rho_{AD}=\frac{\sum_{i=1}^{n}\sum_{j=1}^{n}(d_{ij}-\mu_{\boldsymbol{D}})(a_{ij}-\mu_{\boldsymbol{A}})}{\sqrt{\sum_{i=1}^{n}\sum_{j=1}^{n}(d_{ij}-\mu_{\boldsymbol{D}})^2\sum_{i=1}^{n}\sum_{j=1}^{n}(a_{ij}-\mu_{\boldsymbol{A}})^2}}, \tag{12.53}$$

where $\mu_{\boldsymbol{D}}$ ($\mu_{\boldsymbol{A}}$) denotes the mean value of the elements of matrix $\boldsymbol{D}$ ($\boldsymbol{A}$). The large negative correlation indicates that the clusters are well separated.

The *similarity matrix* provides a visual representation of similarity within the clusters and can be obtained from the distance matrix by grouping rows and columns by clusters.

Cluster cohesion tells us how tightly the cluster is packed, while *cluster separation* describes the average distance between clusters. The distortion parameter, defined by Eq. (12.52), measures cohesion. Alternatively, we could measure cohesion by the average distance between all pairs of points. We have more options to characterize distance separation, including the average distance among centroids, among all points in the clusters, and from centroids to a midpoint.

The *silhouette coefficient* combines cluster cohesion and separation in a single parameter, as illustrated in Fig. 12.13. Let C_i denote the cluster to which data point $\boldsymbol{x}_i$ belongs. Cluster *cohesion* $c(\boldsymbol{x}_i)$ is defined as the average (intracluster) distance between data point x_i and all other points in the same cluster C_i:

$$c(\boldsymbol{x}_i)=\frac{1}{|\{\boldsymbol{y}|\boldsymbol{y}\in C_i,\boldsymbol{y}\neq\boldsymbol{x}_i\}|}\sum_{\substack{\boldsymbol{x}_i\in C_i\\ \boldsymbol{y}\neq\boldsymbol{x}_i}}^{|C_i|}d(\boldsymbol{x}_i,\boldsymbol{y}). \tag{12.54}$$

The cluster separation d_j represents the average intercluster distance between data point $\boldsymbol{x}_i$ and all points in cluster C_j that differ from C_i; that is,

$$d_j(\boldsymbol{x}_i)=\frac{1}{|\{\boldsymbol{y}|\boldsymbol{y}\in C_j\}|}\sum_{\boldsymbol{y}\in C_j}^{|C_j|}d(\boldsymbol{x}_i,\boldsymbol{y}). \tag{12.55}$$

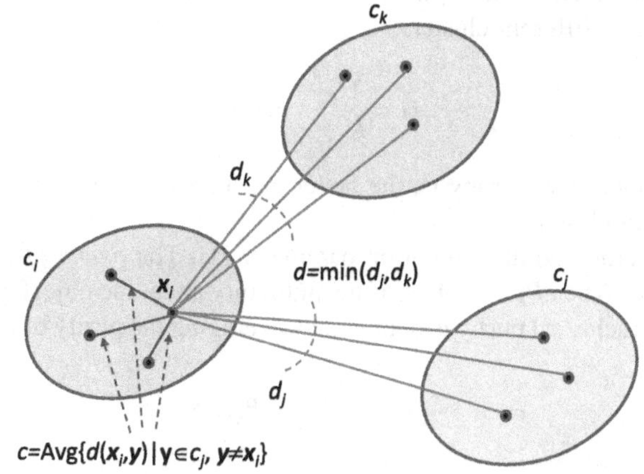

FIGURE 12.13

Defining the silhouette coefficient.

Let the minimum of all d_j's be denoted d, which can be calculated by

$$d(\boldsymbol{x}_i) = \min_{j:\ C_j \neq C_i} d_j(\boldsymbol{x}_i). \tag{12.56}$$

The *silhouette coefficient* for data point $\boldsymbol{x}_i$ is now defined as

$$s(\boldsymbol{x}_i) = \frac{d(\boldsymbol{x}_i) - c(\boldsymbol{x}_i)}{\max(c(\boldsymbol{x}_i), d(\boldsymbol{x}_i))} = 1 - \frac{c(\boldsymbol{x}_i)}{d(\boldsymbol{x}_i)}, \tag{12.57}$$

where we assume that $d > c$; otherwise, the clustering algorithm would be terribly designed. When the silhouette coefficient is close to 1, the points in the cluster are much closer together than the points from any other neighboring cluster. The average *silhouette coefficient* of the whole cluster is defined by

$$s = \frac{1}{n}\sum_{i=1}^{n} s(\boldsymbol{x}_i), \tag{12.58}$$

and this computation can be applied to large datasets.

The popular external measures of clustering quality are entropy and purity. *Entropy* measures uncertainty (randomness)—low intracluster entropy and high intercluster entropy define a good clustering algorithm. External clustering validation requires knowledge of the labels, as mentioned earlier. Let n_{ik} denote the number of labels of type i in cluster C_k and n_k the number of objects in cluster C_k. Then the probability (more precisely, the frequency) of occurrence of object i in cluster C_k is determined by $p_{ik} = n_{ik}/n_k$. The entropy of the k-the cluster is now defined by

$$H_k = -\sum_{i=1}^{L} p_{ik} \log_2 p_{ik}, \tag{12.59}$$

where L is the number of different object types (labels). The intracluster entropy is obtained by averaging the entropies of different clusters:

$$H = \frac{1}{K}\sum_{k=1}^{K} n_k H_k. \tag{12.60}$$

The smaller the intracluster entropy is, the better, because this parameter indicates better uniformity of the labels within clusters.

Purity is another useful external validation parameter [1]. The *purity of cluster k*, denoted P_k, is defined as $P_k = \max_i p_{ik}$. When $P_k = 1$, there is no ambiguity about the object type; all objects have the same label. Overall (intracluster) purity is obtained by averaging the purity of different clusters; that is,

$$P = \frac{1}{K}\sum_{k=1}^{K} n_k P_k. \tag{12.61}$$

12.1.4.3 Expectation-maximization clustering

EM clustering employs the probability distributions in a clustering algorithm instead of the Euclidean distance using K-means and thus represents a soft-clustering algorithm. The Euclidean distance might result in poor clustering results when the clusters are not circularly symmetric, and zero-mean Gaussian distribution does not accurately describe the dataset points. EM clustering comprises (1) an expectation (E) step and (2) a maximization (M) step, summarized as follows:

- *E-step*: Recalculate the probabilities needed in the M-step.
- *M-step*: Recalculate the maximum likelihood estimates for the parameters of the corresponding distributions.

A certain similarity with the K-means can be summarized in two steps: (1) redesign the cluster based on the current centroids, and (2) recompute the new centroids based on current clusters. When the corresponding distributions are circular Gaussian and clustering is hard, EM clustering essentially reduces to the K-means. Namely, the maximum likelihood detector for additive white Gaussian noise channels has the same performance as the Euclidean distance-based detector of Refs. [7,10]. This indicates that K-means is a special case of EM clustering during which a hard-clustering approach is applied. In clustering, we use EM to determine the hidden parameters of the corresponding distributions in K clusters. Each cluster in EM clustering is described by a generative model, such as Gaussian or multivariate. Each cluster corresponds to the probability density function, and the distribution parameters, such as mean values and covariance matrix, must be discovered through EM.

EM clustering provides a dataset $\mathrm{D} = \{\boldsymbol{x}_1, \boldsymbol{x}_2, \ldots, \boldsymbol{x}_n\}$, and we are interested in clustering this dataset in K clusters $\{C_1, \ldots, C_K\}$ that partition D. EM clustering is a soft-clustering algorithm, so we need to associate with the i-th data point the probability of being in the k-th cluster ($k = 1, \ldots, K$), denoted with $p_{k,i}$, that is, $p_{k,i} = P(\boldsymbol{x}_i \in C_k)$; thus, $\sum_k p_{k,i} = 1$. For each cluster k, we associate the density function over R^p, where p is the number of unknown parameters, denoted $f_{\theta_k(\boldsymbol{x})}$. We assume that the density for $(x_i)_{1 \le i \le n}$ can be described by

$$f(\boldsymbol{x}) = \sum_{k=1}^{K} p_k f_{\theta_k}(\boldsymbol{x}), \quad \forall\, \boldsymbol{x} \in R^p, \tag{12.62}$$

where $p_k \geq 0$ ($k = 1,\ldots,K$) denotes the fraction of the k-th distribution from which sample $\boldsymbol{x}$ is drawn, and $\boldsymbol{\theta}_k$ is used to denote the unknown parameters of the distribution corresponding to the k-th cluster. Because the sample must be drawn from one of the K distributions, we find $\sum_k p_k = 1$. Eq. (12.62) describes the mixture model. A very popular model of this type is the *Gaussian mixture model*, for which the parameters are the mean values $\boldsymbol{\mu}_k$ and covariance matrices $\boldsymbol{\Sigma}_k$; that is, we can specify the parameters as $\boldsymbol{\theta}_k=(\boldsymbol{\mu}_k,\boldsymbol{\Sigma}_k)$. The corresponding multivariate Gaussian distribution is given by

$$f_{\boldsymbol{\theta}_k}(\boldsymbol{x}) = \frac{1}{\sqrt{(2\pi)^n \det(\boldsymbol{\Sigma}_k)}} \exp\left[-\frac{1}{2}(\boldsymbol{x}-\boldsymbol{\mu}_k)^T \sum_k^{-1} (\boldsymbol{x}-\boldsymbol{\mu}_k)\right]. \tag{12.63}$$

We can now apply the Bayes rule to determine the degree of membership of point $\boldsymbol{x}$:

$$P(C_k|\boldsymbol{x}) = \frac{P(C_k)P(\boldsymbol{x}|C_k)}{P(\boldsymbol{x})} = \frac{p_k f_{\boldsymbol{\theta}_k}(\boldsymbol{x})}{\sum_{k'=1}^{K} p_{k'} f_{\boldsymbol{\theta}_{k'}}(\boldsymbol{x})}. \tag{12.64}$$

The partition of the dataset is now performed by

$$C_k = \left\{ x \middle| p_k f_{\boldsymbol{\theta}_k}(\boldsymbol{x}) > \max_{k' \neq k} p_{k'} f_{\boldsymbol{\theta}_{k'}}(\boldsymbol{x}) \right\}; \quad k = 1, 2, \cdots, K. \tag{12.65}$$

Let us define the *likelihood function*:

$$\Lambda(\boldsymbol{\theta}|\mathfrak{D}) = P_{\boldsymbol{\theta}}(\mathfrak{D}) = \prod_{n=1}^{n} P_{\boldsymbol{\theta}}(\boldsymbol{x}_i). \tag{12.66}$$

We can formulate the *maximum likelihood estimation* (MLE) as the optimization problem:

$$\widehat{\boldsymbol{\theta}} = \arg\max_{\boldsymbol{\theta}} \Lambda(\boldsymbol{\theta}|\mathfrak{D}). \tag{12.67}$$

To avoid the resolution problem when multiplying many probabilities, we use the log-likelihood function instead:

$$L(\boldsymbol{\theta}|\mathfrak{D}) = \log \Lambda(\boldsymbol{\theta}|\mathfrak{D}) = \log \prod_{i=1}^{n} P_{\boldsymbol{\theta}}(\boldsymbol{x}_i) = \sum_{i=1}^{n} \underbrace{\log P_{\boldsymbol{\theta}}(\boldsymbol{x}_i)}_{L_{\boldsymbol{\theta}}(\boldsymbol{x}_i)} = \sum_{i=1}^{n} L_{\boldsymbol{\theta}}(\boldsymbol{x}_i). \tag{12.68}$$

The MLE can now be reformulated as

$$\widehat{\boldsymbol{\theta}} = \arg\max_{\boldsymbol{\theta}} L(\boldsymbol{\theta}|\mathfrak{D}). \tag{12.69}$$

For the mixture model given by Eq. (12.62), the MLE problem can be represented as

$$\left(\widehat{\boldsymbol{p}}, \widehat{\boldsymbol{\theta}}\right) = \arg\max_{(\boldsymbol{p},\boldsymbol{\theta})} L((\boldsymbol{p},\boldsymbol{\theta})|\mathfrak{D}), \quad L((\boldsymbol{p},\boldsymbol{\theta})|\mathfrak{D}) = \sum_{i=1}^{n} \log\left(\sum_{k=1}^{K} p_k f_{\boldsymbol{\theta}_k}(\boldsymbol{x})\right), \quad \boldsymbol{p} = (p_1, \ldots, p_K). \tag{12.70}$$

This complicated optimization problem can be solved by the gradient ascent in $(\boldsymbol{p},\boldsymbol{\theta})$ provided that the log-likelihood function is convex, which is not necessarily true. An alternative solution is the EM algorithm [13].

By skipping some derivations, we can formulate the *log-domain EM algorithm* as follows:

- *Given*: The dataset D = $\{x_1, x_2, \ldots, x_n\}$, and the parametric model is $f_{\boldsymbol{\theta}}$.
- *Initialization*: Select an initial distribution $\boldsymbol{p}^{(0)}$ and parameter vector $\boldsymbol{\theta}^{(0)}$ randomly in the corresponding state space.
- *E-step*: Recalculate the probabilities that i-th point ($i = 1,\ldots,n$) belongs to k-th cluster ($k = 1,\ldots,K$):

$$p_{k,i}^{(t)} = \frac{p_k^{(t-1)} f_{\boldsymbol{\theta}_k^{(t-1)}}(\boldsymbol{x}_i)}{\sum_{k'=1}^{K} p_{k'}^{(t-1)} f_{\boldsymbol{\theta}_{k'}^{(t-1)}}(\boldsymbol{x}_i)},$$

where (t) denotes the index of the current iteration.

- *M-step*: Recalculate the parameters $p^{(t)}$ and $\boldsymbol{\theta}^{(t)}$ by the following maximizations:

$$\widehat{\boldsymbol{p}} = \arg \max_{p:\ \sum p_k = 1} \sum_{k=1}^{K} \log(p_k) \sum_{i=1}^{n} p_{k,i}^{(t)} \tag{12.71}$$

$$\widehat{\boldsymbol{\theta}}_k^{(t)} = \arg \max_{\boldsymbol{\theta}} \sum_{i=1}^{n} p_{k,i}^{(t)} \log f_{\boldsymbol{\theta}_k}(\boldsymbol{x}_i); \ \ \forall \ k = 1, \cdots, K \tag{12.72}$$

- Repeat the E and M steps until convergence is achieved.

Eq. (12.71) can be simplified by applying the Lagrangian method,

$$\frac{\partial}{\partial p_k}\left[\sum_{k=1}^{K} \log(p_k) \sum_{i=1}^{n} p_{k,i} + \lambda\left(1 - \sum_{k=1}^{n} p_k\right)\right] = 0, \tag{12.73}$$

to obtain

$$p_k = \frac{1}{n} \sum_{i=1}^{n} p_{k,i}. \tag{12.74}$$

For a Gaussian mixture of unknown means and covariance matrix, Eq. (12.72) in the M-step simplifies to

$$\boldsymbol{\mu}_k^{(t)} = \sum_{i=1}^{n} p_{k,i}^{(t)} \boldsymbol{x}_i, \ \ \Sigma_k^{(t)} = \sum_{i=1}^{n} p_{k,i}^{(t)} \left(\boldsymbol{x}_i - \boldsymbol{\mu}_k^{(t)}\right)^{\mathrm{T}} \left(\boldsymbol{x}_i - \boldsymbol{\mu}_k^{(t)}\right). \tag{12.75}$$

12.1.4.4 K-nearest neighbors algorithm

K-NN represents the simplest learning algorithm—it does not require any calculating in training, and all calculations are postponed for the scoring phase. Existing labeled data are used for training. Once the new, unlabeled object is presented to the learner, it searches for the K closest neighbors and labels

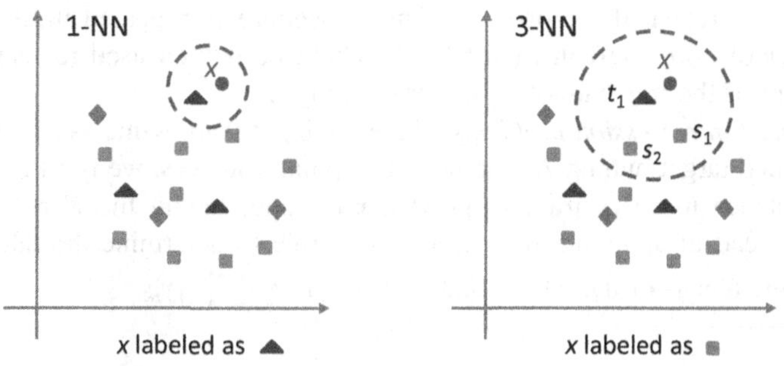

FIGURE 12.14

K-nearest neighbors algorithm for (left) *K* = 1 and (right) *K* = 3. *NN*, nearest neighbors.

the new incoming object by applying the majority rule. As an illustration, in Fig. 12.14, the object denoted in the "circle" (in red color [dark gray in print version]) is presented to the learner. For $K = 1$ (on the left), the learner searches for the closest neighbor and labels the unknown object as "triangle" (in blue color [black in print version]). On the other hand, for $K = 3$ (on the right), the learner searches for the three nearest neighbors and applies the majority rule, thus classifying the unknown object as the "square" (in orange color [light gray in print version]). Given that no training is needed, just tracking of labeled data, this algorithm is often called the *lazy algorithm*.

We can also associate weights for the contributions of different neighbors to account for the importance of their closeness. Thus, it makes sense to use the reciprocal of the distance for this weight. Let us again observe the 3-NN example in Fig. 12.14, taking the distance weight into account. Given that $1/d(x,t_1) > 1/d(x,s_1) + 1/d(x,s_2)$, the unknown object is classified as the triangle (blue [black in print version]). Alternatively, we can use the frequency weight, with the number of squares being $S = 9$, while the number of triangles is $T = 3$. We can weight each square as 1, while the triangle is $R/T = 3$. Going back to the 3-NN example, the "square score" is 2, the triangle score is 3, and the unknown object is classified as the triangle.

Given its simplicity, K-NN can be combined with other MLAs. For instance, PCA can be used to reduce dimensionality first, and then we can apply K-NN for classification in lower-dimensional space.

K-NN has some limitations and is not effective for data with skewed classes. K-NN is also sensitive to all features, including those that may not be relevant. Finally, finding the *K*-closest neighbors can be computationally complex when the Euclidean distance is used. For $K = 1$, this problem is known as the nearest post-office problem, where for a given address, we need to identify the closest post office. For *d*-dimensional data, the complexity of finding the *K* nearest neighbors is $O(nd)$. We can reduce complexity by reducing dimensionality or not comparing all training examples. By employing the K-D-tree construction and searching (described below), the number of operations can be reduced to $O(d \log_2 n)$. Unfortunately, the K-D tree can miss some neighbors and only applies when $d << n$. For HD discrete and sparse datasets, we should use inverted lists, wherein the complexity is in order $O(n'd')$, $n' << n$, $d' << d$. Finally, for HD real-valued or discrete problems that are not sparse, we should use locality-sensitive hashing whose complexity is $O(n'd)$, $n' << n$. In the K-D tree, we select an attribute at random and then find the median (for that dimension) among all instances in the dataset and split it into two equal parts. We select a different feature at random in the next iteration and repeat the procedure. This procedure is repeated until we reach the predetermined number of points left in each branch. This tree is then used to search for the nearest neighbors. The dept of the tree cannot be larger than $\log_2 n$.

K-NN can be used in *regression problems* where the input is the same as a classification problem but with a real-valued target output. As the new data point $\boldsymbol{x}$ arrives, we run the same algorithm by calculating the distance to every training point $d(\boldsymbol{x},\boldsymbol{x}_i)$, and finding the K nearest neighbors, say $\boldsymbol{x}_{i1},\ldots,\boldsymbol{x}_{iK}$. Now, instead of applying the majority vote rule to determine the label of the incoming object, we output the means of the labels, that is, $\hat{y} = (1/K)\sum_{k=1}^{k} y_{ik}$.

12.1.5 Boosting

Different learning algorithms have their weaknesses and strengths. Ensemble learning combines multiple models into one model with better generalization performance. The constituent models are known as *weak* (base) *learners*, while the composite model is commonly referred to as a *strong learner*. The generic procedure in ensemble learning can be described as a two-step procedure. In the first step, we design K weak learners. In the second step, we combine weak learners to obtain a composite, strong learner. Namely, we train K weak classifiers using (the same as before) dataset $\{(\boldsymbol{x}_1,y_1),\ldots,(\boldsymbol{x}_n,y_n)\}$. The subsets used to train weak classifiers may overlap. The weak classifiers should be stronger (they should have better accuracy) than random guessing. Popular ensemble learning methods include bagging, random forests, and boosting. In bagging, which is short for bootstrap aggregating, weak learners simply vote, and decisions are made by majority rule. To improve performance, the training of weak learners is performed by random training subsets.

The boosting represents an ensemble learning method that builds a stronger classifier from weak classifiers complementing each other. To achieve this goal, weak classifiers are trained iteratively and sequentially, so the next weak classifier learns what the previous weak classifier could not do. Thus, the next iteration weak classifier concentrates on misclassified data instances of the previous classifiers. The strong classifier is created by properly selecting and weighting the weak classifiers. Boosting algorithms are different in how they weight classifiers and training data objects.

AdaBoost, short for adaptive boosting, does not need knowledge about the accuracy of previous weak learners but adapts to the accuracies at hand. At each iteration, AdaBoost identifies the key weakness in the composite classifier built in the previous iteration and selects one unused classifier that can fix that weakness. Once selected, it is properly weighted before incorporating it into the composite classifier. Therefore, AdaBoost is an iterative algorithm in which we generate a series of composite classifiers C_t $(t = 1,2,\ldots,T)$, using the K weak classifiers c_k $(k = 1,\ldots,K)$. For iteration index t the composite classifier is created by

$$C_t(\boldsymbol{x}_i) = C_{t-1}(\boldsymbol{x}_i) + \alpha_t k_t(\boldsymbol{x}_i), \quad \alpha_t \geq 0, \tag{12.76}$$

where k_t is a weak classifier selected from the list of unused weak classifiers and α_t denotes the corresponding weight. To determine the "exact role" of the k_t classifier, that is, the corresponding weight, we use the *exponential loss function* [1]:

$$E_t = \sum_{i=1}^{n} \exp\{-y_i[C_{t-1}(\boldsymbol{x}_i) + \alpha_t k_t(\boldsymbol{x}_i)]\}, \tag{12.77}$$

where y_i is the label that corresponds to $\boldsymbol{x}_i$. Any error in the composite classifier is penalized by increasing the loss function significantly. To select the weak classifier to use in iteration t, we need to look at the terms in the equation above, increasing the loss function and selecting the k_t that minimizes the loss function. To simplify the selection of the weak classifier, we can split the exponential loss function into two terms:

$$E_t = e^{-\alpha_t} \underbrace{\sum_{i:y_i=k_t(\boldsymbol{x}_i)} \exp[-y_i C_{t-1}(\boldsymbol{x}_i)]}_{W_1^{(t)}} + e^{\alpha_t} \underbrace{\sum_{i:y_i \neq k_t(\boldsymbol{x}_i)} \exp[-y_i C_{t-1}(\boldsymbol{x}_i)]}_{W_2^{(t)}} \tag{12.78}$$
$$= e^{-\alpha_t} W_1^{(t)} + e^{\alpha_t} W_2^{(t)}.$$

Given that the second term is related to the number of misses from the unused weak classifiers, we select one that minimizes the W_2-term and denote it k_t. Once the weak classifier k_t is selected, W_1 and W_2 are fixed, and the optimum weight can be determined by determining the derivative $dE_t/d\alpha_t$ as follows:

$$\frac{dE_t}{d\alpha_t} = -e^{-\alpha_t} W_1^{(t)} + e^{\alpha_t} W_2^{(t)}, \tag{12.79}$$

and setting it to zero and solving for α_t, we obtain

$$\alpha_t = \frac{1}{2} \log \left[\frac{W_1^{(t)}}{W_2^{(t)}} \right]. \tag{12.80}$$

The AdaBoost algorithm iteration t can be summarized as follows:

- *Step 1*: Of the unused weak classifiers, select one, denoted k_t, that minimizes the number of misses specified by W_2.
- *Step 2*: Once the weak classifier k_t is selected, calculate W_1 and W_2 using the definitions above, then calculate the weight factor α_t.
- Repeat Steps 1 and 2 until all weak classifiers are employed.

12.1.6 Regression analysis

Given the finite training dataset $\{(\boldsymbol{x}_1,y_1),\ldots,(\boldsymbol{x}_n,y_n)\}$, in regression analysis, we are concerned with finding an approximate function $f(\boldsymbol{x}_i)$ that will accurately represent the training instances. It resembles the supervised classification; however, the range of y_i can take any value in the set of real numbers R, while in supervised classification, the labels y_i are discrete. To evaluate the performance of an approximating function, we can use the residual sum of squares (RSS) [14—17]:

$$E = \sum_{i=1}^{n} (y_i - f(\boldsymbol{x}_i))^2. \tag{12.81}$$

The minimization of the error function over arbitrary families of functions yields many solutions, and we need to restrict the problem to a smaller set of functions. Moreover, a zero error does not necessarily mean that the generalization performance is good; instead, we might be running to the overfitting problem. Very often, a good solution could be found by searching in a family of functions characterized by certain parameters, which is known as parametric function optimization.

The method of *linear least-squares* is a parametric model for which the approximation function can be represented as

$$f(\boldsymbol{x},\boldsymbol{\beta})=\boldsymbol{\beta}^T\boldsymbol{x},\ \ \boldsymbol{x}=[x_1\cdots x_d]^T,\ \ x_i\in\Re, \tag{12.82}$$

where $\boldsymbol{\beta}$ is unknown parametric vector. The RSS for this problem becomes

$$E(\boldsymbol{\beta})=\sum_{i=1}^{n}(y_i-f(\boldsymbol{x}_i,\boldsymbol{\beta}))^2. \tag{12.83}$$

Let us now arrange the input data vectors (known as regressors) into a matrix $\boldsymbol{X}$, for which the i-th input vector $\boldsymbol{x}_i$ is placed in the i-th column. We can represent the model output $\boldsymbol{y}$ as follows:

$$\boldsymbol{y}=\boldsymbol{X\beta}+z, \tag{12.84}$$

where z is the noise vector whose components are samples from the zero-mean Gaussian distribution with variance σ^2. In least-squares (LS) estimation, we minimize the squared error between the actual output $\mathbf{y}$ and the estimated output $\boldsymbol{X}\widehat{\boldsymbol{\beta}}$:

$$\widehat{\boldsymbol{\beta}}=\arg\min_{\boldsymbol{\beta}}E(\boldsymbol{\beta})=\arg\min_{\boldsymbol{\beta}}\|\boldsymbol{y}-\boldsymbol{X\beta}\|_2. \tag{12.85}$$

We can determine the optimum parameter vector $\boldsymbol{\beta}$ by solving the equation $\partial E(\boldsymbol{\beta})/\partial\boldsymbol{\beta}=0$. For reasonably high signal-to-noise ratio, the model output approximates $\boldsymbol{y}\cong\boldsymbol{X\beta}$; however, matrix $\boldsymbol{X}$ of size $d\times n$ is not a square matrix and thus cannot be inverted. If we premultiply this equation with $\boldsymbol{X}^T$ from the left we obtain $\boldsymbol{X}^T\boldsymbol{y}\cong\boldsymbol{X}^T\boldsymbol{X\beta}$, the matrix $\boldsymbol{X}^T\boldsymbol{X}$ is of size $d\times d$ and thus invertible. By multiplying the modified model by $(\boldsymbol{X}^T\boldsymbol{X})^{-1}$ from the left, we obtain

$$(\boldsymbol{X}^T\boldsymbol{X})^{-1}\boldsymbol{X}^T\boldsymbol{y}\cong\underbrace{(\boldsymbol{X}^T\boldsymbol{X})^{-1}\boldsymbol{X}^T\boldsymbol{X}}_{\boldsymbol{I}}\boldsymbol{\beta}, \tag{12.86}$$

from which we determine the optimum parameter vector by

$$\widehat{\boldsymbol{\beta}}=(\boldsymbol{X}^T\boldsymbol{X})^{-1}\boldsymbol{X}^T\boldsymbol{y}. \tag{12.87}$$

An alternative method is to use the *pseudoinverse* of $\boldsymbol{X}$, denoted $\boldsymbol{X}^+$. For this purpose, we need to perform SVD of the real-valued X matrix first, that is, $\boldsymbol{X}=\boldsymbol{U\Sigma V}^T$, where $\boldsymbol{U}$ and $\boldsymbol{V}$ are orthogonal matrices, and $\boldsymbol{\Sigma}$ is the diagonal matrix $\boldsymbol{\Sigma}=\mathrm{diag}(\sigma_1,\sigma_2,\ldots)$, with $\sigma_1\geq\sigma_2\geq\ \ldots\ \geq0$ being the *singular values* of $\boldsymbol{X}$. Given that the pseudoinverse of a diagonal matrix $\boldsymbol{\Sigma}$ is simply $\boldsymbol{\Sigma}^+=\mathrm{diag}(\sigma_1^{-1},\cdots,\sigma_r^{-1},0,\cdots,0)$, where r is the number of nonzero singular values, we can define the pseudoinverse of $\boldsymbol{X}$ as follows:

$$\boldsymbol{X}^+=\begin{cases}\boldsymbol{V}\begin{bmatrix}\boldsymbol{\Sigma}^+\\ \mathbf{0}\end{bmatrix}\boldsymbol{U}^T, & d\leq n\\ \boldsymbol{V}\begin{bmatrix}\boldsymbol{\Sigma}^+ & \mathbf{0}\end{bmatrix}\boldsymbol{U}^T, & d>n\end{cases} \tag{12.88}$$

The pseudoinverse is a unique solution to the minimization problem:

$$X^{+} = \arg\min_{X} \|X\beta - I_d\|_2, \tag{12.89}$$

where I_d is the identity matrix of size $d \times d$. When the **0**-matrix is not required in (12.88), the pseudoinverse of X is simply

$$X^{+} = \left(X^T X\right)^{-1} X^T, \tag{12.90}$$

which is easy to verify using SVD of X, namely $X = U\Sigma V^T$. To summarize, the optimum parameter vector can also be determined by

$$\widehat{\beta} = X^{+} y. \tag{12.91}$$

LS estimation is also applicable for higher-order polynomials approximation, which for $x \in R$, has the form

$$f(x, \beta) = \beta_0 + \beta_1 x + \cdots + \beta_p x^p, \; \beta = \left[\beta_0 \cdots \beta_p\right]^T. \tag{12.92}$$

However, for a higher-order polynomial, the orthogonal polynomial might provide more accurate results.

To ensure sparsity or smoothness of the solution, we can regularize the optimization problem as follows:

$$\widehat{\beta} = \arg\min_{\beta} \left[\|y - X\beta\|^2 + \|\Gamma X\|^2\right], \tag{12.93}$$

where Γ is a properly chosen Tikhonov matrix. Very often, we can choose the Tikhonov matrix as δI (δ is a positive constant), and in this case, the minimization problem reduces to

$$\widehat{\beta} = \arg\min_{\beta} \left[\|y - X\beta\|^2 + \delta^2 \|X\|^2\right], \; \delta > 0. \tag{12.94}$$

The optimization problem given by the previous equation is also known as the *ridge regression* problem, which is used to shrink the regression coefficients by imposing a penalty on their size [16]. When we replace the ridge penalty term, defined using L_2-norm, by the corresponding term using the L_1-norm, the optimization problem is known as the *LASSO regression* problem:

$$\widehat{\beta} = \arg\min_{\beta} \left[\|y - X\beta\|_2^2 + \delta \|X\|_1\right], \; \delta > 0, \tag{12.95}$$

(The LASSO is an abbreviation for the least absolute shrinkage and selection operator.) The second constraint now makes the solutions nonlinear in the y_i.

12.1.7 Neural networks

A neural network is a network of nodes called *neurons*. To every input data vector $x = [x_1 \ldots x_d]^T$, the neural network produces an output vector $y = f(x, \theta)$, which is nonlinear for the corresponding parameters θ. The neurons are connected by weighted edges called *synapses*. A neuron receives weighted inputs and generates a multivariate or univariate response, and we refer to this process as the activation of the neuron, while the corresponding function describing the output−input relationship is called the *activation function*. Artificial neural networks (ANNs) are inspired by the mammalian central nervous system; however, the update rule is different from that of natural neural networks. A common characteristic of various ANNs is that the activation function is always nonlinear.

12.1.7.1 Perceptron and activation functions

The neural system composed of only a single neuron is called the *perceptron*, and its operation is illustrated in Fig. 12.15. (Strictly speaking, it is not a network.) Mathematically, the perceptron operation can be described as

$$y = f(\boldsymbol{x}) = \phi\left(\boldsymbol{w}^T\boldsymbol{x} + b\right), \ \boldsymbol{w}^T\boldsymbol{x} = \sum_{i=1}^{d} w_i x_i, \tag{12.96}$$

where $\phi(u)$ is the activation function, $\boldsymbol{w} = [w_1 \ \dots \ w_d]^T$ is the weight vector describing the connections strength, and b is the bias. $\sum$ is the schematic representation of neuron, with the operation being $\Sigma = \boldsymbol{w}^T\boldsymbol{x} + b$. The possible *activation functions* include

- The identity function:

$$\phi(u) = u$$

- The sigmoid (logistic) function:

$$\phi(u) = \frac{1}{1 + e^{-u}}$$

- The hyperbolic tangent function:

$$\phi(u) = \tanh(u) = \frac{e^u - e^{-u}}{e^u + e^{-u}} = \frac{e^{2u} - 1}{e^{2u} + 1}$$

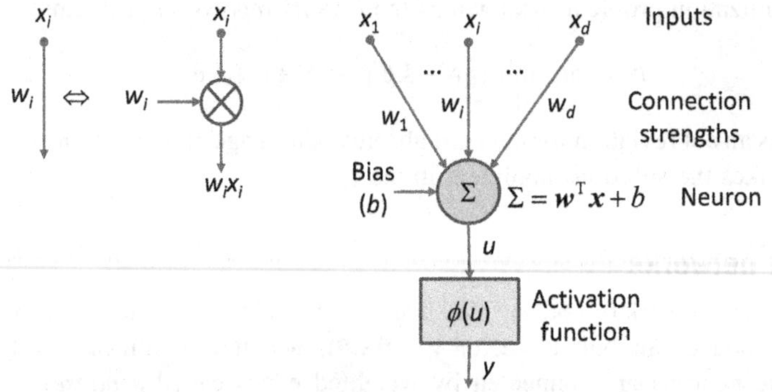

FIGURE 12.15

Operating principle of the perceptron.

- The hard threshold function:

$$\phi_{u_{tsh}}(u) = \mathcal{I}(u \geq u_{tsh})$$

- The rectified linear unit (ReLU) function:

$$\phi(u) = \max(0, x).$$

The activation functions listed above are provided in Fig. 12.16. The sigmoid function, denoted $s(u)$ in Fig. 12.17, has historically been the most frequently used, since it is differentiable and has the range [0,1]. The first derivative of the sigmoid function is simply $s'(u) = s(u)[1-s(u)]$. However, its derivative is close to zero when $|u|$ is larger than 4, as shown in Fig. 12.17, and it causes numerical problems for multilayer feedforward ANNs with many hidden layers. The ReLU function addresses this problem. However, its derivative is zero for $u < 0$, and the following modification should be employed to ensure that the gradient is nonzero for small values of u:

$$\phi(u) = \max(0, x) + \delta\min(0, x), \quad \delta > 0,$$

where δ is a small positive constant.

The perceptron for the threshold (in the hard-threshold function) being set to 0 has the following output–input function:

$$y = f(\boldsymbol{x}) = \phi_0\left(\boldsymbol{w}^T\boldsymbol{x} + b\right) = \begin{cases} 1, & \boldsymbol{w}^T\boldsymbol{x} + b > 0 \\ 0, & \text{otherwise} \end{cases} \tag{12.97}$$

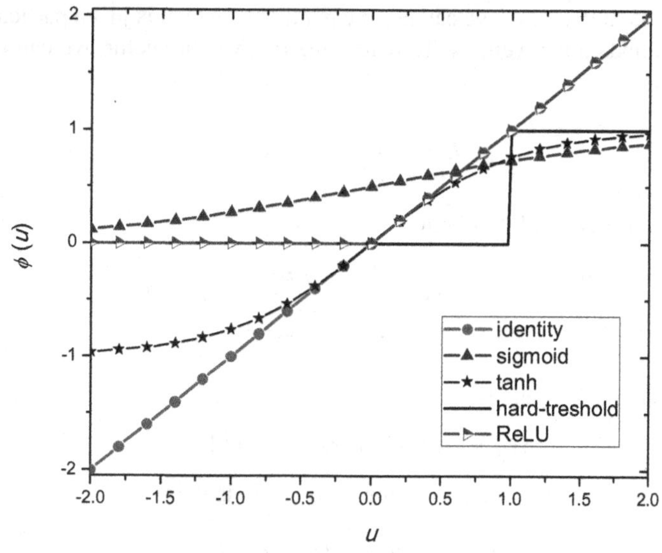

FIGURE 12.16

Possible activation functions.

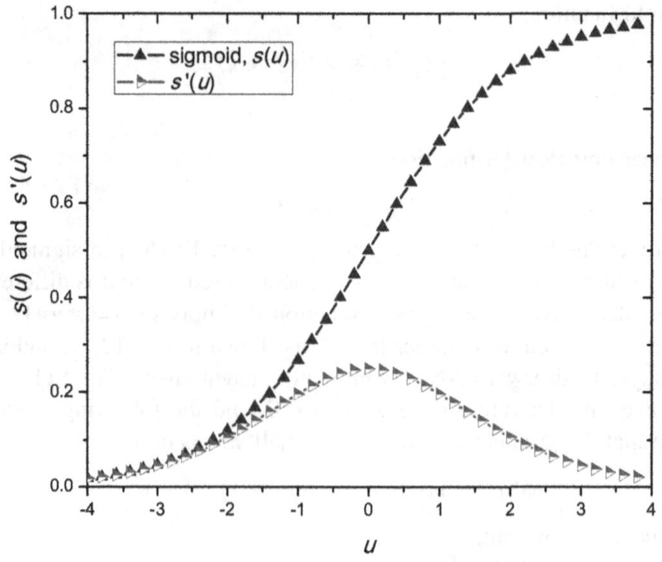

FIGURE 12.17

Sigmoid function and its first derivative.

and can be used as a binary classifier. The purpose of the bias here is to change the decision boundary. The training of the perceptron to determine the weight and bias will converge only if the problem is linearly separable, and in that case, we can use the *delta rule,* which is just a particular instance of the gradient descent, to update the weights. To determine the weight vector, we can minimize the error function, defined as

$$E = \frac{1}{2}\sum_{i=1}^{n}(y_i - f(\boldsymbol{x}_i))^2, \tag{12.98}$$

by taking the first derivative of E to obtain

$$\frac{\partial E}{\partial w_j} = -(y_i - f(\boldsymbol{x}_i))\underbrace{\frac{\partial(\boldsymbol{w}^T\boldsymbol{x}+b)}{\partial w_j}}_{x_j} = -(y_i - f(\boldsymbol{x}_i))x_j. \tag{12.99}$$

We need to update the weights by

$$\Delta w_j = -\mu(y_i - f(\boldsymbol{x}_i))x_j;\ j = 1, \cdots, d, \tag{12.100}$$

where μ is the learning rate. The weight vector update rule is then

$$\boldsymbol{w}^{(i)} = \boldsymbol{w}^{(i-1)} - \mu(y_i - f(\boldsymbol{x}_i))\boldsymbol{x}_i;\ i = 1, \cdots, n. \tag{12.101}$$

12.1.7.2 Feedforward networks

In a *feedforward network*, the neurons do not form a loop; the information flows in one direction from the input layer through one or more intermediate levels, known as *hidden* levels, to the output layer, as illustrated in Fig. 12.18. In other words, each neuron at a given layer is connected to all neurons on the next layer, but there is no connection among neurons within the same layer. ANNs with more than one hidden layer are called "deep" ANNs, and the corresponding learning algorithm is commonly referred to as *deep learning*. Feedforward ANNs are suitable for nonlinear classification and regression problems.

In principle, we can apply different activation functions for different levels. In practice, all hidden layers typically use the same activation function; this is often the sigmoidal activation function. In regression problems, the activation function in the output layer is not needed; in other words, we can use the identity activation function. In binary classification problems, only one output neuron is needed, and the output gives a prediction of $P(y=1|x)$, and since the range for sigmoidal activation function is [0,1], it is suitable for use in the output layer. On the other hand, in multiclass classification problems, we need one output neuron per class c, and the corresponding output gives the prediction of $P_c = P(y=c|x)$. These output probabilities must sum to 1, that is, $\sum_c P_c = 1$, indicating the sigmoidal activation function is not suitable for this problem. The multidimensional *softmax-function* should be used instead, as defined by

$$\text{soft max}(z)_i = \frac{e^{z_i}}{\sum_j e^{z_j}}. \tag{12.102}$$

Feedforward ANNs use numerous training methods, the most popular being *backpropagation*, which represents a generalization of the delta rule for a single perceptron [14–18]. Let us return to Fig. 12.18, with only one hidden layer. We adopted a notation similar to that of Ref. [18] to simplify the explanations. The indices i, j, and k denote a particular neuron at the output, hidden, and input layer, respectively. The weight coefficient related to the edge between the j-th node in the hidden layer and k-th node in the input layer is denoted v_{jk}. Similarly, the weight coefficient related to the edge between

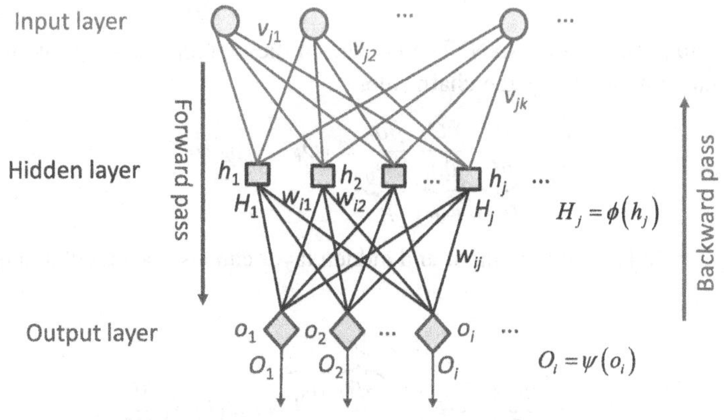

FIGURE 12.18

Feedforward neural network with one hidden layer.

the i-th output node and j-th node in the hidden layer is denoted w_{ij}. Based on Fig. 12.18, we conclude that the input to the j-th node of the hidden layer, denoted h_j, is related to inputs x_k as follows:

$$h_j = \sum_k v_{jk} x_k, \tag{12.103}$$

where to facilitate the explanation, we set the bias to zero. The output from the hidden neuron j, denoted H_j, is given by

$$H_j = \phi(h_j) = \phi\left(\sum_k v_{jk} x_k\right), \tag{12.104}$$

where $\phi(u)$ is the activation function of the hidden neurons. From Fig. 12.18, we conclude that the input to the i-th neuron of the output layer, denoted o_i, is related to the hidden layer outputs H_j:

$$o_i = \sum_j w_{ij} H_j. \tag{12.105}$$

The output from neuron i at the output layer, denoted O_i, is given by

$$O_i = \psi(o_i), \tag{12.106}$$

and $\psi(u)$ is the activation function of the output neurons.

We can use the sum of squares to evaluate the error performance, defined as

$$E = \frac{1}{2} \sum_i (O_i - y_i)^2, \tag{12.107}$$

where y_i is the desired output corresponding to the input data vector $\boldsymbol{x}$ obtained from the training dataset. In the *backpropagation* method representing the steepest descent algorithm, each weight is changed in the direction opposite to the gradient of the error; that is, we can write

$$\Delta w_{ij} = -\mu \frac{\partial E}{\partial w_{ij}}, \quad \Delta v_{jk} = -\mu \frac{\partial E}{\partial v_{jk}}, \tag{12.108}$$

where μ is the learning rate of the ANN. The change in the weights between the hidden and output layer can be calculated by applying the chain rule:

$$\frac{\partial E}{\partial w_{ij}} = \underbrace{\frac{\partial E}{\partial O_i}}_{O_i - y_i} \underbrace{\frac{\partial O_i}{\partial o_i}}_{\psi'(o_i)} \underbrace{\frac{\partial o_i}{\partial w_{ij}}}_{H_j} = (O_i - y_i)\psi'(o_i) H_j. \tag{12.109}$$

The change in weight between the input and hidden layer can also be calculated by again applying the chain rule:

$$\frac{\partial E}{\partial v_{jk}} = \frac{\partial E}{\partial H_j}\frac{\partial H_j}{\partial v_{jk}} = \frac{\partial E}{\partial H_j} \underbrace{\overbrace{\frac{\partial H_j}{\partial h_j}}^{\phi'(h_j)} \overbrace{\frac{\partial h_j}{\partial v_{jk}}}^{x_k}}_{(\partial \backslash \partial v_{jk}) H_j} = \phi'(h_j) x_k \frac{\partial E}{\partial H_j}. \tag{12.110}$$

To calculate $\partial E/\partial H_j$, we also apply the chain rule:

$$\frac{\partial E}{\partial H_j}=\sum_i \frac{\partial E}{\partial o_i}\frac{\partial o_i}{\partial H_j}=\sum_i \underbrace{\frac{\partial E}{\partial o_i}}_{\frac{\partial E}{\partial O_i}\frac{\partial O_i}{\partial o_i}}\underbrace{\frac{\partial o_i}{\partial H_j}}_{w_{ij}}=\sum_i \underbrace{\frac{\partial E}{\partial O_i}}_{O_i-y_i}\underbrace{\frac{\partial O_i}{\partial o_i}}_{\psi'(o_i)}w_{ij}=\sum_i (O_i-y_i)\psi'(o_i)w_{ij}. \tag{12.111}$$

Finally, after substituting (12.111) into (12.110), we obtain

$$\frac{\partial E}{\partial v_{jk}}=\phi'(h_j)x_k\sum_i (O_i-y_i)\psi'(o_i)w_{ij}. \tag{12.112}$$

In the learning stage, we calculate the outputs O_i as well as all intermediate values and update the weights based on the partial derivatives calculated in the previous iteration:

$$w_{ij}^{(\text{new})}=w_{ij}^{(\text{old})}-\mu\frac{\partial E}{\partial w_{ij}},\quad v_{jk}^{(\text{new})}=v_{jk}^{(\text{old})}-\mu\frac{\partial E}{\partial v_{jk}}. \tag{12.113}$$

We refer to this stage as the *forward pass*. We determine the error terms (O_i-y_i) in the *backward pass* and update the partial derivatives based on Eqs. (12.109) and (12.112). These partial derivatives are propagated to the weights for the next iteration. The error is backpropagated as soon as each new training instance is received.

By close inspection of Eqs. (12.109) and (12.112), we conclude that in the backward pass, partial derivatives calculated at a deeper level are needed at the shallower level and should be reused for efficient implementation. From the illustrative example of Fig. 12.18, we provide a sequence of partial derivative calculations (in the backward direction):

$$\frac{\partial E}{\partial o_i}=\underbrace{\frac{\partial E}{\partial O_i}}_{O_i-y_i}\underbrace{\frac{\partial O_i}{\partial o_i}}_{\psi'(o_i)}=(O_i-y_i)\psi'(o_i) \tag{12.114.1}$$

$$\frac{\partial E}{\partial w_{ij}}=\frac{\partial E}{\partial o_i}\underbrace{\frac{\partial o_i}{\partial w_{ij}}}_{H_j}=H_j\frac{\partial E}{\partial o_i} \tag{12.114.2}$$

$$\frac{\partial E}{\partial h_j}=\sum_i \frac{\partial E}{\partial o_i}\underbrace{\frac{\partial o_i}{\partial h_j}}_{\frac{\partial o_i}{\partial H_j}\frac{\partial H_j}{\partial h_j}}=\sum_i \frac{\partial E}{\partial o_i}\phi'(h_j)w_{ij} \tag{12.114.3}$$

$$\frac{\partial E}{\partial v_{jk}}=\frac{\partial E}{\partial h_j}\underbrace{\frac{\partial h_j}{\partial v_{jk}}}_{x_k}=x_k\frac{\partial E}{\partial h_j} \tag{12.114.4}$$

In the forward step, we use the partial derivatives given by Eqs. (12.114.2) and (12.114.4) to update the weights in Eq. (12.113).

In deep learning, the number of hidden layers can be large, so it is important to study the backward pass for feedforward ANNs containing $L>1$ hidden layers, as illustrated in Fig. 12.19, wherein we

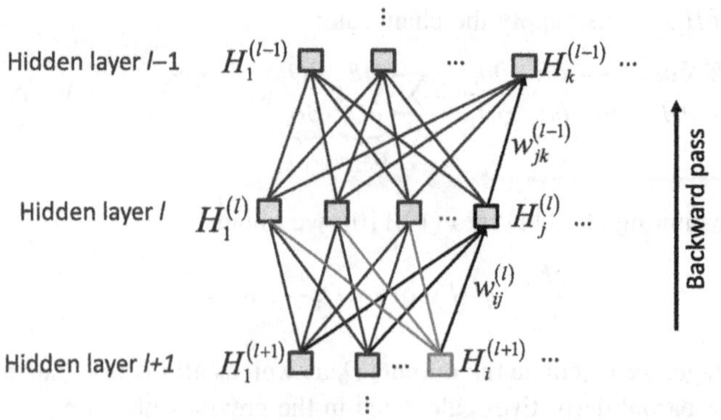

FIGURE 12.19

Backward pass for deep feedforward neural networks.

distinguish the neurons by the corresponding outputs (rather than inputs as used in Fig. 12.18). The input layer can be interpreted as the 0-th hidden level, thus setting $\boldsymbol{H}^{(0)} = \boldsymbol{x}$ and $w_{jk}^{(0)} = v_{jk}$. Based on the previous example and Eq. (12.104), the output of the i-th neuron at layer (l+1) is given by

$$H_j^{(l+1)} = \phi\left(h_j^{(l+1)}\right) = \phi\left(\sum_i w_{ji}^{(l)} H_i^{(l)}\right). \tag{12.115}$$

The partial derivative of E with respect to $H_j^{(l)}$ is given by

$$\frac{\partial E}{\partial H_j^{(l)}} = \sum_i \frac{\partial E}{\partial H_i^{(l+1)}} \underbrace{\frac{\partial H_i^{(l+1)}}{\partial H_j^{(l)}}}_{\phi'\left(h_i^{(l+1)}\right) w_{ij}^{(l)}} = \sum_i \phi'\left(h_i^{(l+1)}\right) w_{ij}^{(l)} \frac{\partial E}{\partial H_i^{(l+1)}}. \tag{12.116}$$

The partial derivative of E with respect to $w_{jk}^{(l-1)}$ is given by

$$\frac{\partial E}{\partial w_{jk}^{(l-1)}} = \frac{\partial E}{\partial H_j^{(l)}} \underbrace{\frac{\partial H_j^{(l)}}{\partial w_{jk}^{(l-1)}}}_{\phi'\left(h_j^{(l)}\right) H_k^{(l-1)}} = \frac{\partial E}{\partial H_j^{(l)}} \phi'\left(h_j^{(l)}\right) H_k^{(l-1)}. \tag{12.117}$$

The weight coefficients can now be updated as

$$w_{jk}^{(l-1,\,\text{new})} = w_{jk}^{(l-1,\,\text{old})} - \mu \frac{\partial E}{\partial w_{jk}^{(l-1)}}. \tag{12.118}$$

For the output layer, the weights can be updated according to (12.109):

$$w_{ij}^{(\text{new})} = w_{ij}^{(\text{old})} - \mu \frac{\partial E}{\partial w_{ij}}, \quad \frac{\partial E}{\partial w_{ij}} = \underbrace{\frac{\partial E}{\partial O_i}}_{O_i - y_i} \underbrace{\frac{\partial O_i}{\partial o_i}}_{\psi'(o_i)} \underbrace{\frac{\partial o_i}{\partial w_{ij}}}_{H_j^{(L)}} = (O_i - y_i)\psi'(o_i) H_j^{(L)}. \tag{12.119}$$

12.2 The Ising model, adiabatic quantum computing, and quantum annealing

Many problems in ML can be mapped to the Ising problem formalism; first, let us describe the *Ising model* [19].

12.2.1 The Ising model

The *Ising model*, named for German physicist Ernst Ising, was introduced to study magnetic dipole moments of atoms in statistical physics. The magnetic dipole moments of atomic spins can be in one of two states, +1 or −1. The spins, represented as discrete variables, are arranged in a graph, typically a *lattice* graph, and this model allows each spin to interact with the closest neighbors. Spins in the neighborhood that share the same value tend to have lower energy; however, heat disturbs this tendency, so various phase transitions exist. Let the individual spin on the lattice be denoted $\sigma_i \in \{-1,1\}$; $i = 1, 2, \ldots, L$, with L as the lattice size. To represent the energy of the Ising model, the following Hamiltonian is typically used:

$$H(\boldsymbol{\sigma}) = -\sum_{\langle i,j\rangle} J_{ij}\sigma_i\sigma_j - \mu\sum_i h_i\sigma_i, \tag{12.120}$$

where J_{ij} denotes the interaction strength between the i-th and j-th sites, h_i denotes the strength of the external magnetic field interacting with the i-th spin site in the lattice, and μ is the magnetic moment. We use the notation $\langle i,j\rangle$ to denote direct nearest neighbors. When $J_{ij}>0$, the interaction is *ferromagnetic*; when $J_{ij}<0$, it is *antiferromagnetic*; and when $J_{ij} = 0$, there is no interaction between spins i and j. On the other hand, when $h_i > 0$, the i-th spin tends to line up with the external field in a positive direction, while for $h_i < 0$, it tends to line up in a negative direction. The *configuration probability* is described by the *Boltzmann distribution*:

$$P_\beta(\boldsymbol{\sigma}) = \frac{1}{Z_\beta}e^{-\beta H(\boldsymbol{\sigma})}, \quad \beta = 1/(k_B T), \tag{12.121}$$

where $\beta = 1/(k_B T)$ (k_B-the Boltzmann constant and T-the absolute temperature) is the *inverse temperature*, while Z_β denotes the *partition function*:

$$Z_\beta = \sum_{\boldsymbol{\sigma}} e^{-\beta H(\boldsymbol{\sigma})}. \tag{12.122}$$

By denoting the interaction matrix by $\boldsymbol{J}=(J_{ij})$, we can determine the *optimum configuration* as follows:

$$\hat{\boldsymbol{\sigma}} = \arg\min_{\boldsymbol{\sigma}}\left(\boldsymbol{\sigma}^{\mathrm{T}}\boldsymbol{J}\boldsymbol{\sigma} + \boldsymbol{h}^{\mathrm{T}}\boldsymbol{\sigma}\right), \tag{12.123}$$

and this problem is similar to the *quadratic unconstrained binary optimization* (QUBO) problem, which is known to be NP-hard and can be formulated by

$$\hat{\boldsymbol{x}} = \arg\min_{\boldsymbol{x}}\left(\sum_{i,j=1}^{N} w_{ij}x_i x_j + \sum_{i=1}^{N} b_i x_i\right), \quad x_i \in \{0,1\}, \tag{12.124}$$

wherein $\sigma_i = 2x_i - 1$, $\mu = 1$, and $b_i = h_i$ is the bias. By introducing the matrix $\boldsymbol{W} = (w_{ij})$, we can reformulate the QUBO problem in a fashion similar to (12.123) by

$$\widehat{\boldsymbol{x}} = \arg\min_{\boldsymbol{x}} \left(\boldsymbol{x}^{\mathrm{T}} \boldsymbol{W} \boldsymbol{x} + \boldsymbol{b}^{\mathrm{T}} \boldsymbol{x}\right), \quad \boldsymbol{x} \in \{0, 1\}^N. \tag{12.125}$$

By replacing the spin variables with corresponding Pauli operators (matrices), we can express the Ising model using a quantum mechanical interpretation:

$$\widehat{H} = -\sum_{\langle i,j \rangle} J_{ij} \sigma_i^Z \sigma_j^Z - \sum_i h_i \sigma_i^Z, \tag{12.126}$$

where the Pauli $\sigma^Z = Z$ matrix is defined as $\sigma^Z = Z = \begin{bmatrix} 1 & 0 \\ 0 & -1 \end{bmatrix}$, and the action on the computation basis vector is $\sigma^Z|0\rangle = |0\rangle$, $\sigma^Z|1\rangle = -|1\rangle$. The expected value of the Ising Hamiltonian is $\langle \widehat{H} \rangle = \langle \psi | \widehat{H} | \psi \rangle$, for arbitrary state $|\psi\rangle$. The evolution is described by the unitary operator U to obtain $U|\psi\rangle$. The corresponding Schrödinger equation is

$$j\hbar \frac{d}{dt} U = HU. \tag{12.127}$$

The solution for the time-invariant Hamiltonian is $U = e^{-jHt/\hbar}$, which is unitary.

When the external field is transversal, the *transverse-filed Hamiltonian* is

$$\widehat{H} = -\sum_{\langle i,j \rangle} J_{ij} \sigma_i^Z \sigma_j^Z - \sum_i h_i \sigma_i^X, \tag{12.128}$$

where the Pauli X-operator is defined by $\sigma^X = X = \begin{bmatrix} 0 & 1 \\ 1 & 0 \end{bmatrix}$. The Pauli X and Z operators do not commute, since $XZ = -ZX$.

12.2.2 Adiabatic quantum computing

An *adiabatic process* is a process that changes slowly (gradually) so the system can adapt its configuration accordingly. If the system was initially in the ground state of an initial Hamiltonian H_0, it would end up in the ground state of the final Hamiltonian H_1 after the adiabatic change. This claim is commonly referred to as the *adiabatic theorem*, proved by Born and Fock in 1928 [20]. We typically choose an initial Hamiltonian H_0 whose ground state is easy to obtain, and we choose a Hamiltonian H_1 for the ground state we are interested in. To determine its ground state, we consider the following Hamiltonian:

$$H(t) = (1 - t)H_0 + tH_1, \quad t \in [0, 1], \tag{12.129}$$

and start with the ground state of $H(0) = H_0$ and then gradually increase the parameter t, representing the time normalized with the duration of the whole process. According to the quantum adiabatic theorem, the system will gradually evolve to the ground state of H_1, provided there is no degeneracy for the ground state energy. This transition depends on the time Δt during which the adiabatic change occurs and depends on the minimum energy gap between the ground state and the first excited state of the adiabatic process. When applied to the Ising model, we select the initial Hamiltonian $H_0 = \sum_i X_i$, describing the interaction of the transverse external field with the spin sites. On the other hand, the final Hamiltonian is given by $H_1 = -\sum_{\langle i,j\rangle} J_{ij} Z_i Z_j - \sum_i h_i X_i$, representing the classical Ising model. The time-dependent Hamiltonian combines two models in Eq. (12.129). We gradually change the parameter t from zero to 1 so that the adiabatic pathway is chosen, and the final state is the ground state of Hamiltonian H_1, which encodes the solution of the problem we are trying to solve. The gap ΔE between the ground state and the first excited state is parameter t dependent. The time scale (the runtime of the algorithm) is reversely proportional to the min $[\Delta E(t)]^2$ [21]. So if the minimum gap during the adiabatic pathway is too small, the runtime of the entire algorithm could be large.

In *adiabatic quantum computing,* the Hamiltonian is defined as follows:

$$H = \underbrace{-\sum_{\langle i,j\rangle} J_{ij} Z_i Z_j - \sum_i h_i Z_i}_{\text{classical Ising model}} - \underbrace{\sum_{\langle i,j\rangle} g_{ij} X_i X_j}_{\text{interaction between transverse fields}}, \tag{12.130}$$

and it can be shown that this model can simulate *universal* quantum computation.

12.2.3 Quantum annealing

In metallurgy, *annealing* represents a process to create crystals without defects. To do this, the crystal is locally heated, and cooling is done in a controlled manner so that internal stress is relieved and imperfectly placed atoms can return to their minimum energy positions. The temperature dictates the rate of diffusion. In *simulated annealing,* we map the optimization problem to the atomic configuration, with random dislocation representing the change in the solution. To avoid being trapped into a local minimum, the simulating annealing accepts the intermediate solution with a certain probability. So the probability here has the same role as a temperature in metallurgy; the higher it is, the more random dislocations are allowed. Therefore, the simulated annealing represents a metaheuristic approach to an approximate global optimization problem in a large search space. It is suitable when the search space is discrete.

Because thermal annealing has a higher barrier than quantum tunneling, *quantum annealing* applies quantum tunneling to control acceptance probabilities [22,23]. Therefore, the quantum tunneling field strength has the same role as the temperature in metallurgy. Quantum annealing is a metaheuristic method of finding a global minimum of an objective function over a given set of candidate solutions (states) by employing quantum-fluctuation-based computation. In adiabatic quantum computing, the minimum energy gap can be hard to calculate. To solve this problem, we can apply a similar approach to adiabatic quantum computing but repeatedly anneal while probably violating the adiabatic condition and select the result with the lowest energy. By employing quantum annealing, we can bypass the higher-level models for quantum computing.

Following the problem definition, we need to map the problem to the Ising model, which is essentially the QUBO problem. The next step is to perform the *minor embedding* [24], which is similar to quantum compilation. In quantum hardware, not all pairs of qubits are entangled, and the connectivity is typically sparse. We can often use the arrangement of nodes known as the *Chimera*

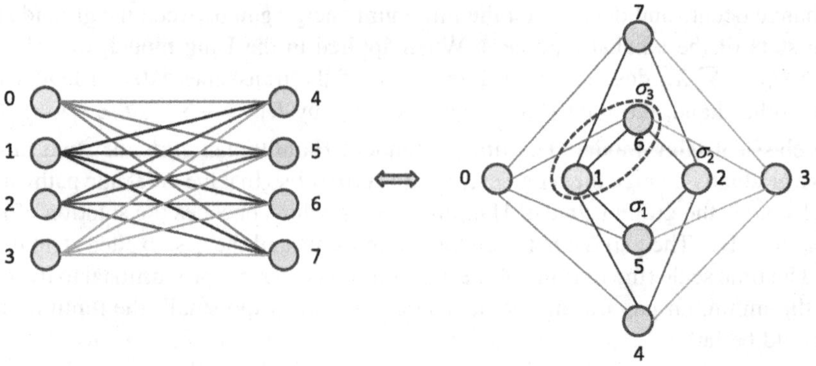

FIGURE 12.20

Eight-node cluster in Chimera graph.

graph [25,26]. In the Chimera graph, we create an eight-node cluster, representing the full bipartite graph, as illustrated in Fig. 12.20. The nodes can be further connected to the north and south clusters in the grid. Due to fabrication imperfections, some of the qubits might not be operational. Let us now consider the minimization problem min $(\sigma_1\sigma_2+\sigma_2\sigma_3+\sigma_1\sigma_3)$, which is the triangle graph of three nodes. However, there are no three nodes in the graph on the right, creating a triangle. We can group the physical qubits one and six into a logical qubit σ_3. We can interpret qubits 1 and 6 as strongly coupled to create the logical qubit. After that, we execute the quantum annealing algorithm on the quantum processing unit (QPU).

12.3 Quantum approximate optimization algorithm and variational quantum eigensolver

Quantum circuits are imperfect, which prevents us from running famous quantum algorithms using the gate-based quantum computing approach. To overcome this problem, a new breed of quantum algorithms has recently been introduced [27–29] that employs parametrized shallow quantum circuits, which can be called *variational (quantum) circuits*. The variational circuits are suitable for noisy and imperfect quantum computers, also called noisy intermediate-scale quantum devices [30]. They employ the *variational method* to find the approximation to the lowest energy eigenket or ground state, illustrated in Fig. 12.21. The QPU is embedded in the environment (bath), and the interaction between the QPU and the environment causes decoherence. The decoherence problem limits circuit control, and therefore, quantum circuit depth. To avoid this problem, we can run a short burst of calculation on QPU, extract the results for the classical central processing unit, which optimizes the QPU parameters to reduce the system's energy consumption and feedback a new set of parameters to the QPU for the next iteration. We iterate this hybrid quantum-classical algorithm until the minimum global energy of the system is achieved. Any Hamiltonian H can be decomposed in terms of the Hamiltonians of corresponding subsystems, that is, $H = \sum_i H_i$, where

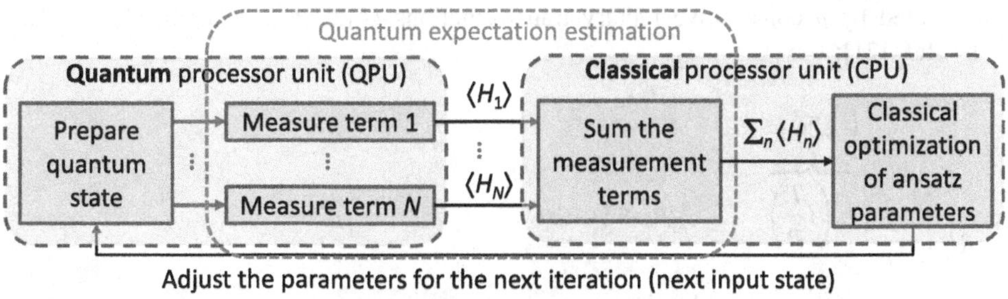

FIGURE 12.21

Variational quantum eigensolver principle.

H_i is the Hamiltonian of the i-th subsystem. The expected value of Hamiltonian can be then determined as

$$\langle H\rangle = \langle\psi|H|\psi\rangle = \langle\psi|\sum_i H_i|\psi\rangle = \sum_i \langle\psi|H_i|\psi\rangle = \sum\nolimits_i \langle H_i\rangle, \tag{12.131}$$

where we exploit the linearity of quantum observables and $|\psi\rangle$ denotes the arbitrary input state. By applying this approach, we effectively replaced the long coherent evolution of QPU with multiple short coherent evolutions of the subsystems. The expected value of Hamiltonian can also be decomposed in terms of corresponding eigenkets of Hamiltonian $|\lambda_i\rangle$ as follows:

$$\langle H\rangle = \langle\psi|H|\psi\rangle = \sum_{\lambda_i,\ \lambda_j \in \mathrm{Spectrum}(H)} \langle\psi|(|\lambda_i\rangle\langle\lambda_i|)H(|\lambda_j\rangle\langle\lambda_j|)|\psi\rangle = \sum_{\lambda_i,\ \lambda_j} \langle\psi|\lambda_i\rangle\langle\lambda_i|H|\lambda_j\rangle\langle\lambda_j|\psi\rangle$$

$$= \sum_{\lambda_i} |\langle\psi|\lambda_i\rangle|^2 \underbrace{\langle\lambda_i|H|\lambda_i\rangle}_{\lambda_i} \geq \sum_{\lambda_i} |\langle\psi|\lambda_i\rangle|^2 E_0 = E_0, \tag{12.132}$$

where E_0 is the lowest energy eigenket of the Hamiltonian. Varying over the entire Hilbert space is too complicated; we can choose a subspace of the entire Hilbert space, parametrized by some real-valued parameters p_i ($i = 1, 2, \ldots, N$), which is known as the ansatz. The choice of the ansatz is relevant, as some choices of ansatzes can lead to better approximations.

The key idea behind the *quantum approximate optimization algorithm (QAOA)* is to approximate the quantum adiabatic pathway [27,31]. This algorithm is suitable for maximum satisfiability (MAX-SAT), MAX-CUT, and number partitioning problems; however, it can potentially be used to solve various NP-complete and NP-hard problems. As before, the optimization problem should be mapped to the Ising model with desired (final) Hamiltonian being $H_f = -\sum_{\langle i,j\rangle} J_{ij} Z_i Z_j - \sum_i h_i X_i$, while initial Hamiltonian being $H_0 = \sum_i X_i$. As before, we apply the adiabatic quantum computing approach and introduce the combined Hamiltonian by $H(t) = (1-t/T)H_0 + (t/T)H_f$, where T is the duration of the computation and $t \in [0,T]$. The corresponding unitary evolution operator is denoted by $U(t)$. The authors of Ref. [21] have proved the Lemma that unitary evolution operator at the end of the process $U(T)$ can be approximated by $U'(T)$ if the distance between Hamiltonians is limited by $||H(t) - H'(t)|| \leq \delta$ for every t. The same authors proved the theorem claiming that the unitary transformation $U(T)$ can be approximated by p consecutive unitary transformations $U'_1,\ldots,U'_p$, wherein U'_i $(i = 1,\ldots,p)$ is defined by Ref. [21]:

$$
\begin{aligned}
U'_i &= e^{-j\frac{T}{p}\overbrace{H\left(i\frac{T}{p}\right)}^{\left(1-\frac{i}{p}\right)H_0+\frac{i}{p}H_f}} = e^{-j\frac{T}{p}\frac{i}{p}H_f-j\frac{T}{p}\left(1-\frac{i}{p}\right)H_0} \\
&\cong \underbrace{e^{-j\frac{T}{p}\frac{i}{p}H_f}}_{U\left(H_f,\frac{i}{p}\frac{T}{p}\right)}\ \underbrace{e^{-j\frac{T}{p}\left(1-\frac{i}{p}\right)H_0}}_{U\left(H_0,\left(1-\frac{i}{p}\right)\frac{T}{p}\right)} = U\left(H_f,\frac{i}{p}\frac{T}{p}\right)U\left(H_0,\left(1-\frac{i}{p}\right)\frac{T}{p}\right).
\end{aligned}
\tag{12.133}
$$

wherein we apply the second line to the Campbell–Baker–Hausdorff (CBH) theorem, claiming that we can approximate parallel Hamiltonians with consecutive ones—in other words, $\left\|e^{A+B} - e^{A}e^{B}\right\|_2 \cong O(\|AB\|_2)$ [32]. So the key idea is to discretize the computation time T into p intervals of duration T/p and then apply H_0 and H_f of duration $(1-i/p)T/p$ and $(i/p)T/p$, respectively, p times. The discretized time steps do not need to be uniformly spaced, as illustrated in Fig. 12.22. In other words, the i-th step can be represented as

$$U'_i = U(H_f,\gamma_i)U(H_0,\beta_i),\ \ \beta_i+\gamma_i = \Delta T_i,\ \ \sum_i \Delta T_i = T;\ i = 1,\cdots,p \tag{12.134}$$

Therefore, we apply H_0 and H_f Hamiltonians of duration β_i and γ_i as horizontal and vertical approximations of the adiabatic pathway in an alternative fashion as follows:

$$U(T) \cong U(H_f,\gamma_p)U(H_0,\beta_p)\cdots U(H_f,\gamma_1)U(H_0,\beta_1). \tag{12.135}$$

Overall, the final ground state in both adiabatic and approximating pathways would be the same. This approximation is known as Trotterization (Trotter–Suzuki decomposition). The accuracy can be improved by increasing the number of steps p. The Hamiltonian H_0 for the ground state has the tensor

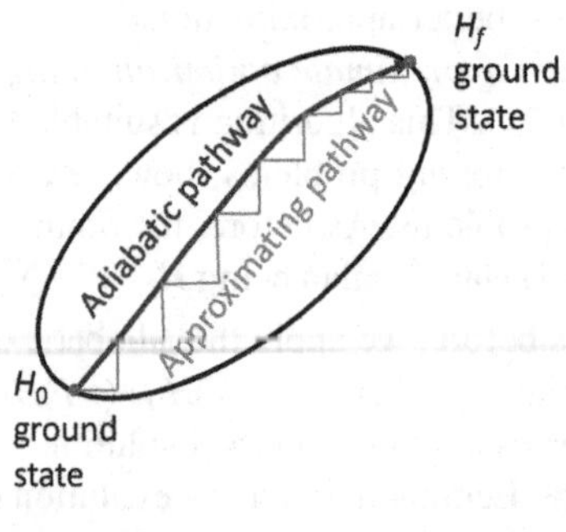

FIGURE 12.22

Approximating the adiabatic pathway.

product of diagonal states $|+\rangle=(|0\rangle+|1\rangle)/\sqrt{2}$ (lowest energy state), which is one of two eigenkets of the Pauli-X operator; that is,

$$|\psi_0\rangle = |+\rangle^{\otimes N} = |+\rangle_{N-1} \otimes \cdots \otimes |+\rangle_0 = \frac{1}{\sqrt{2^n}} \sum\nolimits_z |z\rangle, \quad z \in \{0,1\}^N. \tag{12.136}$$

The final state in p-step QAOA is

$$|\psi_f(\boldsymbol{\beta},\boldsymbol{\gamma})\rangle = |\boldsymbol{\beta},\boldsymbol{\gamma}\rangle = U(H_f,\gamma_p)U(H_0,\beta_p)\cdots U(H_f,\gamma_1)U(H_0,\beta_1)|\psi_0\rangle. \tag{12.137}$$

As an illustration, the QAOA can be used to straightforwardly solve the following optimization problem:

$$C_{\max} = \max_z C(z) = \max_z \sum_{k=1}^{m} C_k(z), \quad z \in \{0,1\}^n, \tag{12.138}$$

where $C_k: \{0,1\}^N \to \{0,1\}$ for Boolean constraints (with "1" meaning that the constraint is satisfied) and $C_k: \{0,1\}^N \to \mathfrak{R}$ ($\mathfrak{R}$—the set of real numbers) for weighted constraints. The objective function can be represented as an operator:

$$C|z\rangle = \sum_{k=1}^{m} C_k(z)|z\rangle = f(z)|z\rangle. \tag{12.139}$$

Based on the previous equation, we conclude that the operator C has the eigenvalues $f(z) = \sum_z C_k(z)$. The largest eigenvalue of $f(z)$ can be denoted $C_{\max} = f(z_{\max})$. The expectation of operator C for an arbitrary state $|\psi\rangle = \sum_z \alpha_z |z\rangle$ is given by

$$\langle\psi|C|\psi\rangle = \sum_{z\in\{0,1\}^N} f(z)|\alpha_z|^2 \le \sum_{z\in\{0,1\}^N} f(z_{\max})|\alpha_z|^2 = f(z_{\max}) \underbrace{\sum\nolimits_{z\in\{0,1\}^N}|\alpha_z|^2}_{1} = f(z_{\max}). \tag{12.140}$$

For the superposition state given by (12.136), the expectation of C becomes

$$\langle\psi_0|C|\psi_0\rangle = \sum_{z\in\{0,1\}^N} f(z)\frac{1}{2^N} = \frac{1}{2^N}\sum_{z\in\{0,1\}^N} f(z). \tag{12.141}$$

The probability of obtaining $|z_{\max}\rangle$ from each measurement is 2^{-N}. The corresponding unitary evolution operator for C is given by

$$\begin{aligned} U(C,\gamma) &= e^{-j\gamma C} = e^{-j\gamma \sum_{k=1}^{m} C_k} \\ &= \prod_{k=1}^{m} e^{-j\gamma C_k}, \end{aligned} \tag{12.142}$$

where the second line comes from the CBH theorem, since $[C_i,C_j] = 0$. Let us now observe the action of $U(C,\gamma)$ on the arbitrary state $|\psi\rangle$:

$$\underbrace{U(C,\gamma)}_{\prod_{k=1}^{m} e^{-j\gamma C_k}} \underbrace{|\psi\rangle}_{\sum_z \alpha_z|z\rangle} = \sum_z \alpha_z \prod_{k=1}^{m} e^{-j\gamma C_k}|z\rangle, \quad e^{-j\gamma C_k} = \begin{cases} I, & C_k = 0 \\ e^{-j\gamma} I, & C_k = 1 \end{cases}, \tag{12.143}$$

indicating that for any $|z\rangle$, the phase term $\exp(-j\gamma)$ is added when condition C_k is satisfied.

The evolution of the initial Hamiltonian $B=H_0=\sum_i X_i$ is given by

$$\begin{aligned} U(H_0,\beta) &= e^{-j\beta H_0} = e^{-j\beta\sum_{i=1}^{N} X_i} \\ &= \prod_{i=1}^{N} e^{-j\beta X_i} = \prod_{i=1}^{N} U(B_i,\beta), \quad U(B_i,\beta) = e^{-j\beta X_i}. \end{aligned} \tag{12.144}$$

where the second line is obtained from the commutativity of X_i and X_j; that is, $[X_i,X_j]=0$. $U(B_i,\beta)$ is just the rotation operator for 2β for the x-axis [33]; that is, $R_x(2\beta)$. Now by applying (12.137), the final state is given by

$$|\beta,\gamma\rangle = U(C,\gamma_p)U(B,\beta_p)\cdots U(C,\gamma_1)U(B,\beta_1)|\psi_0\rangle. \tag{12.145}$$

Now we must choose the parameters β and γ to maximize the expectation of C with respect to state $|\beta,\gamma\rangle$; in other words,

$$C_p = \max_{(\beta,\gamma)} \langle\beta,\gamma|C|\beta,\gamma\rangle. \tag{12.146}$$

As mentioned earlier, the parameters must be carefully chosen to increase accuracy as we increase p; that is, $C_p \geq C_{p-1}$. As the number of steps tends to infinity, we find that

$$\lim_{p\to\infty} C_p = \max_z C(z) = C_{\max}. \tag{12.147}$$

The QAOA is summarized in Fig. 12.23.

Let us now apply the QAOA to the MAX-CUT problem, which represents the problem of partitioning the nodes (vertices) of a given undirected graph $G=\{V,E\}$ (V is the set of vertices and E is the set of edges connecting them), into two sets, say S and $\underline{S}$, such that the number of edges connecting the

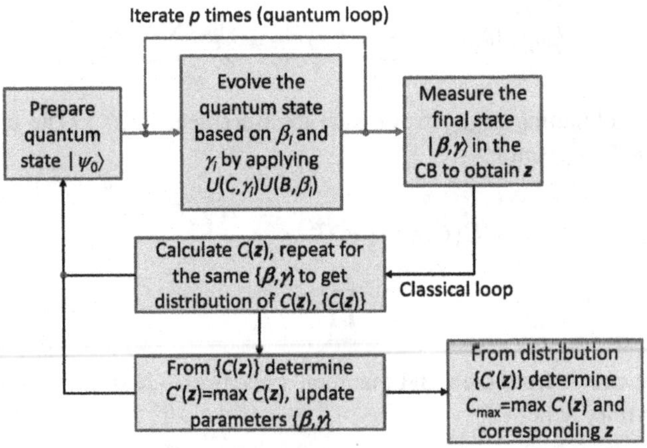

FIGURE 12.23

Quantum approximate optimization algorithm. *CB*, computational basis.

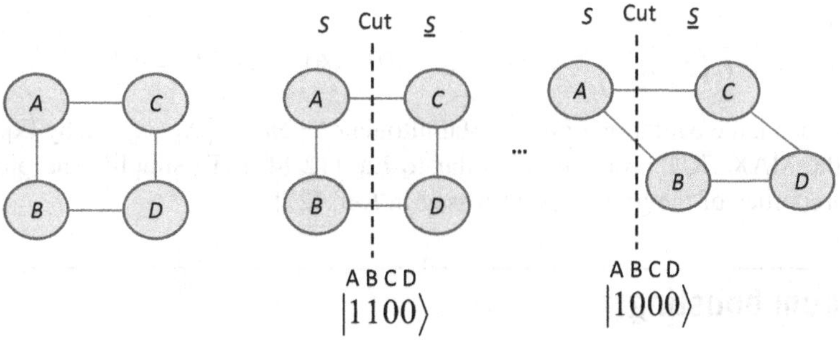

FIGURE 12.24

The MAX-CUT problem.

nodes in opposite sets is maximized. The *cut* δ is defined as the set of edges in the graph separating S and $\underline{S}$. As an illustration, let us consider the square ring graph in Fig. 12.24(left). Since there are $N=4$ nodes, 2^N representations of the graph partitioning exist, with two of these illustrated in Fig. 12.24. We can use 4-bit strings to represent different partitions. If a particular bit position belongs to the set S, we assign the corresponding bit 1, otherwise 0.

Any undirected graph is described by N vertices (nodes) and the edge set $\{\langle j,k\rangle\}$. Let the $C_{\langle j,k\rangle}$ function return one when the edge spanned two nodes j and k in different sets, and 0 otherwise. The corresponding cost function is then

$$C=\sum_{\langle j,k\rangle} C_{\langle j,k\rangle},\ C_{\langle j,k\rangle}=\frac{1}{2}\left(1-z_j z_k\right), \tag{12.148}$$

where $z_i=+1$ if the i-th ($i\in\{j,k\}$) node is located in the set S, otherwise $z_i=-1$. In other words, if the edge $\langle j,k\rangle$ is cut, we obtain $z_j z_k=-1$; if the same edge is not cut, the result is $z_j z_k=1$. The objective function can be represented as an operator:

$$C|z\rangle=\sum_{\langle i,k\rangle} C_{\langle i,k\rangle}(z)|z\rangle=C(z)|z\rangle, \tag{12.149}$$

where the action of operator $C_{\langle j,k\rangle}$ on state $|z\rangle$ is given by

$$C_{\langle i,k\rangle}|z\rangle=\frac{1}{2}(I-Z_i Z_k)|z\rangle=\begin{cases}|z\rangle, & \text{if the edge}\langle i,k\rangle\text{is cut, that is } Z_i Z_k|z\rangle=-|z\rangle\\ 0, & \text{if the edge}\langle i,k\rangle\text{is not cut, that is } Z_i Z_k|z\rangle=|z\rangle\end{cases} \tag{12.150}$$

The corresponding unitary evolution operator for C is given by

$$U(C,\gamma)=e^{-j\gamma C}=e^{-j\gamma\sum_{\langle i,k\rangle} C_{\langle i,k\rangle}}=\prod_{\langle i,k\rangle}\underbrace{e^{-j\gamma C_{\langle i,k\rangle}}}_{U\left(C_{\langle i,k\rangle},\lambda\right)}=\prod_{\langle i,k\rangle}U\left(C_{\langle i,k\rangle},\lambda\right), \tag{12.151}$$

where

$$U\left(C_{\langle i,k\rangle},\lambda\right)=e^{-j\gamma C_{\langle i,k\rangle}}=e^{-j\frac{\gamma}{2}(I-Z_iZ_k)}=e^{-j\frac{-\gamma}{2}Z_iZ_k}e^{-j\frac{\gamma}{2}I}. \tag{12.152}$$

On the other hand, the evolution of initial Hamiltonian $B=H_0=\sum_i X_i$ is given by Eq. (12.144). The workflow for the MAX-CUT problem is similar to Eq. (12.145). To simplify the problem, we can exploit some properties of the graphs as discussed in Ref. [27].

12.4 Quantum boosting

In quantum boosting (QBoost), we use different loss functions than the exponential loss function used in the AdaBoost, described in Section 12.1.1. Similarly as before, for a given dataset $\{(\boldsymbol{x}_1,y_1),\ldots,(\boldsymbol{x}_n,y_n)\}$ we are also given K weak learners, say $\{h_k(\boldsymbol{x})\text{: } R^d\rightarrow\{-1,1\};\, k=1,\ldots,K\}$. Now we need to combine those weak learners appropriately to obtain a composite, strong learner C_m $(m\leq K)$ as follows:

$$C_m(\boldsymbol{w},\boldsymbol{x}_i)=C_{m-1}(\boldsymbol{w},\boldsymbol{x}_i)+w_m h_m(\boldsymbol{x}_i), \tag{12.153}$$

where w_m is the weight derived by minimizing the corresponding cost function. The boosting problem can be interpreted as the following optimization problem [26,34]:

$$\widehat{\boldsymbol{w}}=\arg\min_{\boldsymbol{w}}\left[\frac{1}{N}\sum_{n=1}^{N}\left(\sum_{k=1}^{K}w_k h_k(\boldsymbol{x}_n)-y_n\right)^2+\lambda\|\boldsymbol{w}\|_0\right], \tag{12.154}$$

where the second term is the regularization term with the zero-norm counting the number of nonzero entries. The zero-weight means that a particular weak learner is not included in the model. The smaller hyperparameter λ indicates that the penalty of including more weak learners is lower. This model is continuous since the weights are real-valued; however, what matters is whether a particular weak learner is included. So we can discretize the model by representing the weight with a certain number of bits. We need to modify the minimization problem to make it more suitable for the Hamiltonian encoding, the Ising model representation. So we can take the square of terms in bracket to obtain

$$\widehat{\boldsymbol{w}}=\arg\min_{\boldsymbol{w}}\left\{\frac{1}{N}\sum_{n=1}^{N}\left[\left(\sum_{k=1}^{K}w_k h_k(\boldsymbol{x}_n)\right)^2-2y_n\sum_{k=1}^{K}w_k h_k(\boldsymbol{x}_n)+y_n^2\right]+\lambda\|\boldsymbol{w}\|_0\right\}. \tag{12.155}$$

Given that y_n^2 is just the offset, not w-dependent, we can omit it, and after squaring the summation term and rearranging the summation operations, we obtain

$$\widehat{\boldsymbol{w}}=\arg\min_{\boldsymbol{w}}\left\{\frac{1}{N}\sum_{k=1}^{K}\sum_{l=1}^{K}w_k w_l\left(\sum_{n=1}^{N}h_k(\boldsymbol{x}_n)h_l(\boldsymbol{x}_n)\right)-\frac{2}{N}\sum_{k=1}^{K}w_k\sum_{n=1}^{N}h_k(\boldsymbol{x}_n)y_n+\lambda\|\boldsymbol{w}\|_0\right\}. \tag{12.156}$$

The second term is the bias term, which corresponds to the $\sum_k b_k\sigma_k$-term (the bias term) in the Ising model, with b_k given by $b_k=\sum_n h_k(\boldsymbol{x}_n)y_n$. The b_k is related to the correlation between h_k and $\boldsymbol{y}$. The first term can be related to the Ising model interaction term $\sum\sigma_k\sigma_l J_{kl}$, wherein the coupling strength is given by $J_{kl}=\sum_n h_k(\boldsymbol{x}_n)h_l(\boldsymbol{x}_n)$, representing, therefore, the correlation between the k-th and

l-th weak learners. Upon the quantization of weights, the boosting problem becomes the QUBO problem. If we use the superscript (Q) to denote the quantization terms, we can rewrite the previous optimization problem as follows:

$$\widehat{\boldsymbol{w}} = \arg\min_{\boldsymbol{w}} \left\{ \frac{s}{N} \sum_{k=1}^{K} \sum_{l=1}^{K} w_k^{(Q)} w_l^{(Q)} \left(\sum_{n=1}^{N} h_k(\boldsymbol{x}_n) h_l(\boldsymbol{x}_n) \right) - \frac{2s}{N} \sum_{k=1}^{K} w_k^{(Q)} \sum_{n=1}^{N} h_k(\boldsymbol{x}_n) y_n + \lambda \sum_{k=1}^{K} w_k^{(Q)} \right\}, \tag{12.157}$$

where we introduced the scaling factor s to maintain the desired margins to ensure the mappings h_k: $\boldsymbol{x}_n \rightarrow \{-1/N, 1/N\}$. For one-bit quantization, the minimization problem becomes the QUBO problem. By setting $\sigma_k = 2w_k^{(Q)} - 1$ and using the definitions for b_k and J_{kl} from above, we converted this QUBO problem to the Ising model. Now we can employ either quantum annealing or the QAOA to efficiently solve this problem. For more than one quantization bit, we need to introduce an axillary variable $w_{k,\mathrm{aux}}$ for each weight w_k to enforce the regularization condition by Refs. [26,34]:

$$\lambda \left\| \boldsymbol{w}^{(Q)} \right\| = \sum_{k=1}^{K} \kappa w_k^{(Q)} \left(1 - w_{k,\mathrm{aux}}\right) + \lambda w_{k,\mathrm{aux}}, \tag{12.158}$$

where $w_{k,\mathrm{aux}}$ serves as the indicator having value 1 when $w_k > 0$, and 0 otherwise. The normalization term κ needs to be sufficiently large for proper regularization.

12.5 Quantum random access memory

Classical random-access memory (RAM) allows access to a memory unit with a unique address. With an n-bit address, we can access 2^n memory units. On the other hand, quantum RAM (QRAM) allows us to address a superposition of memory cells with the help of a quantum index (address) register [35]. In other words, the address register is a superposition of addresses $\sum_i p_i |i\rangle_a$, while the output register is a superposition of information in data register d, entangled to the address register as follows [35]:

$$\sum_i p_i |i\rangle_a \xrightarrow{QRAM} \sum_i p_i |i\rangle_a |m_i\rangle_d, \tag{12.159}$$

where m_i is the content of the i-th memory cell. By employing the bucket-brigade architecture, the authors of Ref. [35] have shown that to retrieve an item in QRAM with the address register composed of n qubits, we will need $O(\log 2^n)$ switching cells instead of the 2^n needed in conventional (classical or quantum) RAM. The operating principle of conventional RAM is illustrated by the bifurcation graph in Fig. 12.25. We assume that RAM is composed of 2^n memory cells, placed at the bottom of the bifurcation graph. The address register is composed n flip-flops, as shown in Fig. 12.25, for $n = 4$. The content of the i-th flip-flop corresponds to the i-th level of the graph. If the i-th component of address is 0, we move to the next level using the left edge, otherwise the right edge. The binary representation of the third memory location (with 4 bits) is 0011, and from the root node, we follow the sequence of edges left, left, right, and right to reach the correct memory unit. The same architecture can be applied at the quantum level, as suggested in Ref. [36]; however, the requirement for coherence is too stringent. In the bucket-brigade, the authors of Ref. [35] proposed to place "trit" memory elements at the tree nodes in one of three states: *waiting*, *left*, and *right*. They are all initialized to the low-energy waiting

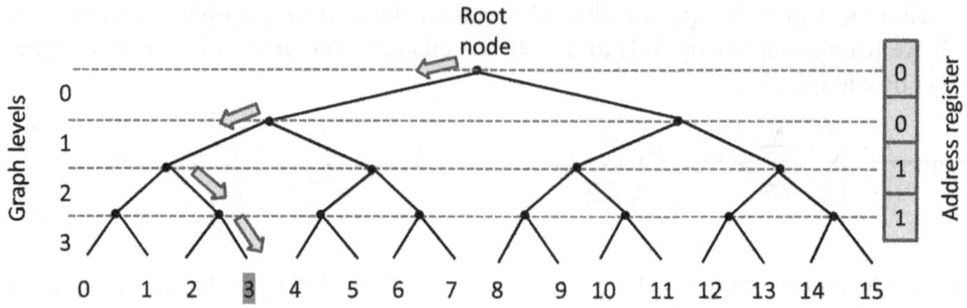

FIGURE 12.25

Operating principle of conventional RAM.

state, as illustrated in Fig. 12.26. The address bits are sent to the graph in a serial fashion. When an incoming bit encounters a waiting state, 0-bit transforms it to the left and 1-bit to the right. If a trit memory unit on a given level is not waiting, it just routs the current bit. When all address bits are sent, the corresponding route is carved into the tree as illustrated in Fig. 12.27 for address 011.

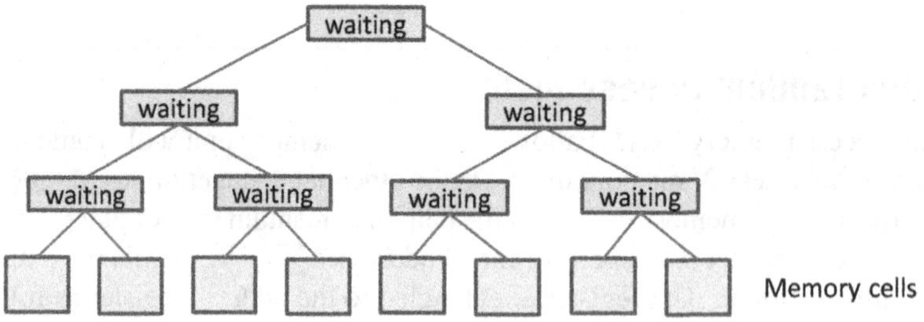

FIGURE 12.26

Initialization of bifurcation graph for bucket-brigade architecture.

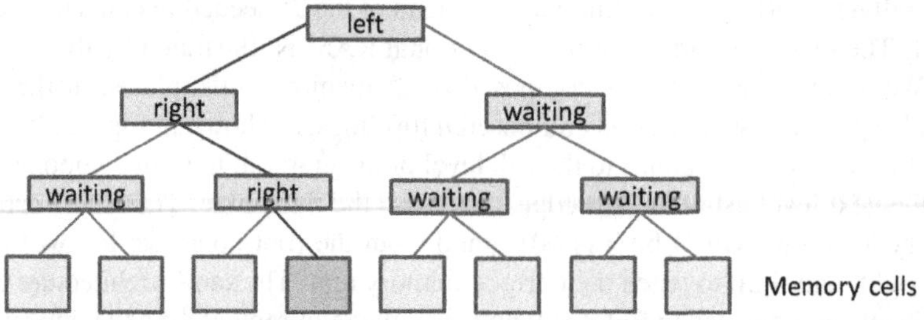

FIGURE 12.27

Addressing bifurcation graph for bucket-brigade architecture for address 011.

In QRAM, the trits are replaced by qutrits in one of three possible quantum states: $|\text{waiting}\rangle$, $|\text{left}\rangle$, and $|\text{right}\rangle$. The address qubits are sent similarly to the manner described above. However, the key difference is that when the route is carved through the tree, the quantum superposition of addresses will carve the superpositions of routes through the graph. The bus qubit is then injected to interact with the memory cells along the routes being activated. The interaction result is then sent back through the graph to write it into the output register through the root node. Finally, the reverse evolution of states is performed on the qutrits to reset all states to the $|\text{waiting}\rangle$ state.

The key advantage of the bucket-brigade procedure is the small number of qutrits needed to retrieve information, which is $\log 2^n$. Based on ion traps, one possible realization is proposed in Ref. [35].

12.6 Quantum matrix inversion

The quantum matrix inversion algorithm, also known as the *Harrow–Hassidim–Lloyd (HHL) algorithm* [37], has often been used as an ingredient of other QML algorithms, including quantum vector support machine [38] and quantum principal component analysis (QPCA) [39–41], and as such, it shall be briefly described here. We are interested in solving the linear system of equations $\boldsymbol{Ax}=\boldsymbol{b}$, wherein $\boldsymbol{A}$ is an $N \times N$ Hermitian matrix, and $\boldsymbol{b}$ is a column vector $\boldsymbol{b}=[b_1 \ldots b_N]^{\mathrm{T}}$. In other words, we are interested in finding $\boldsymbol{x}=\boldsymbol{A}^{-1}\boldsymbol{b}$. If matrix $\boldsymbol{A}$ is not Hermitian or square matrix, we can do matrix inversion of $\boldsymbol{A}'$ matrix instead:

$$\boldsymbol{A}' = \begin{bmatrix} \boldsymbol{0} & \boldsymbol{A}^{\dagger} \\ \boldsymbol{A} & \boldsymbol{0} \end{bmatrix}, \tag{12.160}$$

which is Hermitian and satisfies the following linear system of equations:

$$\begin{bmatrix} \boldsymbol{0} & \boldsymbol{A}^{\dagger} \\ \boldsymbol{A} & \boldsymbol{0} \end{bmatrix} \begin{bmatrix} \boldsymbol{x} \\ \boldsymbol{0} \end{bmatrix} = \begin{bmatrix} \boldsymbol{0} \\ \boldsymbol{b} \end{bmatrix}. \tag{12.161}$$

We map this problem to the corresponding quantum problem as follows:

$$A|x\rangle = |b\rangle, \quad |b\rangle = \begin{bmatrix} b_1 \\ \vdots \\ b_N \end{bmatrix} = \sum_{i=1}^{N} b_i |i\rangle, \tag{12.162}$$

where the matrix representation of Hermitian operator A is given by matrix $\boldsymbol{A}$. To do so, we need to amplitude encode $\boldsymbol{b}$ by $|b\rangle = \sum_i b_i |i\rangle$ and apply Hermitian encoding for $\boldsymbol{A}$. The corresponding evolution operator is defined by $U = \exp(jAt)$, a unitary operator with eigenvalues $\exp(j\lambda_i t)$, where λ_i are eigenvalues of matrix $\boldsymbol{A}$ with corresponding eigenkets being $|u_i\rangle$. The spectral decomposition of A is $A = \sum_i \lambda_i \, |u_i\rangle\langle u_i|$, so the goal is finding $|x\rangle = A^{-1}|b\rangle$. If the matrix $\boldsymbol{A}$ is either sparse or well-conditioned, the run time of the quantum matrix inversion algorithm is $O(\log N \, \kappa^2)$ (κ-the condition number) [37], which represents the exponential speedup over corresponding classical algorithm whose runtime is $O(N\kappa)$. In QML, we are often not interested in finding the $\boldsymbol{x}$ values themselves but rather the result of applying the operator M to $|x\rangle$ or expectation of M, $\langle M\rangle$. The condition number is defined as the ratio of the maximum and minimum eigenvalues of $\boldsymbol{A}$.

When $\boldsymbol{A}$ matrix is diagonal, that is, $\boldsymbol{A} = \text{diag}(\lambda_1,\ldots,\lambda_N)$, we need to perform the transformation $|i\rangle \rightarrow \lambda_i^{-1}|i\rangle$, which can be done probabilistically assuming that the normalization constant c is chosen such that $c/|\lambda| \leq 1$. Now by performing the mapping:

$$|i\rangle \rightarrow \left(\sqrt{1-\frac{c^2}{|\lambda_i|^2}}|0\rangle + \frac{c}{\lambda_i}|1\rangle\right) \otimes |i\rangle, \tag{12.163}$$

and measuring the first qubit until we obtain 1, the resulting state is $A^{-1}|b\rangle$. By applying the unitary operator $U = \exp(jAt)$ to $|b\rangle$ for a superposition of different time instances, we can decompose $|b\rangle$ into eigenkets of $\boldsymbol{A}$, denoted by $|u_i\rangle$, while at the same time we can determine the corresponding eigenvalue λ_i with the help of quantum phase estimation (QPE) algorithm as illustrated in Fig. 12.28. After that decomposition, the system output can be represented as follows:

$$|b\rangle|\mathbf{0}\rangle \rightarrow \sum_i \beta_i |u_i\rangle|\lambda_i\rangle. \tag{12.164}$$

We then perform the conditional rotation on ancilla $|0\rangle$:

$$\sum_i \beta_i |u_i\rangle|\lambda_i\rangle \otimes |0\rangle \rightarrow \sum_i \beta_i |u_i\rangle|\lambda_i\rangle \left(\sqrt{1-\frac{c^2}{|\lambda_i|^2}}|0\rangle + \frac{c}{\lambda_i}|1\rangle\right), \tag{12.165}$$

which is clearly not unitary. Now we need to uncompute the eigenvalue register (by applying QPE^{-1}) to uncorrelate it from the data register ($|b\rangle$-register). Given that the mapping above is not unitary, we need to apply rejection sampling on ancilla (by repeating) until we obtain 1, which effectively performs the following mapping on the data register:

$$\sum_i \beta_i |u_i\rangle \rightarrow \sum_i \beta_i \lambda_i^{-1} |u_i\rangle = A^{-1}|b\rangle. \tag{12.166}$$

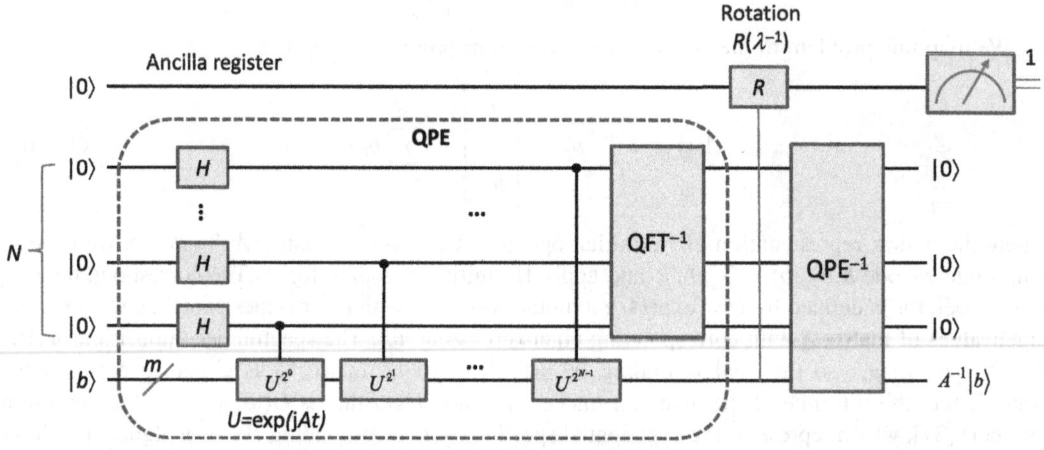

FIGURE 12.28

Quantum matrix inversion.

12.7 Quantum principal component analysis

QPCA is based on the spectral decomposition of a matrix, which is equivalent to simulating the Hamiltonian [26,39]. The quantum simulator approximates the evolution operator exp(jHt), where the Hamiltonian H can be expressed as the simpler Hamiltonians $H=\sum_i H_i$. The goal is to simulate the evolution operator exp(jHt) by a sequence of exponentials exp($-jH_it$) such that the error of the final state does not exceed certain $\varepsilon>0$. The approximation of the evolution operator by a sequence of exponentials is based on the CBH formula [32]:

$$e^X e^Y = e^{X+Y+\frac{1}{2}[X,Y]+\frac{1}{12}[X,[X,Y]]-\frac{1}{12}[Y,[X,Y]]+\cdots} \tag{12.167}$$

When X and Y are sufficiently small in Lie algebra, the series on the right side of exponential is convergent, and we can apply the Lie product formula:

$$e^X e^Y = \lim_{n\to\infty}\left(e^{X/n}e^{Y/n}\right)^n. \tag{12.168}$$

To simulate the evolution operator, we now use the following Trotter decomposition [43]:

$$e^{-jHt} = \left(e^{-jH_1\Delta t}e^{-jH_2\Delta t}\cdots e^{-jH_l\Delta t}\right)^n + \frac{t^2}{2n}\sum_{i>k}[H_i,H_k] + \sum_{p=3}^{\infty}\varepsilon(p),\quad \Delta t = t/n, \tag{12.169}$$

wherein we have split the total simulation time t into n short time intervals $\Delta t = t/n$. When the dimensionality of Hilbert space H_i is d_i, the number of operations needed to simulate exp(jH_it) is $\sim d_i^2$, according to Lloyd [42]. When the simpler Hamiltonians H_i act on local qubits only, the evaluation of exp(jH_it) is more efficient. Based on Lloyd's interpretation, the number of operations needed to simulate exp($-jHt$) is

$$n\left(\sum_{i=1}^{l} d_i^2\right) \le nld^2,\quad d = \max d_i. \tag{12.170}$$

The higher-order terms in Eq. (12.169) are bounded by Ref. [43]:

$$\|\varepsilon(p)\|_{\sup} \le \frac{n}{p!}\|H\Delta t\|_{\sup}^p, \tag{12.171}$$

where we use $\|\bullet\|_{\sup}$ to denote the supremum (maximum expectation of a given operator over all possible states). The total error for approximating exp($-jHt$) with the first term in (12.169) is bounded by $\left\|n\left(e^{-jH\Delta t}-1-jH\Delta t\right)\right\|_{\sup}$. For a given threshold error ε, from the second term in (12.169), we conclude that $\varepsilon\sim t^2/n$. Therefore, the total number of operations required to simulate the evolution operator is O(t^2ld^2/ε). If every Hamiltonian acts on no more than k local qubits, the parameter l is bounded by $l_{\max} = \binom{n}{k} < n^k/k!$, indicating that the number of operations is polynomial in the problem size.

By applying the density operator ρ as a Hamiltonian on another density operator σ, the complexity needed for the component Hamiltonian can be reduced down to O(log d) [41]. By taking the partial trace of the first variable and the swap operator S, we obtain the following [41]:

$$\mathrm{Tr}_P\left(e^{-jS\Delta t}\rho\otimes\sigma e^{-jS\Delta t}\right) = \left(\cos^2\Delta t\right)\sigma + \left(\sin^2\Delta t\right)\rho - j\sin\Delta t\cos\Delta t[\rho,\sigma] \cong \sigma - j\Delta t[\rho,\sigma] + O\left(\Delta t^2\right). \tag{12.172}$$

Given that the swap matrix S is sparse, $\exp(-jS\Delta t)$ can be computed efficiently. Assuming that n copies of ρ are available, we repeat the swap operations on $\rho\otimes\sigma$ to construct the unitary operators $\exp(-jn\rho\Delta t)$ and thus simulate unitary time evolution $\exp(-jn\rho\Delta t)\sigma\exp(jn\rho\Delta t)$. Based on Suzuki-Trotter theory, to simulate $\exp(-j\rho t)$ to accuracy ε, we need $O(t^2\varepsilon^{-1}\,\|\rho-\sigma\|^2)\leq O(t^2\varepsilon^{-1})$ steps, where $t=n\Delta t$. To retrieve ρ using QRAM, we need $O(\log d)$ operations. To implement $\exp(-j\rho t)$, we need $O(t^2\varepsilon^{-1})$ copies of ρ. Therefore, to implement $\exp(-j\rho t)$ with accuracy ε, we will need $O(t^2\varepsilon^{-1}\log d)$ operations.

Let us now apply the *QPE algorithm* for $U=\exp(-j\rho t)$ to determine the eigenkets and eigenvalues of the unknown density matrix. The QPE conditionally executes U for different time intervals and maps the initial state $|\psi\rangle|0\rangle$ to $\sum_i \psi_i|\lambda_i\rangle|\widetilde{r}_i\rangle$, where $|\lambda_i\rangle$ are eigenkets of ρ, $\psi_i=\langle\lambda_i|\psi\rangle$, and $\widetilde{r}_i$ are estimates of corresponding eigenvalues. By applying the QPE on ρ itself, we essentially perform the following mapping [41]:

$$\rho|0\rangle\langle 0|\rightarrow\sum_i r_i|\lambda_i\rangle\langle\lambda_i|\otimes|\widetilde{r}_i\rangle\langle\widetilde{r}_i|. \tag{12.173}$$

By sampling from this state, we can reveal the features of the eigenkets and eigenvalues of ρ. Using multiple copies of ρ to construct eigenvalues and eigenkets is commonly referred to as QPCA.

QPCA is suitable for state discrimination and assignment problems, relevant to supervised and unsupervised learning, particularly classification and clustering. As an illustration, let us assume that two sets of data vectors are available represented by $\{|\psi_i\rangle\}$ $(i=1,\ldots,m)$ and $\{|\phi_i\rangle\}$ $(i=1,\ldots,n)$, which can also be represented using density matrix formalism as follows $\rho=(1/m)\sum_i|\psi_i\rangle\langle\psi_i|$ and $\sigma=(1/n)\sum_i|\phi_i\rangle\langle\phi_i|$. To classify the incoming state $|\zeta\rangle$ we can represent it on the eigenbasis of $\rho-\sigma$. For positive eigenvalues, we assign it to the first set; otherwise, to the second set.

12.8 Quantum optimization-based clustering

In clustering, for a given dataset $\mathfrak{D}=\{\boldsymbol{x}_i\}$ $(i=1,\ldots,N)$, $\boldsymbol{x}_i\in\mathfrak{R}^d$, we need to assign the data points to clusters; in other words, to assign labels based on their similarity. We can use the distance measure between two points $\boldsymbol{x}_i$ and $\boldsymbol{x}_j$ to measure similarity/dissimilarity, and we can create the Gram matrix $\boldsymbol{K}=(K_{ij})$, $K_{ij}=d(\boldsymbol{x}_i,\boldsymbol{x}_j)$. The Euclidean distance is traditionally used for this purpose; however, other similarity measures can be used, such as the scalar product $\boldsymbol{x}_i^T\boldsymbol{x}_j$. Now we can introduce the weighted undirected graph $G=(V,E)$, where V is the set of vertices and E is the set of the edges connecting the vertices. The set of the vertices is represented by the points from the dataset. We can straightforwardly create the fully connected graph, such as the one provided in Fig. 12.29 for $N=4$. In clustering, the distant data points are assigned to different clusters; therefore, maximizing the sum of distances of points with different labels can be used as a clustering approach. Clearly, this problem is equivalent to the MAX-CUT problem discussed in Section 12.3. The weight of the edge between points i and j, denoted as w_{ij}, corresponds to the distance $d(\boldsymbol{x}_i,\boldsymbol{x}_j)$. In the MAX-CUT problem, the goal is to partition the vertices of the graph G into two sets S and S, such that the number of edges connecting the nodes in opposite sets is maximized. The *cut* δ is defined as the set of the edges in the graph separating S and S, which is illustrated in Fig. 12.29. The *cost of the cut* can be defined as the sum of the weight of the edges connecting the vertices in S with vertices in $\underline{S}$; that is, we can write

$$w(\delta(S))=\sum\nolimits_{i\in S,\ j\in\underline{S}}w_{ij}=\sum\nolimits_{i\in S,\ j\in\underline{S}}d(\boldsymbol{x}_i,\boldsymbol{x}_j). \tag{12.174}$$

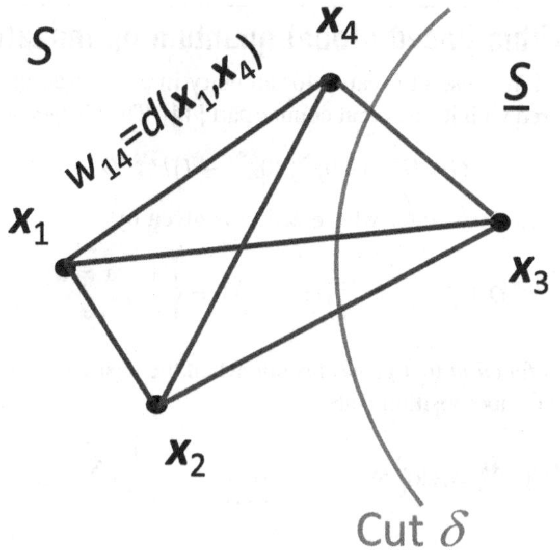

FIGURE 12.29

Relating clustering to MAX-CUT problem.

The MAX-CUT optimization problem can now be defined as

$$\max_{S \subset V} w(\delta(S)) = \sum\nolimits_{i \in S,\ j \in \underline{S}} d(\boldsymbol{x}_i, \boldsymbol{x}_j). \tag{12.175}$$

Let us now map this optimization problem to the Ising model by introducing the Ising spin variables $\sigma_i \in \{-1,1\}$, wherein $\sigma_i = +1$ when $\boldsymbol{x}_i \in S$, otherwise $\sigma_i = -1$ (when $\boldsymbol{x}_i \in \underline{S}$). Given the symmetry of the distances, $w_{ij} = w_{ji}$, we can redefine the cost of the cut as follows:

$$w(\delta(S)) = \sum\nolimits_{i \in S,\ j \in \underline{S}} w_{ij} = \frac{1}{4}\sum_{i,j \in V} w_{ij} - \frac{1}{4}\sum_{i,j \in V} w_{ij}\sigma_i\sigma_j = \frac{1}{4}\sum_{i,j \in V} w_{ij}\left(1 - \sigma_i\sigma_j\right). \tag{12.176}$$

Clearly, when both data points belong to the same cluster (be S or $\underline{S}$), their contribution to the cut cost is 0. By replacing the spin variables with spin operators, we obtain the following Hamiltonian of the corresponding Ising model:

$$H_S = -\frac{1}{4}\sum_{i,j \in V} w_{ij}\left(1 - \sigma_i^z \sigma_j^z\right), \tag{12.177}$$

wherein we introduced minus-sign to ensure that the optimal solution is the minimum energy state. Regarding the summation term, in the Ising model, the nearest neighbors are considered in summation, denoted as $\langle i,j \rangle$. Given that the graph G here is fully connected $\sum_{\langle i,j \rangle}$ is the same as $\sum_{i,j}$. Now we can use the QAOA to determine the minimum energy state as it was done in Section 12.3. Alternatively, quantum annealing can be used.

12.9 Grover algorithm-based global quantum optimization

The Grover search algorithm is used to search for an entry in an unstructured database and provides quadratic speedup compared with its classical counterpart [44]. The Grover search operator is given by

$$G=H^{\otimes n}(2|0\rangle^{\otimes n}\langle 0|^{\otimes n}-I)H^{\otimes n}O, \tag{12.178}$$

where O is the quantum *oracle* operator whose action is given by

$$O_a|x\rangle=(-1)^{f(x)}|x\rangle,\quad f(x)=\begin{cases}1, & x=a\\ 0, & x\neq a\end{cases} \tag{12.179}$$

with $f(x)$ being the *search function* that generates one when the desired item a is found. The action of $H^{\otimes n}$ on $|0\rangle^{\otimes n}$ states is the superposition state:

$$H^{\otimes n}|0\rangle^{\otimes n}=H|0\rangle\otimes\cdots\otimes\underbrace{H|0\rangle}_{(|0\rangle+|1\rangle)/\sqrt{2}}=\frac{1}{\sqrt{2^n}}\sum_{x=0}^{2^n}|x\rangle=|s\rangle, \tag{12.180}$$

so the Grover search operator can also be represented by

$$G=\underbrace{(2|s\rangle\langle s|-I)}_{G_d}O_a=G_dO_a,\quad G_d=2|s\rangle\langle s|-I, \tag{12.181}$$

where G_d is known as the *Grover diffusion operator.* An alternative representation of oracle operator is to consider it a conditional operator with an ancillary qubit being the target qubit, and the corresponding action of the oracle is then

$$O_a|x\rangle|y\rangle=|x\rangle|y\oplus f(x)\rangle, \tag{12.182}$$

and for $|y\rangle=|-\rangle=H|1\rangle=(|0\rangle-|1\rangle)/\sqrt{2}$, we obtain

$$O_a|x\rangle|-\rangle=\frac{1}{\sqrt{2}}|x\rangle|(|0\rangle-|1\rangle)\oplus f(x)\rangle=\begin{cases}\frac{1}{\sqrt{2}}|x\rangle|(|1\rangle-|0\rangle)\rangle=-|x\rangle|-\rangle, & f(x)=1\\ \frac{1}{\sqrt{2}}|x\rangle|(|0\rangle-|1\rangle)\rangle=|x\rangle|-\rangle, & f(x)=0\end{cases}, \tag{12.183}$$

The simplified version of the Grover search algorithm is provided in Fig. 12.30. To obtain the desired entry, we need to apply the Grover search operator $\sim(\pi/4)2^{n/2}$ times. The measurement at the end of the algorithm will reveal the searched item a with a very high probability.

By redefining the oracle operator, the Grover search algorithm can also be used in the optimization of objective function $f(x)$: $\{0,1\}^n\rightarrow R$, as proposed in Refs. [45,46], and the corresponding algorithm can be called the *Grover algorithm-based global optimization.* The corresponding *minimization oracle* operator is given by

$$O_y|x\rangle=(-1)^{h(x)}|x\rangle,\quad h(x)=\begin{cases}1, & f(x)<y\\ 0, & f(x)\geq y\end{cases} \tag{12.184}$$

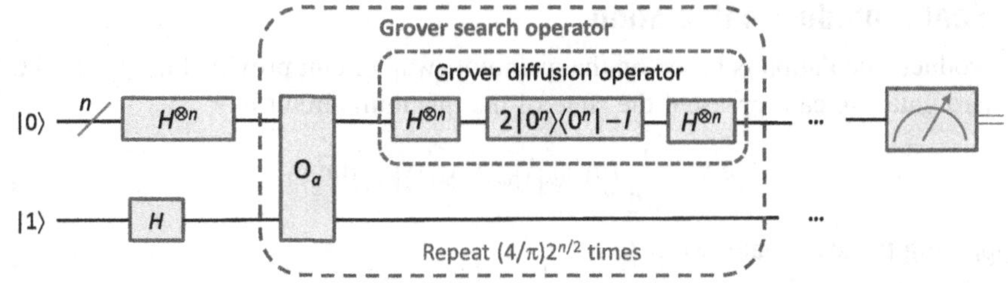

FIGURE 12.30

Quantum circuit to perform Grover search algorithm.

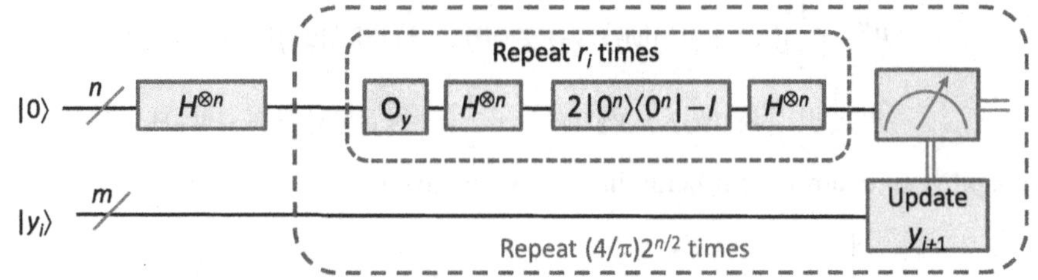

FIGURE 12.31

Quantum circuit to perform Grover search-based minimization algorithm.

To initialize the oracle operator, we randomly select input x_0 and set $y_0 = f(x_0)$ to be a starting point in the algorithm. We also create the superposition state as described above. In the i-th iteration, we apply the Grover search operator G r_i times. The output of the Grover algorithm is state $|x\rangle$; we then calculate $y = f(x)$ and compare the result with y_i. If $y < y_i$, we set $x_{i+1} = x$ and $y_{i+1} = y$, which represents the classical step. We then update the threshold in the oracle function and move to the next iteration, as illustrated in Fig. 12.31. After approximately $O(2^{n/2})$ Grover algorithm iterations, we measure the upper register to determine $x_{\min}$. The number of inner loops r_i is significantly lower than the number of outer loops 2^n. The inner loops are needed to avoid the system becoming trapped in a local minimum.

12.10 Quantum K-means

Many quantum clustering algorithms employ the Grover search algorithm [47]; however, they do not provide exponential speedup over the corresponding classical clustering algorithm. In particular, finding the distance of vectors to the centroids in large dimensional space is computationally complex on a classical computer, and the corresponding quantum algorithm for distance calculation can provide an exponential speedup. To describe the quantum K-means algorithm, we need first to explain how to calculate the dot product efficiently and how to calculate the distance between vectors using the quantum procedure [40].

12.10.1 Scalar product calculation

The scalar product calculation is based on the quantum swap circuit provided in Fig. 12.32. After the first Hadamard gate, we can represent the state of the quantum circuit by

$$|\Psi\rangle = \frac{1}{\sqrt{2}}(|0\rangle|\psi_1\rangle|\psi_2\rangle + |1\rangle|\psi_1\rangle|\psi_2\rangle). \tag{12.185}$$

Upon applying the swap gate, we obtain

$$|\Psi'\rangle = \frac{1}{\sqrt{2}}(|0\rangle|\psi_1\rangle|\psi_2\rangle + |1\rangle|\psi_2\rangle|\psi_1\rangle). \tag{12.186}$$

After the second Hadamard gate, the overall state becomes

$$\begin{aligned} |\Psi''\rangle &= \frac{1}{2}[(|0\rangle + |1\rangle)|\psi_1\rangle|\psi_2\rangle + (|0\rangle - |1\rangle)|\psi_2\rangle|\psi_1\rangle] \\ &= \frac{1}{2}|0\rangle(|\psi_1\rangle|\psi_2\rangle + |\psi_2\rangle|\psi_1\rangle) + \frac{1}{2}|1\rangle(|\psi_1\rangle|\psi_2\rangle - |\psi_2\rangle|\psi_1\rangle). \end{aligned} \tag{12.187}$$

The probability of control qubit being in $|0\rangle$-state is given by

$$\begin{aligned} P(|0\rangle) = |\langle 0|\Psi''\rangle|^2 &= \left|\frac{1}{2}\underbrace{\langle 0|0\rangle}_{1}(|\psi_1\rangle|\psi_2\rangle + |\psi_2\rangle|\psi_1\rangle) + \frac{1}{2}\underbrace{\langle 0|1\rangle}_{0}|1\rangle(|\psi_1\rangle|\psi_2\rangle - |\psi_2\rangle|\psi_1\rangle)\right|^2 \\ &= \frac{1}{4}(\langle\psi_1|\langle\psi_2| + \langle\psi_2|\langle\psi_1|)(|\psi_1\rangle|\psi_2\rangle + |\psi_2\rangle|\psi_1\rangle) = \frac{1}{4}[2\langle\psi_1|\psi_1\rangle\langle\psi_2|\psi_2\rangle + 2\langle\psi_1|\psi_2\rangle\langle\psi_2|\psi_1\rangle] \\ &= \frac{1}{2} + \frac{1}{2}|\langle\psi_1|\psi_2\rangle|^2. \end{aligned} \tag{12.188}$$

$P(|0\rangle) = 0.5$ means that corresponding states are orthogonal, while $P(|0\rangle) = 1$ means that the states are identical. By repeating the measurement multiple times, we can obtain an accurate estimate of $P(|0\rangle)$, and we can determine the magnitude squared of the scalar product by

$$|\langle\psi_1|\psi_2\rangle|^2 = 2P(|0\rangle) - 1. \tag{12.189}$$

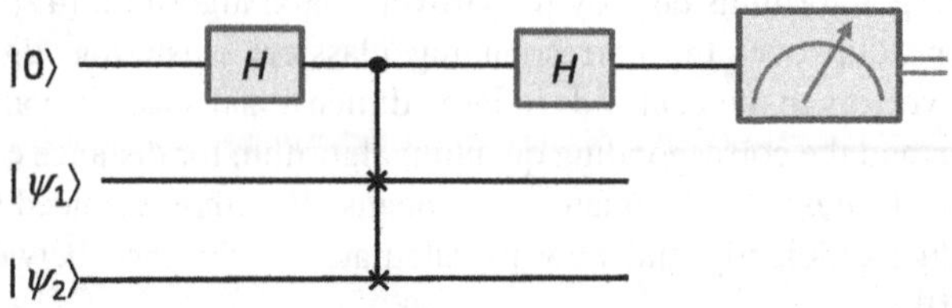

FIGURE 12.32

Quantum swap testing circuit.

12.10.2 Quantum distance calculation

The classical information contained in the vector $\boldsymbol{x}$ can be represented as a quantum state by employing the amplitude encoding, as introduced earlier:

$$\|\boldsymbol{x}\|^{-1}\boldsymbol{x} \rightarrow |x\rangle = \sum_{i=0}^{d-1} \|\boldsymbol{x}\|^{-1} x_i |x_i\rangle, \tag{12.190}$$

and to reconstruct the state, we need to query QRAM $\log_2 d$ times. To determine the distance between two data points $\boldsymbol{x}_i$ and $\boldsymbol{x}_j$, we initialize the following two states [40]:

$$\begin{aligned} |\psi\rangle &= \frac{1}{\sqrt{2}}(|0\rangle|x_i\rangle + |1\rangle|x_j\rangle), \\ |\phi\rangle &= Z^{-1/2}(\|\boldsymbol{x}_i\||x_i\rangle + \|\boldsymbol{x}_j\||x_j\rangle), \; Z = \|\boldsymbol{x}_i\|^2 + \|\boldsymbol{x}_j\|^2. \end{aligned} \tag{12.191}$$

To create the first state, we use QRAM, while for the second state, we create the state $H|1\rangle|0\rangle$ and let it evolve by applying the following Hamiltonian:

$$\hat{H} = (\|\boldsymbol{x}_i\||0\rangle\langle 0| + \|\boldsymbol{x}_j\||1\rangle\langle 1|) \otimes \sigma_X, \tag{12.192}$$

to obtain

$$\begin{aligned} e^{j\hat{H}t} \underbrace{H|1\rangle}_{(|0\rangle-|1\rangle)/\sqrt{2}} |0\rangle &= 2^{-1/2}[\cos(\|\boldsymbol{x}_i\|t)|0\rangle - \cos(\|\boldsymbol{x}_j\|t)|1\rangle]|0\rangle \\ &\quad - j2^{-1/2}[\sin(\|\boldsymbol{x}_i\|t)|0\rangle - \sin(\|\boldsymbol{x}_j\|t)|1\rangle]|1\rangle \end{aligned} \tag{12.193}$$

For a sufficiently small t, by measuring the ancilla qubit, we obtain the desired state $|\phi\rangle$ with a probability of $0.5Z^2t^2$. To determine the Euclidean distance squared between $\boldsymbol{x}_i$ and $\boldsymbol{x}_j$, we first need to run the quantum swap test from Fig. 12.32, by setting $|\psi_1\rangle = |\psi\rangle$ and $|\psi_2\rangle = |\phi\rangle$ to determine the $|\langle\phi|\psi\rangle|^2 = 2P(|0\rangle) - 1$. The Euclidean distance squared can now be determined by

$$\|\boldsymbol{x}_i - \boldsymbol{x}_j\|^2 = 2Z|\langle\phi|\psi\rangle|^2. \tag{12.194}$$

The dot product between $\boldsymbol{x}_i$ and $\boldsymbol{x}_j$ can now be calculated simply by $\left(Z - \|\boldsymbol{x}_i - \boldsymbol{x}_j\|^2\right)/2 = \boldsymbol{x}_i^{\mathrm{T}}\boldsymbol{x}_j$.

12.10.3 Grover algorithm-based *K*-means

Now we are in a position to formulate the Grover algorithm-based quantum K-means algorithm. We randomly initialize the cluster centroids to $\boldsymbol{c}_k$ ($k = 1,\ldots,K$), such as in the K-means++ algorithm [48]. We iterate two inner loops below until convergence is achieved.

- ***Determine the closest cluster centroid.*** Loop over training samples and for each training sample $\boldsymbol{x}_i$ ($i = 1,\ldots,N$) use the *quantum distance calculation* procedure, described in Section 12.10.2, to calculate the distances to each cluster centroids $\|\boldsymbol{x}_i - \boldsymbol{c}_k\|$ ($k = 1,\ldots,K$). Use the *Grover algorithm-based optimization* procedure, described in Section 12.9 (see Fig. 12.31), to determine the index of cluster centroid $c^{(k)}$ that minimizes the distance between training sample and cluster centroid:

$$c^{(\hat{k})} = \arg\min_{k} \|\boldsymbol{x}_i - \boldsymbol{c}_k\|^2. \tag{12.195}$$

- ***Calculate new cluster centroids.*** Loop over clusters C_k ($k = 1,\ldots,K$) and calculate the means of the data points assigned to each cluster as follows:

$$\boldsymbol{c}_k = \frac{1}{|C_k|}\sum_{j\in C_k} \boldsymbol{x}_j, \tag{12.196}$$

where $|C_k|$ denotes the number of vectors assigned to the cluster C_k. Clearly, this step is purely classical.

The convergence is achieved when the distances between cluster centroids at the current and previous iterations are lower than a given threshold ζ:

$$\left\|\boldsymbol{c}_k^{(\text{current})} - \boldsymbol{c}_k^{(\text{previous})}\right\|^2 < \zeta. \tag{12.197}$$

The complexity of classical Lloyd's K-means algorithm is O(N log(Nd)) [40], with factor N coming from the fact that every vector is tested for assignment. To reduce the complexity further, we can formulate the K-means problem as the QUBO problem and apply quantum annealing or the QAOA in a similar fashion as we did in Section 12.8. The complexity can be reduced down to O(Klog(Nd)) or even possibly O(log(KNd)) according to Ref. [40]. Namely, to reduce the complexity from O(N log$N$) to O(log$N$), the output of quantum K-means should not be the list of N vectors and their assignments to clusters, but instead the following superposition clustering state [40]:

$$|\psi_c\rangle = N^{-1/2}\sum_i |c_i\rangle|i\rangle = N^{-1/2}\sum_c\sum_{i\in c_i}|c\rangle|i\rangle, \tag{12.198}$$

wherein the state denotes the superposition of clusters, with each cluster being the superposition of vectors assigned to it. To obtain the statistical picture of the clustering algorithm, we need to sample from this state. The procedure to construct this clustering state by the adiabatic algorithm is provided in the supplementary material of ref. [40], and constructing the clustering state with accuracy ε has O(ε^{-1} K log(KNd)) complexity. When the clusters are well separated, so the adiabatic gap is constant during the adiabatic computing, the complexity can be further reduced down to O(ε^{-1} log(KNd)).

12.11 Quantum support vector machines

The simplest quantum SVM discretizes the cost function and performs the Grover algorithm-based optimization [49] in a fashion similar to that described in Section 12.9. The real strength of such an algorithm is the ability to optimize the nonconvex objective function; however, the speedup over the corresponding classical soft-margin SVM is not exponential.

For the exponential speedup, we should use a quantum SVM with the LS formulation proposed in Ref. [50] and experimentally studied in Ref. [38]. The key idea behind classical LS SVM is to modify the soft-margin optimization as follows [51]:

$$(\widehat{w}, \widehat{b}) = \underset{w,b}{\arg\min}\left(\frac{1}{2}\|w\|^2 + \frac{\gamma}{2}\sum_{j=1}^{N}\zeta_j^2\right), \text{ subject to: } y_j(w^{\mathrm{T}} \cdot \varphi(x_j) - b) = 1 - \zeta_j,\ \zeta_j \geq 0;\ \forall$$
$$j = 1, 2, \cdots, n \tag{12.199}$$

where $\{x_j, y_j\}$ ($x_j \in \Re^d$, $y_j \in \{-1,1\}$) represents the dataset, while the nonlinear mapping $\varphi(x_j)$ is used in the kernel function $k(x_j, x_k) = \varphi(x_j)^{\mathrm{T}} \cdot \varphi(x_k)$. We can reformulate the optimization problem using the Lagrangian method as follows:

$$L(w, b, \alpha, \zeta) = \frac{1}{2}w^{\mathrm{T}}w + \frac{\gamma}{2}\sum_{j=1}^{N}\zeta_j^2 - \sum_{j=1}^{N}\alpha_i\left\{\left[w^{\mathrm{T}} \cdot \varphi(x_j) - b\right] - y_j + \zeta_j\right\},\quad \alpha = [\alpha_1 \cdots \alpha_N]^{\mathrm{T}}. \tag{12.200}$$

Now by taking the first derivatives of L with respect to w, b, α_i, ζ_i and setting them to zero we obtain the following optimality conditions:

$$\begin{aligned}
&\frac{\partial L}{\partial w} = 0 \Rightarrow w = \sum_{j=1}^{N}\alpha_j\varphi(x_j),\\
&\frac{\partial L}{\partial b} = 0 \Rightarrow \sum_{j=1}^{N}\alpha_j = 0,\\
&\frac{\partial L}{\partial \alpha_j} = 0 \Rightarrow y_j = w^{\mathrm{T}} \cdot \varphi(x_j) + b + \zeta_j,\\
&\frac{\partial L}{\partial \zeta_j} = 0 \Rightarrow \alpha_j = \gamma\zeta_j;\ j = 1, \cdots, N.
\end{aligned} \tag{12.201}$$

After eliminating w and ζ_j we obtain the following system of linear equations to solve, instead of the QUBO problem, as follows:

$$\begin{bmatrix} 0 & \mathbf{1}^T \\ 1 & K + \gamma^{-1}I \end{bmatrix}\begin{bmatrix} b \\ \alpha \end{bmatrix} = \begin{bmatrix} 0 \\ y \end{bmatrix},\ y = [y_1 \cdots y_N]^T,\ K = [K_{ij}],\ K_{ij} = \varphi(x_i)^T\varphi(x_j), \tag{12.202}$$

where K is the kernel matrix of size $N \times N$ and I is the identity matrix (of the same size).

To solve for $[b\ \alpha]^T$, we need to invert the matrix:

$$F = \begin{bmatrix} 0 & \mathbf{1}^T \\ 1 & K + \gamma^{-1}I \end{bmatrix} = \underbrace{\begin{bmatrix} 0 & \mathbf{1}^T \\ 1 & 0 \end{bmatrix}}_{J} + \underbrace{\begin{bmatrix} 0 & 0 \\ 1 & K + \gamma^{-1}I \end{bmatrix}}_{K_\gamma} = J + K_\gamma, \tag{12.203}$$

whose complexity on the classical computer is O[$(N+1)^3$]. On the other hand, using quantum matrix inversion (the HHL algorithm), complexity can be significantly reduced. We map this linear system equation solving problem to the corresponding quantum problem as follows:

$$\widehat{F}|b,\boldsymbol{\alpha}\rangle = |\boldsymbol{y}\rangle, \quad \widehat{F} = \boldsymbol{F}/\mathrm{Tr}(\boldsymbol{F}), \quad \|\widehat{F}\| < 1. \tag{12.204}$$

To perform the quantum matrix inversion, we first apply the Lie product by

$$e^{-j\hat{F}\Delta t} = e^{-j\Delta t\boldsymbol{J}/\mathrm{Tr}(\boldsymbol{K}_\gamma)}e^{-j\Delta t\gamma^{-1}\boldsymbol{I}/\mathrm{Tr}(\boldsymbol{K}_\gamma)}e^{-j\Delta t\boldsymbol{K}/\mathrm{Tr}(\boldsymbol{K}_\gamma)} + O(\Delta t^2) \tag{12.205}$$

Given that the matrix $\boldsymbol{J}$ is sparse, we can apply the method due to Berry *et al.* [39] However, the matrix $\boldsymbol{K}$ is not sparse; we can apply QPCA by creating several copies of density matrix ρ and perform the $\exp(-j\rho t)$ as explained in Section 12.7. To create the normalized kernel matrix, we encode the training points by employing Eq. (12.190) and then determine the dot products between $\boldsymbol{x}_i$ and $\boldsymbol{x}_j$ as explained in Section 12.10. The training data oracle performs the following mapping [50]:

$$N^{-1/2}\sum_{i=1}^{N}|i\rangle \to |\chi\rangle = C_\chi^{-1/2}\sum_{i=1}^{N}\|\boldsymbol{x}_i\||i\rangle|\boldsymbol{x}_i\rangle, \quad C_\chi = \sum_{i=1}^{N}\|\boldsymbol{x}_i\|^2, \tag{12.206}$$

with complexity O(log(Nd)). By disregarding the dataset register, we obtain the desired kernel matrix:

$$\mathrm{Tr}_2(|\chi\rangle\langle\chi|) = \sum_{j=1}^{N}\sum_{i=1}^{N}\langle\boldsymbol{x}_j|\boldsymbol{x}_i\rangle\|\boldsymbol{x}_i\|\|\boldsymbol{x}_i\||i\rangle\langle j| = \frac{\boldsymbol{K}}{\mathrm{Tr}(\boldsymbol{K})} = \widehat{K}. \tag{12.207}$$

Now we perform the exponentiation in O(log$N$) steps by Ref. [50]:

$$\begin{aligned} e^{-jL_{\hat{K}}\Delta t}(\rho) &= e^{-j\hat{K}\Delta t}\rho e^{j\hat{K}\Delta t} \cong \mathrm{Tr}_1\left(e^{-jS\Delta t}\widehat{K}\rho e^{jS\Delta t}\right) = \rho - j\Delta t\left[\widehat{K},\rho\right] + O(\Delta t^2), \\ S &= \sum_{m=1}^{N}\sum_{n=1}^{N}|m\rangle\langle n|\otimes|n\rangle\langle m|, \end{aligned} \tag{12.208}$$

with S being the swap matrix. Regarding the vector $|\boldsymbol{y}\rangle$, it can be expanded into eigenkets $|u_k\rangle$ of F, with corresponding eigenvalues being λ_k, with the help of PCE, described in Section 12.8, as follows $|\tilde{y}\rangle = \sum_{k=1}^{N+1}\langle u_k|\tilde{y}\rangle|u_k\rangle$. Now we follow the steps from Section 12.8, including controlled rotation of ancilla qubit and eigenvalue register uncomputation, effectively performing the following mappings [50]:

$$|\tilde{y}\rangle|0\rangle \to \sum_{k=1}^{N+1}\langle u_k|\tilde{y}\rangle|u_k\rangle|\lambda_k\rangle \to \sum_{k=1}^{N+1}\frac{\langle u_k|\tilde{y}\rangle}{\lambda_k}|u_k\rangle. \tag{12.209}$$

The $|b,\,\boldsymbol{\alpha}\rangle$-state can be expressed in the training set labels basis by Ref. [50]:

$$|b,\boldsymbol{\alpha}\rangle = C^{-1/2}\left(b|0\rangle + \sum_{k=1}^{N}\alpha_k|k\rangle\right), \quad C = b^2 + \sum_{k=1}^{N}\alpha_k^2. \tag{12.210}$$

Based on the parameters of the $|b,\,\boldsymbol{\alpha}\rangle$-state, we construct the state

$$|\tilde{u}\rangle = C_{\tilde{u}}^{-1/2}\left(b|0\rangle|0\rangle + \sum_{i=1}^{N}\alpha_i\|\boldsymbol{x}_i\||i\rangle|\boldsymbol{x}_i\rangle\right), \quad C_{\tilde{u}} = b^2 + \sum_{i=1}^{N}\alpha_i^2\|\boldsymbol{x}_i\|^2, \tag{12.211}$$

which is used to *classify the query state* $|\boldsymbol{x}\rangle$, which can be constructed in QRAM as

$$|\widetilde{x}\rangle = C_{\widetilde{x}}^{-1/2}\left(|0\rangle|0\rangle + \sum_{i=1}^{N}\|\boldsymbol{x}\||i\rangle|\boldsymbol{x}\rangle\right), \quad C_{\widetilde{u}} = N\|\boldsymbol{x}\|^2 + 1. \tag{12.212}$$

For classification purpose, we perform the swap text as explained in Section 12.10. The key idea is to construct the state $|\psi\rangle = 2^{-1/2}\left(|0\rangle|\widetilde{u}\rangle + |1\rangle\left|\widetilde{x}\right\rangle\right)$ and measure the overlap with $|\phi\rangle = 2^{-1/2}(|0\rangle + |0\rangle)$. The success probability is

$$P = |\langle\phi|\psi\rangle|^2 = \frac{1}{2}(1 - \langle\widetilde{u}|\widetilde{x}\rangle), \quad \langle\widetilde{u}|\widetilde{x}\rangle = C_{\widetilde{x}}^{-1/2}\, C_{\widetilde{u}}^{-1/2}\left(b + \sum_{i=1}^{N}\alpha_i\|\boldsymbol{x}_i\|\|\boldsymbol{x}\|\langle\boldsymbol{x}_i|\boldsymbol{x}\rangle\right). \tag{12.213}$$

When $P < 1/2$, we classify $|\boldsymbol{x}\rangle$ as $+1$; otherwise we classify it as -1.

12.12 Quantum neural networks

In this section, we describe several representative examples of *quantum neural networks* (QNNs), including feedforward QNNs, the perceptron, and *quantum convolutional neural networks* (QCNNs).

12.12.1 Feedforward quantum neural networks

Feedforward QNNs represent a straightforward generalization of the classical ANNs [52,53]. However, there are two key differences with respect to classical neural networks: (1) quantum operations are reversible and unitary, while the classical activation function is not reversible, and (2) the weighting of the quantum states has no practical importance given that quantum states are of unit length. However, if we interpret the neurons as qubits, we can use the Hamiltonian to describe the QNN, with weights now representing the interaction strength. To solve the reversibility problem, we can add extra inputs denoted as 0-inputs and propagate the data points to the output layer as illustrated in Fig. 12.33.

In this particular example, we assume that the numbers of neurons (of irreversible neural network) at the input, hidden, and output layers are I, J, and K, respectively. For a neural network to be reversible, we need to propagate the inputs to the output layer.

To represent the classical data input x_i as a quantum state, we can use the binary representation $x_i = [x_{i,0}, \ldots, x_{i,L-1}]^T$ and encode it to the quantum state as $|x_i\rangle = |x_{i,0}\rangle \otimes \ldots \otimes |x_{i,L-1}\rangle$. In a similar fashion, the k-th output quantum state is $|y_k\rangle = |y_{k,0}\rangle \otimes \ldots \otimes |y_{k,L-1}\rangle$. The quantum feedforward neural network can now be obtained by replacing the nodes in the hidden layer with corresponding unitary quantum gates U_j ($j = 1,\ldots,J$) as illustrated in Fig. 12.34. The action of the j-th unitary operator on $M = I + K$ qubits can be represented by

$$U_i = \exp\left[j\sum_{m_1,\cdots,m_M=0\cdots0}^{3\cdots3}\alpha_{m_1,\cdots,m_N}^{(i)}\,\sigma_{m_1}^{(i)}\otimes\cdots\otimes\sigma_{m_N}^{(i)}\right], \quad \sigma_{m_i}\in\{I,X,Y,Z\}, \tag{12.214}$$

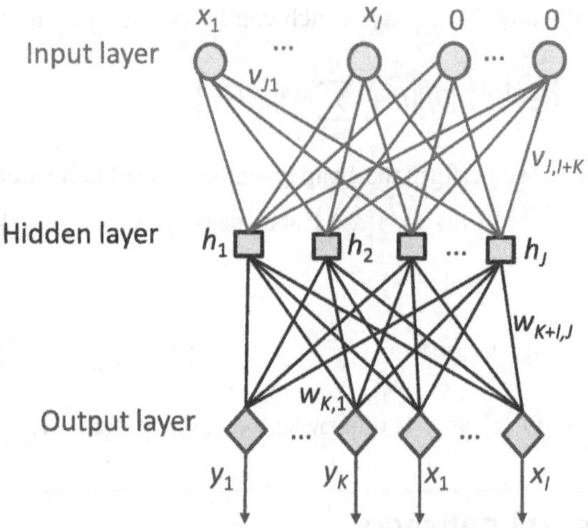

FIGURE 12.33

Reversible feedforward neural network example.

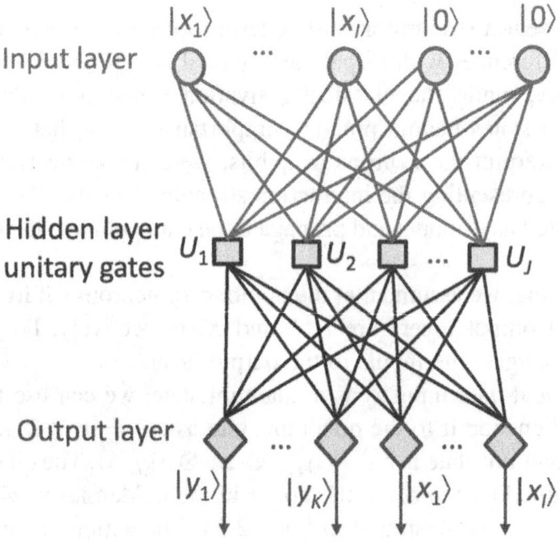

FIGURE 12.34

Quantum feedforward neural network example.

where σ_{m_l} $(l = 1,\ldots,M)$ represent the Pauli matrices. The cost function can be defined as follows:

$$C = -\sum_{n=1}^{N} \langle y_U^{(n)} | y^{(n)} \rangle, \tag{12.215}$$

where $y_U^{(n)}$ is dependent on the unitary operator parameters $\alpha_{m_1,\cdots,m_N}^{(i)}$ and $y^{(n)}$ represents the labels at the output neurons for the n-th training instance $(n = 1,\ldots,N)$. We can use the gradient descent to and update the α-parameters in the i-th unitary matrix as follows:

$$\Delta\alpha_{m_1,\cdots,m_N}^{(i)} = -\eta \frac{\partial C}{\partial \alpha_{m_1,\cdots,m_N}^{(i)}}, \tag{12.216}$$

where η dictates the rate of the update. Quantum gradient descent algorithms would be more adequate for this purpose [54,55]. Another version of the cost function was proposed in Ref. [52]:

$$C = \sum_{n=1}^{N} \sum_{i=1}^{3} f_{i,n}[\langle \sigma_i \rangle_{\text{actual}} - \langle \sigma_i \rangle], \quad f_{i,n} \geq 0, \tag{12.217}$$

where the $\langle \sigma_i \rangle$ are expected values of Pauli operators at individual outputs $y^{(n)}$ (assuming that there exists a single output node $|y\rangle$), while $\langle \sigma_i \rangle_{\text{actual}}$ represent actual expected values of Pauli operators (obtained by the measurements on QNN outputs).

The state of the QNN composed of N quantum neurons can be represented as the multiparticle state living in 2^N-dimensional Hilbert space [56]:

$$|\Psi\rangle = \sum_{i=1}^{2^N} \alpha_i |i_1 i_2 \cdots i_N\rangle, \tag{12.218}$$

where α_i represents the complex amplitude of the i-th QNN basis state. Zak and Williams introduced quantum formalism to capture the nonlinearity and dissipation properties of the QNN by replacing the activation function with a quantum measurement [57]. They were concerned with the unitary quantum walk between computational basis states rather than the evolution of quantum neurons. The evolution, in their formalism, maps the amplitude vector $\boldsymbol{\alpha} = (\alpha_1 \cdots \alpha_i \ldots \alpha_{2^N})$ to $\boldsymbol{\alpha}'$ by applying the unitary operator U; that is, we can write

$$\boldsymbol{\alpha}' = U\boldsymbol{\alpha}. \tag{12.219}$$

The unitary matrix U is of size 2×2^N and describes the transitions among QNN basis states $|i_1 \ldots i_N\rangle$. The projective measurement σ collapses the QNN's superposition state (12.218) into one of the QNN basis states; that is,

$$\boldsymbol{\alpha} = (\alpha_1 \cdots \alpha_i \ldots \alpha_{2^N}) \rightarrow (0_1 \cdots 1_i \; 0_{i+1} \cdots 0_{2^N}), \tag{12.220}$$

which happens with the probability $|\alpha_i|^2$ and represents the nonlinear, dissipative, and irreversible mapping, thus playing the role of the quantum activation function. QNN dynamics are represented by the QNN amplitude vector update from t to $(t+1)$-time instances as follows:

$$\boldsymbol{\alpha}^{(t+1)} = \sigma(U\boldsymbol{\alpha}^{(t)}). \tag{12.221}$$

12.12.2 Quantum perceptron

The concept of the *quantum perceptron* was introduced by Altaisky in 2001 [58] and modeled by the following quantum updating function:

$$|y^{(t)}\rangle = \widehat{F}\sum_{i=1}^{m} \widehat{w}_{iy}^{(t)}|x_i\rangle, \tag{12.222}$$

where $\widehat{F}$ is an arbitrary quantum operator, while $\widehat{w}_{iy}$ operators represent synaptic weight operators acting on m input qubits $|x_i\rangle$. The quantum perceptron training rule is a generalization of the corresponding classical learning rule:

$$\widehat{w}_{jy}^{(t+1)} = \widehat{w}_{jy}^{(t)} + \eta(|d\rangle - |y^{(t)}\rangle)\langle x_i|, \tag{12.223}$$

where $|d\rangle$ is the desired target state, while $|y^{(t)}\rangle$ the state of a quantum neuron at discrete time instance t (and η dictates the learning rate). Unfortunately, the learning rule above is not a unitary operator. To solve this problem, the author in Ref. [59] proposed to use the projector instead of the QNN updated function, defined as $P = |\psi\rangle\langle\psi|$, wherein $\langle\psi|x_1 \ldots, x_N\rangle = |d|$, with d being the desired target output. As an illustration, according to the author, the XOR operation can be performed with the help of the two-input quantum perceptron employing the projector $P = \underbrace{|00\rangle\langle 00| + |11\rangle\langle 11|}_{P_-} + \underbrace{|01\rangle\langle 01| + |10\rangle\langle 10|}_{P_+} = P_- + P_+$. Given that P_- and P_+ are complete and orthogonal, all input states can be unambiguously classified.

12.12.3 Quantum convolutional neural networks

In the rest of this section, we describe the recently proposed QCNNs [60], which can be used in quantum phase recognition and *quantum error correction* (QEC). Classical convolutional neural networks (CNNs) are often used in image recognition problems (see, for example, [61]). The CNN comprises an input layer, multiple hidden layers, and an output layer. The hidden layers further consist of a sequence of interleaved layers known as convolutional layers, pooling layers, and fully connected layers, which is illustrated in Fig. 12.35. The mapping between different layers is known as feature mapping. Convolutional layers "convolve" the input and forward the corresponding results to the next layer. The operation that convolution layers apply is not convolution at all, but rather the sliding scalar product calculation. In the l-th convolution layer, the new pixel values $x_{i,j}^{(l)}$ are calculated from the corresponding ones in the neighborhood, but from the previous layer, so we can write $x_{i,j}^{(l)} = \sum_{m,n}^{w} w_{m,n} x_{i+m,j+n}^{(l-1)}$, wherein $w_{m,n}$ are corresponding weights from a window of $w \times w$ pixels. The purpose of the pooling layers is to reduce the dimensions of the hidden layer by combining the outputs of neuron clusters at the previous layer into a single neuron in the next layer. Local pooling combines small clusters of typical sizes 2×2. On the other hand, global pooling acts on all the neurons of the previous, hidden layer. The pooling operation may compute either a max or an average operation of

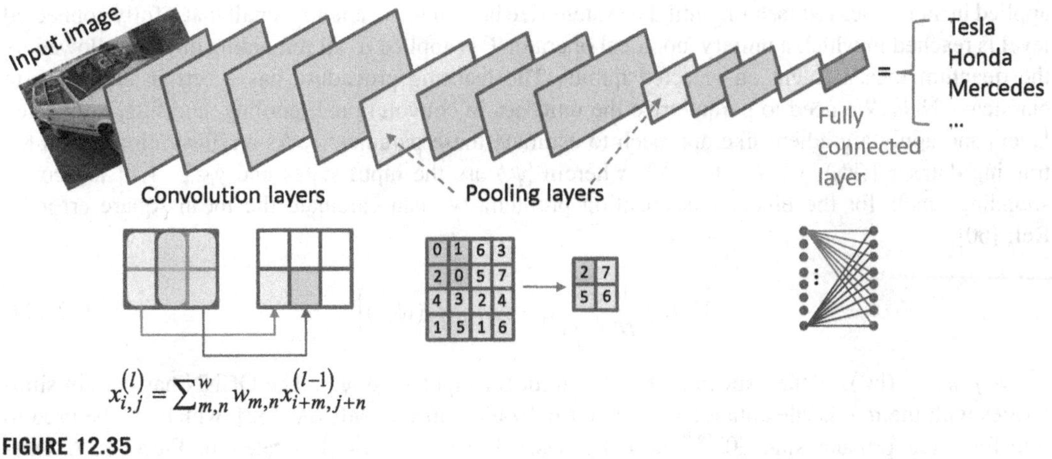

FIGURE 12.35

Simplified classical convolutional neural network example.

small neuron clusters in the previous layer. Max pooling calculates the maximum value from each of a cluster of neurons at the prior layer. On the other hand, the average pooling calculates the average value from each cluster of neurons at the prior layer. Fully connected layers connect every neuron in the previous layer to every neuron in the next layer. Namely, once the size of the feature map becomes reasonably small, the final output gets computed based on all remaining neurons in the previous layer. The weights of the fully connected map are optimized based on training by a large dataset, while the weights on convolutional and pooling layers are typically fixed.

The QCNN introduced in Ref. [60] applies the CNN concept but on quantum states rather than images, which is illustrated in Fig. 12.36. The input to the QCNN is a quantum state $|\psi_{\text{in}}\rangle$. The convolution layer is composed of quasi-local unitary operators U_i, which are translationally invariant. In poling level, some of the qubits undergo quantum measurement, and their classical output is used to control unitary rotation operators V_i. Quantum measurements introduce nonlinearity to the system and simultaneously allow us to reduce the number of qubits. The convolution and pooling layers are

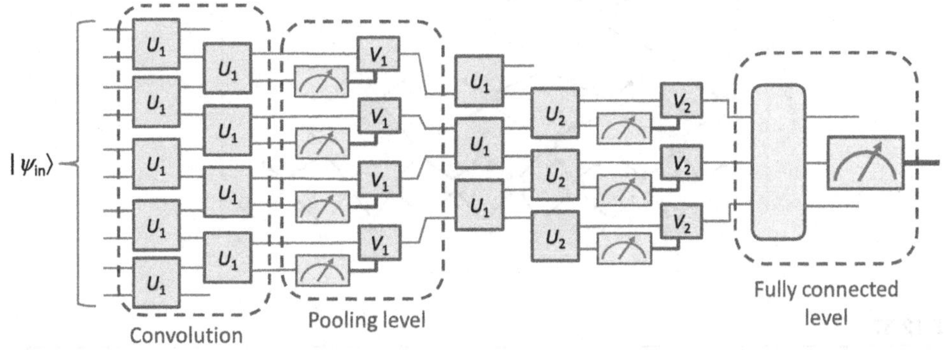

FIGURE 12.36

Quantum convolutional neural network example.

applied in an interleaved fashion until the system size becomes sufficiently small that a fully connected level is reached in which a unitary, nonlocal operator F is applied to all remaining qubits, followed by the quantum measurement on selected qubits. The learning procedure has a certain similarity to classical CNNs. We need to parametrize the unitaries in convolutional, pooling, and fully connected layers and apply a gradient-like approach to estimate these parameters. As an illustration, given the training dataset $\{(|\psi_i\rangle, y_i)\ |i=1,\ldots,N\}$, wherein $|\psi_i\rangle$ are the input states and $y_i \in \{-1,1\}$ are corresponding labels for the binary classification problem, we can calculate the mean square error by Ref. [60]:

$$MSE = \frac{1}{N}\sum_{i=1}^{N}\left[y_i - f_{(U_j,V_j,F)}(|\psi_i\rangle)\right], \tag{12.224}$$

where $f_{(Uj,Vj,F)}(|\psi_i\rangle)$ is the expected QCNN output for input state $|\psi_i\rangle$. The QCNN has certain similarities with the multiscale entanglement renormalization ansatz (MERA) [62], which can be used to transform the product state $|0\rangle^{\otimes N}$ into the desired state $|\psi\rangle$, as illustrated in Fig. 12.37. One-dimensional MERA requires the use of $2N-1$ gates that are organized in O(log$N$) layers. The MERA is composed of three types of tensors [62]: (1) top tensor $t=u|0\rangle|0\rangle$, (2) isometric tensors $w=u|0\rangle$, and (3) unitary gates. The MERA can also be used to transform an arbitrary product state $\otimes_{n=1}^{N}|\phi_n^{[n]}\rangle$ into a different entangled state (compared with the transformation of initial state $|0\rangle^{\otimes N}$), denoted as $\left|\psi_{\{\phi_n\}}\right\rangle$. Therefore, the QCNN can be interpreted as a particular realization of MERA run in the opposite direction. Some of the measurements in the QCNN can be related to quantum syndrome measurements, relevant in QEC, thus improving the reliability of the QCNN.

The authors of Ref. [60] have shown that the QCNN can be successfully used to solve the quantum phase recognition problem, that is, to verify whether a given quantum many-body system in ground state $|\psi_g\rangle$ belongs to the quantum phase P. If yes, the output of the final measurement, after fully connected layer, will give the result 1.

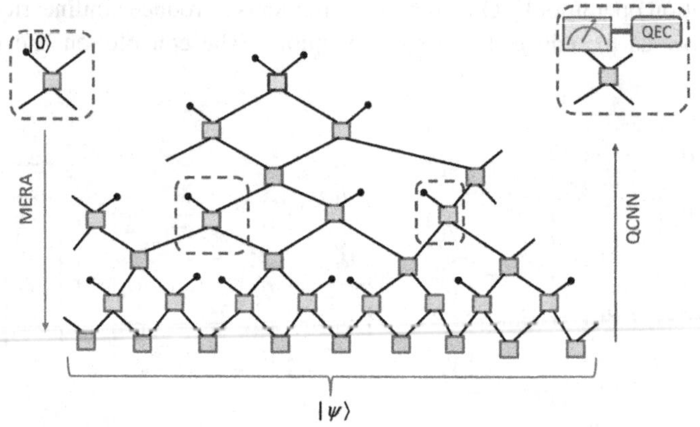

FIGURE 12.37

Relationship between quantum convolutional neural network (QCNN) and multiscale entanglement renormalization ansatz (MERA).

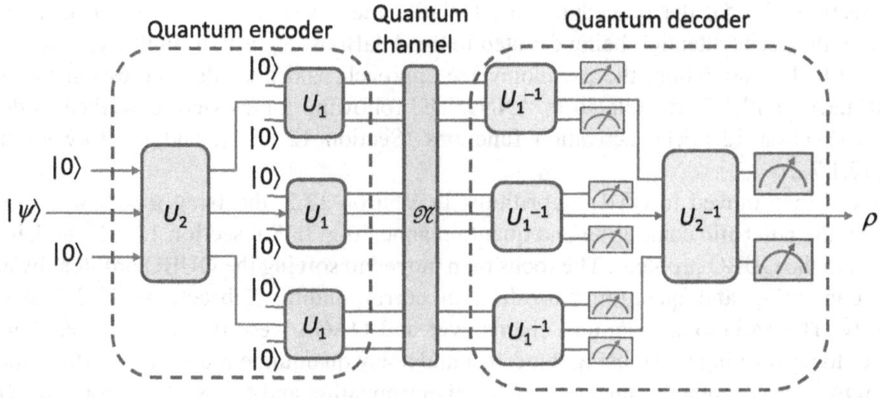

FIGURE 12.38

Quantum convolutional neural network (QCNN)-based [9,1] quantum error correction scheme encoding single information qubit into a nine-qubit codeword. The goal is to design the QCNN to maximize $\langle\psi|\rho|\psi\rangle$.

The QCNN can also be used in QEC, as suggested by the authors of Ref. [60]. Given the quantum error channel model N, the goal is determining the encoding E_q and decoding D_q quantum circuits such the recovery fidelity is maximized:

$$F_q = \sum_{|\psi\rangle} \langle\psi|\mathfrak{D}_q\{N[E_q(|\psi\rangle\langle\psi|)]\}|\psi\rangle. \tag{12.225}$$

To demonstrate the efficiency of the QCNN in QEC, they have designed a [9,1] QEC, with configuration provided in Fig. 12.38, significantly outperforming the corresponding Shor's code in a correlated quantum channel model. Another interesting QNN example, suitable for classification problems, is provided in Ref. [63].

12.13 Summary

This chapter was devoted to the relevant both classical and QML algorithms. In Section 12.1.1, the relevant classical MLAs, categorized as supervised, unsupervised, and RL algorithms, were briefly introduced. In the same section, the concept of the feature space was introduced as well as the corresponding generalization performance of MLAs. The following topics from classical ML were described: PCA, SVMs, clustering, boosting, regression analysis, and neural networks. The PCA section (Section 12.1.2) described how to determine the principal components from the correlation matrix, followed by a description of SVD-based PCA and the PCA scoring phase. Section 12.1.3 was related to SVM-based algorithms, which were described from both geometric and Lagrangian method-based points of view. For classification problems that are not linearly separable, the concept of soft margins was introduced. For classification problems with nonlinear decision boundaries, the kernel method was described. Section 12.1.4 was devoted to the clustering problems. The following clustering algorithms were described: K-means (Section 12.1.4.1), EM (12.1.4.3), and K-NN (12.1.4.4). In the same section, the clustering quality evaluation (Section 12.1.4.2) was discussed. In boosting section (Section 12.1.5), the procedure how to build the strong learner from weak learners was described, with special attention being devoted to the AdaBoost algorithm. In the regression analysis Section 12.1.6, LS estimation, the pseudoinverse approach, and the ridge regression method were described. Section 12.1.7 was related to ANNs. The following topics were described in detail: the perceptron (Section 12.1.7.1), activation functions (Section 12.1.7.1), and feedforward networks (Section 12.1.7.2).

The focus then moved to QML algorithms. In Section 12.2, the Ising model was introduced, followed by the adiabatic computing and quantum annealing. In Subsection 12.2.1, the Ising model was related to the QUBO problem. The focus then moved to solving the QUBO problem by adiabatic quantum computing and quantum annealing in corresponding Subsections 12.2.2 and 12.2.3, respectively. The variational quantum eigensolver and QAOA were discussed in Section 12.3 as approaches to performing QML using imperfect and noisy quantum circuits. To illustrate the impact of the QAOA, it was applied to the combinatorial optimization and MAX-CUT problems. Quantum boosting was discussed in Section 12.4, and it was related to the QUBO problem. Section 12.5 was devoted to QRAM, allowing us to address the superposition of memory cells with the help of a quantum register. Quantum matrix inversion (the HHL algorithm) was described in Section 12.6, which can be used as a basic ingredient for other QML algorithms such as QPCA, which was described in Section 12.7. In the quantum optimization-based clustering Section 12.8, we have described how the MAX-CUT problem can be related to the clustering and thus be solved by adiabatic computing, quantum annealing and the QAOA. In Section 12.9, the Grover algorithm-based quantum optimization was discussed. In the quantum K-means section (Section 12.10), we have described how to calculate the dot product and the quantum distance and how to use them in the Grover search-based K-means algorithm. In the quantum SVM Section 12.11, we formulated the SVM problem using LS and described how to solve it with the help of the HHL algorithm. In the QNN Section 12.12, we described feedforward QNNs, quantum perceptron, and quantum convolutional networks.

References

[1] M. Stamp, Introduction to Machine Learning with Applications in Information Security, CRC Press, Boca Raton, 2018.

[2] S. Raschka, V. Mirjalili, Python Machine Learning, third ed., Packt Publishing, Birmingham-Mumbai, 2019.

[3] J. Unpingco, Python for Probability, Statistics, and Machine Learning, second ed., Springer Nature Switzerland AG, Cham, Switzerland, 2019.

[4] V. Feldman, V. Guruswami, P. Raghavendra, Y. Wu, Agnostic learning of monomials by halfspaces is hard, in: Proc. 2009 50th Annual IEEE Symposium on Foundations of Computer Science, Atlanta, GA, 2009, pp. 385–394.

[5] V. Feldman, V. Guruswami, P. Raghavendra, Y. Wu, Agnostic learning of monomials by halfspaces is hard, SIAM J. Comput. 41 (6) (2012) 1558–1590.

[6] J. Shlens, A Tutorial on Principal Component Analysis, April 2014. https://arxiv.org/abs/1404.1100.

[7] I.B. Djordjevic, Advanced Optical and Wireless Communications Systems, Springer International Publishing, Cham: Switzerland, December 2017.

[8] E. Biglieri, Coding for Wireless Channels, Springer, 2005.

[9] A. Goldsmith, Wireless Communications, Cambridge University Press, Cambridge, 2005.

[10] M. Cvijetic, I.B. Djordjevic, Advanced Optical Communications and Networks, Artech House, January 2013.

[11] V. Vapnik, A. Lerner, Pattern recognition using generalized portrait method, Autom. Rem. Contr. 24 (1963) 774–780.

[12] V. Vapnik, A.Y. Chervonenkis, A class of algorithms for pattern recognition learning, Avtomat. i Telemekh. 25 (6) (1964) 937–945.

[13] A.P. Dempster, N.M. Laird, D.B. Rubin, Maximum likelihood from incomplete data via the EM algorithm, J. Roy. Stat. Soc. B 39 (1) (1977) 1–38.

[14] D. Forsyth, Applied Machine Learning, Springer Nature Switzerland, Cham, Switzerland, 2019.

[15] P. Wittek, Quantum Machine Learning, Elsevier/Academic, Amsterdam-Boston, 2014.

[16] T. Hastie, R. Tibshirani, J. Friedman, The Elements of Statistical Learning: Data Mining, Inference, and Prediction, second ed., Springer Science+Business Media, New York, NY, 2009.

[17] C. Giraud, Introduction to High-Dimensional Statistics, CRC Press, Boca Raton, FL, 2015.

[18] O. Lund, M. Neilsen, C. Lundergaard, C. Kesmir, S. Brunak, Biological Bioinformatics, MIT Press, Cambridge, MA, 2005.

[19] G.H. Wannier, Statistical Physics, Dover Publications, Inc., New York, 1987.

[20] M. Born, V. Fock, Beweis des adiabatensatzes, Z. Phys. 51 (1–3) (1928) 165–180.

[21] W. van Dam, M. Mosca, U. Vazirani, How powerful is adiabatic quantum computation?, in: Proc. the 42nd Annual Symposium on Foundations of Computer Science Newport Beach, , CA, USA, 2001, pp. 279–287.

[22] T. Kadowaki, H. Nishimori, Quantum annealing in the transverse Ising model, Phys. Rev. E 58 (5) (1998) 5355–5363.

[23] A.B. Finnila, et al., Quantum annealing: a new method for minimizing multidimensional functions, Chem. Phys. Lett. 219 (5–6) (1994) 343–348.

[24] W. Vinci, T. Albash, G. Paz-Silva, I. Hen, D.A. Lidar, Quantum annealing correction with minor embedding, Phys. Rev. A 92 (4) (2015) 042310.

[25] C.C. McGeoch, C. Wang, Experimental evaluation of an adiabatic quantum system for combinatorial optimization, in: Proc. ACM International Conference on Computing Frontiers (CF 13), 2013, pp. 23: 1–23:11.

[26] P. Wittek, Quantum Machine Learning: What Quantum Computing Means to Data Mining, Elsevier/Academic Press, Amsterdam-Boston, 2014.

[27] E. Farhi, J. Goldstone, S. Gutmann, A Quantum Approximate Optimization Algorithm, 2014, pp. 1–16. Available at: https://arxiv.org/abs/1411.4028.

[28] A. Peruzzo, et al., A variational eigenvalue solver on a photonic quantum processor, Nat. Commun. 5 (2014) 4213.

[29] J. R McClean, J. Romero, R. Babbush, A. Aspuru-Guzik, The theory of variational hybrid quantum-classical algorithms, New J. Phys. 18 (2016) 023023.

[30] J. Preskill, Quantum computing in the NISQ era and beyond, Quantum 2 (2018) 79.

[31] J. Choi, J. Kim, A tutorial on quantum approximate optimization algorithm (QAOA): fundamentals and applications, in: Proc. 2019 International Conference on Information and Communication Technology Convergence (ICTC), Jeju Island, S. Korea, 2019, pp. 138–142.

[32] R. Bhatia, Matrix Analysis, Graduate Texts in Mathematics, vol. 169, Springer-Verlag, New York, 1997.

[33] I.B. Djordjevic, Quantum Information Processing: An Engineering Approach, Elsevier/Academic Press, Amsterdam-Boston, 2012.

[34] H. Neven, V.S. Denchev, G. Rose, W.G. Macready, Training a Binary Classifier with the Quantum Adiabatic Algorithm, November 4, 2008. Available at: https://arxiv.org/abs/0811.0416.

[35] V. Giovannetti, S. Lloyd, L. Maccone, Quantum random access memory, Phys. Rev. Lett. 100 (16) (2008) 160501.

[36] M.A. Nielsen, I.L. Chuang, Quantum Computation and Quantum Information, Cambridge University Press, Cambridge, 2000.
[37] A.W. Harrow, A. Hassidim, S. Lloyd, Quantum algorithm for solving linear systems of equations, Phys. Rev. Lett. 103 (15) (2009) 150502.
[38] Z. Li, X. Liu, N. Xu, J. Du, Experimental realization of a quantum support vector machine, Phys. Rev. Lett. 114 (2015) 140504.
[39] D.W. Berry, G. Ahokas, R. Cleve, B.C. Sanders, Efficient quantum algorithms for simulating sparse Hamiltonians, Commun. Math. Phys. 270 (2) (2007) 359−371.
[40] S. Lloyd, M. Mohseni, P. Rebentrost, Quantum Algorithms for Supervised and Unsupervised Machine Learning, November 4, 2013. Available at: https://arxiv.org/abs/1307.0411.
[41] S. Lloyd, M. Mohseni, P. Rebentrost, Quantum principal component analysis, Nat. Phys. 10 (2014) 631−633.
[42] S. Lloyd, Universal quantum simulators, Science 273 (5278) (1996) 1073−1078.
[43] G. Benenti, G. Casati, D. Rossini, G. Strini, Principles of Quantum Computation and Information: A Comprehensive Textbook, World Scientific, Singapore, 2019.
[44] L.K. Grover, A Fast Quantum Mechanical Algorithm for Database Search, November 19, 1996. Available at: https://arxiv.org/abs/quant-ph/9605043.
[45] C. Dürr, P. Høyer, A Quantum Algorithm for Finding the Minimum, July 18, 1996. Available at: https://arxiv.org/abs/quant-ph/9607014.
[46] A. Ahuja, S. Kapoor, A Quantum Algorithm for Finding the Maximum, November 18, 1999. Available at: https://arxiv.org/abs/quant-ph/9911082.
[47] E. Aïmeur, G. Brassard, S. Gambs, Quantum speed-up for unsupervised learning, Mach. Learn. 90 (2) (2013) 261−287.
[48] D. Arthur, S. Vassilvitskii, k-means++: the advantages of careful seeding, in: Proc. 18th Annual ACM-SIAM Symposium on Discrete Algorithms, Society for Industrial and Applied Mathematics, Philadelphia, PA, USA, 2007, pp. 1027−1035.
[49] D. Anguita, S. Ridella, F. Rivieccio, R. Zunino, Quantum optimization for training support vector machines, Neural Netw. 16 (5−6) (2003) 763−770. Special issue: Advances in neural networks research.
[50] P. Rebentrost, M. Mohseni, S. Lloyd, Quantum Support Vector Machine for Big Data Classification, July 10, 2014. Available at: https://arxiv.org/abs/1307.0471v3.
[51] J.A. Suykens, J. Vandewalle, Least squares support vector machine classifiers, Neural Process. Lett. 9 (3) (1999) 293−300.
[52] K.H. Wan, O. Dahlsten, H. Kristjnsson, R. Gardner, M.S. Kim, Quantum generalisation of feedforward neural networks, npj Quant. Inf. 3 (36) (2017).
[53] G. Verdon, M. Broughton, J. Biamonte, A Quantum Algorithm to Train Neural Networks Using Low-Depth Circuits, 2017. Available at: https://arxiv.org/abs/1712.05304.
[54] I. Kerenidis, A. Prakash, Quantum Gradient Descent for Linear Systems and Least Squares, April 17, 2017. Available at: https://arxiv.org/abs/1704.04992.
[55] P. Rebentrost, M. Schuld, L. Wossnig, F. Petruccione, S. Lloyd, Quantum Gradient Descent and Newton's Method for Constrained Polynomial Optimization, December 6, 2016. Available at: https://arxiv.org/abs/1612.01789.
[56] M. Schuld, I. Sinayskiy, F. Petruccione, The quest for a quantum neural network, Quant. Inf. Process. 13 (11) (2014) 2567−2586.
[57] M. Zak, C.P. Williams, Quantum neural nets, Int. J. Theor. Phys. 3 (2) (1998) 651−684.
[58] M.V. Altaisky, Quantum Neural Network, July 5, 2001. Available at: https://arxiv.org/abs/quant-ph/0107012.
[59] M. Siomau, A quantum model for autonomous learning automata, Quant. Inf. Process. 13 (2014) 1211−1221.

[60] I. Cong, S. Choi, M.D. Lukin, Quantum convolutional neural networks, Nat. Phys. 15 (2019) 1273–1278.
[61] A. Krizhevsky, I. Sutskever, G.E. Hinton, ImageNet classification with deep convolutional neural networks, Commun. ACM 60 (6) (2017) 84–90.
[62] G. Vidal, Class of quantum many-body states that can be efficiently simulated, Phys. Rev. Lett. 101 (2008) 110501.
[63] E. Farhi, H. Neven, Classification with Quantum Neural Networks on Near Term Processors, February 16, 2018. Available at: https://arxiv.org/abs/1802.06002.

Further reading

[1] J.S. Otterbach, et al., Unsupervised Machine Learning on a Hybrid Quantum Computer, December 15, 2017. Available at: https://arxiv.org/abs/1712.05771.
[2] C.G.J. Jacobi, Über ein leichtes Verfahren, die in der Theorie der Säkularstörungen vorkommenden Gleichungen numerisch aufzulösen, Crelle's J. 30 (1846) 51–94 (in German).
[3] G.H. Golub, H.A. van der Vorst, Eigenvalue computation in the 20th century, J. Comput. Appl. Math. 123 (1–2) (2000) 35–65.

CHAPTER

Fault-tolerant quantum error correction

13

Chapter outline

13.1 Fault-tolerance basics

One of the most powerful applications of quantum error correction is protecting quantum information as it dynamically undergoes quantum computation [1–6]. Imperfect quantum gates affect quantum computation by introducing errors in computed data. Moreover, imperfect control gates introduce errors into the processing sequence because the wrong operations are applied. As a result the quantum error correction coding (QECC) scheme now must deal with errors introduced by not only the quantum channel but also imperfect quantum gates during the encoding/decoding process. Because of this, the reliability of data processed by a quantum computer is not a priori guaranteed by QECC. The reason is threefold: (1) the gates used for encoders and decoders are composed of imperfect gates, including imperfect control gates; (2) syndrome extraction applies unitary operators to entangle ancilla qubits

Quantum Communication, Quantum Networks, and Quantum Sensing. https://doi.org/10.1016/B978-0-12-822942-2.00009-1

with the code block; and (3) the error recovery action requires controlled operation to correct errors. Nevertheless, it can be shown that arbitrary good quantum error protection can be achieved even with imperfect gates, provided that the error probability per gate is below a certain *threshold*; this claim is known as the accuracy threshold theorem [2–4].

From the discussion above, it is clear that QECC does not a priori improve the reliability of a quantum information processing (QIP) module or a quantum communication system. So the purpose of *fault-tolerant design* is to ensure the reliability of the QIP module given the threshold by properly implementing fault-tolerant quantum gates. As in previous chapters, we observe the error model having the following properties: (1) qubit errors occur independently; (2) X, Y, and Z qubit errors occur with the same probability; (3) the error probability per qubit is the same for all qubits; and (4) errors introduced by the imperfect quantum gate only affect qubits acted on by that gate. The last condition (property) is added to ensure that errors introduced by imperfect gates do not propagate and cause catastrophic errors. Based on this error mode, we define fault-tolerant operation as follows:

Definition 1. We say that a quantum operation is fault-tolerant if the occurrence of a single gate/storage error during the operation does not produce more than one error on each encoded qubit block.

In other words, an operation is fault-tolerant if up to the first order in single-error probability p (i.e., $O(p)$), the operation generates no more than one error per block. The key idea of fault-tolerant QIP is constructing fault-tolerant operations to perform the logic on encoded states. Note that someone may argue that if the QECC we use is sufficiently strong, we could allow for more than one error per code block. However, the QECC scheme is typically designed to deal with random errors introduced by quantum channels or storage devices. Allowing for too many errors due to imperfect gates may exceed the error correction capability of QECC when both random and imperfect gate errors are present. We simplify the fault-tolerant design by restricting the number of faulty gate errors to one.

Example. Let us observe the SWAP gate introduced in Chapter 3, which interchanges the states of two qubits: $U_{\text{SWAP}}^{(12)}|\psi_1\rangle|\psi_2\rangle = |\psi_2\rangle|\psi_1\rangle$. If the SWAP gate is imperfect, it can introduce errors on both qubits, and based on definition 1 above, this gate is faulty. However, if we introduce an additional auxiliary state, we can fault-tolerantly perform the SWAP operation as follows: $U_{\text{SWAP}}^{(23)}U_{\text{SWAP}}^{(12)}U_{\text{SWAP}}^{(13)}|\psi_1\rangle|\psi_2\rangle|\psi_3\rangle = |\psi_2\rangle|\psi_1\rangle|\psi_3\rangle$. For example, let us assume that a storage error occurs on qubit 1, leading to $|\psi_1\rangle \rightarrow |\psi_1'\rangle$, before the SWAP operation takes place. The resulting state will be $|\psi_2\rangle|\psi_1'\rangle|\psi_3\rangle$, thus, a single error occurring during the SWAP operation results in only one error per block, and therefore, this modified gate has fault-tolerant operation. Another related definition commonly used in fault-tolerant QIP, namely *transversal operation,* can be shown to be fault-tolerant.

Definition 2. An operation that satisfies one of the following two conditions is transversal: (i) it employs only one-qubit gates to the qubits in a code block, and (ii) the ith qubit in one code block interacts only with the ith qubit in a different code block or a block of ancilla qubits.

Condition (i) is consistent with the fault-tolerance definition. Let us now observe two-qubit operations, particularly the operation of a gate involving the ith qubits of two different blocks. Based on the error model above, only these two qubits can be affected by this faulty gate. The transversal operation is fault-tolerant since these two interacting qubits belong to two different blocks. The transversal operation definition is quite useful, as it is much easier to verify than the fault-tolerant definition. In the upcoming section, to familiarize the reader with fault-tolerant QIP concepts, we will develop fault-tolerant gates assuming that the QECC scheme is based on Steane code [6].

13.2 Fault-tolerant quantum information processing concepts

In this section, we introduce basic fault-tolerant computation concepts. This section aims to gradually introduce the reader to fault-tolerant concepts before moving to a more rigorous exposition in Section 13.3.

13.2.1 Fault-tolerant Pauli gates

The key idea of fault-tolerant QIP is to construct fault-tolerant operations to perform logic on encoded states. The encoded Pauli operator for Pauli X and Z operators for Steane [7,1] code is as follows:

$$\overline{Z} = Z_1Z_2Z_3Z_4Z_5Z_6Z_7 \quad \overline{X} = X_1X_2X_3X_4X_5X_6X_7 \tag{13.1}$$

The encoded Pauli gate Y can be easily implemented as

$$\overline{Y} = \overline{X}\,\overline{Z} \tag{13.2}$$

Let us now determine whether the encoded Pauli gates are transversal. Based on the quantum channel model from the previous section, it is clear that the single-qubit error does not propagate. Because every encoded gate is implemented in a bitwise fashion, the encoded Pauli gates are transversal and consequently fault-tolerant. The implementation of fault-tolerant Pauli gates is shown in Fig. 13.1.

13.2.2 Fault-tolerant Hadamard gate

An encoded Hadamard gate H should interchange X and Z under conjugation the same way an unencoded Hadamard gate interchanges Z and X:

$$\overline{H} = H_1H_2H_3H_4H_5H_6H_7 \tag{13.3}$$

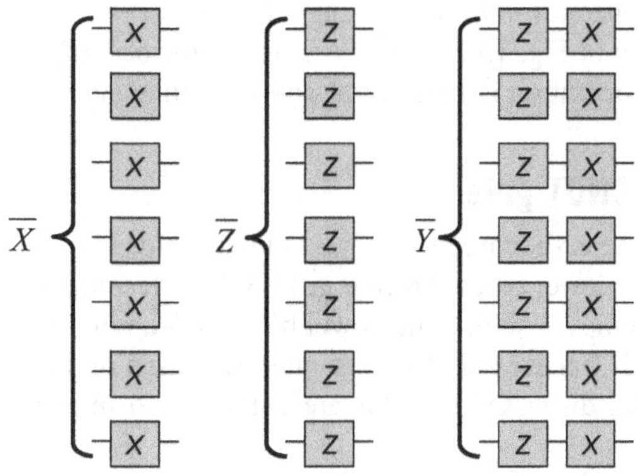

FIGURE 13.1

Transversal Pauli gates on a qubit encoded using Steane [7,1] code.

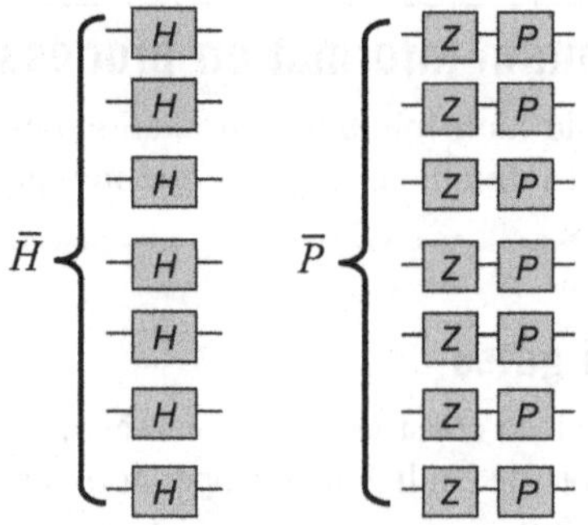

FIGURE 13.2

Transversal Hadamard and phase gates on a qubit encoded using Steane [7,1] code.

A fault-tolerant Hadamard gate implementation is shown in Fig. 13.2 (left). A single error occurring on one qubit propagates no further, indicating that the encoded Hadamard gate is transversal and consequently fault-tolerant.

13.2.3 Fault-tolerant phase gate (*P*)

Under conjugation, the phase gate P maps Z to Z and X to Y as follows:

$$PZP^{\dagger} = Z \quad PXP^{\dagger} = Y \tag{13.4}$$

The fault-tolerant phase gate $\overline{P}$ should perform the same action on encoded Pauli gates. The bitwise operation of $\overline{S} = S_1S_2S_3S_4S_5S_6S_7$ on encoded Pauli Z and X gates gives

$$\overline{P}\,\overline{Z}\,\overline{P}^{\dagger} = \overline{Z} \quad \overline{P}\,\overline{X}\,\overline{P}^{\dagger} = -\overline{Y}, \tag{13.5}$$

indicating that the sign in the conjugation operation of the encoded Pauli-X gate is incorrect, which can be fixed by inserting the Z operator in front of a single P operator so that the fault-tolerant phase gates can be represented as shown in Fig. 13.2 (right). A single error occurring on one qubit propagates no further, indicating that the operation is transversal and fault-tolerant.

13.2.4 Fault-tolerant CNOT gate

The fault-tolerant CNOT gate is shown in Fig. 13.3. Let us assume that the ith qubit in the upper block interacts with ith qubit of the lower block. From Fig. 13.3, it is clear that the error in the ith qubit of the upper block will only affect the ith qubit of the lower block, so condition (ii) of definition 2 is satisfied, so the gate from Fig. 13.3 is transversal and consequently (based on definition 1) fault-tolerant. The fault-tolerant quantum gates discussed thus far are sufficient to implement arbitrary encoders and decoders fault-tolerantly.

13.2.5 Gate fault-tolerant π/8 (*T*) gate

For a complete set of universal gates, a nontransversal gate such as the $\pi/8$ (T) gate must be implemented. To implement the T-gate fault-tolerantly, we must apply the following three-step procedure [3,6]:

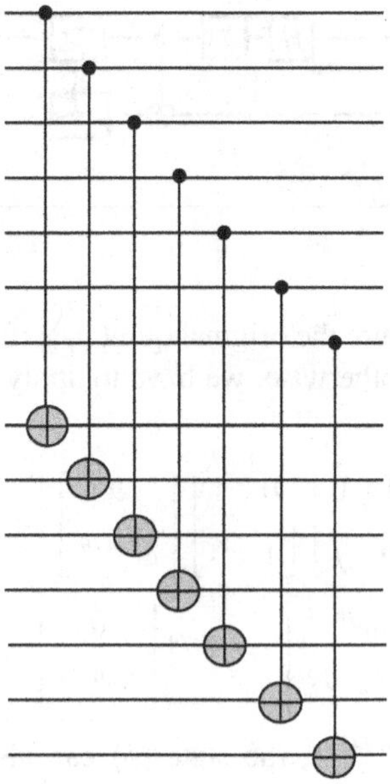

FIGURE 13.3

Fault-tolerant CNOT gate implementation.

1. Prepare an ancilla in the base state and apply a Hadamard followed by a non-fault-tolerant T gate:

$$|\phi\rangle = TH|0\rangle = T\frac{|0\rangle + |1\rangle}{\sqrt{2}} = \frac{|0\rangle + e^{j\pi/4}|1\rangle}{\sqrt{2}} \tag{13.6}$$

2. Apply a CNOT gate on the prepared state $|\phi\rangle$ as the control qubit; and on state $|\psi\rangle$, we want to transform as the target qubit:

$$\begin{aligned} U_{\text{CNOT}}^{(12)}|\phi\rangle|\psi\rangle &= U_{\text{CNOT}}^{(12)}\frac{|0\rangle + e^{j\pi/4}|1\rangle}{\sqrt{2}}\otimes(a|0\rangle + b|1\rangle) \\ &= \frac{1}{\sqrt{2}}U_{\text{CNOT}}^{(12)}\left[|0\rangle(a|0\rangle + b|1\rangle) + e^{j\pi/4}|1\rangle(a|0\rangle + b|1\rangle)\right] \\ &= \frac{1}{\sqrt{2}}(a|0\rangle|0\rangle + b|0\rangle|1\rangle) + e^{j\pi/4}(a|1\rangle|1\rangle + b|1\rangle|0\rangle) \\ &= \frac{1}{\sqrt{2}}\left[\left(a|0\rangle + be^{j\pi/4}|1\rangle\right)|0\rangle + \left(b|0\rangle + ae^{j\pi/4}|1\rangle\right)|1\rangle\right] \end{aligned} \tag{13.7}$$

where we use superscript (12) to denote the action of the CNOT gate from qubit 1 (control-qubit) to qubit 2 (target-qubit).

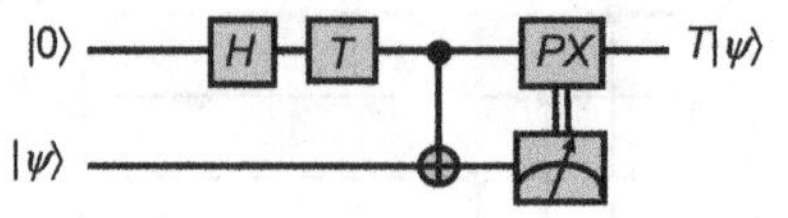

FIGURE 13.4

Fault-tolerant *T*-gate implementation.

3. In the third step, we need to measure the original qubit $|\psi\rangle$: if the result of a measurement is zero, the final state is $a|0\rangle + be^{j\pi/4}|1\rangle$; otherwise, we have to apply the *PX* gate on the original ancilla qubit as follows:

$$PX\begin{bmatrix} b \\ ae^{j\pi/4} \end{bmatrix} = \begin{bmatrix} 1 & 0 \\ 0 & j \end{bmatrix}\begin{bmatrix} 0 & 1 \\ 1 & 0 \end{bmatrix}\begin{bmatrix} b \\ ae^{j\pi/4} \end{bmatrix} = \begin{bmatrix} 0 & 1 \\ j & 0 \end{bmatrix}\begin{bmatrix} b \\ ae^{j\pi/4} \end{bmatrix}$$
$$= \begin{bmatrix} ae^{j\pi/4} \\ be^{j\pi/2} \end{bmatrix} = e^{j\pi/4}\begin{bmatrix} a \\ be^{j\pi/4} \end{bmatrix} = e^{j\pi/4}T|\psi\rangle \quad (13.8)$$

Because $THZHT^{+} = TXT^{+} = e^{-j\pi/4}PX$, the state $|\phi\rangle$ can also be fault-tolerantly prepared from $|0\rangle$by measuring *PX:* if the result is $+1$, the preparation was successful; otherwise, the procedure must be repeated. Fault-tolerant *T* gate implementation is shown in Fig. 13.4. In this implementation, we assume the measurement procedure is perfect. The fault-tolerant measurement procedure is discussed in the upcoming section.

13.2.6 Fault-tolerant measurement

A quantum circuit for measuring a single qubit operator *U* is shown in Fig. 13.5A. Its operation was described in Chapter 3. Clearly, this circuit is faulty. The first idea of performing the fault-tolerant measurement is to put *U* in the transversal form *U*′, as illustrated in Fig. 13.5B, and then perform the measurement in a bitwise fashion with the same control ancilla. Unfortunately, such a measurement is not fault-tolerant because an error in the ancilla qubit would affect several data qubits. We can make the circuit shown in Fig. 13.5B fault-tolerant using three ancilla qubits as shown in Fig. 13.6.

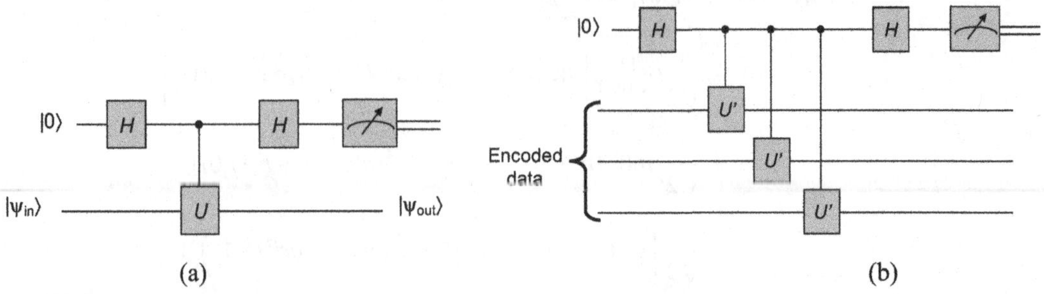

FIGURE 13.5

Faulty measurement circuits: (A) original circuit and (B) modified circuit obtained by putting *U* in transversal form.

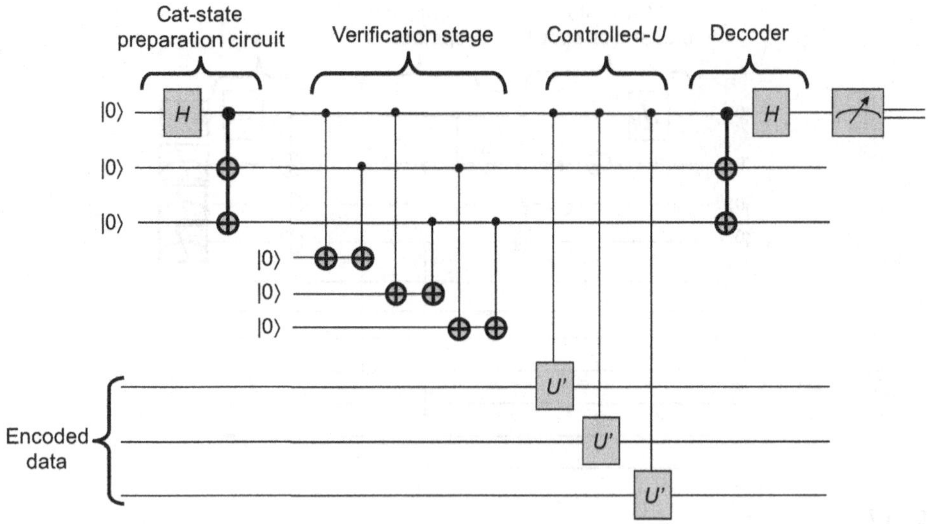

FIGURE 13.6

Fault-tolerant measurement circuit.

This new circuit has several stages. The first stage is the cat-state $(|000\rangle+|111\rangle)/\sqrt{2}$ preparation circuit. The encoder in this stage is very similar to the three-qubit flip code encoder described in Chapter 7. The second stage is the verification stage, which is similar to the three-qubit flip code error correction circuit. Namely, this stage is used to verify whether the state after the first stage is indeed the cat-state. The third stage is a controlled U stage similar to that in Fig. 13.5B. The fourth stage is the decoder stage, which returns ancilla states to their original states. As expected, this circuit has similarities with the three-qubit code decoder. The final measurement stage can be faulty. To reduce its importance on qubit error probability, we perform the whole measurement procedure several times and apply majority rule for the final measurement result.

13.2.7 Fault-tolerant state preparation

The fault-tolerant state preparation procedure can be described as follows. We prepare the desired state and apply the verification procedure measurement section above. If the verification step is successful, that state is used for further quantum computation; otherwise, we restart the verification procedure with new fresh ancilla qubits.

13.2.8 Fault-tolerant measurement of stabilizer generators

A fault-tolerant measurement procedure has been introduced for an observable corresponding to a single qubit. A similar procedure can be applied to an arbitrary stabilizer generator. For example, the fault-tolerant measurement circuit of the first stabilizer generator of Steane code $g_1 = X_1X_5X_6X_7$ is shown in Fig. 13.7. After this elementary introduction of fault-tolerant concepts, gates, and procedures, we turn our attention to a more rigorous description.

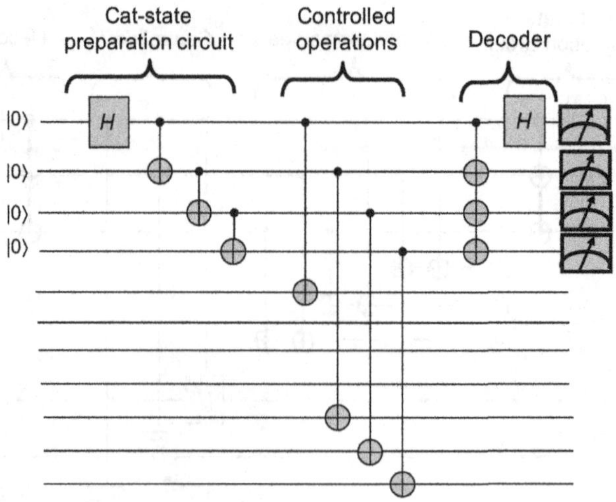

FIGURE 13.7

Fault-tolerant measurement for the generator $X_1X_5X_6X_7$.

13.3 Fault-tolerant quantum error correction

This section is devoted to fault-tolerant quantum error correction. Before we proceed with this topic, we will review some basic stabilizer code concepts introduced in Chapters 7 and 8. Let S be the largest abelian subgroup of G_N (the N-qubit Pauli multiplicative error group) that fixes all elements $|c\rangle$ from quantum code C_Q, called the stabilizer group. The $[N,K]$ stabilizer code C_Q is defined as the K-dimensional subspace of the N-qubit Hilbert space H_2^N, as follows:

$$C_Q = \bigcap_{s\in S}\{|c\rangle \in H_2^N \,|\, s|c\rangle = |c\rangle\}. \tag{13.9}$$

A very important concept of quantum stabilizer codes is the syndrome of an error E from G_N. Let C_Q be a quantum stabilizer code with generators $g_1, g_2,\ldots,g_{N-K}$; the error syndrome of $E \in G_N$ can then be defined as the bit string $S(E) = [\lambda_1\lambda_2 \ldots \lambda_{N-K}]^{\mathrm{T}}$ with the component bits determined by

$$\lambda_i = \begin{cases} 0, [E, g_i] = 0 \\ 1, \{E, g_i\} = 0 \end{cases} \quad (i = 1,\ldots,N-K) \tag{13.10}$$

It was shown in Chapter 8 that the corrupted codeword $E|c\rangle$ is a simultaneous eigenket of generators g_i $(i = 1,\ldots, N-K)$, so we can write

$$g_i(E|c\rangle) = (-1)^{\lambda_i}E(g_i|c\rangle) = (-1)^{\lambda_i}E|c\rangle. \tag{13.11}$$

By inspection of Eqs. (13.10) and (13.11), it can be concluded that eigenvalues $(-1)^{\lambda_i}$ (and consequently the syndrome components λ_i) can be determined by measuring g_i when codeword $|c\rangle$ is corrupted by an error E. By representing error E and generator g_i as products of corresponding X- and

Z-containing operators, as was done in Chapter 8, namely, $E = X(\boldsymbol{a}_E)Z(\boldsymbol{b}_E), g_i = X(\boldsymbol{a}_i)Z(\boldsymbol{b}_i)$, where the nonzero positions in $\boldsymbol{a}$ ($\boldsymbol{b}$) give the locations of X- (Z-) operators. Thus, based on Eq. (13.11), we can write

$$g_i E = (-1)^{\boldsymbol{a}_E \cdot \boldsymbol{b}_i + \boldsymbol{b}_E \cdot \boldsymbol{a}_i} E g_i, \tag{13.12}$$

where $\boldsymbol{a}_E \cdot \boldsymbol{b}_i + \boldsymbol{b}_E \cdot \boldsymbol{a}_i$ denotes the symplectic product (the bitwise addition is mod 2). From Eqs. (13.11) and (13.12), the ith component of the syndrome can be obtained by

$$\lambda_i = \boldsymbol{a}_E \cdot \boldsymbol{b}_i + \boldsymbol{b}_E \cdot \boldsymbol{a}_i \bmod 2; i = 1, \cdots, N - K, \tag{13.13}$$

By inserting identity operators represented in terms of a unitary operator U as $I = U^\dagger U$, into Eq. (13.11), the following is obtained:

$$g_i \overbrace{U^\dagger U}^{I} E \overbrace{U^\dagger U}^{I} |c\rangle = (-1)^{\lambda_i} E \overbrace{U^\dagger U}^{I} |c\rangle \overset{\cdot U(\text{from the left})}{\Leftrightarrow} (U g_i U^\dagger)(U E U^\dagger) U|c\rangle = (-1)^{\lambda_i} (U E U^\dagger) U|c\rangle \tag{13.14}$$

By introducing the notation $\overline{g}_i = U g_i U^\dagger, \overline{E} = U E U^\dagger$, and $|\overline{c}\rangle = U|c\rangle$, the right side of Eq. (13.14) can be rewritten as

$$\overline{g}_i \overline{E} |\overline{c}\rangle = (-1)^{\lambda_i} \overline{E} |\overline{c}\rangle. \tag{13.15}$$

Therefore, the corrupted image of the codeword $\overline{E}|\overline{c}\rangle$ is a simultaneous eigenket of the images of generators $\overline{g}_i$. Similarly, as in Ref. Eq. (13.14), we can show that

$$g_i|c\rangle = |c\rangle \Leftrightarrow g_i \overbrace{U^\dagger U}^{I} |c\rangle = |c\rangle \overset{\cdot U(\text{from the left})}{\Leftrightarrow} U g_i \overbrace{U^\dagger U}^{I} |c\rangle = U|c\rangle \Leftrightarrow \overline{g}_i |\overline{c}\rangle = |\overline{c}\rangle, \tag{13.16}$$

which will be used in the next section.

13.3.1 Fault-tolerant syndrome extraction

To further simplify the exposition of fault-tolerant quantum error correction, we will assume for the moment that generators $g_i = X(\boldsymbol{a}_i)Z(\boldsymbol{b}_i)$ do not contain Y-operators; in other words, $a_{ij} b_{ij} = 0 \;\; \forall j$. In syndrome extraction, the ancilla qubits interact with a codeword to encode the syndrome $S(E)$ information into ancilla qubits and then perform appropriate measurements on ancilla qubits to reveal the syndrome vector $S(E)$. We will show that the syndrome information $S(E)$ can be incorporated by applying the transversal Hadamard $\overline{H}$ and transversal CNOT gate $\overline{U}_{\text{CNOT}}$ on the corrupted codeword $E|c\rangle$ as follows:

$$\overline{H}_i \overline{U}_{\text{CNOT}}(\boldsymbol{a}_i + \boldsymbol{b}_i) \overline{H}_i E|c\rangle \otimes |\boldsymbol{A}\rangle = E|c\rangle \otimes \sum_{\boldsymbol{A}_{\text{even}}} |\boldsymbol{A}_{\text{even}} + \boldsymbol{b}_E \cdot \boldsymbol{a}_i + \boldsymbol{a}_E \cdot \boldsymbol{b}_i\rangle, \tag{13.17}$$

where $|\boldsymbol{A}\rangle$ is the *Shor state*, defined by

$$|\boldsymbol{A}\rangle = \frac{1}{2^{(\text{wt}(g_i)-1)/2}} \sum_{\boldsymbol{A}_{\text{even}}} |\boldsymbol{A}_{\text{even}}\rangle, \tag{13.18}$$

with summation performed for all even-parity bit strings of length $\mathrm{wt}(g_i)$, denoted $\boldsymbol{A}_{\text{even}}$. The Shor state is obtained by first applying the cat-state preparation circuit on $|\boldsymbol{0}\rangle_{\mathrm{wt}(g_i)} = \underbrace{|0\rangle \otimes \cdots \otimes |0\rangle}_{\mathrm{wt}(g_i)\ \text{times}}$ ancilla qubits, as explained in the corresponding text of Fig. 13.7, followed by the set of $\mathrm{wt}(g_i)$ Hadamard gates, as shown in Fig. 13.8. Therefore, we can represent the Shor state in terms of cat-state $|\psi_{\text{cat}}\rangle$ as follows:

$$|\boldsymbol{A}\rangle = \prod_{j=1}^{\mathrm{wt}(g_i)} H_j|\psi_{\text{cat}}\rangle, \ |\psi_{\text{cat}}\rangle = \frac{1}{\sqrt{2}}[|0\cdots0\rangle + |1\cdots1\rangle]. \tag{13.19}$$

It is evident from Fig. 13.8 that the stage of Hadamard gates is transversal, as the ith error in the observed block introduces errors only on the ith qubit of the corresponding interacting block. On the other hand, the error introduced in an arbitrary inner CNOT gate of a cat-state preparation circuit propagates, and the qubits below are affected, indicating that the cat-state preparation circuit is not fault-tolerant. We will return to this problem and Shor state verification later in this section after first introducing the transversal Hadamard gate $\overline{H}$ and transversal CNOT gate $\overline{U}_{\text{CNOT}}$. The transversal Hadamard gate $\overline{H}$ and transversal CNOT gate $\overline{U}_{\text{CNOT}}$, related to the generator

$$g_i = X(\boldsymbol{a}_i)Z(\boldsymbol{b}_i); \boldsymbol{a}_i = [a_{i1}\cdots a_{iN}]^{\mathrm{T}}, \boldsymbol{b}_i = [b_{i1}\cdots b_{iN}]^{\mathrm{T}} (a_{ij}b_{ij} = 0 \ \forall j),$$

are respectively defined as

$$\overline{H}_i \doteq \prod_{j=1}^{N} (H_j)^{a_{ij}}, \overline{U}_{\text{CNOT}}(\boldsymbol{a}_i + \boldsymbol{b}_i) \doteq \prod_{j=1}^{N} (U_{\text{CNOT}})^{a_{ij}+b_{ij}}. \tag{13.20}$$

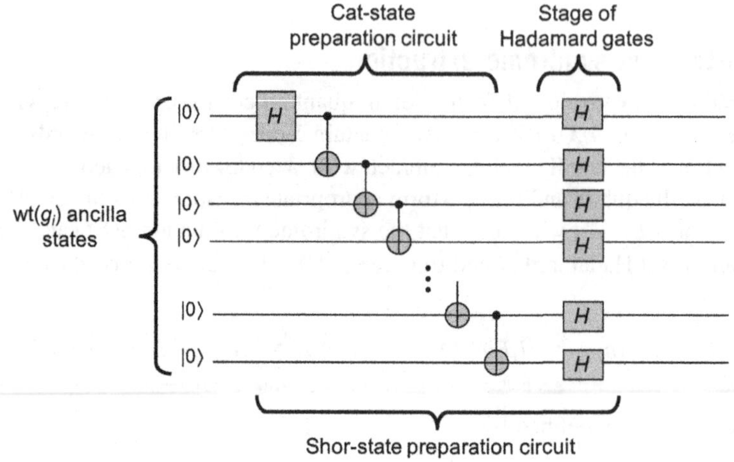

FIGURE 13.8

Shor state preparation circuit.

In Eq. (13.20), the transversal CNOT gate $\overline{U}_{\mathrm{CNOT}}$ is controlled by wt(g_i) qubits, for which $a_{ij}+b_{ij}=1$. From Eq. (13.17), it is evident that the encoded and ancilla blocks are nonentangled, and by measuring the ancilla block, we will not affect the encoded block (to collapse into one of the eigenkets). By measuring the ancilla qubits, the modified Shor state will collapse into one of the eigenkets, say $|\boldsymbol{A}'_{\mathrm{even}}+\boldsymbol{b}_E\cdot\boldsymbol{a}_i+\boldsymbol{a}_E\cdot\boldsymbol{b}_i\rangle$. Since $\mathrm{wt}(\boldsymbol{A}'_{\mathrm{even}})\bmod 2=0$, the result of the measurement will be

$$(-1)^{\boldsymbol{b}_E\cdot\boldsymbol{a}_i+\boldsymbol{a}_E\cdot\boldsymbol{b}_i}=(-1)^{\lambda_i}, \tag{13.21}$$

and the ith syndrome component $\lambda_i=\boldsymbol{b}_E\cdot\boldsymbol{a}_i+\boldsymbol{a}_E\cdot\boldsymbol{b}_i \bmod 2$ will be revealed. By repeating a similar procedure for all syndrome components, we can identify the most probable error E based on syndrome $S(E)=[\lambda_1\ \ldots\ \lambda_{N\text{-}K}]^{\mathrm{T}}$ and perform the corresponding error recovery action $\boldsymbol{R}=R_1\ \ldots\ R_N$.

As a reminder, the action of the Hadamard gate on the jth qubit performs the transformation $H_jX_jH_j=Z_j$. Applying the transversal H-gate on generator g_i yields

$$\overline{g}_i=\overline{H}_i g_i\overline{H}_i^{\dagger}\overset{\overline{H}_i^{\dagger}=\overline{H}_i}{=}\overline{H}_iX(\boldsymbol{a}_i)Z(\boldsymbol{b}_i)\overline{H}_i=\overline{H}_iX(\boldsymbol{a}_i)\overline{H}_iZ(\boldsymbol{b}_i)\overset{H_jX_jH_j=Z_j}{=}Z(\boldsymbol{a}_i)Z(\boldsymbol{b}_i)=Z(\boldsymbol{a}_i+\boldsymbol{b}_i). \tag{13.22}$$

Similarly, we can show that the action of the transversal H-gate on error E gives

$$\begin{aligned}
\overline{E}=\overline{H}_iE\overline{H}_i^{\dagger}&\overset{\overline{H}_i^{\dagger}=\overline{H}_i}{=}\overline{H}_iX(\boldsymbol{a}_E)Z(\boldsymbol{b}_E)\overline{H}_i\overset{\overline{\boldsymbol{a}}_i+\boldsymbol{a}_i=(1\cdots1)}{=}\overline{H}_iX(\boldsymbol{a}_E\cdot(\overline{\boldsymbol{a}}_i+\boldsymbol{a}_i))Z(\boldsymbol{b}_E\cdot(\overline{\boldsymbol{a}}_i+\boldsymbol{a}_i))\overline{H}_i\\
&=\overline{H}_i\underbrace{X(\boldsymbol{a}_E\cdot\overline{\boldsymbol{a}}_i+\boldsymbol{a}_E\cdot\boldsymbol{a}_i)}_{X(\boldsymbol{a}_E\cdot\overline{\boldsymbol{a}}_i)X(\boldsymbol{a}_E\cdot\boldsymbol{a}_i)}\underbrace{Z(\boldsymbol{b}_E\cdot\overline{\boldsymbol{a}}_i+\boldsymbol{b}_E\cdot\boldsymbol{a}_i)}_{Z(\boldsymbol{b}_E\cdot\overline{\boldsymbol{a}}_i)Z(\boldsymbol{b}_E\cdot\boldsymbol{a}_i)}\overline{H}_i\\
&=\overline{H}_iX(\boldsymbol{a}_E\cdot\overline{\boldsymbol{a}}_i)\overline{H}_i\underbrace{\overline{H}_iX(\boldsymbol{a}_E\cdot\boldsymbol{a}_i)\overline{H}_i}_{Z(\boldsymbol{a}_E\cdot\boldsymbol{a}_i)}\overline{H}_iZ(\boldsymbol{b}_E\cdot\overline{\boldsymbol{a}}_i)\overline{H}_i\underbrace{\overline{H}_iZ(\boldsymbol{b}_E\cdot\boldsymbol{a}_i)\overline{H}_i}_{X(\boldsymbol{b}_E\cdot\boldsymbol{a}_i)}\\
&=X(\boldsymbol{a}_E\cdot\overline{\boldsymbol{a}}_i)Z(\boldsymbol{a}_E\cdot\boldsymbol{a}_i)Z(\boldsymbol{b}_E\cdot\overline{\boldsymbol{a}}_i)X(\boldsymbol{b}_E\cdot\boldsymbol{a}_i),
\end{aligned} \tag{13.23}$$

where the "•" operation is defined by $\boldsymbol{a}\cdot\boldsymbol{b}=(a_1b_1,\cdots,a_Nb_N)$, and $\overline{\boldsymbol{a}}_i$ is the bitwise complement $\overline{\boldsymbol{a}}_i=(a_{i1}+1,\cdots,a_{iN}+1)$ In the derivation above, we used the property $\overline{\boldsymbol{a}}_i+\boldsymbol{a}_i=(1,\cdots,1)$. The left-hand side of Eq. (13.17) can be rewritten based on Eq. (13.23) as follows:

$$\begin{aligned}
\overline{H}_i\overline{U}_{\mathrm{CNOT}}(\boldsymbol{a}_i+\boldsymbol{b}_i)\overline{H}_iE|c\rangle\otimes|\boldsymbol{A}\rangle&=\overline{H}_i\overline{U}_{\mathrm{CNOT}}(\boldsymbol{a}_i+\boldsymbol{b}_i)\overbrace{\overline{H}_iE\overline{H}_i^{\dagger}}^{\overline{E}}\overbrace{\overline{H}_ig_i\overline{H}_ig_i\overline{H}_i|c\rangle}^{\overline{g}_i^2|\overline{c}\rangle=|\overline{c}\rangle}\otimes|\boldsymbol{A}\rangle\\
&=\overline{H}_i\overline{U}_{\mathrm{CNOT}}(\boldsymbol{a}_i+\boldsymbol{b}_i)\overline{E}|\overline{c}\rangle\otimes|\boldsymbol{A}\rangle.
\end{aligned} \tag{13.24}$$

Substituting the last line of Eq. (13.23) into Eq. (13.24), we obtain

$$\overline{H}_i\overline{U}_{\mathrm{CNOT}}(\boldsymbol{a}_i+\boldsymbol{b}_i)\overline{H}_iE|c\rangle\otimes|\boldsymbol{A}\rangle \overset{|\overline{c}\rangle=\sum_m\overline{c}(\boldsymbol{m})|\boldsymbol{m}\rangle}{=} \overline{H}_i\overline{U}_{\mathrm{CNOT}}(\boldsymbol{a}_i+\boldsymbol{b}_i)X(\boldsymbol{a}_E\cdot\overline{\boldsymbol{a}}_i) Z(\boldsymbol{a}_E\cdot\boldsymbol{a}_i+\boldsymbol{b}_E\cdot\overline{\boldsymbol{a}}_i)X(\boldsymbol{b}_E\cdot\boldsymbol{a}_i)\sum_m\overline{c}(\boldsymbol{m})|\boldsymbol{m}\rangle\otimes|\boldsymbol{A}\rangle, \tag{13.25}$$

where we use the expansion $|\overline{c}\rangle=\sum_m\overline{c}(\boldsymbol{m})|\boldsymbol{m}\rangle$ in terms of computational basis kets. Applying the X-containing operators, we obtain

$$\begin{aligned}\overline{H}_i\overline{U}_{\mathrm{CNOT}}(\boldsymbol{a}_i+\boldsymbol{b}_i)\overline{H}_iE|c\rangle\otimes|\boldsymbol{A}\rangle &= \overline{H}_i\overline{U}_{\mathrm{CNOT}}(\boldsymbol{a}_i+\boldsymbol{b}_i)X(\boldsymbol{a}_E\cdot\overline{\boldsymbol{a}}_i)Z(\boldsymbol{a}_E\cdot\boldsymbol{a}_i+\boldsymbol{b}_E\cdot\overline{\boldsymbol{a}}_i)\sum_m\overline{c}(\boldsymbol{m})|\boldsymbol{m}+\boldsymbol{b}_E\cdot\boldsymbol{a}_i\rangle\otimes|\boldsymbol{A}\rangle\\ &=(-1)^{\lambda}\overline{H}_i\overline{U}_{\mathrm{CNOT}}(\boldsymbol{a}_i+\boldsymbol{b}_i)Z(\boldsymbol{a}_E\cdot\boldsymbol{a}_i+\boldsymbol{b}_E\cdot\overline{\boldsymbol{a}}_i)X(\boldsymbol{a}_E\cdot\overline{\boldsymbol{a}}_i)\sum_m\overline{c}(\boldsymbol{m})|\boldsymbol{m}+\boldsymbol{b}_E\cdot\boldsymbol{a}_i\rangle\otimes|\boldsymbol{A}\rangle\\ &=(-1)^{\lambda}\overline{H}_i\overline{U}_{\mathrm{CNOT}}(\boldsymbol{a}_i+\boldsymbol{b}_i)Z(\boldsymbol{a}_E\cdot\boldsymbol{a}_i+\boldsymbol{b}_E\cdot\overline{\boldsymbol{a}}_i)\sum_m\overline{c}(\boldsymbol{m})|\boldsymbol{m}+\boldsymbol{b}_E\cdot\boldsymbol{a}_i+\boldsymbol{a}_E\cdot\overline{\boldsymbol{a}}_i\rangle\otimes|\boldsymbol{A}\rangle,\end{aligned} \tag{13.26}$$

where $\lambda=(\boldsymbol{b}_E\cdot\overline{\boldsymbol{a}}_i)\cdot(\boldsymbol{a}_E\cdot\overline{\boldsymbol{a}}_i)$. (Since $X(\boldsymbol{a}_E\cdot\overline{\boldsymbol{a}}_i)$ and $Z(\boldsymbol{a}_E\cdot\boldsymbol{a}_i)$ act on different qubits they commute.) By applying the transversal-CNOT gate on the Shor state from Ref. Eq. (13.26), we obtain

$$\overline{H}_i\overline{U}_{\mathrm{CNOT}}(\boldsymbol{a}_i+\boldsymbol{b}_i)\overline{H}_iE|c\rangle\otimes|\boldsymbol{A}\rangle=(-1)^{\lambda}\overline{H}_iZ(\boldsymbol{a}_E\cdot\boldsymbol{a}_i+\boldsymbol{b}_E\cdot\overline{\boldsymbol{a}}_i)\sum_m\overline{c}(\boldsymbol{m})|\boldsymbol{m}+\boldsymbol{b}_E\cdot\boldsymbol{a}_i+\boldsymbol{a}_E\cdot\overline{\boldsymbol{a}}_i\rangle\otimes$$

$$\sum_{\boldsymbol{A}_{\mathrm{even}}}\Bigg|\boldsymbol{A}_{\mathrm{even}}+\underbrace{(\boldsymbol{a}_i+\boldsymbol{b}_i)(\boldsymbol{b}_E\cdot\boldsymbol{a}_i+\boldsymbol{a}_E\cdot\boldsymbol{b}_i)}_{\boldsymbol{a}_E\cdot\boldsymbol{b}_i+\boldsymbol{b}_E\cdot\boldsymbol{a}_i}\Bigg\rangle=E|c\rangle\otimes\sum_{\boldsymbol{A}_{\mathrm{even}}}|\boldsymbol{A}_{\mathrm{even}}+\boldsymbol{b}_E\cdot\boldsymbol{a}_i+\boldsymbol{a}_E\cdot\boldsymbol{b}_i\rangle, \tag{13.27}$$

therefore proving Eq. (13.17).

Example. The [5,1,3] code (revisited). The generators of this code are

$$g_1=X_1Z_2Z_3X_4,\ g_2=X_2Z_3Z_4X_5,\ g_3=X_1X_3Z_4Z_5,\ \text{and}\ g_4=Z_1X_2X_4Z_5.$$

Based on Eq. (13.20) and the generators above, we obtain the transversal Hadamard gates as follows:

$$\overline{H}_1=H_1H_4,\ \overline{H}_2=H_2H_5,\ \overline{H}_3=H_1H_3,\ \overline{H}_4=H_2H_4.$$

Based on Eq. (13.22), applying the transversal H-gate on the generators g_i ($i=1,\ldots,4$) yields

$$\overline{g}_1=Z(\boldsymbol{a}_1+\boldsymbol{b}_1)=Z_1Z_2Z_3Z_4,\ \overline{g}_2=Z(\boldsymbol{a}_2+\boldsymbol{b}_2)=Z_2Z_3Z_4Z_5,\ \overline{g}_3=Z(\boldsymbol{a}_3+\boldsymbol{b}_3)=Z_1Z_3Z_4Z_5,$$
$$\overline{g}_4=Z_1Z_2Z_4Z_5.$$

The corresponding circuit for syndrome extraction is shown in Fig. 13.9. The upper block corresponds to action on codeword qubits, and the lower block corresponds to action on ancilla qubits (Shor state). A single qubit error in the codeword block clearly affects only one qubit in the ancilla block,

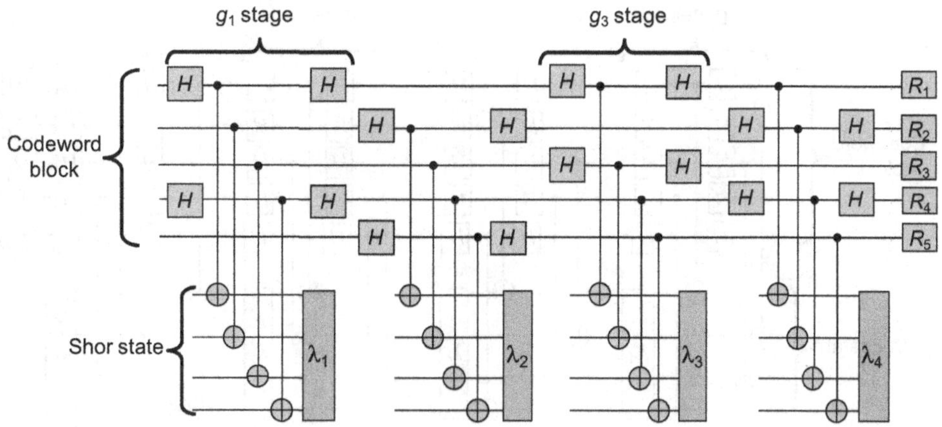

FIGURE 13.9

Syndrome extraction circuit for [5,1,3] stabilizer code.

suggesting that this implementation is transversal and consequently fault-tolerant. Let us observe the stage corresponding to generator g_1. Based on the transparent Hadamard gate $\overline{H}_1 = H_1 H_4$, we must place two Hadamard gates taking action on codeword block qubits H_1 and H_4, respectively. The corresponding generator image $\overline{g}_1 = Z_1 Z_2 Z_3 Z_4$ indicates that qubits 1, 2, 3, and 4 in the codeword block are control qubits, while the qubits in the Shor state block are target qubits. The purpose of this manipulation is to encode syndrome information into ancilla qubits. We perform further measurements on the modified Shor state; the measurement result provides information about syndrome component λ_1. The other g_i $(i = 2,3,4)$ stages operate similarly. Once the syndrome vector is determined, we determine the most probable error from the corresponding lookup table and apply the corresponding recovery action $\boldsymbol{R} = R_1 R_2 R_3 R_4 R_5$ to undo the actions of imperfect gates.

As an alternative, Plenio et al. [7] proposed rearranging the sequence of CNOT gates in Eq. (13.17) as follows:

$$\overline{H}_i \overline{U}_{\text{CNOT}}(\boldsymbol{a}_i) \overline{H}_i \overline{U}_{\text{CNOT}}(\boldsymbol{b}_i) E|c\rangle \otimes |\boldsymbol{A}\rangle, \tag{13.28}$$

where the transversal Hadamard gate is now applied to all qubits:

$$\overline{H}_i = \prod_{j=1}^{N} H_j. \tag{13.29}$$

Based on the generators above and Eqs. (13.28) and (13.29), the corresponding syndrome extraction circuit is shown in Fig. 13.10. The operating principle of this circuit version is similar to the one shown in Fig. 13.9.

In the discussion above, we assumed that generators do not contain Y-operators. In cases where generators contain Y-operators, we must find the gate that performs the transformation $\widetilde{H}_k Y_k \widetilde{H}_k^{\dagger} = Z_k$.

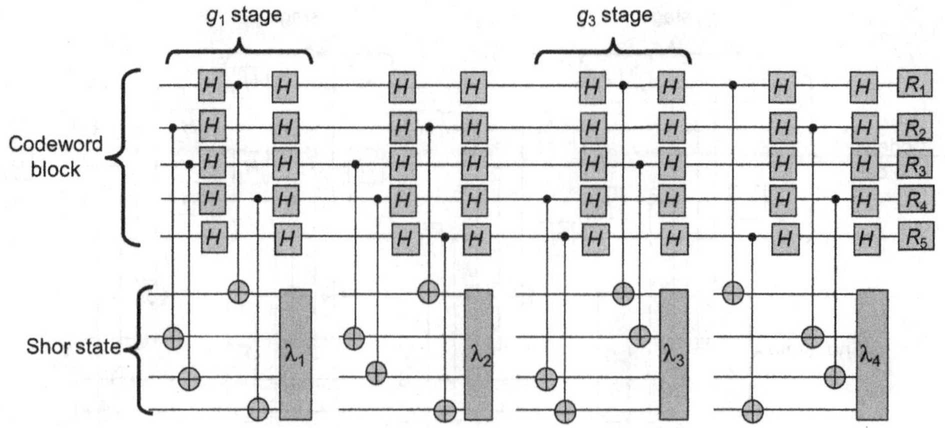

FIGURE 13.10

Syndrome extraction circuit for [5,1,3] stabilizer code based on the approach of Plenio et al.

when constructing images of generators $\overline{g}_i$ rather than applying Hadamard gate H_k to the kth qubit. Matrix multiplication can easily verify that the following gate performs this transformation:

$$\widetilde{H}_k = \frac{1}{\sqrt{2}}\begin{bmatrix} 1 & -j \\ -j & 1 \end{bmatrix}. \tag{13.30}$$

The transversal Hadamard gate is now based on Eqs. (13.20) and (13.30), defined as follows:

$$\overline{H}_i \doteq \prod_{j=1}^{N} \left(\widetilde{H}_j\right)^{(a_i \cdot b_i)_j} \left(H_j\right)^{(a_i \cdot \overline{b}_i)_j}, \overline{b}_{ij} = b_{ij} + 1, \tag{13.31}$$

where we use the notation $(a_i \cdot b_i)_j$ to denote the jth component of $a_i \cdot b_i$. The images of the generators now become

$$\overline{g}_i = \overline{H}_i g_i \overline{H}_i^\dagger = \overline{H}_i X(a_i) Z(b_i) \overline{H}_i^\dagger = \overline{H}_i Y(a_i \cdot b_i) X(a_i \cdot \overline{b}_i) Z(\overline{a}_i \cdot b_i) \overline{H}_i^\dagger$$

$$= \underbrace{\prod_{j=1}^{N} \left(\widetilde{H}_j\right)^{(a_i \cdot b_i)_j} Y(a_i \cdot b_i) \left(\widetilde{H}_j^\dagger\right)^{(a_i \cdot b_i)_j}}_{Z(a_i \cdot b_i)} \underbrace{\prod_{j=1}^{N} \left(H_j\right)^{(a_i \cdot \overline{b})_j} X(a_i \cdot \overline{b}_i) \left(H_j\right)^{(a_i \cdot \overline{b}_i)_j}}_{Z(a_i \cdot \overline{b})} Z(\overline{a}_i \cdot b_i)$$

$$= Z\left(a_i \cdot b_i + a_i \cdot \overline{b}_i + \overline{a}_i \cdot b_i\right) = Z\left(a_i \cdot b_i + a_i + a_i \cdot b_i + b_i + a_i \cdot b_i\right) = Z(a_i + b_i + a_i \cdot b_i). \tag{13.32}$$

Similarly, the image of error E can be found by

$$\overline{E} = \overline{H}_i E \overline{H}_i^\dagger = (-1)^\lambda Z\left(a_E \cdot (a_i \cdot \overline{b}_i) + b_E \cdot \overline{a_i \cdot \overline{b}_i}\right) X\left(a_E \cdot a_i \cdot \overline{b}_i + b_E \cdot a_i\right), \tag{13.33}$$

where $\lambda = \left(a_E \cdot \overline{a_i \cdot \overline{b}_i}\right) \cdot \left(b_E \cdot \overline{a_i \cdot \overline{b}_i}\right)$ By applying the transversal-CNOT gate on $E|c\rangle \otimes |A\rangle$, we obtain

$$\overline{U}_{CNOT}(a_i + b_i + a_i \cdot b_i)E|c\rangle \otimes |A\rangle = \overline{E}|\overline{c}\rangle \otimes \sum_{A_{\text{even}}} |A_{\text{even}} + b_E \cdot a_i + a_E \cdot b_i\rangle. \tag{13.34}$$

Finally, by applying the transversal Hadamard gate given by Eq. (13.31) on Eq. (13.34), we derive

$$\overline{H}_i \overline{U}_{CNOT}(a_i + b_i + a_i \cdot b_i)\overline{H}_i E|c\rangle \otimes |A\rangle = E|c\rangle \otimes \sum_{A_{\text{even}}} |A_{\text{even}} + b_E \cdot a_i + a_E \cdot b_i\rangle, \tag{13.35}$$

which has a form similar to that of Eq. (13.17).

Example. The [4,2,2] code (revisited). The starting point in this example is the generators of the code

$$g_1 = X_1 Z_2 Z_3 X_4 \text{ and } g_2 = Y_1 X_2 X_3 Y_4.$$

Based on Eq. (13.31), we obtain the following transversal Hadamard gates corresponding to generators g_i $(i = 1,2)$:

$$\overline{H}_1 = H_1 H_4 \text{ and } H_2 = \widetilde{H}_1 H_2 H_3 \widetilde{H}_4.$$

Based on the last line of Eq. (13.32), we obtain the images of the generators as follows:

$$\overline{g}_1 = Z_1 Z_2 Z_3 Z_4 \text{ and } \overline{g}_2 = Z_1 Z_2 Z_3 Z_4.$$

Finally, based on Eq. (13.35), the syndrome extraction circuit for the [4,2,2] code is shown in Fig. 13.11.

The syndrome extraction circuits discussed above only employ transversal operations provided that the Shor state was prepared fault-tolerantly. Closer inspection of the Shor preparation circuit (see Fig. 13.8) reveals that errors introduced by a wrong qubit or faulty CNOT gate can cause error propagation. Fig. 13.8 shows that the first and last qubits have even parity. We can verify whether the

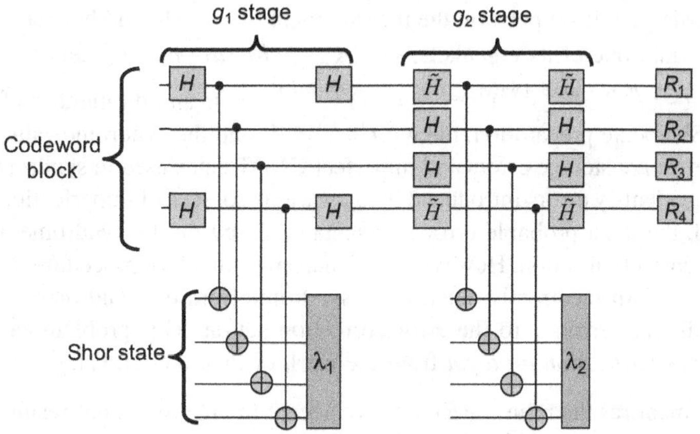

FIGURE 13.11

Syndrome extraction circuit for [4,2,2] stabilizer code.

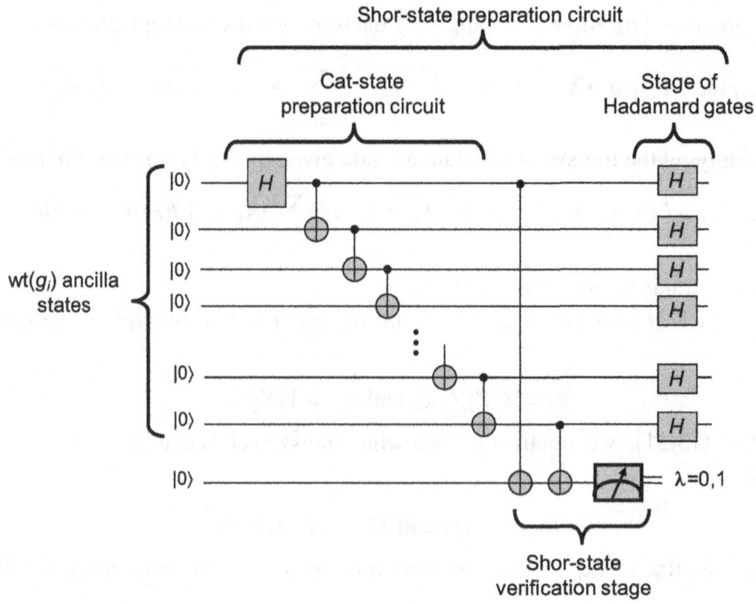

FIGURE 13.12

Modified Shor-state preparation circuit with verification stage.

even parity condition has been satisfied by adding a parity check on the first and last qubit. If even parity has been satisfied, we know that up to $O(p)$, no flip-error has occurred. If the measurement of Z on the additional ancilla qubit is 1, we know that an odd number of errors has occurred, and we must repeat the verification procedure. The modified Shor preparation circuit enabling this verification is shown in Fig. 13.12.

We learned earlier that if we perform the measurement on a modified Shor state, the modified Shor state will collapse into one of its eigenkets, say $|A'_{\text{even}} + \boldsymbol{b}_E \cdot \boldsymbol{a}_i + \boldsymbol{a}_E \cdot \boldsymbol{b}_i\rangle$, and the result of the measurement will be $(-1)^{\boldsymbol{b}_E \cdot \boldsymbol{a}_i + \boldsymbol{a}_E \cdot \boldsymbol{b}_i + \text{wt}(A'_{\text{even}})} = (-1)^{\lambda_i + \text{wt}(A'_{\text{even}})}$. If an odd number of bit-flip errors has occurred during Shor state preparation, the wt($\boldsymbol{A}'_{\text{even}}$) = 1, and the syndrome value will be λ_i+1. The other sources of error are storage errors and imperfect CNOT gates used in syndrome calculation. The syndrome $S(E)$ can identify errors introduced by the various sources of imperfections discussed above. In an ideal world, the most probable error is detectable based on the syndrome and can be used to perform the error correction action. However, the syndrome extraction procedure, as mentioned above, can be faulty as well. An incorrectly determined syndrome from the syndrome extraction procedure can introduce additional errors into the error correction action. This problem can be solved by the following *syndrome verification protocol* from the work of Plenio et al. [7]:

1. If the measurement results return $S(E) = 0$, we accept the measurement result as correct and assume that no recovery operation is needed.

2. On the other hand, if the measurement results return $S(E) \neq 0$, rather than using the syndrome to identify the most probable error followed by error correction action, we repeat the whole procedure until the measurement returns $S(E) = 0$.

This protocol might be time-consuming but ensures that a wrongly determined syndrome cannot introduce error propagation. Note that even the $S(E) = 0$ result does not guarantee that a storage error has not occurred after the syndrome was determined—for example, in the last transversal gates of Figs. 13.9–13.11. However, this undetectable error will be present in the codeword block after error correction, indicating that the protocol is still fault-tolerant.

13.3.2 Fault-tolerant encoding operations

We defined encoded gates in Chapter 8 as unitary operators that perform the mapping of one codeword to another codeword as given below:

$$\left(UsU^{\dagger}\right)U|c\rangle = Us|c\rangle \overset{s|c\rangle=|c\rangle}{=} U|c\rangle; \quad \forall |c\rangle \in C_Q, s \in S \tag{13.36}$$

If $UsU^{\dagger} = s \ \forall s \in S$, that is, if stabilizer S is fixed by U under conjugation ($USU^{\dagger} = S$), $U|c\rangle$ will be a codeword based on Eq. (13.36) that is fixed for any element from S. The operator U, therefore, represents an *encoded operation* (the operation that maps one codeword to another). As shown in Chapter 8, the set of operators U that fix S represents a subgroup of the *unitary group* $U(N)$ (the set of all unitary matrices of dimensions $N \times N$). This subgroup is called the *normalizer* of S with respect to $U(N)$ and denoted here as $N_U(S)$. Namely, the normalizer $N_U(S)$ of the stabilizer S is defined as the set of operators $U \in U(N)$ that fix S under conjugation; in other words, $N_U(S) = \{U \in U(N) | USU^{\dagger} = S\}$. In Chapter 8, we introduced an equivalent concept, namely the centralizer of S with respect to $U(N)$, denoted $C_U(S)$, as the set of all operators U (from $U(N)$) that commute for all stabilizer elements; in other words, $C_U(S) = \{U \in U(N) | US = SU\}$. The centralizer condition $US = SU$ can be converted to the normalizer by multiplying from the right by $U^{\dagger}$ to obtain $USU^{\dagger} = S$. Note that the key difference between this section and Section 8.1 is that here we are concerned with the action of U from $U(N)$ on S, while in Section 8.1, we were concerned with the action of an operator from the Pauli multiplicative group G_N. As $G_N \subset U(N)$, it is obvious that $N_G(S) \subset N_U(S)$.

An important fault-tolerant quantum error correction group is the *Clifford* group, which is defined as the normalizer of the Pauli multiplicative group G_N with respect to $U(N)$ and denoted $N_U(G_N)$. The *Clifford operator U*, therefore, is an operator that preserves the elements of the Pauli group under conjugation, namely $\forall O \in G_N : UOU^{\dagger} \in G_N$. Clifford group elements can be generated based on a simple set of gates: Hadamard, phase, and CNOT gates or equivalently, Hadamard, phase, and controlled-Z gates. In this section, we are concerned with encoded gates $U \in N(G_N) \cap N(S)$ that can be constructed using these gates and encoded operations that are implemented in transversal fashion and are consequently fault-tolerant.

We learned in Chapter 8 that the quotient group $C(S)/S$ could be represented by

$$C(S)/S = \cup\{\overline{X}_m S, \overline{Z}_n S : m, n = 1, ..., K\} \wedge \{j^l : l = 0, 1, 2, 3\} \tag{13.37}$$

where $\{\overline{X}_m, \overline{Z}_n | m, n = 1, \cdots, K\}$ denote the Pauli-encoded operators. The elements of $C(S)/S$ (cosets), denoted $\overline{O}S$, can be represented as follows:

$$\overline{O}S = j^l \overline{X}(\boldsymbol{a}) \overline{Z}(\boldsymbol{b}) S = j^l \left(\overline{X}_1 S\right)^{a_1} \cdots \left(\overline{X}_k S\right)^{a_K} \left(\overline{Z}_1 S\right)^{b_1} \cdots \left(\overline{Z}_1 S\right)^{b_K}, \tag{13.38}$$

as proved in Chapter 8. By denoting the conjugation mapping image of encoded Pauli operators by $\widetilde{X}_m = U\overline{X}_m U^\dagger, \widetilde{Z}_n = U\overline{Z}_n U^\dagger$, we can show that

$$\begin{aligned}\left[\widetilde{X}_m, g_n\right] &= \widetilde{X}_m g_n - g_n \widetilde{X}_m = U\overline{X}_m U^\dagger g_n - g_n U\overline{X}_m U^\dagger \overset{g_n = Us_n U^\dagger}{=} U\overline{X}_m U^\dagger U s_n U^\dagger - U s_n U^\dagger U\overline{X}_m U^\dagger \\ &= U\overline{X}_m s_n U^\dagger - U s_n \overline{X}_m U^\dagger \overset{s_n\overline{X}_m = \overline{X}_m s_n}{=} U\overline{X}_m s_n U^\dagger - U\overline{X}_m s_n U^\dagger = 0\end{aligned} \tag{13.39}$$

where we represent the generators g_n in terms of some stabilizer element s_n by $g_n = U s_n U^\dagger$ (namely, as U fixes the stabilizer under conjugation, any element $s \in S$, including generators, can be represented by $UsU^\dagger$). The image of Pauli operators also commutes with generators g_n and therefore belongs to $C(S)$ as well. Because operator U fixes stabilizer S under conjugation based on Eq. (13.38) and the commutation relation of Eq. (13.39), operator U maps $C(S)/S$ to $C(S)/S$, as coset $\overline{O}S$ is mapped to

$$\widetilde{O}S = j^l\left(\widetilde{X}_1 S\right)^{a_1} \cdots \left(\widetilde{X}_k S\right)^{a_K} \left(\widetilde{Z}_1 S\right)^{b_1} \cdots \left(\widetilde{Z}_1 S\right)^{b_K}. \tag{13.40}$$

Homomorphism is the mapping f from group G onto group H if the image of a product ab equals the product of the images of a and b. When f is a homomorphism and bijective (one-to-one and onto), f is said to be an *isomorphism* from G to H and is commonly denoted $G \cong H$. Finally, if $G = H$, we call the isomorphism an *automorphism*. Let us now observe the mapping f from G_N to G_N that acts on elements from G_N by conjugation, namely $f(g) = UgU^\dagger \ \forall \ g \in G_N$, $U \in N(G_N)$. Below, this mapping will be shown to be an automorphism. We first need to show that the mapping is one-to-one. Let us observe two *different* elements $a, b \in G_N$ and assume that $f(a) = f(b)$. The conjugation definition ($f(g) = UgU^\dagger$) function indicates that $f(a) = UaU^\dagger = UbU^\dagger = f(b)$. Multiplying by U from the right, we obtain $Ua = Ub$, and multiplying by $U^\dagger$ from the left, we obtain $a = b$, which is a contradiction. Therefore, mapping $f(g)$ is a one-to-one mapping. In the second step, we must prove that mapping $f(g)$ is an onto mapping. We showed in the previous paragraph that the range of mapping $f(g)$, denoted as $f(G_N)$, is contained in G_N. Since mapping $f(g)$ is one-to-one, it must be $f(G_N) = G_N$, indicating that the conjugation mapping is indeed onto mapping. In the final stage, we need to prove that the image of the product ab, where $a,b \in G_N$ must equal the product of images $f(a)f(b)$. This claim can easily be proved from the definition of conjugation mapping as follows:

$$f(ab) = UabU^\dagger = Ua\overbrace{U^\dagger U}^{I} bU^\dagger = \overbrace{\left(UaU^\dagger\right)}^{f(a)}\overbrace{\left(UbU^\dagger\right)}^{f(b)} = f(a)f(b). \tag{13.41}$$

Therefore, the mapping $f(g) = UgU^\dagger$ is an automorphism of G_N.

Because every Clifford operator U from $N(G_N)$ is an automorphism of G_N, the action of $N(G_N)$ generators on G_N generators is sufficient to determine the action of U on G_N. We also know that every Clifford generator can be represented by Hadamard, phase, and CNOT (or controlled-Z) gates. It is also essential that we study the action of single-qubit Hadamard and phase gates on the generators of G_1 and the action of two-qubit CNOT gates on the generators of G_2.

The matrix representation of the Hadamard gate in the computational bases from Chapter 3 is given by

$$H = \frac{1}{\sqrt{2}}\begin{bmatrix} 1 & 1 \\ 1 & -1 \end{bmatrix}. \tag{13.42}$$

The Pauli gates X, Y, and Z are under conjugation operation $HOH^{\dagger}$ ($O \in \{X,Y,Z\}$) mapped to

$$\begin{aligned} X \to HXH^{\dagger} &= \frac{1}{\sqrt{2}}\begin{bmatrix} 1 & 1 \\ 1 & -1 \end{bmatrix}\begin{bmatrix} 0 & 1 \\ 1 & 0 \end{bmatrix}\frac{1}{\sqrt{2}}\begin{bmatrix} 1 & 1 \\ 1 & -1 \end{bmatrix} = Z, \\ Z \to HZH^{\dagger} &= \frac{1}{\sqrt{2}}\begin{bmatrix} 1 & 1 \\ 1 & -1 \end{bmatrix}\begin{bmatrix} 1 & 0 \\ 0 & -1 \end{bmatrix}\frac{1}{\sqrt{2}}\begin{bmatrix} 1 & 1 \\ 1 & -1 \end{bmatrix} = X, \\ Y \to HYH^{\dagger} &= \frac{1}{\sqrt{2}}\begin{bmatrix} 1 & 1 \\ 1 & -1 \end{bmatrix}\begin{bmatrix} 0 & -j \\ j & 0 \end{bmatrix}\frac{1}{\sqrt{2}}\begin{bmatrix} 1 & 1 \\ 1 & -1 \end{bmatrix} = -Y. \end{aligned} \tag{13.43}$$

The matrix representation of phase gate in the computational basis is given by

$$P = \begin{bmatrix} 1 & 0 \\ 0 & j \end{bmatrix} \tag{13.44}$$

The Pauli gates X, Y, and Z are under conjugation operation $POP^{\dagger}$ ($O \in \{X,Y,Z\}$) mapped to

$$\begin{aligned} X \to PXP^{\dagger} &= \begin{bmatrix} 1 & 0 \\ 0 & j \end{bmatrix}\begin{bmatrix} 0 & 1 \\ 1 & 0 \end{bmatrix}\begin{bmatrix} 1 & 0 \\ 0 & -j \end{bmatrix} = Y, \\ Z \to PZP^{\dagger} &= \begin{bmatrix} 1 & 0 \\ 0 & j \end{bmatrix}\begin{bmatrix} 1 & 0 \\ 0 & -1 \end{bmatrix}\begin{bmatrix} 1 & 0 \\ 0 & -j \end{bmatrix} = Z, \\ Y \to PYP^{\dagger} &= \begin{bmatrix} 1 & 0 \\ 0 & j \end{bmatrix}\begin{bmatrix} 0 & -j \\ j & 0 \end{bmatrix}\begin{bmatrix} 1 & 0 \\ 0 & -j \end{bmatrix} = -X. \end{aligned} \tag{13.45}$$

Eqs. (13.43) and (13.45) show that both operators (H and P) permute the generators of G_1 and that the phase operator performs the rotation around the z-axis for 90°.

The last gate needed to implement any Clifford operator is the CNOT gate. Since the CNOT gate is a two-qubit gate, we need to study its action on the generators from G_2—namely, $X_1 = XI$,

X_2, Z_1, and Z_2. On a computational basis, the CNOT gate U_{CNOT} and generators from G_2 can be represented as

$$
\begin{aligned}
U_{\mathrm{CNOT}} &= \begin{bmatrix} I & 0 \\ 0 & X \end{bmatrix}, \; X_1 = X \otimes I = \begin{bmatrix} 0 & 1 \\ 1 & 0 \end{bmatrix} \otimes I = \begin{bmatrix} 0 & I \\ I & 0 \end{bmatrix}, \\
X_2 &= I \otimes X = \begin{bmatrix} 1 & 0 \\ 0 & 1 \end{bmatrix} \otimes X = \begin{bmatrix} X & 0 \\ 0 & X \end{bmatrix}, \\
Z_1 &= Z \otimes I = \begin{bmatrix} 1 & 0 \\ 0 & -1 \end{bmatrix} \otimes I = \begin{bmatrix} I & 0 \\ 0 & -I \end{bmatrix}, \\
Z_2 &= I \otimes Z = \begin{bmatrix} 1 & 0 \\ 0 & 1 \end{bmatrix} \otimes Z = \begin{bmatrix} Z & 0 \\ 0 & Z \end{bmatrix}.
\end{aligned}
\tag{13.46}
$$

The generators of G_2—X_1, X_2, Z_1, and Z_2—under conjugation operation $U_{CNOT} O U_{CNOT}^{\dagger}$ ($O \in \{X_1, X_2, Z_1, Z_2\}$), are mapped to

$$
\begin{aligned}
X_1 \rightarrow U_{\mathrm{CNOT}} X_1 U_{\mathrm{CNOT}}^{\dagger} &= \begin{bmatrix} I & 0 \\ 0 & X \end{bmatrix} \begin{bmatrix} 0 & I \\ I & 0 \end{bmatrix} \begin{bmatrix} I & 0 \\ 0 & X \end{bmatrix} = \begin{bmatrix} 0 & X \\ X & 0 \end{bmatrix} = X \otimes X = X_1 X_2, \\
X_2 \rightarrow U_{\mathrm{CNOT}} X_2 U_{\mathrm{CNOT}}^{\dagger} &= \begin{bmatrix} I & 0 \\ 0 & X \end{bmatrix} \begin{bmatrix} X & 0 \\ 0 & X \end{bmatrix} \begin{bmatrix} I & 0 \\ 0 & X \end{bmatrix} = I \otimes X = X_2, \\
Z_1 \rightarrow U_{\mathrm{CNOT}} Z_1 U_{\mathrm{CNOT}}^{\dagger} &= \begin{bmatrix} I & 0 \\ 0 & X \end{bmatrix} \begin{bmatrix} I & 0 \\ 0 & -I \end{bmatrix} \begin{bmatrix} I & 0 \\ 0 & X \end{bmatrix} = Z \otimes I = Z_1, \\
Z_2 \rightarrow U_{\mathrm{CNOT}} Z_2 U_{\mathrm{CNOT}}^{\dagger} &= \begin{bmatrix} I & 0 \\ 0 & X \end{bmatrix} \begin{bmatrix} Z & 0 \\ 0 & Z \end{bmatrix} \begin{bmatrix} I & 0 \\ 0 & X \end{bmatrix} = Z \otimes Z = Z_1 Z_2.
\end{aligned}
\tag{13.47}
$$

Eq. (13.47) makes it clear that an error introduced on the control qubit affects the target qubit, while a phase error introduced on a target qubit affects the control qubit.

For certain implementations, such as cavity quantum electrodynamics, the controlled-Z U_{CZ} gate is more appropriate than the CNOT gate [8]. The CNOT gate can be implemented with the help of a controlled-Z gate, as shown in Fig. 13.13. The gate shown in Fig. 13.13 can be expressed on a computation basis as

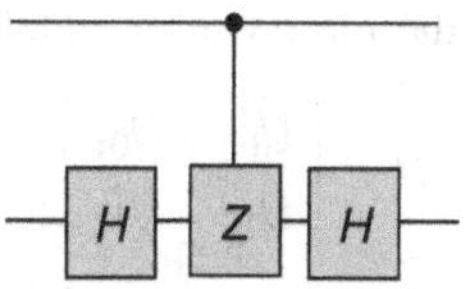

FIGURE 13.13

CNOT gate expressed in terms of a controlled-Z gate.

$$\begin{bmatrix} I & 0 \\ 0 & HZH \end{bmatrix} = \begin{bmatrix} I & 0 \\ 0 & X \end{bmatrix} = U_{\text{CNOT}}, \tag{13.48}$$

indicating that a controlled-Z gate can indeed be used instead of a CNOT gate. We now study the action of controlled-Z gate U_{CZ} on generators of G_2. The generators of G_2—X_1, X_2, Z_1, and Z_2—under conjugation operation $U_{CZ}OU_{CZ}^{\dagger}$ ($O\in\{X_1,X_2,Z_1,Z_2\}$) are mapped to

$$\begin{aligned} X_1 \rightarrow U_{CZ}X_1U_{CZ}^{\dagger} &= \begin{bmatrix} I & 0 \\ 0 & Z \end{bmatrix}\begin{bmatrix} 0 & I \\ I & 0 \end{bmatrix}\begin{bmatrix} I & 0 \\ 0 & Z \end{bmatrix} = X\otimes Z = X_1Z_2, \\ X_2 \rightarrow U_{CZ}X_2U_{CZ}^{\dagger} &= \begin{bmatrix} I & 0 \\ 0 & Z \end{bmatrix}\begin{bmatrix} X & 0 \\ 0 & X \end{bmatrix}\begin{bmatrix} I & 0 \\ 0 & Z \end{bmatrix} = Z\otimes X = Z_1X_2, \\ Z_1 \rightarrow U_{CZ}Z_1U_{CZ}^{\dagger} &= \begin{bmatrix} I & 0 \\ 0 & Z \end{bmatrix}\begin{bmatrix} I & 0 \\ 0 & -I \end{bmatrix}\begin{bmatrix} I & 0 \\ 0 & Z \end{bmatrix} = Z\otimes I = Z_1, \\ Z_2 \rightarrow U_{CNOT}Z_2U_{CNOT}^{\dagger} &= \begin{bmatrix} I & 0 \\ 0 & X \end{bmatrix}\begin{bmatrix} Z & 0 \\ 0 & Z \end{bmatrix}\begin{bmatrix} I & 0 \\ 0 & X \end{bmatrix} = I\otimes Z = Z_2. \end{aligned} \tag{13.49}$$

From Eq. (13.49), it is evident that the bit-flip error on one qubit introduces the phase error on the other qubit, while the phase error on either qubit does not propagate.

We now turn our attention to determining the unitary operator when the corresponding automorphism is known, using an example from the work of Gottesman [9]. The automorphism on G_1 in this example is given by the cyclic permutation of Pauli gates [9]:

$$f_U: X \rightarrow Y \rightarrow Z \rightarrow X. \tag{13.50}$$

We know from Chapter 3 that the computation basis ket $|0\rangle$ is an eigenket of Pauli operator Z with an eigenvalue of $+1$. Let us now apply the unknown gate U on the basis ket $|0\rangle$:

$$U|0\rangle \overset{Z|0\rangle=|0\rangle}{=} UZ|0\rangle = UZ\overbrace{U^{\dagger}U}^{I}|0\rangle = \left(UZU^{\dagger}\right)U|0\rangle = X(U|0\rangle), \tag{13.51}$$

where we use the mapping property of Eq. (13.50). Comparing the first and the last portions of Eq. (13.51), we conclude that $U|0\rangle$ is an eigenket of Pauli operator X with an eigenvalue of $+1$, which is given by

$$U|0\rangle = \frac{1}{\sqrt{2}}(|0\rangle + |1\rangle). \tag{13.52}$$

The elements of the first column in the matrix representation of U,

$$U = \begin{bmatrix} U_{00} & U_{01} \\ U_{10} & U_{11} \end{bmatrix}, \tag{13.53}$$

can be determined from Eq. (13.52) with $U_{00} = \langle 0|U|0\rangle = 1/\sqrt{2}$, and $U_{10} = \langle 1|U|0\rangle = 1/\sqrt{2}$. The computational ket $|1\rangle$ can be obtained as $X|0\rangle = |1\rangle$, and using an approach similar to that of Eq. (13.51), we obtain

$$\begin{aligned} U|1\rangle &\overset{|1\rangle = X|0\rangle}{=} UX|0\rangle = UX\overbrace{U^{\dagger}U}^{I}|0\rangle = \left(UXU^{\dagger}\right)U|0\rangle \\ &= Y(U|0\rangle) \overset{U|0\rangle = \frac{1}{\sqrt{2}}(|0\rangle + |1\rangle)}{=} \frac{1}{\sqrt{2}}Y(|0\rangle + |1\rangle) = \frac{-j}{\sqrt{2}}(|0\rangle - |1\rangle). \end{aligned} \tag{13.54}$$

The elements of the second column of Eq. (13.53) can be found from Eq. (13.54) with $U_{01} = \langle 0|U|1\rangle = -\mathrm{j}/\sqrt{2}$, and $U_{11} = \langle 1|U|1\rangle = \mathrm{j}/\sqrt{2}$, so the matrix representation of U is given by

$$U_T = \frac{1}{\sqrt{2}}\begin{bmatrix} 1 & -j \\ 1 & j \end{bmatrix}, \tag{13.55}$$

The generators of G_1; X, Y, and Z; are under conjugation operation $U_T O U_T^{\dagger}$ ($O \in \{X, Y, Z\}$) mapped to

$$\begin{aligned} X \rightarrow U_T X U_T^{\dagger} &= \begin{bmatrix} 1 & -j \\ 1 & j \end{bmatrix}\begin{bmatrix} 1 & 0 \\ 0 & 1 \end{bmatrix}\begin{bmatrix} 1 & 1 \\ j & -j \end{bmatrix} = Y, \\ Y \rightarrow U_T Y U_T^{\dagger} &= \begin{bmatrix} 1 & -j \\ 1 & j \end{bmatrix}\begin{bmatrix} 0 & -j \\ j & 0 \end{bmatrix}\begin{bmatrix} 1 & 1 \\ j & -j \end{bmatrix} = Z, \\ Z \rightarrow U_T Z U_T^{\dagger} &= \begin{bmatrix} 1 & -j \\ 1 & j \end{bmatrix}\begin{bmatrix} 1 & 0 \\ 0 & -1 \end{bmatrix}\begin{bmatrix} 1 & 1 \\ j & -j \end{bmatrix} = X, \end{aligned} \tag{13.56}$$

which is consistent with Eq. (13.50).

13.3.3 Measurement protocol

In this section, we are concerned with applying the quantum gate on a quantum register through a measurement. Transversally applied, this method will be used in the upcoming sections to enable fault-tolerant QIP using arbitrary quantum stabilizer code. DiVincenzo and Shor [10] have described a method that can be used to perform the measurements using any operator O from the multiplicative Pauli group G_N (see also [1,10]). There are three possible options involving the relationship of this operator O and stabilizer group S: (1) it can belong to S when measuring O will not provide any useful information about the system state, as the result will always be $+1$ for a valid codeword; (2) it can commute with all elements from S, indicating that O belongs to $N(S)/S$; and (3) it can anticommute with certain elements from S, which represents the practical case discussed in Chapters 7 and 8. Let us create a set of generators $S_0 = \{g_1, g_2, \ldots, g_M\}$ such that the operator O anticommutes with g_1 and commutes with the remaining generators. If the mth generator g_m anticommutes with O, we can replace it with $g_1 g_m$, which will commute with O. Because the operator O commutes with generators g_m ($m = 2,3, \ldots, M$), the measurement of O will not affect them (they can be simultaneously measured with complete precision). On the other hand, since g_1 anticommutes with O, the measurement of O will affect g_1. The measurement of O will perform a projection of an arbitrary quantum state $|\psi_0\rangle$ as $P_\pm |\psi_0\rangle$, with the projection operators ($P_\pm^2 = P_\pm$) given by

$$P_\pm = \frac{1}{2}(I \pm O), \tag{13.57}$$

with Chapter 2 providing a justification of this representation. Because operator O has a second order ($O^2 = I$, see Chapter 8), its eigenvalues are ± 1. Therefore, the projection operators $P_\pm$ will project state $|\psi_0\rangle$ onto ± 1 eigenkets. The action of generators g_m on state $|\psi_0\rangle$ is given by $g_m|\psi_0\rangle = |\psi_0\rangle$, as expected.

We have reached the place in our discussion where we can introduce the **measurement protocol**, which consists of three steps:

1. Perform the measurement of O fault-tolerantly, as described in Section 13.2.
2. If the measurement outcome is $+1$, do not take any further action. The final state will be $P_+|\psi_0\rangle = |\psi_+\rangle$.
3. If the measurement outcome is -1, apply the g_1-operator on the postmeasurement state to obtain

$$g_1 P_- |\psi_0\rangle = g_1 \frac{1}{2}(I - O)|\psi_0\rangle = \frac{1}{2}(I + O) g_1 |\psi_0\rangle = P_+ g_1 |\psi_0\rangle = P_+ |\psi_0\rangle = |\psi_+\rangle. \tag{13.58}$$

In both cases, the measurement of O performs the mapping $|\psi_0\rangle \rightarrow |\psi_+\rangle$. The final state is fixed by stabilizer $S_1 = \{O, g'_1, \ldots, g'_M\}$, with the generators given by

$$g'_m = \begin{cases} g_m, [g_m, O] = 0 \\ g_1 g_m, \{g_m, O\} = 0 \end{cases}; m = 2, \cdots, M \tag{13.59}$$

By observing the generators from S_0, it can be shown that the measurement procedure above also performs $S_0 \rightarrow S_1$. The centralizer of S_i ($i = 0,1$) is the set of operators from G_N that commute with all elements from S_i. Below, we show that the measurement protocol performs the mapping $C(S_0)/S_0 \rightarrow C(S_1)/S_1$. To prove this claim, let us observe an element n from $C(S_0)$ and a state $|\psi_0\rangle$ fixed by S_0. The action of operator n on state $P_+|\psi_0\rangle$ during the measurement of operator O must be the same as $P_+ n' |\psi_0\rangle$, where n' is the image of n (under measurement of O). If n commutes with O, then $n' = n$. On the other hand, if n anticommutes with O, $g_1 n$ will commute with O, and the image of n will be $n' = g_1 n$. Therefore, the measurement of O in the first case maps $n \rightarrow n$ and in the second case maps $n \rightarrow g_1 n$. Since in the first case, n commutes with O and $n' = n$, n' must belong to the cosset nS_1. In the second case, $g_1 n$ commutes with O, and n' belongs to the coset $(g_1 n)S_1$. In conclusion, in the first case, the measurement of O performs the mapping $nS_0 \rightarrow nS_1$, while in the second case, $nS_0 \rightarrow g_1 n S_1$, proving the claim above. Let us denote the encoded Pauli operators for S_0 by $\widehat{X}_i, \widehat{Z}_i$. To summarize, to determine the action of measurement of O on initial state $|\psi_0\rangle$, we must perform the following procedure [2]:

1. Identify a generator g_1 from S_0 that anticommutes with O. Substitute any generator g_m ($m > 1$) that anticommutes with O with $g_1 g_m$.

2. Create the stabilizer S_1 from S_0 by replacing g_1 (from S_0) with operator O. The remaining generators for S_1 are created according to the following rule:

$$g'_m = \begin{cases} g_m, [g_m, O] = 0 \\ g_1 g_m, \{g_m, O\} = 0 \end{cases}; m = 2, \cdots, M.$$

3. As the final stage, substitute any $\widehat{X}_i(\widehat{Z}_i)$ that anticommutes with operator O by $g_1\widehat{X}_i(g_1\widehat{Z}_i)$, to ensure that $\widehat{X}_i(\widehat{Z}_i)$ belongs to C(S_1).

The following example from Gottesman [9] (see also [2]) will be used to illustrate this measurement procedure. Consider a two-qubit system $|\psi\rangle \otimes |0\rangle$, where $|\psi\rangle$ is an arbitrary state of qubit 1. Let us now apply the CNOT gate using the rules we derived by Eq. (13.47). The stabilizer for $|\psi\rangle \otimes |0\rangle$ has a generator g given by $I \otimes Z$. The operators $\widehat{X} = X \otimes I, \widehat{Z} = Z \otimes I$ both commute with generator g and mutually anticommute, satisfying the properties of Pauli-encoded operators. The CNOT gate with qubit 1 serving as the control qubit and qubit 2 as the target will perform the following mapping based on Eq. (13.47):

$$\begin{aligned} \widehat{X} &= X_1 \rightarrow \widehat{X}' = X \otimes X = X_1 X_2, \\ \widehat{Z} &= Z_1 \rightarrow \widehat{Z}' = Z \otimes I = Z_1, \\ g &= Z_2 \rightarrow g' = Z \otimes Z = Z_1 Z_2. \end{aligned} \tag{13.60}$$

We now study the measurement of operator $O = I \otimes Y$. Clearly, the operators $\widehat{X}$ and g anticommute with O (as both differ from O in one position). The measurement of O will perform the following mapping based on the three-step procedure given above:

$$\begin{aligned} \widehat{X}' &= X \otimes X = X_1 X_2 \rightarrow \widehat{X}'' = g'\widehat{X}' = (Z \otimes Z)(X \otimes X) = -Y \otimes Y = -Y_1 Y_2, \\ \widehat{Z}' &= Z \otimes I = Z_1 \rightarrow \widehat{Z}'' = \widehat{Z}' = Z_1, \\ g' &= Z \otimes Z = Z_1 Z_2 \rightarrow g'' = O = Y_2. \end{aligned} \tag{13.61}$$

The measurement protocol leaves the second qubit in a +1 eigenket of Y_2, and the second portions of $g'', \widehat{X}''$ and $\widehat{Z}''$ fix it. As qubits 1 and 2 stay nonentangled after the measurement, qubit 2 can be discarded from further consideration. The left-portions of operators $\widehat{X}, \widehat{X}', \widehat{Z}, \widehat{Z}'$ will determine the operation carried out on qubit 1 during the measurement. From Eq. (13.61), it is evident that during the measurement of O, X is mapped to $-Y$, and Z is mapped to Z, which is the same as the action of $P^\dagger$. Therefore, the measurement of O and the CNOT gate can be used to implement the phase gate.

The Hadamard gate can be represented in terms of the phase gate as follows:

$$PR_x(-\pi/2)^\dagger P = \begin{bmatrix} 1 & 0 \\ 0 & j \end{bmatrix} \frac{1}{\sqrt{2}} \begin{bmatrix} 1 & j \\ j & 1 \end{bmatrix}^\dagger \begin{bmatrix} 1 & 0 \\ 0 & j \end{bmatrix} = \frac{1}{\sqrt{2}} \begin{bmatrix} 1 & 1 \\ 1 & -1 \end{bmatrix} = H, \tag{13.62}$$

where with $R_x(-\pi/2)$, we denote the rotation about the x-axis by $-\pi/2$, with the matrix representation given by

$$R_x(-\pi/2) = \frac{1}{\sqrt{2}} \begin{bmatrix} 1 & j \\ j & 1 \end{bmatrix}.$$

This maps the Pauli operators X and Z to X and Y, respectively. What remains is to show that this rotation operator can be implemented using the measurement protocol and CNOT gate. The initial two-qubit system will now be $|\psi\rangle\otimes(|0\rangle+|1\rangle)$, and the corresponding generator is $g=X_2$. The encoded Pauli operator analogs are given by $\widehat{X}=X_1, \widehat{Z}=Z_1$, and operator O is given as X_2. Let us now apply the CNOT gate, where qubit 2 is the control qubit with qubit 1 as the target qubit. The CNOT gate will perform the following mapping based on Eq. (13.47):

$$\begin{aligned}
&\widehat{X}=X_1\rightarrow\widehat{X}'=X\otimes I=X_1,\\
&\widehat{Z}=Z_1\rightarrow\widehat{Z}'=Z\otimes Z=Z_1Z_2,\\
&g=X_2\rightarrow g'=X\otimes X=X_1X_2.
\end{aligned}\tag{13.63}$$

We then apply the phase gate on qubit 2, which performs the following mapping:

$$\begin{aligned}
&\widehat{X}'=X_1\rightarrow\widehat{X}''=X_1,\\
&\widehat{Z}'=Z_1Z_2\rightarrow\widehat{Z}''=Z_1Z_2,\\
&g'=X_1X_2\rightarrow g''=X_1Y_2.
\end{aligned}\tag{13.64}$$

In the final stage, we perform the measurement on operator O. It is clear from Eq. (13.64) that $\widehat{Z}''=Z_1Z_2$ and g'' anticommute with operator O (because they both differ from O in one position). By applying the three-step measurement procedure above, we obtain the following mapping:

$$\begin{aligned}
&\widehat{X}''\rightarrow\widehat{X}'''=X_1,\\
&\widehat{Z}''\rightarrow\widehat{Z}'''=g''\widehat{Z}''=(X_1Y_2)(Z_1Z_2)=Y_1X_2,\\
&g''\rightarrow g'''=O=X_2.
\end{aligned}\tag{13.65}$$

By disregarding qubit 2, we conclude that the measurement procedure maps X_1 to X_1 and Z_1 to Y_1, which is the same as the action of the rotation operator $R_x(-\pi/2)$. In conclusion, we have just shown that any gate from the Clifford group can be implemented based on the CNOT gate, the measurement procedure described above, and a properly prepared initial state.

13.3.4 Fault-tolerant stabilizer codes

We start our description of fault-tolerant stabilizer codes with a four-qubit unitary operation from Gottesman [9] that can generate an encoded version of itself for all $[N,1]$ stabilizer codes when applied transversally. This four-qubit operation performs the following mapping [9]:

$$\begin{aligned}
&X\otimes I\otimes I\otimes I\rightarrow X\otimes X\otimes X\otimes I\\
&I\otimes X\otimes I\otimes I\rightarrow I\otimes X\otimes X\otimes X\\
&I\otimes I\otimes X\otimes I\rightarrow X\otimes I\otimes X\otimes X\\
&I\otimes I\otimes I\otimes X\rightarrow X\otimes X\otimes I\otimes X\\
&Z\otimes I\otimes I\otimes I\rightarrow Z\otimes Z\otimes Z\otimes I\\
&I\otimes Z\otimes I\otimes I\rightarrow I\otimes Z\otimes Z\otimes Z\\
&I\otimes I\otimes Z\otimes I\rightarrow Z\otimes I\otimes Z\otimes Z\\
&I\otimes I\otimes I\otimes Z\rightarrow Z\otimes Z\otimes I\otimes Z.
\end{aligned}\tag{13.66}$$

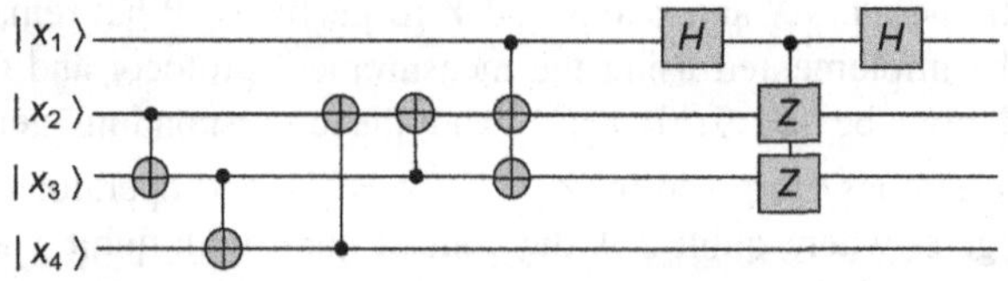

FIGURE 13.14

Quantum circuit to perform mapping as in the work of Gottesman [9], given by Eq. (13.66).

The ith row ($i = 2,3,4$) is obtained by cyclic permutation of the ($i-1$)th row one position to the right. Similarly, the jth row ($j = 6,7,8$) is obtained by cyclic permutation of the ($j-1$)th row one position to the right. The corresponding quantum circuit to perform this manipulation is shown in Fig. 13.14. For any element $s = e^{jl\pi/2}X(\boldsymbol{a}_s)X(\boldsymbol{b}_s) \in S$, where the phase factor $\exp(jl\pi/2)$ is employed to ensure the Hermitian properties of s, this quantum circuit maps $s \otimes I \otimes I \otimes I \rightarrow s \otimes s \otimes s \otimes I$ and its cyclic permutations as given below:

$$\begin{aligned} s\otimes I\otimes I\otimes I &\rightarrow s\otimes s\otimes s\otimes I \\ I\otimes s\otimes I\otimes I &\rightarrow I\otimes s\otimes s\otimes s \\ I\otimes I\otimes s\otimes I &\rightarrow s\otimes I\otimes X\otimes s \\ I\otimes I\otimes I\otimes s &\rightarrow s\otimes s\otimes I\otimes s. \end{aligned} \tag{13.67}$$

It can be easily verified that images of generators are themselves the generators of stabilizers $S\otimes S\otimes S\otimes S$. The encoded Pauli operators $\overline{X}_m = e^{jl_m\pi/2}X(\boldsymbol{a}_m)Z(\boldsymbol{b}_m), \overline{Z}_m = e^{jl'_m\pi/2}X(\boldsymbol{c}_m)Z(\boldsymbol{d}_m)$ are mapped similarly:

$$\begin{aligned} \overline{X}_m\otimes I\otimes I\otimes I &\rightarrow \overline{X}_m\otimes \overline{X}_m\otimes \overline{X}_m\otimes I \\ I\otimes \overline{X}_m\otimes I\otimes I &\rightarrow I\otimes \overline{X}_m\otimes \overline{X}_m\otimes \overline{X}_m \\ I\otimes I\otimes \overline{X}_m\otimes I &\rightarrow \overline{X}_m\otimes I\otimes \overline{X}_m\otimes \overline{X}_m \\ I\otimes I\otimes I\otimes \overline{X}_m &\rightarrow \overline{X}_m\otimes \overline{X}_m\otimes I\otimes \overline{X}_m \\ \overline{Z}_m\otimes I\otimes I\otimes I &\rightarrow \overline{Z}_m\otimes \overline{Z}_m\otimes \overline{Z}_m\otimes I \\ I\otimes \overline{Z}_m\otimes I\otimes I &\rightarrow I\otimes \overline{Z}_m\otimes \overline{Z}_m\otimes \overline{Z}_m \\ I\otimes I\otimes \overline{Z}_m\otimes I &\rightarrow \overline{Z}_m\otimes I\otimes \overline{Z}_m\otimes \overline{Z}_m \\ I\otimes I\otimes I\otimes \overline{Z}_m &\rightarrow \overline{Z}_m\otimes \overline{Z}_m\otimes I\otimes \overline{Z}_m. \end{aligned} \tag{13.68}$$

The transversal application of the four-qubit operation provides an encoded version of the operation itself, which is true for all [N,1] stabilizer codes. On the other hand, for [$N,K > 1$] stabilizer codes, the four-qubit operation maps the encoded Pauli operators $\overline{X}_m, \overline{Z}_m$ for all m-positions simultaneously. Certain modifications are needed to make it suitable for mapping a subset of encoded Pauli operators, as described later in this section.

Let us now provide a simple example from Gottesman [9] (see also [2]). We are concerned with mapping the four-qubit state $|\psi_1\psi_2\rangle\otimes|00\rangle$, where $|\psi_1\psi_2\rangle$ is an arbitrary two-qubit state. The stabilizer S_0 is described by the two generators $g_1 = Z_3,\ g_2 = Z_4$. The encoded Pauli operators are given by

$$\tilde{X}_1 = X_1, \tilde{X}_2 = X_2, \tilde{Z}_1 = Z_1, \tilde{Z}_2 = Z_2. \tag{13.69}$$

Applying the four-qubit operation given by Eq. (13.66), we obtain the following mapping:

$$g_1 \to g_1' = Z_1 Z_3 Z_4, g_2 \to g_1' = Z_1 Z_2 Z_4, \widetilde{X}_1 \to \widetilde{X}_1' = X_1 X_2 X_3, \widetilde{X}_2 \to \widetilde{X}_2' = X_2 X_3 X_4,$$
$$\widetilde{Z}_1 \to \widetilde{Z}_1' = Z_1 Z_2 Z_3, \widetilde{Z}_2 \to \widetilde{Z}_2' = Z_2 Z_3 Z_4.$$

We now perform the measurements using operators $O_1 = X_3$ and $O_2 = X_4$. Applying the measurement protocol from the previous section, we obtain the following mapping as the result of measurements:

$$\begin{aligned} &g_1'' \to O_1, g_1' = Z_1 Z_3 Z_4, g_2 \to O_2, g_2' = Z_1 Z_2 Z_4, \widetilde{X}_1' \to \widetilde{X}_1'' = X_1 X_2 X_3, \widetilde{X}_2' \to \widetilde{X}_2'' = X_2 X_3 X_4, \\ &\widetilde{Z}_1' \to \widetilde{Z}_1'' = g_1' g_2' \widetilde{Z}_1' = Z_1, \widetilde{Z}_2' \to \widetilde{Z}_2'' = g_1' g_2' \widetilde{Z}_2' = Z_4. \end{aligned} \tag{13.70}$$

By disregarding the measured ancilla qubits in Eqs. (13.69) and (13.70) (qubits 3 and 4), we conclude that we have effectively performed the following mapping on the first two qubits:

$$\begin{aligned} &X_1 \to X_1 X_2, X_2 \to X_2, \\ &Z_1 \to Z_1, Z_2 \to Z_1 Z_2, \end{aligned} \tag{13.71}$$

which is the same as the CNOT gate given by Eq. (13.47). Therefore, using the four-qubit operation and measurements, we can implement the CNOT gate. In the previous section, we learned how to implement an arbitrary Clifford operator using a CNOT gate and measurements. If we apply a similar procedure to four code blocks and disregard the ancilla code blocks, we can perform a fault-tolerant encoding operation that maps (on the first two qubits)

$$\begin{aligned} &\overline{X}_m \otimes I \to \overline{X}_m \otimes \overline{X}_m, I \otimes \overline{X}_m \to I \otimes \overline{X}_m, \\ &\overline{Z}_m \otimes I \to \overline{Z}_m \otimes I, I \otimes \overline{Z}_m \to \overline{Z}_m \otimes \overline{Z}_m, \end{aligned} \tag{13.72}$$

Eq. (13.72) is simply a block-encoded CNOT gate. For $[N,K > 1]$ stabilizer codes with this procedure, the encoded CNOT gate is applied to all *m*th encoded qubits in two code blocks simultaneously. Another difficulty is that this procedure cannot be applied to two encoded qubits lying in the same code block or one lying in the first block and the other in the second block.

We turn our attention to describing a procedure that can be applied to $[N,K > 1]$ stabilizer codes as well. The first problem can be solved by moving the *m*th qubit to two ancilla blocks employing the procedure described below. During this transfer, the *m*th encoded qubit is transferred to two ancilla qubit blocks, while other qubits in ancilla blocks are initialized to $|\overline{0}\rangle$ We then apply the block-encoded CNOT gate given by Eq. (13.71) and the description above it. Because we transferred the *m*th encoded qubit to ancilla blocks, the other coded- locks and ancilla block will not be affected. Once we apply the desired operation to the *m*th encoded qubit, we can transfer it back to the original code block. Let us assume that the data qubit is an arbitrary state $|\psi\rangle$ and ancilla qubit prepared in the $+1$ eigenket of $X^{(a)}$ $((|0\rangle+|1\rangle)/\sqrt{2}$-state in computation basis), where we use superscript (a) to denote the ancilla qubit. The stabilizer for this two-qubit state is described by the generator $g = I^{(d)}X^{(a)}$, where we now use the superscript (d) to denote the data qubit. The encoded Pauli operator analogs are given by $\widetilde{X} = X^{(d)}I^{(a)}, \widetilde{Z} = Z^{(d)}I^{(a)}$. The CNOT gate is further applied, assuming that the ancilla qubit is a control qubit, and the data qubit is a target qubit. The CNOT gate performs the following mapping (see Eq. 13.47):

$$g = I^{(d)}X^{(a)} \to g' = X^{(d)}X^{(a)}, \tilde{X} = X^{(d)}I^{(a)} \to \tilde{X}' = X^{(d)}I^{(a)}, \tilde{Z} = Z^{(d)}I^{(a)} \to \tilde{Z}' = Z^{(d)}Z^{(a)}. \quad (13.73)$$

We then perform the measurement on operator $O = Z^{(d)}I^{(a)}$, and by applying the measurement procedure from the previous section, we perform the following mapping:

$$g' = X^{(d)}X^{(a)} \to g'' = O = Z^{(d)}I^{(a)}, \tilde{X} \to \tilde{X}'' = g'\tilde{X}' = I^{(d)}X^{(a)}, \tilde{Z}' = Z^{(d)}I^{(a)} \to \tilde{Z}'' = Z^{(d)}Z^{(a)}. \quad (13.74)$$

By disregarding the measurement (data) qubit, we conclude that we effectively performed the mapping

$$X^{(d)} \to X^{(a)}, Z^{(d)} \to Z^{(a)}, \quad (13.75)$$

indicating that the data qubit is transferred to the ancilla qubit. On the other hand, if the initial ancilla qubit was in a $|0\rangle$-state, and the CNOT gate application had an ancilla qubit serve as the control qubit, the target (data) qubit would be unaffected. The procedure discussed in this paragraph can therefore be used to transfer an encoded qubit to an ancilla block, with other qubits being initialized to $|\overline{0}\rangle$ To summarize, the procedure can be described as follows. The ancilla block is prepared in such a way that all encoded qubits n different from the mth qubit are initialized to the $+1$ eigenket $|\overline{0}\rangle$ of $\overline{Z}_n^{(a)}$. The encoded qubit m is, on the other hand, initialized to the $+1$ eigenket of $\overline{X}_m^{(a)}$. The block-encoded CNOT gate described by Eq. (13.72) is then applied using the ancilla block as the control block. We further perform the measurement on $\overline{Z}_m^{(d)}$. With this procedure, the mth encoded qubit is transferred to the ancilla block, while the other encoded ancilla qubits remain unaffected. Upon the measurement, the mth qubit is left of the $+1$ eigenket $|\overline{0}\rangle$ of $\overline{Z}_m^{(d)}$, wherein the remaining encoded data qubits remain unaffected. Through the foregoing procedure, we can transfer an encoded qubit from one block to another. What remains now is to transfer the mth encoded qubit from the ancilla-encoded block back to the original data block, which can be performed by the procedure described in the next paragraph.

Let us assume now that our two-qubit data-ancilla system is initialized as $|0\rangle^{(d)} \otimes |\psi\rangle^{(a)}$, with the generator of the stabilizer given by $g = Z^{(d)}I^{(a)}$. The corresponding encoded Pauli generator analogs are given by $\tilde{X} = I^{(d)}X^{(a)}, \tilde{Z} = I^{(d)}Z^{(a)}$. We then apply the CNOT gate using the ancilla qubit as a control qubit and the data qubit as a target qubit, which performs the following mapping (again, see Eq. 13.47):

$$g = Z^{(d)}I^{(a)} \to g' = Z^{(d)}Z^{(a)}, \tilde{X} = I^{(d)}X^{(a)} \to \tilde{X}' = X^{(d)}X^{(a)}, \tilde{Z} = I^{(d)}Z^{(a)} \to \tilde{Z}' = I^{(d)}Z^{(a)}. \quad (13.76)$$

We further perform the measurement on operator $O = I^{(d)}X^{(a)}$, which performs the mapping given by

$$g' = Z^{(d)}Z^{(a)} \to g'' = O = I^{(d)}X^{(a)}, \tilde{X}' = X^{(d)}X^{(a)} \to \tilde{X}'' = X^{(d)}X^{(a)}, \tilde{Z}' = I^{(d)}Z^{(a)} \to \tilde{Z}'' = g'\tilde{Z}' = Z^{(d)}I^{(a)}. \quad (13.77)$$

By disregarding the measurement (ancilla) qubit, we conclude that we have effectively performed the mapping

$$X^{(a)} \rightarrow X^{(d)}, Z^{(a)} \rightarrow Z^{(d)}, \tag{13.78}$$

which is just the opposite of the mapping given by Eq. (13.75). With this procedure, we can therefore move the mth encoded qubit back to the original encoded data block.

We now turn our attention to the second problem, applying the CNOT gate from the mth to the nth qubit within the same block or between two different blocks. This problem can be solved by transferring the qubits under consideration to two ancilla blocks with all other positions being in a $|\overline{0}\rangle$ state. We further apply the block-encoded CNOT gate to the corresponding ancilla blocks in a procedure that will leave other ancilla qubits unaffected. After performing the desired CNOT operation, we will need to transfer the corresponding qubits back to the original data positions. The key device for this manipulation is the fault-tolerant SWAP gate. The SWAP gate was described in Chapter 3; the corresponding quantum circuit is provided in Fig. 13.15A to facilitate the description of fault-tolerant SWAP gate implementation. In Fig. 13.15B, we describe a convenient method to implement the CNOT gate operation from qubit 2 to qubit 1. Although it is trivial to prove the correctness of the equivalences given in Fig. 13.15, it is interesting to prove the equivalences using the procedure we introduced in Section 13.3.2. To prove the equivalence given in Fig. 13.15A, we apply the CNOT gate from qubit 1 to qubit 2, which performs the following mapping:

$$XI \rightarrow XX, IX \rightarrow IX, ZI \rightarrow ZI, IZ \rightarrow ZZ. \tag{13.79}$$

In the second stage, we apply the CNOT gate using qubit 2 as the control qubit, and based on Eq. (13.47), we perform the following mapping:

$$XX = (XI)(IX) \rightarrow (XI)(XX) \overset{X^2=I}{=} IX, IX \rightarrow XX, ZI \rightarrow ZZ, ZZ = (ZI)(IZ) \rightarrow (ZZ)(IZ) \overset{Z^2=I}{=} ZI. \tag{13.80}$$

In the third stage, we again apply the CNOT gate, using qubit 1 as the control qubit:

$$IX \rightarrow IX, XX = (XI)(IX) \rightarrow (XX)(IX) = XI, ZZ = (ZI)(IZ) \rightarrow (ZI)(ZZ) = IZ, ZI \rightarrow ZI. \tag{13.81}$$

The overall mapping can be written as

$$XI \rightarrow IX, IX \rightarrow XI, ZI \rightarrow IZ, IZ \rightarrow ZI, \tag{13.82}$$

which is the SWAP-ing action of qubits 1 and 2. To prove the equivalence given in Fig. 13.15B, we first apply the Hadamard gates on both qubits, and based on Eq. (13.43), we can write

$$XI \rightarrow ZI, IX \rightarrow IZ, ZI \rightarrow XI, IZ \rightarrow IX. \tag{13.83}$$

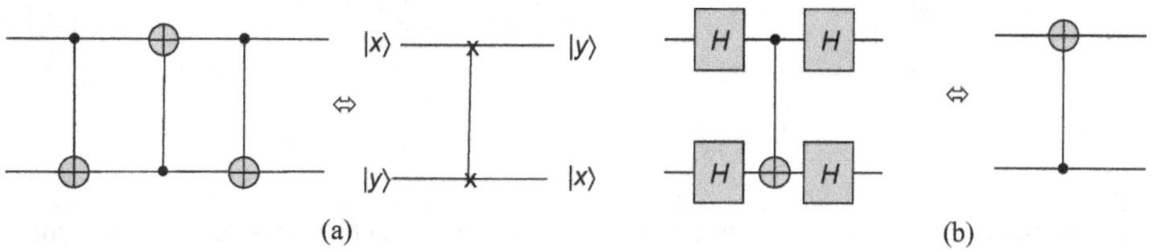

FIGURE 13.15

Quantum circuits to perform (A) SWAP operation, and (B) CNOT gate action from qubit 2 to qubit 1.

We further apply the CNOT gate using qubit 1 as the control qubit, and based on Eq. (13.47), we perform the following mapping:

$$ZI \rightarrow ZI, IZ \rightarrow ZZ, XI \rightarrow XX, IX \rightarrow IX. \tag{13.84}$$

Finally, we again apply the Hadamard gates on both qubits to obtain

$$ZI \rightarrow XI, ZZ \rightarrow XX, XX \rightarrow ZZ, IX \rightarrow IZ. \tag{13.85}$$

The overall mapping can be written as

$$XI \rightarrow XI, IX \rightarrow XX, ZI \rightarrow ZZ, IZ \rightarrow IZ, \tag{13.86}$$

which is the same as the action of the CNOT gate from qubit 2 to qubit 1.

In certain implementations, it is difficult to implement the *automorphism group* of stabilizer S, denoted Aut(S) and defined as the group of functions performing the mapping $G_N \rightarrow G_N$. (Note that the automorphism group of group G_N preserves the multiplication table.) In these situations, it is instructive to design a SWAP gate based on one of its unitary elements, say U, that is possible to implement. This operator is an automorphism of stabilizer S that fixes it, maps codewords to codewords as an encoded operator, and maps C(S)/S to C(S)/S. Based on the fundamental isomorphism theorem, we know that $C(S)/S \cong G_K$, and $U \in$ N(S), indicating that the action of U can be determined by studying its action on generators of G_K. The automorphism U must act on the first and nth encoded qubits, and by measuring other encoded qubits upon the action of U, we come up with an encoding operation in N(G_2) that acts only on the first and nth qubits. Gottesman has shown in Ref. [10] that as far as single-qubit operations are concerned, the only two-qubit gates in $N(G_2)$ are the identity, CNOT gate, SWAP gate, and CNOT gate followed by the SWAP gate. If the automorphism U yields to the action of the SWAP gate, our design is complete. On the other hand, if the automorphism U yields to the action of the CNOT gate, using the equivalency 15(a), we can implement the SWAP gate. Finally, if the automorphism U yields to the action of CNOT gate followed by SWAP gate, the desired SWAP gate that swaps the first qubit being in encoded state $|\overline{0}\rangle$ and nth encoded qubit can be implemented based on circuit shown in Fig. 13.16A. Note that the SWAP of two qubits in the block is not a transversal operation. An error introduced during swapping can cause errors on both qubits being swapped. This problem can be solved using the circuit shown in Fig. 13.16B, in which two qubits to be swapped do not interact directly but by means of an ancilla qubit.

There are certain implementations, such as linear optics, in which implementing the CNOT gate is challenging, so we should keep the number of CNOT gates low. Typically, the CNOT gate in linear

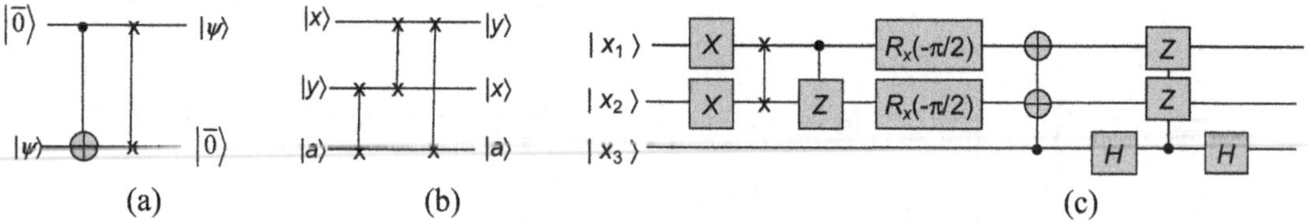

FIGURE 13.16

(A) Quantum circuit to swap encoded qubits 1 and n, (B) quantum circuit to swap kets $|x\rangle$ and $|y\rangle$ employing auxiliary ket $|a\rangle$, and (C) quantum circuit to perform mapping as in the work of Gottesman [9].

optics is implemented as a probabilistic gate [11–13] that performs a desired action with a certain probability. For such technologies, we can implement the SWAP gate through *quantum teleportation* [14], which was introduced in Chapter 3. Alternatively, highly nonlinear optical fibers can be combined with the four-wave mixing effect for quantum teleportation. Namely, in quantum teleportation, entanglement in the Bell state, also known as the Einstein–Podolsky–Rosen pair, is used to transport arbitrary quantum state $|\psi\rangle$ between two distant observers A and B (often called Alice and Bob). The quantum teleportation system employs three qubits: qubit 1 is an arbitrary state to be transferred, while qubits 2 and 3 are in the Bell state $(|00\rangle+|11\rangle)/\sqrt{2}$. Observer A has access to qubits 1 and 2, while observer B has access to qubit 3. The observers also share an auxiliary classical communication channel. Therefore, the initial three-qubit state is $|\psi\rangle\otimes(|00\rangle+|11\rangle)/\sqrt{2}$. The corresponding generators of the stabilizer are $g_1 = IXX$ and $g_2 = IZZ$, while the encoded Pauli operator analogs are $\tilde{X} = XII, \tilde{Z} = ZII$. Observer A first applies the CNOT gate using qubit 1 as the control qubit and qubit 2 as the target. Based on Eq. (13.47), we know that the generators and encoded Pauli operator analogs will be mapped to

$$g_1 = IXX \rightarrow g_1' = IXX, g_2 = IZZ \rightarrow g_2' = ZZZ, \tilde{X} = XII \rightarrow \tilde{X}' = XXI, \tilde{Z} = ZII \rightarrow \tilde{Z}' = ZII. \quad (13.87)$$

Observer A further applies the Hadamard gate on qubit 1, which performs the following mapping:

$$g_1' = IXX \rightarrow g_1' = IXX, g_2' = ZZZ \rightarrow g_2' = XZZ, \tilde{X}' = XXI \rightarrow \tilde{X}'' = ZXI, \tilde{Z}' = ZII \rightarrow \tilde{Z}'' = XII. \quad (13.88)$$

Observer A then performs measurement on operators $O_1 = Z_1$ and $O_2 = Z_2$ with the measurement procedure (protocol) described in Section 13.3.3. This measurement protocol performs the following mapping:

$$\begin{aligned} g_1'' = IXX \rightarrow g_1'' = O_2, g_2'' = XZZ \rightarrow g_2''' = O_1, \tilde{X}'' = ZXI \rightarrow \tilde{X}''' = g_1''\tilde{X}'' = ZIX, \tilde{Z}'' \\ = XII \rightarrow \tilde{Z}''' = g_2''\tilde{Z}'' = IZZ. \end{aligned} \quad (13.89)$$

By discarding the measured qubits, we conclude that we have effectively performed the mapping $X_1 \rightarrow X_3, Z_1 \rightarrow Z_3$, meaning that qubit 1 has been transferred to qubit 3, which is equivalent to the SWAP operation. We now explain using the quantum teleportation procedure to perform encoded qubit swapping. We first prepare an encoded block in which all qubits are in state $|\overline{0}\rangle$ except the nth encoded qubit, which is an arbitrary state $|\psi\rangle$. We further prepare another block in which all qubits are in state $|\overline{0}\rangle$ except the first and nth encoded qubits, which are prepared into the encoded Bell state $(|\overline{00}\rangle+|\overline{11}\rangle)/\sqrt{2}$. Quantum teleportation is then performed between the nth encoded qubits in two blocks, which finishes by transferring the nth encoded qubit from the first block into the first encoded qubit in the second block while leaving the nth encoded qubit in the second block in the state $|\overline{0}\rangle$. By discarding the first coded block, we effectively perform the encoded SWAP operation. This procedure requires a block-encoded CNOT gate, block-Hadamard gate, and $O_n = Z_n$ measurements, which is doable for the arbitrary stabilizer code. Therefore, we have just completely described how to fault-tolerantly apply encoded operations in the Clifford group $N(G_N)$ for arbitrary stabilizer code. Before concluding this section, in the upcoming subsection, we provide an example of [5,1,3] stabilizer code from the work of Gottesman [9] (see also [2]).

13.3.5 The [5,1,3] fault-tolerant stabilizer code

The [5,1,3] stabilizer code can correct a single error (because the minimum distance is 3), and it is also a perfect quantum code, as it satisfies the quantum Hamming bound with equality sign (see Section 7.4). This quantum code is generated by the four generators of the stabilizer $g_1 = XZZXI$, $g_2 = IXZZX$, $g_3 = XIXZZ$, and $g_4 = ZXIXZ$, and the corresponding quantum-check matrix is given by

$$A_Q = \begin{bmatrix} \quad X \quad\quad Z \\ 10010|01100 \\ 01001|00110 \\ 10100|00011 \\ 01010|10001 \end{bmatrix}. \tag{13.90}$$

Since every next row is a cyclic shift of the previous row, this code is a cyclic quantum code. The corresponding encoded Pauli operators, using standard formalism, are obtained as follows:

$$\overline{X} = X_1X_2X_3X_4X_5, \quad \overline{Z} = Z_1Z_2Z_3Z_4Z_5. \tag{13.91}$$

In Section 13.3.2, we introduced the following Pauli gates cyclic permutation operator:

$$U: X \to Y \to Z \to X. \tag{13.92}$$

When this operator is applied transversally to the codeword of the [5,1,3] code, it leaves the stabilizer S invariant shown below:

$$\begin{aligned} g_1 &= XZZXI \to YXXYI = g_3g_4, g_2 = IXZZX \to IYXXY = g_1g_2g_3, \\ g_3 &= XIXZZ \to YIYXX = g_2g_3g_4, g_4 = ZXIXZ \to XYIYX = g_1g_2. \end{aligned} \tag{13.93}$$

Since operator U fixes stabilizer S, when applied transversally, it can serve as an encoded fault-tolerant operation. To verify this claim, let us apply U to the Pauli-encoded operators:

$$\begin{aligned} \overline{X} = X_1X_2X_3X_4X_5 \to Y_1Y_2Y_3Y_4Y_5 = \overline{Y}, \quad \overline{Z} &= Z_1Z_2Z_3Z_4Z_5 \to X_1X_2X_3X_4X_5 = \overline{X}, \\ \overline{Y} &= Y_1Y_2Y_3Y_4Y_5 \to Z_1Z_2Z_3Z_4Z_5 = \overline{Z}. \end{aligned} \tag{13.94}$$

In Section 13.3.4, we introduced a four-qubit operation (see Eq. 13.66) that can be applied transversally to generate an encoded version of itself for all stabilizer codes. Alternatively, the following three-qubit operation can serve the same role for [5,1,3] code [9]:

$$T_3 \doteq \begin{bmatrix} 1 & 0 & j & 0 & j & 0 & 1 & 0 \\ 0 & -1 & 0 & j & 0 & j & 0 & -1 \\ 0 & j & 0 & 1 & 0 & -1 & 0 & -j \\ j & 0 & -1 & 0 & 1 & 0 & -j & 0 \\ 0 & j & 0 & -1 & 0 & 1 & 0 & -j \\ j & 0 & 1 & 0 & -1 & 0 & -j & 0 \\ -1 & 0 & j & 0 & j & 0 & -1 & 0 \\ 0 & 1 & 0 & j & 0 & j & 0 & 1 \end{bmatrix}. \tag{13.95}$$

This three-qubit operation performs, with the corresponding quantum circuit shown in Fig. 13.16C, the following mapping on unencoded qubits:

$$XII \rightarrow XYZ, ZII \rightarrow ZXY, IXI \rightarrow YXZ, IZI \rightarrow XZY, IIX \rightarrow XXX, IIZ \rightarrow ZZZ. \tag{13.96}$$

The transversal application of T_3 performs the following mapping on the generators of [5,1,3] code:

$$\begin{aligned}
&g_1 \otimes I \otimes I = XZZXI \otimes IIIII \otimes IIIII \rightarrow XZZXI \otimes \underbrace{YXXYI}_{g_3g_4} \otimes \underbrace{ZYYZI}_{g_1g_3g_4} = g_1 \otimes g_3g_4 \otimes g_1g_3g_4 \\
&g_2 \otimes I \otimes I = IXZZX \otimes IIIII \otimes IIIII \rightarrow IXZZX \otimes \underbrace{IYXXY}_{g_1g_2g_3} \otimes \underbrace{IZYYZ}_{g_1g_3} = g_2 \otimes g_1g_2g_3 \otimes g_1g_3 \\
&g_3 \otimes I \otimes I \rightarrow g_3 \otimes g_2g_3g_4 \otimes g_2g_4, g_4 \otimes I \otimes I \rightarrow g_4 \otimes g_1g_2 \otimes g_1g_2g_4 \\
&I \otimes g_1 \otimes I \rightarrow g_3g_4 \otimes g_1 \otimes g_1g_3g_4, I \otimes g_2 \otimes I \rightarrow g_1g_2g_3 \otimes g_2 \otimes g_1g_3 \\
&I \otimes g_3 \otimes I \rightarrow g_2g_3g_4 \otimes g_3 \otimes g_2g_4, I \otimes g_4 \otimes I \rightarrow g_1g_2 \otimes g_4 \otimes g_1g_2g_4 \\
&I \otimes I \otimes g_n \rightarrow g_n \otimes g_n \otimes g_n; n = 1,2,3,4
\end{aligned} \tag{13.97}$$

By closer inspection of Eq. (13.97), we conclude that the T_3 operator applies U and U^2 on the other two blocks, except the last line. Since both operators fix S, it turns out that T_3 fixes $S \otimes S \otimes S$ as well for all lines in Eq. (13.97) except the last one. On the other hand, by inspection of the last line in Eq. (13.97), we conclude that the T_3 operator also fixes $S \otimes S \otimes S$ in this case. Therefore, the T_3-operator applied transversally represents an encoded fault-tolerant operation for the [5,1,3] code. In a similar fashion, by applying U and U^2 on the other two blocks in encoded Pauli operators, we obtain the following mapping:

$$\begin{aligned}
&\overline{X} \otimes I \otimes I \rightarrow \overline{X} \otimes \overline{Y} \otimes \overline{Z}, \overline{Z} \otimes I \otimes I \rightarrow \overline{Z} \otimes \overline{X} \otimes \overline{Y}, I \otimes \overline{X} \otimes I \rightarrow \overline{Y} \otimes \overline{X} \otimes \overline{Z}, \\
&I \otimes \overline{Z} \otimes I \rightarrow \overline{X} \otimes \overline{Z} \otimes \overline{Y}, I \otimes I \otimes \overline{X} \rightarrow \overline{X} \otimes \overline{X} \otimes \overline{X}, I \otimes I \otimes \overline{Z} \rightarrow \overline{Z} \otimes \overline{Z} \otimes \overline{Z}.
\end{aligned} \tag{13.98}$$

Therefore, the T_3-operator, applied transversally, has performed the mapping to the encoded version of the operation itself. Now by combining the operations U and T_3 with the measurement protocol, we are in a position to fault-tolerantly implement all operators in N(G_N) for [5,1,3] stabilizer code. Because of the equivalence of Eqs. (13.96) and (13.98), it is sufficient to describe the unencoded design procedures. We start our description with the phase-gate implementation description. Our starting point is the three-qubit system $|00\rangle \otimes |\psi\rangle$. The generators of the stabilizer are $g_1 = ZII$ and $g_2 = IZI$, while the corresponding encoded Pauli operator analogs are $\widetilde{X} = IIX, \widetilde{Z} = IIZ$. The application of T_3-operation results in the following mapping:

$$g_1 = ZII \rightarrow g_1' = ZXY, g_2 = IZI \rightarrow g_2' = XZY, \widetilde{X} = IIX \rightarrow \widetilde{X}' = XXX, \widetilde{Z} = IIZ \rightarrow \widetilde{Z}' = ZZZ. \tag{13.99}$$

Next, we perform the measurement of operators O_1=IZI and O_2=IIZ. We see that g'_1 and $\widetilde{X}'$ anticommute with O_1, while g'_1, g'_2 and $\widetilde{X}'$ anticommute with O_2. Based on measurement protocol, we perform the following mapping:

$$g_1' = ZXY \rightarrow g_1'' = O_1, g_2' \rightarrow g_2'' = O_2, \widetilde{X}' = XXX \rightarrow \widetilde{X}'' = g_1'\widetilde{X}' = YIZ, \widetilde{Z}' \rightarrow \widetilde{Z}'' = ZZZ. \tag{13.100}$$

By disregarding the measurement qubits (qubits 2 and 3), we effectively perform the mapping $X_3 \rightarrow Y_1$, and $Z_3 \rightarrow Z_1$, which is the phase-gate combined with transfer from qubit 3 to qubit 1. Since

$$U^{\dagger}P = \frac{1}{\sqrt{2}}\begin{bmatrix} 1 & j \\ j & 1 \end{bmatrix} = R_x(-\pi/2),$$

and the Hadamard gate can be represented as $H=PR_x(-\pi/2)P^\dagger$, we can implement an arbitrary single-qubit gate in the Clifford group. For arbitrary Clifford operator implementation, the CNOT gate is needed. The starting point is the three-qubit system $|\psi_1\rangle \otimes |\psi_2\rangle \otimes |0\rangle$. The generator of the stabilizer is given by $g = IIZ$, while the encoded Pauli operator analogs are given by $\widetilde{X}_1 = XII, \widetilde{X}_2 = IXI, \widetilde{Z}_1 = ZII, \widetilde{Z}_2 = IZI$. We then apply the T_3-operator, which performs the following mapping:

$$g = IIZ \rightarrow g' = ZZZ, \widetilde{X}_1 = XII \rightarrow \widetilde{X}_1' = XYZ, \widetilde{X}_2 = IXI \rightarrow \widetilde{X}_2' = YXZ, \widetilde{Z}_1 = ZII \rightarrow \widetilde{Z}_1' = ZXY, \widetilde{Z}_2 = IZI \rightarrow \widetilde{Z}_2' = XZY. \tag{13.101}$$

We further perform the measurement of $O=IXI$. Clearly, g', $\widetilde{X}_1'$, and $\widetilde{Z}_2$ anticommute with O, and the measurement protocol performs the mapping

$$g' \rightarrow g'' = O, \widetilde{X}_1' = XYZ \rightarrow \widetilde{X}_1'' = g'\widetilde{X}_1' = YXI, \widetilde{X}_2' \rightarrow \widetilde{X}_2'' = YXZ, \widetilde{Z}_1' \rightarrow \widetilde{Z}_1'' = ZXY, \widetilde{Z}_2' = XZY \rightarrow \widetilde{Z}_2'' = g'\widetilde{Z}_2' = YIX. \tag{13.102}$$

By disregarding the measurement qubit (namely qubit 2), we effectively perform the following mapping:

$$X_1 I_2 \rightarrow Y_1 I_3, \widetilde{X}_2 = I_1 X_2 \rightarrow Y_1 Z_3, \widetilde{Z}_1 = Z_1 I_2 \rightarrow Z_1 Y_3, I_1 Z_2 \rightarrow Y_1 X_3. \tag{13.103}$$

By relabeling qubit 3 in the images with qubit 2, the mapping of Eq. (13.103) can be described by the equivalent gate U_e [9] (see also [1]):

$$U_e = (I \otimes U^2)(P \otimes I)U_{\text{CNOT}}^{(21)}(I \otimes R_x(-\pi/2)), \tag{13.104}$$

where with $U_{\text{CNOT}}^{(21)}$, we denote a CNOT gate with qubit 2 as the control qubit and qubit 1 as the target qubit. Using the fact that U gate is of order 3 ($U^3 = I \Leftrightarrow (U^2)^\dagger = U$), the $U_{\text{CNOT}}^{(21)}$ gate can be expressed in terms of U_e gate as

$$U_{\text{CNOT}}^{(21)} = (P^\dagger \otimes I)(I \otimes U)U_e\left(I \otimes (R_x(-\pi/2))^\dagger\right). \tag{13.105}$$

The CNOT gate $U_{\text{CNOT}}^{(21)}$ can be implemented based on $U_{\text{CNOT}}^{(21)}$ and four Hadamard gates in a fashion similar to that shown in Fig. 13.15B. This description of CNOT gate implementation completes the description of all gates needed to implement an arbitrary Clifford operator.

13.4 Summary

This chapter was devoted to fault-tolerant quantum error correction. After introducing fault-tolerance basics and traversal operations in Section 13.1, we provided the basics of fault-tolerant QIP concepts and procedures in Section 13.2, employing Steane code as an example. Section 13.3 was devoted to a rigorous description of fault-tolerant quantum error correction. After a short review of some basic stabilizer code concepts, fault-tolerant syndrome extraction is described in Subsection 13.3.1, followed by fault-tolerant encoding operations in Subsection 13.3.2. Subsection 13.3.3 was concerned with quantum gates applied on a quantum register through a measurement protocol. This method, when applied transversally, was used in Subsection 13.3.4 to enable fault-tolerant error correction based on an arbitrary quantum stabilizer code. Fault-tolerant stabilizer codes were described in Subsection 13.3.4. The [5,1,3] fault-tolerant stabilizer code was provided as an example in Subsection 13.3.5.

References

[1] D. Gottesman, I.L. Chuang, Demonstrating the viability of universal quantum computation using teleportation and single-qubit operations, Nature 402 (1999) 390–393.

[2] F. Gaitan, Quantum Error Correction and Fault Tolerant Quantum Computing, CRC Press, 2008.

[3] I.B. Djordjevic, Quantum Information Processing, Quantum Computing, and Quantum Error Correction: An Engineering Approach, second ed., Elsevier/Academic Press, 2021.

[4] M.A. Neilsen, I.L. Chuang, Quantum Computation and Quantum Information, Cambridge University Press, Cambridge, 2000.

[5] P.W. Shor, Fault-tolerant quantum computation, in: Proc. 37nd Annual Symposium on Foundations of Computer Science, IEEE Computer Society Press, 1996, pp. 56–65.

[6] X. Zhou, D.W. Leung, I.L. Chuang, Methodology for quantum logic gate construction, Phys. Rev. A 62 (5) (2000). Paper 052316.

[7] M.B. Plenio, V. Vedral, P.L. Knight, Conditional generation of error syndromes in fault-tolerant error correction, Phys. Rev. A 55 (6) (1997) 4593–4596.

[8] I.B. Djordjevic, Cavity quantum electrodynamics based quantum low-density parity-check encoders and decoders, Paper no. 7948-38, in: Proc. SPIE Photonics West 2011, Advances in Photonics of Quantum Computing, Memory, and Communication IV, The Moscone Center, San Francisco, California, USA, January 22–27, 2011.

[9] D. Gottesman, Theory of fault-tolerant quantum computation, Phys. Rev. A 57 (1) (1998) 127–137.

[10] D.P. DiVincenzo, P.W. Shor, Fault-tolerant error correction with efficient quantum codes, Phys. Rev. Lett. 77 (15) (1996) 3260–3263.

[11] T.C. Ralph, N.K. Langford, T.B. Bell, A.G. White, Linear optical Controlled-NOT gate in the coincidence basis, Phys. Rev. A 65 (2002) 062324-1–062324-5.

[12] E. Knill, R. Laflamme, G.J. Milburn, A scheme for efficient quantum computation with linear optics, Nature 409 (2001) 46–52.

[13] A. Politi, M. Cryan, J. Rarity, S. Yu, J.L. O'Brien, Silica-on-silicon waveguide quantum circuits, Science 320 (5876) (2008) 646–649.

[14] C.H. Bennett, G. Brassard, C. Crépeau, R. Jozsa, A. Peres, W.K. Wootters, Teleporting an unknown quantum state via dual classical and Einstein-Podolsky-Rosen channels, Phys. Rev. Lett. 70 (13) (1993) 1895–1899.

Further reading

[1] A.Y. Kitaev, Quantum computations: algorithms and error correction, Russ. Math. Surv. 52 (6) (1997) 1191–1249.

[2] D. Gottesman, Stabilizer Codes and Quantum Error Correction (PhD Dissertation), California Institute of Technology, Pasadena, CA, 1997.

[3] A. Barenco, A universal two-bit quantum computation, Proc. *R. Soc. Lond. A*, Math. Phys. Sci. 449 (1937) (1995) 679–683.
[4] D. Deutsch, Quantum computational networks, Proc. R. Soc. Lond. A, Math. Phys. Sci. 425 (1868) (1989) 73–90.
[5] E. Knill, R. Laflamme, Concatenated Quantum Codes, 1996. Available at: http://arxiv.org/abs/quant-ph/9608012.
[6] E. Knill, R. Laflamme, W.H. Zurek, Resilient quantum computation, Science 279 (5349) (1998) 342–345.
[7] C. Zalka, Threshold Estimate for Fault Tolerant Quantum Computation, 1996. Available at: http://arxiv.org/abs/quant-ph/9612028.
[8] M. Grassl, P. Shor, G. Smith, J. Smolin, B. Zeng, Generalized concatenated quantum codes, Phys. Rev. A 79 (5) (2009). Paper 050306(R).
[9] A.Y. Kitaev, Fault-tolerant quantum computation by anyons, Ann. Phys. 303 (1) (2003) 2–30. Available at: http://arxiv.org/abs/quant-ph/9707021.
[10] S. B. Bravyi, A. Y. Kitaev, Quantum codes on a lattice with boundary, Available at: http://arxiv.org/abs/quant-ph/9811052.
[11] A.G. Fowler, M. Mariantoni, J.M. Martinis, A.N. Cleland, Surface codes: towards practical large-scale quantum computation, Phys. Rev. A 86 (3) (2012) 032324.
[12] D. Litinski, A Game of Surface codes: large-scale quantum computing with lattice surgery, Quantum 3 (2019) 128.
[13] A.J. Landahl, C. Ryan-Anderson, Quantum Computing by Color-Code Lattice Surgery, 2014. Available at: https://arxiv.org/abs/1407.5103.
[14] C. Horsman, A.G. Fowler, S. Devitt, R.V. Meter, Surface code quantum computing by lattice surgery, New J. Phys. 14 (2012) 123011.
[15] A.G. Fowler, C. Gidney, Low Overhead Quantum Computation Using Lattice Surgery, 2018. Available at: https://arxiv.org/abs/1808.06709.
[16] R. Raussendorf, D.E. Browne, H.J. Briegel, Measurement-based quantum computation with cluster states, Phys. Rev. A 68 (2003) 022312.
[17] S. Bravyi, A. Kitaev, Universal quantum computation with ideal Clifford gates and noisy ancillas, Phys. Rev. A 71 (2) (2005) 022316.

Index

'*Note:* Page numbers followed by "f" indicate figures and "t" indicate tables.'

D

E

F

G

M

N

R

S

T

U

V

W

Z

Printed in the United States
by Baker & Taylor Publisher Services